Biomarkers for Psychiatric Disorders

Christoph W. Turck
Editor

Biomarkers for Psychiatric Disorders

Springer

Editor
Christoph W. Turck
Max Planck Institute of Psychiatry
Kraepelinstrasse 2-10
D-80804 Munich, Germany
turck@mpipsykl.mpg.de

ISSN: xxx-x-xxx-xxxxx-x e-ISSN: xxx-x-xxx-xxxxx-x
ISBN: 978-1-387-79250-7 e-ISBN: 978-1-387-79251-4
DOI: 10.1007/978-1-387-79251-4

Library of Congress Control Number: 2008930767

Printed on acid-free paper

springer.com

Preface

Biomarkers are receiving a great deal of attention in the life sciences and the global market for disease - specific biomarkers is expected to see a significant increase in the years to come. Among the different disease areas, psychiatric disorders are without doubt the most challenging in terms of understanding their pathophysiology, drug development and patient treatment. The limited knowledge of etiology and pathogenesis, the great clinical heterogeneity, uncertain phenotype boundaries, genetic overlap between psychiatric disorders and the great influence of non-genetic factors all contribute to this situation.

Although a detailed pathobiology of psychiatric disorders remains elusive, it is now believed that several neural circuits located in more than one area of the brain are involved. There are presumably many ways in which these circuits can be disrupted through the effects of several genes that code for protein products, which in turn have an impact on metabolic and signaling pathways. As a consequence, biomarkers for an apparently similar disease phenotype may vary. Only through an improved understanding of the neural circuitries will it be possible to better stratify different disease phenotypes and identify the relevant biomarkers. Good animal models representing distinct features of a psychiatric disorder phenotype are one way to come up with specific biomarkers. Here a good representation of the relevant psychiatric disease mechanism in an animal disease model is mandatory but challenging, to enable a pre-clinical to clinical study translation.

Not surprising is the fact that because the diagnostic tools used now for psychiatric disorders, which are restricted to the evaluation of behavioral and clinical phenotypes, many scientific studies are compromised. What the field therefore needs more than anything else are specific and sensitive biomarkers instead of the highly variable and subjective clinical parameters currently used to track a disease. This includes biomarkers that are of sufficient sensitivity at the pre-symptomatic stage.

Major challenges for the development of novel psychiatric drugs in the course of clinical trials are the placebo response observed in many patients, side effects and tolerance. Here biomarkers could have a major impact for stratifying placebo responders, an objective which cannot be achieved through knowledge of the genotype alone but which requires markers for acute state assessments. The hope is that biomarkers will benefit drug trials by predicting clinical outcome and reducing development costs through smaller trial size, a concept that will also eventually enable a true personalized-medicine approach. Ultimately biomarkers will be judged by their clinical utility, in other words whether the added benefit justifies the cost involved. It is now widely believed that only a panel of biomarkers will achieve the necessary specificity and sensitivity for a reliable clinical assay. However, with an increasing complexity of biomarker panels, both measurement and data analysis will become rate-limiting with consequences for regulatory compliance and associated cost. Still, it is my believe that even elaborate diagnostic procedures can be justified if the patient lives longer or feels better.

The road to a successful biomarker consists of several steps that range from discovery, assay development, validation to clinical implementation. The apparent dearth of new biomarkers for psychiatric disorders is on one hand caused by the earlier mentioned patient heterogeneity. At the same time it is apparent that current technologies are still insufficiently sensitive for low abundant marker detection. It therefore may be critical to come up with markers from an array of different platforms including "omics" technologies, *in vivo* imaging, clinical data and others. For this to come about, the science culture has to adapt accordingly. Not only will it be mandatory for specialists of different technology platforms to share and understand each other's data, but also the various stakeholders including academia, pharma and biotech industries, diagnostic industries, advocacy groups and regulatory agencies need to come up with a concerted effort. Programs have indeed been initiated to bring all the stakeholders to the table: to mention are the *fNIH Biomarkers Consortium* that consists of steering committees, each representing a different disease area including psychiatric disorders. In a related attempt the *Innovative Medicines Initiative* of the *European Union* aims at establishing public-private partnerships between the pharmaceutical industry and academia to support a more efficient discovery and development of better medicines, including the use of biomarkers.

The present book attempts to give an overview of current efforts for psychiatric disorder biomarker discovery. The methods employed for the identification of "wet" biomarkers from brain tissue, body fluids and cells are mainly based on "omics" technologies, including genomics, transcriptomics, proteomics and metabolomics. At the same time neuroimaging and sleep analysis methods are capable of producing "dry" biomarkers. After diagnostic applications, biomarkers used in drug development efforts and the realization of personalized medicine are discussed. Finally, an approach for modeling of psychiatric disorders based on biomarker information and other data sources is submitted, a method that will undoubtedly play an increasingly important role in the understanding of the pathobiology, identification of new drug targets and patient care.

I would like to close by pointing out that the quest for biomarkers for psychiatric disorders is not a novel idea, but was already pursued by the founding director of the *Max Planck Institute of Psychiatry*, Emil Kraepelin (1856-1926). In the late nineteenth century Kraepelin designed a writing scale which was used to measure writing pressure curves of patients suffering from psychiatric disorders. The figure below shows the results of such writing pressure analyses, which Kraepelin used for the classification of mood disorders long before the *Diagnostic and Statistical Manual of Mental Disorders* was published.

Munich, 2008
Christoph W. Turck
Max Planck Institute of Psychiatry

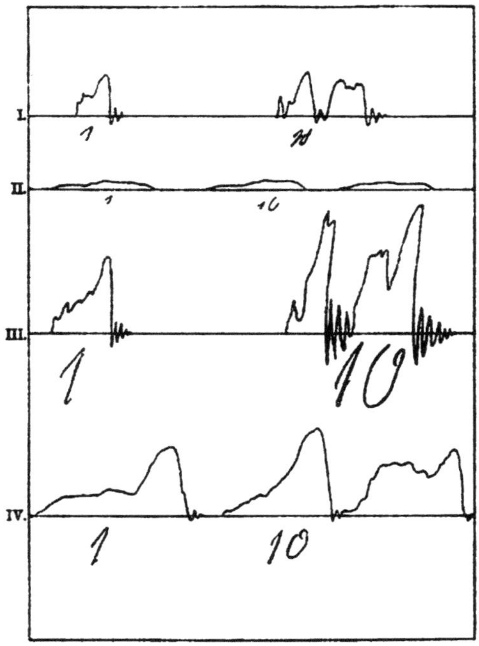

Pressure curves recorded by a writing balance. From top to bottom: (I) healthy control, (II) patient with depressed episode, (III) patient with manic episode, (IV) same patient as in (III), after clinical improvement (adapted from E. Kraepelin, Psychiatry Textbook, Vol. II, 6th Edition, Leipzig, Verlag von Johann Ambrosius Barth, 1899, p. 373).

Contents

Chapter 1
Blood and Brain Gene Expression in Major Psychiatric Disorders: A Search for Biomarkers

Gursharan Chana, Stephen J. Glatt, Ian P. Everall, and Ming T. Tsuang(✉)

Abstract Microarray investigations in psychiatry have so far implicated a number of genes to be associated with major psychiatric disorders, in particular schizophrenia. In postmortem brain studies, alterations in the expression of transcripts encoding for oligodendrocyte functioning and myelination, mitochondrial-related genes, and energy metabolism, as well as synaptic structure and transmission, have been demonstrated. To a certain extent, these alterations reflect changes in associated mRNAs and proteins previously seen in neuropathological investigations of major psychiatric disorders and hence are of great interest. Assessing gene expression changes in the blood of patients with psychiatric disorders will aid in the characterization of their genetic profile. This, in turn, may eventually allow us to relate these findings to brain-related changes, and hence to potentially identify biomarkers for detection, intervention, and treatment. However, while microarray technology has opened the way for high-throughput gene expression analysis, a significant amount of methodological and technical variability still exists in their application. Therefore, it is necessary that a stringent approach be adopted by researchers in designing such experiments and precaution taken in the final analysis and interpretation of results. The aim of this chapter is to provide a balanced view of microarray investigations in the blood and brain in major psychiatric disorders by highlighting the strengths and weaknesses of such studies in identifying candidate genes. Strategies to overcome these weaknesses will be discussed in the context of advancing and improving future microarray investigations in psychiatry.

Abbreviations ACHR: Acetylcholine receptor; AMPA1-2: 2-Aminomethyl phenylacetic acid; Apo: Apolipoprotein; BDNF: Brain-derived neurotrophic

---

M.T. Tsuang
Behavioral Genomics Endowed Chair and University Professor, Distinguished Professor of Psychiatry and Director, Center for Behavioral Genomics, Department of Psychiatry, University of California, San Diego, MTF 453, MC 0603, 9500 Gilman Drive, La Jolla, CA 92093-0603
mtsuang@ucsd.edu

C.W. Turck (ed.) *Biomarkers for Psychiatric Disorders*,
doi: 10.1007/978-0-387-79251-4_1,

factor; BTG1: B-cell translocation gene 1; CCK: Cholecystokinin; CD14: Cell differentiation 14; CNP: Cyclic nucleotide phosphodiesterase; dlPFC: Dorsolateral prefrontal cortex; DRD2: Dopamine receptor D2; EDGF: Epidermal-derived growth factor; ER: Endoplasmic reticulum; erbB2: Erythroblastic leukemia viral oncogene homolog 2; FGF: Fibroblast growth factor; GABA: Gamma-amino butyric acid; GAP-43: Growth associated protein-43; GAD67: Glutamate decarboxylase-67; GluR1-2: Glutamate receptor 1–2; GNA01: Guanine nucleotide binding protein alpha 1; GO: Gene ontology; GSK3A: Glycogen synthase kinase 3 alpha; HERC2: Heat domain and rcc1-like domain 2) Giα1 (g-protein inhibitory alpha1); HLA-DRB1: Major histocompatibility complex DR beta 1; HER3: Heregulin-3; HML-2: Human macrophage lectin 2; HNRPA3: Heterogeneous nuclear ribonucleoprotein A3; IFITM1: Interferon-induced protein with tetratricopeptide repeats 1; HSPB1: Heat shock protein beta 1; Kir2.3: Potassium channel inward rectifying 2.3; LARS2: Leucyl-tRNA synthetase, mitochondrial; LCM: Laser capture microdissection; MAG: Myelin-associated glycoprotein; MAL: Myelin and lymphocyte protein; MARCKS: Myristolated alanine-rich C-kinase substrate; MBP: Myelin basic protein; MDD: Major depressive disorder; MDH1: Malate dehydrogenase 1; MM: Mismatch; MOBP: Myelin-oligodendrocyte basic protein; MOG: Myelin-oligodendrocytic protein; MRI: Magnetic resonance imaging; MT2A: Metallothionein 2A; Neurod1: Neurogenic differentiation 1; NMDA1: *N*-methyl-D-aspartate 1; NPY: Neuropeptide Y; NRG-1: Neuregulin-1; NSF: *N*-ethylmaleimide sensitive fusion protein; PDE4D: Phosphodiesterase 4D; PEA-15: Phosphoprotein enriched in astrocytes 15; PFC: Prefrontal cortex; PLP: Proteolipid protein; PM: Perfect match; PMI: Postmortem interval; PMP22: Peripheral myelin protein 22; qRT-PCR: Quantitative real time polymerase chain reaction; RGS4: Regulator of g-protein signalling-4; RMA: Robust multichip algorithm; S100: Calcium-binding protein A1; SELENBP1: Selenium binding protein-1; SERPINA3: Alpha-1-anti-chymotrypsin A3; SFRS1: Splicing factor, arginine/serine rich 1; SMDF: Sensory and motor neuron derived factor; SNP: Single nucleotide polymorphism; SOX10: SRY-related homeobox gene 10; SPR: Sepiaterin reductase; TGF-α: Transforming growth factor-alpha; TGF-β 1: Transforming growth factor beta 1; TRAF4: Tumor necrosis factor receptor associated factor 4; XBP1: X box binding protein 1

1.1 Introduction

The use of microarray technology to assess gene expression changes in major psychiatric disorders has increased steadily over the past two decades. This increase has been fueled largely by a huge body of literature demonstrating inconsistent neuropathological changes in the brains of patients with these disorders. Therefore, the premise behind gene expression investigations has been to identify candidate mRNAs and genes in the brains of patients, which can then be validated and related at the protein level to provide us with clues to the etiology of these disorders. This reversal of the traditional scientific method is logical in principle, as it allows us to take a more global view of brain changes in these complex disorders before refining to a more mechanistic picture. While some interesting findings have emerged from microarray investigations, they have unfortunately not always been replicated by independent research groups when using different brain cohorts. This discrepancy in results may be partially attributed to inconsistencies in methodologies and platforms used for analysis; however, it is also largely a consequence of the heterogeneous and polygenic profile of these disorders. Nevertheless, as microarray technology and study designs become more consistent we are now beginning to build up gene expression profiles for major psychiatric disorders. Furthermore, the construction of such profiles may eventually help in the separation of overlapping clinical profiles between disorders.

One recent area of interest in gene expression studies in psychiatry has been to assess changes in mRNA transcripts in the blood cells of patients and relate them to brain levels. This avenue of blood-based gene expression studies is potentially very powerful, as it may enable the identification of potential biomarkers for these disorders, which could then ultimately be used as diagnostic and prognostic indicators. While the discovery of such markers would greatly advance psychiatry, the realization of this may take some time. We begin by briefly discussing the study design of microarray investigations prior to covering current gene expression findings in the blood and brain in psychiatric disorders. Finally, we look at problems associated with gene expression studies and how solving these will get us closer to the goal of identifying biomarkers for psychiatric disorders.

1.2 Designing Gene Expression Investigations

While the design of postmortem gene expression investigations may be limited by the availability of samples, many steps can still be taken to reduce variability in findings through careful planning prior to experimentation. This is paramount for blood-based investigations, in which more control can be achieved but more external variables can easily influence gene expression. The most obvious criterion that can add power to any gene expression study is the number of samples available for investigation. As mentioned, this is often limited in postmortem studies by a limited amount of available tissue, with competition for samples among researchers. Also, for postmortem investigations, matching of clinical and demographic variables that can affect RNA quality, and hence gene expression findings, should be attempted. When this is not achievable, the variable should be included as a covariate in the final analysis. These variables include factors such as age, gender, cause of death, medication histories, history of drug abuse, pH, and postmortem interval (PMI) (for a review see Bunney, Bunney et al., 2003). Ideal matching of all or even half of these parameters is, however, rarely achieved. For blood-based gene expression studies, there are even more parameters that can substantially affect gene expression including diet, genetics, time to last meal, time of day, any medications used, and frequency of exercise (Radich, Mao et al., 2004). These variables to a large extent can be controlled by the researcher, i.e., patients could be asked to give a blood sample first thing in the morning before eating. Additional questionnaires pertaining to the patient's lifestyle can also be administered to glean as much information as possible regarding potential influences on gene expression changes. Finally, for blood-based gene expression studies, it is worthwhile attempting the recruitment of first-degree relatives into the study. This has the potential for differentiating disease-related genes from medication-related genes or genes that change in response to chronic mental illness.

1.3 Choosing a Platform

Selecting a suitable platform for conducting a gene expression study depends ultimately on the aims of the investigation. For major psychiatric disorders, most investigators to date have employed high-throughput microarray gene chips that encompass the whole human genome. This approach has been adopted because of the lack of information regarding etiological mechanisms for major psychiatric disorders. Nevertheless, a fairly broad choice still exists when selecting a genome-wide chip for screening patient brain or blood samples. While a detailed technical dissection of these platforms is beyond the scope of this review, it is worthwhile considering the major strengths and weaknesses of the various platforms in order to make an informed decision. Many of the microarray studies in psychiatry have utilized gene chips produced by Affymetrix (Santa Clara, USA). However, other competitors for the whole human gene chip market include Codelink, Agilent, Applied Biosystems, and Illumina.

The primary factor influencing the early adoption of Affymetrix gene chips has been the highly automated process by which they are manufactured. The Affymetrix gene chip manufacturing process comprises the use of highly specific photolithographic masks and a solid-phase DNA synthesis to construct 25mer probes on the surface of the chip (Hardiman, 2004). Other platforms such as Agilent, Codelink, and Illumina utilize 60–70mer probe sets which are more sensitive in detecting mRNA signal than the 25mer probe sets used by Affymetrix. Affymetrix, however, increases its specificity of detection by including 11 probe sets, which are then averaged following analysis of the chip posthybridization. Also contained on the chip are mismatch (MM) probe sets. These probe sets are identical to the perfect match (PM) probes with the exception of a single base difference located in a central position. The MM probes serve as controls for specific hybridization and facilitate background correction due to crosshybridization signals (Hardiman, 2004). The newer Affymetrix U133 plus 2.0 platform has been a more recently used gene chip in a number of different gene expression investigations in psychiatry. This gene chip comprises probes for analysis of over 47,000 transcripts, including the entire human genome, and provides a comprehensive way to assess global gene expression changes. This chip has recently been superseded by the Affymetrix Exon arrays, which are capable in addition of assessing gene expression of splice variants of transcripts. This novel technology will help to better understand the contribution of individual variants to overall gene expression and hence better relate to gene function.

However, with the further development of microarray technology, Affymetrix's competitors are also beginning to produce highly reproducible arrays for analyzing whole human genome expression. For instance Ilumina's latest chip sets are also produced in a highly automated way, utilizing a randomly self-assembled silica bead pool with attached oligonucleotide probes that are recorded on specific locations on a patterned substrate (Fan, Chee et al., 2006). Owing to the small size of the bead used for this procedure, the density of these arrays can be up to 40,000 times greater than that of spotted arrays (Michael, Taylor et al., 1998).

Another deciding factor for choosing a microarray platform is how well the gene expression changes detected can be validated by quantitative real-time polymerase chain reaction (qRT-PCR). This is of major significance, given the need to validate microarray findings by this much more sensitive molecular technique. The final and perhaps most important reason for choosing a microarray platform is the ability of the researchers to compare their data on gene expression changes with that generated by other groups. A number of studies to date, looking at cross-platform comparisons, have unfortunately seen a very low level of correlation in gene expression (Tan, Downey et al., 2003; Hollingshead, Lewis et al., 2005), even with mRNA samples derived from homogenous tissue sources such as cell lines. This has led to the conclusion that gene expression data cannot be combined reliably between platforms. Knowledge of this result has often led researchers to choose platforms on the basis of what past investigations have utilized. While the logic behind this choice is concrete, it does to a certain extent preclude an informed decision based on the merits of individual platforms. Furthermore, recent comparisons of cross-platform correlations have yielded more concordant data (Shippy, Sendera et al., 2004; Schlingemann, Habtemichael et al., 2005; Wang, Barbacioru et al., 2006; Bosotti, Locatelli et al., 2007), suggesting that either the technology or the scientists' ability to harness it is improving.

1.4 Postmortem Gene Expression in Major Psychiatric Disorders

Early postmortem gene expression studies of the brain used pooled samples of RNA to identify differences in expression between control and patient groups. Although these studies were cost efficient, they masked many of the significant gene expression differences that exist between individuals in both control and patient groups. Therefore, current postmortem gene expression investigations tend to use one gene chip per sample when carrying out their analysis. Mirnics et al. (2000), were the first to conduct a high-throughput gene expression study for schizophrenia, comparing 250 functional gene groups between matched pairs of a patient and a control. While differences in specific genes were not consistent between the pairs, they did observe downregulation of functional gene groups. The most consistent of these findings related to a downregulation of genes encoding presynaptic proteins (Mirnics, Middleton et al., 2000). Included in this list were the genes for *N*-ethylmaleimide-sensitive factor and synapsin II, which were validated and confirmed by in situ hybridization. Interestingly, the regulator of G protein signaling 4 (RGS4), which has subsequently received much attention from gene association studies in schizophrenia, was also highlighted as a possible candidate for schizophrenia. Since then a number of different postmortem microarray investigations for major psychiatric disorders have been carried out in a number of different brain regions, but have overall tended to focus on schizophrenia and gene expression changes in the dorsolateral prefrontal cortex (dlPFC). The main reason for the dlPFC being chosen as a candidate region is the relatively consistent cognitive and functional deficits seen in patients with schizophrenia (Goldberg, Weinberger et al., 1987; Goldman-Rakic, 1994; Wible, Anderson et al., 2001; Wolf, Gur et al., 2007), and the involvement of dlPFC in the execution of these abilities. The study design and main findings of these investigations are summarized in Table 1.1 .

Looking at the overall picture of candidate genes for major psychiatric disorders uncovered by gene expression studies, some trends can be seen, especially for schizophrenia, on which most microarray investigations have focused. One of the most consistent findings has been a reduction in the genes related to myelin structure and oligodendrocyte functioning in the prefrontal cortex in schizophrenia (Hakak, Walker et al., 2001; Tkachev, Mimmack et al., 2003; Aston, Jiang et al., 2004; Tkachev, Mimmack et al., 2007), with reductions also being demonstrated in bipolar disorder and major depressive disorder (Tkachev, Mimmack et al., 2003; Aston, Jiang et al., 2005; Sun, Wang et al., 2006).

While demyelination has not been observed in schizophrenia or bipolar disorder, alterations in oligodendrocyte numbers and associated proteins have been demonstrated histopathologically (Uranova, Orlovskaya et al., 2001; Hof, Haroutunian et al., 2002; Flynn, Lang et al., 2003; Uranova, Vostrikov et al., 2004) together with alterations in white matter visualized via magnetic resonance imaging (MRI) (Davis, Stewart et al., 2003; Stewart and Davis, 2004). Given the intimate relationship between oligodendrocytes and axons and their role in speeding neurotransmission, it seems logical that alterations in oligodendrocytes or myelin proteins could cause or contribute to some of the symptoms of schizophrenia.

Genes related to metabolism and mitochondrial pathways are another ontological group that have been found by a number of studies to be reduced in expression in schizophrenia (Pongrac, Middleton et al., 2002; Tkachev, Mimmack et al., 2003; Prabakaran, Swatton et al., 2004; Iwamoto, Bundo et al., 2005; Konradi, 2005) and by some in bipolar disorder (Iwamoto, Bundo et al., 2005). Changes in the expression of mitochondrial genes involved in energy pathways from the study by Middleton et al. (2002) demonstrated reductions in transcripts associated with the regulation of ornithine, polyamine metabolism, the mitochondrial malate shuttle system, the tricarboxyclic acid cycle, and amino acid and ubiquitin metabolism. While the studies that followed this initial finding also showed alterations in similar mitochondrial- and metabolism-related genes, alterations in identical genes were not necessarily seen. Nevertheless, this picture of reduction in metabolism, especially in the prefrontal cortex of schizophrenics, fits in well with reductions in

Table 1.1 Postmortem microarray investigations in major psychiatric disorders

Brain region	Sample	Platform	Main gene expression findings	Reference
Prefrontal cortex (BA9)	6(10) C; 6(10) S	cDNA array (7000)	Reduced Presynaptic (NSF, Synapsin II); Reduced GABA and glutamate neurotransmission (GAD67, AMPA1-2, GluR1-2)	Mirnics et al. (2000)
Prefrontal cortex (BA46)	12C; 12S	HuGeneFL (6000)	Reduced Myelin Related (MAG, CNP, MAL, HER3); Increased developmental (GAP-43, MARCKS) and increased GABA neurotransmission (GAD67, GABA-A)	Hakak et al. (2001)
Prefrontal cortex (BA10)	15C; 15S; 15B; 15MDD	HuGeneFL (6000)	Reduced neuropeptide Y in schizophrenia and bipolar disorder	Kuromitsu et al. (2001)
Prefrontal cortex	10C; 10B	cDNA array (1200)	Reduced TGF-beta 1; Increased Caspase 8 and erbB2	Bezchlibnyk et al. (2001)
Prefrontal cortex	15C; 15S; *15B; 15MDD*	Custom array (300);	Increased apoL1; QPCR (increased ApoL2 and L4 in schizophrenia)	Mimmack et al. (2002)
Prefrontal cortex (BA9)	11(50)C; 11(54)S; 14B	U133A	Reduced myelin and oligodendrocytic genes (MOBP, MOG, MAG, PLP, MBP) in schizophrenia and bipolar disorder; Reduced metabolic and increased oxidative stress genes	Tkachev et al. (2003); Prabakaran et al. (2004))
Prefrontal cortex (BA10)	15C; 13S; 11B; 11MDD	U95Av2	Reduced genes for receptors, channels, or transporters in bipolar disorder; Increased genes for stress response proteins or molecular chaperones in bipolar disorder. Reduced *SOX10* in schizophrenia; Increased *LARS2* in schizophrenia and bipolar disorder	Iwamoto et al. (2004); Yamada et al. (2005); Munakata et al. (2005)
Prefrontal Cortex/ Ant. Cingulate	7C; 6B; 9MDD	U133A	Reduced expression of FGF1 and 2, FGFR2 and 3 in MDD in DLPFC and ACC as well as increased expression of FGF9.	Evans et al. (2004)
Prefrontal cortex (BA47)	6C; 6S	cDNA array (1,373)	Reduced astrocytic and oligodendrocyte genes (PEA-15, S100, MAL, MBP, MOBP); Reduced growth/neurotrophic factors (BDNF, EDGF, TGF-a, trkB, erbB1)	Sugai et al. (2004)
Prefrontal cortex	19C; 19MDD	U133A	No detectable differences between depression and suicide and controls	Sibille et al. (2004)
Prefrontal cortex (BA46)	35C; 33S; 34B	U133A	Reduced mitochondrial-related genes in schizophrenia and bipolar disorder; medication-free bipolar patients showed increased expression	Iwamoto et al. (2005)

(continued)

Table 1.1 (continued)

Brain region	Sample	Platform	Main gene expression findings	Reference
Prefrontal cortex	27C; 19S	U133A	Increased expression of SELENBP1, BTG1, HNRPA3, and SFRS1; Reduced expression of GSK3A, HLA-DRB1.	Glatt et al. (2005)
Prefrontal cortex (BA46)	35S; 33S; 34B	Custom array	Increased expression of HML-2 in bipolar disorder and schizophrenia.	Frank et al. (2005)
Prefrontal cortex	35B, 35C	Custom array (19K)	Reduced mitochondrial/electron transport genes (ETC complex I, ETC Complex IV, ETC Complex V)	Sun et al. (2006)
Prefrontal cortex	14C; 14S	Custom array	Increased expression of genes related to immune and chaperone function (SERPINA3, IFITM1, IFITM2, IFITM3, CHI3L1, MT2A, CD14, HSPB1, HSPA1B, HSPA1A).	Arion et al. (2007)
Prefrontal cortex	14C; 14S	Custom array	Reduced expression of GABA related genes (GAD67, GABAT1, NPY, SST, CCK, GABAR subunits a1, a2, b3, g2, d)	Hashimoto et al. (2007)
Temporal cortex (BA21)	14C;12S; 12MDD	U95Av2	Reduced expression of Oligodendrocyte Genes (MAG, PLLP, PLP1, ErbB3); Reduced expression of neurodevelopmental genes (TRAF4, Neurod1, histone deacetylase-3) in schizophrenia; Reduced expression of oligodendrocytes related genes (CNP, MAG, MAL, MOG, MOBP, PMP22, PLLP, PLP1) in major depressive disorder.	(Aston et al., 2004, 2005)
Hippocampus	2C; 2S	cDNA array (8600)	Increased expression of chondrex, histamine releasing factor, HERC2 and HSP 70.	Chung et al. (2003)
Hippocampus	10C; 8S; 9B	U95Av2	Reduced expression of mictochondrial and energy metabolism related genes; Reduced expression of GAD67, SST	Konradi et al. (2004)
Entorhinal cortex	9C; 8S	cDNA array	Reduced expression of Gi 1, GluR3, NMDA1, synaptophysin, phospholemman	Hemby et al. (2002)
Multiple region	8,10C; 8,10S	Custom array (1127)	Reduced expression of synaptic and proteolytic related genes in the cerebellum and prefrontal cortex.	Vawter et al. (2001)

C Controls, *B* Bipolar Disorder, *MDD* Major Depressive Disorder

Abbreviations: *NSF*, *N*-ethylmaleimide sensitive fusion protein; *GABA*, gamma-amino butyric acid; *GAD67*, glutamate decarboxylase-67; *AMPA1-2*, 2-(aminomethyl phenylacetic acid; *GluR1-2*, glutamate receptor 1–2; *MAG*, myelin associated glycoprotein; *CNP*, cyclic nucleotide phosphodiesterase; *MAL*, myelin and lymphocyte protein; *HER3*, heregulin-3; *GAP-43* growth associated protein-43; *MARCKS*, myristolated alanine-rich C-kinase substrate; *TGF-beta 1*, Transforming growth factor beta 1;*erbB2*, erythroblastic leukemia viral oncogene homolog 2; *Apo*, apolipoprotein; *MOBP*, myelin-oligodendrocyte basic protein; *MOG*, myelin-oligodendrocytic protein; *PLP*, proteolipid protein; *MBP*, myelin basic protein; *SOX10*, SRY-related homeobox gene 10; *LARS2*, leucyl-tRNA synthetase, mitochondrial; *FGF*, fibroblast growth factor; *PEA-15*, phosphoprotein enriched in astrocytes 15; *S100*, calcium binding protein A1; *BDNF*, brain derived neurotrophic factor; *EDGF*, epidermal derived growth factor; *TGF-a*, transforming growth factor-alpha; *SELENBP1*, selenium binding protein-1; *BTG1*, B-cell translocation gene 1; *HNRPA3*, heterogeneous nuclear ribonucleoprotein A3; *SFRS1*, splicing factor, arginine/serine rich 1; *GSK3A*, glycogen synthase kinase 3 alpha; *HLA-DRB1*, major histocompatibility complex DR beta 1; *HML-2*, human macrophage lectin 2; *SERPINA3*, alpha-1-antichymotrypsin A3; *IFITM1*, interferon induced protein with tetratricopeptide repeats 1; *CHI3L1*, chitinase 3 like 1; *MT2A*, metallothionein 2A; *CD14*, cell differentiation 14; *HSPB1*, heat shock protein beta 1; *CCK*, cholecystokinin; *TRAF4*, tumor necrosis factor receptor associated factor 4; *Neurod1*, neurogenic differentiation 1; *PMP22*, peripheral myelin protein 22; *HERC2*, heat domain and rcc1-like domain 2; *Giα1*, g-protein inhibitory alpha1; *NMDA1*, *N*-methyl-D-aspartate 1.

activity demonstrated in this brain region in patients performing working memory tasks (Wible, Anderson et al., 2001; Achim and Lepage, 2005). This pattern suggests that, like the weak link in a chain, perhaps only one of several genes in a complex biological pathway needs to be "broken" (in this case, dysregulated in expression) to cause the entire pathway or system to malfunction. Yet, a potentially strong confounder of the mitochondrial gene expression findings observed in both bipolar disorder and schizophrenia may be agonal pH state. In a recent study, Vawter and colleagues demonstrated that 28% of genes related to mitochondrial function were differentially expressed between control brains at low and high pH, respectively (Vawter, Tomita et al., 2006).

Other interesting changes that have been observed in postmortem gene expression investigations in schizophrenia and bipolar disorder have been alterations in synaptic and synaptic-related genes as well as in genes encoding neurotransmitters and their associated proteins (Mirnics, Middleton et al., 2000; Hakak, Walker et al., 2001; Vawter, Barrett et al., 2001; Hemby, Ginsberg et al., 2002; Konradi, Eaton et al., 2004; Hashimoto, Arion et al., 2007). Of interest, these findings relate to reductions in synaptic proteins such as SNAP-25 and synaptobrevin-1 (VAMP-1) observed in postmortem PFC of schizophrenics (Young, Arima et al., 1998; Honer, Falkai et al., 1999; Fatemi, Earle et al., 2001) as well as reductions in GAD-67 (Guidotti, Auta et al., 2000; Volk, Austin et al., 2000; Woo, Walsh et al., 2004), a key enzyme involved in the synthesis of gamma aminobutyric acid (GABA). In relation to correlating changes in mRNA expression to previous protein alterations in postmortem neuropathological investigations, however, although translation of mRNA changes to protein adds credence to microarray investigations, inconsistencies in postmortem neuropathological data (including synaptic and/or neurotransmitter markers) are one of the main reasons behind investigators turning to microarray technology. Therefore, at the moment this could be a "catch 22" situation for data interpretation.

Finally, postmortem microarray investigations for major depressive disorders have been less numerous than for schizophrenia and bipolar disorder, but are beginning to increase in number. One interesting finding was that by Evans et al., demonstrating dysregulation of fibroblast growth factor genes (Evans, Choudary et al., 2004). Of direct correlation to these findings, we observed reductions in FGF2 and an upregulation of FGF9 in in vitro human fetal brain aggregates exposed to cortisol (Salaria, Chana et al., 2006). As hypercortisolemia is a defining feature in MDD, these findings may implicate a role for FGFs in MDD etiology. Further in vitro correlation to postmortem gene expression changes are necessary to better validate microarray findings.

1.5 Blood Gene Expression in Major Psychiatric Disorders

To date, the number of blood-based microarray investigations for major psychiatric disorders has been limited, as summarized in Table 1.2 . Half of these studies have focused on assessing gene expression changes in lymphoblastoid cells cultured from patients with schizophrenia and bipolar disorder. The earliest one of these

Table 1.2 Blood-based microarray investigations in major psychiatric disorders

Source	Sample	Platform	Main gene expression findings	Reference
Lymphoblastoid cells	1C; 2B	U95Av2	Reduced expression of ER stress response gene XBP1 in affected twins.	Kakiuchi et al. (2003)
Lymphoblastoid cells	9C; 5S	Custom Array (1127)	Reduced expression of NPY, GNA01, MDH1 in schizophrenia	Vawter et al. (2004)
Lymphoblastoid cells	2B	cDNA array (2400)	Increased expression of alpha1B-AR in bipolar prior to lithium treatment. Chronic lithium treatment reduced alpha1B-AR, ACHR, PDE4D, SPR in lithium responders.	Sun et al. (2006)
Peripheral blood sample	33S; 5B	U133A	Increased expression of NRG-1 variant SMDF in patients with schizophrenia.	Petryshen et al. (2005); Middleton et al. (2005)
Peripheral blood sample	10C; 13S	Custom array (3200)	Increased expression of DRD2 and Kir2.3 in drug-naïve schizophrenic subjects.	(Zvara et al., 2005)
Peripheral blood sample	24(17) C; 30 (30S); (7)B	Custom array (3200)	Increased expression of SELENBP1, GSK3alpha in schizophrenia; Reduced expression of BTG1, HLADRB1, HNRPA3, SFRS1 in schizophrenia.	Glatt et al. (2005); Tsuang et al., (2005)
Peripheral blood sample	21 C; 21B	U133A 2.0	Reduced expression of electron transfer chain genes in bipolar cultured lymphocytes following glucose deprivation vs. controls with increased expression.	Naydenov et al. (2007)

Abbreviations: *ER*, endoplasmic reticulum; *XBP1*, X box binding protein 1; *NPY*, neuropeptide Y; *GNA01*, guanine nucleotide binding protein alpha 1; *MDH1*, malate dehydrogenase 1; *ACHR*, acetylcholine receptor; *PDE4D*, phosphodiesterase 4D; *SPR*, sepiaterin reductase; *NRG-1*, neuregulin-1; *SMDF*, sensory and motor neuron derived factor; *DRD2*, dopamine receptor D2; *Kir2.3*, potassium channel inward rectifying 2.3

studies was that by Kakiuchi et al. in 2003. In this investigation they used the Affymetrix U95Av2 chip to assess gene expression in a pair of controls vs. two pairs of twins discordant for bipolar disorder. They demonstrated reduction in expression of XBP1 in the affected member of a twin pair discordant for bipolar disorder. XBP1 is a pivotal gene involved in the endoplasmic reticulum (ER) stress response. They further found that a polymorphism substitution at position 116 (C–G) in the promoter region of XBP1 lymphoblasts derived from Japanese bipolar patients conferred a reduced ER stress response that was rescued by treatment with the mood stabilizer valproate (Kakiuchi, Iwamoto et al., 2003). More recently, this group extended this finding by demonstrating that lithium treatment was more effective in patients with the 116C allele as opposed to patients homozygous for 116G (Kakiuchi, Ishiwata et al., 2006). While this finding is of interest as it demonstrates a polymorphism and expression change linked to treatment response, it requires validation in separate, independent patient cohorts. In the largest microarray study of lymphoblastoid cells so far by Vawter et al. (2004), it was found that the expression of NPY and GNA01 were reduced in cells derived from patients with schizophrenia. They also demonstrated an increase in the mitochondrial-related gene MDH1 in schizophrenics (Vawter, Ferran et al., 2004). Changes in NPY have been observed by some postmortem investigations in schizophrenia and bipolar disorder (Frederiksen, Ekman et al., 1991; Iritani, Kuroki et al., 2000; Kuromitsu, Yokoi et al., 2001). Nevertheless, sample numbers were relatively small in this blood-based investigation with only five schizophrenic and nine control subjects and therefore these results require replication in larger cohorts.

Recent peripheral blood microarray findings have implicated a number of genes in schizophrenia and bipolar disorder. Of the genes implicated, SMDF, a splice variant of NRG-1 has the strongest correlation with previous genetic or neuropathological research. Numerous genetic association studies have associated the gene for NRG1 with schizophrenia susceptibility; for a review see Harrison and Weinberger, 2005. More recently, evidence for reduced NRG1 levels in the PFC of brains of patients with schizophrenia and unipolar depression has also emerged (Bertram, Bernstein et al., 2007). While some studies have failed to demonstrate (Thiselton, Webb et al., 2004; Duan, Martinez et al., 2005; Rosa, Gardner et al., 2007) association of NRG1 with schizophrenia susceptibility, recent meta-analyses have supported a relationship (Li, Collier et al., 2006; Munafo, Thiselton et al., 2006). In a comparative blood and brain gene expression study carried out by the authors of this chapter, it was demonstrated that the gene for SELENBP1 was upregulated in both compartments in schizophrenia (Glatt, Everall et al., 2005). In a follow-up QPCR study looking at a separate brain cohort, our group also observed increased SELENBP1 expression in DLPFC of patients with schizophrenia. SELENBP1 mRNA was significantly increased by 12% in psychosis; the clinical group found a 11% increase in schizophrenia and a 14% increase in psychotic bipolar cases compared to nonpsychotic bipolar subjects and controls which did not differ. However, while increased expression in SELENBP1 in the DLPFC in schizophrenia may be present, the direct functional significance of this up-regulation remains to be elucidated. One potential explanation may be related to its role as a neurogenic factor that has been evidenced by two investigations demonstrating its ability to promote neurite outgrowth in the rat cerebral cortex (Zhao, Nair et al., 2000) as well as colocalization with actin in growing tips of human-derived neuronal SY5Y cells (Miyaguchi, 2004). Given that reductions in synaptic and dendritic arbors have been observed in schizophrenia, elevated SELENBP1 may play a compensatory role in restoring neuronal connectivity and functioning. Further work to better define this tentative mechanism is required.

Finally, a recent study by Naydenov et al. demonstrated a reduction in expression of electron transport chain genes in lymphocytes isolated from patients with bipolar disorder following glucose deprivation (Naydenov, MacDonald et al., 2007). This reduction was in contrast to an upregulation of these components seen in controls. This finding may be of significance, given the reductions in mitochondrial and energy-related genes in postmortem microarray investigations of schizophrenia and bipolar disorder.

1.6 Methodological Considerations for Microarray Investigations in Psychiatry

An inherent difficulty in conducting microarray investigations in the brain is the phenotypic diversity in cell types and between and across various brain regions (Mirnics, Levitt et al., 2006). While many of the studies listed in Table 1.1 define Brodmann's areas for sampling, the proportion of neurons and glia that make up the final constituent cells for RNA extraction remain largely unknown. As gene expression between neurons and glial cells is likely to be different and related to their various functional roles, differential sampling of cells in postmortem microarray investigations would lead to gene expression differences associated with the various proportions of cells present. A way around this problem may be via the use of laser capture microdissection (LCM) to cut out individual cell types and analyze their gene expression separately. However, the use of LCM to harvest a significant number of cells for this purpose is labor intensive. Furthermore, dissection of individual cells from brain tissue will lead to sampling of some of the neuropil, which will also confound findings. Nevertheless, with the development of more automated techniques for LCM and increased precision, this method presents a way to assess gene expression changes in specific brain cell populations in major psychiatric disorders. This may also extend to assessment of subpopulations of neurons and glia, i.e., assessment of gene expression changes in interneuronal populations in BA9. Other challenges that face brain microarray studies include (1) variability introduced by genetic diversity, (2) effects of disease treatment on gene expression, (3) differential diagnoses, (4) comorbidity with other disorders, (5) variation in age, PMI, pH, and drug abuse between groups in a cohort, (6) limited sample sizes with a limited number of samples yielding high-quality RNA for investigation, and (7) variability in platform types and methods for hybridization (Mirnics, Levitt et al., 2006).

Apart from (4) and (5) all these challenges are also faced by blood-based gene expression studies. In addition, lifestyle factors also significantly impact blood gene expression. Factors such as diet, exercise, smoking, and time of last meal can all affect gene expression in the blood and hence matching and normalization for these factors where possible should be a standard consideration along with the other factors mentioned earlier when designing such studies. Just as with brain microarray investigations, blood contains several different cell types as its constituent components. These cell types fall primarily into three categories: erythrocytes, leukocytes and thrombocytes. With leukocytes making up the immune component of blood, focus has been cast on assessing gene expression changes within this subcategory of blood cells. However, even within this category, several cell types exist: neutrophils, eosinophils, basophils, lymphocytes, monocytes, and macrophages. This inherent variability and difference in the functional roles of leukocytes is no doubt reflected in differences in their gene expression profiles. Furthermore, on the basis of the immune status of an individual, different proportions of these cell types may be present. While this significant variability in cell types of the blood paints a rather bleak picture for gleaning information on gene expression changes in patients and identifying biomarkers for disease, specific patterns of gene expression can be seen in blood samples of patients with schizophrenia and bipolar disorder and controls (Tsuang, Nossova et al., 2005; Bowden, Weidenhofer et al., 2006). In an attempt to circumnavigate problems with cell variability in blood samples, some researchers have chosen to transform and culture B-lymphocytes into lymphoblastoid cell lines using Epstein Barr virus prior to microarray analysis (see Table 1.2). While these studies have the advantage of assessing gene expression effects in a relatively homogenous cell population free from the temporal state of individuals, effects of exposure to virus and chromosomal alterations during culture must be considered as a confounder of these results (Iwamoto and Kato, 2006).

1.6.1 Chips and Fish?: Cross-Platform Variability and Data Analysis

A significant problem with microarrays may lie in cross-platform variability and more so in the variability in the tools for analysis used by investigators. While more recent studies have demonstrated better correlation between platforms, even a 10% divergence can lead to differences in thousands of transcripts between platforms. Linked with this drift between platforms is the relative insensitivity of microarray platforms to detect changes in the expression of genes that are present at low abundance. A good illustration of this is given by Mirnics et al. in a recent review of microarrays in psychiatry: if a cross-hybridization of 2% exists between gene X and gene Y, and X is expressed at 10 copies/cell and Y at 1000 copies/cell then 65% of the observed target X will originate from Y. This example illustrates the need for validation of microarray results by techniques such as QPCR and in situ hybridization, with QPCR having a significantly increased sensitivity over microarrays. However, the ability to detect subtle changes in transcripts of low abundance will also depend on the length of the oligonucleotide probes present on microarray chips. This is a potential consideration in RNA extraction methodology as well as in choosing a platform for analysis, as some, such as Codelink, use longer 60–70mer probes as opposed to Affymetrix who uses shorter 20–30mer probes on their chips.

Data analysis and data mining of microarrays also represent areas for potential error in generating false-positive and false-negative results. Analyses of chip image files are generally completed with programs such as robust multichip analysis algorithim (RMA) and then subjected to ontological analysis by a variety of different bioinformatics tools all utilizing the gene ontology (GO) database (http://www.geneontology.org). An area of subjectivity in microarray analysis comes from the thresholds used to filter gene lists. To date, many studies have utilized fold change cut-offs to filter gene expression lists. Unfortunately, consistency between selecting fold changes has not been seen between studies, with selection of cut-offs often being based on the data generated or potential genes of interest. Given that major psychiatric disorders are likely to have complex genetic etiologies and with a lack of reliable candidate genes identified so far, this method of arbitrary filtering may lead to generation of false-positive and false-negative results. Ontological programs such as GoSurfer, GoMiner, OntoExpress, DAVID, and Metacore essentially carry out similar analyses, separating gene lists according to biological, cellular, and molecular classification. However, variations in filtering and in generating networks of association also make this step of analysis open to a level of subjectivity.

1.7 Future Directions for Gene Expression Analysis in Psychiatry

Over the last decade microarray investigations assessing gene expression for major psychiatric disorders have greatly increased our power to detect candidate genes for the etiologies of these illnesses. Nevertheless, significant problems exist in terms of methodological and technical issues with microarrays which require amendment of current protocols. Probably, the biggest problem facing microarray studies is the need for standardization. This includes standardization in diagnosis of disorders, collection procedures of samples (for both postmortem and blood-based studies), platform design, laboratory techniques, and tools used for analysis. Many problems exist within these individual categories, the details of which have been elucidated in previous sections and elsewhere (Bunney, Bunney et al., 2003; Iwamoto and Kato, 2006; Mirnics, Levitt et al., 2006). Problems associated with the differential diagnosis of major psychiatric disorders are a consequence of overlapping phenotypes between these disorders. With the lack of consistent biomarkers for major psychiatric disorders, separation of these disorders will be a difficult task for the psychiatrist. Therefore, the solution to this problem itself may come at the end of the evolution of microarray investigations in the blood, whereby enough data has been generated to reliably identify biomarkers and hence aid differentiation of disorders. This process may also involve several iterations, with present diagnoses facilitating the identification of genes of relevance to select traits, and novel classifications of traits emerging from unsupervised classifications of similar gene expression patterns.

With regard to collection of samples for microarray investigations, approaches for the collection of brain material for postmortem study and RNA extraction are well established and researchers should look to mine the existing databases prior to processing samples. For blood-based investigations, a number of different methods exist for the isolation of leukocytes and extraction of RNA, with some of these methodologies being validated in terms of RNA quantity and quality extracted. The significant contribution of gene expression differences in subtypes of cells requires addressing. Isolation of specific lymphocyte types may be achievable by the use of techniques such as immunolabeling with beads for specific cell-surface markers and separation through electromagnetic columns. However, such techniques are relatively inefficient ways of yielding sufficient numbers of layers for a representative microarray investigation. Better techniques for isolating these individual cell populations will lead to greater sensitivity to detect gene expression changes in the blood and brain of patients with major psychiatric disorders.

Another confounding variable touched upon earlier is the effect of medications on gene expression changes. These effects manifest in both blood and brain compartments and require consideration and inclusion of relevant controls in microarray investigations in psychiatry. In studies generating lymphoblastoid cells from patients, treatment of these cells with atypical antipsychotics followed by assessment of gene expression effects will no doubt help in separating disease-related and medication-related genes. An alternative approach to assessing changes in gene expression associated with antipsychotic medications is via in vitro and in vivo investigation. Indeed, some studies have already demonstrated alteration in transcripts following exposure of rats and cell lines to typical and atypical antipsychotics (Kontkanen, Lakso et al., 2000; Kontkanen, Lakso et al., 2002; Takahashi, Kumanishi et al., 2004; Chen and Chen, 2005). Using an in vitro human brain aggregate model exposed to haloperidol and clozapine, we have recently demonstrated that clozapine has a much greater ability to alter gene expression than haloperidol. While similar neurotrophic and synaptic genes were positively regulated by both compounds, the effects were more pronounced for clozapine. With reductions in synaptic proteins appearing as some of the most consistent postmortem brain changes observed in schizophrenia, increases in the genes encoding these proteins by clozapine provides a potential long-term mechanism of therapeutic action for the atypical antipsychotics (Chana, 2007). Further in vitro and blood-based expression studies are needed to determine the influence of current atypicals that affect many neurotransmitter systems.

1.8 Conclusion

Microarray studies in psychiatry have great potential for elucidating etiological mechanisms. With the advent of blood-based investigations to assess peripheral correlates of brain dysfunction in patients, we have moved a step closer to identifying potential biomarkers. However, care must be taken in the initial stages of these studies to avoid overinterpretation of such generated data, with potential confounders and problems still being numerous. By implementing careful study designs, many of these potential confounders can be limited. Further still, separation of specific cell populations for study, including both blood and brain investigations, will allow us to glean more knowledge as to the mechanism of gene dysregulation within major psychiatric disorders. The field of microarray platform design and development is also moving very fast. With the availability of newer platform types the ability to detect splice variants of transcripts as well as incorporating single nucleotide polymorphism (SNP) analysis, a more detailed understanding of gene-based biomarkers can be achieved. This can also then be related to other epigenetic mechanisms with the potential to influence gene expression changes. This increase in data retrieval from both blood and brain compartments will allow us to not only understand gene expression changes but also what in-built mechanisms may be regulating them. The ability to mine and extract useful data, however, requires standardization in analysis techniques and the free distribution of data to all researchers. An example of this can be seen in the Stanley Foundation's online genomics database (https://www.stanleygenomics.org/), where study results from numerous investigators and platforms have been summarized and made available, or the NCBI's Gene Expression Omnibus repositories. Lastly, analyzing the transcriptome using microarrays must be coupled to other advancing fields such as proteomics and metabolomics and findings related across these disciplines. This strategy will no doubt greatly facilitate our search for biomarkers in psychiatry.

Acknowledgments Thanks are due to Dr Sharon Doc Chandler for her assistance in revising this chapter.

Reference

Achim, A. M. and M. Lepage (2005). "Episodic memory-related activation in schizophrenia: meta-analysis." Br J Psychiatr 187: 500–9.

Aston, C., L. Jiang, et al. (2004). "Microrray analysis of postmortem temporal cortex from patients with schizophrenia." J Neurosci Res 77(6): 858–66.

Aston, C., L. Jiang, et al. (2005). "Transcriptional profiling reveals evidence for signaling and oligodendroglial abnormalities in the temporal cortex from patients with major depressive disorder." Mol Psychiatr 10(3): 309–22.

Bezchlibnyk YB, Wang JF, McQueen GM, Young LT. Gene expression differences in bipolar disorder revealed by cDNA array analysis of post-mortem frontal cortex. J Neurochem. 2001 Nov;79(4):826–34.

Bertram, I., H. G. Bernstein, et al. (2007). "Immunohistochemical evidence for impaired neuregulin-1 signaling in the prefrontal cortex in schizophrenia and in unipolar depression." Ann N Y Acad Sci 1096: 147–56.

Bosotti, R., G. Locatelli, et al. (2007). "Cross platform microarray analysis for robust identification of differentially expressed genes." BMC Bioinformatics 8(Suppl 1): S5.

Bowden, N. A., J. Weidenhofer, et al. (2006). "Preliminary investigation of gene expression profiles in peripheral blood lymphocytes in schizophrenia." Schizophr Res 82(2–3): 175–83.

Bunney, W. E., B. G. Bunney, et al. (2003). "Microarray technology: a review of new strategies to discover candidate vulnerability genes in psychiatric disorders." Am J Psychiatr 160(4): 657–66.
Chana, G. L. J. E. I. P. (2007). "In vitro Gene Expression Changes Associated with Haloperidol and Clozapine in Human Fetal Brain Aggregates." Biol Psychiatr 61(1S–266S): 205.
Chen, M. L. and C. H. Chen (2005). "Microarray analysis of differentially expressed genes in rat frontal cortex under chronic risperidone treatment." Neuropsychopharmacology 30(2): 268–77.
Chung C, Tallerico T, Seeman P. Schizophrenia hippocampus has elevated expression of chondrex glycoprotein gene. Synapse. 2003 Oct;50(1)29–34.
Davis, K. L., D. G. Stewart, et al. (2003). "White matter changes in schizophrenia: evidence for myelin-related dysfunction." Arch Gen Psychiatr 60(5): 443–56.
Duan, J., M. Martinez, et al. (2005). "Neuregulin 1 (NRG1) and schizophrenia: analysis of a US family sample and the evidence in the balance." Psychol Med 35(11): 1599–610.
Evans, S. J., P. V. Choudary, et al. (2004). "Dysregulation of the fibroblast growth factor system in major depression." Proc Natl Acad Sci U S A 101(43): 15506–11.
Fan, J. B., M. S. Chee, et al. (2006). "Highly parallel genomic assays." Nat Rev Genet 7(8): 632–44.
Fatemi, S. H., J. A. Earle, et al. (2001). "Altered levels of the synaptosomal associated protein SNAP-25 in hippocampus of subjects with mood disorders and schizophrenia." Neuroreport 12(15): 3257–62.
Flynn, S. W., D. J. Lang, et al. (2003). "Abnormalities of myelination in schizophrenia detected in vivo with MRI, and post-mortem with analysis of oligodendrocyte proteins." Mol Psychiatr 8(9): 811–20.
Frank O, Giehl M, Zheng C, Hehlmann R, Leib-Mösch C, Seifarth W. Human endogenous retrovirus expression profiles in samples from brains of patients with schizophrenia and bipolar disorders. J Virol. 2005 Sep;79(17)10890–901.
Frederiksen, S. O., R. Ekman, et al. (1991). "Reduced concentrations of galanin, arginine vasopressin, neuropeptide Y and peptide YY in the temporal cortex but not in the hypothalamus of brains from schizophrenics." Acta Psychiatr Scand 83(4): 273–7.
Glatt, S. J., I. P. Everall, et al. (2005). "Comparative gene expression analysis of blood and brain provides concurrent validation of SELENBP1 up-regulation in schizophrenia." Proc Natl Acad Sci U S A 102(43): 15533–8.
Goldberg, T. E., D. R. Weinberger, et al. (1987). "Further evidence for dementia of the prefrontal type in schizophrenia? A controlled study of teaching the Wisconsin Card Sorting Test." Arch Gen Psychiatr 44(11): 1008–14.
Goldman-Rakic, P. S. (1994). "Working memory dysfunction in schizophrenia." J Neuropsychiatr Clin Neurosci 6(4): 348–57.
Guidotti, A., J. Auta, et al. (2000). "Decrease in reelin and glutamic acid decarboxylase67 (GAD67) expression in schizophrenia and bipolar disorder: a postmortem brain study." Arch Gen Psychiatr 57(11): 1061–9.
Hakak, Y., J. R. Walker, et al. (2001). "Genome-wide expression analysis reveals dysregulation of myelination-related genes in chronic schizophrenia." Proc Natl Acad Sci U S A 98(8): 4746–51.
Hardiman, G. (2004). "Microarray platforms--comparisons and contrasts." Pharmacogenomics 5(5): 487–502.
Harrison, P. J. and D. R. Weinberger (2005). "Schizophrenia genes, gene expression, and neuropathology: on the matter of their convergence." Mol Psychiatr 10(1): 40–68; image 5.
Hashimoto, T., D. Arion, et al. (2007). "Alterations in GABA-related transcriptome in the dorsolateral prefrontal cortex of subjects with schizophrenia." Mol Psychiatr.
Hemby, S. E., S. D. Ginsberg, et al. (2002). "Gene expression profile for schizophrenia: discrete neuron transcription patterns in the entorhinal cortex." Arch Gen Psychiatr 59(7): 631–40.
Hof, P. R., V. Haroutunian, et al. (2002). "Molecular and cellular evidence for an oligodendrocyte abnormality in schizophrenia." Neurochem Res 27(10): 1193–200.

Hollingshead, D., D. A. Lewis, et al. (2005). “Platform influence on DNA microarray data in postmortem brain research.” Neurobiol Dis 18(3): 649–55.

Honer, W. G., P. Falkai, et al. (1999). “Synaptic and plasticity-associated proteins in anterior frontal cortex in severe mental illness.” Neuroscience 91(4): 1247–55.

Iritani, S., N. Kuroki, et al. (2000). “Morphological changes in neuropeptide Y-positive fiber in the hippocampal formation of schizophrenics.” Prog Neuropsychopharmacol Biol Psychiatr 24(2): 241–9.

Iwamoto, K. and T. Kato (2006). “Gene expression profiling in schizophrenia and related mental disorders.” Neuroscientist 12(4): 349–61.

Iwamoto, K., M. Bundo, et al. (2005). “Altered expression of mitochondria-related genes in postmortem brains of patients with bipolar disorder or schizophrenia, as revealed by large-scale DNA microarray analysis.” Hum Mol Genet 14(2): 241–53.

Iwamoto K, Bundo M, Yamada K, Takao H, Iwayama-Shigeno Y, Yoshikawa T, Kato T. DNA methylation status of SOX10 correlates with its downregulation and oligodendrocyte dysfunction in schizophrenia. J Neurosci. 2005 Jun 1;25(22):5376–81.

Kakiuchi, C., K. Iwamoto, et al. (2003). “Impaired feedback regulation of XBP1 as a genetic risk factor for bipolar disorder.” Nat Genet 35(2): 171–5.

Kakiuchi, C., M. Ishiwata, et al. (2006). “XBP1 induces WFS1 through an endoplasmic reticulum stress response element-like motif in SH-SY5Y cells.” J Neurochem 97(2): 545–55.

Konradi, C. (2005). “Gene expression microarray studies in polygenic psychiatric disorders: applications and data analysis.” Brain Res Brain Res Rev 50(1): 142–55.

Konradi, C., M. Eaton, et al. (2004). “Molecular evidence for mitochondrial dysfunction in bipolar disorder.” Arch Gen Psychiatr 61(3): 300–8.

Kontkanen, O., M. Lakso, et al. (2000). “A functional genomic study of the effects of ****antipsychotic agent chlorpromazine in PC12 cells.” 38(9): 911–915.

Kontkanen, O., M. Lakso, et al. (2002). “Chronic antipsychotic drug treatment induces long-lasting expression of fos and jun family genes and activator protein 1 complex in the rat prefrontal cortex.” Neuropsychopharmacology 27(2): 152–162.

Kuromitsu, J., A. Yokoi, et al. (2001). “Reduced neuropeptide Y mRNA levels in the frontal cortex of people with schizophrenia and bipolar disorder.” Brain Res Gene Expr Patterns 1(1): 17–21.

Li, D., D. A. Collier, et al. (2006). “Meta-analysis shows strong positive association of the neuregulin 1 (NRG1) gene with schizophrenia.” Hum Mol Genet 15(12): 1995–2002.

Michael, K. L., L. C. Taylor, et al. (1998). “Randomly ordered addressable high-density optical sensor arrays.” Anal Chem 70(7): 1242–8.

Middleton FA, Peng L, Lewis DA, Levitt P, Mirnics K. Altered expression of 14-3-3 genes in the prefrontal cortex of subjects with schizophrenia. Neuropsychopharmacology. 2005 May;30(5) 974–83.

Mirnics, K., F. A. Middleton, et al. (2000). “Molecular characterization of schizophrenia viewed by microarray analysis of gene expression in prefrontal cortex.” Neuron 28(1): 53–67.

Mirnics, K., P. Levitt, et al. (2006). “Critical appraisal of DNA microarrays in psychiatric genomics.” Biol Psychiatr 60(2): 163–76.

Miyaguchi, K. (2004). “Localization of selenium-binding protein at the tips of rapidly extending protrusions.” Histochem Cell Biol 121(5): 371–6.

Munafo, M. R., D. L. Thiselton, et al. (2006). “Association of the NRG1 gene and schizophrenia: a meta-analysis.” Mol Psychiatr 11(6): 539–46.

Munakata K, Iwamoto K, Bundo M, Kato T. Mitochondrial DNA 3243A>G mutation and increased expression of LARS2 gene in the brains of patients with bipolar disorder and schizophrenia. Biol Psychiatry. 2005 Mar 1;57(5)525–32.

Naydenov, A. V., M. L. MacDonald, et al. (2007). “Differences in lymphocyte electron transport gene expression levels between subjects with bipolar disorder and normal controls in response to glucose deprivation stress.” Arch Gen Psychiatr 64(5): 555–64.

Petryshen TL, Middleton FA, Kirby A, Aldinger KA, Purcell S, Tahl AR, Morley CP, McGann L, Gentile KL, Rockwell GN, Medeiros HM, Carvalho C, Macedo A, Dourado A, Valente J,

Ferreira CP, Patterson NJ, Azevedo MH, Daly MJ, Pato CN, Pato MT, Sklar P. Support for involvement of neuregulin 1 in schizophrenia pathophysiology. Mol Psychiatry. 2005 Apr;10(4) 366–74, 328.

Pongrac, J., F. A. Middleton, et al. (2002). "Gene expression profiling with DNA microarrays: advancing our understanding of psychiatric disorders." Neurochem Res 27(10): 1049–63.

Prabakaran, S., J. E. Swatton, et al. (2004). "Mitochondrial dysfunction in schizophrenia: evidence for compromised brain metabolism and oxidative stress." Mol Psychiatr 9(7): 684–97, 643.

Radich, J. P., M. Mao, et al. (2004). "Individual-specific variation of gene expression in peripheral blood leukocytes." Genomics 83(6): 980–8.

Rosa, A., M. Gardner, et al. (2007). "Family-based association study of neuregulin-1 gene and psychosis in a Spanish sample." Am J Med Genet B Neuropsychiatr Genet.

Salaria, S., G. Chana, et al. (2006). "Microarray analysis of cultured human brain aggregates following cortisol exposure: implications for cellular functions relevant to mood disorders." Neurobiol Dis 23(3): 630–6.

Schlingemann, J., N. Habtemichael, et al. (2005). "Patient-based cross-platform comparison of oligonucleotide microarray expression profiles." Lab Invest 85(8): 1024–39.

Shippy, R., T. J. Sendera, et al. (2004). "Performance evaluation of commercial short-oligonucleotide microarrays and the impact of noise in making cross-platform correlations." BMC Genom 5(1): 61.

Sibille E, Arango V, Galfalvy HC, Pavlidis P, Erraji-Benchekroun L, Ellis SP, John Mann J. Gene expression profiling of depression and suicide in human prefrontal cortex. Neuropsychopharmacology. 2004 Feb;29(2)351–61.

Stewart, D. G. and K. L. Davis (2004). "Possible contributions of myelin and oligodendrocyte dysfunction to schizophrenia." Int Rev Neurobiol 59: 381–424.

Sugai T, Kawamura M, Iritani S, Araki K, Makifuchi T, Imai C, Nakamura R, Kakita A, Takahashi H, Nawa H. Prefrontal abnormality of schizophrenia revealed by DNA microarray impact on glial and neurotrophic gene expression. Ann N Y Acad Sci. 2004 Oct;1025: 84–91.

Sun, X., J. F. Wang, et al. (2006). "Downregulation in components of the mitochondrial electron transport chain in the postmortem frontal cortex of subjects with bipolar disorder." J Psychiatry Neurosci 31(3): 189–96.

Takahashi, Y., T. Kumanishi, et al. (2004). "Using a DNA microarray method to examine gene expression in brain from clozapine-injected mice." Ann N Y Acad Sci 1025: 561–9.

Tan, P. K., T. J. Downey, et al. (2003). "Evaluation of gene expression measurements from commercial microarray platforms." Nucleic Acids Res 31(19): 5676–84.

Thiselton, D. L., B. T. Webb, et al. (2004). "No evidence for linkage or association of neuregulin-1 (NRG1) with disease in the Irish study of high-density schizophrenia families (ISHDSF)." Mol Psychiatr 9(8): 777–83; image 729.

Tkachev, D., M. L. Mimmack, et al. (2003). "Oligodendrocyte dysfunction in schizophrenia and bipolar disorder." Lancet 362(9386): 798–805.

Tkachev, D., M. L. Mimmack, et al. (2007). "Further evidence for altered myelin biosynthesis and glutamatergic dysfunction in schizophrenia." Int J Neuropsychopharmacol 10(4): 557–63.

Tsuang, M. T., N. Nossova, et al. (2005). "Assessing the validity of blood-based gene expression profiles for the classification of schizophrenia and bipolar disorder: a preliminary report." Am J Med Genet B Neuropsychiatr Genet 133(1): 1–5.

Uranova, N., D. Orlovskaya, et al. (2001). "Electron microscopy of oligodendroglia in severe mental illness." Brain Res Bull 55(5): 597–610.

Uranova, N. A., V. M. Vostrikov, et al. (2004). "Oligodendroglial density in the prefrontal cortex in schizophrenia and mood disorders: a study from the Stanley Neuropathology Consortium." Schizophr Res 67(2–3): 269–75.

Vawter, M. P., T. Barrett, et al. (2001). "Application of cDNA microarrays to examine gene expression differences in schizophrenia." Brain Res Bull 55(5): 641–50.

Vawter, M. P., E. Ferran, et al. (2004). "Microarray screening of lymphocyte gene expression differences in a multiplex schizophrenia pedigree." Schizophr Res 67(1): 41–52.

Vawter, M. P., H. Tomita, et al. (2006). "Mitochondrial-related gene expression changes are sensitive to agonal-pH state: implications for brain disorders." Mol Psychiatr 11(7): 615, 663–79.

Volk, D. W., M. C. Austin, et al. (2000). "Decreased glutamic acid decarboxylase67 messenger RNA expression in a subset of prefrontal cortical gamma-aminobutyric acid neurons in subjects with schizophrenia." Arch Gen Psychiatr 57(3): 237–45.

Wang, Y., C. Barbacioru, et al. (2006). "Large scale real-time PCR validation on gene expression measurements from two commercial long-oligonucleotide microarrays." BMC Genom 7: 59.

Wible, C. G., J. Anderson, et al. (2001). "Prefrontal cortex, negative symptoms, and schizophrenia: an MRI study." Psychiatr Res 108(2): 65–78.

Wolf, D. H., R. C. Gur, et al. (2007). "Alterations of fronto-temporal connectivity during word encoding in schizophrenia." Psychiatr Res 154(3): 221–32.

Woo, T. U., J. P. Walsh, et al. (2004). "Density of glutamic acid decarboxylase 67 messenger RNA-containing neurons that express the N-methyl-D-aspartate receptor subunit NR2A in the anterior cingulate cortex in schizophrenia and bipolar disorder." Arch Gen Psychiatr 61(7): 649–57.

Yamada K. (2005). "Algorithm for the treatment of rapid cycling. "Psychiatry Clin Neurosci 53 Suppl S73–5.

Young, C. E., K. Arima, et al. (1998). "SNAP-25 deficit and hippocampal connectivity in schizophrenia." Cereb Cortex 8(3): 261–8.

Zhao, Z., S. M. Nair, et al. (2000). "Expression and role of sulfoglucuronyl (HNK-1) carbohydrate and its binding protein SBP-1 in developing rat cerebral cortex." J Neurosci Res 62(2): 186–205.

Zvara A, Szekeres G, Janka Z, Kelemen JZ, Cimmer C, Sántha M, Puskás LG. Over-expression of dopamine D2 receptor and inwardly rectifying potassium channel genes in drug-naive schizophrenic peripheral blood lymphocytes as potential diagnostic markers. Dis Markers. 2005;21(2):61-9.

Chapter 2
Biomarkers in Schizophrenia

Laura A. Feldcamp and Albert H.C. Wong(✉)

Abstract Schizophrenia is a heterogeneous disorder presenting as episodes of psychosis against a background of cognitive, social, and functional impairments. Schizophrenia has been studied extensively, and a large number of biological abnormalities associated with the disorder have been described. Many of these abnormalities have been proposed as biomarkers, some of which may represent useful endophenotypes for dissecting the etiology or pathophysiology of schizophrenia. The main molecular, electrophysiological, imaging, and psychological features of schizophrenia are described, with a critical evaluation of their utility as diagnostic and endophenotypic biomarkers. While none of these biomarkers are useful at present for clinical diagnosis, they may identify subgroups of schizophrenia or represent dimensions of the illness that can be subject to further study. With the recent identification of several promising candidate susceptibility genes, models of pathophysiology can now be generated to integrate this diverse collection of abnormalities from both molecular and abnormal psychology perspectives.

Abbreviations AS: Antisaccade; BDNF: Brain derived neurotrophic factor; CAPON: Also known as nitric oxide synthase 1(neuronal) adaptor protein (NOS1AP); CHRNA7: Cholinergic receptor, nicotinic, alpha 7; CI: Cortical inhibition; CNS: Central nervous system; COMT: Catechol-*o*-methyltransferase; CSF: Cerebrospinal fluid; CT: Catscan; DISC1: Disrupted in schizophrenia 1; DNA: Deoxyribonucleic acid; DPFC: Dorsolateral prefrontal cortex; DTNBP1: Dysbindin-1, dystrobrevin-binding protein 1; EGF: Epidermal growth factor; EPN4: Epsin-related protein, clathrin interactor 1; ErbB: Epidermal growth factor receptor; ERP: Event-related potential; fMRI: Functional MRI; G72/DOA/DAOA: D-amino acid oxidase,

A. H. C. Wong
Centre for Addiction and Mental Health, 250 College Street, Room 711, Toronto, ON, M5T 1R8, Canada
albert.wong@utoronto.ca

C.W. Turck (ed.) *Biomarkers for Psychiatric Disorders*,
doi: 10.1007/978-0-387-79251-4_2, © Springer Science+Business Media, LLC 2008

D-amino acid oxidase activator; GABA: γ-Aminobutyrate; GAD67: 67-kDa isoform of glutamic acid decarboxylase; GRM3: Glutamate receptor, metabotropic 3; GS: Gamma synchrony; IBZM: Iodobenzamide; IL: Interleukin; INFγ: Interferon gamma; MEG: Magnetoencephalography; MMN: Mismatch negativity; MRI: Magnetic resonance imaging; mRNA: Messenger ribonucleic acid; NRG1: Neuregulin 1; PET: Positron emission tomography; PFC: Prefrontal cortex; PPI: Prepulse inhibition; PPP3C: Protein phosphatase 3 (formerly 2B), catalytic subunit, gamma isoform; PRODH: Proline dehydrogenase (proline oxidase); RGS4: Regulator of G-protein signaling 4; SELENBP1: Selenium binding protein 1; SNP: Single nucleotide polymorphism; SPECT: Single positron emission computed tomography; TAAR6: Trace amine associated receptor 6; TMS: Transcranial magnetic stimulation; TNFα: Tumour necrosis factor alpha; TOM: Theory of mind

2.1 Introduction

2.1.1 Schizophrenia

Schizophrenia is a chronic brain disorder characterized by symptoms including hallucinations, delusions, thought disorder, and blunted effect, as well as impaired cognition, insight and social interaction (Wong and Van Tol, 2003). Although the prevalence has typically been cited as being 0.5–1% of the population worldwide (Sartorius et al., 1986), more recent evidence reveals substantial geographical variation, with incidence ranging from 7.7–43 per 100,000 (median 15.2/100,000) (McGrath et al., 2004), and a median lifetime prevalence of 4 per 1000 (Saha et al., 1994). Diagnostic clinical features typically appear in the late second to third decade of life, with average age of onset ~5 years earlier in males than in females, and an overall male:female ratio of 1.4 (McGrath, 2007). Schizophrenia is a complex disease with a polygenic, non-Mendelian pattern of inheritance, with heritability estimates as high as 80% (Cardno and Gottesman, 2000). The rate of concordance among monozygotic twins is approximately 50% (Gottesman, 1991), suggesting that a combination of environmental, developmental, stochastic, and epigenetic factors may contribute to the disease susceptibility in addition to DNA sequence variation (Wong et al., 2005a). The neurodevelopmental hypothesis for schizophrenia has strong support from neuropathological observations such as the relative absence of neurodegeneration, the effect of prenatal factors on disease risk, and the emergence of subtle abnormalities in childhood (Lewis and Levitt, 2002).

Some of the difficulties in defining schizophrenia stem from this range of hypothesized contributing factors and the well-established clinical and neurobiological heterogeneity of the illness (Tamminga and Holcomb, 2005). Many clinical and molecular findings have not been replicated consistently, including genetic linkage and association results and the impact of particular environmental risks (Sawa and Snyder, 2002). These inconsistencies have been attributed to population heterogeneity in race, geography, environment, clinical diagnostic classification, and the inherent variability in an outbred species such as humans. There are clear differences between patients with schizophrenia and unaffected controls on neuropsychological test performance, structural and functional brain imaging, postmortem histology, gene transcription, and genetic polymorphisms (Lewis and Lieberman, 2000; Wong and Van Tol, 2003). However, an important observation is that the effect size separating cases and controls diminishes as the measured parameter becomes more distal from the symptoms. For example, neuropsychological test abnormalities have greater effect size than brain imaging, which in turn has a larger effect size than gene expression differences. The resulting overlap between schizophrenic patients and controls on virtually all reported abnormalities means that no diagnostic test is able to supersede clinical diagnosis based on symptoms (Heinrichs, 2001).

The search for definitive biomarkers in schizophrenia has been the focus of much research over the last century, but an objective marker that could be used for diagnosis, sub-typing, or predicting treatment response has not emerged. Although a clinically useful biomarker for schizophrenia is not available, the many and varied abnormalities that affect schizophrenia patients as a group have provided a great deal of information on the disease process, and hold the promise of eventually leading to the identification of novel treatment targets or preventative strategies. In subsequent sections, a more detailed discussion of these issues is followed by a survey of the main findings on biomarkers in schizophrenia, and finally an attempt is made to synthesize these results into some general conclusions.

2.1.2 Endophenotypes in Schizophrenia

The heterogeneity of schizophrenia in its clinical manifestations and etiology has been a significant obstacle to gene discovery, despite the relatively high heritability estimates (Ross et al., 2006). For this reason, biological markers that can segregate pathological subtypes of schizophrenia would be useful for both reducing heterogeneity in research cohorts and in guiding clinical treatment decisions. Defining endophenotypes for schizophrenia represents a major strategy for finding a component or dimension of the illness that may be more closely tied to a molecular or genetic etiology (Gottesman and Gould, 2003).

The term endophenotype refers to a discrete and heritable phenotype based on a biochemical test, microscopic examination, or other laboratory test that is consistently associated with the disorder. They should also cosegregate with the illness in families, and also be observable in unaffected relatives at a higher rate than the disorder itself (Gottesman and Gould, 2003). Endophenotypes can be used as a surrogate for a clinical diagnostic category or a subgroup of patients within that category, i.e., schizophrenia. Endophenotypes should be relatively easy to obtain, show high penetrance, be more reliable than clinical diagnosis, and represent a simpler phenotype for analysis than the overall disorder (Pulver, 2000). Genetic analysis of an endophenotype can specify heritability and may eventually result in gene identification (Adler et al., 1999), which can help to understand disease risk if the locus is linked to the disease or contains a causal gene (Price et al., 2006). Endophenotypes are also known as intermediate phenotypes, and can be neurophysiological, biochemical, endocrinological, neuroanatomical, or cognitive, and can be useful for bridging the gap between a particular gene and the disease (Gottesman and Gould, 2003).

Kendler has suggested that endophenotypes should act as a refinement filter for a study sample before genetic analysis is pursued (Kendler and Gardner, 1997). Endophenotypes may improve on "hard" clinical indicators of disease for investigation in that a multivariate phenotype might "have the greatest likelihood of assisting in the search for schizophrenia related genes" (Iacono, 1998). They may provide much-needed clarity to meta-analysis by breaking population samples into more homogeneous subcategories that permit more precise diagnosis and therefore reduce noise inherent in the more broadly defined umbrella diagnosis (Price et al., 2006). In the search for psychiatric disease genes, endophenotypes may be more easily measured in animal models than the symptom clusters that define the disease in humans (Turetsky et al., 2007). Endophenotypes can facilitate a reductionistic approach to understanding components of the overall disease phenotype, which would be impossible to achieve by studying only the clinical syndrome as a whole. Multiple endophenotypes may represent different dimensions of schizophrenia that may be expressed together in different combinations to produce the full clinical presentation.

2.1.3 Biomarkers in Psychiatric Disease

An ideal biomarker should detect a fundamental feature of the underlying pathophysiology of the disease and distinguish the illness from other conditions with acceptable negative and positive predictive values (Sunderland et al., 2005). Tests for biomarkers should be simple to perform, relatively noninvasive, inexpensive, and reliably reproducible between laboratories (Sunderland et al., 2005). The clinical value of biomarkers lies in their utility — whether a single biomarker or a combination of markers is used — as accurate diagnostic tests, predictors of disease course and progression, or treatment response.

The search for biomarkers for schizophrenia in particular, and for psychiatric disease in general, raises questions about the nature and classification of mental disorders. The original concept of schizophrenia is usually attributed to Kraepelin and Bleuer, where the distinction from bipolar and other mood disorders is central (Kraepelin, 1971). However, recent evidence of significant genetic overlap between schizophrenia and bipolar disorder suggests that the current nosology is inadequate (Craddock et al., 2007; Craddock and Owen, 2007). This is only one example of a wider problem with behavioral disorders in which known individual etiologies do not map neatly onto clinical symptoms or presentation (Prusiner, 1998). This issue is also present in general medical disorders, where two patients with the same diagnosis (e.g., lupus or atherosclerotic disease) may have different and nonoverlapping clinical manifestations. Without knowledge of etiology, the diagnosis of psychiatric diseases is subject to circular reasoning in which symptom clusters are used to define conditions with the hope of finding etiologies that in turn may not generate the same symptoms in every patient (McHugh, 1995).

An alternative model is to reject the traditional disease classifications, at least for biomarker research, and to approach pathological behavior from a dimensional rather than categorical perspective. This perspective has a long and robust history in the psychological and social sciences, but is less widely accepted in medicine (Plomin et al., 1994). It is possible that specific genetic or other molecular lesions will generate dysfunction in one or more components of brain function, and that these clusters of functional impairments together produce the symptoms eventually labeled as disease (Porteous et al., 2006). Endophenotypes represent, at least implicitly, this philosophical stance, in that subcomponents of schizophrenia may be more closely tied to a particular cognitive, electrophysiological or molecular brain system than the disease concept as a whole (Hennah et al., 2006).

An additional complication that affects biomarker research in schizophrenia is the complexity of behavior itself and the nonlinear path from structural brain changes and experiential influences to individual behavior at any given time point. Consistent with the other natural sciences, there are phenomena in schizophrenia research that are nondeterministic (Prigogine, 1996). Thinking, feeling, and behaving certainly fall in this category of emergent properties of complex systems, and so it may be unrealistic to expect that simple lab or functional tests will predict behavior with sufficient accuracy to be clinically useful (Koch and Laurent, 1999). More useful though, will be an understanding of how basic brain systems affect complex psychological functions, which can be considered as quantitative traits rather than disease categories (Andreasen, 1997).

2.2 Nonmolecular Markers in Schizophrenia

2.2.1 Gross Cognitive Deficits in Schizophrenia

Cognitive dysfunction may be the best predictor of long-term functional outcome for schizophrenic patients (Bowie and Harvey, 2005). Cognitive abnormalities are present throughout the lifespan of individuals with schizophrenia, may be noted during childhood and adolescence, and are present at the initial onset of psychosis (Tamminga, 2006). Cognitive problems in schizophrenia affect attention, declarative memory, and higher-order problem solving (Harvey et al., 2001). These deficits appear at the onset of illness and persist throughout the course of illness, with evidence that cognitive impairment may also be progressive (Lewis and Lieberman, 2000). Specific types of cognitive deficits in schizophrenia include domains such as cortical inhibition (CI), working memory, and mentalization or theory of mind (TOM). These are described in more detail in the following sections.

2.2.2 Cortical Inhibition

A deficit in CI has been suggested as a component of the pathophysiology of schizophrenia (Daskalakis et al., 2002; Freedman et al., 1983), although the exact nature of the deficit in schizophrenia is not clear. CI can be assessed with transcranial magnetic stimulation (TMS) (Haraldsson et al., 2004), and may be dependent on clinical state (Saka et al., 2005). It is certainly affected by antipsychotic medication (Fitzgerald et al., 2002). CI refers to inhibition of dorsolatoral prefrontal cortical pyramidal neurons by disruption of their coordinated and sustained firing between a stimulus and the initiation of a behavioral response. γ-Aminobutyrate (GABA) is the major inhibitory neurotransmitter involved in this pathway, and disrupting the GABAergic system through use of antagonists has been shown to disrupt working memory (Lewis et al., 2005).

CI deficits may be due to gene expression differences in cortical GABAergic neurons. Glutamic acid decarboxylase (67-kDa isoform of glutamic acid decarboxylase, GAD67), calcium-binding proteins calbindin and paravalbumin, and GABA membrane transporter-1 (responsible for GABA reuptake) messenger ribonucleic acid (mRNA) are examples of gene products that appear to be modified in schizophrenic patients, and which may change firing patterns depending on the GABA neuron subpopulations affected. Changes in GABA neuron function may also affect the development of neuronal projections (Lewis et al., 2005). For example, individuals with schizophrenia have been reported to have double the number of GABA ergic receptors with the α2 subunit in pyramidal neuron axon initial segments in comparison with the α1 subunit normally seen in unaffected controls (Lewis et al., 2005). CI therefore represents a putative combination of functional and molecular biomarkers for schizophrenia.

2.2.3 Cognitive and Working Memory Impairment

The impairment of cognition and working memory significantly decreases quality of life for schizophrenia patients (Green, 2006), and some have argued that it is the core deficit in schizophrenia (Elvevag and Goldberg, 2000). Cognitive impairment is clearly a prominent feature of the illness, and may serve as a biomarker for severity of disease (Tamminga, 2006). A number of measurement tools are commonly used, such as the Wisconsin Card Sorting Test (Everett et al., 2001), or tests for executive functioning, verbal fluency, alertness, working memory, and optical vigilance (Hofer et al., 2007). Cognitive impairment may be considered an objective marker of prefrontal cortex (PFC) function for areas innervated by cortical and subcortical dopamine projections (Piskulic et al., 2007).

Working memory is a short-term retention of information that is manipulated during more complex learning and decision-making tasks, and is a crucial component of normal cognition (Baddeley, 2003). Schizophrenia is associated with a significant impairment in working memory ability and this may be a trait marker for schizophrenia since family members and those at high risk for schizophrenia show similar impairment (Goldman-Rakic, 1994). Working memory is regulated by dopamine, specifically D1 receptor subtypes in the frontal cortex (Castner et al., 2000), and can be affected by dopamine levels and metabolism (Meyer-Lindenberg et al., 2005). Schizophrenic patients show reduced dorsolateral prefrontal cortex (DPFC) activation during task performance (Lewis et al., 2005), and have difficulty with tasks involving storage and manipulation of information, short-term apprehension, and impairment in the ability to update working memory, all of which are associated with poverty of environmental engagement in schizophrenia (Galletly et al., 2005). This characteristic is tested in part by the auditory oddball paradigm (P300, discussed later) and is one of the most robust biological abnormalities in schizophrenia (Bramon et al., 2004). It is also assessed by N100 latency, which is significantly shorter in schizophrenic patients than in controls (McCarley et al., 1991). This impairment is also related to consistent GABA deficiencies in the DPFC of schizophrenic patients, where tissue concentrations of mRNA encoding GAD67, a GABA synthetic enzyme, are consistently reduced (Lewis et al., 2005). Taken together, impairment of cognitive capacity and working memory provide a characteristic and prominent biomarker for schizophrenia in susceptible individuals.

2.2.4 Theory of Mind

TOM, also known as mentalization, is the ability to attribute mental states, beliefs, desires, and thoughts to explain and predict the behavior of others (Premack and Woodruff, 1978). In essence, it describes thinking about thoughts. Many symptoms of schizophrenia may originate psychologically from TOM deficits including hallucinations, delusions of reference, external control, and persecutory paranoia (Brune, 2005). TOM deficits may also affect social withdrawal and dysfunction, and may be related to disruption in willed action, self and other monitoring, and an inability to accurately perceive ones own intentions/behaviors (Frith and Frith, 1999). Other behaviors that may arise from disordered TOM include apathy and bizarre behavior, a reduced awareness of thoughts arising from within (delusions and hallucinations), and a lack of awareness of the thoughts and intentions of others, resulting in persecutory paranoia, misidentification, and disorganized communication (Harrington et al., 2005).

Some tests used to assess TOM include assessment of first- and second-order false beliefs, visual mental-state jokes, comprehension of linguistic devices such as metaphors and irony, and completion of verbal and nonverbal intention stories (Harrington et al., 2005). It is not clear whether some of these tasks reflect a state or a trait, but overall, the evidence supports the trait hypothesis (Harrington et al., 2005). TOM dysfunction is a recognizable characteristic of schizophrenia and includes a spectrum of related abnormalities that may contribute to the clinical symptoms.

2.3 Electrophysiological Endophenotypes as Biomarkers

Electrophysiological endophenotypes serve as easily measurable divergences from the norm in schizophrenia. They meet the criteria required of an endophenotype for validation, and are noninvasive. Most importantly, when used in conjunction with one another they can characterize deficiencies in responses that may not otherwise be well described, since they objectively measure sensory gating and information integration. The event-related potential (ERP) is the starting point for acquisition of these data.

2.3.1 Event-related Potentials

The ERP measures brain electrical activity in response to a variety of stimulus events in combination, such as pictures, words, and sounds, and can be reliably measured using scalp electroencephalography. It is used to determine a range of electrophysiological abnormalities that are potential endophenotypes in schizophrenic subjects. The ERP can be a measure of attention and information processing through ERP-related events (Braff and Geyer, 1990) such as P50 suppression, P300 peak measurements, antisaccades, and mismatch negativity (Turetsky et al., 2007) (Fig. 2.1). Disturbances in attention and information processing have been proposed as a potential mechanism for generating the symptoms of schizophrenia. Electrophysiological measures seem integral among endophenotypes for schizophrenia, and may be of use in conjunction with other biomarkers to predict phenotype.

2.3.1.1 P50 Suppression (P50)

P50 suppression refers to the early positive deflection in the P50 component of the ERP. It is induced by two clicks presented 500 ms apart, and the P50 wave is generated in the second (test) click. The P50 is normally suppressed relative to the P50 wave generated in the first (conditioning) click (Freedman et al., 1996; Freedman et al., 1991). The reduction of this waveform is present in patients with chronic schizophrenia (Potter et al., 2006) and their unaffected first-degree relatives (Clementz et al., 1998). P50 suppression appears to be unaffected by typical antipsychotic medication (Freedman et al., 1983); however, atypical antipsychotics modify the P50 component (Light et al., 2000). The suppression of the P50 component is attributed to the activation of inhibitory neural circuitry by the first conditioning click (Freedman et al., 1996) and can be interpreted as a form of sensory gating to filter against information overload (Braff and Light, 2005). There is no effect of illness duration on the P50, so it can be considered a "trait" marker of schizophrenia (Bramon et al., 2004). P50 suppression has equivalents in other mammals such as

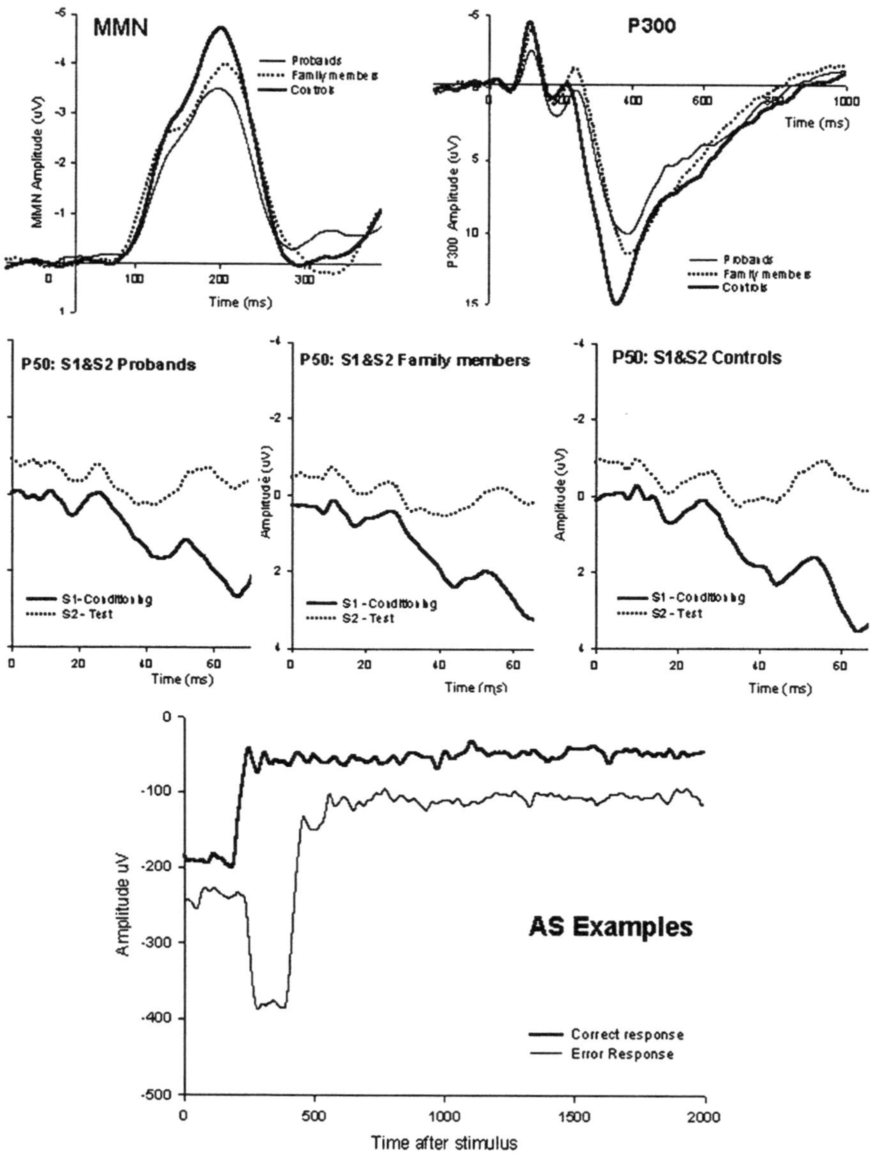

Fig. 2.1 This figure, from Price et al. (2006) gives representative waveforms of grand averages from MMN, P300, P50 conditioning (S1) and test (S2) stimuli in addition to AS.

rats, cats, and monkeys, indicating that animal modeling of this particular endophenotype may be a useful tool for the discovery of complementary molecular biomarkers.

2.3.1.2 Auditory Oddball Target P300 Peak

This endophenotype, the most studied of ERP components, is conceptualized as being the physiological correlate of a working memory update of changes in the environment and the allocation of attentional resources (Coull, 1998). The P300 is generated during an auditory oddball paradigm in response to infrequent target stimuli that require an overt response (Price et al., 2006). Reduction in amplitude and increased latency are seen in chronic schizophrenia (Salisbury et al., 1998) and at first episode (Kumari et al., 2005a). Diminished P300 auditory-evoked potentials are related to a decreased ability to distinguish two tones (Freedman et al., 1991). These differences are also seen in unaffected first-degree relatives (Turetsky et al., 2007), siblings (Karoumi et al., 2000), and children at high risk for schizophrenia (Schreiber et al., 1992). This endophenotype appears to be unaffected by medication (Pfefferbaum et al., 1989), symptom state (Blackwood, 2000), or duration of illness. This stability over time suggests that it is also a trait marker of schizophrenia (Bramon et al., 2004).

2.3.1.3 Antisaccade

AS error describes an oculomotor paradigm designed to evaluate the inhibitory capacity of the brain. A fixation cue appears unpredictably in an eccentric location, and the subject is asked to make an eye movement (saccade) in the opposite direction (Hallett, 1978). Errors are recorded if the saccade is in the wrong direction. An increased error rate in AS associated with schizophrenia was proposed as an endophenotype, and is found in individuals with chronic schizophrenia (Curtis et al., 2001), at the first episode of psychosis (Hutton et al., 1998), and in first degree relatives (Calkins et al., 2003). The AS error measures both a deficit in frontally mediated inhibition (failure to correct errors) and a deficit in response generation (Hutton and Ettinger, 2006). Developmental profiling of AS errors is broadly consistent with the known development of the PFC (Hutton and Ettinger, 2006). The rate of error appears to be unaffected by state (Curtis et al., 2001), but the effects of medication are uncertain (Price et al., 2006). Significant interlaboratory differences have been found in the measurement of this endophenotype (Hutton and Ettinger, 2006).

2.3.1.4 Mismatch Negativity

Mismatch negativity (MMN) is the negative component of the ERP produced in response to low-probability deviant sounds in a sequence of standard sound stimuli (Todd et al., 2007), while the subject's attention is directed elsewhere (i.e., by reading a book, carrying out a visual task, or watching a video). The mismatch is generated by any discriminable change in a repetitive background of auditory stimuli (Winkler et al., 1996), and occurs when the auditory system detects a deviation from an established pattern of acoustic stimuli (Price et al., 2006). This is considered an automatic/preattentive brain response and has been reported for deviants on a variety of auditory stimuli, including intensity, frequency, duration, spatial location, phonemes, and spectrotemporal patterns (Michie, 2001). MMN amplitude reduction has been proposed as an endophenotype for schizophrenia (Shelley et al, 1991), and is present in patients with chronic schizophrenia (Javitt et al., 1993), their first-degree relatives (Jessen et al., 2001) and high-risk children (Schreiber et al., 1992), but not necessarily in first-episode patients (Salisbury et al., 2002). MMN reduction is unaffected by medication (Korostenskaja et al., 2005) or by clinical state (Shinozaki et al., 2002) and is relatively specific to schizophrenia (Catts et al., 1995), indicating that it may be a good biomarker when combined with other markers of the disease.

2.3.2 Prepulse Inhibition

The startle reflex is a whole body response to intense, sudden stimulus, and is normally attenuated in human and animal subjects when the startle stimulus is preceded by a weaker warning stimulus – the prepulse (Hsieh et al., 2006). The reflex, as well as its prepulse inhibition (PPI), are highly conserved evolutionarily across many species and can be measured by eyeblink, as it often is in humans, or by whole-body movement, as it is in rodents. PPI is quantified as the reduction of startle response magnitude when a prepulse is given and occurs when the interval between the prepulse and startle stimulus is 30–500 ms (Hsieh et al., 2006). Schizophrenic patients usually exhibit PPI deficits regardless of the stimulus category: auditory, tactile, or electrocutaneous (Braff et al., 2001). A deficit in PPI in schizophrenic patients has been suggested to be a strong endophenotype by several groups (Braff and Freedman, 2002; Kumari et al., 2005b); however, direct comparisons between studies can be challenging, as PPI can be affected by background noise and prepulse duration, frequency, and interval (Hsieh et al., 2006). Abnormal PPI has been found in the unaffected relatives of patients with schizophrenia (Cadenhead et al., 2000). There is also evidence of significant heritability of PPI in healthy subjects (Anokhin et al., 2003). PPI is disrupted by increased noradrenergic activity, higher blood glycine levels (Hammer et al., 2007), and amphetamine (Swerdlow et al., 2003), and is improved with antipsychotics (Swerdlow et al., 2006). Disrupted PPI, when combined with other endophenotypes, is a relatively robust indicator of schizophrenia (Price et al., 2006; Swerdlow et al., 2007).

2.3.3 Gamma Synchrony

Gamma synchrony (GS) describes a “binding mechanism” for integrating disparate brain networks mediating perception, cognition, and memory (Uhlhaas and Singer, 2006). It is related to the visual perception of gestalt-like or meaningful stimuli including illusory figures and attention (Tallon-Baudry et al., 1996). GS is observed in healthy humans responding to sensory stimuli and in cognitive tasks using both visual and auditory stimulation and learned associations and is measured via electroencephalography (EEG) or magnetoencephalography (MEG) (Symond et al., 2005). GS manifests as synchronous high-frequency oscillations in brain electrical activity from 30–100 Hz, which occurs across several brain regions. GS may therefore serve to integrate brain activity across separate neuronal networks (Lee et al., 2003). The functional significance of GS in the perception of both auditory and visual stimuli is in the variation in gamma band amplitude, for which a difference is seen between controls and individuals with schizophrenia, notably in that GS is disrupted in schizophrenics (Lee et al., 2003). These disruptions are characterized by global delays and decreased temporal connectivity in neural activity during sensory responses to task-relevant stimuli (Symond et al., 2005), and may be related to the disorganization syndrome in these patients (Bressler, 2003). Impairments in frontal lobe synchrony may be particularly relevant to disordered cognitive control or flexibility in schizophrenia (Cho et al., 2006).

2.4 Brain Imaging

Brain imaging studies in schizophrenia have demonstrated differences in brain volumes, glucose metabolism, and blood flow, both at rest and during the performance of cognitive tasks, the experience of symptoms, or with pharmacological manipulation (Ross et al., 2006). Imaging techniques have also permitted the examination of antipsychotic action and negative symptoms. Perhaps unique among the repertoire of research methods, functional brain imaging allows real-time visualization of brain function and allows comparisons of anatomy, regional activation, neurotransmitter occupancy, and metabolism between patients with schizophrenia and various control groups. Imaging studies have been crucial in demonstrating that the symptoms of schizophrenia originate in brain changes, and have been instrumental in stimulating research using other methods on the neurobiology of schizophrenia. However, as with all human clinical studies, confounding factors such as genetic heterogeneity and the impossibility of controlling for environmental factors may hamper the interpretation of some results. Analytical and technical methods are also diverse, and have changed with time, making direct comparisons between studies complex at best.

2.4.1 Structural Magnetic Resonance Imaging (MRI)/Computed Tomography (CT)

Changes in brain structure are one of the most robust biological abnormalities in schizophrenia, with many replicated studies showing enlarged ventricles and reduced cortical volume (Reveley et al., 1982; Suddath et al., 1990; Zipursky et al., 1992). These abnormalities are present at the first episode of psychosis (Steen et al., 2006; Zipursky et al., 1998), and are progressive (Cahn et al., 2002; Lieberman et al., 2001). A recent meta-analysis of structural MRI concluded that whole-brain and hippocampal volume reduction at first episode is subtle, and close to the detection limit of current MRI (Steen et al., 2006). Gray matter volume loss, especially in the frontal cortex, is correlated with poor outcome, while reduced thalamic volume is not (Gur et al., 1998; Staal et al., 2000). The temporal lobes, especially the left medial and superior temporal gyrus, show the greatest reduction in volume with schizophrenia (Honea et al., 2005). Both typical and atypical antipsychotics are associated with changes in grey matter volumes, though this does not nullify the primary effect of the illness (Dazzan et al., 2005). The type and stage of psychosis also have an effect on hippocampal and amygdala volumes (Velakoulis et al., 2006), and overall outcome is loosely related to ventricular enlargement (Staal et al., 1999; van Haren et al., 2003). It has been suggested that white matter abnormalities may disrupt connectivity between brain regions and contribute to the functional impairment, and diffusion tensor imaging has been applied to specifically examine fiber tracts to verify this hypothesis. Consistent findings, however, have yet to emerge (Kanaan et al., 2005; Kubicki et al., 2007). Overall, the strongest finding from structural brain imaging in schizophrenia is ventricular enlargement, and regional volumetric studies point to volume loss in the temporal lobes as being of the greatest magnitude (Harrison, 1999).

2.4.2 Functional Imaging: Positron Emission Tomography (PET), Single Photon Emission Computed Tomography (SPECT), and Functional MRI (fMRI)

Functional imaging allows a real-time assessment of brain state at rest, during particular tasks, or with pharmacological interventions. These methods permit measurement of regional activation in association with cognitive processing, symptoms such as hallucinations, or specific subject experiences. Therefore links can be established between brain metabolism, blood flow or radiotracer distribution and symptoms, task deficits, emotional states, or drug binding in schizophrenia. Functional imaging studies in schizophrenia have been mostly aimed at addressing three main questions. The first is whether there are differences in regional activation, blood flow, or metabolism. The second is whether abnormalities in the dopamine system can be demonstrated, in support of the dopamine hypothesis of schizophrenia, and the related issue of how antipsychotic medication occupancy of dopamine receptors affects psychotic symptom control, side effects, and mental state (Zipursky et al., 2007). While the latter topic is of great clinical interest, it does not serve as a biomarker for the illness, and will not be covered in this chapter. Hypofrontality and dopamine system changes will be discussed in the following sections.

2.4.2.1 Hypofrontality

There have been hundreds of functional imaging studies in schizophrenia, using all the available techniques of SPECT, PET, and fMRI. As with other neurobiological studies on this heterogeneous illness, the findings of different studies have often been inconsistent, generating considerable debate. Some authors have argued strongly that the experimental support for hypofrontality in schizophrenia is weak (Gur and Gur, 1995), and a review of the literature reports that only one-third of papers show this finding (Chua and McKenna, 1995). In addition to the disease heterogeneity that confounds all clinical research in schizophrenia, the dynamic nature of functional imaging and the effects of task or other state variables on imaging measures make direct comparisons among disparate studies difficult. Nevertheless, several broad conclusions may be synthesized from this large amount of published data.

The first report of decreased relative blood flow or metabolism in frontal brain regions (Ingvar and Franzen, 1974) has not been consistently replicated, but a thorough meta-analysis on this issue has concluded that hypofrontality at rest and with task activation is a feature of schizophrenia, however evidence for altered regional functional architecture is weak (Hill et al., 2004). Studies can be roughly divided into those that measured whole-brain activation and those that measured relative or absolute hypofrontality at rest or during task activation. For resting hypofrontality, the effect size for relative differences (–0.32) was less than for absolute differences (–0.55), which might be accounted for by decreased whole-brain activation in schizophrenia. In contrast, both relative and absolute activation hypofrontality have similar effect sizes (–0.37 and –0.42, respectively). A further important conclusion is that methodology did not have a large impact on the degree of hypofrontality (Hill et al., 2004).

An unresolved question is whether task-activated hypofrontality is a core deficit in schizophrenia, or whether it is simply a correlate of decreased task performance (Weinberger and Berman, 1996). Decreased perfusion or metabolism of specific brain regions is correlated with known regional anatomy of specific tasks, for example, decreased activation of amygdala and hippocampus during facial emotion processing by schizophrenic patients (Gur et al., 2002), or reduced dorsolateral prefrontal cortical activation during maintenance and manipulation of working memory tasks (Cannon et al., 2005). However, reduced right-sided temporal and ventral prefrontal cortex activation is also correlated with more global deficits such as negative symptoms that may affect motivation or capacity for task performance across many domains of cognitive function (Potkin et al., 2002).

Although activation hypofrontality is a general finding in schizophrenia, this global observation greatly oversimplifies the complexity of brain regions recruited to tackle even simple cognitive tasks. For example, patterns of both over- or under-activation are observed depending on working memory load, which may act as an index of schizophrenia affecting ability to allocate neural resources for specific tasks (Johnson et al., 2006). As well, thalamic activation is correlated with performance on working memory tasks in schizophrenia (Andrews et al., 2006), highlighting that frontal lobe activity may have dynamic and differential impacts on specific tasks or components of tasks that are not captured by the idea of a generalized or global hypoactivation. In conclusion, functional imaging methods have been useful in understanding the pathophysiology of symptoms and cognitive dysfunction in schizophrenia rather than as a global biomarker of the diagnosis.

2.4.2.2 Dopamine System Imaging in Schizophrenia

The dopamine hypothesis of schizophrenia posits that the symptoms of this illness arise from overactivity of the dopamine system. While antipsychotics clearly exert their therapeutic effects through dopamine D2 receptors (Seeman and Lee, 1975), it has been controversial whether primary dysfunction of the dopamine system is inherent to the pathophysiology of schizophrenia. Several possibilities have been investigated through functional brain imaging: the amount of synaptic dopamine itself could be elevated, either through increased release or reduced metabolism or uptake; the number of postsynaptic dopamine receptors could be elevated; or downstream signal transduction through the receptors could be augmented (Zipursky et al., 2007). Technically, the most straightforward experiment is to use PET or SPECT to image radioligands for dopamine D2 receptors, and this was the first kind of study published. These studies have not produced consistent results (Frankle and Laruelle, 2002), which could be explained by differences in radioligand affinity for the D2 receptor, resulting in differential competitive displacement by endogenous dopamine (Seeman et al., 1989; Seeman and Tallerico, 1998).

To overcome this problem, an alternative approach has been to compare receptor occupancy before and after depletion of presynaptic dopamine stores, reducing competition between the dopamine D2 receptor radioligand and dopamine, thereby permitting a more accurate estimate of receptor density. Using the radiotracer [^{123}I]IBZM and SPECT, a greater difference between pre- and postdepletion D2 occupancy was demonstrated, suggesting that synaptic dopamine is elevated in schizophrenia (Abi-Dargham et al., 2000). This was present in both first-episode never-treated patients and previously treated chronic patients, and elevated synaptic dopamine was predictive for antipsychotic efficacy in treating psychotic symptoms. In a related experiment, the same group found higher striatal amphetamine-induced dopamine release in drug-free schizophrenia patients. Furthermore, the amount of dopamine release was correlated with the emergence or exacerbation of psychotic symptoms (Laruelle et al., 1996). This was confirmed by another group using PET and the radiotracer [^{11}C]raclopride (Breier et al., 1997). Together, these findings support the contention that a hyperdopaminergic state is present in schizophrenia when psychotic symptoms are present, but not in periods of remission (Laruelle et al., 1999).

2.5 Blood-based and Molecular Biomarkers

Developing diagnostic tests for schizophrenia using biomarkers obtained from peripheral blood would be useful for diagnosis or predicting prognosis in patients with complex symptoms or for individuals at high risk of developing schizophrenia.

There is evidence for a substantial genetic contribution to schizophrenia from twin, adoption, and other family studies (Gottesman, 1991), and delineating the risk genes and alleles would provide a definitive biomarker. Although the mode of transmission remains uncertain, it is clear that schizophrenia is not a Mendelian single-gene disorder. It remains unclear whether some subgroups of schizophrenia may be caused by variation in single genes that differ from family to family (Porteous et al., 2006), or whether a number of genes contribute to schizophrenia in most cases, consistent with a multiplicative (interacting genes) or additive (cumulative effect) multilocus model (Risch, 1990). It seems likely that even when specific genes are implicated in the etiology, individual vulnerability alleles may have different effects on gene function, expression, or protein interactions that add further heterogeneity to study samples. For example, it is possible that some cases or families may contain mutations in a single gene or a small number of genes with large effects. In other patients, a relatively large number of genetic variants, each with subtle individual effects, may be required to generate the illness.

An alternative view is that specific symptoms or cognitive dysfunctions map onto molecular and anatomic networks that do not assort neatly into current diagnostic criteria. Thus, genes that confer susceptibility to psychosis may have variants important for many diagnoses in which psychotic symptoms may be present, such as schizophrenia plus the affective psychoses (bipolar disorder with psychotic features and psychotic depression) (Badner and Gershon, 2002; Craddock et al., 2007). In this dimensional rather than categorical framework, genetic or other molecular biomarkers would be useful for predicting and understanding the basis of specific symptom types, rather than for classification according to current nosology (Craddock and Owen, 2007).

Because of these complications, no single methodology has been able to generate consistently replicated definitive evidence of a specific schizophrenia susceptibility gene or allele. Instead, the convergence of different sources of data is the most parsimonious approach to assessing the importance of particular candidate genes and risk alleles (Harrison and Weinberger, 2005). Ideally, this convergence should include results from whole-genome genetic linkage scans or chromosomal lesions, candidate gene association analysis, gene expression or functional effects of the genetic variants, and mechanistic models with experimental verification in animal or in vitro model systems (Arguello and Gogos, 2006; DeLisi, 2000). It is likely that even when definitive risk alleles are established, they will have to act in combination with other genes and environmental risk factors, and will be applicable only to a subset of patients with a clinical diagnosis of schizophrenia. As the genes conferring the greatest disease risk are identified, it may be possible to generate genetic subtypes of the disorder to further refine the molecular diagnosis with additional genetic, blood-based, or other biomarkers.

Traditional linkage studies with large multiplex families have found a number of chromosomal loci linked to schizophrenia, and those that have reached genome-wide significance or been replicated in independent samples include 1q21–22, 1q42, 5q21–33, 6p24–22, 6q21–25, 8p22–21, 10p15–11, 10q25–26, 13q32–34, 17p11-q25, and 22q11–22 (Owen et al., 2005; Pulver, 2000). Chromosomal abnormalities such as deletions or translocations that result in psychosis and other psychiatric symptoms have provided another source of positional candidates for schizophrenia. Examples include the 22q11 region deleted in DiGeorge or velocardiofacial syndrome (Bassett and Chow, 1999; Bassett et al., 2003), and the 1q42.1–11q14.3 balanced translocation that led to the identification of DISC1 as a candidate gene for schizophrenia (Porteous et al., 2006). Located within these linkage regions are a number of strong positional candidate genes, including *CAPON* (1q22), *RGS4* (1q23), *DISC1* (1q42.1), *EPN4* (5q33), *DTNBP1* or *dysbindin* (6p22), *TAAR6* (6q23), *NRG1* or *neuregulin 1* (8p12), *PPP3CC* (8p21), G*72* or *DAOA* (13q34), *PRODH* and *COMT* (22q11) (Gogos and Gerber 2006). Other candidate genes that are not located in these linkage regions, but that show altered expression, strong genetic association, and biologically plausible relationships to schizophrenia include *GRM3* (7q21–22), *Akt1* (14q22–32), and *CHRNA7* (15q13–14) (Harrison and Weinberger, 2005). Research into these genes is quite active, and any attempt to decide which genes are most important before more data are available is likely to become obsolete quickly. A summary is provided in Table 2.1.

Both *DTNBP1* and *NRG1* have demonstrated genetic association with schizophrenia in many, though not all, studies to date, and are located within linkage regions (Kirov et al., 2004; Stefansson et al., 2003; Stefansson et al., 2002; Straub et al., 2002; Williams et al., 2004; Williams et al., 2003). However, the allelic variants of both these genes associated with schizophrenia are not consistent across different study populations, suggesting that altered expression or RNA processing may be important (Owen et al., 2005). This is supported by evidence of altered gene expression in schizophrenia (Law et al., 2006; Talbot et al., 2004). *DTNBP1*

Table 2.1 Genes and gene products implicated as susceptibility genes in schizophrenia

Gene	Contributing evidence	References
BDNF	Brain-derived neurotrophic factor	
	met allele at val66met polymorphism	
	Associated with working memory performance impairment in patients with schizophrenia	(Rybakowski et al., 2006)
	Associated with age of onset, clinical symptoms that remain after long term treatment	(Numata et al., 2006)
	Appears to be population specific	
	No correlation between polymorphism and schizophrenia in Asian populations	(Naoe et al., 2007)
	Risk correlation in Scottish population	(Neves-Pereira et al., 2005)
	Located at chr 11p.13	
DISC-1	Disrupted in schizophrenia-1	
	Associated with working memory performance impairment and hippocampal structure	(Callicott et al. 2005)
	Implicated in bipolar disorder	(Porteous et al. 2006)
	DISC-1 binding partner expression abnormalities (decreases) may be responsible contribution	(Lipska et al. 2006)
	Located at 1q42.1	

(continued)

Table 2.1 (continued)

Gene	Contributing evidence	References
COMT	Catechol-*o*-methyltransferase	
	met108/158met scz patients have improved PFC efficiency vs. val108/158met patients	(Meyer-Lindenberg et al. 2005)
	met variant results in 40% reduction in enzymatic activity of COMT	(Woodward et al., 2007)
	May increase availability of dopamine and norepinephrine in PFC	
	Located at 22q11.21–23	
NRG1	Neuregulin 1	
	Association with schizophrenia confirmed in multiple populations	(Harrison and Law 2006)
	Two haplotypes at least implicated in the pathogenesis of schizophrenia and in bipolar disorder	(Blackwood et al., 2007)
	Functions in neural migration, differentiation, oligodendrocyte survival, impaired in schizophrenia	(Sei et al., 2007)
	Codes 15 different isoforms, at least	
	Abnormal mRNA expression patterns found in PFC of schizophrenia patients	(Bertram et al., 2007)
	Located at chromosome 8p12–21	
DTNBP1	Dysbindin-1, dystrobrevin-binding protein 1	
	Involved in NMDA receptor-mediated glutamatergic transmission	(Vilella et al. 2008)
	Nonfunctional SNP variant associated with schizophrenia, confirmed in Han Chinese population	(Liao and Chen 2004) (Straub et al. 2002) (Owen et al. 2005)
	May confer risk of schizophrenia by mediating effects of postsynaptic structure and function	
	Significant associations between DTNBP1 and schizophrenia in at least ten samples	
	Located at 6p22.3	
DOA and DAOA	D-Amino acid oxidase, D-amino acid oxidase activator	
	both show significant associations with schizophrenia	(Detera-Wadleigh and McMahon 2006)
	Gene interaction between the two increases schizophrenia risk	
	Located at 12q24 and 13q33.2–3.4 respectively	(Chumakov et al., 2002)
RGS4	Regulator of G-protein signaling 4	(Levitt et al., 2006)
	Decreased expression in schizophrenia	(Mirnics et al., 2000)
	Controls duration/timing of synaptic responsiveness for Gi/Go/Gq couple neurotransmitter systems	
	Protein implicated both in pathophysiology and in pharmacological treatment	
	mRNA levels decreased across highly divergent cortical regions in schizophrenic patients	(Brzustowicz et al., 2000)
	Located at 1q23.3, highly associated with famililal schizophrenia susceptability	

is located in synaptic nerve terminals but its function is not well understood (Owen et al., 2004). More is known about *NRG1*; interactions with ErbB receptors are important for radial glia formation and differentiation as part of the scaffold for cortical neuron migration during development (Corfas et al., 2004). These interactions also regulate synaptic plasticity in conjunction with GABA neurotransmission at inhibitory synapses in the cortex, with a plausible impact on thought disorder and electrophysiological endophenotypes as discussed earlier (Role and Talmage, 2007).

DISC1 is located within a linkage peak for schizophrenia, and was originally discovered through the presence of a balanced translocation that severs the gene in a family with a high prevalence of mental illness (Millar et al., 2000). As with *DTNBP1* and *NRG1*, there are multiple allelic variants associated with schizophrenia, as well as evidence that susceptibility SNPs alter binding to functional partners such as *Fez1, Lis1*, and *NUDEL*, which together regulate neuronal migration in the developing cortex (Lipska et al., 2006; Ozeki et al., 2003). *DISC1* also interacts with *PDE4B* (Millar et al., 2005); together, they regulate cAMP signaling with obvious relevance to the memory and cognitive deficits in schizophrenia (Porteous and Millar, 2006). Two recent mouse models have emerged, with mutations in the *DISC1* gene, both with behavioral abnormalities consistent with some of the endophentoypes discussed earlier in this chapter (Clapcote et al., 2007; Hikida et al., 2007). The precise relationship between DISC1 gene sequence variations and schizophrenia, however, remain unresolved, such as whether a haploinsufficiency or dominant negative mechanism is at work. Although *DISC1* is an important schizophrenia candidate gene, many of the carriers of the 1q42.1–11q14.3 translocation have other types of mental illness, ranging from mood to learning problems, lending further support to the dimensional view of mental illness.

2.5.1 Gene Expression Biomarkers in Postmortem Brain and Peripheral Tissues

Many investigators have used quantification of mRNA and protein levels from both postmortem brain and peripheral blood samples in schizophrenia to identify biomarkers. These studies can be divided into hypothesis-driven measurements of a specific candidate gene product, and screening strategies that cover the genome with microarray or other high-throughput methods. Genome-wide mRNA expression profiling using microarrays has become cheaper and more accessible and allows rapid screening of tens of thousands of gene transcripts simultaneously from small quantities of biological samples. This powerful technique has been successfully applied to understanding disease biology in a number of important areas, including cancer (Wang et al., 2005) and immunology (Calvano et al., 2005).

In schizophrenia in particular, gene expression studies are likely to identify further associated molecular pathways, or interactions with existing candidate genes, and have the potential to specify biomarkers that may be downstream from the etiology.

By enabling expression-level measurement of so many genes at once, microarray analysis can identify multiple genes whose collective pattern of expression may be diagnostic for schizophrenia. The advantage of this empirical strategy is that no individual gene expression level is critical to the diagnosis, which is important in a complex genetic disorder in which many susceptibility genes are likely to be involved and significant genetic heterogeneity between individual patients is to be expected (Wong and Van Tol, 2003).

Gene expression studies of postmortem brain tissue in schizophrenia (Iwamoto and Kato, 2006) have indicated altered transcription of synaptic (Mirnics et al., 2000; Vawter et al., 2001; Vawter et al., 2002), metabolic (Altar et al., 2005; Iwamoto et al., 2005; Middleton et al., 2002), oligodendrocyte (Aston et al., 2004; Hakak et al., 2001; Katsel et al., 2005), and neuropeptide Y and p72 genes (Kuromitsu et al., 2001). Despite some successful replications, brain tissue microarray findings have inconsistencies, likely due to the use of different array platforms (Hollingshead et al., 2005; Jurata et al., 2004) and the limitations associated with postmortem tissue (e.g., scarce samples, heterogeneity, tissue conditions) (Iwamoto and Kato, 2006). Attempts to assess the effects of antipsychotic medication on the brain transcriptome have included administering antipsychotics to nonhuman primates (Mirnics et al., 2000) and rodents (Wong et al., 2005b; Wong et al., 2003), but since these animal models are not afflicted with schizophrenia, the effects of antipsychotics on gene expression in schizophrenia remain unclear. The public availability of online databases containing expression microarray data now allows researchers to compare their own data and mine for possible relationships between gene expression and other disease variables (Higgs et al., 2006).

Peripheral blood is a far more accessible tissue that may be obtained from living subjects. Comparisons of gene expression in peripheral blood and multiple CNS tissues show significant similarities between transcriptomes (Sullivan et al., 2006), with a 0.5 median nonparametric correlation between transcripts present in both blood and brain. In addition, 50% of known candidate genes for schizophrenia are expressed in both peripheral blood and cortex, and the expression levels of many genes relevant to schizophrenia biology are not significantly different between whole blood and PFC (Sullivan et al., 2006).

There are several examples in which peripheral transcription profiling has successfully identified relevant disease biomarkers. In fibroblasts, a unique gene expression signature can distinguish familial Alzheimer's disease mutation carriers from their unaffected siblings (Nagasaka et al., 2005). In olfactory epithelium, differences in transcription are seen between schizophrenic patients and controls (McCurdy et al., 2006). In peripheral blood, interindividual differences in leukocyte gene expression have been shown to indicate changes in response to environmental and genetic differences (Radich et al., 2004). Gene expression analysis of peripheral blood leukocytes from discordant sib-pairs with schizophrenia and bipolar disorder has also shown some convergence in molecular pathways identified by either genetic or genomic approaches (Middleton et al., 2005). Peripheral blood gene expression profiling is able to distinguish schizophrenia or bipolar disorder from unaffected controls (Tsuang et al., 2005). Furthermore, a recent study identified

SELENBP1 upregulation in schizophrenia peripheral blood samples, consistent with differential expression of this gene in frontal cortex, confirming that peripheral blood samples can yield useful information about brain (Glatt et al., 2005). mRNA levels of genes associated with integral brain functions such as neuronal myelination, neurodevelopment, neuroprotection, and cellular signaling show modified expression patterns in lymphocytes of schizophrenic subjects compared to matched, healthy controls (Bowden et al., 2006). Bowden et al. (2006) found that age in relation to clinical diagnosis can subdivide schizophrenia into groups showing altered gene expression profiles, though not without taking into account differences in expression that may relate to duration of illness.

2.5.2 Immune Markers (Cytokines)

There is evidence suggesting that the immune system is perturbed in schizophrenia, with changes seen in the levels of cytokines associated with both antipsychotic treatment and the diagnosis itself. The immune, endocrine, and nervous system interact with each other through cytokines, hormones, and neurotransmitters, as reviewed in Drzyzga et al. (2006). In the brain, cytokines are regulated in a cascade that includes positive and negative feedback regulation (Kronfol and Remick, 2000). Certain cytokines, including interleukin (IL)-2, IL-6, and tumor necrosis factor-α (TNF-α) are important mediators of interactions between the immune system and the CNS (Pollmacher et al., 1997), and abnormal levels of IL-1, -2, and -4 have been found in the cerebrospinal fluid of schizophrenic patients (Yao and van Kammen, 2004). Blood levels of IL-1, -2, -6, -18, and TNF-α are changed in schizophrenia in a state-dependent fashion and related to the duration of the disorder and antipsychotic medication (Nawa and Takei, 2006; Yao et al., 2003). Interferon gamma (INF-γ) production is significantly reduced during acute phases of paranoia but not during periods of residual symptoms (Wilke et al., 1996). Additionally, cerebrospinal fluid (CSF) IL-2 levels are elevated during relapse and associated with risk of relapse in schizophrenia (Yao et al., 2003).

There is evidence for low-level constitutive expression of cytokines in blood vessels in the brain (Kronfol and Remick, 2000). Excluding CNS synthesis and excretion of cytokines by astrocytes and/or microglia, and in some cases neurons, cytokines cannot cross the blood–brain barrier under physiological conditions. Cytokines may cross the blood–brain barrier passively at regions without a blood–brain barrier; through carrier-mediated transport across the blood–brain barrier; by activation at peripheral afferent nerve terminals where cytokines are released; and by binding to cerebral vascular endothelium, inducing second messenger cascades (Kronfol and Remick, 2000). The cytokines implicated in the pathophysiology of schizophrenia belong to a family that consists of molecular homologs including neuregulin-1, that can produce neurobehavioral changes in schizophrenic patients when modified, as well as neurodevelopmental changes when exposure occurs in utero (Nawa et al., 2000). Cytokine genes that have been implicated in the risk for schizophrenia include IL-1 gene complex, IL-10, neuregulin-1, epidermal growth factor (EGF), and TNF-α, most of which have been suggested through single nucleotide polymorphism (SNP) or linkage studies and some of which may affect levels of glutamate receptors and dopamine (Nawa et al., 2000). Despite these reports, the literature is far from consistent in this area, with a large body of conflicting evidence (Drzyzga et al., 2006). Therefore, although the immune system may play a role in the pathophysiology of schizophrenia, there are no clear patterns of easily measured cytokine levels that are pathogenomic.

2.6 Summary and Perspectives

Biomarkers for schizophrenia include abnormalities at all levels of description, from whole organism behavior, to electrophysiology and molecular markers. The current challenge is to unify these findings into a coherent model of pathophysiology, of which there are several possible scenarios. The simplest way to integrate all of the myriad biomarkers discussed above would be to discover a final common pathway for the symptoms of schizophrenia, where a number of different genetic and environmental insults converge. However, it is more likely that the neurobiology of particular symptoms or endophenotypes will be characterized individually, and their combined presence in a given patient will determine the clinical presentation. In other words, neurobiology will map onto dimensions of the illness rather than the diagnosis itself.

Research on schizophrenia has made rapid progress, and with the emergence of a short list of candidate genes, investigators can now move beyond cataloguing abnormalities to characterizing disease mechanisms. There are several competing, but not necessarily exclusive, pathophysiological models that may connect molecular lesions to physiology and ultimately behavioral manifestations of schizophrenia. Disruption of cortical neurodevelopment due to a combination of genetic and prenatal environmental insults is likely to be important, but the specific etiology may be quite heterogeneous between individual patients. Both structural and neurochemical changes in the adult could result from developmental derailment and from the primary effects of genetic variants, which in turn disrupt electrophysiology, plasticity, and cognition, and ultimately produce psychotic and negative symptoms.

Balancing this optimistic view is the possibility that complex mental phenomena are not deterministic, and that even a clear account of the pathophysiology of schizophrenia covering the main etiologies will not result in reliable diagnostic or prognostic biomarkers. Many complex systems exist in which emergent properties can only be described in probabilistic terms. How the brain regulates behavior is certainly complex enough that a simple set of biomarkers may be insufficient to improve diagnostic accuracy beyond clinical assessment. More important though, is that understanding disease mechanisms may reveal novel treatment targets, especially since all existing antipsychotics work through dopamine D2 receptors. Novel targets are likely to emerge from understanding of specific genetic subtypes of schizophrenia and ongoing research into schizophrenia biomarkers.

Acknowledgments AHCW is supported by a Clinician-Scientist Fellowship from the Canadian Institutes for Health Research (CIHR).

References

Abi-Dargham A, Rodenhiser J, Printz D, Zea-Ponce Y, Gil R, Kegeles LS, Weiss R, Cooper TB, Mann JJ, Van Heertum RL, Gorman JM, Laruelle M (2000) Increased baseline occupancy of D2 receptors by dopamine in schizophrenia. Proc Natl Acad Sci USA 97: 8104–9
Adler LE, Freedman R, Ross RG, Olincy A, Waldo MC (1999) Elementary phenotypes in the neurobiological and genetic study of schizophrenia. Biol Psychiatr 46: 8–18
Altar CA, Jurata LW, Charles V, Lemire A, Liu P, Bukhman Y, Young TA, Bullard J, Yokoe H, Webster MJ, Knable MB, Brockman JA, Palfreyman MG, Hook DJ, Klimczak LJ, Brockman JA, Evans DM, Altar CA, Jordan S, Koprivica V, Chen R, Tottori K, Kikuchi T, Altar CA (2005) Deficient hippocampal neuron expression of proteasome, ubiquitin, and mitochondrial genes in multiple schizophrenia cohorts. Biol Psychiatr 58: 85–96
Andreasen NC (1997) Linking mind and brain in the study of mental illnesses: a project for a scientific psychopathology. Science 275: 1586–93
Andrews J, Wang L, Csernansky JG, Gado MH, Barch DM (2006) Abnormalities of thalamic activation and cognition in schizophrenia. Am J Psychiatr 163: 463–9
Anokhin AP, Heath AC, Myers E, Ralano A, Wood S (2003) Genetic influences on prepulse inhibition of startle reflex in humans. Neurosci Lett 353: 45–8
Arguello PA, Gogos JA (2006) Modeling madness in mice: one piece at a time. Neuron 52: 179–96
Aston C, Jiang L, Sokolov BP (2004) Microarray analysis of postmortem temporal cortex from patients with schizophrenia. J Neurosci Res 77: 858–66
Baddeley A (2003) Working memory: looking back and looking forward. Nat Rev Neurosci 4: 829–39
Badner JA, Gershon ES (2002) Meta-analysis of whole-genome linkage scans of bipolar disorder and schizophrenia. Mol Psychiatr 7: 405–11
Bassett AS, Chow EW (1999) 22q11 deletion syndrome: a genetic subtype of schizophrenia. Biol Psychiatr 46: 882–91
Bassett AS, Chow EW, AbdelMalik P, Gheorghiu M, Husted J, Weksberg R (2003) The schizophrenia phenotype in 22q11 deletion syndrome. Am J Psychiatr 160: 1580–6
Bertram I, Bernstein HG, Lendeckel U, Bukowska A, Dobrowolny H, Keilhoff G, Kanakis D, Mawrin C, Bielau H, Falkai P, Bogerts B (2007) Immunohistochemical evidance for impaired neuregulin-1 signaling in the prefrontal cortex in schizophrenia and in unipolor depression Ann Ny Acad Sci. 1096: 147–56
Blackwood D H, B J. Pickard et al. (2007) Are some genetic risk factors common to schizophrenia, bipolar disorder and depression? Evidence from DISC1, GRIK4 and NRG1. Neurotox Res 11: 73–83
Blackwood D (2000) P300, a state and a trait marker in schizophrenia. Lancet 355: 771–2
Bowden, N. A., J. Weidenhofer, et al. (2006) Preliminary investigation of gene expression profiles in peripheral blood lymphocytes in schizophrenia. Schizophr Res 82: 175–83
Bowie CR, Harvey PD (2005) Cognition in schizophrenia: impairments, determinants, and functional importance. Psychiatr Clin North Am 28: 613–33; 626
Braff, D. L. and M. A. Geyer (1990) Sensorimotor gating and schizophrenia. Human and animal model studies. Arch Gen Psychiatry 47: 181–8
Braff DL, Freedman R (2002) Endophenotypes in studies of the genetics of schizophrenia. In: Davis KL, Charney D, Coyle JT (eds) Neuropsychopharmacology: the fifth generation of progress. American College of Neuropsychopharmacology, p 703
Braff DL, Light GA (2005) The use of neurophysiological endophenotypes to understand the genetic basis of schizophrenia. Dialogues Clin Neurosci 7: 125–35

Braff DL, Geyer MA, Swerdlow NR (2001) Human studies of prepulse inhibition of startle: normal subjects, patient groups, and pharmacological studies. Psychopharmacology (Berl) 156: 234–58

Bramon E, Rabe-Hesketh S, Sham P, Murray RM, Frangou S (2004) Meta-analysis of the P300 and P50 waveforms in schizophrenia. Schizophr Res 70: 315–29

Breier A, Su TP, Saunders R, Carson RE, Kolachana BS, de Bartolomeis A, Weinberger DR, Weisenfeld N, Malhotra AK, Eckelman WC, Pickar D (1997) Schizophrenia is associated with elevated amphetamine-induced synaptic dopamine concentrations: evidence from a novel positron emission tomography method. Proc Natl Acad Sci U S A 94: 2569–74

Bressler SL (2003) Cortical coordination dynamics and the disorganization syndrome in schizophrenia. Neuropsychopharmacology 28 (Suppl 1): S35–9

Brune M (2005) "Theory of mind" in schizophrenia: a review of the literature. Schizophr Bull 31: 21–42

Brzustowicz, L. M., K. A. Hodgkinson, et al. (2000) Location of a major susceptibility locus for familial schizophrenia on chromosome 1q21-q22. Science 288: 678–82

Cadenhead KS, Swerdlow NR, Shafer KM, Diaz M, Braff DL (2000) Modulation of the startle response and startle laterality in relatives of schizophrenic patients and in subjects with schizotypal personality disorder: evidence of inhibitory deficits. Am J Psychiatr 157: 1660–8

Cahn W, Hulshoff Pol HE, Lems EB, van Haren NE, Schnack HG, van der Linden JA, Schothorst PF, van Engeland H, Kahn RS (2002) Brain volume changes in first-episode schizophrenia: a 1-year follow-up study. Arch Gen Psychiatr 59: 1002–1010

Callicott, J. H., R. E. Straub, et al. (2005) Variation in DISC1 affects hippocampal structure and function and increases risk for schizophrenia. Proc Natl Acad Sci U S A 102: 8627–32

Calkins ME, Iacono WG, Curtis CE (2003) Smooth pursuit and antisaccade performance evidence trait stability in schizophrenia patients and their relatives. Int J Psychophysiol 49: 139–146

Calvano SE, Xiao W, Richards DR, Felciano RM, Baker HV, Cho RJ, Chen RO, Brownstein BH, Cobb JP, Tschoeke SK, Miller-Graziano C, Moldawer LL, Mindrinos MN, Davis RW, Tompkins RG, Lowry SF (2005) A network-based analysis of systemic inflammation in humans. Nature 437: 1032–1037

Cannon TD, Glahn DC, Kim J, Van Erp TG, Karlsgodt K, Cohen MS, Nuechterlein KH, Bava S, Shirinyan D (2005) Dorsolateral prefrontal cortex activity during maintenance and manipulation of information in working memory in patients with schizophrenia. Arch Gen Psychiatr 62: 1071–1080

Cardno AG, Gottesman, II (2000) Twin studies of schizophrenia: from bow-and-arrow concordances to star wars Mx and functional genomics. Am J Med Genet 97: 12–17

Castner SA, Williams GV, Goldman-Rakic PS (2000) Reversal of antipsychotic-induced working memory deficits by short-term dopamine D1 receptor stimulation. Science 287: 2020–2022

Catts SV, Shelley AM, Ward PB, Liebert B, McConaghy N, Andrews S, Michie PT (1995) Brain potential evidence for an auditory sensory memory deficit in schizophrenia. Am J Psychiatr 152: 213–219

Cho RY, Konecky RO, Carter CS (2006) Impairments in frontal cortical gamma synchrony and cognitive control in schizophrenia. Proc Natl Acad Sci U S A 103: 19878–19883

Chua SE, McKenna PJ (1995) Schizophrenia—a brain disease? A critical review of structural and functional cerebral abnormality in the disorder. Br J Psychiatr 166: 563–582

Chumakov I, Blumenfeld M, Guerassimenko O, Cavarec L, Palicio M, Abderrahim H, Bougueleret L, Barry C, Tanaka H, La Rosa P, Puech A, Tahri N, Cohen-Akenine A, Delabrosse S, Lissarrague S, Picard FP, Maurice K, Essioux L, Millasseau P, Grel P, Debailleul V, Simon AM, Caterina D, Dufaure I, Malekzadeh K, Belova M, Luan JJ, Bouillot M, Sambucy JL, Primas G, Saumier M, Boubkiri N, Martin-Saumier S, Nasroune M, Peixoto H, Delaye A, Pinchot V, Bastucci M, Guillou S, Chevillon M, Sainz-Fuertes R, Meguenni S, Aurich-Costa J, Cherif D, Gimalac A, Van Duijn C, Gauvreau D, Ouellette G, Fortier I, Raelson J, Sherbatich T, Riazanskaia N, Rogaev E, Raeymaekers P, Aerssens J, Konings F, Luyten W, Macciardi F, Sham PC, Straub RE, Weinberger DR, Cohen N, Cohen D. (2004) Genetic and physiological data implicating the new human gene G72 and the gene for D-amino acid oxidase in schizophrenia. Proc Natl Acad Sci 99:13675–13680

Clapcote SJ, Lipina TV, Millar JK, Mackie S, Christie S, Ogawa F, Lerch JP, Trimble K, Uchiyama M, Sakuraba Y, Kaneda H, Shiroishi T, Houslay MD, Henkelman RM, Sled JG, Gondo Y, Porteous DJ, Roder JC (2007) Behavioral phenotypes of Disc1 missense mutations in mice. Neuron 54: 387–402
Clementz BA, Geyer MA, Braff DL (1998) Poor P50 suppression among schizophrenia patients and their first-degree biological relatives. Am J Psychiatr 155: 1691–1694
Corfas G, Roy K, Buxbaum JD (2004) Neuregulin 1-erbB signaling and the molecular/cellular basis of schizophrenia. Nat Neurosci 7: 575–580
Coull JT (1998) Neural correlates of attention and arousal: insights from electrophysiology, functional neuroimaging and psychopharmacology. Prog Neurobiol 55: 343–361
Craddock N, Owen MJ (2007) Rethinking psychosis: the disadvantages of a dichotomous classification now outweigh the advantages. World Psychiatr 6: 20–27
Craddock N, O'Donovan MC, Owen MJ (2007) Symptom dimensions and the Kraepelinian dichotomy. Br J Psychiatr 190: 361; author reply 361–362
Curtis CE, Calkins ME, Grove WM, Feil KJ, Iacono WG (2001) Saccadic disinhibition in patients with acute and remitted schizophrenia and their first-degree biological relatives. Am J Psychiatr 158: 100–106
Daskalakis ZJ, Christensen BK, Chen R, Fitzgerald PB, Zipursky RB, Kapur S (2002) Evidence for impaired cortical inhibition in schizophrenia using transcranial magnetic stimulation. Arch Gen Psychiatr 59: 347–354
Dazzan P, Morgan KD, Orr K, Hutchinson G, Chitnis X, Suckling J, Fearon P, McGuire PK, Mallett RM, Jones PB, Leff J, Murray RM (2005) Different effects of typical and atypical antipsychotics on grey matter in first episode psychosis: the AESOP study. Neuropsychopharmacology 30: 765–774
DeLisi LE (2000) Critical overview of current approaches to genetic mechanisms in schizophrenia research. Brain Res Brain Res Rev 31: 187–192
Detera-Wadleigh, S. D. and F. J. McMahon (2006) G72/G30 in schizophrenia and bipolar disorder: review and meta-analysis. Biol Psychiatry 60: 106–14
Drzyzga L, Obuchowicz E, Marcinowska A, Herman ZS (2006) Cytokines in schizophrenia and the effects of antipsychotic drugs. Brain Behav Immun 20: 532–345
Elvevag B, Goldberg TE (2000) Cognitive impairment in schizophrenia is the core of the disorder. Crit Rev Neurobiol 14: 1–21
Everett J, Lavoie K, Gagnon JF, Gosselin N (2001) Performance of patients with schizophrenia on the Wisconsin Card Sorting Test (WCST). J Psychiatr Neurosci 26: 123–130
Fitzgerald PB, Brown TL, Daskalakis ZJ, Kulkarni J (2002) A transcranial magnetic stimulation study of the effects of olanzapine and risperidone on motor cortical excitability in patients with schizophrenia. Psychopharmacology (Berl) 162: 74–81
Frankle WG, Laruelle M (2002) Neuroreceptor imaging in psychiatric disorders. Ann Nucl Med 16: 437–446
Freedman R, Adler LE, Waldo MC, Pachtman E, Franks RD (1983) Neurophysiological evidence for a defect in inhibitory pathways in schizophrenia: comparison of medicated and drug-free patients. Biol Psychiatr 18: 537–551
Freedman R, Waldo M, Bickford-Wimer P, Nagamoto H (1991) Elementary neuronal dysfunctions in schizophrenia. Schizophr Res 4: 233–343
Freedman R, Adler LE, Myles-Worsley M, Nagamoto HT, Miller C, Kisley M, McRae K, Cawthra E, Waldo M (1996) Inhibitory gating of an evoked response to repeated auditory stimuli in schizophrenic and normal subjects. Human recordings, computer simulation, and an animal model. Arch Gen Psychiatr 53: 1114–1121
Frith CD, Frith U (1999) Interacting minds—a biological basis. Science 286: 1692–1695
Galletly CA, Clark CR, McFarlane AC (2005) Clozapine improves working memory updating in schizophrenia. Eur Neuropsychopharmacol 15: 601–608
Glatt SJ, Everall IP, Kremen WS, Corbeil J, Sasik R, Khanlou N, Han M, Liew CC, Tsuang MT (2005) Comparative gene expression analysis of blood and brain provides concurrent validation of SELENBP1 up-regulation in schizophrenia. Proc Natl Acad Sci U S A 102(43): 15533–15538

Gogos JA, Gerber DJ (2006) Schizophrenia susceptibility genes: emergence of positional candidates and future directions. Trends Pharmacol Sci 27: 226–233
Goldman-Rakic PS (1994) Working memory dysfunction in schizophrenia. J Neuropsychiatr Clin Neurosci 6: 348–357
Gottesman II (1991) Schizophrenia Genesis: The Origins of Madness. W.H. Freeman
Gottesman, II, Gould TD (2003) The endophenotype concept in psychiatry: etymology and strategic intentions. Am J Psychiatr 160: 636–45
Green MF (2006) Cognitive impairment and functional outcome in schizophrenia and bipolar disorder. J Clin Psychiatr 67 (Suppl 9): 3–8; discussion 36–42
Gur RC, Gur RE (1995) Hypofrontality in schizophrenia: RIP. Lancet 345: 1383–4
Gur RE, Cowell P, Turetsky BI, Gallacher F, Cannon T, Bilker W, Gur RC (1998) A follow-up magnetic resonance imaging study of schizophrenia. Relationship of neuroanatomical changes to clinical and neurobehavioral measures. Arch Gen Psychiatr 55: 145–52
Gur RE, McGrath C, Chan RM, Schroeder L, Turner T, Turetsky BI, Kohler C, Alsop D, Maldjian J, Ragland JD, Gur RC (2002) An fMRI study of facial emotion processing in patients with schizophrenia. Am J Psychiatr 159: 1992–9
Hakak Y, Walker JR, Li C, Wong WH, Davis KL, Buxbaum JD, Haroutunian V, Fienberg AA (2001) Genome-wide expression analysis reveals dysregulation of myelination-related genes in chronic schizophrenia. Proc Natl Acad Sci U S A 98: 4746–51
Hallett PE (1978) Primary and secondary saccades to goals defined by instructions. Vision Res 18: 1279–96
Hammer TB, Oranje B, Glenthoj BY (2007) The effects of imipramine on P50 suppression, prepulse inhibition and habituation of the startle response in humans. Int J Neuropsychopharmacol 25: 1–9
Haraldsson HM, Ferrarelli F, Kalin NH, Tononi G (2004) Transcranial Magnetic Stimulation in the investigation and treatment of schizophrenia: a review. Schizophr Res 71: 1–16
Harrington L, Siegert RJ, McClure J (2005) Theory of mind in schizophrenia: a critical review. Cognit Neuropsychiatr 10: 249–86
Harrison PJ (1999) The neuropathology of schizophrenia. A critical review of the data and their interpretation. Brain 122: 593–624
Harrison, P. J. and A. J. Law (2006) Neuregulin 1 and schizophrenia: genetics, gene expression, and neurobiology. Biol Psychiatry 60: 132–40
Harrison PJ, Weinberger DR (2005) Schizophrenia genes, gene expression, and neuropathology: on the matter of their convergence. Mol Psychiatr 10: 40–68; image 5
Harvey PD, Bowie CR, Friedman JI (2001) Cognition in schizophrenia. Curr Psychiatr Rep 3: 423–8
Heinrichs RW (2001) In Search of Madness: Schizophrenia and Neuroscience. Oxford University Press, Oxford
Hennah W, Thomson P, Peltonen L, Porteous D (2006) Genes and schizophrenia: beyond schizophrenia: the role of DISC1 in major mental illness. Schizophr Bull 32: 409–16
Higgs BW, Elashoff M, Richman S, Barci B (2006) An online database for brain disease research. BMC Genomics 7: 70
Hikida T, Jaaro-Peled H, Seshadri S, Oishi K, Hookway C, Kong S, Wu D, Xue R, Andrade M, Tankou S, Mori S, Gallagher M, Ishizuka K, Pletnikov M, Kida S, Sawa A (2007) Dominant-negative DISC1 transgenic mice display schizophrenia-associated phenotypes detected by measures translatable to humans. Proc Natl Acad Sci U S A 3: 3
Hill K, Mann L, Laws KR, Stephenson CM, Nimmo-Smith I, McKenna PJ (2004) Hypofrontality in schizophrenia: a meta-analysis of functional imaging studies. Acta Psychiatr Scand 110: 243–256
Hofer A, Niedermayer B, Kemmler G, Rettenbacher MA, Trebo E, Widschwendter CG, Fleischhacker WW (2007) Cognitive impairment in schizophrenia: clinical ratings are not a suitable alternative to neuropsychological testing. Schizophr Res 92: 126–131
Hollingshead D, Lewis DA, Mirnics K (2005) Platform influence on DNA microarray data in postmortem brain research. Neurobiol Dis 18: 649–655
Honea R, Crow TJ, Passingham D, Mackay CE (2005) Regional deficits in brain volume in schizophrenia: a meta-analysis of voxel-based morphometry studies. Am J Psychiatr 162: 2233–2245

Hsieh MH, Swerdlow NR, Braff DL (2006) Effects of background and prepulse characteristics on prepulse inhibition and facilitation: implications for neuropsychiatric research. Biol Psychiatr 59: 555–559
Hutton SB, Ettinger U (2006) The antisaccade task as a research tool in psychopathology: a critical review. Psychophysiology 43: 302–313
Hutton SB, Crawford TJ, Puri BK, Duncan LJ, Chapman M, Kennard C, Barnes TR, Joyce EM (1998) Smooth pursuit and saccadic abnormalities in first-episode schizophrenia. Psychol Med 28: 685–692
Iacono WG (1998) Identifying psychophysiological risk for psychopathology: examples from substance abuse and schizophrenia research. Psychophysiology 35: 621–637
Ingvar DH, Franzen G (1974) Abnormalities of cerebral blood flow distribution in patients with chronic schizophrenia. Acta Psychiatr Scand 50: 425–462
Iwamoto K, Kato T (2006) Gene expression profiling in schizophrenia and related mental disorders. Neuroscientist 12: 349–361
Iwamoto K, Bundo M, Kato T (2005) Altered expression of mitochondria-related genes in postmortem brains of patients with bipolar disorder or schizophrenia, as revealed by large-scale DNA microarray analysis. Hum Mol Genet 14: 241–253
Javitt DC, Doneshka P, Zylberman I, Ritter W, Vaughan HG, Jr. (1993) Impairment of early cortical processing in schizophrenia: an event-related potential confirmation study. Biol Psychiatr 33: 513–519
Jessen F, Fries T, Kucharski C, Nishimura T, Hoenig K, Maier W, Falkai P, Heun R (2001) Amplitude reduction of the mismatch negativity in first-degree relatives of patients with schizophrenia. Neurosci Lett 309: 185–188
Johnson MR, Morris NA, Astur RS, Calhoun VD, Mathalon DH, Kiehl KA, Pearlson GD (2006) A functional magnetic resonance imaging study of working memory abnormalities in schizophrenia. Biol Psychiatr 60: 11–21
Jurata LW, Bukhman YV, Charles V, Capriglione F, Bullard J, Lemire AL, Mohammed A, Pham Q, Laeng P, Brockman JA, Altar CA (2004) Comparison of microarray-based mRNA profiling technologies for identification of psychiatric disease and drug signatures. J Neurosci Methods 138: 173–188
Kanaan RA, Kim JS, Kaufmann WE, Pearlson GD, Barker GJ, McGuire PK (2005) Diffusion tensor imaging in schizophrenia. Biol Psychiatr 58: 921–929
Karoumi, B., M. Saoud, et al. (2001) Poor performance in smooth pursuit and antisaccadic eye-movement tasks in healthy siblings of patients with schizophrenia. Psychiatry Res 101: 209–19
Katsel P, Davis KL, Haroutunian V (2005) Variations in myelin and oligodendrocyte-related gene expression across multiple brain regions in schizophrenia: a gene ontology study. Schizophr Res 79: 157–173
Kendler KS, Gardner CO (1997) The risk for psychiatric disorders in relatives of schizophrenic and control probands: a comparison of three independent studies. Psychol Med 27: 411–419
Kirov G, Ivanov D, Williams NM, Preece A, Nikolov I, Milev R, Koleva S, Dimitrova A, Toncheva D, O'Donovan MC, Owen MJ (2004) Strong evidence for association between the dystrobrevin binding protein 1 gene (DTNBP1) and schizophrenia in 488 parent-offspring trios from Bulgaria. Biol Psychiatr 55: 971–975
Koch C, Laurent G (1999) Complexity and the nervous system. Science 284: 96–98
Korostenskaja M, Dapsys K, Siurkute A, Maciulis V, Ruksenas O, Kahkonen S (2005) Effects of olanzapine on auditory P300 and mismatch negativity (MMN) in schizophrenia spectrum disorders. Prog Neuropsychopharmacol Biol Psychiatr 29: 543–548
Kraepelin E (1971) Dementia Praecox and Paraphrenia. Robert R. Krieger, New York
Kronfol Z, Remick DG (2000) Cytokines and the brain: implications for clinical psychiatry. Am J Psychiatr 157: 683–694
Kubicki M, McCarley R, Westin CF, Park HJ, Maier S, Kikinis R, Jolesz FA, Shenton ME (2007) A review of diffusion tensor imaging studies in schizophrenia. J Psychiatr Res 41: 15–30
Kumari V, Das M, Zachariah E, Ettinger U, Sharma T (2005a) Reduced prepulse inhibition in unaffected siblings of schizophrenia patients. Psychophysiology 42: 588–594

Kumari V, Ettinger U, Crawford TJ, Zachariah E, Sharma T (2005b) Lack of association between prepulse inhibition and antisaccadic deficits in chronic schizophrenia: implications for identification of schizophrenia endophenotypes. J Psychiatr Res 39: 227–240
Kuromitsu J, Yokoi A, Kawai T, Nagasu T, Aizawa T, Haga S, Ikeda K (2001) Reduced neuropeptide Y mRNA levels in the frontal cortex of people with schizophrenia and bipolar disorder. Brain Res Gene Expr Patterns 1: 17–21
Laruelle M, Abi-Dargham A, Gil R, Kegeles L, Innis R (1999) Increased dopamine transmission in schizophrenia: relationship to illness phases. Biol Psychiatr 46: 56–72
Laruelle M, Abi-Dargham A, van Dyck CH, Gil R, D'Souza CD, Erdos J, McCance E, Rosenblatt W, Fingado C, Zoghbi SS, Baldwin RM, Seibyl JP, Krystal JH, Charney DS, Innis RB (1996) Single photon emission computerized tomography imaging of amphetamine- induced dopamine release in drug-free schizophrenic subjects. Proc Natl Acad Sci U S A 93: 9235–9240
Law AJ, Lipska BK, Weickert CS, Hyde TM, Straub RE, Hashimoto R, Harrison PJ, Kleinman JE, Weinberger DR (2006) Neuregulin 1 transcripts are differentially expressed in schizophrenia and regulated by 5' SNPs associated with the disease. Proc Natl Acad Sci U S A 103: 6747–6752
Lee KH, Williams LM, Breakspear M, Gordon E (2003) Synchronous gamma activity: a review and contribution to an integrative neuroscience model of schizophrenia. Brain Res Brain Res Rev 41: 57–78
Levitt, P., P. Ebert, et al. (2006) Making the case for a candidate vulnerability gene in schizophrenia: Convergent evidence for regulator of G-protein signaling 4 (RGS4). Biol Psychiatry 60: 534–7
Lewis DA, Lieberman JA (2000) Catching up on schizophrenia: natural history and neurobiology. Neuron 28: 325–334
Lewis DA, Levitt P (2002) Schizophrenia as a disorder of neurodevelopment. Annu Rev Neurosci 25: 409–432. Epub 2002 Mar 22
Lewis DA, Hashimoto T, Volk DW (2005) Cortical inhibitory neurons and schizophrenia. Nat Rev Neurosci 6: 312–324
Liao, H. M. and C. H. Chen (2004) Mutation analysis of the human dystrobrevin-binding protein 1 gene in schizophrenic patients. Schizophr Res 71: 185–9
Lieberman J, Chakos M, Wu H, Alvir J, Hoffman E, Robinson D, Bilder R (2001) Longitudinal study of brain morphology in first episode schizophrenia. Biol Psychiatr 49: 487–499
Light GA, Geyer MA, Clementz BA, Cadenhead KS, Braff DL (2000) Normal P50 suppression in schizophrenia patients treated with atypical antipsychotic medications. Am J Psychiatr 157: 767–771
Lipska BK, Peters T, Hyde TM, Halim N, Horowitz C, Mitkus S, Weickert CS, Matsumoto M, Sawa A, Straub RE, Vakkalanka R, Herman MM, Weinberger DR, Kleinman JE (2006) Expression of DISC1 binding partners is reduced in schizophrenia and associated with DISC1 SNPs. Hum Mol Genet 15: 1245–1258
McCarley RW, Faux SF, Shenton ME, Nestor PG, Adams J (1991) Event-related potentials in schizophrenia: their biological and clinical correlates and a new model of schizophrenic pathophysiology. Schizophr Res 4: 209–231
McCurdy RD, Feron F, Perry C, Chant DC, McLean D, Matigian N, Hayward NK, McGrath JJ, Mackay-Sim A (2006) Cell cycle alterations in biopsied olfactory neuroepithelium in schizophrenia and bipolar I disorder using cell culture and gene expression analyses. Schizophr Res 6: 6
McGrath JJ (2007) The surprisingly rich contours of schizophrenia epidemiology. Arch Gen Psychiatr 64: 14–16
McGrath J, Saha S, Welham J, El Saadi O, MacCauley C, Chant D (2004) A systematic review of the incidence of schizophrenia: the distribution of rates and the influence of sex, urbanicity, migrant status and methodology. BMC Med 2: 13
McHugh PR (1995) Witches, multiple personalities, and other psychiatric artifacts. Nat Med 1: 110–114
Meyer-Lindenberg A, Kohn PD, Kolachana B, Kippenhan S, McInerney-Leo A, Nussbaum R, Weinberger DR, Berman KF (2005) Midbrain dopamine and prefrontal function in humans: interaction and modulation by COMT genotype. Nat Neurosci 8: 594–596
Michie PT (2001) What has MMN revealed about the auditory system in schizophrenia? Int J Psychophysiol 42: 177–194

Middleton FA, Mirnics K, Pierri JN, Lewis DA, Levitt P (2002) Gene expression profiling reveals alterations of specific metabolic pathways in schizophrenia. J Neurosci 22: 2718–2729

Middleton FA, Pato CN, Gentile KL, McGann L, Brown AM, Trauzzi M, Diab H, Morley CP, Medeiros H, Macedo A, Azevedo MH, Pato MT (2005) Gene expression analysis of peripheral blood leukocytes from discordant sib-pairs with schizophrenia and bipolar disorder reveals points of convergence between genetic and functional genomic approaches. Am J Med Genet B Neuropsychiatr Genet 136: 12–25

Millar JK, Wilson-Annan JC, Anderson S, Christie S, Taylor MS, Semple CA, Devon RS, Clair DM, Muir WJ, Blackwood DH, Porteous DJ (2000) Disruption of two novel genes by a translocation co-segregating with schizophrenia. Hum Mol Genet 9: 1415–1423

Millar JK, Pickard BS, Mackie S, James R, Christie S, Buchanan SR, Malloy MP, Chubb JE, Huston E, Baillie GS, Thomson PA, Hill EV, Brandon NJ, Rain JC, Camargo LM, Whiting PJ, Houslay MD, Blackwood DH, Muir WJ, Porteous DJ (2005) DISC1 and PDE4B are interacting genetic factors in schizophrenia that regulate cAMP signaling. Science 310: 1187–1191

Mirnics K, Middleton FA, Marquez A, Lewis DA, Levitt P (2000) Molecular characterization of schizophrenia viewed by microarray analysis of gene expression in prefrontal cortex. Neuron 28: 53–67

Nagasaka Y, Dillner K, Ebise H, Teramoto R, Nakagawa H, Lilius L, Axelman K, Forsell C, Ito A, Winblad B, Kimura T, Graff C (2005) A unique gene expression signature discriminates familial Alzheimer's disease mutation carriers from their wild-type siblings. Proc Natl Acad Sci U S A 102: 14854–14859

Naoe, Y., T. Shinkai, et al. (2007) No association between the brain-derived neurotrophic factor (BDNF) Val66Met polymorphism and schizophrenia in Asian populations: Evidence from a case-control study and meta-analysis. Neurosci Lett 415: 108–12

Nawa H, Takei N (2006) Recent progress in animal modeling of immune inflammatory processes in schizophrenia: implication of specific cytokines. Neurosci Res 56: 2–13

Nawa H, Takahashi M, Patterson PH (2000) Cytokine and growth factor involvement in schizophrenia—support for the developmental model. Mol Psychiatr 5: 594–603

Neves-Pereira, M., J. K. Cheung, et al. (2005) BDNF gene is a risk factor for schizophrenia in a Scottish population. Mol Psychiatry 10: 208–12

Numata, S., S. Ueno, et al. (2006) Brain-derived neurotrophic factor (BDNF) Val66Met polymorphism in schizophrenia is associated with age at onset and symptoms. Neurosci Lett 401: 1–5

Owen MJ, Williams NM, O'Donovan MC (2004) Dysbindin-1 and schizophrenia: from genetics to neuropathology. J Clin Investig 113: 1255–1257

Owen MJ, Craddock N, O'Donovan MC (2005) Schizophrenia: genes at last? Trends Genet 21: 518–525

Ozeki Y, Tomoda T, Kleiderlein J, Kamiya A, Bord L, Fujii K, Okawa M, Yamada N, Hatten ME, Snyder SH, Ross CA, Sawa A (2003) Disrupted-in-Schizophrenia-1 (DISC-1): mutant truncation prevents binding to NudE-like (NUDEL) and inhibits neurite outgrowth. Proc Natl Acad Sci U S A 100: 289–294

Pfefferbaum A, Ford JM, White PM, Roth WT (1989) P3 in schizophrenia is affected by stimulus modality, response requirements, medication status, and negative symptoms. Arch Gen Psychiatr 46: 1035–1044

Piskulic D, Olver JS, Norman TR, Maruff P (2007) Behavioural studies of spatial working memory dysfunction in schizophrenia: a quantitative literature review. Psychiatr Res 150: 111–121

Plomin R, Owen MJ, McGuffin P (1994) The genetic basis of complex human behaviors. Science 264: 1733–1739

Pollmacher T, Hinze-Selch D, Fenzel T, Kraus T, Schuld A, Mullington J (1997) Plasma levels of cytokines and soluble cytokine receptors during treatment with haloperidol. Am J Psychiatr 154: 1763–1765

Porteous DJ, Millar JK (2006) Disrupted in schizophrenia 1: building brains and memories. Trends Mol Med 12: 255–261

Porteous DJ, Thomson P, Brandon NJ, Millar JK (2006) The genetics and biology of DISC1—an emerging role in psychosis and cognition. Biol Psychiatr 60: 123–131

Potkin SG, Alva G, Fleming K, Anand R, Keator D, Carreon D, Doo M, Jin Y, Wu JC, Fallon JH (2002) A PET study of the pathophysiology of negative symptoms in schizophrenia. Positron emission tomography. Am J Psychiatr 159: 227–237

Potter D, Summerfelt A, Gold J, Buchanan RW (2006) Review of clinical correlates of P50 sensory gating abnormalities in patients with schizophrenia. Schizophr Bull 32: 692–700

Premack DG, Woodruff G (1978) Does the chimpanzee have a theory of mind? Behav Brain Sci 1: 515–526

Price GW, Michie PT, Johnston J, Innes-Brown H, Kent A, Clissa P, Jablensky AV (2006) A multivariate electrophysiological endophenotype, from a unitary cohort, shows greater research utility than any single feature in the Western Australian family study of schizophrenia. Biol Psychiatr 60: 1–10

Prigogine I (1996) The End of Certainty. The Free Press, New York

Prusiner SB (1998) Prions. Proc Natl Acad Sci U S A 95: 13363–13383

Pulver AE (2000) Search for schizophrenia susceptibility genes. Biol Psychiatr 47: 221–230

Radich JP, Mao M, Stepaniants S, Biery M, Castle J, Ward T, Schimmack G, Kobayashi S, Carleton M, Lampe J, Linsley PS (2004) Individual-specific variation of gene expression in peripheral blood leukocytes. Genomics 83: 980–988

Reveley AM, Reveley MA, Clifford CA, Murray RM (1982) Cerebral ventricular size in twins discordant for schizophrenia. Lancet 1: 540–541

Risch N (1990) Linkage strategies for genetically complex traits. I. Multilocus models. Am J Hum Genet 46: 222–228

Role LW, Talmage DA (2007) Neurobiology: new order for thought disorders. Nature 448: 263–265

Ross CA, Margolis RL, Reading SA, Pletnikov M, Coyle JT (2006) Neurobiology of schizophrenia. Neuron 52: 139–153

Rybakowski, J. K., A. Borkowska, et al. (2006) Prefrontal cognition in schizophrenia and bipolar illness in relation to Val66Met polymorphism of the brain-derived neurotrophic factor gene. Psychiatry Clin Neurosci 60: 70-6

Saha N, Tsoi WF, Low PS, Basair J, Tay JS (1994) Lack of association of the dopamine D3 receptor gene polymorphism (BalI) in Chinese schizophrenic males. Psychiatr Genet 4: 201–204

Saka MC, Atbasoglu EC, Ozguven HD, Sener HO, Ozay E (2005) Cortical inhibition in first-degree relatives of schizophrenic patients assessed with transcranial magnetic stimulation. Int J Neuropsychopharmacol 8: 595–599

Salisbury DF, Shenton ME, Sherwood AR, Fischer IA, Yurgelun-Todd DA, Tohen M, McCarley RW (1998) First-episode schizophrenic psychosis differs from first-episode affective psychosis and controls in P300 amplitude over left temporal lobe. Arch Gen Psychiatr 55: 173–180

Salisbury DF, Shenton ME, Griggs CB, Bonner-Jackson A, McCarley RW (2002) Mismatch negativity in chronic schizophrenia and first-episode schizophrenia. Arch Gen Psychiatr 59: 686–694

Sartorius N, Jablensky A, Korten A (1986) Early manifestations and first contact incidence of schizophrenia in different cultures: a preliminary report on the inital evaluation phase of the WHO collaborative study on determinants of outcome of severe mental disorders. Psychol Med 16: 909–928

Sawa A, Snyder SH (2002) Schizophrenia: diverse approaches to a complex disease. Science 296: 692–695

Schreiber H, Stolz-Born G, Kornhuber HH, Born J (1992) Event-related potential correlates of impaired selective attention in children at high risk for schizophrenia. Biol Psychiatr 32: 634–651

Seeman P, Lee T (1975) Antipsychotic drugs: direct correlation between clinical potency and presynaptic action on dopamine neurons. Science 188: 1217–1219

Seeman P, Tallerico T (1998) Antipsychotic drugs which elicit little or no parkinsonism bind more loosely than dopamine to brain D2 receptors, yet occupy high levels of these receptors. Mol Psychiatr 3: 123–134

Seeman P, Guan HC, Niznik HB (1989) Endogenous dopamine lowers the dopamine D2 receptor density as measured by [3H]raclopride: implications for positron emission tomography of the human brain. Synapse 3: 96–97

Sei, Y., R. Ren-Patterson, et al. (2007) Neuregulin1-induced cell migration is impaired in schizophrenia: association with neuregulin1 and catechol-o-methyltransferase gene polymorphisms. Mol Psychiatry 12: 946–57

Shinozaki N, Yabe H, Sato Y, Hiruma T, Sutoh T, Nashida T, Matsuoka T, Kaneko S (2002) The difference in Mismatch negativity between the acute and post-acute phase of schizophrenia. Biol Psychol 59: 105–119

Staal WG, Hulshoff Pol HE, Kahn RS (1999) Outcome of schizophrenia in relation to brain abnormalities. Schizophr Bull 25: 337–348

Staal WG, Hulshoff Pol HE, Schnack HG, Hoogendoorn ML, Jellema K, Kahn RS (2000) Structural brain abnormalities in patients with schizophrenia and their healthy siblings. Am J Psychiatr 157: 416–421

Steen RG, Mull C, McClure R, Hamer RM, Lieberman JA (2006) Brain volume in first-episode schizophrenia: systematic review and meta-analysis of magnetic resonance imaging studies. Br J Psychiatr 188: 510–518

Stefansson H, Sigurdsson E, Steinthorsdottir V, Bjornsdottir S, Sigmundsson T, Ghosh S, Brynjolfsson J, Gunnarsdottir S, Ivarsson O, Chou TT, Hjaltason O, Birgisdottir B, Jonsson H, Gudnadottir VG, Gudmundsdottir E, Bjornsson A, Ingvarsson B, Ingason A, Sigfusson S, Hardardottir H, Harvey RP, Lai D, Zhou M, Brunner D, Mutel V, Gonzalo A, Lemke G, Sainz J, Johannesson G, Andresson T, Gudbjartsson D, Manolescu A, Frigge ML, Gurney ME, Kong A, Gulcher JR, Petursson H, Stefansson K (2002) Neuregulin 1 and susceptibility to schizophrenia. Am J Hum Genet 71: 877–92

Stefansson H, Sarginson J, Kong A, Yates P, Steinthorsdottir V, Gudfinnsson E, Gunnarsdottir S, Walker N, Petursson H, Crombie C, Ingason A, Gulcher JR, Stefansson K, Clair DS (2003) Association of neuregulin 1 with schizophrenia confirmed in a Scottish population. Am J Hum Genet 72: 83–87

Straub RE, Jiang Y, MacLean CJ, Ma Y, Webb BT, Myakishev MV, Harris-Kerr C, Wormley B, Sadek H, Kadambi B, Cesare AJ, Gibberman A, Wang X, O'Neill FA, Walsh D, Kendler KS (2002) Genetic variation in the 6p22.3 gene DTNBP1, the human ortholog of the mouse dysbindin gene, is associated with schizophrenia. Am J Hum Genet 71: 337–348

Suddath RL, Christison GW, Torrey EF, Casanova MF, Weinberger DR (1990) Anatomical abnormalities in the brains of monozygotic twins discordant for schizophrenia. N Engl J Med 322: 789–794

Sullivan PF, Fan C, Perou CM (2006) Evaluating the comparability of gene expression in blood and brain. Am J Med Genet B Neuropsychiatr Genet 141: 261–268

Sunderland T, Gur RE, Arnold SE (2005) The use of biomarkers in the elderly: current and future challenges. Biol Psychiatr 58: 272–276

Swerdlow NR, Stephany N, Wasserman LC, Talledo J, Shoemaker J, Auerbach PP (2003) Amphetamine effects on prepulse inhibition across-species: replication and parametric extension. Neuropsychopharmacology 28: 640–650

Swerdlow NR, Light GA, Cadenhead KS, Sprock J, Hsieh MH, Braff DL (2006) Startle gating deficits in a large cohort of patients with schizophrenia: relationship to medications, symptoms, neurocognition, and level of function. Arch Gen Psychiatr 63: 1325–1335

Swerdlow NR, Sprock J, Light GA, Cadenhead K, Calkins ME, Dobie DJ, Freedman R, Green MF, Greenwood TA, Gur RE, Mintz J, Olincy A, Nuechterlein KH, Radant AD, Schork NJ, Seidman LJ, Siever LJ, Silverman JM, Stone WS, Tsuang DW, Tsuang MT, Turetsky BI, Braff DL (2007) Multi-site studies of acoustic startle and prepulse inhibition in humans: initial experience and methodological considerations based on studies by the Consortium on the Genetics of Schizophrenia. Schizophr Res 92: 237–251

Symond MP, Harris AW, Gordon E, Williams LM (2005) "Gamma synchrony" in first-episode schizophrenia: a disorder of temporal connectivity? Am J Psychiatr 162: 459–65

Talbot K, Eidem WL, Tinsley CL, Benson MA, Thompson EW, Smith RJ, Hahn CG, Siegel SJ, Trojanowski JQ, Gur RE, Blake DJ, Arnold SE (2004) Dysbindin-1 is reduced in intrinsic, glutamatergic terminals of the hippocampal formation in schizophrenia. J Clin Invest 113: 1353–1363

Tallon-Baudry C, Bertrand O, Delpuech C, Pernier J (1996) Stimulus specificity of phase-locked and non-phase-locked 40 Hz visual responses in human. J Neurosci 16: 4240–4249

Tamminga CA (2006) The neurobiology of cognition in schizophrenia. J Clin Psychiatr 67: e11
Tamminga CA, Holcomb HH (2005) Phenotype of schizophrenia: a review and formulation. Mol Psychiatr 10: 27–39
Todd J, Michie PT, Schall U, Karayanidis F, Yabe H, Naatanen R (2007) Deviant Matters: Duration, Frequency, and Intensity Deviants Reveal Different Patterns of Mismatch Negativity Reduction in Early and Late Schizophrenia. Biol Psychiatr 20: 20
Tsuang MT, Nossova N, Yager T, Tsuang MM, Guo SC, Shyu KG, Glatt SJ, Liew CC (2005) Assessing the validity of blood-based gene expression profiles for the classification of schizophrenia and bipolar disorder: a preliminary report. Am J Med Genet B Neuropsychiatr Genet 133: 1–5
Turetsky BI, Calkins ME, Light GA, Olincy A, Radant AD, Swerdlow NR (2007) Neurophysiological endophenotypes of schizophrenia: the viability of selected candidate measures. Schizophr Bull 33: 69–94
Uhlhaas PJ, Singer W (2006) Neural synchrony in brain disorders: relevance for cognitive dysfunctions and pathophysiology. Neuron 52: 155–168
van Haren NE, Cahn W, Hulshoff Pol HE, Schnack HG, Caspers E, Lemstra A, Sitskoorn MM, Wiersma D, van den Bosch RJ, Dingemans PM, Schene AH, Kahn RS (2003) Brain volumes as predictor of outcome in recent-onset schizophrenia: a multi-center MRI study. Schizophr Res 64: 41–52
Vawter MP, Barrett T, Cheadle C, Sokolov BP, Wood WH III, Donovan DM, Webster M, Freed WJ, Becker KG (2001) Application of cDNA microarrays to examine gene expression differences in schizophrenia. Brain Res Bull 55: 641–650
Vawter MP, Thatcher L, Usen N, Hyde TM, Kleinman JE, Freed WJ (2002) Reduction of synapsin in the hippocampus of patients with bipolar disorder and schizophrenia. Mol Psychiatr 7: 571–578
Velakoulis D, Wood SJ, Wong MT, McGorry PD, Yung A, Phillips L, Smith D, Brewer W, Proffitt T, Desmond P, Pantelis C (2006) Hippocampal and amygdala volumes according to psychosis stage and diagnosis: a magnetic resonance imaging study of chronic schizophrenia, first-episode psychosis, and ultra-high-risk individuals. Arch Gen Psychiatr 63: 139–149
Vilella, E., J. Costas, et al. (2008) Association of schizophrenia with DTNBP1 but not with DAO, DAOA, NRG1 and RGS4 nor their genetic interaction. J Psychiatr Res 42: 278–88
Wang Y, Klijn JG, Zhang Y, Sieuwerts AM, Look MP, Yang F, Talantov D, Timmermans M, Meijer-van Gelder ME, Yu J, Jatkoe T, Berns EM, Atkins D, Foekens JA (2005) Gene-expression profiles to predict distant metastasis of lymph-node-negative primary breast cancer. Lancet 365: 671–679
Weinberger DR, Berman KF (1996) Prefrontal function in schizophrenia: confounds and controversies. Philos Trans R Soc Lond B Biol Sci 351: 1495–1503
Wilke I, Arolt V, Rothermundt M, Weitzsch C, Hornberg M, Kirchner H (1996) Investigations of cytokine production in whole blood cultures of paranoid and residual schizophrenic patients. Eur Arch Psychiatr Clin Neurosci 246: 279–284
Williams NM, Preece A, Spurlock G, Norton N, Williams HJ, Zammit S, O'Donovan MC, Owen MJ (2003) Support for genetic variation in neuregulin 1 and susceptibility to schizophrenia. Mol Psychiatr 8: 485–487
Williams NM, Preece A, Morris DW, Spurlock G, Bray NJ, Stephens M, Norton N, Williams H, Clement M, Dwyer S, Curran C, Wilkinson J, Moskvina V, Waddington JL, Gill M, Corvin AP, Zammit S, Kirov G, Owen MJ, O'Donovan MC (2004) Identification in 2 independent samples of a novel schizophrenia risk haplotype of the dystrobrevin binding protein gene (DTNBP1). Arch Gen Psychiatr 61: 336–344
Winkler I, Karmos G, Naatanen R (1996) Adaptive modeling of the unattended acoustic environment reflected in the mismatch negativity event-related potential. Brain Res 742: 239–252
Wong AH, Van Tol HH (2003) Schizophrenia: from phenomenology to neurobiology. Neurosci Biobehav Rev 27: 269–306
Wong AH, Macciardi F, Klempan T, Kawczynski W, Barr CL, Lakatoo S, Wong M, Buckle C, Trakalo J, Boffa E, Oak J, Azevedo MH, Dourado A, Coelho I, Macedo A, Vicente A, Valente J, Ferreira CP, Pato MT, Pato CN, Kennedy JL, Van Tol HH (2003) Identification of candidate

genes for psychosis in rat models, and possible association between schizophrenia and the 14-3-3eta gene. Mol Psychiatr 8: 156–166
Wong AH, Gottesman, II, Petronis A (2005a) Phenotypic differences in genetically identical organisms: the epigenetic perspective. Hum Mol Genet 14 (Spec No 1): R11–8
Wong AH, Lipska BK, Likhodi O, Boffa E, Weinberger DR, Kennedy JL, Van Tol HH (2005b) Cortical gene expression in the neonatal ventral-hippocampal lesion rat model. Schizophr Res 77: 261–70
Woodward, N. D., K. Jayathilake, et al. (2007) COMT val108/158met genotype, cognitive function, and cognitive improvement with clozapine in schizophrenia. Schizophr Res 90: 86–96
Yao JK, van Kammen DP (2004) Membrane phospholipids and cytokine interaction in schizophrenia. Int Rev Neurobiol 59: 297–326
Yao JK, Sistilli CG, van Kammen DP (2003) Membrane polyunsaturated fatty acids and CSF cytokines in patients with schizophrenia. Prostaglandins Leukot Essent Fatty Acids 69: 429–436
Zipursky RB, Lim KO, Sullivan EV, Brown BW, Pfefferbaum A (1992) Widespread cerebral gray matter volume deficits in schizophrenia. Arch Gen Psychiatr 49: 195–205
Zipursky RB, Lambe EK, Kapur S, Mikulis DJ (1998) Cerebral gray matter volume deficits in first episode psychosis. Arch Gen Psychiatr 55: 540–546
Zipursky RB, Meyer JH, Verhoeff NP (2007) PET and SPECT imaging in psychiatric disorders. Can J Psychiatr 52: 146–157

Chapter 3
Proteomic Strategies for Biomarker Discovery: From Differential Expression to Isoforms to Pathways

Christoph W. Turck(✉), Claudia Ditzen, and Eser Sayan-Ayata

Abstract Proteomics technologies are improving at a great pace with the ultimate goal to allow the sensitive and comprehensive analysis of proteins in tissues and body fluids. The comparison of proteomes from two or more sources is still a challenging task but has already made significant inroads into biomarker discovery. This chapter describes strategies and platforms for biomarker discovery in mouse models and human specimens. We submit that the discovery of proteins that are differentially expressed will ultimately provide valuable information about the pathways involved in the pathobiology of psychiatric disorders.

Abbreviations BPD: Bipolar disorder; CSF, Cerebrospinal fluid; Dkk3: Dickkopf-3 related protein; DSM-IV: Diagnostic and Statistical Manual of Mental Disorders IV; EP: Enolase phosphatase; Glx1: Glyoxalase 1; HAB: High anxiety-related behavior; ICD: International Classification of Diseases; iTRAQ: Isobaric tags for relative and absolute quantitation; LAB: Low anxiety-related behavior; NAB: Normal anxiety-related behavior; NMDAR: *N*-methyl-D-aspartate-type excitatory amino acid receptor; PEDF: Pigment epithelium-derived protein; SAM: *S*-adenosylmethionine; SDS: Sodium dodecyl sulfate; SNP: Single nucleotide polymorphism; UPD: Unipolar depression; 2DPAGE: Two-dimensional polyacrylamide gel electrophoresis

3.1 Introduction

Unlike other major debilitating diseases including cancer and ischaemic heart disease, where we have seen a significant decrease in mortality over the past 30 years, a similar trend is not observed for psychiatric disorders. Also, the prevalence for psychiatric illness has not shown any appreciable decline. It therefore does not

C.W. Turck
Max Planck Institute of Psychiatry Kraepelinstrasse 2–10, Munich, D-80804 Germany
turck@mpipsykl.mpg.de

C.W. Turck (ed.) *Biomarkers for Psychiatric Disorders*,
doi: 10.1007/978-0-387-79251-4_3, © Springer Science+Business Media, LLC 2008

come as a surprise that the "World Health Organization 2002 Report" has labeled psychiatric disorders the major cause for disability in the Western world today (Murray and Lopez, 1997; Kessler et al., 2003).

Genome-wide association studies strongly suggest that a combination of several genes accounts for the various psychiatric disorder phenotypes, with each gene contributing to a small extent. Genetic variations include single nucleotide polymorphisms (SNPs), as well as deletions and duplications of genetic material (Steffanson et al., 2008; The International Schizophrenia Consortium, 2008). As a consequence, it is now widely accepted that only gene interaction analysis will make possible a comprehensive understanding of the genetic contribution to psychiatric disorders. Furthermore, because of the low effect sizes of the candidate genes that have resulted from poorly powered studies and low rates of replication, a validation by meta-analyses is mandatory in order to validate candidate genes (Sullivan et al., 2000; Camp and Cannon-Albright, 2005). Further complicating the interpretation of genotyping results is the fact that many SNPs do not reside inside genes, where they could give rise to altered protein expression, structure, and function, but are found in intergenic regions, the so-called dark matter of the genome, where their impact on the phenotype remains obscure. Finally, epigenetic factors caused by the environment, especially those with effects on critical aspects of developmental experience and stressful life events, seem to greatly contribute to psychiatric disorder pathobiology (Mill and Petronis, 2007). Specific environmental factors will ultimately trigger a genetic risk background to develop into a psychiatric disorder and different patient phenotypes for each genotype. A quotation from Leo Tolstoy's novel Anna Karenina "all happy families resemble one another, but each unhappy family is unhappy in its own way" sums up the heterogeneous presentation of psychiatric disorders that results from the complex interplay between genes and the environment.

Unlike other diseases, psychiatric disorders do not reveal themselves through a lesion and in all likelihood originate from an abnormal processing and/or activity in neural networks that involve several brain areas (Insel and Quirion, 2005). The characterization of these altered networks and neural circuits will be critical to get a better understanding of psychiatric disorders and in turn make possible the definition of targets for the development of more specific medicines. Since current medications act on targets that are in all likelihood quite remote from pathways that are relevant for the pathobiology of psychiatric disorders, they are characterized by limited efficiency and a number of side effects (Hyman, 2007).

Another complicating factor in the area of psychiatric disorders is the imprecise diagnosis that is not based on a molecular pathophysiology but instead relies on clinical observations of symptom clusters (Hyman, 2007). Clinical criteria for the diagnosis of psychiatric disorders are based on the *Diagnostic and Statistical Manual of Mental Disorders IV* (DSM-IV) and the *International Classification of Diseases* (ICD-10) and to a large extent include verbal communication between psychiatrist and patient, which renders them very subjective. In addition, an overlap of the general symptoms makes it particularly difficult to deal with the great diversity of psychiatric disorders. In this regard it is also increasingly questioned whether the traditional boundary between psychiatric and neurological disorders is still appropriate despite the fact that the latter are characterized by well-defined lesions (Insel and Quirion, 2005). Even though psychiatric disorders in all likelihood involve dysfunctional neural circuits, there is increasing evidence that certain aspects of their pathobiology are related to neurological diseases, a finding that is also supported by the often-observed comorbidity of neurological and psychiatric disorders. Owing to the late detection of psychiatric disorder onset, their prevention is not strategic and treatment is by and large based on trial and error (Hyman, 2007). New drug candidates are therefore needed, making the obligatory response studies as important as disease susceptibility investigations.

Here is where the field of biomarker discovery can make important contributions. For both types of investigation, drug response and disease susceptibility, biomarkers are needed to move the area of psychiatric disorders into the rest of medicine (Hyman, 2007). Only through a combined interrogation of genetic variations and biomarkers will it be possible to achieve this goal. The obscure etiology and pathogenesis of psychiatric disorders, combined with the fact that today's treatments are empirical and at best symptomatic, also provide a great incentive for psychiatric disorder biomarker discovery efforts. Ultimately, it is hoped that biomarkers will assist in stratifying patient groups with similar clinical features and at the same time help in the identification of neural circuitries that are responsible for disease etiology. This will enable the complementation of the presently applied DSM-IV and ICD-10 criteria with specific sets of biomarkers and result in a more precise nosological framework for psychiatric disorders (Hyman, 2007). In addition, biomarkers will make possible a predictive pharmacogenomics approach for newly developed medicines with new target sites that will result in more specific medicines with fewer side effects.

A number of different assays and methods can produce biomarkers for psychiatric disorders (Fig. 3.1). In this chapter we will focus on the identification of proteins, the functional molecules in cells and organisms, and their roles in biomarker discovery.

3.2 Strategies

The 'Proteomics and Biomarkers' research group at the Max Planck Institute of Psychiatry is engaged in the identification of biomarkers for psychiatric and neurological disorders. Several global proteomics approaches with human and rodent tissue and body fluid specimens are used for this purpose (Turck et al., 2005) (Fig. 3.2). In addition, the examination of autoantibodies in patient cerebrospinal fluid (CSF) constitutes an exploratory project that is based on the hypothesis of a dysfunctional immune response in the pathobiology of psychiatric disorders. Finally, we also attempt to exploit the great wealth of information on psychiatric disorders in the public domain through an *in silico* interrogation of experimental data deposited in various databases and article texts.

A particular focus of our biomarker discovery efforts is the analysis of mouse models that represent features relevant for psychiatric disorders. The rationale

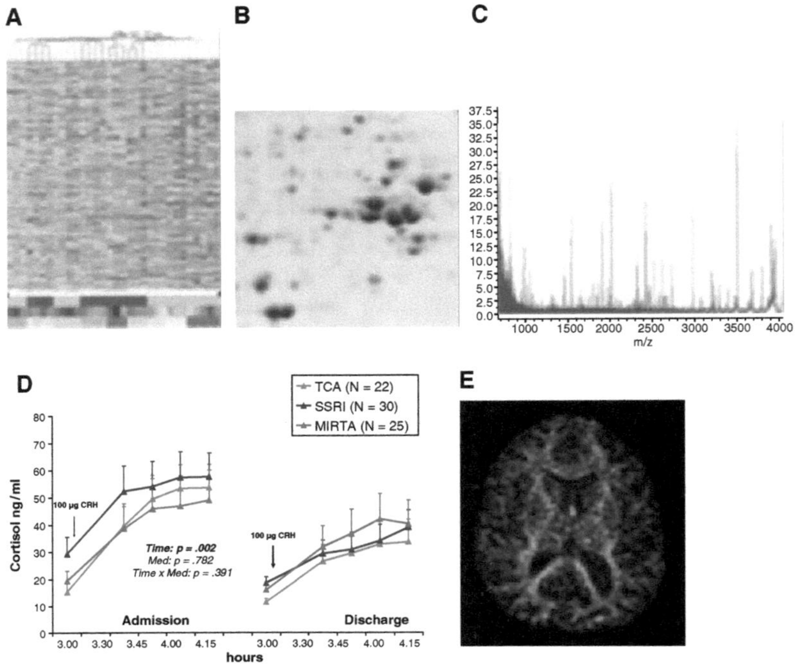

Fig. 3.1 A biomarker is any characteristic that can be objectively measured to reflect physiological, pharmacological, or disease processes in animals or humans (De Gruttola et al., 2001). Various platforms for different types of biomarkers can be used for psychiatric disorders. (**a**) RNA microarray, (**b**) Protein expression analysis by gel electrophoresis, (**c**) Mass spectrometry-based protein and metabolite profiling, (**d**) Neuroendocrine assay, (**e**) Brain MRI scan (*See Color Plates*)

behind the use of animal models is that the analysis of patient specimens is plagued by several limiting factors. Foremost, as discussed in the Introduction, psychiatric disorders present themselves in a rather heterogeneous manner owing to diverse sets of disease-causing genes and different environmental influences varying from patient to patient. Also, since each gene contributes to the disease to a rather small extent one can expect only minor effects on particular biochemical pathways, resulting in a low disease marker signal-to-noise ratio. Other interindividual differences that are not related to the disease phenotype further complicate this issue. Finally, human specimens relevant for the proteomic analysis of psychiatric disorders are available in limited amounts and their retrieval is difficult to control in a consistent manner. This is true for both types of patient specimen that are commonly used for proteomic analyses of psychiatric disorders: *postmortem* brain tissue and CSF. In the case of *postmortem* brain tissue, prolonged times between death and tissue preservation can lead to protein degradation and/or modification, which will produce artifacts picked up during proteomics analysis (Sköld et al., 2007). Similar effects are also seen in the case of CSF specimens obtained by lumbar puncture, a procedure that is difficult to control with regard to timing and patient constitution.

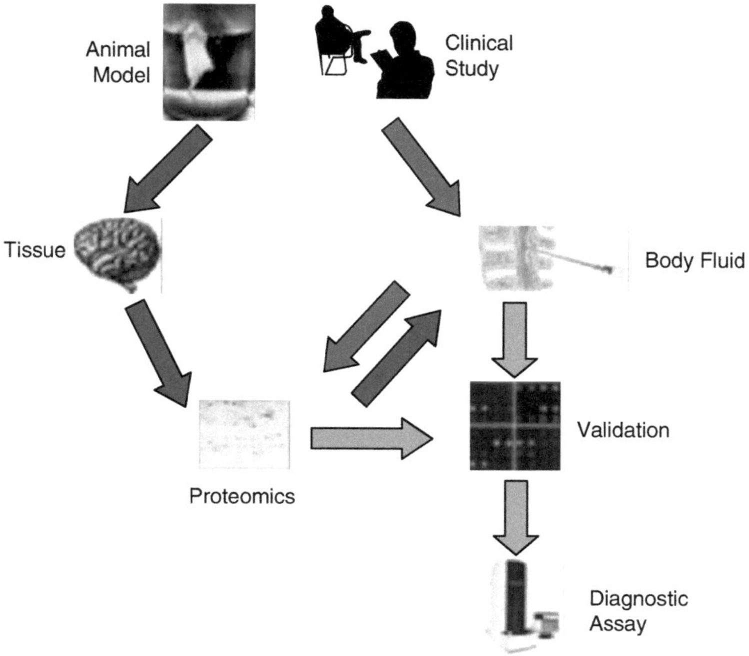

Fig. 3.2 Roads to protein biomarker discovery. On one hand, brains from animal models that reflect certain aspects of the disease phenotype are investigated by the proteomics methods. Once candidate markers have been identified from the animal model, they need to be validated in patient specimens. Alternatively, clinical specimens are used directly for biomarker discovery, validated in a great number of samples, and ultimately used for establishing a diagnostic assay platform (*See Color Plates*)

Patient studies are also often compromised by medication and other treatments which will affect the proteome constituents.

Despite these challenges, we have not excluded human specimens from our biomarker discovery studies (Turck, 2005). Experienced psychiatrists in our clinic ensure that CSF specimens are obtained from carefully phenotyped patient populations in a consistent manner and include detailed clinical and neuroendocrine data, which make the grouping of samples to be used for proteomic analysis more viable.

In addition to a global proteome analysis of CSF, we are also using a semitargeted approach for the identification of novel disease markers that exploits the great specificity of the body's immune system. For this purpose we are interrogating the antibody pool that is present in CSF. The basis for this approach stands on the hypothesis that several factors may trigger an autoimmune response within the central nervous system (CNS) especially in individuals with psychiatric disorders that are dispositioned to immune system dysfunction and impaired blood—brain barriers. We therefore hypothesize that the presence of markers for psychiatric diseases is reflected by the appearance of autoantibodies in CSF. These autoantibodies can be used for the identification of their respective autoantigen targets.

The mining of already existing data in the public domain is a recent approach that we have begun to explore. It is based on the assumption that a great amount of information pertinent to biomarker discovery has already been acquired in the numerous genetic, proteomic, and clinical studies and deposited in databases or text documents where it awaits exploitation with the right bioinformatic tools. Obviously, not all the publicly deposited data is relevant and of sufficient quality, but the sheer volume of the data warrants a thorough interrogation with regard to their significance as biomarkers.

3.3 Trait Anxiety Mouse Model

The utilization of animal models for psychiatric disorders comes with the realization that these models represent only certain aspects of the disorder and not the disorder itself (Insel, 2007). Even if one had the knowledge of most, if not all, genes causing a particular psychiatric disorder, it is at present not technically feasible to create models by manipulating a great number of genes at a time in the same animal. The generation of animal models based on the manipulation of single genes, on the other hand, has a poor chance of achieving a penetrating level that is sufficient to reflect the disease phenotype observed in patients.

The symptom anxiety is the normal action to danger but becomes abnormal when the response is unproportional. Clinical and epidemiological data have shown that the symptom anxiety is often found in other psychiatric disorders including depression, obsessive-compulsive disorders, and post-traumatic stress disorder (Gross and Hen, 2004).

A number of animal models reflecting the anxiety phenotype have been generated through either genetic manipulation, exposure to trauma or social stress, or maternal separation in early developmental stages. Alternatively, inbred mouse strains that inherently differ in their natural anxiety levels are used to study the phenotype in greater detail (Finn et al., 2003; Cryan and Mombereau, 2004; Gordon and Hen, 2004).

Dr. Rainer Landgraf, head of the 'Behavioral Neuroendocrinology' research group at the Max Planck Institute of Psychiatry, has established a robust mouse model of trait anxiety using a selective breeding protocol, based on the animal's behavior in a commonly used assay for anxiety, the elevated plus maze (Krömer et al., 2005). This model does not have the disadvantage inherent to studies dealing with a comparison of unselected inbred or outbred mouse lines, which in addition to anxiety also differ in other phenotypes. The intrastrain breeding approach has the benefit that it centers on only a limited set of traits related to anxiety. Consequently, studies with this type of mouse model will increase the likelihood of identifying trait-relevant parameters no matter whether they are genetic, proteomic, metabolomic, or otherwise in nature. On the basis of a bidirectional breeding approach, mouse lines of hyperanxious (HAB) and hypoanxious (LAB) phenotypes were established and validated. These lines were derived from CD1 mice, which, due to their outbred nature, differ in their anxiety-related behavior using the elevated plus maze and other behavioral assays that reveal symptoms indicative of psychiatric disorders.

Our biomarker discovery efforts with the above mouse lines initially used a classical proteomics platform, two-dimensional polyacrylamide gel electrophoresis (2DPAGE) (Klose, 1975; O'Farrell, 1975), for differential protein expression analysis (Krömer et al., 2005). Two proteins were identified that showed quantitative and qualitative differences, respectively, between HAB and LAB mice. The quantitative difference was identified as glyoxalase 1 (Glx1), a protein expressed in the cytosol of cells and tissues of many organisms (Hayes et al., 1989; Thornalley, 1993). The enzyme plays a major role in the detoxification of methylglyoxal, which is a potent cytotoxic metabolite. Glx1 catalyzes the transformation of methylglyoxal and glutathione to *S*-lactoylglutathione, which is then converted to D-lactic acid by glyoxalase 2. Owing to its ubiquitous expression, Glx1 is believed to be of fundamental importance for cellular metabolism. The fact that the enzyme uses glutathione as a cosubstrate points to a functional role of Glx1 in oxidative stress mechanisms. Studies have implicated Glx1 in several brain disorders including Alzheimer's disease (Chen et al., 2004) and autism (Junaid et al., 2004). A possible connection between Glx1 and unipolar affective disease has been found in a linkage study (Tanna et al., 1989). Interestingly, in an analysis using different inbred strains of mice it was also found that Glx1 and glutathione reductase 1, another enzyme with a function in oxidative stress mechanisms, play a causal role in anxiety (Hovatta et al., 2005). Still, the question whether Glx1 represents a risk marker or a risk factor for the anxiety-related phenotype in mice remains unknown at the present time (Thornalley, 2006).

In the HAB/LAB animals, Glx1 is present in many cell types in which its expression level reflects the one found in the brain and makes its determination in blood cells feasible. On the basis of this finding we have used a western blot assay specific for Glx1 to screen red blood cell specimens from patients afflicted with anxiety disorders and depression (Ditzen et al., 2006). These studies are presently expanded to a great number of samples to find out whether Glx1 can be used as a biomarker for anxiety in the clinical laboratory. In this regard it has been recently demonstrated that reduced levels of Glx1 mRNA can be found in mood disorder patients in a current depressive state compared to healthy control subjects (Fujimoto et al., 2008).

The other difference found during proteome analysis of HAB and LAB brains also represents an enzyme, enolase phosphatase (EP), which is expressed as different isoforms in the two lines (Ditzen et al., 2006). Caused by two SNPs that result in amino acid changes, the protein isoforms migrate at different positions during 2DPAGE. Apart from its altered mobility in SDS gels, the HAB EP isoform has also a lower enzymatic activity compared to the LAB/NAB isoform. This activity difference probably affects the methionine salvage pathway, of which EP is a member. The metabolic pathway is of interest in the area of psychiatric disorders as it includes a metabolite, *S*-adenosylmethionine (SAM), which is reported to be a natural mood stabilizer. Furthermore, antidepressant activities of SAM have been demonstrated in clinical trials (Bressa, 1994; Silveri et al., 2003). Aside from SAM, the methionine salvage pathway has an interesting connection to another pathway that is relevant for psychiatric disorders. The polyamine pathway includes putrescine, spermidine, and spermine, which have been shown to affect neurotransmission caused by their ability to modulate ion channels such as the *N*-methyl-D-aspartate-type excitatory amino acid receptor (NMDAR) (Bernstein and Müller, 1999). The latter is involved in glutamatergic neurotransmission and associated with long-term potentiation, neuronal development and neuronal plasticity (Williams, 1997), as well as affective disorder pathobiology (Skolnik, 1999). Furthermore, animal models for depression have shown that levels of the three polyamines are altered in specific brain areas compared to control animals and that SAM administration has the ability to modulate polyamine levels (Genedani et al., 2001).

The proteomic analysis of a robust and valid animal model for trait anxiety has resulted in two marker candidates that are part of metabolic pathways pertinent to the disease phenotype. Although not necessarily causative for the anxiety phenotype, these proteins provide valuable information with regard to the pathways that are involved. As a consequence, we submit that genetic and proteomic differences that are identified in animal models are not only useful as biomarkers themselves but at the same time open the gate for an extended interrogation of metabolites that are part of the affected pathways and may represent biomarkers in their own right. The elucidated pathways can thus provide valuable information for metabolic assays of specimens from patients afflicted with anxiety and affective disorders. Studies are in progress that will assess metabolites from the methionine salvage and polyamine pathways to examine their potential use in clinical assays for anxiety (Fig. 3.3).

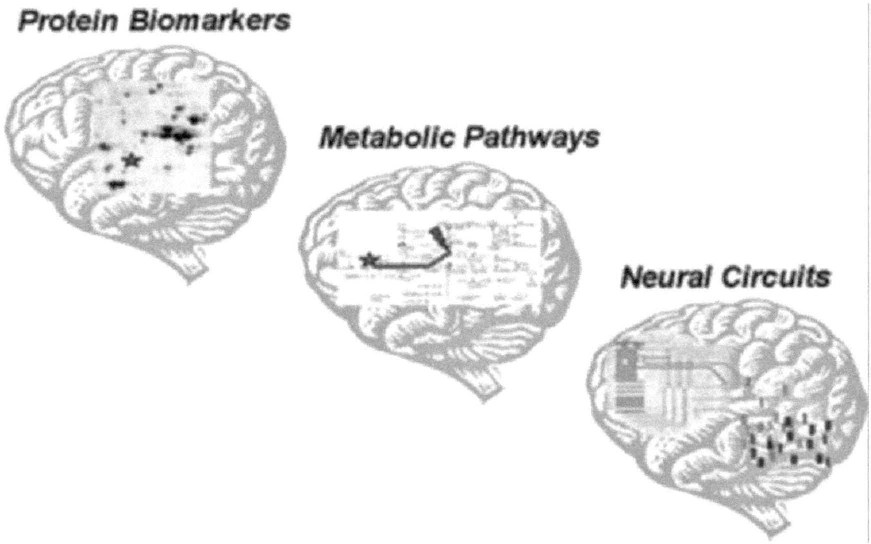

Fig. 3.3 From protein biomarkers to metabolic pathways to neural circuits. Phenotype-related protein expression differences provide information on metabolic pathways that are affected in psychiatric disorders. A combination of altered metabolic pathways may ultimately precipitate disease by distressing critical neural circuits (*See Color Plates*)

Despite our success in biomarker discovery using the 2DPAGE method, we seek to extend the list of trait anxiety markers using more sophisticated proteomic methods. For complex diseases such as psychiatric disorders, a great number of markers will be needed for a reliable diagnosis characterized by high specificity and sensitivity. To achieve this goal, we have established a more sensitive, comprehensive, and consistent proteomics platform that omits the inherent limitations of the 2DPAGE method and allows quantitative mass spectrometry with great precision and sensitivity. This method involves metabolic labeling of mouse models with stable isotopes. Using this method, tissues and body fluids from metabolically labeled case and nonlabeled control mice can be combined before any sample work-up steps such as lysis, protein fractionation, and digestion are performed. The metabolic labeling procedure therefore avoids the introduction of any artificial differences caused by an inconsistent sample preparation prior to and during the proteomic comparison. The method was initially restricted to lower organisms such as bacteria and yeast, but was later also applied to mammalian cells in culture. More recently, the metabolic labeling approach has been used in mammals when rats were differentially grown on ^{15}N-enriched and -depleted diets (Wu et al., 2004).

To achieve optimal sensitivity, mice are labeled with the ^{15}N isotope-enriched diet *in utero* with continued feeding after birth. Mice are then sacrificed after assessing their phenotype with several behavioral assays, and brain tissue and blood is quick-frozen in liquid nitrogen. Applying the method to the HAB/LAB mouse model, we have found that ^{15}N-enrichment in brain tissue and blood is 93 and 95%, respectively, which is a sufficiently high incorporation rate for sensitive protein analysis and quantitation by tandem mass spectrometry. For a thorough interrogation of the mouse proteomes, we are using brain sections and organellar fractions (cytosolic, membrane, nuclear). Peptide ratios from tryptic digests are determined on the basis of the ion current ratios of each light (^{14}N) and heavy (^{15}N) peptide pair. With the help of a mass spectrometry data quantitation software that we have developed, changes in protein expression can be estimated by using multiple peptide pairs for each protein (Fig. 3.4).

3.4 Cerebrospinal Fluid

Because of the close proximity and perfusion of the brain, CSF contains mediators that reflect metabolic processes in the CNS. CSF therefore represents the most appropriate biomarker source for patients afflicted with psychiatric disorders. Proteins that are secreted or shed from brain cells are present in CSF and, owing to an exchange, also found in blood, albeit at reduced levels. This is an important aspect for a biomarker as, owing to its easier availability, blood will ultimately be the body fluid of choice for a routine clinical diagnostic assay for psychiatric disorders (LaBaer, 2005). We therefore propose to use CSF for biomarker discovery and blood for eventual routine screening in the clinical laboratory.

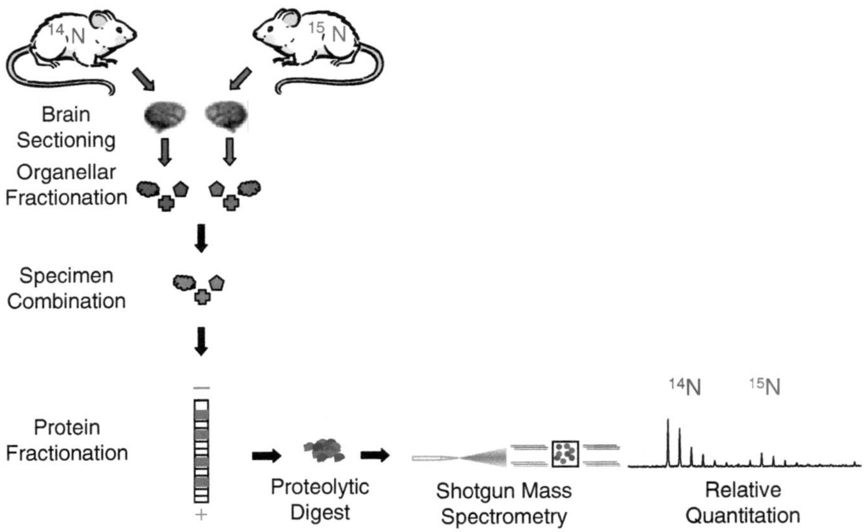

Fig. 3.4 Metabolic labeling approach for sensitive and comprehensive protein analysis and comparison. Mice are fed with either regular ^{14}N or labeled ^{15}N diets. Organ or body fluid specimens are then combined and processed for relative protein quantitation by mass spectrometry (*See Color Plates*)

The proteomic analysis of body fluids such as CSF remains a challenge (Anderson and Anderson, 1998). An important prerequisite for biomarker discovery is that retrieval of CSF by lumbar puncture is carried out in a controlled fashion to minimize variability. The limiting amounts of starting material and the large dynamic range of protein concentrations in CSF (up to 12 orders of magnitude) between the highest and lowest expressed proteins make any proteomic-based analysis difficult (Anderson and Anderson, 1998). As current technologies are limited by the amount of protein they can reliably detect and the large dynamic range, only a fraction of the CSF proteome is interrogated. In addition, owing to a leaky blood—brain barrier that is especially pronounced in patients with brain disorders, many serum proteins can infiltrate CSF. This makes it difficult to know in many cases whether a CSF protein is derived from the brain or serum.

To improve the depth of CSF proteome coverage, abundant proteins are first depleted and the remaining less abundant proteins further fractionated prior to mass spectrometry analysis. As is always the case in protein analysis, the more one fractionates, the more proteins can be identified. However, because of limited sample amounts, this can be done only to a certain extent.

In the first set of studies, we have carried out proteome mining experiments to get a feel of the complexity of the CSF protein constituents (Maccarrone et al., 2004a, b). For this purpose we employed the shotgun mass spectrometry approach after depletion of the abundant proteins followed by an extensive fractionation at the protein or tryptic peptide level. This was achieved by protein anion exchange chromatography, SDS-PAGE and tryptic peptide isoelectric focusing. We have been able to identify over 1,000 CSF proteins that belong to several groups with regard to molecular function, biological process, and cellular compartment distribution. On the basis of a comparison with two human serum protein databases, approximately half of the proteins are derived from brain tissue (Anderson et al., 2004; Chan et al., 2004).

The logical next step in our CSF proteomics studies is to move from a mining to a scoring mode. In other words, we compare CSF proteomes from different patient groups to identify biomarkers for specific psychiatric phenotypes. These endeavors have been met with only limited success. In retrospect, we were probably overly optimistic in assuming that the identification of biomarkers with a small number of CSF specimens was feasible. The problems associated with the proteomic analysis of human specimens have been discussed in the Introduction. In addition, it is now widely accepted in the area of genome-wide association studies of complex disorders, in which individual genes have a rather small effect on the phenotype, that a great number of samples have to be interrogated in order to come up with significant hits. This situation in all likelihood is not much different for protein biomarker discovery efforts. Comparative proteomic analyses that we have attempted with samples from patients with unipolar depression (UPD) and normal controls, using either 2DPAGE or the iTRAQ stable isotope labeling method (Ross et al., 2004) in combination with quantitative mass spectrometry, have resulted in a limited set of differences that were not consistently found for all patients. This is, of course, not surprising in light of the fact that not every patient is predicted to have the same set of biomarkers. Instead, it is more likely that each patient will have different sets of markers with a certain degree of overlap between patients.

Analysis and comparison by 2DPAGE of CSF from UPD patients and controls revealed interesting differences in protein isoform expression. One example of such an isoform variation in UPD patients is the pigment epithelium derived protein (PEDF) (Fig. 3.5). PEDF is a glycoprotein with neuroprotective effects by inducing prosurvival genes in neurons (Yabe et al., 2005). If and how these isoform discrepancies contribute to the pathomechanism of depressive disorders is unclear at this point. Since PEDF was also found as a potential marker for early diagnosis of Alzheimer's disease (Yamagishi et al., 2004), it might be indicative of a possible compensatory mechanism of the brain to fight against neuronal cell injury.

The other difference of interest observed by 2DPAGE is the Dickkopf-3 related protein (Dkk3), which is expressed at lower levels in CSF of UPD patients. Dkk3 is a secreted glycoprotein that suppresses the neurodevelopmental wingless cascade and is critical for embryonic head development and synaptic function in the adult brain. A decrease in Dkk3 mRNA in *postmortem* brains of schizophrenic individuals has been reported (Ftouh et al., 2005). We are presently assessing to what extent the two protein differences can contribute to a biomarker list for UPD.

On the basis of the CSF protein mining results (Maccarrone et al., 2004a), we are now establishing a different type of proteomic screen that involves an antibody array platform. A project proposal to the 'Human Proteome Resource' in Sweden

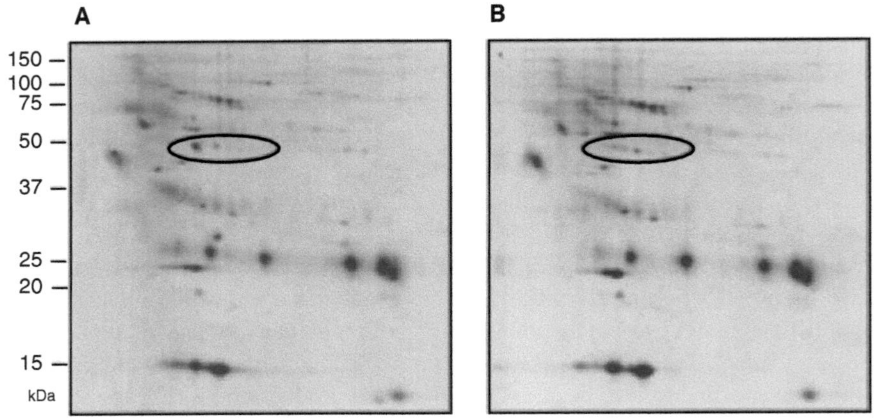

Fig. 3.5 2DPAGE comparision of CSF from controls (**a**) and UPD patients (**b**). The pigment epithelium-derived protein isoforms are circled. Positions of molecular weight standards are indicated on the left

(Uhlen, 2007) was approved and will provide us with 200 antibodies specific for proteins that we have previously identified in human CSF. This method has also the advantage that it consumes a significantly smaller amount of CSF sample and therefore will permit screening with a greater number of CSF specimens. As explained above, we expect to find different but overlapping sets of biomarkers between patients of the same disease group. It remains to be seen whether different patient groups will also result in such an overlap, which is not inconceivable given the fact that different psychiatric disorders also show a genetic overlap.

3.5 Autoantibodies

Autoantibodies against brain proteins in serum and CSF of patients afflicted with psychiatric disorders have been reported (Wang et al., 2003; Hornig et al., 1999). In one such study, it was found that an increased immunoglobulin G titer was present in CSF and serum during a state of depression (Hornig et al., 1999). Besides an immune dysfunction, there is also evidence for an impaired blood–brain barrier in patients with psychiatric disorder, which results in an increased risk for an autoimmune response (Wang et al., 2003). The autoimmune response targets, the autoantigens, may represent a valuable class of biomarkers for psychiatric disorders. Similar approaches have already provided important information in other disease areas. These include Type I autoimmune diabetes mellitus, rheumatoid arthritis, and multiple sclerosis (Mathey et al., 2007).

In preliminary studies we were able to demonstrate, by western blot immunoassays, that autoantibodies against brain proteins are indeed present in CSF from patients afflicted with bipolar disorder (BPD). In these experiments, human brain protein extracts were fractionated by SDS gel electrophoresis and transferred to a membrane. The brain protein blot was then probed with different patient CSF specimens, which after blot development resulted in the detection of discrete bands that represent brain proteins specifically recognized by CSF autoantibodies. The limited amount of CSF specimen available prompted us to explore sensitive phage display and protein array screens for the identification of the detected brain autoantigens. Unlike the western blot, these methods allow not only the detection but also the subsequent identification of protein antigens with the small amounts of autoantibodies that are present in CSF.

Phage display screens with CSF from BPD patients resulted in 64 positive clones, with some derived from the same protein family or sharing a common metabolic pathway. After *in silico* interrogation by pathway analysis and text mining, 20 candidate biomarkers were selected for further validation studies. In a complementary approach, we used a protein array in order to monitor the CSF autoantibody pool. The protein array consisted of over 27,000 proteins derived from expression-verified, full-length, as well as shorter cDNA clones. Proteins printed in duplicate onto 22 × 22 cm membranes were overlaid with CSF and developed, which resulted in a number of positive hits that were also found during the phage display screens (Fig. 3.6).

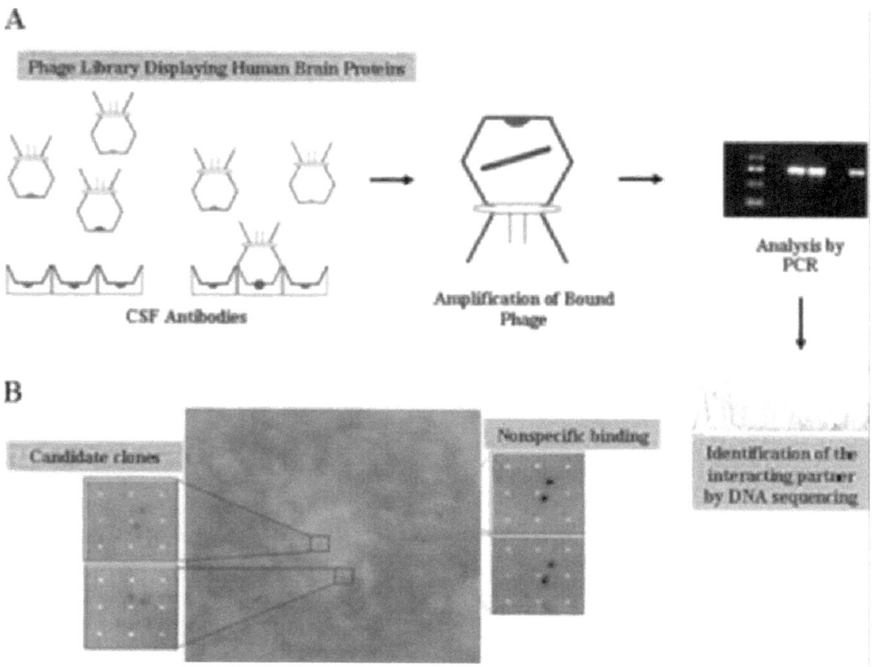

Fig. 3.6 Phage display (**a**) and protein array (**b**) screens for CSF autoantibodies. Positive clones are candidate brain autoantigen biomarkers (*See Color Plates*)

3.6 Validation

With the great flood of data resulting from today's "omics" technologies, criteria need to be developed for the subsequent validation of the most promising set of candidates before translating them into the clinical laboratory. As mentioned previously, we believe that the protein markers identified with the various experimental platforms may not necessarily be the best biomarkers themselves. However, they may have great potential in identifying disease-pertinent pathways. Other proteins or metabolites that are part of these pathways may be more applicable as biomarkers or could even serve as drug targets. For this reason, we are using pathway programs and protein interaction databases as well as text mining software to establish a list of marker targets by starting with the proteins identified during the proteomic interrogation of mouse model and patient specimens.

As a first step in the validation of a biomarker candidate, we are typically using immunoassay-based screens with the same material that was used as a source in the proteomics discovery process. This assay eliminates any candidate markers that result from artifacts. In the second step of validation, patient body fluid specimens, including serum and CSF, are used. Here it is of paramount importance that the patients are carefully phenotyped by the physicians and grouped according to the results of the clinical phenotyping data. Only then can one expect meaningful results from the validation studies. As explained in the Introduction, it should also be kept in mind that not every patient of a group will have the same set of biomarkers. A thorough statistical analysis for biomarker patterns is therefore critical and in all likelihood will result in biomarker overlaps between patients. We take advantage of a large CSF specimen bank that has been established at our institute through the associated Psychiatry and Neurology Clinics. This bank now contains over 1,000 samples that have been carefully prepared and stored and are derived from groups of patients who have been thoroughly characterized by physicians in our clinic. As is the case for all polygenic diseases, we do not anticipate that a single marker will be able to unequivocally distinguish different clinical phenotypes. Only through a combination of a great number of markers will it be possible to gain statistical significance to differentiate complex traits and establish a sensitive and selective diagnostic assay.

3.7 Conclusion

From the above discussion of our biomarker discovery strategies, it is clear that only a combination of technologies will result in a valid list of markers that can ultimately be used for a clinical assay. Once this goal has been achieved, patient stratification and diagnosis of psychiatric disorders will greatly improve. On the basis of our own data, there is a good chance that a number of different brain disorders including neurological diseases such as Alzheimer's disease, Parkinson disease and multiple sclerosis will share certain biomarkers with psychiatric disorders.

Data that we have obtained from biomarker analyses of the *Experimental Autoimmune Encephalomyelitis* mouse model for multiple sclerosis indicate that this is indeed the case (Jastorff et al., submitted).

Looking ahead, biomarker information will not only be a critical requirement for the establishment of more reliable clinical diagnostic assays but at the same

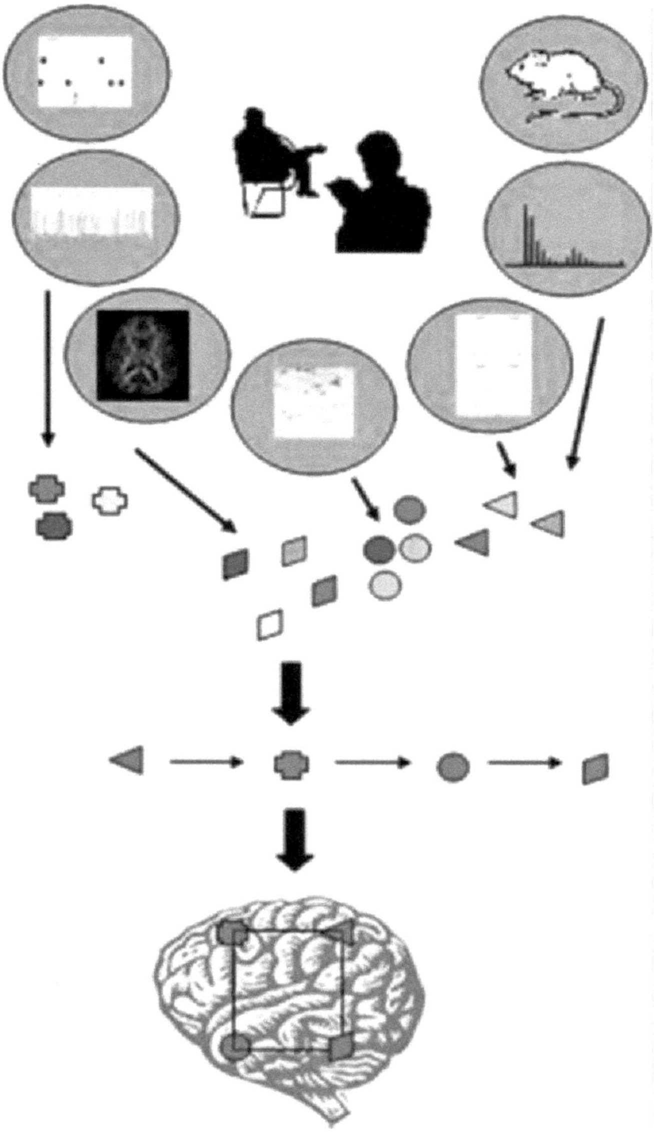

Fig. 3.7 Data integration for insights into psychiatric disorder pathophysiology. Experimental and clinical data obtained from diverse platforms are consolidated with the help of bioinformatics tools in order to identify affected neural pathways (*See Color Plates*)

time reveal pathologic mechanisms for psychiatric disorders, which in all likelihood involve dysfunctional neural pathways. A prerequisite, however, will be the consolidation and integration of a number of different data sets resulting from disciplines such as epidemiology, statistical genetics, proteomics, metabolomics, and others (Fig. 3.7).

References

Anderson NL, Anderson NG (1998) Proteome and proteomics: new technologies, new concepts, and new words. Electrophoresis 19:1853–1861

Anderson NL, Polanski M, Pieper R et al (2004) The human plasma proteome: a nonredundant list developed by combination of four separate sources. Mol Cell Proteomics 3:311–326

Bernstein H-G, Müller M (1999) The cellular localization of the L-ornithine decarboxylase/ polyamine system in normal and diseased central nervous systems. Prog Neurobiol 57:485–505

Bressa, GM (1994) *S*-adenosyl-l-methionine (SAMe) as antidepressant: meta-analysis of clinical studies. Acta Neurol Scand 154 (Suppl):7–14

Camp NJ, Cannon-Albright LA (2005) Dissecting the genetic aetiology of major depressive disorder using linkage analysis. Trends Mol Med 11:138–144

Chan KC, Lucas DA, Hise D et al (2004) Analysis of the human serum proteome. Clin Proteomics 1:101–225

Chen F, Wollmer MA, Hoerndli F et al (2004) Role for glyoxalase I in Alzheimer's disease. Proc Natl Acad Sci U S A 101:7687–7692

Cryan JF, Mombereau C (2004) In search of a depressed mouse: utility of models for studying depression-related behavior in genetically modified mice. Mol Psychiatr 9:326–357

De Gruttola VG, Clax P, DeMets DL et al (2001) Considerations in the evaluation of surrogate endpoints in clinical trials. Summary of a National Institutes of Health workshop. Control Clin Trials 22:485–502

Ditzen C, Jastorff AM, Kessler MS et al (2006) Protein biomarkers in a mouse model of extremes in trait anxiety. Mol Cell Proteomics 5:1914–1920

Finn DA, Rutledge-Gorman MT, Crabbe JC (2003) Genetic animal models of anxiety. Neurogenetics 4:109–135

Ftouh S, Akbar MT, Hirsch SR, de Belleroche JS (2005) Down-regulation of Dickkopf 3, a regulator of the Wnt signalling pathway, in elderly schizophrenic subjects. J Neurochem 94:520–530

Fujimoto M, Uchida S, Watanuki, T et al (2008) Reduced expression of glyoxalase-1 mRNA in mood disorder patients. Neurosci Let 438:196–199

Genedani S, Saltini S, Benelli A et al (2001) Influence of SAMe on the modifications of brain polyamine levels in an animal model of depression. Neuroreport 12:3939–3942

Gordon JA, Hen R (2004) Genetic approaches to the study of anxiety. Annu Rev Neurosci 27:193–222

Gross C, Hen R (2004) The developmental origins of anxiety. Nat Rev Neurosci 5:545–552

Hayes JD, Milner SW, Walker SW (1989) Expression of glyoxalase, gluthatione peroxidase and gluthatione S-transferase isoenzymes in different bovine tissues. Biochim Biophys Acta 994:21–29

Hornig M, Amsterdam JD, Kamoun M, Goodman DBP (1999) Autoantibody disturbances in affective disorders: a function of age and gender? J Affect Disord 55:29–37

Hovatta I, Tennant RS, Helton R et al (2005) Glyoxalase 1 and glutathione reductase 1 regulate anxiety in mice. Nature 438:662–666

Hyman SE (2007) Can neuroscience be integrated into the DSM-V? Nat Rev Neurosci 8:725–732

Insel TR (2007) From animal models to model animals. Biol Psychiatry 62:1337–1339
Insel TR, Quirion R (2005) Psychiatry as a clinical neuroscience discipline. J Am Med Assoc 294:2221–2224
Jastorff AM et al. (submitted)
Junaid MA, Kowal D, Barua M et al (2004) Proteomic studies identified a single nucleotide polymorphism in glyoxalase I as autism susceptibility factor. Am J Med Genet 131:11–17
Kessler RC, Berglund P, Demler O et al (2003) The epidemiology of major depressive disorder: results from the national comorbidity survey replication (NCS-R). JAMA 289:3095–3105
Klose J (1975) Protein mapping by combined isoelectric focusing and electrophoresis of mouse tissues. A novel approach to testing for induced point mutations in mammals. Humangenetik 26:231–243
Krömer SA, Kessler MS, Milfay D et al (2005) Identification of glyoxalase-I as a protein marker in a mouse model of extremes in trait anxiety. J Neurosci 25:4375–4384
LaBaer J (2005) So, you want to look for biomarkers. J Proteome Res 4:1053–1059
Maccarrone G, Birg I, Malisch E et al (2004a) In-depth analysis of the human CSF proteome using protein prefractionation. Clin Proteomics J 1:333–364
Maccarrone G, Milfay D, Birg I (2004b) Mining the human CSF proteome by immunodepletion and shotgun mass spectrometry. Electrophoresis 25:2402–2412
Mathey EK, Derfuss T, Storch MK et al (2007) Neurofascin as a novel target for autoantibody-mediated axonal injury. J Exp Med 204:2363–2372
Mill J, Petronis A (2007) Molecular studies of major depressive disorder: the epigenetic perspective. Mol Psychiatr 12:799–814
Murray CJ, Lopez AD (1997) Global mortality, disability, and the contribution of risk factors: global burden of disease study. Lancet 349:1436–1442
O'Farrell PH (1975) High resolution two-dimensional electrophoresis of proteins. J Biol Chem 250:4007–4021
Ross PL, Huang YN, Marchese JN et al (2004) Multiplexed protein quantitation in Saccharomyces cerevisiae using amine-reactive isobaric tagging reagents. Mol Cell Proteomics 3:1154–1169
Silveri MM, Parow AM, Villafuerte RA et al (2003) S-adenosyl-L-methionine: effects on brain bioenergetic status and transverse relaxation time in healthy subjects. Biol Psychiatr 54:833–839
Sköld K, Svensson M, Norrman M et al (2007) The significance of biochemical and molecular sample integrity in brain proteomics and peptidomics: Stathmin 2–20 and peptides as sample quality indicators. Proteomics 7:4445–4456
Skolnik P (1999) Antidepressants for the new millenium. Eur J Pharmacol 375:31–40
Stefansson H et al. (2008) Large recurrent microdeletions associated with schizophrenia. Nature Jul 30, Epub ahead of print
Sullivan PF, Neale MC, Kendler KS (2000) Genetic epidemiology of major depression: review and meta-analysis. Am J Psychiatr 157:1552–1562
Tanna VL, Wilson AF, Winokur G, Elston RC (1989) Linkage analysis of pure depressive disease. J Psychiatr Res 23:99–107
The International Schizophrenia Consortium (2008) Rare chromosomal deletions and duplications increase risk of schizophrenia. Nature Jul 30, Epub ahead of print
Thornalley PJ (1993) The glyoxalase system in health and disease. Mol Aspects Med 14:287–371
Thornalley PJ (2006) Unease on the role of glyoxalase 1 in high-anxiety-related behaviour. Trends Mol Med 12:195–199
Turck CW (2005) Understanding and treating psychiatric and neurological disorders: the benefits of cerebrospinal fluid analysis. Eur Biopharm Rev 46–49
Turck CW, Maccarrone G, Sayan-Ayata E et al (2005) The quest for brain disorder biomarkers. J Med Invest 52 (Suppl):231–235
Uhlen M (2007) Mapping the human proteome using antibodies. Mol Cell Proteomics 6:1455–1456
Wang X-F, Wang D, Zhu W et al (2003). Studies characterizing 60 kDa autoantibodies in subjects with schizophrenia. Biol Psychiatr 53:361–375
Williams K (1997) Interaction of polyamines with ion channels. Biochem J 325:289–297

World Health Organization: International Statistical Classification of Diseases, 2nd edn, 10th Revision

Wu CC, MacCoss MJ, Howell KE et al (2004) Metabolic labeling of mammalian organisms with stable isotopes for quantitative proteomic analysis. Anal Chem 76:4951–4959

Yabe T, Kanemitsu K, Sanagi T et al (2005) Pigment epithelium-derived factor induces pro-survival genes through cyclic AMP-responsive element binding protein and nuclear factor kappa B activation in rat cultured cerebellar granule cells: implication for its neuroprotective effect. Neuroscience 133:691–700

Yamagishi S, Inagaki Y, Takeuchi M, Sasaki N (2004) Is pigment epithelium-derived factor level in cerebrospinal fluid a promising biomarker for early diagnosis of Alzheimer's disease? Med Hypotheses 63:115–117

Chapter 4
Schizophrenia Biomarkers: A Means to Advance Disease Understanding, Diagnosis and Treatment

Emanuel Schwarz(✉) and Sabine Bahn(✉)

Abstract Schizophrenia is a multifaceted neuropsychiatric disorder affecting approximately 1% of the population worldwide. Its onset is the result of a complex interplay of genetic predisposition and environmental factors. The heterogeneity inherent to schizophrenia has so far been a great obstacle, hindering advances to elucidate disease mechanisms and the discovery of useful biomarkers. Recently, technological advances have been implemented that allow a search for molecular alterations with unprecedented sensitivity. In this chapter, we describe the findings generated using multi-omics approaches. The accessibility, complexity and heterogeneity of samples are some of the greatest challenges connected to biomarker discovery and other proteomic approaches for schizophrenia. We illustrate how far these challenges can be addressed using current technology and describe promising emerging techniques, which have yet to be applied to research in the schizophrenia field. Finally, we address challenges faced when analyzing large scale datasets and present methods that have already been successfully applied to discover biomarkers for this devastating disorder.

Abbreviations CNS: Central Nervous System; CSF: Cerebrospinal Fluid; FEW: Family Wise Error rate; LDL: Low Density Lipoprotein; MALDI: Matrix Assisted Lased Desorption Ionisation; NMR: Nuclear Magnetic Resonance; PLS-DA: Partial Least Squares Discriminant Analysis; pFDR: Positive False Discovery Rate; PTM: Post Translational Modification; PCA: Principal Component Analysis; SELDI: Surface Enhanced Laser Desorption Ionisation; VLDL: Very Low Density Lipoprotein

E. Schwarz, S. Bahn
Institute of Biotechnology, University of Cambridge, Tennis Court Road,
CB2 1QT, Cambridge, United Kingdom
es404@cam.ac.uk, sb209@cam.ac.uk

C.W. Turck (ed.) *Biomarkers for Psychiatric Disorders*,
doi: 10.1007/978-0-387-79251-4_4,

4.1 Introduction

4.1.1 Schizophrenia and the Importance of BioMarkers

Schizophrenia is a complex psychiatric disorder affecting approximately 1% of the population with equal risk across genders. It is a debilitating disease with disastrous consequences for affected individuals who often suffer from lifelong impairments. Being the seventh most expensive medical illness, schizophrenia furthermore represents a serious burden to the health care system (Freedman, 2003). This is partly due to the early onset, typically in mid to late adolescence or early adulthood, the frequently chronic course and the limited treatment success with current drug regimes. Within the last decades, considerable progress has been made in the search for factors contributing to the development of schizophrenia; however, the etiology of the disorder is still far from being understood. The heterogeneity of the disorder is a major factor slowing down scientific progress. One aspect of the heterogeneity is the contribution of multiple factors to the development of the disorder. It is well established that schizophrenia has a strong genetic contribution with heritability estimates reaching up to 80–85% (reviewed in Cardno and Gottesman, 2000). However, this estimation is controversial and other reports suggest a much lower heritability of as low as 30%. The genetic contribution to an increased susceptibility of being affected is also supported by studies on monozygotic twins, where the concordance rates reach almost 50% (Cardno and Gottesman, 2000). However, the increased risk of developing the disorder is not accounted for by a single gene, but a multitude of different genes. Furthermore, the inheritance patterns of schizophrenia do not follow simple Mendelian laws and single genes have only a small effect towards the overall risk of being affected. Even for the most promising gene polymorphisms (such as Neuregulin 1), the additional risk is low, with relative risks of approximately 2% instead of the 1% in the general population (reviewed in Tosato et al., 2005). In light of these facts it is unlikely that genetic biomarkers alone will advance the clinical management of schizophrenia, as they do not reflect the environmental disease determinants; and, as exemplified by the twin studies, even a strong genetic predisposition does not inevitably result in disease manifestation. This fact is often not fully appreciated in the literature. However, genetic determinants for variations in drug metabolism relating to antipsychotic drug efficacy/toxicity are promising. Therefore, genetic information might be valuable in the decision making process with regard to which pharmacotherapy to initiate in specific schizophrenia subtypes.

A second aspect of the heterogeneity is the varied clinical manifestation of schizophrenia, presenting with either predominant positive (delusions, hallucinations and thought disorder) or negative (lack of motivation, poverty of speech or inappropriate display of emotions) symptoms. The symptom spectrum overlaps with other psychiatric and neurological disorders and it remains unclear whether the different manifestations reflect subtypes of schizophrenia with different etiologies or whether indeed diverse clinical syndromes may have overlapping pathologies.

As no objective diagnostic tests exist at present, the diagnosis of schizophrenia is based on a subjective interview. The diagnostic criteria were first defined by Emil Kraepelin over 100 years ago. This is a major obstacle for the discovery of valid diagnostic biomarkers, as the biological underpinnings of the symptoms may not align with the current descriptive diagnostic approach. This conundrum impacts on multiple steps of the biomarker discovery process, such as sample choice or data analysis strategy. Obviously, in terms of the development of a novel diagnostic tool, biomarkers would be of limited use if they do not outperform the interview process used at the moment. Therefore, the ultimate goal of biomarker research in schizophrenia is to redefine disease entities with common pathophysiologies, which in turn will help to treat the underlying disease mechanisms more effectively.

If biomarkers were found to be predictive of the onset of schizophrenia, early intervention would be feasible and treatment could be commenced long before symptoms are apparent. Early intervention results in an improved prognosis of many diseases and increasing lines of evidence suggest that this is also the case for schizophrenia (McGorry et al., 2002; Holmes et al., 2006; McGlashan et al., 2006). Biomarkers could be used to monitor treatment response and compliance. If the markers showed an early response to treatment and were predictive of treatment success, ineffective treatment attempts could be reduced to a minimum and the efficacy of interventions could be assessed at an unprecedented speed. Even if the biological mechanism of how altered markers contribute to the development of schizophrenia is not apparent, biomarkers would be priceless for the development of novel therapeutics, foster the development of appropriate animal models and would help to generate new hypotheses about the etiology of schizophrenia.

Ideally, diagnostic markers should meet several criteria (adapted from Vitzthum et al., 2005). First of all, they should provide definite and reliable diagnostic information. This means that psychiatrists or even general physicians should be provided with a tool that unequivocally diagnoses a person as having schizophrenia under varying external conditions. This is closely connected to the need that biomarkers have a high sensitivity and specificity. The sensitivity is a measurement of how many schizophrenic patients can be identified among all people who actually have schizophrenia and take the test. Obviously, this ratio should be as high as possible but at the same time a person who does not have schizophrenia should not be diagnosed as being schizophrenic, which is measured as the specificity of the test. The test should furthermore not only be specific for schizophrenia alone but should differentiate schizophrenia from other disorders with similar clinical manifestations. The sensitivity and specificity of a diagnostic test is the measurement of its applicability and is very important if treatment is commenced based on the outcome of the test. As there is often a trade-off between sensitivity and specificity, it is crucial to know the probability of an incorrect diagnosis. There are obvious ethical questions, e.g., if the outcome of the test is positive, is it appropriate to start treatment if the treatment has severe side effects? In the context of schizophrenia the correct identification of possible subgroups of the disease is especially important because certain medications might only be effective for a specific subtype of schizophrenia.

A diagnostic test should be easy to use, so that ideally health professionals without expert knowledge concerning the biochemical mechanism of the test should be able to administer and interpret the test results correctly. One aspect of this would be the automation of the process where the sample of a person is used as an input and the diagnosis is directly given as an output by the diagnostic tool. Therefore, variability caused by the user would be minimized. Another very important aspect is the cost of the diagnostic tool. Ideally, the test should be inexpensive so that a broad distribution of the test – ideally to be used by the general practitioner – is feasible.

4.1.2 Current Knowledge About Biomarkers in Schizophrenia

In 1980 a 'niacin test' was suggested by Horrobin for the diagnosis of schizophrenia (Horrobin, 1980). It was proposed that the niacin test measures reduced membrane arachidonic acid levels in an indirect manner. Niacin induces a variety of reactions and it remains unclear whether the differences detected with the niacin test ultimately reflect differences in membrane arachidonic acid. However, reduced arachidonic acid levels have been repeatedly reported in schizophrenia (Berger et al., 2006). Niacin (methyl nicotinate) acts via niacin receptors on immunocompetent cells which results in an on-demand production of prostaglandins (e.g. PGD2) (Morrow et al., 1992). The skin flushing is then observed as a consequence of vasodilatation. Although the finding of hyporeaction to the niacin test in schizophrenia has been repeatedly replicated (Puri et al., 2001; Puri et al., 2002; Tavares et al., 2003; Bosveld-van Haandel et al., 2006), the sensitivity varies greatly between 23 and 90% and the response in schizophrenia patients was observed to be heterogeneous (Bosveld-van Haandel et al., 2006). The differential reaction to niacin in some patients is biologically interesting, but has not found its way into clinical use due to low sensitivity and specificity.

Neuroimaging techniques like magnetic resonance imaging (MRI) or computer tomography (CT) have consistently revealed structural abnormalities of some brain regions in schizophrenia patients (reviewed in Lawrie and Abukmeil, 1998). It has been reported that structural abnormalities are already present in individuals with high genetic risk and a possible application of these findings to identify people with high liability to develop schizophrenia was suggested (Lawrie et al., 1999; Lawrie et al., 2001). Together with functional imaging techniques like functional MRI (fMRI), positron emission tomography (PET) and single-photon emission computerized tomography (SPECT), these methods have provided very interesting insights into the pathology of schizophrenia (reviewed in Abou-Saleh, 2006). However, to our knowledge, to date, none of the imaging approaches offers a high enough sensitivity and specificity to be directly applicable as a diagnostic 'biomarker'.

However, the combination of imaging studies in conjunction with other biomarkers could be very fruitful; especially the correlation of CSF, serum and genetic biomarkers with imaging data holds promise to impact substantially on the field of schizophrenia. Thus interdisciplinary collaborations will be required, where clinical, experimental and operational procedures are clearly defined and samples are being collected from investigated patients and controls with the view of an integrated systems-based analysis.

The complexity and heterogeneity of schizophrenia as well as the fact that no hypothesis driven approach has to date been able to come up with a usable biomarker (or novel treatment approach), underlines the great need for a more global investigation of the disease pathology and associated biomarker signatures. The advances in systems biology approaches hold particular promise to impact on the clinical management of schizophrenia and related disorders. It is conceivable (and even likely) that single molecules with sufficient sensitivity and specificity might not exist and a ‘biomarker’ has to be based on a panel of different biochemical analytes. Non-hypothesis driven profiling approaches are especially suited for capturing the complexity of biological systems like the blood proteome and analyzing the alteration of the system in the disease state as well as its response to drug treatment. Before we look at different profiling techniques, their advantages and disadvantages and how they were implemented in the research of schizophrenia, we want to address the problems that relate to the important issue of sample choice and preparation with respect to biomarker discovery in schizophrenia.

4.2 Methods for Biomarker Discovery in Schizophrenia

4.2.1 Challenges of ‘Sampling the Disease’

In cancer, diagnosis can often be established by investigating tissue samples/biopsies directly taken from the diseased tissue. For other diseases, like cardiovascular disease and diabetes, the pathology is systemic and thus alterations are directly reflected in most body fluids, especially blood. For this reason useful biomarkers, for e.g., myocardial infarction like the widely used troponin have been identified. The situation is more complicated in schizophrenia as it has traditionally been regarded as a brain disorder. A major limiting factor in schizophrenia research is that it is not possible to sample the brain of a living patient directly. Brain samples are available *post mortem* but their usefulness is limited for biomarker discovery per se, though such studies are essential in enhancing our understanding of the pathophysiology of schizophrenia. Therefore, biomarker research has to rely on the analysis of body fluids or peripheral tissues. It is easily imaginable that this restriction comes with various problems. Biomarkers that directly relate to brain dysfunction may be very dilute or not detectable in peripheral body fluids. However, the question whether schizophrenia is a systemic disorder is an issue of debate. It is conceivable that similar to Creutzfeldt-Jakob Disease the disease progresses from the periphery to the brain and might even result from a systemic biochemical disturbance in schizophrenic patients. Therefore, it is very important to look for altered biochemical analytes in as many accessible body fluids and tissues as possible.

Cerebrospinal fluid (CSF) occupies the ventricular system of the brain and surrounds the spinal cord. CSF is clinically accessible by lumbar puncture but although this intervention is not commonly associated with serious adverse effects, the procedure can be painful and has to be performed by a skilled clinician. CSF samples are very valuable for biomarker research into CNS disorders as the detection of altered markers provides unique insights into pathological disease mechanisms in the brain and as the CSF is in dynamic continuum with the blood, such markers can potentially be traced in other, more easily accessible body fluids. With a volume of 150 mL and a daily production of approximately 500 mL, CSF has a high turnover. It is mainly produced by the choroid plexus of the ventricles and has important functions including the mechanical protection of the brain and the transport and distribution of neuroendocrine factors. Due to its proximity to the brain, alterations in CSF could directly relate to brain disturbances. Although CSF contains various molecules like peptides, sugars, lipids and proteins, its molecular complexity is manageable. It is advantageous that extensive sample prefractionation is usually not necessary and possible variation introduced by sample preparation can thus be kept to a minimum. However, the protein concentration is low at 0.18–0.58 mg ml^{-1} (Seehusen et al., 2003) which is two to three orders of magnitude lower than serum protein concentrations and embodies a serious challenge for the sensitivity of analytical techniques.

Serum and plasma are obviously more easily accessible than CSF and the choice between these two body fluids is an important decision during sample preparation (reviewed in Villar-Garea et al., 2006). These blood-derived samples are very interesting to analyze because they are the medium to transport a vast variety of different molecules and are in direct contact with all body tissues. Furthermore, tissue specific molecules are likely to be released to the bloodstream as a result of cell damage or death. Therefore, blood samples can give a comprehensive picture of the physiological state of an individual and are often the first point of call to look for potential biomarkers. However, blood sample investigations come with several inherent difficulties. First of all, the dynamic range of the serum proteome exceeds ten orders of magnitude and 97% of the proteins belong to one of the seven most abundant protein groups (Schulte et al., 2005). This is a challenge for instrumentation as no existing technique comes close to covering the dynamic range of the entire proteome in a blood sample. However, novel techniques do have an impressive sensitivity. This becomes clear if one imagines dropping a 1 cm^3 sugar cube into an Olympic swimming pool and injecting a sample into a mass spectrometer; advanced instruments reach sensitivity limits of attomoles (10^{-18} mol) per microliter (Steen et al., 2001) which would allow us to measure the concentration of the sugar in the swimming pool correctly, determine which sugar was added and how pure it is. However, it has to be taken into consideration that instrument sensitivity is established by using purified compounds and the situation is clearly different when analyzing complex samples like blood. The concentration of the plasma proteome ranges from

50 mg ml^{-1} (serum albumin) to approximately 5 pg ml^{-1} (Interleukin-6) (Anderson and Anderson, 2002). The analytic techniques available at the moment have a dynamic range of 10^2–10^4. Therefore, they only allow looking at a very narrow window depicting only a minor fraction of the full dynamic range of the serum or plasma proteomes. If a crude serum or plasma sample is analyzed one can only detect the most abundant proteins, while medium and low abundance proteins are obscured. It is not possible to just inject more sample volume to increase the amount of lower abundant proteins as the detector will be saturated with the higher abundant proteins. Therefore, the sample has to be fractionated in a way that allows looking at subsets of the proteome with different concentration ranges. These samples can be appropriately concentrated and their independent analysis allows looking through different windows and thereby 'zooming in' on different ranges of protein abundance.

One of these fractionation methods is the use of antibody- based depletion columns that remove the most abundant protein species. Whereas this procedure allows the detection of a higher number of lower abundant proteins, proteins bound to albumin, especially, might be lost. Another approach involves the precipitation of more abundant proteins using organic solvents like acetonitrile which was reported to simultaneously dissociate small proteins from carrier molecules like albumin (Merrell et al., 2004). However, other authors criticize depletion methods based on organic solvent precipitation due to their poor selectivity (Fu et al., 2005).

The serum proteome has a very high complexity; 1,175 non redundant proteins were reported to be present in serum (Anderson, 2005) but the real number is likely to be far higher. One reason for this estimation is the presence of a considerable amount of protein isoforms due to multiple post-translational modifications (PTM). A large proportion of proteins are supposedly phosphorylated by one of more than 500 protein kinases (Manning et al., 2002) and the complexity is further increased by glycosylation and other PTMs. For this reason a promising approach for reducing the complexity of serum is its sub fractionation and the creation of 'sub proteomes'. This approach has been pursued for glycosylated (reviewed in Geyer and Geyer, 2006), and phosphorylated proteins (reviewed in Gafken and Lampe, 2006), but, to our knowledge, has not found its way into research in schizophrenia. Differential analysis of PTM is an exciting area of research, promises to provide insight into schizophrenia etiology and bears the potential for biomarker discovery.

For mass spectrometry-based profiling techniques the problem of sample complexity is further increased if digestion with trypsin is necessary. Mass spectrometric methods suffer from the physical phenomenon of ion suppression that occurs when a dominant peptide peak suppresses the intensity measurement of a lower abundant peak injected into the MS at the same time. To address the problem of sample complexity, techniques have been developed where only the N-terminal fragment of each protein is selectively isolated and injected into the MS which greatly reduces the complexity (McDonald et al., 2005). However, for the peptide intensity to reliably reflect the protein concentration, a separation strategy has to be developed to avoid ion suppression.

Biomarker discovery approaches are mostly non hypothesis driven and call for profiling techniques that are able to quantitatively or semi-quantitatively measure as many biochemical analytes as possible. The above section outlines problems associated with sampling biochemical analytes from body fluids or tissues. No technique is currently available that can analyze all the molecules contained in body fluids like blood or even CSF in one run. This gives rise to the necessity to fractionate complex samples followed by the separate analysis of each fraction. The two most obvious subfractions of a blood sample are the proteins and the small molecules like lipids, sugars and small peptides, called metabolites. Starting from these two large domains there are various subfractions of each of them which can be analyzed separately. For these tasks several analytical platforms are available that have their own strengths and weaknesses. In the following section we look at the most promising platforms and discuss their potential application for biomarker discovery in schizophrenia.

4.2.2 Proteomics

4.2.2.1 Two Dimensional Gel Electrophoresis Mass Spectrometry

One of the oldest and still widely used proteomics technology is two dimensional gel electrophoresis MS (2D-GE – MS). Proteins are first separated based on their isoelectric point using immobilized pH gradient (IPG) strips. These gradients are commercially available in a variety of pH ranges. In the second dimension, proteins are separated based on their molecular weight using denaturing polyacrylamide gels. Gels are stained and scanned to quantify proteins based on the density of the protein spots. Afterwards, the spots are automatically excised, the proteins destained in gel digested with trypsin and analyzed in a mass spectrometer. Despite the long history of 2D-GE – MS the technique has not been extensively used for the discovery of schizophrenia biomarkers. Wan C. et al used 2D-GE – MS to profile plasma proteins of schizophrenia patients and healthy volunteers and found upregulation of four acute phase proteins in schizophrenic patients (Wan et al., 2007). The same group reported alterations of transthyretin and apolipoprotein E in the CSF of schizophrenic patients (Wan et al., 2006). Jiang L. et al used 2D-GE – MS to profile the CSF of schizophrenia patients identifying apolipoprotein A IV as being significantly altered (Jiang et al., 2003). One advantage of 2D-GE – MS is its capability of simultaneously analyzing different isoforms of the same protein, differing either in isoelectric point or molecular weight. However, the property that products of single genes can migrate to multiple spots due to factors including posttranslational or artifactual modifications as well as differential protein processing is a major obstacle for correctly quantifying protein concentration (Gygi et al., 2000). This problem is further increased by the possibility of comigration when proteins with similar properties do not migrate to separate spots. A further development that increases the resolving power is the use of narrow range IPG strips (nrIPG) but it has been shown that they also fail to solve the problems of overcrowded gels (reviewed in Corthals et al., 2000). In addition, high and low molecular weight proteins are not displayed on two dimensional gels.

A further drawback of 2D-GE is its low dynamic range. Silver staining of plasma proteins in 2D-GE cannot display more than four orders of magnitude, which is only a small proportion of the total dynamic range of plasma. The detectability of low abundance/ low copy number proteins can be improved by increasing the amount of protein loaded on the gel (Corthals et al., 2000; Gygi et al., 2000) but in many cases samples are precious and even if more protein is loaded, 2D-GE – MS fails to detect a large proportion of the lower abundant proteins as distinct spots (Corthals et al., 2000). Indeed, in the case of CSF, samples have to be pooled to obtain a sufficient protein concentration for 2D-GE – analysis which makes statistical evaluation difficult and leaves one with analyzing fold changes of proteins.

A noteworthy addition to the 2D-GE methodology is difference in gel electrophoresis (DIGE) (Tonge et al., 2001; Shaw et al., 2003). In this technique up to three samples are independently labeled with different fluorescent reagents, pooled and run on the same gel. Therefore, comparisons between gels can be avoided. Prabakaran et al. used this technique on liver and red blood cells to find protein alterations in schizophrenia, which supported the involvement of oxidative stress in schizophrenia (Prabakaran et al., 2007). However, this extension of 2D-GE does not solve the major problems of twodimensional gel electrophoresis.

4.2.2.2 Isotope Coded Affinity Tags

For two-dimensional gel electrophoresis the role of the mass spectrometer is the identification of proteins. The quantification is performed by determining the density of the stained protein spots. Newer approaches try to directly quantify proteins based on the signal detected in the mass spectrometer. It was argued that the signal intensity measured in a mass spectrometer is not a reliable measurement of protein concentration (Patterson and Aebersold, 2003). This is because measured intensities depend on chemical compositions of peptides as well as other properties which are only partly understood. However, as identical peptides with different isotopic composition could be reliably compared in a mass spectrum, relative quantification between those peaks offered a solution to the quantification problem. The first technique using this rationale was based on isotope coded affinity tags (ICAT) (Gygi et al., 1999). This group of reagents has an alkylation group that binds covalently to cysteine residues and a biotin affinity tag to allow for selective isolation. These two groups are connected with a linker containing either eight hydrogen or eight deuterium atoms generating two isotopic forms of the reagent. To compare protein concentrations in two samples, the samples are labeled with different isotopic forms of the reagent and pooled afterwards. The labeled proteins can be enriched optionally using avidin affinity chromatography and can be enzymatically digested. The relative difference of protein concentration between the two samples can now be easily determined by comparing the intensity of the differently labelled peptides (for a more extensive description of the method see Patterson and Aebersold, 2003).

An advantage of this approach is that cysteine-containing peptides can be selectively isolated, which reduces the complexity of the sample dramatically. Furthermore, samples can be analyzed simultaneously, which reduces the technical variation. However, the specificity to cysteine residues can be a drawback as only a subset of the proteome is being analyzed. The more severe disadvantage, probably, is that only two samples can be analyzed simultaneously due to the limited number of different labeling reagents.

These drawbacks have been slightly alleviated by the introduction of new labeling procedures which comprise labeling reagents with a higher number of different isotopes as well as other binding properties. Isobaric tagging for relative and absolute protein quantification (iTRAQ) is an example with four labeling reagents that bind to the N-terminal end of each peptide and can therefore label all peptides (for more information see Zieske, 2006).

These stable isotope dilution techniques have not found their way into research in schizophrenia as yet. A particular disadvantage is that only a limited number of samples can be investigated as at present no labeling technique exists that offers more than ten different reagents. Due to the inherent heterogeneity in schizophrenia it is unlikely that profiling approaches based on small sample numbers will be fruitful. However, the possibility of simultaneously analyzing several samples seems very attractive, owing to lower experimental variation. Therefore, methods that enable efficient normalization between different batches of labeled samples could be a valuable asset and make labeling techniques in principle very attractive for the discovery of schizophrenia biomarkers.

4.2.2.3 SELDI – TOF MS

Surface enhanced laser desorption ionization (SELDI) is an extension of matrix assisted lased desorption ionization (MALDI). Similar to MALDI, a laser transfers energy onto an energy absorbing surface and the energy is then transferred to the analytes embedded in the matrix. The analytes heat up and are desorbed from the surface (Merchant and Weinberger, 2000). SELDI allows for a certain reduction of the sample complexity as the surfaces used typically bind to analytes with special chemical properties (Issaq et al., 2002). One advantage of this method is the huge variety of experiments that can be carried out using different chip surfaces. The limitation of possibilities depends on the creativity of the scientist as reactive chip surfaces can be manipulated individually according to the experimental design. Besides profiling, SELDI is especially suited for antibody capturing studies and the analysis of posttranslational modifications like glycosylation or phosphorylation (Issaq et al., 2002). Further advantages are its ease of use, the relatively inexpensive instrumentation and its high throughput. Huang et al. used this technology to screen CSF of drug naïve, paranoid, first onset schizophrenia patients for biomarkers and identified VGF and transthyretin as potential biomarkers (Huang et al., 2006).

However, the SELDI TOF MS technology has been criticized repeatedly. One of its greatest limitations is that direct identification of potential biomarkers is not possible. Therefore a purification strategy has to be developed which can be very difficult and time consuming. Low mass accuracy and resolution are further drawbacks of the SELDI technology. Furthermore, variability between spectra was reported to result from chip or laser variability (Rogers et al., 2003; Xiao et al., 2005; Villar-Garea et al., 2007) and other confounding factors. It is also conceivable that competition phenomena between higher and lower abundant proteins occur for binding sites on the chip surface as well as in term of ion suppression. Suggestions of how to overcome these problems are reviewed in 2006).

4.2.2.4 Quantitative Label Free Shotgun Proteomics

Despite the notion that mass spectrometry is inherently not a quantitative technique, recent advances in mass spectrometric technology provide examples of how the direct analysis of ion chromatograms can be used as a relative quantification method to detect alterations in complex samples. Several studies based on measurements of standards showed a high correlation between mass spectral peak intensities of peptide ions and the respective protein abundance in the sample (Bondarenko et al., 2002; Chelius and Bondarenko, 2002; Wang et al., 2003; Liu et al., 2004). These findings were reproduced using complex samples from human cells (Old et al., 2005). So- called sample statistics are generally used to quantify protein abundance, which includes spectral count (number of MS/MS spectra recorded for the peptides of a given protein), peptide count (number of peptides corresponding to a protein identified per analysis) and sequence coverage (coverage of the protein sequence by the peptides identified per analysis) (Zhang et al., 2006). Besides these, the integration of peak areas per peptide is also a common way to determine peptide abundance (Higgs et al., 2005). The technology is still in its infancy, especially in terms of its application for large-scale profiling studies. However, it bears great potential for discovering biomarkers as the sample size is not limited a priori. Furthermore, labeling procedures are not necessary, which reduces the experimental variation.

4.2.2.5 Functional Proteomics

While profiling approaches usually include mapping and differential expression measurement, functional proteomics comprises techniques to measure protein activity, interactions and the presence of posttranslational modifications (Choudhary and Grant, 2004). In the following subsections we quickly look at activity-based protein profiling and protein microarrays as they have possible applications for biomarker discovery in schizophrenia.

4.2.2.6 Activity-Based Protein Profiling

In activity-based protein profiling, samples are treated with reagents that react covalently with proteins that have a specific type of activity. Such modified proteins can be detected with a second tag (e.g., biotin) that is a part of the reagent (Phizicky et al., 2003). This approach is suitable for biomarker discovery and has previously been used to track the activity of cysteine proteases and to identify different labeling profiles during the progression of skin cancer in mice (Greenbaum et al., 2000). A further development involves SDS – gel separation of selectively isolated enzymes with the targeted activity class (Greenbaum et al., 2002) and subsequent ingel digestion and mass spectrometric analysis (Kidd et al., 2001).

Owing to the selective reagents, classes of active enzymes can be isolated, which promises to give interesting insights into alterations of biochemical pathways and offers multiple applications for the discovery of biomarkers. However, the technique has not found its way into research in schizophrenia as yet.

4.2.2.7 Protein Microarray

Among several different types of protein microarrays (reviewed in Phizicky et al., 2003) analytical microarrays which are still in their infancy probably bear the greatest potential for approaches in the discovery of biomarkers. In this method, biochemical analytes like antibodies are bound to the array surface and the concentrations of proteins contained in a complex sample that bind to these analytes can be quantified. Drawbacks of this technology include possible cross-reactivity of the analytes fixed to the array that results in problems of quantification (Haab et al., 2001). However, this method bears great potential as many analytes can be measured in parallel and further developments of this methodology might be a valuable asset for biomarker discovery in schizophrenia.

4.2.3 Metabolomics

Alterations in metabolite concentrations are often regarded as consequences of regulatory or metabolic processes and embody the downstream elements in the gene – protein – metabolite cascade. As the metabolome is highly dynamic it gives a molecular picture of a complex system at a very specific point in time and reflects a systems reaction to exterior or internal (e.g. pathological) effects. Therefore, metabolic profiling is a promising approach for the discovery of biomarkers, with previous studies focusing on nuclear magnetic resonance (NMR) and increasingly on LC-MS methods. However, metabolomics experiments have rarely been carried out in research in schizophrenia as yet. This is probably due to the several limitations of the current metabolomics technology. One of the drawbacks common to all methods is that most research groups with access to schizophrenia samples analyze samples from patients treated with drugs. The drug treatment does not only obscure the true biological differences, but might also lead to the discovery of false positive hits if drug effects cannot be excluded as a reason for the observed metabolic alteration. The property of the metabolome being highly dynamic also introduces problems, such as being prone to confounding factors including diet, smoking, alcohol abuse and others. In the case of schizophrenia it is especially worthwhile mentioning metabolic alterations caused by the use of cannabis and smoking; cannabis consumption and smoking are highly prevalent among schizophrenia patients. Here we look at NMR, LC-MS and GC-MS methods as biomarker discovery tools for schizophrenia.

4.2.3.1 NMR

Most of the metabolic profiling work in other diseases has been carried out using NMR spectroscopy. Indeed, Tsang et al. pioneered NMR-based metabolomics experiments in schizophrenia by profiling plasma from identical twins discordant for schizophrenia, finding lipids from LDL and VLDL as the most discriminating factors (Tsang et al., 2006). Later, Holmes et al. profiled the metabolome of 54 drug naïve, paranoid, first onset and 28 minimally treated schizophrenia patients and found altered levels of glucose, acetate and lactate as well as a shift from glutamate to glutamine due to a pH decrease of 0.1 in schizophrenia patients (Holmes et al., 2006). This study further demonstrated that the metabolic alterations normalize in approximately half of the treated patients and data suggesting that treatment commenced after the first psychotic episode supports this normalization process. From our viewpoint NMR is a very useful technique for the discovery of a biomarker in schizophrenia as the identification of potential biomarkers is quite straightforward. However, the sensitivity of current NMR technology allows investigation of only the more abundant metabolites. So only dozens of metabolites can be profiled and relatively large volumes of the sample are needed, which is especially problematic in the case of CSF.

4.2.3.2 LC-MS – GC-MS

An interesting alternative to NMR spectroscopy are mass spectrometry based metabolic profiling methods. They offer a greater sensitivity but unfortunately, to date, do not allow for a straightforward identification. Usually the identification is based on the high mass accuracy of the instrument, which in combination with the isotopic pattern of a metabolite, allow the determination of the elemental composition. Obviously, the number of suggested formulae increases with increasing mass and if no target class of molecules is searched for, the identification of the elemental composition is very difficult. One solution to this problem lies in the establishment of databases containing MS-libraries of standard compounds against which the detected molecules can be searched. This would be especially helpful in the case of LC-MS as currently no publicly accessible database is available. Owing to the high reproducibility of the ionization process used in gas chromatographic mass spectrometry, databases are in place for GC-MS data and identification can often be achieved without too much trouble. MS/MS further helps in the identification of metabolites, but often sufficient fragmentation is either not observed or not informative enough to determine the identity of metabolites. In contrast to LC-MS, GC-MS relies on the ability of the molecules to be analyzed to transit from the liquid phase into the gas phase, which depends on the volatility of the compounds. For polar components the volatility is rather low, which makes derivatization necessary. Despite these difficulties impressive recent developments make the MS- technology highly attractive for the discovery of biomarkers in schizophrenia. O'Hagan et al. recently applied a closed loop optimization of mass spectrometric parameters using genetic algorithms to increase the number of detected peaks on a GCGC-MS instrument in comparison to a one-dimensional GC-MS by a factor of three to approximately 1,800 metabolite peaks in human serum (O'Hagan et al., 2007). Considering the scope of these further developments and the potential of the technology, mass spectrometry-based metabolic profiling promises to become an important tool for the analysis of metabolites in schizophrenia on a system level.

4.3 Reported Peripheral Alterations in Schizophrenia

See Table 4.1.

Table 4.1 Reported peripheral markers in schizophrenia

Molecule	Location of detection	Affected pathway	References
Proteins/Peptides:			
Acute phase proteins	Plasma	Immune function	Wan et al. (2007)
Interleukin 2 (IL-2)	blood	Immune function	Zhang et al. (2002)
Apolipoprotein E (Apo E)	CSF	Lipid metabolism	Wan et al. (2006)
Apolipoprotein A1 (Apo A1)	Plasma	Lipid metabolism	La et al. (2007)
Apolipoprotein A4 (Apo A4)	CSF	Lipid metabolism	Jiang et al. (2003)
Transthyretin (TTR)	CSF	Thyroxine transport	Huang et al. (2006); Wan et al. (2006)
Oxidative stress proteins	liver/RBC	Oxidative stress	Prabakaran et al. (2007)
Glutathione	CSF	Oxidative stress	Do et al. (1995)
VGF	CSF	unknown	Huang et al. (2006)
GSK-3beta	CSF	Energy metabolism	Kozlovsky et al. (2004)
mitochondrial complex I activity	platelets	Energy metabolism	Dror et al. (2002)
Reelin	blood	Neuronal function	Fatemi et al. (2001)
Metabolites:			
LDL and VLDL lipids	plasma	Lipid transport	Tsang et al. (2006)
glucose	CSF	Energy metabolism	Holmes et al. (2006)
acetate	CSF	Energy metabolism	Holmes et al. (2006)
lactate	CSF	Energy metabolism	Holmes et al. (2006)
homovanillic acid	plasma	Dopamine metabolism	Friedhoff amin (1997) review
RNA:			
SELENBP1	PBC	Selen binding	Glatt et al. (2005)
D3 receptor	BL	Signal transduction	Ilani et al. (2001)
GCLM gene	fibroblast	Oxidative stress	Tosic et al. (2006)

RBC red blood cells; *PBC* peripheral blood cells; *BL* Blood lymphocytes

4.4 Data Analysis

4.4.1 Challenges of Data Analysis

Data analysis for biomarker discovery is a challenging task. Besides the difficulties connected to the processing of the raw data, researchers can choose from a huge variety of different ways of analyzing processed data and user-friendly software makes the construction of complex mathematical models quite easy. Because of the many possible ways of looking at datasets with different mathematical methods, care has to be taken when choosing from the available tools and interpreting the results. Data analysis methods in the field of biomarker discovery can be divided into univariate methods considering single biochemical analytes one at a time and multivariate methods considering several or many molecules simultaneously. It is beyond the scope of this chapter to comment on the controversial issue of the application of frequentist as compared to Bayesian approaches and we will focus on describing the methods commonly used in the field. Datasets generated using molecular profiling approaches typically are output in the matrix format with the different samples being in the rows and the detected molecules in the columns of the matrix. Each value in the matrix represents the measured quantity for a certain molecule in a certain sample. Because many molecules are analyzed in parallel, these datasets show high dimensionality, which introduces several challenges to the strategies in data analysis. One problem that is common to all studies analyzing only a few samples is that data are often not normally distributed. Many statistical tests like student's T-test or analysis of variance (ANOVA) rely on a normal distribution of the data. Normal distributions are often not present for lower sample numbers which can lead to erroneous results. Second, the high number of measured molecules can easily be the reason for the discovery of false positive findings. If independent statistical tests are carried out to determine the significance of altered levels between the disease and control group for each molecule, the likelihood of discovering significantly changing molecules increases with the number of tests carried out. Therefore, it is advisable to employ methods to correct for this multiple comparison effect. The most stringent correction and therefore the most conservative approach is the Bonferroni correction. In this method the probability of falsely rejecting even one of the true null hypotheses, the family-wise error rate (FWE), is controlled. Therefore, the control of the FWE at a level α has to be carried out at α/n for each individual of n tests performed. If for example the original 'significance level' α was set to be 0.05 and 1,000 molecules are tested for differences in a disease and a control group individually, the 'significance level' for each single test would be 0.00005 to control the FWE.

A less stringent method to address the multiple comparison problem is the control of the false discovery rate as suggested by Benjamini and Hochberg in 1995 (Benjamini and Hochberg, 1995).The false discovery rate is defined as the expected proportion of false positive findings among all rejected hypotheses. Therefore, the proportion of errors is controlled. This means that if a high proportion of hypotheses is rejected, the error of rejecting one incorrectly is small and the applied correction less stringent. Storey introduced the positive false discovery rate (pFDR) as a modified error measure to use (Storey, 2001). This quantity can be directly used to determine q-values, the pFDR analogue of p-values. One further point that should be taken into consideration is that the datasets generated with profiling methods comprise molecules/variables that are often not independent of each other. Methods of how multiple comparison correction procedures can be applied in the case of dependency are beyond the scope of this chapter and are mentioned elsewhere (Reiner et al., 2003).

Besides the univariate methods, multivariate statistics are very common methods employed for the analysis of high dimensional datasets. In contrast to univariate methods, multivariate statistics take the changes of all measured analytes simultaneously into consideration. Therefore, they are useful tools to get an overview of the structure of the dataset. Multivariate methods are typically classified as unsupervised and supervised methods. Unsupervised methods are applied without the prior knowledge of which class an observation belongs to. A typical example of an unsupervised method is principal component analysis (PCA). In this method, the multi-dimensional dataset is projected to a few (mostly two) dimensions that account for the greatest variance in the dataset. The variance is the amount of information that is contained in the data.

The axis along the greatest variance of the dataset is called principal component one, the axis along the second greatest variance and orthogonal to the first axis is called principal component two and so on. Therefore, the dataset can be visualized in a two or three dimensional plot depicting the observations in the space of the principal components while conserving most of the information in the data (scores plot). If on this plot differences are observed between two or more sample groups, the variables accounting for this difference can be identified with the so called loadings plot. The loadings plot represents the correlation of the original variables with the principal components. In other words, the loadings plot shows how the variables account for the information contained in the dataset. Therefore, if a separation between sample groups is observed on the scores plot, the variables with the highest loading in the direction of the observed separation are the variables accounting for most of the difference between the groups.

For very high dimensional datasets or datasets with a high noise level, possible differences between sample groups can easily be obscured by a bigger variance in the dataset, such as differences in sensitivity of the measurement along the experiment. However, as PCA depicts inherent similarities of observations with similar observations being closely together on the scores plot, the method is very useful to assess the quality of the dataset. For example, replicates of an experiment carried out on the same sample are supposed to be very similar. Or differences between samples that are simply caused by a sensitivity change along an MS-run can be assessed.

For supervised methods the group membership of observations is considered a priori. Very common supervised methods are partial least squares (PLS) regression and PLS – discriminant analysis (PLS-DA) (Vong et al., 1988; Barker et al., 2003). In these regression-based methods, the multidimensional dataset is also projected to lower dimensions but this time along the dimensions in which the difference between the predefined groups is the greatest. Therefore, similar to PCA, potential biomarkers can be identified that account the most for the observed separation. However, PLS-DA is a very sensitive technique and even larger sample groups between 50 and 100 observations can be separated completely using a set of a couple of hundred random variables. Although software packages try to cope with this problem by automatically performing a cross-validation on the dataset, false positive results are often observed, especially if certain scaling methods are used. This especially holds true for analyzing body fluids of schizophrenic patients which are very dilute (e.g., serum) and where high fold changes are uncommon. For further examples regarding the application of multivariate data analysis see Goodacre et al. (2004).

If we are able to determine the differences between schizophrenic patients and healthy volunteers we are often interested in how other factors influence these differences. Generally, if we observe differences we are not able to make a decision about whether these differences are causative or a consequence of the disease. While the information might be sufficient for a diagnostic tool, especially in schizophrenia an important feature of a biomarker would be its predictivity, its ability to measure if a person is likely to develop full-blown schizophrenia. Another factor is the effect of drug treatment, as we want to know if the biomarker can be used to monitor the drug response and more importantly if it has predictive validity regarding whether the treatment will be successful.

Multivariate methods are very useful to address these questions. First, they allow looking at the effects of multiple potential biomarkers at the same time, thus, they allow us to examine whether those prodromal patients who develop a psychotic state already show the same molecular alterations as full-blown schizophrenic patients. Furthermore, we can look at how the molecular profile of schizophrenic patients changes after drug treatment and especially if a potential normalization of the profile correlates with a positive treatment outcome.

Data analysis in the context of schizophrenia is especially challenging, as it is not known if the patients have been diagnosed correctly or if subgroups with different etiologies exist. Therefore, methods showing inherent similarities between the observations like the ones mentioned above are very useful but might be prone to the detection of false positives.

4.5 Summary and Perspectives

In this chapter we have outlined several problems connected to the approaches of discovering biomarkers for the complex psychiatric disorder of schizophrenia. Major challenges are the heterogeneity of the disease as well as the high complexity and dynamic range of proteins and metabolites. As a consequence, sample preparation techniques and analytical instrumentation have to be chosen carefully. Downstream of the acquisition of the raw data, the analysis of processed data requires serious thought to address the right questions and simultaneously avoid false discoveries. Furthermore, we have illustrated how profiling techniques have been used for the discovery of biomarkers in schizophrenia. These findings will have to be reproduced in larger independent sample sets to assess their utility as diagnostic tools. From our viewpoint it is likely that panels of multiple biomarkers in combination with genetic information and imaging studies will serve as surrogate clinical outcomes to identify schizophrenia subtypes and could be the basis for early and more successful interventions as is currently achieved.

Many of the recently developed techniques like labeling or label-free proteomics approaches have not yet found their way into research in schizophrenia. However, they hold great potential to provide deeper insights into the complexity of the proteome and its alteration in schizophrenia and we hope that future studies will translate to major improvements of the clinical management of this most devastating disorder.

Acknowledgments This research was kindly supported by the Stanley Medical Research Institute (SMRI). We thank all members of the Bahn Laboratory, especially Yishai Levin, for discussions, help and encouragement. E.S. holds a Cambridge European Trust scholarship and S. B. holds a NARSAD Essel Independent Investigator Fellowship.

References

Abou-Saleh, M. T. (2006). Neuroimaging in psychiatry: An update. J Psychosom Res 61(3): 289–93.

Anderson, L. (2005). Candidate-based proteomics in the search for biomarkers of cardiovascular disease. J Physiol 563(Pt 1): 23–60.

Anderson, N. L. and N. G. Anderson (2002). The human plasma proteome: history, character, and diagnostic prospects. Mol Cell Proteomics 1(11): 845–67.

Barker, M. and W. Rayens (2003). PLS for discrimination. J. Chemometrics 17: 166–73.

Benjamini, Y. and Y. Hochberg (1995). Controlling the false discovery rate: a practical and powerful approach to multiple testing. J. R. Statist. Soc. B (57): 289–300.

Berger, G. E., S. Smesny and G. P. Amminger (2006). Bioactive lipids in schizophrenia. Int Rev Psychiatr 18(2): 85–98.

Bondarenko, P. V., D. Chelius and T. A. Shaler (2002). Identification and relative quantitation of protein mixtures by enzymatic digestion followed by capillary reversed-phase liquid chromatography-tandem mass spectrometry. Anal Chem 74(18): 4741–9.

Bosveld-van Haandel, L., R. Knegtering, H. Kluiter and R. J. van den Bosch (2006). Niacin skin flushing in schizophrenic and depressed patients and healthy controls. Psychiatry Res 143 (2–3): 303–6.

Cardno, A. G. and Gottesman, II (2000). Twin studies of schizophrenia: from bow-and-arrow concordances to star wars Mx and functional genomics. Am J Med Genet 97(1): 12–7.

Chelius, D. and P. V. Bondarenko (2002). Quantitative profiling of proteins in complex mixtures using liquid chromatography and mass spectrometry. J Proteome Res 1(4): 317–23.

Choudhary, J. and S. G. Grant (2004). Proteomics in postgenomic neuroscience: the end of the beginning. Nat Neurosci 7(5): 440–5.

Corthals, G. L., V. C. Wasinger, D. F. Hochstrasser and J. C. Sanchez (2000). The dynamic range of protein expression: a challenge for proteomic research. Electrophoresis 21(6): 1104–15.

Do, K. Q., C. J. Lauer, W. Schreiber, M. Zollinger, U. Gutteck-Amsler, M. Cuenod and F. Holsboer (1995). Gamma-Glutamylglutamine and taurine concentrations are decreased in the cerebrospinal fluid of drug-naive patients with schizophrenic disorders. J Neurochem 65(6): 2652–62.

Dror, N., E. Klein, R. Karry, A. Sheinkman, Z. Kirsh, M. Mazor, M. Tzukerman and D. Ben-Shachar (2002). State-dependent alterations in mitochondrial complex I activity in platelets: a potential peripheral marker for schizophrenia. Mol Psychiatr 7(9): 995–1001.

Fatemi, S. H., J. L. Kroll and J. M. Stary (2001). Altered levels of Reelin and its isoforms in schizophrenia and mood disorders. Neuroreport 12(15): 3209–15.

Freedman, R. (2003). Schizophrenia. N Engl J Med 349(18): 1738–49.

Friedhoff AJ, Amin F (1997). Plasma Homovanillic Acid in Schizophrenia: Implication for Presynaptic Dopamine Dysfunction. Washington, DC, American Psychiatric Press.

Fu, Q., C. P. Garnham, S. T. Elliott, D. E. Bovenkamp and J. E. Van Eyk (2005). A robust, streamlined, and reproducible method for proteomic analysis of serum by delipidation, albumin and IgG depletion, and two-dimensional gel electrophoresis. Proteomics 5(10): 2656–64.

Gafken, P. R. and P. D. Lampe (2006). Methodologies for characterizing phosphoproteins by mass spectrometry. Cell Commun Adhes 13(5–6): 249–62.

Geyer, H. and R. Geyer (2006). Strategies for analysis of glycoprotein glycosylation. Biochim Biophys Acta 1764(12): 1853–69.

Glatt, S. J., I. P. Everall, W. S. Kremen, J. Corbeil, R. Sasik, N. Khanlou, M. Han, C. C. Liew and M. T. Tsuang (2005). Comparative gene expression analysis of blood and brain provides concurrent validation of SELENBP1 up-regulation in schizophrenia. Proc Natl Acad Sci U S A 102(43): 15533–8.

Goodacre, R., S. Vaidyanathan, W. B. Dunn, G. G. Harrigan and D. B. Kell (2004). Metabolomics by numbers: acquiring and understanding global metabolite data. Trends Biotechnol 22(5): 245–52.

Greenbaum, D., A. Baruch, L. Hayrapetian, Z. Darula, A. Burlingame, K. F. Medzihradszky and M. Bogyo (2002). Chemical approaches for functionally probing the proteome. Mol Cell Proteomics 1(1): 60–8.

Greenbaum, D., K. F. Medzihradszky, A. Burlingame and M. Bogyo (2000). Epoxide electrophiles as activity-dependent cysteine protease profiling and discovery tools. Chem Biol 7(8): 569–81.

Gygi, S. P., G. L. Corthals, Y. Zhang, Y. Rochon and R. Aebersold (2000). Evaluation of two-dimensional gel electrophoresis-based proteome analysis technology. Proc Natl Acad Sci USA 97(17): 9390–5.

Gygi, S. P., B. Rist, S. A. Gerber, F. Turecek, M. H. Gelb and R. Aebersold (1999). Quantitative analysis of complex protein mixtures using isotope-coded affinity tags. Nat Biotechnol 17(10): 994–9.

Haab, B. B., M. J. Dunham and P. O. Brown (2001). Protein microarrays for highly parallel detection and quantitation of specific proteins and antibodies in complex solutions. Genome Biol 2(2): RESEARCH0004.

Higgs, R. E., M. D. Knierman, V. Gelfanova, J. P. Butler and J. E. Hale (2005). Comprehensive label-free method for the relative quantification of proteins from biological samples. J Proteome Res 4(4): 1442–50.

Holmes, E., T. M. Tsang, J. T. Huang, F. M. Leweke, D. Koethe, C. W. Gerth, B. M. Nolden, S. Gross, D. Schreiber, J. K. Nicholson and S. Bahn (2006). Metabolic profiling of CSF: evidence that early intervention may impact on disease progression and outcome in schizophrenia. PLoS Med 3(8): e327.

Horrobin, D. F. (1980). Schizophrenia: a biochemical disorder? Biomedicine 32(2): 54–5.

Huang, J. T., F. M. Leweke, D. Oxley, L. Wang, N. Harris, D. Koethe, C. W. Gerth, B. M. Nolden, S. Gross, D. Schreiber, B. Reed and S. Bahn (2006). Disease biomarkers in cerebrospinal fluid of patients with first-onset psychosis. PLoS Med 3(11): e428.

Ilani, T., D. Ben-Shachar, R. D. Strous, M. Mazor, A. Sheinkman, M. Kotler and S. Fuchs (2001). A peripheral marker for schizophrenia: Increased levels of D3 dopamine receptor mRNA in blood lymphocytes. Proc Natl Acad Sci U S A 98(2): 625–8.

Issaq, H. J., T. D. Veenstra, T. P. Conrads and D. Felschow (2002). The SELDI-TOF MS approach to proteomics: protein profiling and biomarker identification. Biochem Biophys Res Commun 292(3): 587–92.

Jiang, L., K. Lindpaintner, H. F. Li, N. F. Gu, H. Langen, L. He and M. Fountoulakis (2003). Proteomic analysis of the cerebrospinal fluid of patients with schizophrenia. Amino Acids 25(1): 49–57.

Kidd, D., Y. Liu and B. F. Cravatt (2001). Profiling serine hydrolase activities in complex proteomes. Biochemistry 40(13): 4005–15.

Kozlovsky, N., W. T. Regenold, J. Levine, A. Rapoport, R. H. Belmaker and G. Agam (2004). GSK-3beta in cerebrospinal fluid of schizophrenia patients. J Neural Transm 111(8): 1093–8.

La, Y. J., C. L. Wan, H. Zhu, Y. F. Yang, Y. S. Chen, Y. X. Pan, G. Y. Feng and L. He (2007). Decreased levels of apolipoprotein A-I in plasma of schizophrenic patients. J Neural Transm 114(5): 657–663.

Lawrie, S. M. and S. S. Abukmeil (1998). Brain abnormality in schizophrenia. A systematic and quantitative review of volumetric magnetic resonance imaging studies. Br J Psychiatr 172: 110–20.

Lawrie, S. M., H. Whalley, J. N. Kestelman, S. S. Abukmeil, M. Byrne, A. Hodges, J. E. Rimmington, J. J. Best, D. G. Owens and E. C. Johnstone (1999). Magnetic resonance imaging of brain in people at high risk of developing schizophrenia. Lancet 353(9146): 30–3.

Lawrie, S. M., H. C. Whalley, S. S. Abukmeil, J. N. Kestelman, L. Donnelly, P. Miller, J. J. Best, D. G. Owens and E. C. Johnstone (2001). Brain structure, genetic liability, and psychotic symptoms in subjects at high risk of developing schizophrenia. Biol Psychiatr 49(10): 811–23.

Liu, H., R. G. Sadygov and J. R. Yates, III (2004). A model for random sampling and estimation of relative protein abundance in shotgun proteomics. Anal Chem 76(14): 4193–201.

Manning, G., D. B. Whyte, R. Martinez, T. Hunter and S. Sudarsanam (2002). The protein kinase complement of the human genome. Science 298(5600): 1912–34.

McDonald, L., D. H. Robertson, J. L. Hurst and R. J. Beynon (2005). Positional proteomics: selective recovery and analysis of N-terminal proteolytic peptides. Nat Methods 2(12): 955–7.

McGlashan, T. H., R. B. Zipursky, D. Perkins, J. Addington, T. Miller, S. W. Woods, K. A. Hawkins, R. E. Hoffman, A. Preda, I. Epstein, D. Addington, S. Lindborg, Q. Trzaskoma, M. Tohen and A. Breier (2006). Randomized, double-blind trial of olanzapine versus placebo in patients prodromally symptomatic for psychosis. Am J Psychiatr 163(5): 790–9.

McGorry, P. D., A. R. Yung, L. J. Phillips, H. P. Yuen, S. Francey, E. M. Cosgrave, D. Germano, J. Bravin, T. McDonald, A. Blair, S. Adlard and H. Jackson (2002). Randomized controlled trial of interventions designed to reduce the risk of progression to first-episode psychosis in a clinical sample with subthreshold symptoms. Arch Gen Psychiatr 59(10): 921–8.

Merchant, M. and S. R. Weinberger (2000). Recent advancements in surface-enhanced laser desorption/ionization-time of flightmass spectrometry. Electrophoresis 21(6): 1164–77.

Merrell, K., K. Southwick, S. W. Graves, M. S. Esplin, N. E. Lewis and C. D. Thulin (2004). Analysis of low-abundance, low-molecular-weight serum proteins using mass spectrometry. J Biomol Tech 15(4): 238–48.

Morrow, J. D., J. A. Awad, J. A. Oates and L. J. Roberts, II (1992). Identification of skin as a major site of prostaglandin D2 release following oral administration of niacin in humans. J Invest Dermatol 98(5): 812–5.

O'Hagan, S., W. B. Dunn, J. D. Knowles, D. Broadhurst, R. Williams, J. J. Ashworth, M. Cameron and D. B. Kell (2007). Closed-loop, multiobjective optimization of two-dimensional gas chromatography/mass spectrometry for serum metabolomics. Anal Chem 79(2): 464–76.

Old, W. M., K. Meyer-Arendt, L. Aveline-Wolf, K. G. Pierce, A. Mendoza, J. R. Sevinsky, K. A. Resing and N. G. Ahn (2005). Comparison of label-free methods for quantifying human proteins by shotgun proteomics. Mol Cell Proteomics 4(10): 1487–502.

Patterson, S. D. and R. H. Aebersold (2003). Proteomics: the first decade and beyond. Nat Genet 33 (Suppl): 311–23.

Phizicky, E., P. I. Bastiaens, H. Zhu, M. Snyder and S. Fields (2003). Protein analysis on a proteomic scale. Nature 422(6928): 208–15.

Prabakaran, S., M. Wengenroth, H. E. Lockstone, K. Lilley, F. M. Leweke and S. Bahn (2007). 2-D DIGE analysis of liver and red blood cells provides further evidence for oxidative stress in schizophrenia. J Proteome Res 6(1): 141–9.

Puri, B. K., T. Easton, I. Das, L. Kidane and A. J. Richardson (2001). The niacin skin flush test in schizophrenia: a replication study. Int J Clin Pract 55(6): 368–70.

Puri, B. K., S. R. Hirsch, T. Easton and A. J. Richardson (2002). A volumetric biochemical niacin flush-based index that noninvasively detects fatty acid deficiency in schizophrenia. Progr Neuro Psychopharmacol Biol Psychiatr 26(1): 49–52.

Reiner, A., D. Yekutieli and Y. Benjamini (2003). Identifying differentially expressed genes using false discovery rate controlling procedures. Bioinformatics 19(3): 368–75.

Rogers, M. A., P. Clarke, J. Noble, N. P. Munro, A. Paul, P. J. Selby and R. E. Banks (2003). Proteomic profiling of urinary proteins in renal cancer by surface enhanced laser desorption ionization and neural-network analysis: identification of key issues affecting potential clinical utility. Cancer Res 63(20): 6971–83.

Schulte, I., H. Tammen, H. Selle and P. Schulz-Knappe (2005). Peptides in body fluids and tissues as markers of disease. Expert Rev Mol Diagn 5(2): 145–57.

Seehusen, D. A., M. M. Reeves and D. A. Fomin (2003). Cerebrospinal fluid analysis. Am Fam Physician 68(6): 1103–8.

Shaw, J., R. Rowlinson, J. Nickson, T. Stone, A. Sweet, K. Williams and R. Tonge (2003). Evaluation of saturation labelling two-dimensional difference gel electrophoresis fluorescent dyes. Proteomics 3(7): 1181–95.

Steen, H., B. Kuster and M. Mann (2001). Quadrupole time-of-flight versus triple-quadrupole mass spectrometry for the determination of phosphopeptides by precursor ion scanning. J Mass Spectrom 36(7): 782–90.

Storey, J. D. (2001). The positive false discovery rate: A Bayesian interpretation and the q-value. Technical Report 2001–12, Department of Statistics, Stanford University

Tavares, H., J. Yacubian, L. L. Talib, N. R. Barbosa and W. F. Gattaz (2003). Increased phospholipase A2 activity in schizophrenia with absent response to niacin. Schizophr Res 61(1): 1–6.

Tonge, R., J. Shaw, B. Middleton, R. Rowlinson, S. Rayner, J. Young, F. Pognan, E. Hawkins, I. Currie and M. Davison (2001). Validation and development of fluorescence two-dimensional differential gel electrophoresis proteomics technology. Proteomics 1(3): 377–96.

Tosato, S., P. Dazzan and D. Collier (2005). Association between the neuregulin 1 gene and schizophrenia: a systematic review. Schizophr Bull 31(3): 613–7.

Tosic, M., J. Ott, S. Barral, P. Bovet, P. Deppen, F. Gheorghita, M. L. Matthey, J. Parnas, M. Preisig, M. Saraga, A. Solida, S. Timm, A. G. Wang, T. Werge, M. Cuenod and K. Q. Do (2006). Schizophrenia and oxidative stress: glutamate cysteine ligase modifier as a susceptibility gene. Am J Hum Genet 79(3): 586–92.

Tsang, T. M., J. T. Huang, E. Holmes and S. Bahn (2006). Metabolic profiling of plasma from discordant schizophrenia twins: correlation between lipid signals and global functioning in female schizophrenia patients. J Proteome Res 5(4): 756–60.

Villar-Garea, A., M. Griese and A. Imhof (2007). Biomarker discovery from body fluids using mass spectrometry. J Chromatogr B Analyt Technol Biomed Life Sci 849(1–2): 105–114.

Vitzthum, F., F. Behrens, N. L. Anderson and J. H. Shaw (2005). Proteomics: from basic research to diagnostic application. A review of requirements & needs. J Proteome Res 4(4): 1086–97.

Vong, R., P. Geladi, S. Wold and K. Esbensen (1988). Source contributions to ambient aerosol calculated by discriminant partial least squares regression (PLS). J. Chemometrics 2: 281–6.

Wan, C., Y. La, H. Zhu, Y. Yang, L. Jiang, Y. Chen, G. Feng, H. Li, H. Sang, X. Hao, G. Zhang and L. He (2007). Abnormal changes of plasma acute phase proteins in schizophrenia and the relation between schizophrenia and haptoglobin (Hp) gene. Amino Acids 32(1): 101–8.

Wan, C., Y. Yang, H. Li, Y. La, H. Zhu, L. Jiang, Y. Chen, G. Feng and L. He (2006). Dysregulation of retinoid transporters expression in body fluids of schizophrenia patients. J Proteome Res 5(11): 3213–6.

Wang, W., H. Zhou, H. Lin, S. Roy, T. A. Shaler, L. R. Hill, S. Norton, P. Kumar, M. Anderle and C. H. Becker (2003). Quantification of proteins and metabolites by mass spectrometry without isotopic labeling or spiked standards. Anal Chem 75(18): 4818–26.

Xiao, Z., D. Prieto, T. P. Conrads, T. D. Veenstra and H. J. Issaq (2005). Proteomic patterns: their potential for disease diagnosis. Mol Cell Endocrinol 230(1–2): 95–106.

Zhang, B., N. C. VerBerkmoes, M. A. Langston, E. Uberbacher, R. L. Hettich and N. F. Samatova (2006). Detecting differential and correlated protein expression in label-free shotgun proteomics. J Proteome Res 5(11): 2909–18.

Zhang, X. Y., D. F. Zhou, L. Y. Cao, P. Y. Zhang and G. Y. Wu (2002). Decreased production of interleukin-2 (IL-2), IL-2 secreting cells and CD4+ cells in medication-free patients with schizophrenia. J Psychiatr Res 36(5): 331–6.

Zieske, L. R. (2006). A perspective on the use of iTRAQ reagent technology for protein complex and profiling studies. J Exp Bot 57(7): 1501–8.

Chapter 5
RNA Biomarkers in Schizophrenia

Daniel Martins-de-Souza and Emmanuel Dias-Neto

Abstract The deciphering of the human genome and the advances in transcriptome interrogation approaches have turned RNA biomarkers into important promises for the understanding and management of psychiatric diseases. In this chapter we describe the techniques more widely used for gene expression analysis and present the main findings in the search for RNA biomarkers in schizophrenia such as the recurrent observation of alterations in genes that encode proteins involved in pathways related to myelinization, synapses, and energy metabolism. We also discuss the main findings resultant from peripheral blood cell studies and present new techniques and new sources of RNA biomarkers for the future of research in schizophrenia.

Abbreviations AS: Alternative splicing; D2: Dopamine receptor 2; D2L: Long isoform of dopamine receptor 2; D2S: Short isoform of dopamine receptor 2; DPFC: Dorsolateral prefrontal cortex; dT: Deoxythymidine; EST: Expressed sequence tag; FDA: United States Food and Drug Administration; GABAA: Gamma-amino butyric acid – A; GNAO1: Guanine nucleotide-binding regulatory protein Go-alpha; mGluR3: Metabotropic glutamate receptor 3; miRNA: MicroRNA; NCAM1: Neural cell adhesion molecule 1; ncRNA: Noncoding RNA; NMDA: *N*-methyl-D-aspartic acid; PFC: Prefrontal cortex; PSYN: Pre-synaptic function; qPCR: Quantitative polymerase chain reaction; SAGE: Serial analysis of gene expression; SCZ: Schizophrenia
Genes ACADS: acyl-Coenzyme A dehydrogenase, C-2 to C-3 short chain; ACADL: acyl-Coenzyme A dehydrogenase, long chain; ACAT2: acetyl-coenzyme

D. Martins-de-Souza
Laboratório de Neurociências (LIM-27), Instituto de Psiquiatria, Faculdade de Medicina, Universidade de São Paulo (USP), Brazil
Laboratório de Proteômica, Dept. de Bioquímica, Instituto de Biologia, Universidade Estadual de Campinas (UNICAMP), Brazil

E. Dias-Neto
Laboratório de Neurociências (LIM-27), Instituto de Psiquiatria, Faculdade de Medicina, Universidade de São Paulo (USP), Brazil
Present address: University of Texas, MD Anderson Cancer Center, Houston, TX USA

C.W. Turck (ed.) *Biomarkers for Psychiatric Disorders*,
doi: 10.1007/978-0-387-79251-4_5,

A acyltransferase 2; ACO: aconitase; ADSSL1: adenylosuccinate synthetase; AGPS: alkylglycerone phosphate synthase; AMPA2: glutamate receptor ionotropic AMPA2; APOBEC3B: catalytic polypeptide-like apolipoprotein B mRNA editing enzyme 3B; ATM: ataxia telangiectasia mutated; ATP5A1: ATP synthase mitochondrial F1 complex alpha; Azin1: antizyme inhibitor; BTG1: B-cell translocation gene 1, anti-proliferative; CALM3: calmodulin 3; CHRNA7: alpha7-nicotinic-acetylcholine-receptor; CLC: Charcot-Leyden crystal protein; CLDN11: claudin 11; CNP: $2',3'$-cyclic nucleotide $3'$ phosphodiesterase; CPT1: carnitine palmitoyltransferases 1; CPT2: carnitine palmitoyltransferases 2; CRYM: crystallin; CTNNA1: alpha catenin; CXCR1: chemokine C-X-C motif ligand 1; CYP27B1: cytochrome P450 family 1, subfamily B, polypeptide 1; Datf1: death-associated transcription factor 1; DUSP6: dual specificity phosphatase 6; EBP50: ezrin-radixin-moesin phosphoprotein 50; Edg-2: endothelial differentiation gene 2; ENO: enolase; ERBB2: v-erb-b2 erythroblastic leukemia viral oncogene homolog 2, neuro/glioblastoma derived oncogene homolog (avian); ERBB3: v-erb-b2 erythroblastic leukemia viral oncogene; GAP-43: growth-associated protein-43; GAPDH: glyceraldehyde-3-phosphate dehydrogenase; GOT2: glutamic-oxaloacetic transaminase 2; GSK3a: glycogen synthase kinase 3 α; GSN: gelsolin; HLA-DRB1:m ajorh istocompatibilityc omplex,c lassI I,D R β 1; HNRPA3:h eterogeneous nuclear ribonucleoprotein A3; MAG: myelin-associated glycoprotein; MAL: T-lymphocyte maturation-associated protein; MARCKS: myristoylated alanine-rich C kinase substrate; MAZ: myc-associated zinc finger protein; MDH1: malate dehydrogenase 1; MOG: myelin oligodendrocyte glycoprotein; NDUFS1: mitochondrial complex I 75-kDa subunit; NPY: neuropeptide Y; NPY1R: neuropeptide Y receptor Y1 gene; NRG1: neuregulin 1; NSF: N-ethylmaleimide sensitive factor; OAT: ornithine aminotransferase; OXCT1: 3-oxoacid CoA transferase; PDH: pyruvate dehydrogenase; PDLIM5: PDZ and LIM domain 5; PKM1: muscle pyruvate kinase; PLLP/TM4SF11: plasmolipin or transmembrane 4 superfamily 11; PLP: proteolipid protein; QKI: quaking homolog; RAB3C: RAB3C, member RAS oncogene family; S100A9: S100 calcium binding protein A9; SCG-10: superior cervical ganglia-10; SELENBP1: selenium-binding protein 1; SERPINI1: neuroserpin; SFRS1: splicing factor, arginine/serine-rich 1; SMDF: sensory motor neuron derived factor; SYN2: synapsin 2; SYNJ1: synaptojanin 1; TCFL4: MAX-like protein X; TF: transferrin; TIMM17A: translocase of inner mitochondrial membrane 17; TNFR2: tumor necrosis factor receptor 2; UCHL1: ubiquitin C-terminal esterase L1; USP14: ubiquitin-specific protease 14; YWHAH: tyrosine 3-monooxygenase/tryptophan 5-monooxygenase activation protein, eta polypeptide

5.1 Introduction

The United States Food and Drug Administration (FDA) defines a biomarker as "a characteristic that is objectively measured and evaluated as an indicator of normal biologic or pathogenic processes or pharmacological responses to a therapeutic intervention" (Chakravarty, 2003). The utility of biomarkers in clinical medicine is vast, and includes disease diagnosis, indicators of disease status, or targets to monitor and predict response to therapeutics or disease outcome. In this context, RNAs constitute a central class of markers that can be directly influenced by DNA alterations but may also exert a straight influence on protein markers in many diseases.

The primary functional link of RNAs as biomarkers is their obvious role as intermediate between the DNA that encodes the information, and the proteins that are the ultimate effectors. However, as the RNA complexity continues to be revealed by large-scale transcriptome analysis, new RNA categories continue to emerge bringing new classes of potential markers, including regulatory noncoding RNAs such as miRNAs. The recent discovery of these new RNA molecules will certainly bring dramatic changes in the way we see and understand protein regulation, and will also increase the number of RNA markers available for a number of diseases.

The category of RNA biomarkers gained a dramatic impulse after the deciphering of the human genome, and the diverse projects that contributed to catalogue our transcriptomes. The dynamic nature of the transcriptome reveals the complexity of gene regulation and together with other molecular markers should contribute to the understanding of relevant clinical aspects such as disease mechanisms and reaction to drugs, pointing to less ambiguous prognosis, and more precise markers for treatment response.

RNA alterations are certainly involved in the biological basis of a number of diseases. In psychiatry, RNA markers are central, and are certainly implicated in diverse facets such as distinct clinical presentations, neuroimaging alterations, and endophenotypes. Consistent patterns of abnormalities in proteomes can be a result of alternative mRNA splicing, quantitative fluctuations in mRNA, or translational regulation, which can be determined by noncoding regulatory RNAs. These proteomic alterations can affect diverse disease-related aspects, including neural system alterations that may underlie emotional processing and cognitive control, characteristic for psychiatric diseases. Such abnormalities, present at protein and RNA levels, may constitute valuable disease biomarkers, and may potentially help in, disease diagnosis, shedding light on the molecular basis of psychiatric illnesses. Together, genome and transcriptome analysis provide the basis upon which these markers could be investigated and identified by a series of distinct and complementary approaches. In this chapter we will present the most frequently used techniques to interrogate the transcriptome in large-scale and will focus on the most relevant biomarkers that have been revealed in schizophrenia (SCZ), a disease uniquely human that affects 1% of the population world wide.

5.2 *The Main Approaches used for RNA Biomarkers Discovery*

The identification of reliable biomarkers is of utmost relevance for psychiatric diseases. The availability of these markers would be of great value for all psychiatric diseases, and of particular importance in SCZ, a complex disease with pathogenesis mechanisms yet to be defined, despite the various efforts in order to its progressive understanding (Freedman, 2003). Besides some macroscopic alterations observed in brain scans of SCZ patients, the disease diagnosis is essentially clinical, with no molecular basis nor reliable biochemical markers identified up to now and a still unpredictable response to treatment (Frances et al., 1991). The complexity of SCZ is a driving force in the search for reliable markers, at DNA, RNA, protein and brain morphometry levels. Besides the recent exploration of the transcriptome in the postgenomic era, the search for reliable SCZ markers is still an ongoing process.

In its early days, the search of RNA markers in SCZ was focused on the evaluation of the expression of specific genes of interest related to what was known in SCZ. These genes had their cDNA sequenced in samples of patients, such as it has been done in 1986 with the analysis of the coding region of gamma-endorphin gene derived from postmortem brains and pituitary glands of schizophrenic patients (Bovenberg et al., 1986). Nowadays, a number of large-scale techniques allow the interrogation of the entire transcriptome, mainly using hybridization platforms known as gene chips, or cDNA microarrays. The gene arrays available today not only cover the full set of human genes, but also include a large set of noncoding regions and noncoding transcripts, enabling the interrogation of a significant fraction of the whole genome. These comprehensive genome and transcriptome analyses promise a revolution in the comprehension of gene regulation and its effects on human diseases.

The most commonly used techniques for high-throughput transcriptome analysis are cDNA microarrays, Serial Analysis of Gene Expression (SAGE), and the analyses of Expressed Sequence Tag (EST). The discrete capabilities and distinct aspects of these different approaches are important to reveal the diversity of gene regulation aspects. These large-scale methods are absolutely distinct from each other, and not only differ in their throughput and cost, but also in the type of data obtained and in the depth of information provided (Mirnics et al., 2001). These techniques not only allow an analysis of quantitative alterations, an aspect that is the subject of frequent investigation by the majority of RNA biomarker projects, but also help to unravel qualitative aspects of gene expression, such as alternative splicing or alternative polyadenylation events, that also affect protein composition and RNA stability, with direct phenotypic consequences. These aspects make complementary, rather than redundant, the distinct transcriptome interrogation approaches. The combination of large-scale transcriptome interrogation tools will facilitate the investigation of the most complex questions related to diseases of the nervous system (reviewed in Mirnics et al., 2001) and together with new and large-scale DNA sequencing approaches available today, have the potential to transfigure the importance of RNA markers for psychiatric diseases.

5.2.1 Expressed Sequence Tags Analysis (EST)

One of the early approaches used for large-scale gene discovery and analysis was based on the partial sequencing of clones derived from cDNA libraries as proposed by the group of J. Craig Venter, in 1992 (Adams et al., 1992). Due to its capacity for generating relatively long reads of cDNA sequences, this approach is still very useful for gene discovery, alternative splicing, alternative poly-adenylation, polymorphism, and mutation analysis, being extremely rich sources of transcribed biomarkers.

EST analysis starts with the extraction of the RNA of interest (mRNA or total RNA, depending on the approach used) that is going to be cloned into sequencing vectors. The standard approach consists of using oligo-dT primers for the synthesis of the first cDNA strand, followed by the second strand synthesis and the cloning of double stranded cDNA molecules in the sequencing vectors, usually with the help of adaptors for a directional cloning. In this regard, there is a major difference between EST and SAGE (to be described later); in the case of ESTs one usually aims to clone the entire, or the most complete cDNA fragment, while for SAGE the focus is on very small fragments from a defined portion of the transcript. The cDNA clones are then sequenced, from one or both ends. In non-normalized libraries, the number of times a certain gene is sequenced usually corresponds to its frequency in the original RNA pool. Also, as the fragment sequenced is usually long (500–800 nt), sequence analysis provides not only an unambiguous gene identification, but also provides data useful for the observation of transcriptional diversity, such as polymorphisms, mutations or alternative splicing and alternative polyadenylation events.

The major limitations of the EST approach are related to sequencing costs required for broad transcriptome coverage. Each clone sequenced allows the study of a single transcript and due to the high level of expression of some transcripts, sequencing redundancy of the most abundant transcripts could be a limitation. Another drawback is the biased distribution of the ESTs towards the ends of the transcripts, which reduces the coverage of the whole gene. While the reduction on DNA sequencing costs is now a reality, the alternative approaches that have been described to circumvent sequencing redundancy and positional biases of ESTs inevitably abolish its use in the detection of relevant fluctuations in gene expression (Soares et al., 1994; Dias-Neto et al., 2000; Camargo et al., 2001).

In neuropsychiatry, ESTs have been used in the study of Parkinson's disease (Lu et al., 2005a; Kim et al., 2006) and epilepsy (Avedissian et al., 2007). ESTs have also been used in the study of the brains of schizophrenic patients and a few thousand sequences have been generated and are publicly available in the EST database at the NIH (www.ncbi.nlm.nih.gov/dbEST). Most brains used for cDNA library construction were obtained from the Stanley Neuropathology Consortium, and include the frontal lobe of a suicide schizophrenic, male, 34 years old (3,699 ESTs available), and a hippocampus (a pool of three schizophrenic patients), subtracted from a pool of three mentally normal individuals. All sequences are available in public databases, and can be used to point to a number of transcriptional aspects present in these brain samples. In SCZ, ESTs were also useful for the analysis of new gene polymorphisms in the 14-3-3 eta chain gene (Bell et al., 2000).

5.2.2 cDNA Microarrays

The development of cDNA microarrays has provided one of the most powerful tools for the large-scale gene expression investigation in human diseases. Through the use of these arrays, global gene expression analysis could be used as a valuable tool to obtain major insights into diagnosis, progression, prognosis, and response to therapy for a number of human diseases. The basic technologies used for cDNA microarrays were first developed to evaluate gene expression. However, improvements of the initial methods now permits more intricate and widespread uses, including mutation analysis in expressed genes (Klevering et al., 2004; Tennis et al., 2006; Van Bogaert et al., 2007), gene sequencing and polymorphism analysis (Kozal et al., 1996; Günthard et al., 1998) and analysis of splicing events and copy number variations (Cuperlovic-Culf et al., 2006; Hughes et al., 2006; Blencowe, 2006; Cowell and Hawthorn 2007). The analysis of gene expression using cDNA microarrays is one of the most used applications of the gene-chip technology as it permits the rapid, simultaneous and sensitive analysis of a large number of biological samples for the concurrent expression of thousands of genes (Mirnics et al., 2001; Magic et al., 2007). Recent developments allowed the investigation of larger fractions of the transcriptome, increasing the accuracy of cDNA microarrays, and reducing its costs, permitting microarrays to be used as a common tool in the search for markers of human diseases. These studies are often carried out in conjunction with other methods that confirm the differential expression detected with microarrays, including Northern blots and quantitative real-time PCR (qPCR).

The concepts that lead to the development of cDNA microarrays were realized soon after the first description of the double helix by Watson and Crick, in 1953. After the depiction of the DNA structure, it was realized that the two DNA strands could be separated by heat or alkali treatment, in a reversible process that underlies all the methods based on DNA hybridization. It was also observed at that time, that some degree of sequence complementarity was required during the hybridization of two sequences involved in the duplex formation. Double-strand denaturation, sequence complementarity, and renaturation capability were concepts that emerged during this period and allowed the development of analytical methods based on DNA hybridization, which were quickly incorporated into a range of biological investigations. Later, in the mid-1970s the potential of the recombinant DNA technology was realized due to a number of factors, including the capability of detecting specific clones in genomic or cDNA libraries (Grunstein and Hogness 1975; Benton and Davis 1977), using hybridization. Thus, bacterial or phage clones carrying plasmids with different inserts, could be screened and selected on membranes, exploring the concept of anchoring nucleic acids to a solid support for analysis by hybridization with radioactively-labeled probes.

Modern microarrays were developed from these key basic concepts, and now employ microscopic dots, spotted on glass slides, revealed with fluorescent probes. The most used platforms consist of a set of predefined arrays of DNA molecules (such as oligonucleotides, cDNA clones or PCR products) that are usually spotted onto glass slides, as well as labeled cDNAs, derived from the RNAs of the samples of interest. The labeled cDNA molecules are hybridized against the elements on the arrays and the detection is usually made using fluorophores such as Cy-3 or Cy-5. Once the hybridization step is completed, the glass slides are scanned to obtain digital images of the experimental results. These digital images contain hundreds or thousands of points or "spots." Each spot represents a probe, and usually, many probes are available for each transcript. Gene expression levels are then quantified with the help of software designed for image-analysis. After processing the hybridization results of thousands of probes, representing thousands of genes, the typical goal is to find statistically significant up- or down-regulated transcripts.

The technique is sensitive and can detect gene expression fluctuations for most of the transcripts. Commercially available cDNA microarrays can detect as few as one in 250,000 mRNA copies (Mirnics et al., 2001). This level of sensitivity should allow the detection of rare mRNA species in a pool of transcripts found in a typical tissue sample (Lockhart et al., 1996; Bertucci et al., 1999; Kane et al., 2000). However, even with this high sensitivity, many low-abundance transcripts remain undetected. Unfortunately, increasing the absolute amount of the hybridized target usually will not increase the sensitivity of the microarray – as the relative abundance of the transcript within the RNA pool, coupled with microarray probe characteristics, will influence the detection limit for each individual transcript. Thus, transcripts whose expression is restricted to a small subpopulation of neurons are effectively diluted by the most abundant ones and remain undetectable. Consequently, even if one had access to the entire transcriptome on a microarray, the complexity of gene expression in the brain would preclude real global gene expression profiling.

Advantages of microarrays include its high degree of automation, the requirement of relatively small RNA amounts, as well as the capability of simultaneous analysis of thousands of genes in a single experiment, at a relatively low cost (around US$500/slide). However, as each probe is capable of evaluating only its corresponding transcript, the distinct probe properties (such as its GC content, self-complementarity, and its location in the transcript) can affect its hybridization capabilities. Thus, the expression data derived this way is relative rather than absolute, and gene-expression measurements made by microarrays still need to be confirmed by other methods. Due to the relative low cost, and the availability of a number of commercial platforms, cDNA microarrays are one of the most popular approaches to investigate gene expression.

In psychiatry research, microarray analysis has provided a wealth of information to help uncover complex biological processes, to better understand the pathogenesis of many diseases, and to discover novel potential biomarker panels, to mention a few. Transcriptomes of different brain areas have been investigated by many groups, and the results have contributed to the definition of the set of genes expressed in diseased brains, pointing to a potentially relevant set of RNA markers. Most of the markers described in this chapter have been discovered by using microarray analysis.

5.2.3 Serial Analysis of Gene Expression (SAGE)

Gene expression patterns in the human brain exceed the complexity of many other organ systems. The degree of difficulty in the analysis of such patterns is magnified in the investigation of psychiatric disorders, which appear to result from the interplay of polygenic and epigenetic factors on multiple brain circuits. Thus, in many situations of psychiatric research, it would be important to use techniques that do not require an *a priori* definition of the genes that are going to be investigated. This is one of the most interesting aspects of SAGE.

SAGE is a DNA sequencing-based technique, described by Velculescu et al., in 1995 (Velculescu et al., 1995), based on the sampling of short cDNA sequences (called tags) ideally gene-specific, from a population of cells or tissues. The presence and the sequence of these gene tags are determined by the occurrence and location of digestion sites of frequent cutter restriction enzymes. SAGE theory rests on three basic principles: First, considering a traditional SAGE experiment, a theoretical tag of 14 – 4 nt corresponding to the restriction enzyme site, followed by 10 nucleotides downstream – can generate 4^{10} (1,048,576) different tags, which in theory are capable to discriminate most of human transcripts (Patino et al., 2002). Second, tag concatenation allows a faster and less-expensive determination of the tag-sequence *via* cDNA sequencing, allowing the analysis of thousands of genes in a few sequencing reactions; and third, the number of times each tag is sequenced should reflect the expression level of its corresponding gene (Knox and Skuce, 2005), allowing the generation of gene abundance lists, categorizing the expression of each transcript in an absolute fashion.

SAGE requires the extraction of the RNA samples to be compared, followed by the construction and sequencing of SAGE libraries containing tag concatemers. After sequencing a few thousand clones of the SAGE libraries, the frequency of the tags (which are derived from the collection of genes expressed in the studied samples) corresponds to the frequency of the respective gene in the original sample. Due to the concatenation of small fragments for serial sequencing, tag redundancy is used to estimate transcript abundance, and does not represent a limitation as seen in the EST approach.

The construction of SAGE-libraries is based on a chain of enzymatic steps. Briefly, the process starts with the synthesis of cDNA molecules, primed by a biotinylated oligo dT coupled to magnetic beads. The cDNAs are magnetically captured and digested with a frequent (4-bp) cutter restriction enzyme, named anchoring enzyme (usually *NlaIII*). After this digestion, the digested fragments that are not coupled to the oligo-dT in the magnetic columns are washed away, and only the most 3 end fragments remain. These are subjected to the ligation of adapters containing the restriction site of other enzymes, used as tagging enzymes. A tagging enzyme is a type IIS restriction endonuclease (*BsmfI* is typically used), an enzyme that cuts the DNA fragment at a defined distance of its recognition site (up to 20 base pairs). To illustrate, the restriction enzyme *BsmfI* recognizes the sequence GGGAC and will cleave the DNA strand 10 nt downstream. As the DNA is cleaved, it is released from the bead, and the tags are ligated to each other to create ditags. These tags are then concatenated and cloned in sequencing vectors, constituting a SAGE library, ready to be used for serial DNA sequencing. The detailed protocol of SAGE as well as more information can be found at the SAGEnet website (http://www.sagenet.org/protocol/MANUAL1e.pdf – accessed on 30 Dec 2007) as well as in a recent review by Hu and Polyak (Hu and Polyak, 2006).

A typical SAGE-library contains thousands of clones, each one containing a few dozens of small transcript tags (14 bp) cloned in tandem. In a single sequencing run, as many as 25–30 gene tags can be obtained from a clone, and 2,000–3,000 tags can be produced after sequencing a single 96 well plate. After extracting and counting the individual tags, a SAGE-tag list is produced. The tags are mapped to their genes, and their frequency corresponds to the abundance of the corresponding gene in the original RNA pool. Thus, SAGE is an open platform that does not require an *a priori* list of genes to be investigated.

Another important aspect of SAGE is with regard to its quantitative power. A typical SAGE experiment involves the construction and sequencing of a pair of SAGE libraries, allowing transcriptome comparison in two different situations, giving statistical significance to each compared transcript. This type of data permits a direct comparison between different experiments, laboratories, experimental systems, and data types; a crucial aspect of the database dependent analysis of biological systems that is required by functional genomics approaches. With the advent of the new DNA sequencing technologies (Margulies et al., 2005; Nielsen et al., 2006) SAGE lists with millions of tags can be generated in a few days, and data from different laboratories or experiments can be directly compared.

Maybe the most significant problem of SAGE is the generation of ambiguous tags, which could make difficult the interpretation of data. It is estimated that around 30% of the regular SAGE-tags can be mapped to multiple human transcripts, and the correct interpretation of gene regulation in the sample requires tag-extension approaches or alternative SAGE methods (reviewed in Wang, 2007). One way to overcome this limitation is to generate longer and thus, more specific gene tags, as described (Saha et al., 2002). Another limitation of SAGE is the inability of tagging some genes, due to the absence of required sites for the anchoring enzyme in a fraction of the transcripts. It has been calculated that this latter problem would prohibit the proper analysis of expression of 3–5% of the human genes. Due to the high costs, and many days required for library construction and sequencing SAGE libraries, this technique has been poorly used in the study of transcriptome alterations in psychiatric diseases.

The published gene expression analyses using SAGE in models to study neuropsychiatry are limited to a few papers evaluating drug-response and gene expression (Ouchi et al., 2004; Cai et al., 2005), or models for epilepsy (Hendriksen et al., 2001; Arai et al., 2003) or Parkinson's disease (Ryu et al., 2002; Ryu et al., 2005). Few studies have been published with human samples, and are currently limited to epilepsy (Ozbas-Gerçeker et al., 2006), bipolar disorder (Sun et al., 2001), Parkinson's (Noureddine et al., 2005) or Alzheimer's disease (Xu et al., 2007). After the recent development of new DNA sequencing technologies, such as the sequencing by synthesis using pyrosequencing protocols optimized for solid supports (Margulies et al., 2005) we can expect to see a rebirth of sequencing-based RNA biomarker discovery in the coming years. These technologies dramatically reduce sequencing costs, and allow the generation of more than 400,000 reads in a single sequencing run. Less expensive sequencing will make possible the sequencing of hundreds of thousands of ESTs, and millions of SAGE tags in a single day, with a strong impact on the discovery of RNA-biomarkers.

Table 5.1 Advantages and disadvantages of three large-scale transcriptome investigation approaches

Technique	Main advantages	Main disadvantages
EST analysis	– The higher informational content (quantitative/ qualitative)	– cDNA library preparation is a complex step
	– More precise gene identification	– Large amounts of RNA are required
	– Excellent to provide qualitative alterations in gene expression (such as alternative splicing)	– High cost for large-scale sequencing when a broad transcriptome coverage is required
	– Allows the analysis of gene polymorphisms and mutations	
	– Excellent tool for gene finding.	– Low-throughput method
cDNA microarrays	– Low cost after device installation	– Low quantification range
	– Large-scale evaluation of thousands of transcripts.	– Allows the evaluation only of the transcripts present in the array.
	– Relatively small RNA amounts	– Do not analyze unknown genes
		– cDNA may hybridize to similar probes (adjustable problem)
	– Method highly automated	– Require special device to array analysis
	– High-throughput	– Requires complex statistics analysis.
	– The best characterized method	
SAGE	– Allows the analysis of unknown genes	– SAGE library preparation is a complex step
	– Directly estimates gene abundance level	– Tags can be ambiguous
	– Expression data is absolute	– Some transcripts may lack the anchoring enzyme site
	– Excellent quantification range	– Poor detection of rare transcripts
	– High Throughput Method	– Expensive and cumbersome DNA sequencing (applying traditional SAGE method).
	– Do not require special devices	– Results from any new experiment are directly comparable to other SAGE data

A comparative overview of the most important advantages and disadvantages of cDNA microarrays, SAGE and EST analyses is presented in Table 5.1.

5.3 *Searching for RNA Markers in Schizophrenia Brain Samples*

Transcriptome studies in neuropsychiatric diseases indicated consistent alterations in the expression of many genes. In many cases, independent studies, dealing with different samples and distinct experimental platforms, have consistently found alterations in the same genes, or in the same pathways, strongly suggesting the implication of discrete physiological clusters in the pathophysiology of certain diseases. This is a strong suggestion that transcriptional alterations may be very relevant not only to propose appropriate biomarkers, but also to contribute to the comprehension of the biological basis of neuropsychiatric diseases.

On the next pages we will briefly present the regulatory clusters more consistently identified in gene expression studies in SCZ, which constitute the most promising RNA markers for this disease.

5.3.1 Oligodendrocyte-related Genes

The main function of the oligodendrocytes is the formation of myelin sheath units around the axons of the neurons in the central nervous system. Myelin is an electrically-insulating phospholipid layer that surrounds the axons of many neurons and greatly facilitates axonal signal by increasing the speed at which the electrical impulses are propagated and by preventing the electrical current from leaving the axon. Due to the key role of myelin, oligodendrocytes have been an important focus in the study of many neurological and neuropsychiatric diseases. In the grey-matter, oligodendrocytes are more numerous than astrocytes and microglia (Dai et al., 2003), and it has been estimated that oligodendrocytes constitute about 51% of cells around the soma of large neurons in the human cortex (Polak et al., 1982). The atrophy of pyramidal neurons in the hippocampus, which has been reported in the prefrontal cortex (PFC) in SCZ, can be caused by the loss of these important cells (Benes et al., 1991; Arnold et al., 1995; Zaidel et al., 1997; Rajkowska et al., 2001). Whereas the abnormal distribution and decreased density of oligodendrocytes in frontal regions of SCZ brains was observed in histological studies (Uranova et al., 2001), the search for RNA markers in different regions of SCZ brains revealed the altered regulation of many oligodendrocyte-related genes, most involved with myelin-homeostasis, suggesting the important role of myelinization and oligodendrocytes in SCZ. Alterations in the metabolism of oligodendrocytes are among the most consistent SCZ biomarkers.

One of the first publications that reported the altered regulation of oligodendrocytes in SCZ was Hakak et al. (Hakak et al., 2001), using cDNA microarrays in the analysis of dorsolateral prefrontal cortex (DPFC) pools of controls and medicated chronic SCZ patients. The findings of this report suggest that oligodendrocytes can be a specific cell type functionally deficient in SCZ. This suggestion was subsequently reinforced by independent analysis conducted by another four groups, also using cDNA microarrays to compare RNAs from SCZ and controls, using RNA samples extracted from DPFC or other brain regions (Tkachev et al., 2003; Prabakaran et al., 2004; Aston et al., 2004; Katsel et al., 2005; Arion et al., 2007), and by another study that evaluated the expression of selected oligodendrocyte-related genes using qPCR (Dracheva et al., 2006). Some of the oligodendrocyte-related genes found by these groups were recently confirmed by McCullumsmith et al. (McCullumsmith et al., 2007) using *in situ* hybridization. Some of the most relevant papers published on RNA analysis in SCZ, as well as the brain regions and platforms used in the study are presented in Table 5.2. The most consistent oligodendrocyte-related gene expression alterations, identified by at least two groups, are listed in Table 5.3.

Many different large-scale studies, using different samples, different brain regions and diverse microarray platforms showed significant alterations in the expression of myelinization pathways. Many of these findings were subsequently confirmed on an individual basis, by other groups, usually with sensitive approaches such as qPCR analysis. Diverse genes related to oligodendrocyte function were confirmed on an individual basis, including 2',3'-cyclic nucleotide 3' phosphodiesterase (CNP) (Peirce et al., 2006), Myelin-associated glycoprotein (MAG) (Aberg et al., 2006; Wan et al., 2005), Gelsolin (GSN) (Xi et al., 2004), Myelin oligodendrocyte glycoprotein (MOG) (Liu et al., 2005), Neuregulin (NRG1) (Tosato et al., 2005), and many others.

It is interesting to note that the myelination of the PFC occurs in adolescence, a period when SCZ onset is more common. It should also be noted that demyelination diseases such as metachromatic leukodystrophy, are associated with schizophrenic-like psychoses (Hyde et al., 1992). As suggested by Hakak et al. (Hakak et al., 2001), alterations in oligodendrocyte—axon interactions may underlie cytoarchitectural changes found in SCZ. When the consistent observation of a differential regulation of myelin-related genes is taken together with clinical and physiological aspects, oligodendrocytes appear as one of the stronger players in SCZ.

5.3.2 Alterations in Synaptic Function and Plasticity

One of the first large-scale transcriptome analyses of SCZ brains was carried out by Mirnics et al. (Mirnics et al., 2000) who used cDNA microarrays to study gene expression in PFC samples derived from 12 SCZ and 10 control subjects. PFC, the anterior region of the frontal lobes, is one of the most explored brain regions in SCZ RNA studies. This neocortical region is most elaborated in primates and provides diverse and flexible behavioral repertoires, including the differentiation of conflicting thoughts, determination of concepts of "good and bad", and perspectives in accordance with determined actions and moderating correct social behavior. Moreover, an important function operated in PFC is personality expression (Miller and Cohen, 2001; Liston et al., 2006). The platform used by Mirnics et al. (Mirnics et al., 2000) contained over 7,000 unique cDNA elements, and the most relevant findings involved the reduced expression of genes involved in presynaptic function (PSYN) which were subsequently verified by *in situ* hybridization. The reduced expression of *PSYN* genes, can lead to an impaired release of synaptic vesicles at nerve terminals, causing a neurotransmission imbalance frequently observed in SCZ. Two genes showed consistent regulation in individual analysis: N-ethylmaleimide sensitive factor (NSF) and synapsin II (SYN2) in all six SCZ patients included in the study. The markers revealed by this study suggest that SCZ can be the consequence of an abnormality in presynaptic function.

Table 5.2 Analysis of gene expression of oligodendrocytes-related genes of postmortem scz brain tissue

Reference	Type of analysis	Subjects	Brain region
Hakak et al. (2001)	Microarrays	12 SCZ; 12 controls	Prefrontal cortex(BA46; left hemisphere; gray matter)
Tkachev et al. (2003)	Microarrays	15 SCZ; 15 Bipolar; 15 controls	Prefrontal cortex(BA9; gray/white matter)
Prabakaran et al. (2004)	Microarrays	54 SCZ; 50 controls	Prefrontal cortex(BA9 gray/white matter)
Aston et al. (2004)	Microarrays	12 SCZ; 14 controls	Temporal Cortex(Middle Temporal Gyrus – BA21)
Katsel et al. (2005)	Microarrays	16 SCZ; 14 controls	Superior frontal gyrus (BA8) Frontal pole (BA10) Insular cortex (BA44) Dorsolateral prefrontal córtex (BA46) Anterior cingulate (BA: 24/32) Posterior cingulate (BA: 23/31), Parietal (BA: 7) Inferior temporal gyrus (BA20) Middle temporal gyrus (BA21) Superior temporal gyrus (BA22) Parahippocampal gyrus (BA 36/28) Occipital (BA: 17) Hippocampus Caudate nucleus Caudate putamen
Dracheva et al. (2006)	qPCR	I- 30 SCZ; 25 Controls II- 24 SCZ; 21 Controls III- 23 SCZ; 20 Controls IV- 24 SCZ; 19 Controls	I- Cingulate Gyrus (Brodmann area 24/32) II- Hippocampus III- Caudate Nucleus IV- Caudate Putamen
McCullumsmith et al. (2007)	*in situ* hybridization	41 SCZ; 34 Controls	Anterior Cingulate Cortex
Arion et al. (2007)	Microarrays and qPCR	12 SCZ; 12 Controls	Prefrontal cortex(BA46; left hemisphere; gray matter)

Table 5.3 Oligodendrocyte-related genes showing altered expression in scz brain tissues

Gene symbol	Gene name	Gene alteration described by
CNP	2′,3′-cyclic nucleotide 3′ phosphodiesterase	Hakak et al. (2001); Tkachev et al. (2003); Prabakaran et al. (2004); Aston et al. (2004); Katsel et al. (2005); Dracheva et al. (2006); McCullumsmith et al. (2007)
MAG	Myelin-associated glycoprotein	Hakak et al. (2001; Tkachev et al. (2003); Aston et al. (2004); Katsel et al. (2005); Aberg et al. (2006b) Dracheva et al. (2006); McCullumsmith et al. (2007)
ERBB3	v-erb-b2 erythroblastic leukemia viral oncogene	Hakak et al. (2001) Tkachev et al. (2003) Aston et al. (2004) Katsel et al. (2005)
TF	Transferrin	Hakak et al. (2001) Tkachev et al. (2003) Prabakaran et al. (2004) Katsel et al. (2005) Aberg et al. (2006b) McCullumsmith et al. (2007) Arion et al. (2007)
GSN	Gelsolin	Hakak et al. (2001); Prabakaran et al. (2004); Katsel et al. (2005)
MAL	T-lymphocyte maturation-associated protein	Hakak et al. (2001); Aston et al. (2004); Katsel et al. (2005)
CLDN11	Claudin 11; Oligodendrocyte specific protein	Tkachev et al. (2003); Katsel et al. (2005); Dracheva et al. (2006);
MOG	Myelin oligodendrocyte glycoprotein	Tkachev et al. (2003); Katsel et al. (2005); Arion et al. (2007)
PLP	Proteolipid protein	Tkachev et al. (2003); Aston et al. (2004); Aberg et al. (2006b)
PLLP/TM4SF11	Plasmolipin or Transmembrane 4 superfamily 11	Aston et al. (2004); Katsel et al. (2005)
QKI	Quaking homolog	Aberg et al. (2006a); McCullumsmith et al. (2007)

Altered states of synaptic functions and plasticity have been implicated as primary suspects in SCZ for many years (reviewed in Owen et al., 2005). One of the major evidences of a synaptic imbalance is the hyper-dopaminergic state presented by SCZ patients, which has been proven after the demonstration that dopaminergic agonists can induce "psychotic-like status" and after the observation that the potency of an antipsychotic drug is directly proportional to its ability to block dopamine receptors (Seeman et al., 1975). Moreover, an abnormal neurodevelopment, including synapse formation, could be one of the main causes of this hyper-dopaminergic state (Raedler et al., 1998). The synaptic defects in SCZ brains may lead to deficits in episodic memory, malfunction of hippocampal circuitry, and anomalies of axonal sprouting (Ben-Shachar and Laifenfeld, 2004).

Transcriptome studies revealed the differential expression of synapse-related proteins in the brains of SCZ patients and have reinforced the involvement of this pathway in SCZ. Myristoylated alanine-rich C kinase substrate, growth-associated protein-43, superior cervical ganglia-10, and neuroserpin are genes involved in neuronal development and the modulation of synaptic plasticity (Aigner et al., 1995; McNamara and Lenox, 1997), which have been found to be upregulated in SCZ brains (Hakak et al., 2001).

Vawter et al. (Vawter et al., 2001), while studying middle temporal gyrus, cerebellum and PFC pools of SCZ patients, showed the alteration of rab3c, glutamate receptor ionotropic AMPA2, neuroserpin, tyrosine 3-monooxygenase/tryptophan 5-monooxygenase activation protein, eta polypeptide, and calmodulin 3. More recently, Aston et al. (Aston et al., 2004) described the downregulation of SYN2 and synaptojanin 1, in SCZ PFC. The list of these synapsis-related genes may also include NRG1 (a protein also involved in oligodendrocyte function) which is released from glutamate terminals and regulates NMDA glutamate receptors. A synthesis of the more solid findings of synapsis-related gene alterations found in SCZ can be found in Table 5.4.

These genes code for proteins that may have an impact on the function of synapses, including glutamate synapses. The identification of these genes implies that synapses might be one of the primary abnormality sites in SCZ, with a series of downstream consequences that might affect the neural circuitry.

Table 5.4 Synapsis-related genes revealed as differentially expressed in transcriptome studies of SCZ brain tissue

Reference	Gene Symbol	Gene name
Hakak et al. (2001)	MARCKS	Myristoylated alanine-rich C kinase substrate
	GAP-43	Growth-associated protein-43
	SCG-10	Superior cervical ganglia-10
	SERPINI1	Neuroserpin
Mirnics et al. (2000)	NSF	N-ethylmaleimide sensitive factor
	SYN2	Synapsin II
Vawter et al. (2001)	RAB3C	RAB3C, member RAS oncogene family
	AMPA2	Glutamate receptor ionotropic AMPA2
	SERPINI1	Neuroserpin
	YWHAH	Tyrosine 3-monooxygenase/tryptophan 5-monooxygenase activation protein, eta polypeptide
	CALM3	Calmodulin 3
Aston et al. (2004)	SYN2	Synapsin II
	SYNJ1	Synaptojanin 1

5.3.3 Energy Metabolism

Alterations in energy metabolism have been extensively described by large-scale transcriptome and proteome analyses in SCZ (Prabakaran et al., 2004; Ben-Shachar and Laifenfeld, 2004; Bubber et al., 2004; Karry et al., 2004; Glatt et al., 2005; Clark et al., 2006; Martorell et al., 2006; Mehler-Wex et al., 2006). Many aspects of the energy metabolism in brain, as well as its close connection to the neuronal plasticity and synapse (reviewed in Ben-Shachar and Laifenfeld, 2004) and evidence of oxidative damage in SCZ brains (reviewed in Yao et al., 2001) strongly suggest that SCZ has an important energetic component in its pathogenesis.

Middleton et al. (Middleton et al., 2002), studying the transcriptome of SCZ PFC cortex using microarray, showed a reduction in the expression of energy-metabolism genes involved in the mitochondrial malate shuttle, mitochondrial translocases, Ketone body metabolism and others as shown in Table 5.5. Prabakaran et al. (Prabakaran et al., 2004) using PFC from 54 SCZ and 50 controls found transcriptional alterations of genes related to energy metabolism and to oxidative stress as shown in Table 5.5.

Ben-Shachar et al. (Ben-Shachar et al., 2004) hypothesize that the mitochondrial dysfunction in SCZ could be caused by an imbalanced dopamine metabolism in SCZ brains. The hyper-dopaminergic state generates many dopamine oxidized metabolites, which consequently inhibits the mitochondrial respiratory system.

Table 5.5 Energy metabolism-related genes revealed as differentially expressed in transcriptome studies of scz brain tissue

Reference	Gene Symbol	Gene name
Middleton et al. (2002)	GOT2	Glutamic-oxaloacetic transaminase 2
	MDH1	Malate dehydrogenase 1
	ATP5A1	ATP synthase mitochondrial F1 complex alpha
	Azin1	Antizyme inhibitor
	CRYM	Crystallin
	OAT	Ornithine aminotransferase
	TIMM17A	Translocase of inner mitochondrial membrane 17
	OXCT1	3-Oxoacid CoA transferase
	USP14	Ubiquitin-specific protease 14
	UCHL1	Ubiquitin C-terminal esterase L1
Prabakaran et al. (2004)	PKM1	Muscle pyruvate kinase
	ACADS & ACADL	Peroxisomal acyl-CoA oxidase – short- and long-chain – dehydrogenases
	ACAT2	acetyl-coenzyme A acyltransferase 2
	AGPS	ATP citrate lyase, alkylglycerone phosphate synthase
	CPT1 & CPT2	carnitine palmitoyltransferases 1 and 2
	ACO	Aconitase
	ENO	Enolase
	PDH	Pyruvate dehydrogenase
	GAPDH	Glyceraldehyde-3-phosphate dehydrogenase

5.3.4 Other Biochemical Pathways

Table 5.6 presents an overall synthesis of large-scale analyses in SCZ brains, showing the sample size and the brain areas studied, and highlighting the most relevant regulated pathways.

Taken together, the results of the gene expression studies presented here suggest that there is both neuronal and glial involvement in the SCZ disease process, including alterations in the transmission and propagation of the electric impulse in the axons as well as in the synapses, together with central energy deficits in some brain areas.

5.4 *Searching for RNA Biomarkers in Peripheral Blood Cells*

Central biomarkers identified in the brains of SCZ patients are fundamental to help to uncover the biological basis of the disease and to provide new targets for the development of novel therapeutic approaches. However, the difficulty to obtain fresh brain tissues, especially for RNA studies, not only complicates the analysis of a large number of patients, but also makes it difficult to have homogeneous groups with less confounding effects (such as age, diet, *causa-mortis*, gender and use of psychotropic medications). Thus, the advantages of identifying SCZ markers in peripheral tissues are obvious and have clear implications for diagnosis and for the clinical management of the patients.

The large-scale analysis of the blood cell transcriptome in SCZ has been done by a few groups and a number of promising biomarkers have been revealed. However, the recurring identification of biomarkers has been scarce, mainly due to the distinct platforms and the heterogeneity of the clinical samples studied. Vawter et al. (Vawter et al., 2004) analyzed the transcriptome of cultured leukocytes from five patients with SCZ and nine controls using an array of brain-expressed genes. These authors found statistically significant differences in the expression of eight transcripts. The differential expression of two of these genes (neuropeptide Y receptor Y1 gene — NPY1R and guanine nucleotide-binding regulatory protein Go-alpha -GNAO1) could be confirmed by qPCR. The circulating blood transcriptome of 30 SCZ and 17 controls was evaluated by Tsuang et al. (Tsuang et al., 2005), while searching for peripheral RNA markers for the diagnosis of SCZ. For this analysis the authors used cDNA microarrays as a platform to suggest differentially expressed genes, and qPCR for validation. The results allowed the discrimination between SCZ and controls through linear and nonlinear combinations of eight putative biomarker genes as shown in Table 5.7 with an overall accuracy of 95–97%. The confirmation of these findings by other groups is fundamental to validate these markers.

Altered gene expression in peripheral blood cells of SCZ patients was also evaluated by Glatt et al. (Glatt et al., 2005) in one of the most austere approaches published up to now. This study used cDNA microarray analysis to investigate RNA alterations in the brain, followed by the analysis of peripheral blood cells in

Table 5.6 Analysis of gene expression of different brain regions of postmortem scz brain****. Not included here are oligodendrocyte genes, which are shown in Tables 5.2 and 5.3

Reference	Samples	Type of analysis	Regulated pathways	Most important regulated genes
Mirnics et al. (2000)	11 SCZ 11Controls Prefrontal cortex (BA9)	Microarray	* Presynaptic function * GABA neurotransmission * Glutamate neurotransmission	* N-ethylmaleimide sensitive factor * synapsin II
Hakak et al. (2001)	12 SCZ 12 Controls Prefrontal cortex (BA46)	Microarray	* Oligodendrocyte metabolism * Synaptic plasticity * Neuronal development * Neurotransmission * Signal transduction	
Vawter et al. (2001)	5 medicated SCZ 3 unmedicated SCZ 3 Controls (Prefrontal cortex, cerebellum and middle temporal gyrus)	Microarray	* Synaptic signaling * Proteolytic functions	* Tyrosine 3-monooxygenase/tryptophan 5-monooxygenase activation protein, eta polypeptide * Sialyltransferase * Proteasome subunit, alpha type 1 * Ubiquitin carboxyl-terminal esterase L1 * Solute carrier family 10, member 1.
Middleton et al. (2002)	10 SCZ 11 Controls Prefrontal cortex (BA9)	Microarray	* Malate shuttle * Transcarboxylic acid cycle * Ornithine—polyamine metabolism * Aspartate—alanine metabolism * Ubiquitin metabolism	* Antizyme inhibitor * Crystallin * Ornithine aminotransferase * Translocase of inner mitochondrial membrane 17 * Ubiquitin-specific protease 14 * Glutamic-oxaloacetic transaminase 2, mitochondrial * 3-Oxoacid CoA * ATP synthase, mitochondrial F1 complex * Malate dehydrogenase 1, NAD (soluble) * Ubiquitin C-terminal esterase L1 (thiolesterase)

Mimmack et al. (2002)	10 SCZ 10 Controls (Japan/NZ — Prefrontal cortex)	Microarray	High-Density Lipoproteins	* Apolipoprotein L1 * Apolipoprotein L2 * Apolipoprotein L4
Vawter et al. (2002)	15 SCZ 15 Controls Prefrontal Cortex (BA9/ BA46)	Microarray		* Histidine triad nucleotide-binding protein * Uubiquitin conjugating enzyme E2N * Glutamate receptor, ionotropic,
Tkachev et al. (2003)	15 SCZ 15 Controls Prefrontal cortex (BA9/ BA46)	Microarray	* Oligodendrocyte metabolism	* Proteolipid protein 1 * Myelin-associated glycoprotein * Myelin basic protein * Myelin oligodendrocyte glycoprotein * Myelin protein zero * V-erb-b2 erythroblastic leukemia viral oncogene homolog 3 * Transferrin
Prabakaran et al. (2004)	54 SCZ 50 Controls Prefrontal cortex (BA9)	Microarray	* Energy metabolism * Oxidative stress	* Metallothionein * Platelet-derived growth factor * Erythropoietin receptors * Enzyme complexes of mitochondria * Superoxide Dismutase
Aston et al. (2004)	12 SCZ 14 Controls Temporal Gyrus (BA21)	Microarray	* Oligodendrocyte metabolism * Neurodevelopment * Circadian rhythms * Signaling mechanisms	* Myelin-associated glycoprotein * Plasma membrane proteolipid * V-erb-b2 erythroblastic leukemia viral oncogene homolog 3 * Proteolipid protein 1 * TNF receptor-associated factor 4 * Histone deacetylase 3 * Neurod1 * Period homolog 1

(continued)

Table 5.6 (continued)

Reference	Samples	Type of analysis	Regulated pathways	Most important regulated genes
Katsel et al. (2005)	13 SCZ 13 Controls (15 different regions related in Table 5.2)	Microarray	* Glial differentiation * Oligodendrocyte metabolism	* Myelin-associated glycoprotein * Transferrin * Quaking Protein
Dracheva et al. (2006)	30 SCZ 25 Controls Cingulate Gyrus (BA24/32) 24 SCZ 21 Controls Hippocampus 23 SCZ 20 Controls Caudate Nucleus 24 SCZ 19 Controls Caudate Putamen	Q-PCR	* Oligodendrocyte and myelin-related genes	* Myelin-associated glycoprotein * Sex determining region Y-box 10 * Claudin11 * Peripheral myelin protein 22 * 2′,3′-cyclic nucleotide 3′ phosphodiesterase
McCullumsmith et al. (2007)	41 SCZ 34 Controls Anterior Cingulate Cortex	in situ hybridization	* Oligodendrocyte and myelin-related genes	* 2′,3′-cyclic nucleotide 3′-phosphodiesterase * myelin-associated glycoprotein * Transferrin * Quaking
Arion et al. (2007)	14 SCZ 14 Controls Prefrontal Cortex (BA46)	Microarray and qPCR	* Synaptic * Oligodendrocyte * Signal transduction * Immune/chaperone	* SERPINA3 * IFITM1 * IFITM2 * IFITM3 * CHI3L1 * MT2A * CD14 * HSPB1 * HSPA1B * HSPA1A

Table 5.7 Analysis of RNA biomarkers found in peripheral blood cells of scz patients

Reference	Method	Regulated gene in blood previously found in brain
Vawter et al. (2004)	Microarray	Neuropeptide Y (NPY) (Found regulated in brain by Hakak et al., 2001) Malate Dehydrogenase (MDH1) (Found regulated in brain by Hakak et al., 2001; Middleton et al., 2002)
Tsuang et al. (2005)	Microarray and qPCR	Catalytic polypeptide-like apolipoprotein B mRNA editing enzyme 3B (APOBEC3B) Adenylosuccinate synthetase (Adssl1) ataxia telangiectasia mutated (ATM) Charcot-Leyden crystal protein (CLC) C-terminal binding protein 1 Death-associated transcription factor 1 (Datf1) Chemokine C-X-C motif ligand 1 (CXCR1) S100 calcium binding protein A9 (S100A9)
(Glatt et al., 2005)	Microarray	B cell translocation gene 1, antiproliferative (BTG1) (upregulated in DLPFC and downregulated in blood) Glycogen synthase kinase 3 α (GSK3a) (downregulated in DLPFC and upregulated in blood) Heterogeneous nuclear ribonucleoprotein A3 (HNRPA3) (upregulated in DLPFC and downregulated in blood) MHC class II, DR β 1 (HLA-DRB1) (downregulated in DLPFC and blood) Selenium-binding protein 1 (SELENBP1) (upregulated in DLPFC and blood) Splicing factor, arginine/serine-rich 1 (splicing factor 2, alternate splicing factor) (SFRS1) (upregulated in DLPFC and downregulated in blood)
Middleton et al. (2005)	Microarray and qPCR	Alpha catenin (CTNNA1) Neuregulin 1 (NRG1) MAX-like protein X (TCFL4) v-erb-b2 erythroblastic leukemia viral oncogene homolog 2, neuro/glioblastoma derived oncogene homolog (avian) (ERBB2) Sensory motor neuron derived factor (SMDF) Cytochrome P450 family 1, subfamily B, polypeptide 1 (CYP27B1) Quaking homolog (QKI) Dual specificity phosphatase 6 (DUSP6).
Bowden et al. (2006)	Microarray and qPCR	Endothelial differentiation gene 2 (Edg-2) Ezrin-radixin-moesin phosphoprotein 50 (EBP50) Myc-associated zinc finger protein (MAZ) Tumor Necrosis Factor Receptor 2 (TNFR2)
Perl et al. (2006)	qPCR	Alpha7-nicotinic-acetylcholine-receptor (CHRNA7)
Mehler-Wex et al. (2006)	qPCR	Mitochondrial complex I 75-kDa subunit (NDUFS1)
Numata et al. (2007)	qPCR	PDZ and LIM domain 5 (PDLIM5)

a second cohort of patients and controls. A total of 177 putative markers were found in brain, and 123 putative markers were seen in the peripheral blood cells. Six of these RNA markers were found in both compartments as shown in Table 5.7. Middleton et al. (Middleton et al., 2005) analyzed by microarray and qPCR the LBP transcriptome from 33 SCZ patients, compared with controls and bipolar patients. They found many altered genes previously described in SCZ analysis as shown in Table 5.7. Differential expression of Alpha7-nicotinic-acetylcholine-receptor could contribute to sensory gating, leading to a marked dysfunction that can impact employability, treatment adherence, and social skills of SCZ patients. qPCR was used by Perl et al. (Perl et al., 2006) to measure the expression of this gene in peripheral blood cells of 44 SCZ patients and compared with 16 controls, and found a downregulation of this gene ($P < 0.05$) in SCZ patients. Analysis of other individual genes have been performed by Mehler-Wex et al. (Mehler-Wex et al., 2006) showing the upregulation of mitochondrial complex I 75-kDa subunit in neuroleptic-naive SCZ patients and by Numata et al. (Numata et al., 2007) who described the upregulation of PDLIM5, a gene whose product regulates intracellular calcium levels and has a role in neurotransmitter synaptic vesicles. The PDLIM5 gene lies on chromosome 4q22, a locus previously reported to be linked with SCZ (Mowry et al., 2000). A synthesis of these findings is described in Table 5.7.

5.5 *New Classes of RNA Markers in Psychiatry*

5.5.1 Alternative Splicing Studies

The vast majority of the transcriptome analyses performed during the search for RNA markers in SCZ were based on the old central dogma of "one gene, one mRNA, one protein," which clearly oversimplifies and underestimates transcriptome complexity. These studies are usually focused on quantitative alterations of gene expression, and don't consider qualitative variations that may occur. While transcriptional regulation plays important roles within a cell, posttranscriptional regulation, such as alternative splicing (AS), dramatically increases transcriptional diversity and may have remarkable consequences for proteome variety. AS plays a critical role in gene expression regulation and human diseases (Kan et al., 2001; Cartegni et al., 2002). Splicing is a cellular mechanism that joins the exons of a precursor immature RNA, removing its intronic sequences. This mechanism occurs in the spliceosome, a complex cellular compartment composed small nuclear RNAs and hundreds of proteins. Since the alternative combination of different gene exons can generate diverse protein isoforms, which can trigger different mechanisms inside the cell, the study of AS as a qualitative gene expression data has become one of the central issues in biomedical sciences (Buratti et al., 2006) including SCZ research.

Studies of AS isoforms, as possible RNA markers in SCZ, have been focused on a few genes usually those traditionally studied in SCZ. An example is the Dopamine receptor D2 which produces, by AS, two receptor isoforms called D2L (for long isoform) and D2S (for short) (Giros et al., 1989). These two isoforms have distinct functions in vivo: D2L acting mainly at postsynaptic sites and D2S serves as a pre-synaptic autoreceptor (Usiello et al., 2000). These subunits differentially contribute to the therapeutic actions and side effects of antipsychotic agents (Xu et al., 2002). Splicing variants are also responsible for the molecular dissimilarity of genes related to a number of important SCZ pathways discussed in this chapter, including genes involved in myelinization and synapses. Good examples are the synapsins, synaptic vesicle-associated phosphoproteins that have been implicated in the control of neurotransmitter release and synaptogenesis. It should be noted that a member of this family (synapsin III) is located on human chromosome 22q, a SCZ-susceptibility locus, and different AS isoforms have been described in the brain (Kao et al., 1998).

Quaking homolog (QKI) is a development-related protein encoded by a gene highly conserved among different species. A deletion of a portion of this gene causes body tremor and severe dysmyelination of the central nervous system, with dysfunction of oligodendrocytes and a reduced expression of myelin components (Sidman et al., 1964; Ebersole et al., 1996; Hardy et al., 1996). Aberg et al. (Aberg et al., 2006a) showed that one of the four AS isoforms of QKI (named *QKI-7kb*) was reduced in the frontal cortex of SCZ patients. The same group showed later (Aberg et al., 2006b) that the reduced expression of this isoform could be correlated to the reduced expression of three tightly regulated myelin-related genes (*PLP1, MAG*, and *TF*) that also had a reduced expression in SCZ brain samples as compared to controls. These results indicate that QKI may be a master regulator of oligodendrocyte differentiation/maturation in the human brain and also suggest that decreased activities of myelin-related genes in SCZ might be caused by a disturbed *QKI* splicing.

Other examples of AS in SCZ include variations in the relative abundance of alternatively spliced isoforms of the gamma2 subunit of the GABA-A receptor (Huntsman et al., 1998; Zhao et al., 2006), NCAM1 (Neural cell adhesion molecule) (Vawter et al., 2000), NMDA receptor (Clinton et al., 2003), glutamate receptor 3 (mGluR3) (Sartorius et al., 2006), and ErbB4 (a Neuregulin 1 receptor) (Silberberg et al., 2006; Law et al., 2007). These examples clearly demonstrate that this class of transcriptional variation also can be relevant in the search of RNA markers for SCZ. In the near future, the analysis of AS RNA markers in SCZ should be reinforced by the use of cDNA microarrays specifically designed to investigate these events in human transcripts, as well as by new sequencing-based approaches that may reveal this kind of event.

5.5.2 microRNAs

A promising class of markers that still needs to be more explored in neuropsychiatry consists of the group of small noncoding regulatory RNA molecules found from plants to viruses and animals (reviewed in Bartel, 2004). The advent of tiling genomic arrays showed that a significant fraction of the noncoding genome sequence is transcribed (Kapranov et al., 2002; Cheng et al., 2005). Among these noncoding RNA molecules we find the microRNAs (miRNAs) that are considered today to be the most important transcriptional regulators of the human genome (reviewed in Zhang and Farwell, 2008). Some papers suggest that miRNA are more consistent regulators than mRNA and that the study of a few hundred of these molecules can derive stronger markers than tens of thousands mRNAs used in cDNA microarray platforms (Lu et al., 2005b).

miRNAs control distinct processes that lead to regulation of protein abundance, such as transcription, mRNA degradation, stability and translation. An increasing number of studies now reveal the important role of miRNAs in biological processes such as the differentiation and specificity of neurons (Vo et al., 2005; Krichevsky et al., 2006), synapse plasticity (Schratt et al., 2006) development and maintenance of the central nervous system (Krichevsky et al., 2003; Miska et al., 2004; Giraldez et al., 2005; Lukiw and Pogue 2007; Makeyev et al., 2007). The importance of miRNAs in the regulation of these processes, as well as their abundance in the brain, makes these molecules attractive RNA biomarkers to be explored in SCZ.

The analysis of miRNA in SCZ is still in its early days. A recent study evaluated for the first time the differential expression of miRNAs in SCZ (Perkins et al., 2007). In this study, with a custom miRNA array, the expression of 264 distinct miRNAs was contrasted in postmortem PFC tissue samples of individuals with SCZ ($n = 13$) or schizoaffective disorder ($n = 2$) and nonpsychiatric patients ($n = 21$). The authors found 16 miRNA to be differentially expressed in PFC of patient subjects, 94% being down-regulated, which suggests an overall up-regulation of their targets. The authors concluded that genes that were commonly targeted by the regulated miRNAs were significantly clustered in 12 pathways. The most significant pathways identified contain proteins involved in synaptic plasticity at the level of dendritic spines.

Today we know that most of the eukaryotic genome is transcribed as noncoding RNAs (ncRNAs), which play important roles in chromatin organization, gene expression, posttranscription regulation, with consequences for normal physiology and disease etiology. The increasing diversity and high expression of ncRNAs, including miRNAs, identified in the eukaryotic genome suggests a critical link between the regulatory potential of ncRNAs and the complexity of genome expression regulation.

5.6 Summary and Perspectives

The continuous identification of alterations in genes related to synapsis, myelinization and energy metabolism strongly suggests the involvement of these elements in the pathogenesis of SCZ. However, larger studies involving distinct populations and more samples are still required to fully demonstrate the importance of markers already revealed, and to point to new markers valuable for more specific endophenotypes.

The future of neuropsychiatric diseases is likely to be profoundly dependent upon the use of biomarkers that shall guide the clinicians at many steps of disease management including early diagnosis, response to therapy and disease outcome. Biomarkers that detect diseases, and predict their outcome or influence treatment choice will have tremendous importance in the future of neuropsychiatric diseases. However, to understand the significance of each biomarker and its importance in pathogenesis, many samples, comprehending diverse ethnical backgrounds, endophenotypes and disease states need to be investigated. Massive amounts of biological information will need to be investigated with a multidisciplinary approach involving clinicians, biologists, statisticians and bioinformatics in a continuous interface of genomics, transcriptomics and proteomics.

It is clear that, together with DNA and protein, RNA biomarkers will be part of an optimistic scenario where more detailed and personalized data will be evaluated together, guiding researchers and clinicians to understand and to treat complex diseases.

Acknowledgments The authors thank ABADHS (Associação Beneficente Alzira Denise Hertzog Silva), FAPESP (Fundação de Amparo à Pesquisa do Estado de São Paulo — Brazil), CNPq (Conselho Nacional de Pesquisas) and DAAD (Deutscher Akademischer Austauschdienst) for their fundamental support of our research.

References

Aberg K , Saetre P, Lindholm E et al. (2006a) Human QKI, a new candidate gene for schizophrenia involved in myelination. Am J Med Genet B Neuropsychiatr Genet. 141(1):84–90.

Aberg K , Saetre P, Jareborg N, Jazin E (2006b) Human QKI, a potential regulator of mRNA expression of human oligodendrocyte-related genes involved in schizophrenia. Proc Natl Acad Sci U S A. 103(19):7482–7487.

Adams MD, Dubnick M, Kerlavage AR et al. (1992) Sequence identification of 2,375 human brain genes. Nature. 355(6361):632–634.

Aigner L, Arber S, Kapfhammer JP et al. (1995) Overexpression of the neural growth-associated protein GAP-43 induces nerve sprouting in the adult nervous system of transgenic mice. Cell. 83:269–278.

Arai M, Amano S, Ryo A et al. (2003) Identification of epilepsy-related genes by gene expression profiling in the hippocampus of genetically epileptic rat. Brain Res Mol Brain Res. 118 (1–2):147–151.

Arion D, Unger T, Lewis DA (2007) Molecular evidence for increased expression of genes related to immune and chaperone function in the prefrontal cortex in schizophrenia. Biol Psychiatr. 62(7):711–721.

Arnold SE, Franz BR, Gur RC et al. (1995) Smaller neuron size in schizophrenia in hippocampal subfields that mediate cortical-hippocampal interactions. Am J Psychiatr. 152(5):738–748.

Aston C, Jiang L, Sokolov BP (2004) Microarray analysis of postmortem temporal cortex from patients with schizophrenia. J Neurosci Res. 77(6):858–866.

Avedissian M, Longo BM, Jaqueta CB et al. (2007) Hippocampal gene expression analysis using the ORESTES methodology shows that homer 1a mRNA is upregulated in the acute period of the pilocarpine epilepsy model. Hippocampus. 17(2):130–136.

Bartel DP (2004) MicroRNAs: genomics, biogenesis, mechanism, and function. Cell. 116(2):281–297.

Bell R, Munro J, Russ C et al. (2000) Systematic screening of the 14-3-3 eta (eta) chain gene for polymorphic variants and case-control analysis in schizophrenia. Am J Med Genet. 96(6):736–743.
Benes FM, McSparren J, Bird ED et al. (1991) Deficits in small interneurons in prefrontal and cingulate cortices of schizophrenic and schizoaffective patients. Arch Gen Psychiatr. 48(11):996–1001.
Ben-Shachar D, Laifenfeld D (2004) Mitochondria, synaptic plasticity, and schizophrenia. Int Rev Neurobiol. 59:273–296.
Ben-Shachar D, Zuk R, Gazawi H, Ljubuncic P (2004) Dopamine toxicity involves mitochondrial complex I inhibition: implications to dopamine-related neuropsychiatric disorders. Biochem Pharmacol. 67(10):1965–1974.
Benton WD, Davis RW (1977) Screening lambdagt recombinant clones by hybridization to single plaques in situ. Science. 196(4286):180–182.
Bertucci F, Bernard K, Loriod B et al. (1999) Sensitivity issues in DNA array-based expression measurements and performance of nylon microarrays for small samples. Hum Mol Genet. 8, 1715–1722.
Blencowe BJ (2006) Alternative splicing: new insights from global analyses. Cell. 126(1):37–47.
Bovenberg RA, Burbach JP, Wiegant VM et al. (1986) Gamma-Endorphin and schizophrenia: amino acid composition of gamma-endorphin and nucleotide sequence of gamma-endorphin cDNA from pituitary glands of schizophrenic patients. Brain Res. 376(1):29–37.
Bowden NA, Weidenhofer J, Scott RJ et al. (2006) Preliminary investigation of gene expression profiles in peripheral blood lymphocytes in schizophrenia. Schizophr Res. 82(2–3):175–183.
Bubber P, Tang J, Haroutunian V et al. (2004) Mitochondrial enzymes in schizophrenia. J Mol Neurosci. 24(2):315–321.
Buratti E, Baralle M, Baralle FE (2006) Defective splicing, disease and therapy: searching for master checkpoints in exon definition. Nucleic Acids Res. 34(12):3494–3510.
Cai J, Ash D, Kotch LE et al. (2005) Gene expression in pharyngeal arch 1 during human embryonic development. Hum Mol Genet. 14(7):903–912.
Camargo AA, Samaia HP, Dias-Neto E et al. (2001) The contribution of 700,000 ORF sequence tags to the definition of the human transcriptome. Proc Natl Acad Sci U S A. 98(21): 12103–12108.
Cartegni L, Chew SL, Krainer AR (2002) Listening to silence and understanding nonsense: exonic mutations that affect splicing. Nat Rev Genet 3:285–298.
Chakravarty A (2003) Surrogate Markers: Their Role in Regulatory Decision Process. Food and Drug Administration. www.fda.gov/cder/Offices/Biostatistics/Chakravarty_376/sld016.htm.
Cheng J, Kapranov P, Drenkow J et al. (2005) Transcriptional maps of 10 human chromosomes at 5-nucleotide resolution. Science. 308(5725):1149–1154.
Clark D, Dedova I, Cordwell S, Matsumoto I (2006) A proteome analysis of the anterior cingulate cortex gray matter in schizophrenia. Mol Psychiatr. 11:459–470.
Clinton SM, Haroutunian V, Davis KL, Meador-Woodruff JH (2003) Altered transcript expression of NMDA receptor-associated postsynaptic proteins in the thalamus of subjects with schizophrenia. Am J Psychiatr. 160(6):1100–1109.
Cowell JK, Hawthorn L (2007) The application of microarray technology to the analysis of the cancer genome. Curr Mol Med. 7(1):103–120.
Cuperlovic-Culf M, Belacel N, Culf AS, Ouellette RJ (2006) Microarray analysis of alternative splicing. OMICS. 10(3):344–357.
Dai X, Lercher LD, Clinton PM et al. (2003) The trophic role of oligodendrocytes in the basal forebrain. J Neurosci. 23(13):5846–5853.
Dias-Neto E, Correa RG, Verjovski-Almeida S et al. (2000) Shotgun sequencing of the human transcriptome with ORF expressed sequence tags. Proc Natl Acad Sci U S A. 97(7): 3491–3496.
Dracheva S, Davis KL, Chin B et al. (2006) Myelin-associated mRNA and protein expression deficits in the anterior cingulate cortex and hippocampus in elderly schizophrenia patients. Neurobiol Dis. 21(3):531–540.

Ebersole TA, Chen Q, Justice MJ, Artzt K (1996) The quaking gene product necessary in embryogenesis and myelination combines features of RNA binding and signal transduction proteins. Nat Genet. 12(3):260–265.

Frances AJ, First MB, Widiger TA et al. (1991) An A to Z guide to DSM-IV conundrums. J Abnorm Psychol 100:407–412.

Freedman R (2003). Schizophrenia. N Engl J Med. 349:1738–1749.

Giraldez AJ, Cinalli RM, Glasner ME et al. (2005) MicroRNAs regulate brain morphogenesis in zebrafish. Science. 308(5723):833–838.

Giros B, Sokoloff P, Martres MP et al. (1989) Alternative splicing directs the expression of two D2 dopamine receptor isoforms. Nature. 342(6252):923–926.

Glatt SJ, Everall IP, Kremen WS et al. (2005) Comparative gene expression analysis of blood and brain provides concurrent validation of SELENBP1 up-regulation in schizophrenia. Proc Natl Acad Sci U S A. 102(43):15533–15538.

Grunstein M, Hogness DS (1975) Colony hybridization: a method for the isolation of cloned DNAs that contain a specific gene. Proc Natl Acad Sci U S A. 72(10):3961–3965.

Günthard HF, Wong JK, Ignacio CC et al. (1998) Comparative performance of high-density oligonucleotide sequencing and dideoxynucleotide sequencing of HIV type 1 pol from clinical samples. AIDS Res Hum Retrovir. 14(10):869–876.

Hakak Y, Walker JR, Li C et al. (2001) Genome-wide expression analysis reveals dysregulation of myelination-related genes in chronic schizophrenia. Proc Natl Acad Sci U S A. 98(8):4746–4751.

Hardy RJ, Loushin CL, Friedrich VL Jr et al. (1996) Neural cell type-specific expression of QKI proteins is altered in quaking viable mutant mice. J Neurosci. 16(24):7941–7949.

Hendriksen H, Datson NA, Ghijsen WE et al. (2001) Altered hippocampal gene expression prior to the onset of spontaneous seizures in the rat post-status epilepticus model. Eur J Neurosci. 14(9):1475–1484.

Hu M, Polyak K (2006) Serial analysis of gene expression. Nat Protoc. 1:1743–1760.

Hughes TR, Hiley SL, Saltzman AL et al. (2006) Microarray analysis of RNA processing and modification. Methods Enzymol. 410:300–16.

Huntsman MM, Tran BV, Potkin SG et al. (1998) Altered ratios of alternatively spliced long and short gamma2 subunit mRNAs of the gamma-amino butyrate type A receptor in prefrontal cortex of schizophrenics. Proc Natl Acad Sci U S A. 95(25):15066–15071.

Hyde TM, Ziegler JC, Weinberger DR (1992) Psychiatric disturbances in metachromatic leukodystrophy. Insights into the neurobiology of psychosis. Arch Neurol. 49(4):401–406.

Kan Z, Rouchka EC, Gish WR, States DJ (2001) Gene Structure Prediction and Alternative Splicing Analysis Using Genomically Aligned ESTs. Genome Res 11(5):889–900.

Kane MD, Jatkoe TA, Stumpf CR et al. (2000) Assessment of the sensitivity and specificity of oligonucleotide (50mer) microarrays. Nucleic Acids Res. 28, 4552–4557

Kao HT, Porton B, Czernik AJ et al. (1998) A third member of the synapsin gene family. Proc Natl Acad Sci U S A. 95(8):4667–4672.

Kapranov P, Cawley SE, Drenkow J et al. (2002) Large-scale transcriptional activity in chromosomes 21 and 22. Science. 296(5569):916–919.

Karry R, Klein E, Ben Shachar D (2004) Mitochondrial complex I subunits expression is altered in schizophrenia: a postmortem study. Biol Psychiatr. 55(7):676–684.

Katsel P, Davis KL, Haroutunian V (2005) Variations in myelin and oligodendrocyte-related gene expression across multiple brain regions in schizophrenia: a gene ontology study. Schizophr Res. 79(2–3):157–173.

Kim JM, Lee KH, Jeon YJ et al. (2006) Identification of genes related to Parkinson's disease using expressed sequence tags. DNA Res. 13(6):275–286.

Klevering BJ, Yzer S, Rohrschneider K et al. (2004) Microarray-based mutation analysis of the ABCA4 (ABCR) gene in autosomal recessive cone-rod dystrophy and retinitis pigmentosa. Eur J Hum Genet. 12(12):1024–1032.

Knox DP, Skuce PJ (2005) SAGE and the quantitative analysis of gene expression in parasites. Trends Parasitol. 21:322–326.

Kozal MJ, Shah N, Shen N et al. (1996) Extensive polymorphisms observed in HIV-1 clade B protease gene using high-density oligonucleotide arrays. Nat Med. 2(7):753–759.
Krichevsky AM, King KS, Donahue CP et al. (2003) A microRNA array reveals extensive regulation of microRNAs during brain development. RNA. 9(10):1274–1281. Erratum in: RNA. 2004 10(3):551.
Krichevsky AM, Sonntag KC, Isacson O, Kosik KS (2006) Specific microRNAs modulate embryonic stem cell-derived neurogenesis. Stem Cells. 24(4):857–864.
Law AJ, Kleinman JE, Weinberger DR, Weickert CS (2007) Disease-associated intronic variants in the ErbB4 gene are related to altered ErbB4 splice-variant expression in the brain in schizophrenia. Hum Mol Genet. 16(2):129–141.
Liston C, Miller MM, Goldwater DS et al. (2006) Stress-induced alterations in prefrontal cortical dendritic morphology predict selective impairments in perceptual attentional set-shifting. J Neurosci. 26(30):7870–7874.
Liu X, Qin W, He G et al. (2005) A family-based association study of the MOG gene with schizophrenia in the Chinese population. Schizophr Res. 73(2–3):275–280.
Lockhart DJ, Dong H, Byrne MC et al. (1996) Expression monitoring by hybridization to high-density oligonucleotide arrays. Nat. Biotechnol. 14:1675–1680.
Lu L, Neff F, Alvarez-Fischer D et al. (2005a) Gene expression profiling of Lewy body-bearing neurons in Parkinson's disease. Exp Neurol. 195(1):27–39.
Lu J, Getz G, Miska EA et al. (2005b) MicroRNA expression profiles classify human cancers. Nature. 435(7043):834–838.
Lukiw WJ, Pogue AI (2007) Induction of specific micro RNA (miRNA) species by ROS-generating metal sulfates in primary human brain cells. J Inorg Biochem. 101(9):1265–1269.
Magic Z, Radulovic S, Brankovic-Magic M (2007) cDNA microarrays: identification of gene signatures and their application in clinical practice. J BUON. 12 (Suppl. 1):S39–S44.
Makeyev EV, Zhang J, Carrasco MA, Maniatis T (2007) The MicroRNA miR-124 promotes neuronal differentiation by triggering brain-specific alternative pre-mRNA splicing. Mol Cell. 27(3):435–448.
Margulies M, Egholm M, Altman WE et al. (2005) Genome sequencing in microfabricated high-density picolitre reactors. Nature. 437:376–380.
Martorell L, Segues T, Folch G et al. (2006) New variants in the mitochondrial genomes of schizophrenic patients. Eur J Hum Genet. 14(5):520–528.
McCullumsmith RE, Gupta D, Beneyto M et al. (2007) Expression of transcripts for myelination-related genes in the anterior cingulate cortex in schizophrenia. Schizophr Res. 90(1–3): 15–27.
McNamara RK, Lenox RH (1997) Comparative distribution of myristoylated alanine-rich C kinase substrate (MARCKS) and F1/GAP-43 gene expression in the adult rat brain. J Comp Neurol. 379(1):48–71.
Mehler-Wex C, Duvigneau JC, Hartl RT et al. (2006) Increased mRNA levels of the mitochondrial complex I 75-kDa subunit. A potential peripheral marker of early onset schizophrenia? Eur Child Adolesc Psychiatr. 15(8):504–507.
Middleton FA, Mirnics K, Pierri JN et al. (2002) Gene expression profiling reveals alterations of specific metabolic pathways in schizophrenia. J Neurosci. 22(7):2718–2729.
Middleton FA, Pato CN, Gentile KL et al. (2005) Gene expression analysis of peripheral blood leukocytes from discordant sib-pairs with schizophrenia and bipolar disorder reveals points of convergence between genetic and functional genomic approaches. Am J Med Genet B Neuropsychiatr Genet. 136(1):12–25.
Miller EK, Cohen JD (2001) An integrative theory of prefrontal cortex function. Annu Rev Neurosci. 24:167–202.
Mimmack ML, Ryan M, Baba H et al. (2002) Gene expression analysis in schizophrenia: reproducible up-regulation of several members of the apolipoprotein L family located in a high-susceptibility locus for schizophrenia on chromosome 22. Proc Natl Acad Sci U S A. 99(7):4680–4685.

Mirnics K, Middleton FA, Marquez A et al. (2000) Molecular characterization of schizophrenia viewed by microarray analysis of gene expression in prefrontal cortex. Neuron. 28(1):53–67.
Mirnics K, Middleton FA, Lewis DA, Levitt P (2001) Analysis of complex brain disorders with gene expression microarrays: schizophrenia as a disease of the synapse. Trends Neurosci. 8:479–486; Review.
Miska EA, Alvarez-Saavedra E, Townsend M et al. (2004) Microarray analysis of microRNA expression in the developing mammalian brain. Genome Biol. 5(9):R68.
Mowry BJ, Ewen KR, Nancarrow DJ et al. (2000) Second stage of a genome scan of schizophrenia: study of five positive regions in an expanded sample. Am J Med Genet. 96(6):864–869.
Nielsen KL, H gh AL, Emmersen J (2006) DeepSAGE—digital transcriptomics with high sensitivity, simple experimental protocol and multiplexing of samples. Nucleic Acids Res. 34(19): e133.
Noureddine MA, Li YJ, van der Walt JM et al. (2005) Genomic convergence to identify candidate genes for Parkinson disease: SAGE analysis of the substantia nigra. Mov Disord. 20(10):1299–1309.
Numata S, Ueno S, Iga J et al. (2007) Gene expression in the peripheral leukocytes and association analysis of PDLIM5 gene in schizophrenia. Neurosci Lett. 415(1):28–33.
Ouchi Y, Kubota Y, Ito C (2004) Serial analysis of gene expression in methamphetamine- and phencyclidine-treated rodent cerebral cortices: are there common mechanisms? Ann N Y Acad Sci. 1025:57–61.
Owen MJ, O'Donovan MC, Harrison PJ (2005) Schizophrenia: a genetic disorder of the synapse? BMJ 330: 158–159.
Ozbas-Gerçeker F, Redeker S, Boer K et al. (2006) Serial analysis of gene expression in the hippocampus of patients with mesial temporal lobe epilepsy. Neuroscience. 138(2):457–474.
Patino WD, Mian OY, Hwang PM (2002) Serial analysis of gene expression: technical considerations and applications to cardiovascular biology. Circ Res. 91(7):565–569.
Peirce TR, Bray NJ, Williams NM et al. (2006) Convergent evidence for 2′,3′-cyclic nucleotide 3′-phosphodiesterase as a possible susceptibility gene for schizophrenia. Arch Gen Psychiatr. 63(1):18–24.
Perkins DO, Jeffries CD, Jarskog LF et al. (2007) microRNA expression in the prefrontal cortex of individuals with schizophrenia and schizoaffective disorder. Genome Biol. 8(2):R27.
Perl O, Strous RD, Dranikov A et al. (2006) Low levels of alpha7-nicotinic acetylcholine receptor mRNA on peripheral blood lymphocytes in schizophrenia and its association with illness severity. Neuropsychobiology. 53(2):88–93.
Polak M, Haymaker W, Johnson JE, D'Amelio F (1982) Neuroglia and their reactions. In: Haymaker W, Adams RD (Eds.) Histology and Histopathology of the Nervous System. Springfield: Charles C. Thomas.
Prabakaran S, Swatton JE, Ryan MM et al. (2004) Mitochondrial dysfunction in schizophrenia: evidence for compromised brain metabolism and oxidative stress. Mol Psychiatr. 9(7):684–697, 643.
Raedler TJ, Knable MB, Weinberger DR (1998) Schizophrenia as a developmental disorder of the cerebral cortex. Curr Opin Neurobiol. 8(1):157–161.
Rajkowska G, Halaris A, Selemon LD (2001) Reductions in neuronal and glial density characterize the dorsolateral prefrontal cortex in bipolar disorder. Biol Psychiatr. 49(9):741–752.
Ryu EJ, Angelastro JM, Greene LA (2005) Analysis of gene expression changes in a cellular model of Parkinson disease. Neurobiol Dis. 18(1):54–74.
Ryu EJ, Harding HP, Angelastro JM et al. (2002) Endoplasmic reticulum stress and the unfolded protein response in cellular models of Parkinson's disease. J Neurosci. 22(24):10690–10698.
Saha S, Sparks AB, Rago C et al. (2002) Using the transcriptome to annotate the genome. Nat Biotechnol. 20(5):508–512.
Sartorius LJ, Nagappan G, Lipska BK et al. (2006) Alternative splicing of human metabotropic glutamate receptor 3. J Neurochem. 96(4):1139–1148.
Schratt GM, Tuebing F, Nigh EA et al. (2006) A brain-specific microRNA regulates dendritic spine development. Nature. 439(7074):283–289; Erratum in: Nature. 2006 441(7095):902.

Seeman P, Chau-Wong M, Tedesco J, Wong K (1975) Brain receptors for antipsychotic drugs and dopamine: direct binding assays. Proc Natl Acad Sci U S A. 72: 4376–4380.
Sidman RL, Dickie MM, Appel SH (1964) Mutant Mice (Quaking and Jimpy) with deficient myelination in the central nervous system. Science. 144:309–311.
Silberberg G, Darvasi A, Pinkas-Kramarski R, Navon R (2006) The involvement of ErbB4 with schizophrenia: association and expression studies. Am J Med Genet B Neuropsychiatr Genet. 141(2):142–148.
Soares MB, Bonaldo MF, Jelene P et al. (1994) Construction and characterization of a normalized cDNA library. Proc Natl Acad Sci U S A. 91(20):9228–9232.
Sun Y, Zhang L, Johnston NL et al. (2001) Serial analysis of gene expression in the frontal cortex of patients with bipolar disorder. Br J Psychiatr (Suppl. 41):s137–s141.
Tennis M, Krishnan S, Bonner M et al. (2006) p53 Mutation analysis in breast tumors by a DNA microarray method. Cancer Epidemiol Biomarkers Prev. 15(1):80–85.
Tkachev D, Mimmack ML, Ryan MM et al. (2003) Oligodendrocyte dysfunction in schizophrenia and bipolar disorder. Lancet. 362(9386):798–805.
Tosato S, Dazzan P, Collier D (2005) Association between the neuregulin 1 gene and schizophrenia: a systematic review. Schizophr Bull. 31(3):613–617.
Tsuang MT, Nossova N, Yager T et al. (2005) Assessing the validity of blood-based gene expression profiles for the classification of schizophrenia and bipolar disorder: a preliminary report. Am J Med Genet B Neuropsychiatr Genet. 133(1):1–5.
Uranova N, Orlovskaya D, Vikhreva O et al. (2001) Electron microscopy of oligodendroglia in severe mental illness. Brain Res. Bull. 55, 597–610.
Usiello A, Baik JH, Rouge-Pont F et al. (2000) Distinct functions of the two isoforms of dopamine D2 receptors. Nature. 408(6809):199–203.
Van Bogaert P, Azizieh R, Désir J et al. (2007) Mutation of a potassium channel-related gene in progressive myoclonic epilepsy. Ann Neurol. 61(6):579–586.
Vawter MP, Barrett T, Cheadle C et al. (2001) Application of cDNA microarrays to examine gene expression differences in schizophrenia. Brain Res Bull. 55(5):641–650.
Vawter MP, Frye MA, Hemperly JJ et al. (2000) Elevated concentration of N-CAM VASE isoforms in schizophrenia. J Psychiatr Res. 34(1):25–34.
Vawter MP, Crook JM, Hyde TM et al. (2002) Microarray analysis of gene expression in the prefrontal cortex in schizophrenia: a preliminary study. Schizophr Res. 58(1):11–20.
Vawter MP, Ferran E, Galke B et al. (2004) Microarray screening of lymphocyte gene expression differences in a multiplex schizophrenia pedigree. Schizophr Res. 67(1):41–52.
Velculescu VE, Zhang L, Vogelstein B, Kinzler KW (1995) Serial analysis of gene expression. Science. 270:484–487.
Vo N, Klein ME, Varlamova O et al. (2005) A cAMP-response element binding protein-induced microRNA regulates neuronal morphogenesis. Proc Natl Acad Sci U S A. 102(45):16426–16431; Erratum in: Proc Natl Acad Sci U S A. 2006 17;103(3):825.
Wan C, Yang Y, Feng G et al. (2005) Polymorphisms of myelin-associated glycoprotein gene are associated with schizophrenia in the Chinese Han population. Neurosci Lett. 388(3): 126–131.
Wang SM (2007) Understanding SAGE data. Trends Genet. 23(1):42–50.
Xi ZR, Qin W, Yang YF et al. (2004) Transmission disequilibrium analysis of the GSN gene in a cohort of family trios with schizophrenia. Neurosci Lett. 372(3):200–203.
Xu R, Hranilovic D, Fetsko LA et al. (2002) Dopamine D2S and D2L receptors may differentially contribute to the actions of antipsychotic and psychotic agents in mice. Mol Psychiatr. 7(10):1075–1082.
Xu PT, Li YJ, Qin XJ et al. (2007) A SAGE study of apolipoprotein E3/3, E3/4 and E4/4 allele-specific gene expression in hippocampus in Alzheimer disease. Mol Cell Neurosci. 36(3):313–331.
Yao JK, Reddy RD, van Kammen DP (2001) Oxidative damage and schizophrenia: an overview of the evidence and its therapeutic implications. CNS Drugs. 15(4):287–310.

Zaidel DW, Esiri MM, Harrison PJ (1997) Size, shape, and orientation of neurons in the left and right hippocampus: investigation of normal asymmetries and alterations in schizophrenia. Am J Psychiatr. 154(6):812–818.

Zhang B, Farwell MA (2007) microRNAs: a new emerging class of players for disease diagnostics and gene therapy. J Cell Mol Med. 12:3–21.

Zhao C, Xu Z, Chen J et al. (2006) Two isoforms of GABA(A) receptor beta2 subunit with different electrophysiological properties: Differential expression and genotypical correlations in schizophrenia. Mol Psychiatr. 11(12):1092–1105.

Chapter 6
Metabolomics: A Global Biochemical Approach to the Discovery of Biomarkers for Psychiatric Disorders

Rima Kaddurah-Daouk(✉), Jair C. Soares, and Marlon P. Quinones

Abstract A biomarker is a characteristic that is objectively measured and evaluated as an indicator of normal or pathogenic processes, as well as responses to therapeutic interventions. The discovery of biomarkers for psychiatric disorders and their incorporation into clinical decision-making could dramatically change the future delivery of health care. Thus, there is great need for the discovery, evaluation, and clinical validation of biomarkers. Abnormalities present in psychiatric illness might be related to changes in cellular metabolism leading to measurable differences in the composition and levels of the universe of all plasma metabolites known as the metabolome. Characterizing these biochemical changes could be very useful in the identification of disease biomarkers. Metabolomics is the study of metabolism at the global level. The concept that a metabolic state is representative of the overall physiologic status of the organism lies at the heart of metabolomics. Metabolomic studies capture global biochemical events by assaying thousands of small molecules in cells, tissues, organs, or biological fluids, followed by the application of informatic techniques to define metabolomic signatures. Metabolomic studies can lead to enhanced understanding of disease mechanisms in psychiatric illnesses, as demonstrated by early work in schizophrenia and mood disorders. This chapter begins with an overview of the principles underlying biomarker research and changes in metabolism associated with psychiatric disorders. Then, it describes the conceptual basis for metabolomics, the analytical and informatic techniques used to define metabolomic signatures, and how to use this information to identify biomarkers for psychiatric disorders.

Abbreviations 5-HIAA: 5-hydroxyindoleacetic acid; 5HT: Serotonin; Ach: Acetylcholine; AODS: Antioxidant defense system; BD: Bipolar Disorder; CSF: Cerebrospinal fluid; DA: Dopamine; DFA: Discriminant function analysis; GABA:

R. Kaddurah-Daouk
Duke University Medical Center,
Department of Psychiatry and Behavioral Sciences,
Durham, NC USA

C.W. Turck (ed.) *Biomarkers for Psychiatric Disorders*,
doi: 10.1007/978-0-387-79251-4_6, © Springer Science+Business Media, LLC 2008

Gamma-aminobutyric acid; GC: Gas chromatography; Glu: Glutamate; HVA: Homovanillic acid; LC: Liquid chromatography; LCECA: Liquid chromatography coupled with electrochemical array detection; LC-MS: Liquid chromatography coupled with mass spectroscopy; Li: Lithium; MDD: Major Depressive Disorder; MHPG: 3-methoxy-4-hydroxyphenylglycol; MS: Mass spectrometry; NE (or NA): Norepinephrine; NMR: Nuclear magnetic resonance spectroscopy; NT: Neurotransmitter pathways; PC: Phosphotidylcholine; PCA: Principal components analysis; PDBs: Psychiatric disorders biomarkers; PE: Phosphotidylethanolamine; PLS: Partial least squares; SQZ: Schizophrenia; SSRI: Selective serotonin reuptake inhibitors; Vlp: Valproate

6.1 Introduction

The final decades of the twentieth century and the beginning of the twenty-first century have witnessed a revolution in biomedical research. It is now clear that understanding disease can be defined as the ability to *classify* a pathological state using phenotypic observations and molecular diagnostics; *explain* how perturbed molecular processes cause the disease state; and *define* the mechanisms underlying these perturbations (e.g., genetic polymorphisms, pathogens, environmental factors, etc.). Reaching this high-degree of sophistication in our understanding of disease, calls for the use of high-throughput readouts to profile many genes, proteins and metabolites. Interpretation of the massive amount of data generated by these approaches and its transformation into knowledge requires an interdisciplinary approach and industry-academia partnerships.

The discovery of biomarkers is key to the process of understanding disease. A biomarker is a characteristic that is objectively measured and evaluated as an indicator of normal or pathogenic processes or a response, as well as responses to therapeutic interventions. The discovery of psychiatric disorders biomarkers (PDBs) and their incorporation into clinical decision-making could dramatically change the future delivery of health care. These PDBs should be more reliable, have more precise predictive ability than at present, and/or could be more informative about disease pathogenesis ("Biomarkers and surrogate endpoints: preferred definitions and conceptual framework, " 2001).

The path for the discovery of PDBs is now characterized by our ability to move from the study of single genes, single mRNA transcripts, single proteins, or single metabolites, to a more *global approach*. Indeed, there has been a gradual transition from research solely based on a reductionistic, *hypothesis-driven* approach, to a more *holistic (systems view), discovery-driven-* approach to research (Nicholson, 2006). This systems approach may facilitate the discovery of PDBs. The notion is that psychiatric disorders emerge from abnormal patterns of gene expressions related to changes in proteins and levels of metabolites linked to disease-perturbed networks. This disease specific molecular fingerprint can be uncovered by combining the use of high-throughput methods at the core of genomics, proteomics and metabolomics.

This chapter outlines general aspects related to biomarker research and how it relates to metabolic abnormalities in psychiatric disorders. It includes an introduction to metabolomics – its conceptual basis, the analytical techniques that are used to perform metabolomic studies, and the informatic tools that are required to analyze metabolomic data – with specific examples to illustrate how metabolomics is used to define initial signatures for psychiatric disorders. This chapter also outlines the mounting evidence, some of which is reviewed in subsequent sections of this chapter, documenting that psychiatric disorders are characterized by changes in the metabolome. The overall sum of metabolites known as the metabolome represents a snapshot of the status of diverse biochemical pathways at a particular point in time. Metabolomics has emerged as a field of great significance, focusing on the identification and quantification of small molecules, or metabolites in cells, tissues, and body fluids. The application of sophisticated biochemical and informatics platforms in metabolomics has enabled the initial identification of signatures for disorders such as Schizophrenia (SCZ) and Major Depressive Disorder (MDD); and it is likely to provide information on Bipolar Disorder (BD). Thus, the comprehensive study of the metabolome could lead to the identification of new disease-specific signature(s) as possible biomarkers and PDBs useful in diagnosis and/or treatment of psychiatric disorders. Such signatures could also shed light on our understanding of the underlying disease processes. Metabolomics promises to have broad implications for both basic biomedical research and medical practice, because it can capture information with regard to mechanisms of disease and of drug action, making it possible to map disease risk or drug action to metabolic pathways.

6.2 Biomarker Discovery Research: An Overview

Biomarkers, particularly as related to metabolomics, can be measured in any biological sample, e.g., blood, urine, or saliva. They can be indicators of disease traits (or risk markers), disease states, or disease rates (progression). Accordingly, biomarkers have been classified as antecedent biomarkers (identifying the risk of developing an illness), screening biomarkers (screening for subclinical disease), diagnostic biomarkers (recognizing overt disease), staging biomarkers (categorizing disease severity), or prognostic biomarkers (predicting future disease course, including recurrence and response to therapy, and monitoring efficacy of therapy) ("Biomarkers and surrogate endpoints: preferred definitions and conceptual framework, " 2001; Vasan, 2006).

Biomarkers may also serve as surrogate endpoints or as biomarkers intended to substitute for clinical endpoints ("Biomarkers and surrogate endpoints: preferred definitions and conceptual framework," 2001). A surrogate endpoint is expected to predict clinical benefit, (or harm, or lack of benefit) based on epidemiologic, therapeutic, pathophysiologic, or other scientific evidence characteristically used as an outcome in clinical trials to evaluate safety and effectiveness of therapies instead of measurement of the true outcome of interest. This is possible whenever the surrogate endpoint closely tracks changes in the outcome of interest. Surrogate end points have the advantage that they may be gathered in a shorter time frame and with less expense than end points such as morbidity and mortality, which require large clinical trials for evaluation. A good example of a commonly used subrogate endpoint is the assessment of CD4+ T lymphocytes in Human immunodeficiency virus (HIV)-infected subjects as a proxy for disease progression (De Milito, Titanji, & Zazzi, 2003).

One of the main advantages of the use of biomarkers as clinical endpoints is that by virtue of possibly being closer to the underlying disease mechanisms (or intervention) of interest, they may be easier to relate causally than more distant clinical events ("Biomarkers and surrogate endpoints: preferred definitions and conceptual framework," 2001). Nevertheless, in the case of psychiatric illnesses, the power of using biomarkers could be compromised by the fact that, in addition to the characteristic measured, commonly studied clinical outcomes (e.g., mood) are influenced by numerous factors. These residual effects may reduce the validity of one biomarker in providing information about an illness.

6.2.1 Characteristics of an Ideal Biomarker

The overall expectation of PDBs is to enhance the ability of the mental health provider to optimally diagnose and manage patients. For instance, in a person with depression, a biomarker may be expected to facilitate the discrimination between patients with MDD and BD. In a patient presenting to the emergency department in a clearly manic state, a biomarker may be able to assess the likelihood of the following: the likelihood of a therapeutic response to lithium, the risk of suicide, the presence of co-morbidities such as anxiety disorders, the risk of recurrence, etc.

Regardless of the intended purpose for its use, the clinical value of a new biomarker will depend on many factors such as: (a) simplicity of interpretation by clinicians; (b) sensitivity – ability of a test to detect a condition of interest when it is truly present; (c) specificity – ability of a test to exclude the condition of interest in patients who do not have the disease; (d) accuracy – sensitivity and its specificity at select cut-points; (e) reproducibility – same results are obtained when the assay is repeated on the same sample in the same or a different laboratory; (f) predictive value – ability of the test to predict the true positives or true negatives for the outcome it is expected to identify; (g) ability to explain a reasonable proportion of the outcome independent of established predictors; and (h) information suggesting that knowledge of biomarker levels changes management (Vasan, 2006). An important performance metric for a biomarker is also to determine the likelihood ratio. The likelihood ratio combines information about the sensitivity and specificity, providing information about how much a positive or negative result changes the likelihood that a patient would have the disease (Deeks and Altman, 2004).

The desirable properties of biomarkers vary with their intended use. For *screening biomarkers*, features such as high sensitivity, specificity, and predictive values at a low cost are important. For diagnostic markers, in addition to the aforementioned characteristics, the biomarkers assays should be able to be easily run at the point-of-sites of delivery of clinical care. For *biomarkers monitoring disease progression or response to therapy* (general prognostic biomarkers), features such as sensitivity and specificity are important but not so critical, given that baseline values are compared with follow-up values, and that the main goal is not making a correct diagnosis or ruling out a disorder. Small intra-individual variation and good correlation with disease outcome or therapy are critical.

6.2.2 Searching for Different Types of Biomarkers

Biomarkers have also been classified as three different types. *Type 0*, are biomarkers of the natural history of a disease. They correlate longitudinally with known clinical indices. *Type I*, are markers that capture the effects of a therapeutic intervention; and *Type II* biomarkers are intended to substitute for a clinical endpoint and predict clinical benefit on the basis of epidemiological, therapeutic, pathophysiological, or other scientific evidence (Frank and Hargreaves, 2003). Establishing the prognostic utility of a biomarker (i.e., type II marker) is more challenging because it requires a larger sample and a prospective design, whereas demonstrating its value as a diagnostic test requires a smaller sample and a cross-sectional design.

6.2.3 Finding the Biomarker's Niche in Clinical Practice

There are many ways in which a biomarker could be of importance to patients and health-care providers. For instance, a negative test result can offer "reassurance value" or serve as a research tool by providing insights into disease mechanisms. However, it is often the case that given the current nature of health care, biomarkers are expected to change disease management or patient outcomes. The so-called disease-centered variables (such as blood pressure, LDL-cholesterol, prostatic-specific antigen, rheumatoid erosions on a radiograph and many others (see (Bakhtiar, 2008)) have obvious clinical relevance or meaning as they can change disease management (Lassere et al., 2007). Their meaning is determined over time after data are collected from laboratory, epidemiological, and clinical settings as mechanisms of biology and pathology are understood.

A naïve expectation is that single biomarkers can capture the complex processes underlying disease processes. In fact, by looking at the disease through the glasses of systems biology, as malfunctioning regulatory networks, it becomes clear that a multi-parameter analysis (panel of markers) may provide better insight into disease diagnosis, prognosis, and treatment. Building on these notions, metabolomics-based approaches survey for global changes in metabolic pathways that are more likely to provide a wealth of information that may be difficult to capture by looking at only one pathway or one biomarker.

Notably, in 2007 the U.S. Food and Drug Administration (FDA) cleared for marketing a multivariate index assay that determines the likelihood of breast cancer returning within 5–10 years of a woman's initial cancer by looking at 70 genes in a sample of a woman's surgically removed breast cancer tumor, and then applying an algorithm to produce a score that determines whether the patient is deemed low risk or high risk for the spread of the cancer to another site ((Fan et al., 2006) and (http://www.fda.gov/bbs/topics/NEWS/2007/NEW01555.html))

There are some important considerations that ought to be taken into account when considering the generation of multi-marker tests (i.e., panels) based on the use of a composite of several biomarkers (measured in parallel) for the purpose of predicting disease risk and patient outcomes. One of these factors is multi-collinearity, or the inherent correlation between biomarkers tracking the same process in one individual. If the correlation between two markers is very high, then measuring one of them is sufficient to capture an event. Another factor is ensuring incremental utility of adding a new biomarker to a panel, meaning that, at some point, measuring additional factors will not improve diagnosis accuracy or change the management or clinical outcome.

6.3 Evidence for Metabolic Changes in Psychiatric Disorders: Rationale for a Metabolomics Approach for Biomarker Discovery

Is there a rationale for looking for metabolomic changes in psychiatric disorders for the purpose of identifying biomarkers? Several lines of evidence suggest that abnormalities in neurotransmitter (NT)-pathways are linked to the pathogenesis of mood disorders, including BD and MDD, as well as schizophrenia (SQZ). Metabolism of biogenic amines (e.g., serotonin and norepinephrine), concentrations of serum proteins (e.g., albumin), enzymes (e.g., alanine transaminase, aspartate transaminase, creatine kinase), arachidonic acid-cascade derivates such as prostaglandins and some electrolytes (e.g., sodium, potassium, calcium, and magnesium) have all been reported to be affected globally in psychiatric patients suffering from mood disorders or SQZ (Coppen, 1969; Ebuehi, Bishop, Fanmuyiwa, Akinwande, & Ladenegan, 2001; Lieb, Karmali, & Horrobin, 1983; Segal et al., 2007a,b; Sublette, Russ, & Smith, 2004; Tandon, Channabasavanna, & Greden, 1988). Much attention has been given to the analysis of abnormalities in single neurotransmitter systems by measuring their levels (or their metabolites) in circulation (e.g., plasma), cerebrospinal fuild (CSF), or urine (excretion). There is also a growing interest in capturing global changes at the level of lipids, (i.e., high-dimensional lipid analysis technologies or lipidomics; Wiest & Watkins, 2007,Watson 2006) and other important metabolic substrates not only aiming to find primary abnormalities, but also due to the recognition of the negative effects of many of the current psychiatric medications, (e.g., atypical antipsychotics used for the treatment of schizophrenia; Casey, 2004; Kaddurah-Daouk, 2007). Additionally, several lipid changes in psychiatric disorders could be linked to abnormalities in membrane structure and function, and hence neurotransmission (e.g., phospholipids from studies in SQZ as described below).

Decades of research provide support to the notion that measurement of NTs – and related products, as well as abnormalities in lipids, provide a window of opportunity to understand central abnormalities associated with psychiatric disorders (see below). Unfortunately, advances in this field have slowed down due to the limited availability of tools to consistently interrogate several interacting biochemical pathways simultaneously. Metabolomics provides powerful tools that enable mapping of global biochemical changes in psychiatric illnesses. Metabolic signatures of psychiatric disorders and of therapies used to treat these disorders could provide novel tools for biomarker discovery.

6.3.1 Evidence for Metabolic Changes in BD

BD is a mental illness involving episodes of mania and depression (Frederick K. Goodwin, Jamison, & Ghaemi, 2007). BD diagnosis and follow-up remains a challenge, as categorization relies only on a clustering of symptoms. The lack of solid diagnostic tools is a consequence of our limited knowledge of disease pathogenesis. BD is thought to result from the interaction between genetic and environmental factors that affect the brain, and might result in perturbations in neurotransmitter systems including Acetylcholine (Ach) (Janowsky, el-Yousef, Davis, Hubbard, & Sekerke, 1972; Janowsky, el-Yousef, Davis, & Sekerke, 1972), GABA (Brambilla, Perez, Barale, Schettini, & Soares, 2003), Norepinephrine (NE) (Schildkraut, 1974), Dopamine (DA) (F. K. Goodwin & Sack, 1974), Glutamate (Glu) (Kugaya & Sanacora, 2005), and Serotonin (5HT) (Mahmood & Silverstone, 2001; Sobczak, Honig, van Duinen, & Riedel, 2002). Abnormalities in lipids and increased inflammation and immune activation may also contribute to the pathogenesis of mood disorders, in part by modulating the metabolism of NT systems (Wichers et al., 2005). Finally, some of the most commonly used and effective drugs in the treatment of BD. i.e., classical mood stabilizers (Lithium (Li), Valproate (Vlp)) or atypical anti-psychotics, have profound metabolic effects that may be related to treatment response and also to common side-effects (Correll, Frederickson, Kane, & Manu, 2007; Garcia-Portilla et al., 2008; Sicras-Mainar, Blanca-Tamayo, Rejas-Gutierrez, & Navarro-Artieda, 2008; Taylor & MacQueen, 2006; Yumru et al., 2007).

6.3.1.1 Neurotransmitter Abnormalities in BD

Norepinephrine (NE)

A multitude of studies have measured single metabolites within the catecholamine family, i.e., catecholamine's, NE, epinephrine, metanephrine, normetanephrine, vanilmandelic acid, and 3-methoxy-4-hydroxyphenyl-glycol (MHPG), in patients with mood disorders (Fortunati, Mazure, Preda, Wahl, & Bowers, 2002; Greenspan et al., 1970; Maas, Dekirmenjian, & Fawcett, 1974; Mooney et al., 1991; Schatzberg et al., 1980; Schildkraut, 1974; Swann et al., 1983; Young, Warsh, Kish, Shannak, & Hornykeiwicz, 1994). The principal metabolite of NE is MHPG. Disturbances in noradrenergic neurotransmission seem to be closely linked to BD pathogenesis (Schatzberg et al., 1982). MHPG levels can be used to predict which patients with mood disorders will require higher doses of medication, higher plasma levels, and longer treatment periods to respond to treatment (Schatzberg et al., 1980). Pretreatment levels of urinary MHPG correlate with improvement in manic syndrome scores (Swann et al., 1999). Unmedicated unipolar and bipolar depressed patients may have a 'hyper-responsive' noradrenergic system (Grossman & Potter, 1999); however, studies comparing urinary NE and its metabolites in unipolar or bipolar depressed patients and healthy volunteers have not yielded consistent findings (Grossman & Potter, 1999). Urinary NE is associated with severity of the current mood (Joyce et al., 1995). MHPG levels are elevated in the CSF of bipolar patients, and they are correlated with mania symptom ratings (Swann et al., 1983). Plasma levels of MHPG decrease 4 weeks after risperidone administration in manic patients (Yoshimura et al., 2006). Indeed, clinical improvement is significantly correlated with levels of plasma MHPG prior to treatment with Li and Olanzapine (Davila et al., 2006).

Plasma NE and adrenaline levels, which are thought to be reliable biochemical indices of emotional arousal, have been found to be increased in patients during mania, but not during depression (Maj, Ariano, Arena, & Kemali, 1984). CSF and plasma monoamines and their metabolites in euthymic bipolar patients are normal (Berrettini et al., 1985). Together, these data are consistent with findings that have shown abnormal, perhaps excessive, central noradrenergic activity in patients with mania (Swann et al., 1983), but normal activity during euthymia. Of special interest is the observation that abnormalities measured in the periphery, e.g., plasma, may be linked to changes in the brain itself, as increased cortical NE turnover has been described in BD (Young et al., 1994).

Dopamine (DA)

Homovanillic acid (HVA) is a commonly measured dopamine metabolite. Increased dopaminergic tone has been consistently reported to be associated with hypomania/mania (Frederick K. Goodwin et al., 2007; Swann et al., 1983). Plasma levels of HVA decrease 4 weeks after risperidone administration in manic patients (Yoshimura et al., 2006). Interestingly, in depressed patients with selective serotonin reuptake inhibitors (SSRI)-induced mania an increased plasma HVA level was observed (Fortunati et al., 2002). Higher urinary dopamine predicts subsequent manic mood (Joyce et al., 1995). Pretreatment HVA levels are correlated with clinical response to typical antipsychotics in manic patients (Chou et al., 2000) and those with manic psychosis (Mazure & Bowers, 1998).

CSF HVA levels are also increased in manic patients (Gerner et al., 1984). Higher dopaminergic activity in mania could be linked to the well-characterized elevated goal/reward-directed activity commonly present during this mood state. Notably, the addition of Li to a neuroleptic antipsychotic agent may enhance the metabolism of dopamine (Bowers, Mazure, Nelson, & Jatlow, 1992) suggesting an additional synergistic mechanism of action for this mood stabilizer in the treatment of BD.

Gamma-Aminobutyric Acid (GABA)

Convergent evidence from many studies suggests that a GABAergic dysfunction is likely to play a key role in mood disorders (Brambilla et al., 2003). Manic patients have significantly higher plasma GABA levels than controls (Petty & Schlesser, 1981), and plasma levels are also correlated with the severity of manic symptoms (Swann et al., 1999). Most notably, Vlp increases plasma and brain levels of GABA, and higher pretreatment plasma GABA levels predict a better clinical response to Vlp (Petty et al., 1996).

Serotonin (5HT)

A reduction in serotoninergic tone has been consistently linked to increased impulsivity, aggression and suicide (Audenaert, Peremans, Goethals, & van Heeringen, 2006; Mann, Oquendo, Underwood, & Arango, 1999). Plasma levels of the 5HT precursor tryptophan are higher in recovered manics than in active manics (Peet, Moody, Worrall, Walker, & Naylor, 1976). Li increases CSF levels of 5-hydroxyindoleacetic acid (HIAA) (Berrettini et al., 1985). Nevertheless, higher levels of CSF 5HIAA have been reported in mania (Tandon et al., 1988). It has been suggested that since lower serotoninergic tone can be seen in both depressed and manic bipolar patients, disturbances of serotoninergic neurotransmission create a "permissive" milieu for the development of mood symptoms in BD (Ellis, Mellsop, Beeston, & Cooke, 1991). Importantly, Li treatment may reverse some of the abnormalities in serotoninergic neurotransmission, as suggested by increments in 5-HIAA concentrations during lithium addition to conventional antidepressant treatment. Also Li increases 5-HT turnover (Birkenhager et al., 2007).

Glutamate/Glutamine (Glu/Gln)

Compared with healthy controls, individuals with BD have been reported to have elevated Glu plasma levels during mania (Hoekstra et al., 2006). However, reduced glutamate plasma levels have also been reported (Palomino et al., 2007). Gln CSF levels are elevated in depressed bipolars (Levine et al., 2000). Of note, Lamotrigine, a drug that is highly effective in the treatment of bipolar depression, may exert/mediate its pharmacological effects in part by reducing excessive central Glu-release (Makatsori et al., 2004).

6.3.1.2 Lipid Abnormalities in BD

Arachidonic Acid (AA)

Inflammatory cytokines activate the metabolism of AA, a molecule that is released from membrane phospholipids via receptor–G protein–initiated activation of phospholipase A2 (PLA2) (Murakami & Kudo, 2002; Nanda et al., 2007). AA is metabolized to eicosanoids by two groups of enzymes: the cyclo-oxygenases (COX), which produce prostaglandins (PGs) and thromboxanes (TXBs), and the lipoxygenases, which catalyze leukotriene and lipoxin synthesis (Murakami & Kudo, 2002; Nanda et al., 2007). Abnormalities in eicosanoids have become increasingly recognized pathogenic factors in mood disorders (Piccirillo et al., 1994; Sublette et al., 2004). For example, compared with healthy controls, individuals with MDD have increased PG levels in CSF, serum/plasma, and in saliva, in addition to elevated TXB-2 levels (Horrobin, 1977; Lieb et al., 1983; Piccirillo et al., 1994). Importantly, randomized double-blind pilot add-on studies in MDD and BD showed a significant therapeutic effect of selective COX-2 inhibition in mood disorders (Muller, Riedel, & Schwarz, 2004; Muller et al., 2006). Likewise, some evidence supports a role for disturbances in AA metabolism in the pathogenesis of BD (Sublette et al., 2004). Notably, the mood stabilizers Li and Vlp in rats decrease AA turnover in the brain and levels of COX-2 in experimental rodent models, and it has been hypothesized (based on preliminary evidence) that some of the effects of these medications might depend on the effects of Li on AA-metabolism (Rapoport & Bosetti, 2002).

Fatty Acids

Polyunsaturated essential fatty acids (PUFA) oppose the effects of the pro-inflammatory of eicosanoid-derivatives. Fish and seafood are the richest dietary sources of the long-chained omega-3, eicosapentaenoic acid (EPA) and docosahexaenoic acid (DHA). DHA is selectively concentrated in synaptic neuronal membranes and contributes to unique biophysical properties that mediate receptor activity and signal transduction. Changes in plasma levels of fatty acids have been seen in depression (Mueller, Davis, Bunney, Weil-Malherbe, & Cardon, 1970). Notably, higher plasma ratios of AA to EPA are correlated with a greater severity of depressive symptoms (Freeman et al., 2006). Recent meta-analyses, meta-regressions, and critical reviews of the literature concluded that the preponderance of epidemiologic and tissue compositional studies supports a protective effect of omega-3 intake, particularly EPA and DHA, in mood disorders (both MDD and BD) (Appleton et al., 2006; Freeman et al., 2006; Lin & Su, 2007; Ross, Seguin, & Sieswerda, 2007).

6.3.2 *Evidence for Metabolic Changes in Major Depressive Disorders*

The lifetime prevalence of MDD in the US is 16% (Kessler et al., 2003). Like many other psychiatric illnesses, there is a need for the identification of disease biomarkers to aid in the diagnosis and prediction of treatment-response. Multiple neurotransmitter systems are thought to be dysregulated in MDD; however, much attention has been given to the role of impairment in the central monoaminergic function. The monoamine hypothesis of depression predicts deficiencies in the absolute concentrations of NE and 5-HT (Leonard, 2000). Nevertheless, support for this notion has been mostly provided by measuring in isolation levels of different neurotransmitters and their metabolites in CSF, urine, and plasma of patients with depression (Leonard, 2000). Unfortunately, the use of conventional approaches to assess the metabolism of these neurotransmitters has led to equivocal results. Nonetheless, the ability to measure simultaneously most metabolites in these neurotransmitter pathways was not available until recently through development of advanced metabolomics platforms.

6.3.2.1 Serotonin (5HT)

Given that SSRIs are commonly used for the treatment of MDD, 5HT abnormities have been traditionally linked to MDD' pathogenesis (Delgado, 2004). Plasma free tryptophan concentration provides an index of CSF tryptophan and 5-HIAA acid concentrations, and hence of 5HT turnover (Curzon, 1979). Moderate changes of tryptophan supply to the brain can modify serotonergic activity (Curzon, 1981). Indeed, tryptophan depletion produces a marked reduction in plasma tryptophan and consequently brain 5-HT synthesis and release. In healthy volunteers, the effects of tryptophan depletion are influenced by the characteristics of the subjects, and include some mood lowering, some memory impairment, and an increase in aggression (Delgado, 2006). In patients with depression, tryptophan depletion tends to result in no worsening of depression in untreated subjects, but a relapse in those who have responded to antidepressants (particularly serotonergic agents) (Bell, Abrams, & Nutt, 2001). Lower L-tryptophan plasma levels have been observed in patients with depression (Maes et al., 1994). Unipolar depressed patients excrete higher amounts of 5-HIAA, the main 5HT metabolite, than controls and bipolar subjects (Garfinkel, Warsh, & Stancer, 1979). Studies in postmortem brain of depressed and suicide subjects, suggest that decreased 5HT metabolite levels and turnover may be common to both MDD and BD (Young et al., 1994). Moreover, the molar ratio of total plasma tryptophan to the sum of other large neutral amino acids (viz., valine, isoleucine, leucine, tyrosine, phenylalanine), thought to reflect brain serotonin formation, have been suggested as potential tools in predicting drug response (Lucini, Lucca, Catalano, & Smeraldi, 1996). Abnormalities in 5HT are also linked to suicide. Indeed, CSF 5-HIAA levels are negatively correlated with the maximum lethality of suicide attempts during the 2 year follow-up period (Sher et al., 2006).

6.3.2.2 Norepinephrine(NA)

Not only serotonergic, but also noradrenergic systems are likely to be involved in antidepressant action, and specific impairments in these systems may also play a role in disease pathogenesis (Delgado & Moreno, 2000). For instance, studies measuring few NE metabolites have suggested that akin to BD, in MDD there is a "hyper-responsive" noradrenergic system (Grossman & Potter, 1999). Patients with MDD, but not BD, were shown to have significantly higher plasma NE levels than age- and sex-matched controls (Roy, Guthrie, Pickar, & Linnoila, 1987) and higher levels of plasma norepinephrine were present along higher pulse rates than healthy control subjects (Lake et al., 1982). Moreover, a study including 151 healthy, MDD, and BD subjects concluded that the pattern of NE and epinephrine measured discriminates among the forms of mood disorders (Koslow et al., 1983). Differences in dopaminergic function may also differentiate between BD and MDD. In contrast with the descriptions of abnormalities in this system in bipolar patients, particularly during mania (see above), depressed patients and controls have comparable plasma levels of HVA suggesting that the dopaminergic function may be spared in MDD (Roy, 1988).

6.3.2.3 Gamma-Aminobutyric Acid (GABA)

The GABAergic neurotransmitter system has become increasingly implicated in the neurobiology of mood disorders (Brambilla et al., 2003; Sanacora & Saricicek, 2007). GABAergic involvement in the pathophysiology and treatment of mood disorders is supported by several lines of evidence spanning from animal studies showing stress-related changes in GABAergic function to the demonstration of GABAergic abnormalities and genetic associations in depressed patients (Brambilla et al., 2003; Sanacora & Saricicek, 2007). Specifically, the association of lower GABA levels with depression is a consistent finding (Petty & Schlesser, 1981). Indeed, reduced levels have been consistently reported in plasma (Petty & Schlesser, 1981), CSF (Gerner et al., 1984; Kasa et al., 1982) and brain (Bhagwagar et al., 2007; Hasler et al., 2007; Sanacora & Saricicek, 2007) of individuals diagnosed with depression. Likewise, GABA agonists and antagonists have the ability to modulate behavioral models of depression in rodents, existing antidepressant medications have GABAergic effects and there is some evidence for a clinical antidepressant efficacy associated with GABAergic drugs.

6.3.2.4 Glutamate (Glu)/Glutamine (Gln)

There is a growing interest in the evidence for a role of Glus in the pathophysiology of mood disorders. Initial work has revealed glutamatergic abnormalities in MDD patients including decreased CSF glutamate and glycine (Frye, Tsai, Huggins, Coyle, & Post, 2007) but increased levels of Gln have been reported (Levine et al., 2000). Furthermore antidepressant and mood stabilizing medications and Glu modulating agents have glutamatergic effects in the treatment of depression.

Indeed, a single intravenous dose of the Glu NMDA receptor antagonist ketamine is sufficient to produce sustained relief from depressive symptoms (Zarate et al., 2006). Animal studies have also shown stress-related changes in glutamatergic function, and its possible relationship to the pathophysiology and pathogenesis of MDD (Maeng et al., 2008).

6.3.3 *Evidence for Metabolic Changes in Schizophrenia (SQZ)*

The underlying pathophysiology of schizophrenia is poorly understood but impairments in neurotransmitter, signal transduction, antioxidant system, membrane composition and immune functions have been noted among several other changes in patients with schizophrenia. We provide below a brief summary of these biochemical changes and pathways.

6.3.3.1 Altered Neurotransmission Systems

There is sufficient evidence to suggest that aberrant neurotransmission might play a role in the pathogenesis of schizophrenia. In particular, aberrant dopaminergic, serotonergic, and glutamatergic systems have been noted (Miyamoto et al., 2003). Some data also exists that supports the involvement of histamine, and acetylcholine neurotransmitter systems (Miyamoto et al., 2003). It is unclear, however, to what extent these neurochemical findings reflect primary rather than secondary pathology, compensatory mechanisms, or environmental influences, and it is not clear how these pathways interact in schizophrenia pathogenesis.

Dopamine (DA)

Despite several limitations, the hypothesis that hyperactivity of dopaminergic transmission at the dopamine D_2 receptor is the key pathogenic factor in schizophrenia remains the most accepted neurochemical theory (Heritch, 1990; Knable & Weinberger, 1997; Meisenzahl, Schmitt, Scheuerecker, & Moller, 2007). The hypothesis is supported by strong correlation between the therapeutic doses of conventional antipsychotic drugs and their affinities for the D_2 receptor (Pani, Pira, & Marchese, 2007). In addition, indirect dopamine agonists (e.g., L -dopa, cocaine, and amphetamines) can induce psychosis in healthy subjects and, at very low doses, provoke psychotic symptoms in schizophrenics (Ellison, 1994). Postmortem and positron emission tomography (PET) indicates increased dopamine D_2 receptor levels in the brains of schizophrenic patients (Wong et al., 1986) but it is not clear if these changes are due to antipsychotic effects.

By looking at DA related metabolites it is clear that DA abnormalities play a more critical role than dysregulation of the other NT-systems. For instance, while increased plasma HVA are associated with relapse, plasma MHPG levels are not (Kelley, Yao, & van Kammen, 1999); and in patients with SQZ, treatment with clozapine but not with haloperidol increases plasma NE levels. However, the increase in NE is not associated with improvement in total or positive symptomatology (Brown et al., 1997). Instead, there is an association between improvement in negative symptoms and increased HVA on clozapine (Brown et al., 1997). Interestingly, lateral ventricle enlargement in SQZ, a commonly reported neuroanatomical abnormality that has been positively correlated with poor pre-morbid competence, negative symptoms, and poor treatment response; is negatively correlated with CSF-concentrations of HVA (Lewine et al., 1991).

Serotonin (5HT)

Recent findings suggest a role for abnormalities in the serotoninergic system in the pathophysiology of schizophrenia (Lieberman et al., 1998). Several lines of evidence support this hypothesis including the observation that serotonin receptors are involved in the psychotomimetic and psychotogenic properties of hallucinogens (e.g., lysergic acid diethylamide- LSD). Notably, 5-HT_{2A} receptor play a role in the therapeutic profile of atypical antipsychotics (e.g., clozapine), the number of cortical 5-HT_{2A} receptors have been found to be altered in schizophrenic brains and 5-HT_{2A} receptor-mediated activation of the prefrontal cortex may be impaired in some schizophrenics. Moreover, genetic evidence shows that certain polymorphisms of the 5-HT_{2A} receptor gene are associated with schizophrenia.

Several investigators have examined CSF levels of 5-HIAA in schizophrenic patients and report conflicting results with lower and unchanged levels compared to control values (Breier, 1995; Ebuehi et al., 2001; Gerner et al., 1984; Lewine et al., 1991). Clearly, this is an area of research that deserves additional investigation.

6.3.3.2 Membrane Phospholipid Defects

Lipids and their constituent fatty acids provide scaffolding for many key functional systems including neurotransmitter receptor binding, signal transduction, transmembrane ion channels, prostanoid synthesis, and mitochondrial electron-transport systems. Evidence has accrued that phospholipids that play a critical role in the structure and function of membranes seem to be impaired in schizophrenia (Berger et al., 2002; Horrobin, 1998; Skosnik & Yao, 2003). There are reports of decrease in phospholipids in red blood cells (RBC) of psychotic patients (Yao, Thomas, Reddy, & Keshavan, 2005) and medication-free schizophrenic patients (Khan et al., 2002); in fibroblasts from neuroleptic-naïve schizophrenic patients (Mahadik et al., 1994); in platelet membranes of drug-naïve patients with schizophrenia (Schmitt et al., 2001) and in postmortem brain tissue from patients with schizophrenia (Horrobin, Manku, Hillman, Iain, & Glen, 1991). In fact, the above findings indicate that schizophrenia may involve a lipid disorder expressing throughout the whole-body (Horrobin, 1996), providing further validity in utilizing peripheral measures of lipid changes in schizophrenia as biomarkers for the disease (see below).

6.3.3.3 Role of Oxidative Stress in Schizophrenia

There is increasing evidence of antioxidant defense system (AODS) deficits in schizophrenia (Yao, Reddy, & van Kammen, 2001). Significant reductions of plasma antioxidants (e.g., albumin, bilirubin and uric acid) are seen early in the course of schizophrenia (Reddy, Keshavan, & Yao, 2003), consistent with previous findings in patients with chronic schizophrenia (Yao, Reddy, McElhinny, & van Kammen, 1998; Yao, Reddy, & van Kammen, 1998; Yao, Reddy, & van Kammen, 1999; Yao, Reddy, & van Kammen, 2000). More importantly, these reductions are not a consequence of treatment since patients were antipsychotic drug-naïve at entry into the study. Moreover, significantly lower levels of glutathione, glutathione peroxidase, and glutathione reductase were also found in the schizophrenia group than in the control groups without any psychiatric disorders (Yao, Leonard, & Reddy, 2006).

6.4 Metabolomics: A Global Biochemical Approach for Biomarker Discovery

6.4.1 Metabolomics Overview

Metabolomics, the study of metabolism at the global or 'omics' level, has the potential to contribute significantly to biomedical research and, ultimately, to clinical medical practice. In contrast to classical biochemical approaches that often focus on single metabolites, single metabolic reactions and their kinetic properties, and/or defined sets of linked reactions and cycles (i.e., precursor/product, intermediary metabolism), metabolomics involves the collection of quantitative data on a broad series of metabolites in an attempt to gain an overall understanding of metabolism and/or metabolic dynamics associated with conditions of interest, including drug exposure (Kaddurah-Daouk, Kristal, & Weinshilboum, 2008; Kristal, Kaddurah-Daouk, Beal, & Matson, 2007; Lindon, Holmes, & Nicholson, 2007). Many names have been used to refer to this new field, including metabonomics, metabolic profiling, metabolic fingerprinting, and metanomics, among others (Harrigan & Goodacre, 2003; Kristal, Kaddurah-Daouk et al., 2007; Lindon et al., 2007). However, metabolomics has been used most often, so we have applied that term throughout this chapter.

Metabolomic tools enable us to study the metabolome, the repertoire of small molecules present in cells and tissue (Kaddurah-Daouk et al., 2008; Kristal, Kaddurah-Daouk et al., 2007; Lindon et al., 2007). The identities, concentrations, and fluxes of these substances are the final product of interactions between gene expression, protein expression, and the cellular environment (Fig. 6.1). Unlike earlier, more rudimentary analytical methods, metabolomics today offers analytical instruments that can simultaneously quantitate thousands of substances present in a biological sample of interest and sophisticated mathematical tools that can find a molecular signal amongst millions of pieces of data (Kell, 2004).

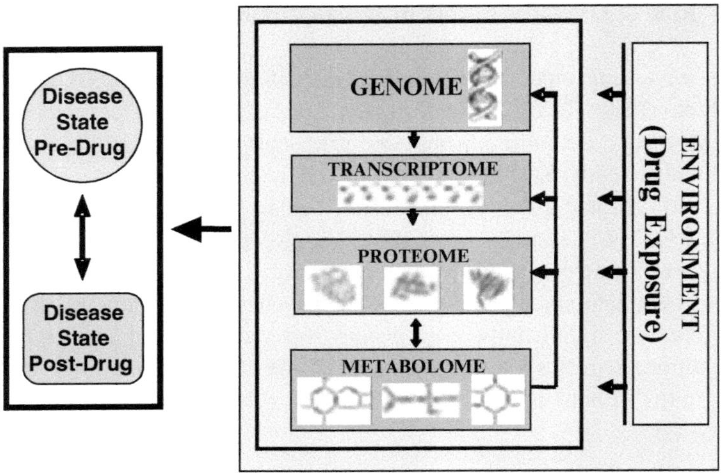

Fig. 6.1 Flow of information from the genetic code (DNA) to proteins, and finally to metabolites. The environment and DNA affect the end products and influence health/disease states. In this case, drug exposure is the environmental alteration being tested (*See Color Plates*)

The size of the metabolome remains a subject of debate and depends strongly on the definition of what should be included and on the analytical platforms and methods used to assess it. Metabolomic information complements data obtained from other fields that comprise the new biology – genomics, transcriptomics, and proteomics – adding a final piece to a systems approach for the study of disease pathophysiology and drug action. Unlike earlier analytical methods, metabolomics utilizes instruments that can simultaneously quantitate thousands of small molecules in a biological sample. This analytical capability must then be joined to sophisticated mathematical tools that can identify a molecular signal among millions of pieces of data.

Ideally, metabolomics will ultimately contribute a detailed map of the regulation of metabolic pathways, and, therefore, of the interaction of proteins encoded by the genome with environmental factors, including drug exposure. Therefore, the metabolome represents a 'state' function for an individual at a particular point in time or after exposure to a specific environmental stimulus, e.g., a specific drug or potentially even a mood state.

Many diseases disrupt metabolism and result in changes that are long lasting and that can be captured as metabolic signatures. Initial metabolomic signatures have already been reported for several disease states, including motor neuron disease (Rozen, Cudkowicz, Bogdanov, & et al., 2005), depression (Paige, Mitchell, Krishnan, Kaddurah-Daouk, & Steffens, 2007), schizophrenia (Holmes et al 2006, Kaddurah-Daouk, 2006; Kaddurah-Daouk et al., 2007), Alzheimer's disease (Han, D, McKeel, Kelley, & Morris, 2002), cardiovascular and coronary artery disease (Brindle et al., 2002; Sabatine et al., 2005), hypertension (Brindle, Nicholson, Schofield, Grainger, & Holmes, 2003), subarachnoid hemorrhage (Dunne,

Bhattachayya, Besser, Rae, & Griffin, 2005), preeclampsia (Crocker, Kenny, Thornton, Szabo, & Baker, 2005), type 2 diabetes (van der Greef et al., 2007; van Doorn et al., 2007; Wang et al., 2005; Yuan, Kong, Guan, Yang, & Xu, 2007), liver cancer (Yang et al., 2004), ovarian cancer (Odunsi et al., 2005), breast cancer (X. Fan, Bai, & Shen, 2005), and Huntington's disease (Underwood et al., 2006). These signatures are made up of tens of metabolites that are deregulated, with concentrations that are modified in the disease state or after drug exposure. As a result, analysis of these signatures and their components can potentially provide information with regard to disease pathophysiology.

The application of metabolomics to the study of the effects of drugs captures signatures representing changes that occur secondary to drug treatment and in which those signatures capture information from pathways that are targets for, or are affected by, drug therapy (Kaddurah-Daouk et al., 2008; Morvan & Demidem, 2007; Portilla et al., 2006; van Doorn et al., 2007; Watkins, Reifsnyder, Pan, German, & Leiter, 2002).

6.4.2 *The Metabolomics Process*

6.4.2.1 From Samples to Databases

An overview of a "typical" metabolomics study is depicted in Fig. 6.2. Samples of interest (e.g., plasma, CSF, tissue biopsy, etc.) are collected. Small molecules are extracted from the sample and are then analyzed using techniques that separate and quantitate molecules of interest. These technologies include, among others, liquid and gas chromatography as well as mass and NMR spectroscopy. Combinations of these techniques can be used to augment separations and/or expand the analyte information collected.

These datasets must be then collected and curated, a process that can take significant time for the overall experiment. After curation, the data are analyzed by one or more software packages designed for studies of datasets that are far too large for human evaluation. A database is then generated for the diseased patients and another for the controls or for the same patients before and after drug therapy. These databases include levels of detectable metabolites and identity (or a description of the properties) of the metabolites, i.e., oxidation-reduction potential, mass/charge ratio, etc. whenever known.

6.4.2.2 Mining Metabolomic Databases

The application of software tools for the analysis of the information contained in a database can: (a) identify signature of a disease (e.g., compounds that highlight a disease state); (b) predict class (e.g., disease or control, pre- or post-drug exposure); (c) identify unrecognized groups in the data (e.g., drug response

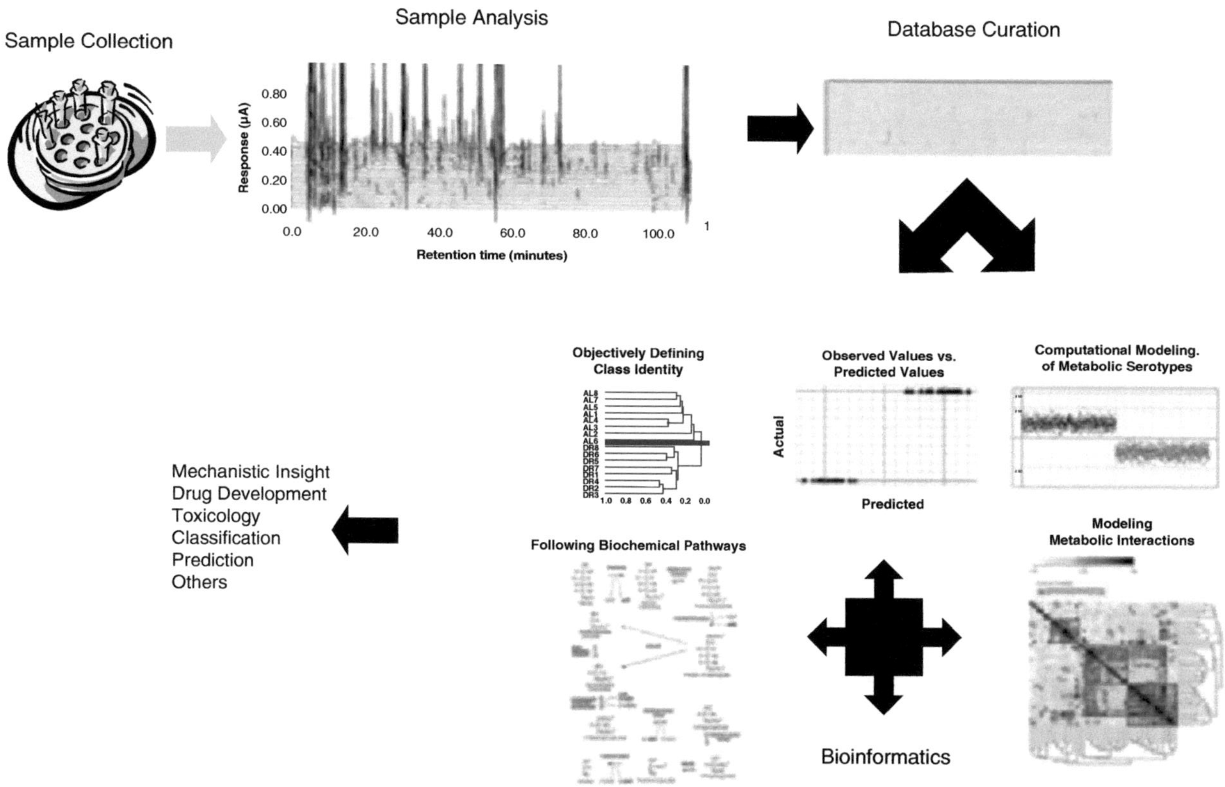

Fig. 6.2 Flow diagram for a typical metabolomics experiment. Samples are collected and individual metabolites are quantitatively analyzed using one or more high data density analytical instruments. These datasets are curated based on the design of the individual experiment, and are then analyzed using a series of high data density informatics approaches. Clockwise from the top left, those shown here are cluster analysis (from Pirouette, infometrix), class prediction (SIMCA-P, Umetrics), principal components analysis of a computationally modeled dataset (SIMCA-P, Umetrics), 2D cluster analysis (GeneLinker Platinum, Improved Outcomes Software), and metabolic pathway analysis (biotech.icmb.utexas.edu) (*See Color Plates*)

subgroups); (d) identify interactions between variables; and (e) place these variables within known biochemical pathways.

A critical conceptual point in metabolomics is that, as discussed earlier, a biomarker that predicts disease, or helps to monitor therapy most often is not a single molecule but rather a pattern of several molecules. This concept determines the need for quantitative precision as well as careful avoidance of artifacts during metabolomic studies. While this is a difficult analytical task in the early stages of detection of such patterns, if the relevant species can be defined and identified, several technologies can be used to develop rapid targeted assays suitable for clinical application.

6.4.2.3 Instrumentation

The choice of metabolomic analytical instrumentation and software is generally goal-specific, since each type of instrument has certain strengths. Liquid chromatography followed by coulometric array detection, for example, has been used in the identification of signatures in amyotrophic lateral sclerosis (Rozen et al., 2005) and most recently other neurodegenerative diseases (Bogdanov et al., 2008). It is excellent for mapping neurotransmitter pathways (e.g., dopamine and serotonin) and pathways involved in oxidative stress. Gas chromatography in conjunction with mass spectroscopy is often used in the analysis of lipid subsets (Kaddurah-Daouk et al., 2007; Watkins, 2004; Watson, 2006). Liquid chromatography with mass spectroscopy is often used to obtain the largest possible subset of biochemical profile information. It is a sensitive tool that can be used to characterize, identify, and quantify a large number of compounds in a biological sample where metabolite concentrations might cover a broad range (Kristal, Shurubor, Kaddurah-Daouk, & Matson, 2007).

In addition to popular high-sample-throughput applications, NMR is particularly powerful for metabolite structural determinations, including the atomic positions of isotopic labels (e.g., ^{13}C, ^{15}N, or ^{2}H) in different isotopomers generated during stable isotope tracer studies (Fan & Lane, 2007). NMR based high throughput analysis has been used successfully in, for example, toxicology studies (Coen et al., 2004; Lindon, Nicholson, Holmes, & Everett, 2000; Lindon et al., 2003). NMR applications provide detailed maps of biochemical pathways or networks, which can also serve as crucial inputs for in silico quantitative flux analysis (Dauner, Bailey, & U., 2001; de Graaf et al., 2000).

6.4.3 *Statistical and Machine-Learning Algorithms for Mining Metabolomics Data*

Metabolomics data can be analyzed with a range of statistical and machine-learning algorithms. These algorithms can be classified into two major classes: unsupervised and supervised (Kell, 2004). They can be useful in the identification of biomarkers (Sajda, 2006; Shin & Markey, 2006). Unsupervised algorithms find patterns in the data without any biases, and are typically driven by the largest changes (variance).

Supervised algorithms require that samples be labeled in groups a priori, and they uncover the features (variables) that best discriminate between those groups. Examples of unsupervised methods that have been routinely used in analyzing molecular fingerprinting data include principal component analysis (PCA) and self-organizing maps (Patterson et al., 2008). These methods are usually guided by the largest average differences between the groups, and are thus very sensitive to outliers. Moreover, because the groupings originate from the data itself, rather than from the analyst, the methods are also very sensitive to how the experiments were carried out. These methods are best used to reveal unknown patterns in the data, but their interpretation needs to be highly connected to the experimental details. Their application is important in the sense that they provide a kind of quality control, by which we will verify which are the most salient features of the data. Supervised methods have also been applied to molecular fingerprinting data, most often ANOVA, partial least squares (PLS), and discriminant function analysis (DFA) (Denkert et al., 2006; Hall, Brown, & Paul, 2007; Steuer, Morgenthal, Weckwerth, & Selbig, 2007; Weljie, Dowlatabadi, Miller, Vogel, & Jirik, 2007).

6.4.4 Translating Findings from Metabolomic Studies into Biomarkers

The complexity of the statistics involved in deciphering the output of metabolomic studies was illustrated in the preceding paragraphs. The next challenge is to translate "PCA patterns" into clinically relevant "panels" or finite set of PDBs, which as described earlier, can be useful for the mental health provider. One approach to achieve this goal could be to pinpoint a few metabolites that can aid in the discrimination between groups (phase 1 studies) and aim for the development of strategies to facilitate their application in the clinic. Another approach could be to rather than select a few metabolites for the analysis, to focus on maintaining the high dimensionality of the data sets, and use this for diagnosis purposes. In this case, the mental health provider would not be responsible for the interpretation of a test result, but rather he or she will be given a final answer extrapolated from a global analysis. The former approach is likely to be more applicable to emergency settings, while the latter is more suitable for clinical scenarios in which diagnosis and management do not have to be defined on the spot. In metabolomics biomarker discovery studies one needs a signature establishing set, a replication set, and a blinded set to estimate the predictive ability of the metabolic biomarkers. Healthy controls and other disease controls are included to define specificity and sensitivity of identified biomarkers.

6.4.4.1 Connecting Central and Peripheral

An obvious challenge for the identification of disease signatures in neuropsychiatric illnesses is the limited access of measure levels of metabolites directly in the brain. Given that CSF can be collected following a lumbar puncture, this fluid is commonly used as a proxy for brain changes. Some evidence seems to suggest that central (CSF) changes in potential biomarkers might be correlated with changes in the periphery (blood; plasma or serum). For instance measurement of inflammatory markers in paired plasma/CSF samples of healthy human volunteers revealed a correlation between central and peripheral levels of pro-inflammatory makers (Maier, Laurer, Rose, Buurman, & Marzi, 2005). This seems also to be the case for some metabolites. Plasma free tryptophan is clearly correlated with brain tryptophan concentration. Indeed, it has been suggested that plasma free tryptophan concentration provides an index of CSF tryptophan and 5-HT turnover in the brain (Curzon, 1979). Likewise, CSF and serum/plasma concentrations of vitamin biomarkers are significantly correlated. Strikingly, the correlation between serum and CSF-folate can be as high 0.69 (Obeid et al., 2007). Another example is the levels of the side chain oxidized oxysterol 24S-hydroxycholesterol, a potential marker of brain injury, which formed almost exclusively in the brain; and whose levels in plasma and CSF are highly correlated (Leoni et al., 2003).

Nonetheless, limited amount of evidence also suggests that abnormalities in the blood level of certain metabolites (plasma or serum) might not be correlated with central abnormalities (Levine, Sela, Osher, & Belmaker, 2005). Thus, additional research using metabolomics-based approaches is needed to define metabolites whose central and peripheral levels are correlated and how these correlations are influenced by neuropsychiatric illnesses. In any case, given the accessibility of human plasma and the vast medical laboratory infrastructure already in place for its analysis, this is likely to remain the preferred diagnostic material for the foreseeable future. Thus, the identification of peripheral metabolomic signatures of neuropsychiatric illnesses is likely to have more potential for translation into the clinical realm. One has yet to show that these peripheral biomarkers are related to central changes and are disease specific.

6.4.5 Initial Findings from Metabolomics Studies in Depression and Schizophrenia

6.4.5.1 Metabolic Signatures in Depression

Given the body of evidence suggesting that abnormalities in the metabolome do exist both centrally and in the periphery, a recently published metabolomics study was conducted in plasma as a first step towards mapping biomarkers for depression. Metabolomic analysis of blood plasma was performed on nine depressed, 11 remitted, and ten never-depressed older adults (Paige, 2007). Hundreds of metabolites were measured using Gas chromatography-mass spectrometry (GC-MS). Metabolite identification was based on a combination of chromatographic properties and mass spectra. A library of 800 commercially available human metabolite standards has been assembled and analyzed on a GC-MS platform that helped in compound identification.

Concentrations of each metabolite were normalized to mean scaled values of two recovery standards that represent hydrophobic and hydrophilic compounds. Univariate

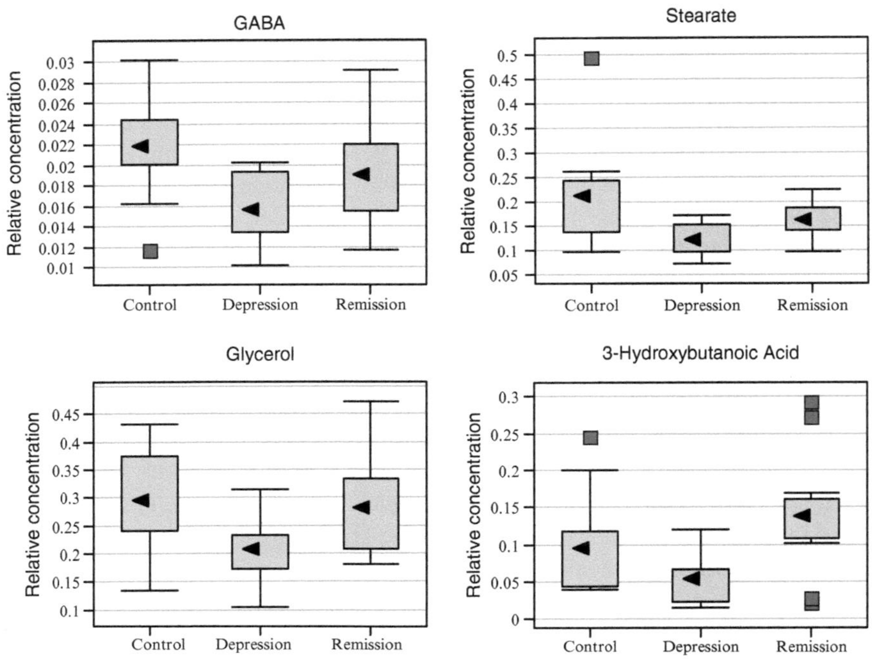

Fig. 6.3 Relative Concentrations of four metabolites in healthy controls and individuals with ongoing or remitted depression. In this box plots, the center line represents the median value, the bottom and top of the box represents the 25th and 75th percentile, respectively, and the dashed lines show the full range for the metabolite in the various groups. When compared with controls and remission groups, the depressed group had overall significant ($p<0.05$) decline in the concentration of GABA and medium chain fatty acids including stearate, palmitate, myristate and oleate. Stearate is shown as a representative fatty acid that is lower in the depressed population (*See Color Plates*)

T-tests (Welch's two-sample t-tests) were performed for each metabolite to determine metabolic differences between healthy controls, and individuals who are depressed or in remission. False discovery rate (FDR) was accounted for, the multiple testing and q-values were estimated for every possible list of "significant" metabolites. Metabolites that were altered in currently depressed patients when compared with controls included several fatty acids, glycerol and gamma-aminobutyric acid (GABA). Analyses comparing concentrations in remitted and currently depressed patients revealed a pattern of metabolite alterations similar to the control vs. currently depressed analyses. One difference observed in the remitted patients relative to the depressed patients was elevation of the concentration of the ketone 3-hydroxybutanoic acid (Fig. 6.3).

These preliminary results will need to be examined and validated in larger longitudinal cohorts. However, these findings suggest that the depressed state may be associated with alterations in the metabolism of lipids and neurotransmitters, and that treatment with antidepressants adjusts some of the aberrant pathways in disease so that the patients in remission have a metabolic profile more similar to controls than to the depressed population. An evaluation of such changes in CSF samples is needed to establish how closely these findings are to central changes.

6.4.5.2 Metabolic Signatures in Schizophrenia and its Treatment

Recently several metabolomic studies have been conducted in an attempt to better define the pathways modified in schizophrenia and its treatment (Holmes et al., 2006; Kaddurah-Daouk et al., 2007; Tkachev, Mimmack, Huffaker, Ryan, & Bahn, 2007; Tsang, Huang, Holmes, & Bahn, 2006). In one study (Kaddurah-Daouk et al., 2007) we used a specialized lipidomics platform and measured more than 300 polar and nonpolar lipid metabolites (structural and energetic lipids) across seven lipid classes to evaluate global lipid changes in schizophrenia before and after treatment with three commonly prescribed atypical antipsychotics: olanzapine, risperidone, and aripiprazole.

As mentioned above, lipidomics is a branch of metabolomics that specifically focuses on comprehensive assessment of lipid biochemistry (German, Gillies, Smilowitz, Zivkovic, & Watkins, 2007; Watson, 2006; Wiest & Watkins, 2007). In this particular study, lipid profiles were obtained for 50 patients with schizophrenia before and after 2–3 weeks of treatment with olanzapine, risperidone, or aripiprazole. At baseline, and prior to drug treatment, major changes were noted in two phospholipid classes, phosphotidylethanolamine (PE) and phosphotidylcholine (PC), suggesting that phospholipids that play a key role in proper membrane structure and function seem to be impaired in patients with schizophrenia [Fig. 6.4 and (Kaddurah-Daouk et al., 2007)]. This confirmed previous observations but established a far more detailed biochemical map for sites of perturbations. Some of the biochemical perturbations were seen within the omega 3 and omega 6 subclasses in PE and PC. Additionally shifts between saturated and ploy unsaturated fatty acids were noted [Fig. 6.4 and (Kaddurah-Daouk et al., 2007)].

The effects of the three antipsychotic drugs, olanzapine, risperidone, and aripiprazole, on lipid biochemical pathways were then evaluated by comparing metabolic profiles at baseline to post treatment (Kaddurah-Daouk et al., 2007). It was of interest that each of the three drugs studied had a unique signature suggesting that while these drugs share some effects, they also have many effects that are unique for each. PE concentrations that were suppressed at baseline in patients with schizophrenia were elevated after treatment with all three drugs. However, olanzapine and risperidone affected a much broader range of lipid classes than did aripiprazole, with approximately 50 lipids that were increased after exposure to these drugs, but not after aripiprazole therapy (Fig. 6.4). On balance, aripiprazole induced minimal changes in the lipidome (Fig. 6.4), consistent with its limited metabolic side effects. There were also increased concentrations of triacylglycerols and decreased free fatty acid concentrations after both olanzapine and risperidone, but not after aripiprazole therapy [Fig. 6.4 and (Kaddurah-Daouk et al., 2007)]. All of these changes suggest peripheral effects that might be related to the metabolic side effects that have been reported for this class of drugs and highlights lipases in the liver as possibly targets for these drugs. Finally, a PCA identified baseline lipid alterations that seemed to correlate with acute treatment response.

Collectively, these results raised the possibility that a more definitive long-term randomized study of these drugs in which global lipid changes would be correlated

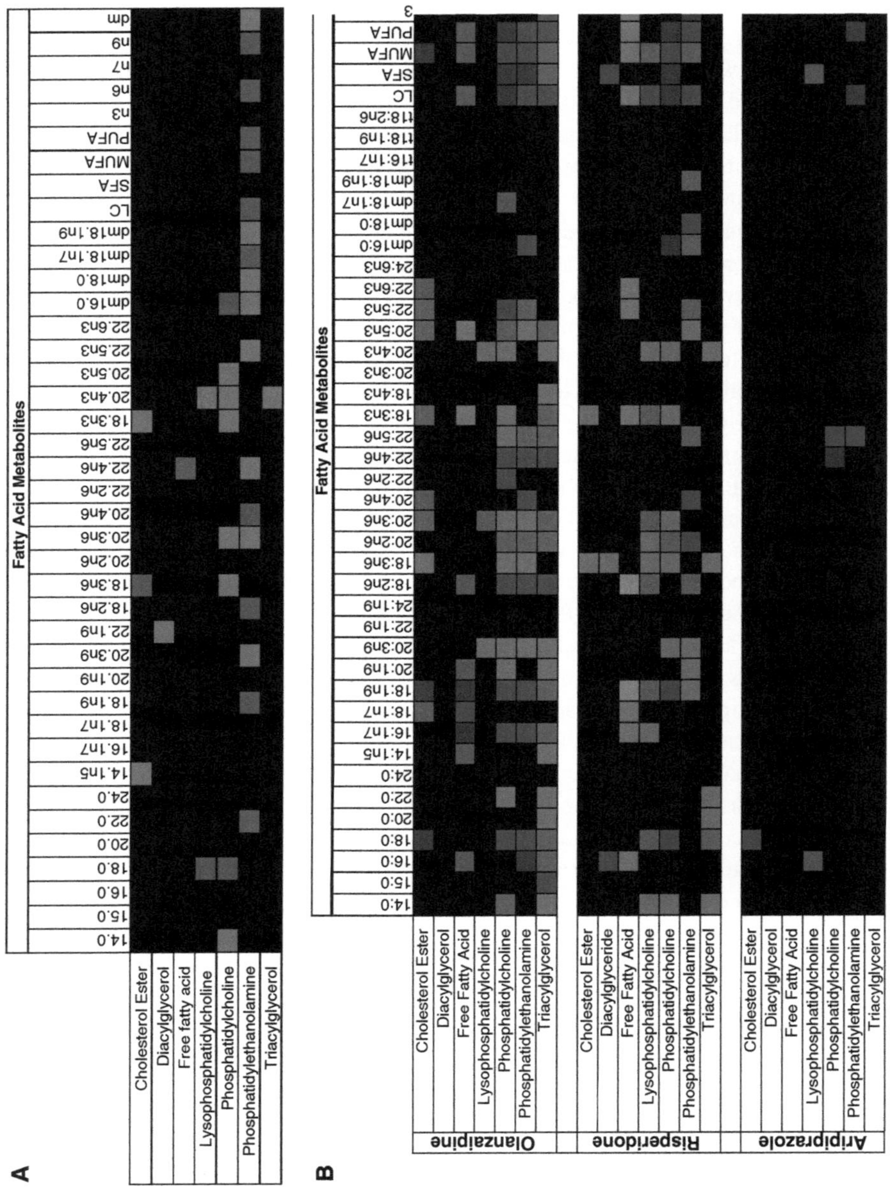
A
Fatty Acid Metabolites
14.0 15.0 16.0 18.0 20.0 22.0 24.0 14.1n5 16.1n7 18.1n7 18.1n9 20.1n9 20.3n9 22.1n9 18.2n6 18.3n6 20.2n6 20.3n6 20.4n6 22.2n6 22.4n6 22.5n6 18.3n3 20.4n3 20.5n3 22.5n3 22.6n3 dm16.0 dm18.0 dm18.1n7 dm18.1n9 LC SFA MUFA PUFA n3 n6 n7 n9 dm
Cholesterol Ester
Diacylglycerol
Free fatty acid
Lysophosphatidylcholine
Phosphatidylcholine
Phosphatidylethanolamine
Triacylglycerol
B
Fatty Acid Metabolites
14:0 15:0 16:0 18:0 20:0 22:0 24:0 14:1n5 16:1n7 18:1n7 18:1n9 20:1n9 20:3n9 22:1n9 24:1n9 18:2n6 18:3n6 20:2n6 20:3n6 20:4n6 22:2n6 22:4n6 22:5n6 18:3n3 18:4n3 20:3n3 20:4n3 20:5n3 22:5n3 22:6n3 24:6n3 dm16:0 dm18:0 dm18:1n7 dm18:1n9 t16:1n7 t18:1n9 t18:2n6 LC SFA MUFA PUFA 3
Olanzapine
Cholesterol Ester
Diacylglycerol
Free Fatty Acid
Lysophosphatidylcholine
Phosphatidylcholine
Phosphatidylethanolamine
Triacylglycerol
Risperidone
Cholesterol Ester
Diacylglyceride
Free Fatty Acid
Lysophosphatidylcholine
Phosphatidylcholine
Phosphatidylethanolamine
Triacylglycerol
Aripiprazole
Cholesterol Ester
Diacylglycerol
Free Fatty Acid
Lysophosphatidylcholine
Phosphatidylcholine
Phosphatidylethanolamine
Triacylglycerol

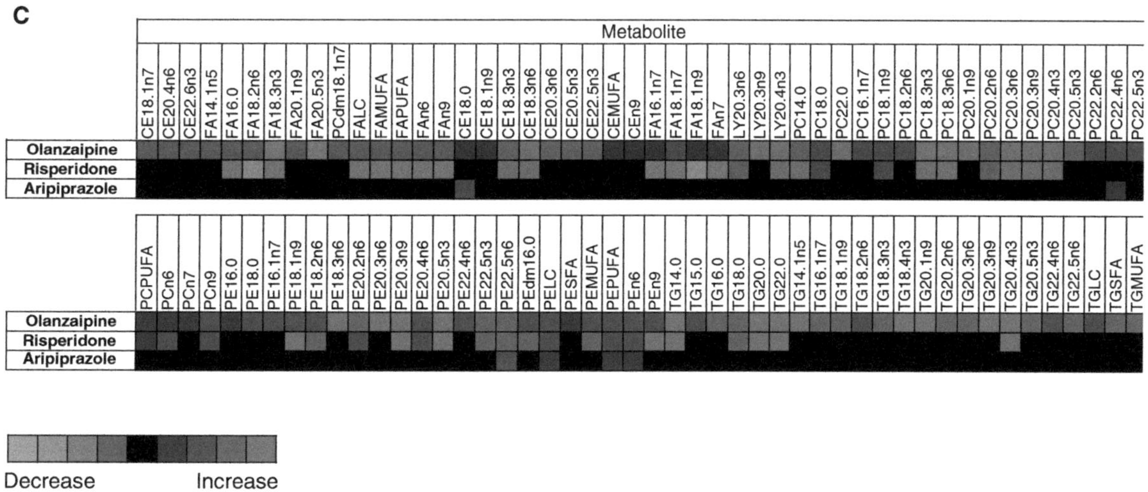

Fig. 6.4 Heat map showing differences in individual lipid metabolites in the plasma of patients with schizophrenia as compared to controls (**a**), in the plasma of schizophrenic patients post treatment as compared to pretreatment (**b**) with olanzapine, risperidone, and aripiprazole. Figure C shows the most significantly modified lipid metabolites in plasma of patients treated with olanzapine and highlights which of these metabolites are also modified upon treatment with risperidone or aripiprazole. The column headers indicate fatty acid metabolites as they appear in each distinct lipid class (rows). In (**a**), lipids whose percent levels were higher in patients vs. controls are shown in red while those with decreased level are shown in green; see methods for details. In (**b**) and (**c**), the percent increase in any lipid upon treatment with drug is shown in red squares and decrease in green squares as described in the methods. The brightness of each color corresponds to the magnitude of the difference in quartiles. The brighter the square is the larger the difference (*See Color Plates*)

with clinical outcomes might yield biomarkers related to response and development of side effects. This study of atypical antipsychotic drugs illustrates the way in which metabolomics might contribute to our understanding of drug response phenotypes and how it provides tools to analyze pathways implicated in variation to response for this class of drugs. Additionally it illustrates how metabolomics could be a very valuable tool for biomarker discovery.

Three additional metabolomics studies in schizophrenia can be highlighted using (1)H NMR spectroscopy-based metabolomic analysis of plasma samples from 21 pairs of monozygotic twins discordant for schizophrenia and eight pairs of age-matched healthy twins demonstrated alterations in lipid profiles of both affected and unaffected schizophrenia twins (Tsang et al., 2006). In another study (1)H NMR profiling of CSF samples from drug-naïve patients with first-onset of schizophrenia, suggested alterations in glucose regulation, an abnormality that seems to get corrected by early treatment with antipsychotics (Holmes et al., 2006). Finally, an interesting metabolomic study on post-mortem tissue provides support to the notion that abnormalities at the level of glutamatergic neurotransmission and myelin synthesis play an important role in schizophrenia (Tkachev et al., 2007).

6.5 Conclusions

There is a growing need for the identification of PDBs to guide clinical decision-making in psychiatry. The discovery of biomarkers relies on sound study design, as well as new technology, to capture disease underpinnings at a more global level. Metabolomics, the study of the complete repertoire of small molecules in cells, tissues, organs, and biological fluids, represents a major and rapidly evolving component of new biology. It is clear that there are numerous metabolic perturbations in psychiatric disorders. Yet the measurement of single metabolites has not enabled the identification of reliable biomarkers. The application of metabolomic technologies to the study of psychiatric disorders will enable simultaneous measurement of many metabolites in key interacting pathways. From these studies numerous new biomarkers will emerge. The development of a series of analytical platforms, NMR, GC-MS, LC-MS, and LCECA, all capable of accurately measuring hundreds or thousands of small molecules in biological samples, promises to substantially advance our understanding of disease pathophysiology making possible the discovery of biomarkers for multiple disorders. An evaluation of global changes in metabolites in the periphery (blood) should be conducted concurrently with analysis of CSF samples to establish how closely peripheral abnormalities are related to central changes.

In-depth knowledge of metabolic perturbations linking psychiatric disorders, multiple biochemical pathways, and treatment effect-response, should provide valuable insights into disease pathophysiology and could provide a novel approach for therapeutic monitoring and outcome. Metabolomics promises to have broad applications in basic research and medical practice that include: (a) information about the mechanisms of disease and pathways involved; (b) development of prognostic, diagnostic and surrogate markers of a disease state; (c) novel ability to sub-classify diseases; (d) better design of clinical trials based on sub-classification of patients and early monitoring of drug effects in each patient (pharmaco-metabolomics); (e) early mapping of the beneficial and side effects of drugs. There can be little doubt that the addition of metabolomic analyses to genomic, transcriptomic and proteomic assays will greatly enhance our understanding of mechanisms underlying the pathogenesis of psychiatric illnesses and drug effects. Collectively, these technologies offer great promise for the identification of clinically relevant PDBs.

Disclosures

Dr. Kaddurah-Daouk is equity holder in Metabolon Inc., a biotechnology company in the metabolomics domain, and she also holds IP interest in this field. The other authors do not have anything to disclose.

Acknowledgments Supported in part by National Institutes of Health grants R24 GM078233, "The Metabolomics Research Network" (R.K.D., B.S.K., R.M.W.); SMRI (R.K.-D.), NARSAD (R.K.-D.), Stanley Medical Research Institute (M.P.Q. and J.C.S.); MH 68766 (J.C.S.), MH 69774 (J.C.S.), MH 068662 (J.C.S.), RR 20571 (J.C.S.), UTHSCSA's GCRC (M01-RR-01346) (J.C.S.) and Department of Psychiatry Friends of Psychiatry Grant (M.P.Q. and J.C.S.), NARSAD (J.C.S.), Veterans Administration (Merit Review) (J.C.S.), and the Krus Endowed Chair in Psychiatry (J.C.S.).

References

Appleton KM, Hayward RC, Gunnell D, et al.: Effects of n-3 long-chain polyunsaturated fatty acids on depressed mood: systematic review of published trials. Am J Clin Nutr 84: 1308–1316, 2006.

Audenaert K, Peremans K, Goethals I, et al.: Functional imaging, serotonin and the suicidal brain. Acta Neurol Belg 106:125–131, 2006.

Bakhtiar R: Biomarkers in drug discovery and development. J Pharmacol Toxicol Meth 57:85–91, 2008.

Bell C, Abrams J, Nutt D: Tryptophan depletion and its implications for psychiatry. Br J Psychiatr 178:399–405, 2001.

Berger GE, Wood SJ, Pantelis C, et al.: Implications of lipid biology for the pathogenesis of schizophrenia. Aust New Zeal J Psychiatr 36:355–366, 2002.

Berrettini WH, Nurnberger JI, Jr., Scheinin M, et al.: Cerebrospinal fluid and plasma monoamines and their metabolites in euthymic bipolar patients. Biol Psychiatr 20:257–269, 1985.

Bhagwagar Z, Wylezinska M, Jezzard P, et al.: Low GABA concentrations in occipital cortex and anterior cingulate cortex in medication-free, recovered depressed patients. Int J Neuropsychopharmacol 1–6, 2007.

Biomarkers and surrogate endpoints: preferred definitions and conceptual framework. Clin Pharmacol Ther 69:89–95, 2001.

Birkenhager TK, van den Broek WW, Fekkes D, et al.: Lithium addition in antidepressant-resistant depression: effects on platelet 5-HT, plasma 5-HT and plasma 5-HIAA concentration. Prog Neuropsychopharmacol Biol Psychiatr 31:1084–1088, 2007.

Bogdanov M, Matson WR, Wang L, Matson T, Saunders-Pullman R, Bressman SS, Flint Beal M. Metabolomic profiling to develop blood biomarkers for Parkinson's disease. Brain 131389–96, 2008.

Bowers MB, Jr., Mazure CM, Nelson JC, et al.: Lithium in combination with perphenazine: effect on plasma monoamine metabolites. Biol Psychiatr 32:1102–1107, 1992.

Brambilla P, Perez J, Barale F, et al.: GABAergic dysfunction in mood disorders. Mol Psychiatr 8:721–737, 715, 2003.

Breier A: Serotonin, schizophrenia and antipsychotic drug action. Schizophr Res 14:187–202, 1995.

Brindle JT, Antti H, Holmes E, et al.: Rapid and noninvasive diagnosis of the presence and severity of coronary heart disease using 1H-NMR-based metabonomics. Nat Med 8:1439–1444, 2002.

Brindle JT, Nicholson JK, Schofield PM, et al.: Application of chemometrics to 1H NMR spectroscopic data to investigate a relationship between human serum metabolic profiles and hypertension. Analyst 128:32–36, 2003.

Brown AS, Gewirtz G, Harkavy-Friedman J, et al.: Effects of clozapine on plasma catecholamines and relation to treatment response in schizophrenia: a within-subject comparison with haloperidol. Neuropsychopharmacology 17:317–325, 1997.

Casey DE: Dyslipidemia and atypical antipsychotic drugs. J Clin Psychiatr 65(Suppl. 18):27–35, 2004.

Chou JC, Czobor P, Tuma I, et al.: Pretreatment plasma HVA and haloperidol response in acute mania. J Affect Disord 59:55–59, 2000.

Coen M, Ruepp SU, Lindon JC, et al.: Integrated application of transcriptomics and metabonomics yields new insight into the toxicity due to paracetamol in the mouse. J Pharm Biomed Anal 35:93–105, 2004.

Coppen A: Electrolytes and mental illness. Proc Annu Meet Am Psychopathol Assoc 58:397–409, 1969.

Correll CU, Frederickson AM, Kane JM, et al.: Does antipsychotic polypharmacy increase the risk for metabolic syndrome? Schizophr Res 89:91–100, 2007.

Crocker IP, Kenny LC, Thornton WA, et al.: Excessive stimulation of poly(ADP-ribosyl)ation contributes to endothelial dysfunction in pre-eclampsia. Br J Pharmacol 144:772–780, 2005.

Curzon G: Relationships between plasma, CSF and brain tryptophan. J Neural Transm (Suppl.): 81–92, 1979.

Curzon G: Influence of plasma tryptophan on brain 5HT synthesis and serotonergic activity. Adv Exp Med Biol 133:207–219, 1981.

Dauner M, Bailey JE, U. S: Metabolic flux analysis with a comprehensive isotopomer model in Bacillus subtilis. Biotechnol Bioeng 76:1440156, 2001.

Davila R, Zumarraga M, Basterreche N, et al.: Influence of the catechol-O-methyltransferase Val108/158Met polymorphism on the plasma concentration of catecholamine metabolites and on clinical features in type I bipolar disorder–a preliminary report. J Affect Disord 92:277–281, 2006.

de Graaf AA, Mahle M, Möllney M, et al.: Determination of full 13C isotopomer distributions for metabolic flux analysis using heteronuclear spin echo difference NMR spectroscopy. J Biotechnol 77:25035, 2000.

De Milito A, Titanji K, Zazzi M: Surrogate markers as a guide to evaluate response to antiretroviral therapy. Curr Med Chem 10:349–365, 2003.

Deeks JJ, Altman DG: Diagnostic tests 4: likelihood ratios. BMJ 329:168–169, 2004.

Delgado PL, Moreno FA: Role of norepinephrine in depression. J Clin Psychiatr 61(Suppl. 1): 5–12, 2000.

Delgado PL: How antidepressants help depression: mechanisms of action and clinical response. J Clin Psychiatr 65(Suppl. 4):25–30, 2004.

Delgado PL: Monoamine depletion studies: implications for antidepressant discontinuation syndrome. J Clin Psychiatr 67 (Suppl. 4):22–26, 2006.

Denkert C, Budczies J, Kind T, et al.: Mass spectrometry-based metabolic profiling reveals different metabolite patterns in invasive ovarian carcinomas and ovarian borderline tumors. Cancer Res 66:10795–10804, 2006.

Dunne VG, Bhattachayya S, Besser M, et al.: Metabolites from cerebrospinal fluid in aneurysmal subarachnoid haemorrhage correlate with vasospasm and clinical outcome: a pattern-recognition 1H NMR study. NMR Biomed 18:24–33, 2005.

Ebuehi OA, Bishop SA, Fanmuyiwa OO, et al.: Biogenic amines metabolism and blood chemistry of psychiatric patients. Afr J Med Med Sci 30:269–273, 2001.

Ellis PM, Mellsop GW, Beeston R, et al.: Platelet tritiated imipramine binding in patients suffering from mania. J Affect Disord 22:105–110, 1991.

Ellison G: Stimulant-induced psychosis, the dopamine theory of schizophrenia, and the habenula. Brain Res Brain Res Rev 19:223–239, 1994..

Fan C, Oh DS, Wessels L, et al.: Concordance among gene-expression-based predictors for breast cancer. N Engl J Med 355:560–569, 2006.

Fan TW-M, Lane AN: Structure-based profiling of metabolites and isotopomers by NMR. Prog NMR Spectrosc, in press.

Fan X, Bai J, Shen P: Diagnosis of breast cancer using HPLC metabonomics fingerprints coupled with computational methods. Conf Proc IEEE Eng Med Biol Soc 6:6081–6084, 2005.

Fortunati F, Mazure C, Preda A, et al.: Plasma catecholamine metabolites in antidepressant-exacerbated mania and psychosis. J Affect Disord 68:331–334, 2002.

Frank R, Hargreaves R: Clinical biomarkers in drug discovery and development. Nat Rev Drug Discov 2:566–580, 2003.

Freeman MP, Hibbeln JR, Wisner KL, et al.: Omega-3 fatty acids: evidence basis for treatment and future research in psychiatry. J Clin Psychiatr 67:1954–1967, 2006.

Frye MA, Tsai GE, Huggins T, et al.: Low cerebrospinal fluid glutamate and glycine in refractory affective disorder. Biol Psychiatr 61:162–166, 2007.

Garcia-Portilla MP, Saiz PA, Benabarre A, et al.: The prevalence of metabolic syndrome in patients with bipolar disorder. J Affect Disord 106:197–201, 2008.

Garfinkel PE, Warsh JJ, Stancer HC: Depression: new evidence in support of biological differentiation. Am J Psychiatr 136:535–539, 1979.

German JB, Gillies LA, Smilowitz JT, et al.: Lipidomics and lipid profiling in metabolomics. Curr Opin Lipidol 18:66–71, 2007.

Gerner RH, Fairbanks L, Anderson GM, et al.: CSF neurochemistry in depressed, manic, and schizophrenic patients compared with that of normal controls. Am J Psychiatr 141: 1533–1540, 1984.

Goodwin FK, Sack RL: Central dopamine function in affective illness: evidence from precursors, enzyme inhibitors, and studies of central dopamine turnover. Adv Biochem Psychopharmacol 12:261–279, 1974.

Goodwin FK, Jamison KR, Ghaemi SN: Manic-depressive illness: bipolar disorders and recurrent depression, 2nd Edition. New York: Oxford University Press, 2007.

Greenspan K, Schildkraut JJ, Gordon EK, et al.: Catecholamine metabolism in affective disorders. 3. MHPG and other catecholamine metabolites in patients treated with lithium carbonate. J Psychiatr Res 7:171–183, 1970.

Grossman F, Potter WZ: Catecholamines in depression: a cumulative study of urinary norepinephrine and its major metabolites in unipolar and bipolar depressed patients versus healthy volunteers at the NIMH. Psychiatr Res 87:21–27, 1999.

Hall JA, Brown R, Paul J: An exploration into study design for biomarker identification: issues and recommendations. Cancer Genomics Proteomics 4:111–119, 2007.

Han X, Holtzman DM, McKeel DW, Jr., et al.: Substantial sulfatide deficiency and ceramide elevation in very early Alzheimer's disease: potential role in disease pathogenesis. J Neurochem 82:809–818, 2002.

Harrigan G, Goodacre R: Metabolic profiling: its role in biomarker discovery and gene function analysis. Boston: Kluwer, 2003.

Hasler G, van der Veen JW, Tumonis T, et al.: Reduced prefrontal glutamate/glutamine and gamma-aminobutyric acid levels in major depression determined using proton magnetic resonance spectroscopy. Arch Gen Psychiatr 64:193–200, 2007.

Heritch AJ: Evidence for reduced and dysregulated turnover of dopamine in schizophrenia. Schizophr Bull 16:605–615, 1990.

Hoekstra R, Fekkes D, Loonen AJ, et al.: Bipolar mania and plasma amino acids: increased levels of glycine. Eur Neuropsychopharmacol 16:71–77, 2006.

Holmes E, Tsang TM, Huang JT, et al.: Metabolic profiling of CSF: evidence that early intervention may impact on disease progression and outcome in schizophrenia. PLoS Med 3:e327, 2006.
Horrobin DF: The roles of prostaglandins and prolactin in depression, mania and schizophrenia. Postgrad Med J 53 (Suppl. 4):160–165, 1977.
Horrobin DF, Manku MS, Hillman H, et al.: Fatty acid levels in the brains of schizophrenics and normal controls. Biol Psychiatr 30:795–805, 1991.
Horrobin DF: Schizophrenia as a membrane lipid disorder which is expressed throughout the body. Prostaglandins Leukot Essent Fatty Acids 55:3–7, 1996.
Horrobin DF: The membrane phospholipid hypothesis as a biochemical basis for the neurodevelopmental concept of schizophrenia. Schizophr Res 30:193–208, 1998.
Janowsky DS, el-Yousef MK, Davis JM, et al.: Cholinergic reversal of manic symptoms. Lancet 1:1236–1237, 1972.
Janowsky DS, el-Yousef MK, Davis JM, et al.: A cholinergic-adrenergic hypothesis of mania and depression. Lancet 2:632–635, 1972.
Joyce PR, Fergusson DM, Woollard G, et al.: Urinary catecholamines and plasma hormones predict mood state in rapid cycling bipolar affective disorder. J Affect Disord 33:233–243, 1995.
Kaddurah-Daouk R: Metabolic profiling of patients with schizophrenia. PLoS Med 3:e363, 2006.
Kaddurah-Daouk R, McEvoy J, Baillie RA, et al.: Metabolomic mapping of atypical antipsychotic effects in schizophrenia. Mol Psychiatr 12:934–945, 2007.
Kaddurah-Daouk R, Kristal BS, Weinshilboum RM: Metabolomics: A global biochemical approach to drug response and disease. Annu Rev Pharmacol Toxicol 48:653–683, 2008.
Kasa K, Otsuki S, Yamamoto M, et al.: Cerebrospinal fluid gamma-aminobutyric acid and homovanillic acid in depressive disorders. Biol Psychiatr 17:877–883, 1982.
Kell DB: Metabolomics and systems biology: making sense of the soup. Curr Opin Micro 7:296–307, 2004.
Kelley ME, Yao JK, van Kammen DP: Plasma catecholamine metabolites as markers for psychosis and antipsychotic response in schizophrenia. Neuropsychopharmacology 20:603–611, 1999.
Kessler RC, Berglund P, Demler O, et al.: The epidemiology of major depressive disorder: results from the National Comorbidity Survey Replication (NCS-R). JAMA 289:3095–3105, 2003.
Khan MM, Evans DR, Gunna V, et al.: Reduced erythrocyte membrane essential fatty acids and increased lipid peroxides in schizophrenia at the never-medicated first-episode of psychosis and after years of treatment with antipsychotics. Schizophr Res 58:1–10, 2002.
Knable MB, Weinberger DR: Dopamine, the prefrontal cortex and schizophrenia. J Psychopharmacol 11:123–131, 1997.
Koslow SH, Maas JW, Bowden CL, et al.: CSF and urinary biogenic amines and metabolites in depression and mania. A controlled, univariate analysis. Arch Gen Psychiatr 40:999–1010, 1983.
Kristal BS, Kaddurah-Daouk R, Beal MF, et al.: Metabolomics: concept and potential neuroscience application, in Handbook of Neurochemistry and Molecular Neurobiology: Brain Energetics. Integration of Molecular and Cellular Processes. Edited by. Berlin Heidelberg New York, Springer, 2007, pp. 889–912.
Kristal BS, Shurubor YI, Kaddurah-Daouk R, et al.: Metabolomics in the study of aging and caloric restriction. Methods Mol Biol 371:393–409, 2007.
Kugaya A, Sanacora G: Beyond monoamines: glutamatergic function in mood disorders. CNS Spectrum 10:808–819, 2005.
Lake CR, Pickar D, Ziegler MG, et al.: High plasma norepinephrine levels in patients with major affective disorder. Am J Psychiatr 139:1315–1318, 1982.
Lassere MN, Johnson KR, Boers M, et al.: Definitions and validation criteria for biomarkers and surrogate endpoints: development and testing of a quantitative hierarchical levels of evidence schema. J Rheumatol 34:607–615, 2007.
Leonard BE: Evidence for a biochemical lesion in depression. J Clin Psychiatr 61 (Suppl. 6):12–17, 2000.
Leoni V, Masterman T, Patel P, et al.: Side chain oxidized oxysterols in cerebrospinal fluid and the integrity of blood-brain and blood-cerebrospinal fluid barriers. J Lipid Res 44:793–799, 2003.

Levine J, Panchalingam K, Rapoport A, et al.: Increased cerebrospinal fluid glutamine levels in depressed patients. Biol Psychiatr 47:586–593, 2000.
Levine J, Sela BA, Osher Y, et al.: High homocysteine serum levels in young male schizophrenia and bipolar patients and in an animal model. Prog Neuropsychopharmacol Biol Psychiatr 29:1181–1191, 2005.
Lewine RR, Risch SC, Risby E, et al.: Lateral ventricle-brain ratio and balance between CSF HVA and 5-HIAA in schizophrenia. Am J Psychiatr 148:1189–1194, 1991.
Lieb J, Karmali R, Horrobin D: Elevated levels of prostaglandin E2 and thromboxane B2 in depression. Prostaglandins Leukot Med 10:361–367, 1983.
Lieberman JA, Mailman RB, Duncan G, et al.: Serotonergic basis of antipsychotic drug effects in schizophrenia. Biol Psychiatr 44:1099–1117, 1998.
Lin PY, Su KP: A meta-analytic review of double-blind, placebo-controlled trials of antidepressant efficacy of omega-3 fatty acids. J Clin Psychiatr 68:1056–1061, 2007.
Lindon JC, Nicholson JK, Holmes E, et al.: Metabonomics: metabolic processes studied by NMR spectroscopy of biofluids. Concepts Magn Reson 12:289–320, 2000.
Lindon JC, Nicholson JK, Holmes E, et al.: Contemporary issues in toxicology the role of metabonomics in toxicology and its evaluation by the COMET project. Toxicol Appl Pharmacol 187:137–146, 2003.
Lindon JC, Holmes E, Nicholson JK: Metabonomics in pharmaceutical R&D. FEBS J 274:1149–1151, 2007.
Lucini V, Lucca A, Catalano M, et al.: Predictive value of tryptophan/large neutral amino acids ratio to antidepressant response. J Affect Disord 36:129–133, 1996.
Maas JW, Dekirmenjian H, Fawcett JA: MHPG excretion by patients with affective disorders. Int Pharmacopsychiatr 9:14–26, 1974.
Maeng S, Zarate CA, Jr., Du J, et al.: Cellular mechanisms underlying the antidepressant effects of ketamine: role of alpha-amino-3-hydroxy-5-methylisoxazole-4-propionic Acid Receptors. Biol Psychiatr 63:349–352, 2008.
Maes M, Scharpe S, Meltzer HY, et al.: Increased neopterin and interferon-gamma secretion and lower availability of L-tryptophan in major depression: further evidence for an immune response. Psychiatr Res 54:143–160, 1994.
Mahadik SP, Mukherjee S, Correnti EE, et al.: Plasma membrane phospholipid and cholesterol distribution of skin fibroblasts from drug-naive patients at the onset of psychosis. Schizophr Res 13:239–247, 1994.
Mahmood T, Silverstone T: Serotonin and bipolar disorder. J Affect Disord 66:1–11, 2001.
Maier B, Laurer HL, Rose S, et al.: Physiological levels of pro- and anti-inflammatory mediators in cerebrospinal fluid and plasma: a normative study. J Neurotrauma 22:822–835, 2005.
Maj M, Ariano MG, Arena F, et al.: Plasma cortisol, catecholamine and cyclic AMP levels, response to dexamethasone suppression test and platelet MAO activity in manic-depressive patients. A longitudinal study. Neuropsychobiology 11:168–173, 1984.
Makatsori A, Duncko R, Moncek F, et al.: Modulation of neuroendocrine response and non-verbal behavior during psychosocial stress in healthy volunteers by the glutamate release-inhibiting drug lamotrigine.
Makatsori A, Duncko R, Moncek F, et al.: Modulation of neuroendocrine response and non-verbal behavior during psychosocial stress in healthy volunteers by the glutamate release-inhibiting drug lamotrigine. Neuroendocrinology 79:34–42, 2004.
Mann JJ, Oquendo M, Underwood MD, et al.: The neurobiology of suicide risk: a review for the clinician. J Clin Psychiatr 60 (Suppl. 2):7–11; discussion 18–20, 113–116, 1999.
Mazure CM, Bowers MB: Pretreatment plasma HVA predicts neuroleptic response in manic psychosis. J Affect Disord 48:83–86, 1998.
Meisenzahl EM, Schmitt GJ, Scheuerecker J, et al.: The role of dopamine for the pathophysiology of schizophrenia. Int Rev Psychiatr 19:337–345, 2007.
Miyamoto S, LaMantia AS, Duncan GE, et al.: Recent advances in the neurobiology of schizophrenia. Mol Interv 3:27–39, 2003.

Mooney JJ, Schatzberg AF, Cole JO, et al.: Urinary 3-methoxy-4-hydroxyphenylglycol and the depression-type score as predictors of differential responses to antidepressants. J Clin Psychopharmacol 11:339–343, 1991.
Morvan D, Demidem A: Metabolomics by proton nuclear magnetic resonance spectroscopy of the response to chloroethylnitrosourea reveals drug efficacy and tumor adaptive metabolic pathways. Cancer Res 67:2150–2159, 2007.
Mueller PS, Davis JM, Bunney WE, Jr., et al.: Plasma free fatty acids concentration in depressive illness. Arch Gen Psychiatr 22:216–221, 1970.
Muller N, Riedel M, Schwarz MJ: Psychotropic effects of COX-2 inhibitors—a possible new approach for the treatment of psychiatric disorders. Pharmacopsychiatry 37:266–269, 2004.
Muller N, Schwarz MJ, Dehning S, et al.: The cyclooxygenase-2 inhibitor celecoxib has therapeutic effects in major depression: results of a double-blind, randomized, placebo controlled, add-on pilot study to reboxetine. Mol Psychiatr 11:680–684, 2006.
Murakami M, Kudo I: Phospholipase A2. J Biochem (Tokyo) 131:285–292, 2002.
Nanda BL, Nataraju A, Rajesh R, et al.: PLA2 mediated arachidonate free radicals: PLA2 inhibition and neutralization of free radicals by anti-oxidants–a new role as anti-inflammatory molecule. Curr Top Med Chem 7:765–777, 2007.
Nicholson JK: Global systems biology, personalized medicine and molecular epidemiology. Mol Syst Biol 2:52, 2006.
Obeid R, Kostopoulos P, Knapp JP, et al.: Biomarkers of folate and vitamin B12 are related in blood and cerebrospinal fluid. Clin Chem 53:326–333, 2007.
Odunsi K, Wollman RM, Ambrosone CB, et al.: Detection of epithelial ovarian cancer using 1H-NMR-based metabonomics. Int J Cancer 113:782–788, 2005.
Paige LA, Mitchell MW, Krishnan KR, et al.: A preliminary metabolomic analysis of older adults with and without depression. Int J Geriatr Psychiatr 22:418–423, 2007.
Palomino A, Gonzalez-Pinto A, Aldama A, et al.: Decreased levels of plasma glutamate in patients with first-episode schizophrenia and bipolar disorder. Schizophr Res 95:174–178, 2007.
Pani L, Pira L, Marchese G: Antipsychotic efficacy: relationship to optimal D2-receptor occupancy. Eur Psychiatr 22:267–275, 2007.
Patterson AD, Li H, Eichler GS, et al.: UPLC-ESI-TOFMS-based metabolomics and gene expression dynamics inspector self-organizing metabolomic maps as tools for understanding the cellular response to ionizing radiation. Anal Chem 80:665–674, 2008.
Peet M, Moody JP, Worrall EP, et al.: Plasma tryptophan concentration in depressive illness and mania. Br J Psychiatr 128:255–258, 1976.
Petty F, Schlesser MA: Plasma GABA in affective illness. A preliminary investigation. J Affect Disord 3:339–343, 1981.
Petty F, Rush AJ, Davis JM, et al.: Plasma GABA predicts acute response to divalproex in mania. Biol Psychiatr 39:278–284, 1996.
Piccirillo G, Fimognari FL, Infantino V, et al.: High plasma concentrations of cortisol and thromboxane B2 in patients with depression. Am J Med Sci 307:228–232, 1994.
Portilla D, Li S, Nagothu KK, et al.: Metabolomic study of cisplatin-induced nephrotoxicity. Kidney Int 69:2194–2204, 2006.
Rapoport SI, Bosetti F: Do lithium and anticonvulsants target the brain arachidonic acid cascade in bipolar disorder? Arch Gen Psychiatr 59:592–596, 2002.
Reddy R, Keshavan M, Yao JK: Reduced plasma antioxidants in first-episode patients with schizophrenia. Schizophr Res 62:205–212, 2003.
Ross BM, Seguin J, Sieswerda LE: Omega-3 fatty acids as treatments for mental illness: which disorder and which fatty acid? Lipids Health Dis 6:21, 2007.
Roy A, Guthrie S, Pickar D, et al.: Plasma norepinephrine responses to cold challenge in depressed patients and normal controls. Psychiatr Res 21:161–168, 1987.
Roy A: Plasma HVA levels in depressed patients and controls. J Affect Disord 14:293–296, 1988.

Rozen S, Cudkowicz ME, Bogdanov M, et al.: Metabolomic analysis and signatures in motor neuron disease. Metabolomics 1:101–108, 2005.
Sabatine MS, Liu E, Morrow DA, et al.: Metabolomic identification of novel biomarkers of myocardial ischemia. Circulation 112:3868–3875, 2005.
Sajda P: Machine learning for detection and diagnosis of disease. Annu Rev Biomed Eng 8:537–565, 2006.
Sanacora G, Saricicek A: GABAergic contributions to the pathophysiology of depression and the mechanism of antidepressant action. CNS Neurol Disord Drug Targets 6:127–140, 2007.
Schatzberg AF, Orsulak PJ, Rosenbaum AH, et al.: Catecholamine measures for diagnosis and treatment of patients with depressive disorders. J Clin Psychiatr 41:35–39, 1980.
Schatzberg AF, Orsulak PJ, Rosenbaum AH, et al.: Toward a biochemical classification of depressive disorders, V: heterogeneity of unipolar depressions. Am J Psychiatr 139:471–475, 1982.
Schildkraut JJ: Biogenic amines and affective disorders. Annu Rev Med 25:333–348, 1974.
Schmitt A, Maras A, Petroianu G, et al.: Effects of antipsychotic treatment on membrane phospholipid metabolism in schizophrenia. J Neural Transm 108:1081–1091, 2001.
Segal M, Avital A, Drobot M, et al.: Serum creatine kinase level in unmedicated nonpsychotic, psychotic, bipolar and schizoaffective depressed patients. Eur Neuropsychopharmacol 17:194–198, 2007a.
Segal M, Avital A, Drobot M, et al.: CK levels in unmedicated bipolar patients. Eur Neuropsychopharmacol 17:763–767, 2007b.
Sher L, Carballo JJ, Grunebaum MF, et al.: A prospective study of the association of cerebrospinal fluid monoamine metabolite levels with lethality of suicide attempts in patients with bipolar disorder. Bipolar Disord 8:543–550, 2006.
Shin H, Markey MK: A machine learning perspective on the development of clinical decision support systems utilizing mass spectra of blood samples. J Biomed Inform 39:227–248, 2006.
Sicras-Mainar A, Blanca-Tamayo M, Rejas-Gutierrez J, et al.: Metabolic syndrome in outpatients receiving antipsychotic therapy in routine clinical practice: a cross-sectional assessment of a primary health care database. Eur Psychiatr 23:100–108, 2008.
Skosnik PD, Yao JK: From membrane phospholipid defects to altered neurotransmission: is arachidonic acid a nexus in the pathophysiology of schizophrenia? Prostaglandins Leukot Essent Fatty Acids 69:367–384, 2003.
Sobczak S, Honig A, van Duinen MA, et al.: Serotonergic dysregulation in bipolar disorders: a literature review of serotonergic challenge studies. Bipolar Disord 4:347–356, 2002.
Steuer R, Morgenthal K, Weckwerth W, et al.: A gentle guide to the analysis of metabolomic data. Methods Mol Biol 358:105–126, 2007.
Sublette ME, Russ MJ, Smith GS: Evidence for a role of the arachidonic acid cascade in affective disorders: a review. Bipolar Disord 6:95–105, 2004.
Swann AC, Secunda S, Davis JM, et al.: CSF monoamine metabolites in mania. Am J Psychiatr 140:396–400, 1983.
Swann AC, Petty F, Bowden CL, et al.: Mania: gender, transmitter function, and response to treatment. Psychiatr Res 88:55–61, 1999.
Tandon R, Channabasavanna SM, Greden JF: CSF biochemical correlates of mixed affective states. Acta Psychiatr Scand 78:289–297, 1988.
Taylor V, MacQueen G: Associations between bipolar disorder and metabolic syndrome: a review. J Clin Psychiatr 67:1034–1041, 2006.
Tkachev D, Mimmack ML, Huffaker SJ, et al.: Further evidence for altered myelin biosynthesis and glutamatergic dysfunction in schizophrenia. Int J Neuropsychopharmacol 10:557–563, 2007.
Tsang TM, Huang JT, Holmes E, et al.: Metabolic profiling of plasma from discordant schizophrenia twins: correlation between lipid signals and global functioning in female schizophrenia patients. J Proteome Res 5:756–760, 2006.
Underwood B, Broadhurst D, Dunn WB, et al.: Huntington disease patients and transgenic mice have similar pro-catabolic serum metabolite profiles. Brain 129:877–886, 2006.

van der Greef J, Martin S, Juhasz P, et al.: The art and practice of systems biology in medicine: mapping patterns of relationships. J Proteome Res 6:1540–1559, 2007.
van Doorn M, Vogels J, Tas A, et al.: Evaluation of metabolite profiles as biomarkers for the pharmacological effects of thiazolidinediones in Type 2 diabetes mellitus patients and healthy volunteers. Br J Clin Pharmacol 63:562–574, 2007.
Vasan RS: Biomarkers of cardiovascular disease: molecular basis and practical considerations. Circulation 113:2335–2362, 2006.
Wang C, Kong H, Guan Y, et al.: Plasma phospholipid metabolic profiling and biomarkers of type 2 diabetes mellitus based on high-performance liquid chromatography/electrospray mass spectrometry and multivariate statistical analysis. Anal Chem 77:4108–4116, 2005.
Watkins SM, Reifsnyder PR, Pan HJ, et al.: Lipid metabolome-wide effects of the PPARgamma agonist rosiglitazone. J Lipid Res 43:1809–1811, 2002.
Watkins SM: Lipomic profiling in drug discovery, development and clinical trial evaluation. Curr Opin Drug Discov Devel 7:112–117, 2004.
Watson AD: Thematic review series: systems biology approaches to metabolic and cardiovascular disorders. Lipidomics: a global approach to lipid analysis in biological systems. J Lipid Res 47:2101–2111, 2006.
Weljie AM, Dowlatabadi R, Miller BJ, et al.: An inflammatory arthritis-associated metabolite biomarker pattern revealed by 1H NMR spectroscopy. J Proteome Res 6:3456–3464, 2007.
Wichers MC, Koek GH, Robaeys G, et al.: IDO and interferon-alpha-induced depressive symptoms: a shift in hypothesis from tryptophan depletion to neurotoxicity. Mol Psychiatr 10:538–544, 2005.
Wiest MM, Watkins SM: Biomarker discovery using high-dimensional lipid analysis. Curr Opin Lipidol 18:181–186, 2007.
Wong DF, Wagner HN, Jr., Tune LE, et al.: Positron emission tomography reveals elevated D2 dopamine receptors in drug-naive schizophrenics. Science 234:1558–1563, 1986.
Yang J, Xu G, Zheng Y, et al.: Diagnosis of liver cancer using HPLC-based metabonomics avoiding false-positive result from hepatitis and hepatocirrhosis diseases. J Chromatogr B Analyt Technol Biomed Life Sci 813:59–65, 2004.
Yao JK, Reddy R, McElhinny LG, et al.: Reduced status of plasma total antioxidant capacity in schizophrenia. Schizophr Res 32:1–8, 1998a.
Yao JK, Reddy R, van Kammen DP: Reduced level of plasma antioxidant uric acid in schizophrenia. Psychiatr Res 80:29–39, 1998b.
Yao JK, Reddy RD, van Kammen DP: Human plasma glutathione peroxidase and symptom severity in schizophrenia. Biol Psychiatr 45:1512–1515, 1999.
Yao JK, Reddy R, van Kammen DP: Abnormal age-related changes of plasma antioxidant proteins in schizophrenia. Psychiatr Res 97:137–151, 2000.
Yao JK, Reddy RD, van Kammen DP: Oxidative damage and schizophrenia: an overview of the evidence and its therapeutic implications. CNS Drugs 15:287–310, 2001.
Yao JK, Thomas EA, Reddy RD, et al.: Association of plasma apolipoproteins D with RBC membrane arachidonic acid levels in schizophrenia. Schizophr Res 72:259–266, 2005.
Yao JK, Leonard S, Reddy R: Altered glutathione redox state in schizophrenia. Dis Markers 22:83–93, 2006.
Yoshimura R, Nakano Y, Hori H, et al.: Effect of risperidone on plasma catecholamine metabolites and brain-derived neurotrophic factor in patients with bipolar disorders. Hum Psychopharmacol 21:433–438, 2006.
Young LT, Warsh JJ, Kish SJ, et al.: Reduced brain 5-HT and elevated NE turnover and metabolites in bipolar affective disorder. Biol Psychiatr 35:121–127, 1994.
Yuan K, Kong H, Guan Y, et al.: A GC-based metabonomics investigation of type 2 diabetes by organic acids metabolic profile. J Chromatogr B Analyt Technol Biomed Life Sci 850:236–240, 2007.
Yumru M, Savas HA, Kurt E, et al.: Atypical antipsychotics related metabolic syndrome in bipolar patients. J Affect Disord 98:247–252, 2007.
Zarate CA, Jr., Singh JB, Carlson PJ, et al.: A randomized trial of an N-methyl-D-aspartate antagonist in treatment-resistant major depression. Arch Gen Psychiatr 63:856–864, 2006.

Chapter 7
Animal Models for Schizophrenia: A Brief Overview

Miyako Furuta and Hiroshi Kunugi(✉)

Abstract Schizophrenia is a devastating disorder affecting approximately 1% of the population worldwide. Since the pathogenesis and pathophysiology of the illness are largely unknown, and current treatment strategies are far from ideal, it is crucial to develop biomarkers, new drugs, and prevention strategies. To this end, the importance of animal models is growing rapidly. The validity of animal models of human diseases should be evaluated by three dimensions: face, constructive, and predictive validity. Behavioral tests to assess the face validity are summarized. Dopamine agonists and *N*-methyl-D-aspartate (NMDA) receptor antagonists have been most extensively studied with respect to their producing schizophrenia-like behavioral abnormalities. Neurodevelopmental animal models of schizophrenia are based on experimentally induced disruption of brain development during pre- or perinatal period that results in altered brain neurochemistry and aberrant schizophrenia-like behaviors. The genetically engineered mouse is a powerful tool to examine the possible mechanisms of genes giving susceptibility to schizophrenia by affecting certain endophenotypes. Here we provide a brief overview of these animal models of schizophrenia, particularly those of rodents.

Abbreviations DISC1: Disrupted-in-schizophrenia 1; LI: Latent inhibition; mGluR: Metabotropic glutamate receptor; NMDA: *N*-methyl-D-aspartate; NPAS: Neuronal PAS domain protein; NR1: NMDA R1 subunit; NRG1: Neuregulin 1; PCP: Phencyclidine; PPI: Prepulse inhibition; VH: Ventral hippocampus

H. Kunugi
Department of Mental Disorder Research, National Institute of Neuroscience, National Center of Neurology and Psychiatry, 4-1-1, Ogawahigashi, Kodaira, Tokyo, 187-8502 Japan
hkunugi@ncnp.go.jp ≤mailto:hkunugi@ncnp.go.jp≥

C.W. Turck (ed.) *Biomarkers for Psychiatric Disorders*,
doi: 10.1007/978-0-387-79251-4_7, © Springer Science+Business Media, LLC 2008

7.1 Introduction

Schizophrenia is a devastating disorder characterized by fundamental disturbances in a wide range of brain functions, including thought, perception, cognition, and emotions. It affects approximately 1% of the population worldwide. The monetary cost of schizophrenia to society is enormous, exceeding that of all cancers in the United States, and the World Health Organization finds this disease the world's fourth leading cause of disability. The illness usually starts in young adulthood and many cases experience chronic courses with exacerbations, and about 10% of patients commit suicide. A half century ago, as many as 50% of all hospital beds were used for patients with schizophrenia in industrialized nations; later, deinstitutionalization, with the advent of antipsychotic drugs, has resulted in an increased number of homeless and jailed populations. Although new antipsychotic (atypical) drugs have been introduced, they are still far from ideal in treating schizophrenia. Therefore it is crucial to understand the pathogenesis and pathophysiology of schizophrenia and develop biomarkers, new drugs, and prevention strategies. To this end, the importance of animal models is growing rapidly in the field of schizophrenia research, which parallels the rapid development of new technology in molecular biology (e.g., molecular genetics, microarray, and genetic engineering). Since brain tissues cannot be obtained from live human subjects with schizophrenia, animal models play a crucial role in elucidating molecular mechanisms of the behavioral abnormalities in schizophrenia. One might suspect, however, that schizophrenia is a human disease and such a model is not feasible. Surely, it might be an unrealistic challenge to make a single ideal model of schizophrenia; however, it is possible to make animal models showing certain characteristics of schizophrenia (schizophrenia-like behaviors), which may be of help to elucidate molecular mechanisms and biomarkers of such characteristics or "endophenotypes." Here we provide a brief overview of animal models of schizophrenia, particularly those of rodents.

7.2 Behavioral Analysis in Animal Models for Schizophrenia

7.2.1 Validity of Animal Models

Validity of animal models of human diseases should be evaluated by three dimensions: face validity, constructive validity, and predictive validity. *Face validity* of an animal model of a disease refers to how well the model resembles symptoms of the human disease. The diagnostic criteria for schizophrenia according to the DSM-IV (American Psychiatric Association, 1994) consist of five characteristic symptoms: delusions, hallucinations, disorganized speech, grossly disorganized or catatonic behavior, and negative symptoms. A diagnosis of schizophrenia requires two or more of these symptoms. In addition to these characteristic symptoms, people with schizophrenia often have a wide range of neurocognitive dysfunctions such as poor performance in memory tasks, low intelligence quotient (IQ), difficulty in sustaining attention, and deficits in sensorimotor gating, all of which are core features disabling such people (Fioravanti et al., 2005; Keefe et al., 2005; Hori et al., 2006). Although it is virtually impossible to see whether animals have delusions, hallucinations, or disorganized speech, it is possible to examine learning and memory, attention, sensorimotor gating, a part of negative symptoms, and psychomotor excitement in animals. To fully examine the face validity, it is ideal to perform a comprehensive test battery of emotional behaviors by a number of behavioral tests as listed in Table 7.1 (Takao et al., 2007). It is essential to assess general health and neurological abnormalities to assess "emotional behaviors", because hypoactivity, a sign of depression-like behavior or lack of energy, for example, may simply result from poor physical condition. Table 7.2 shows observable behaviors in mice which are of potential relevance to signs and symptoms of schizophrenia, as summarized by Powell and Miyakawa (2006).

To establish *construct validity*, pathophysiological and mechanistic characteristics of the animal model should be similar to those of the human disease. In the case of schizophrenia, however, causes of the illness are still unclear, although many

Table 7.1 Comprehensive test battery to assess emotional behaviors (Modified from Takao et al. 2007)

Behavioral test	Measure
General health/neurological screen	Body weight, rectal temperature, whisker, coat, simple reflexes
Wire hang	Muscle strength
Grip strength test	Muscle strength
Light/dark transition	Anxiety-like behavior
Open field activity	Anxiety-like behavior, behavioral sensitization, stereotypy
Elevated plus maze	Anxiety-like behavior
Hot plate	Nociception
Two-bottle choice	Anhedonia
Social interaction (novel environment)	Social behavior
Rotarod	Motor coordination, motor learning
Social interaction (Crawley version)	Social behavior
Prepulse inhibition/startle response	Sensory-motor gating, hearing, startle
Porsolt forced swimming	Depression-like behavior
Radial-arm maze	Working memory, reference memory, perseveration
T-maze	Working memory, reference memory, perseveration
Morris water maze	Reference memory, perseveration, working memory
Barnes maze	Reference memory, perseveration, working memory
Cued and contextual fear conditioning	Context memory, episodic-like memory
Tail suspension test	Depression-like behavior
Gait analysis	Gait, locomotion, motor function
24 h home cage monitoring	24 h activity in home cage, circadian rhythm
24 h home cage monitoring (social interaction)	Social behavior in home cage

Table 7.2 Mouse behaviors of potential relevance to signs and symptoms of schizophrenia (from Powell and Miyakawa, 2006)

I. *Positive Signs/Symptoms:*

A. Psychomotor agitation
- a. Locomotor activity
- b. Locomotor activity in response to novelty

B. Sensitivity to psychotomimetic drugs
- a. Augmented locomotor response to non-competitive NMDA receptor antagonists (MK-801, PCP, ketamine)
- b. Augmented locomotor response to amphetamine
- c. Increased sensitivity of other tests to psychotomimetic drugs (e.g., increased effect of MK-801, PCP, or ketamine on PPI test)

II. *Negative Signs/Symptoms:*

A. Social withdrawal
- a. Decreased interaction with a juvenile conspecific
- b. Decreased place preference for a caged peer conspecific
- c. Decreased preference for social novelty
- d. Altered social dominance on tube test
- e. Altered aggression behavior on resident intruder assay
- f. Decreased nesting behavior
- g. Home-cage social interaction

III. *Cognitive Signs/Symptoms:*

A. Decreased working memory
- a. Impaired alternation in T-maze working memory task
- b. Impaired performance in 8-arm radial maze working memory task

B. Deficits in attention/sensorimotor gating/executive function
- a. Decreased sensorimotor gating (PPI deficits)
- b. Decreased latent inhibition
- c. 5-choice serial reaction time test (5-CSRTT)
- d. Decreased set-shifting ability

C. General cognitive deficits
- a. Decreased spatial learning in Morris water maze
- b. Decreased spatial learning in 8-arm radial maze

NMDA *N*-methyl-D-aspartate; *PCP* phencyclidine; *PPI* prepulse inhibition; *CSRTT* 5-choice serial reaction time task

genetic and environmental risk factors have been raised. In this context, one of the main purposes of animal models of schizophrenia is to see whether a particular etiological or risk-increasing factor produces schizophrenia-like abnormalities in animals. Thus, when a possible environmental factor for schizophrenia is found, one can examine face and predictive validity of animals that are exposed to such an environmental factor. For a possible susceptibility gene for schizophrenia, one can take the advantage of genetically engineered animals to elucidate mechanisms underlying the gene giving susceptibility to schizophrenia.

Predictive validity refers to the efficacy of treatment drugs used for human disease on the phenotype of the model. However, in the case of schizophrenia, currently available drugs (mainly antipsychotic drugs) are limited by their efficacy on its symptoms. They are often effective on positive symptoms and psychomotor excitement in the acute phase; however, they have limited efficacy on negative symptoms and cognitive dysfunctions and sometimes worsen such symptoms. In addition, therapeutic effects of conventional drugs (e.g., haloperidol) are limited to the positive symptoms of the illness and they have substantially less impact on cognitive impairments, whereas atypical antipsychotics (e.g., clozapine, risperidone, olanzapine, quetiapine, and aripiprazole) may ameliorate cognitive deficits to some extent (Keefe et al., 1999; Thornton et al., 2006). Thus even though some antipsychotic drugs do not reinstate such behavioral phenotypes in animals, it does not necessarily indicate invalidation of the animal model. Rather, one should utilize animal models of schizophrenia with face validity in order to develop new drugs that are superior to currently available ones.

7.2.2 Locomotion and Stereotyped Behavior

Alterations in locomotion and stereotyped behavior are assessed in the open field test in which spontaneous activities are observed in an open field for 10–60 min. The principal purposes of the open field test in an animal model of schizophrenia are threefold: to assess anxiety-like behavior, behavioral sensitization, and stereotypy. Parameters of the test include total distance traveled, time spent in the center of the arena, and stereotyped behavior. Since rodents are neophobic and find open spaces aversive, normal animals will prefer to stay close to the walls in a novel open field (thigmotaxis). Therefore the more time spent in the center, the less the animal might be anxious.

Repeated intermittent administration of psychostimulants such as cocaine, amphetamine, and phencyclidine (PCP) produces "behavioral sensitization", which is characterized by a progressive and enduring enhancement in many psychostimulant-induced behaviors. In rodents, sensitized animals will show augmented locomotor response to a subsequent psychostimulant challenge injection (for a review, see Robinson and Becker, 1986; Kalivas et al., 1993; Pierce and Kalivas, 1997). This sensitization is observed in psychostimulant-induced psychosis and chronic schizophrenia in humans as well; these conditions show progressively enhanced susceptibility to abnormal behaviors, psychotic state, and relapse (Sato et al., 1983; Ujike, 2002). Evidence suggests that individuals with schizophrenia have sensitization to environmental stress (Myin-Germeys et al., 2005; Yuii et al., 2007). Thus psychostimulant-induced hyperlocomotion is considered to be a model of chronic schizophrenia with behavioral sensitization.

In schizophrenia, stereotypy is often observed, a symptom characterized by repetitive, functionless motor behavior (Morrens et al., 2006). Stereotypic behaviors in rodents are sniffing, grooming, head bobbing, nail biting, gnawing, and circling. To measure stereotypy, activity in an open field should be assessed with a standardized stereotypy scoring system (Creese and Iversen, 1973). Amphetamine and a nonselective dopamine receptor agonist apomorphine induce stereotyped behaviors in rodents, while antipsychotics (D2 receptor antagonists) suppress such behaviors (Randrup and Munkvad, 1974; Ridley et al., 1994; Ellenbroek and Cools, 2002; Picada et al., 2003). Since excessive release of dopamine is thought to induce positive symptoms of schizophrenia, amphetamine or apomorphine-induced stereotypy is commonly used to evaluate the efficacy of antipsychotic drugs.

7.2.3 Prepulse Inhibition (PPI) of Startle Reflex

PPI refers to the phenomenon in which a weak pre-stimulus or prepulse suppresses the response to a subsequent startling stimulus (Hoffman and Searle, 1968; Graham,1975). This phenomenon is thought to occur because pre-stimulus or prepulse shuts the gate of sensory input, resulting in the reduction of response to subsequent startling stimulus. PPI is therefore a measure of information processing or that of "sensorimotor gating". In such gating works, the interval between pulse and prepulse is within 1000 ms; usually intervals of 30–240 ms are utilized in experiments. Deficits in sensorimotor gating may lead to "sensory flooding" and "cognitive fragmentation". Acoustic stimuli have most often been used for experiments. Reduced PPI in schizophrenia patients was found, for the first time, by Braff et al. (1978), which has been confirmed by a number of subsequent studies (reviewed by Braff et al., 2001), including our Japanese study (Kunugi et al., 2007). Impaired PPI accords with deficits in information processing in schizophrenia patients, revealed and enhanced when high processing loads, multiple tasks, distraction, or other stressors demand the rapid and efficient processing of information (Braff, 1993). All mammalian species exhibit PPI, and the same paradigm is applicable for animals and humans. Swerdlow et al. (1994) demonstrate the face, predictive, and construct validity of disruption of PPI in rats as an animal model of impaired sensorimotor gating in schizophrenia. They showed disruption in PPI by apomorphine in rats, mimicking PPI deficits in schizophrenics (face validity), the ability of antipsychotics to restore PPI in apomorphine-treated rats (predictive validity), and disrupted PPI by infusion of dopamine into the nucleus accumbens (constructive validity). PPI in rodents has repeatedly been shown to be greatly helpful to understand the neurobiology of cognitive deficits in schizophrenia, to predict the antipsychotic efficacy of compounds, and to elucidate the effect of possible susceptibility genes for schizophrenia on sensorimotor gating (reviewed by Swerdlow and Geyer, 1998; Geyer et al., 2001, 2002). Perhaps PPI is now the most important behavioral test in animal studies on schizophrenia.

7.2.4 Latent Inhibition (LI)

LI is a term introduced approximately half a century ago (Lubow and Moore, 1959) to describe a phenomenon in which learning about the consequences of a stimulus is retarded if that stimulus has been experienced without reinforcement.

More simply, a stimulus that is casually familiar enters into new associations more slowly than a novel stimulus (Lubow and Gewirtz, 1995). When an organism is repeatedly exposed to a stimulus that is not followed by a significant consequence, the stimulus subsequently becomes less effective, as compared to a novel stimulus, in the acquisition/performance of a new association, which can be observed in many different species (Lubow et al., 1997). The ubiquitous nature of LI suggests that it has an adaptive significance in biasing an organism to more fully processing new inputs than older and unimportant ones. The experimental procedure in animals, for example, consists of three stages: pre-exposure, in which the to-be-conditioned stimulus, tone, was presented without being followed by reinforcement; acquisition, in which the pre-exposed tone was paired with electric shock; and test, in which LI was indexed by the animals' suppression of licking during tone presentation (Weiner et al., 1984). Deficits in LI are considered to reflect deficits in selective attention in schizophrenia (Lubow and Gewirtz, 1995). In humans, acute positive-symptom schizophrenics, normal volunteers who score high on questionnaire measures of schizotypy and non-patients or animals treated with dopamine agonists show reduced LI (Gray et al., 1992; Kumari et al., 1999). In rats, amphetamine disrupts LI, and this disruption is blocked by antipsychotics (Solomon et al., 1981; Weiner et al., 1984, 1988; Warburton et al., 1994). Moreover, antipsychotics enhance LI and rats' ability to ignore irrelevant stimuli (Feldon and Weiner, 1991; Dunn et al., 1993; Warburton et al., 1994; Weiner et al., 1997). These findings indicate that animal models with disrupted LI might be a good model of schizophrenia, particularly its positive symptoms.

7.2.5 Social Interaction

Schizophrenic patients with negative symptoms suffer from social withdrawal. Few animal models have been described with face validity for schizophrenia's social withdrawal; however, the social interaction test might be a helpful paradigm. In the test, a pair of rodents are placed simultaneously at opposing corners in the open field apparatus and allowed to explore freely. Sociality could be assessed by parameters such as the number of contacts, total duration of contacts, and total distance traveled. Decreased number of contact times and duration resemble social withdrawal of schizophrenia. Sams-Dodd (1998) found repeated administration of both amphetamine and PCP dose-dependently induced stereotyped behavior and locomotor hyperactivity; however, only PCP induced social withdrawal assessed with the social interaction test, which is consistent with the view that PCP, but not amphetamine, induce symptoms analogous to negative symptoms in schizophrenia. An obvious limitation of the use of such a social interaction model in rodents might be the conceptual distance between interactive behavior in rodent pairs and deficits in social skills and social withdrawal exhibited by schizophrenic patients in complex human social contexts.

7.2.6 *Forced Swimming*

In the forced swimming paradigm (Porsolt et al., 1978), animals are put into a chamber filled with water at 27°C. They swim around attempting to escape, which is impossible, and eventually assume an immobile posture of floating. Time spent by the active movement (struggling to escape or swimming) or that of immobility is measured during the test session. Reduced immobility time indicates antidepressant-like behavior. On a subsequent test, the animals more readily adopt the floating behavior, indicating that the animals have learned that they cannot escape, a phenomenon resembling "learned helplessness" of depressive disorder. In mice with repeated treatment of PCP, immobility time in the forced swimming test was enhanced (Noda et al., 1995, 1997). This effect of PCP on immobility appeared to be sensitive to atypical antipsychotic (clozapine and risperidone) treatment, but not to haloperidol or tricyclic antidepressant (Noda et al., 1995, 1997). Since risperidone and clozapine, but not haloperidol and tricyclic antidepressants, improve negative symptoms in schizophrenia, these observations may indicate that immobility in the forced swimming could be an animal model of negative symptoms (e.g., lack of energy or motivation) in schizophrenia.

7.2.7 *Learning and Memory Tasks*

Morris' water maze (Morris, 1984) is a spatial navigation task as a measure for spatial (visual) learning and memory in which the animal swims to find a hidden platform, using a visual cue to mark the position of the platform. Finding a platform to escape from water is the positive reinforcement. The fundamental premise of this task is that rodents are highly motivated to escape from the water environment via the quickest route. Swim speed and swim pathway provide measures of the procedural abilities to perform this task. The performance is dependent on a variety of cognitive substrates including learning and memory and therefore is relevant to a wide range of cognitive deficits in schizophrenia. Administration of noncompetitive *N*-methyl-D-aspartate (NMDA) receptor antagonists such as PCP, ketamine, or MK-801 (dizocilpine) to rats has been shown to impair performance on the Morris water maze task (Heale and Harley, 1990; Jones et al., 1990; Murray and Ridley, 1997).

Radial-arm maze has been proven to be very useful for the assessment of spatial learning and memory in rodents. It measures entries into arms baited with food or water reinforcer. Radial-arm maze tasks often bait only some of the arms (e.g., bait in five out of eight arms), with the same arms baited on successive trials, requiring the animal to learn the location of the baited vs. unbaited arms and to learn not to return to an arm already visited. Errors are scored when a rat or mouse enters an unbaited arm or a previously visited arm. Time elapsed between the start of the test session and finding all food rewards is another measure. Since the disturbed learning and working memory are core features of cognitive deficits in schizophrenia, the radial-arm maze paradigm is very useful to assess the appropriateness of the model for schizophrenia. Indeed, treatment with MK-801 or PCP increased the number of arm revisits in the radial-arm maze (Zhang et al., 2005). On the other hand, some atypical antipsychotics, such as olanzapine and risperidone, have been shown to improve spatial working memory assessed with the radial-arm maze in rats (Wolff and Leander, 2003).

T-maze is an alternative method to assess working memory. Animals are maintained on a food-restricted or water-restricted diet before the test session, habituated to the maze, and then placed to run to the ends of the T-maze to obtain the reward. Each test trial contains two runs: information run and choice run. In a delayed-matching-to-position task, the reward is placed at the same arm for both runs, while in a delayed-nonmatching-to-position task the reward is located in different arms for the two runs. The latter procedure is particularly useful to assess reference memory as well as working memory.

7.3 Pharmacological Animal Models

There have been numerous studies using a variety of psychotropic or psychotomimetic drugs to produce animal models of schizophrenia. Dopamine agonists and NMDA receptor antagonists have been most extensively studied since these drugs induce schizophrenia-like psychosis or corresponding behavioral abnormalities in humans and animals.

7.3.1 Dopamine Agonists

The well-known dopamine hypothesis is based on the discovery that clinical and pharmacological potencies parallel with dopamine receptor binding (Creese et al., 1976) and on the fact that dopamine agonists such as cocaine (dopamine reuptake inhibitor), amphetamine (dopamine releaser and reuptake inhibitor), apomorphine (non-selective dopamine receptor agonist), and L-Dopa (intermediate in dopamine biosynthesis) cause schizophrenia-like symptoms in humans. However, such dopamine agonists-induced symptoms do not incorporate negative symptoms or important features of cognitive dysfunction (i.e., learning and memory). In rodents, administration of amphetamine, for example, induces hyperlocomotion, stereotyped behavior (e.g., Randrup and Munkvad, 1974; Sams-Dodd, 1998), deficits in attention and sensorimotor gating assessed by PPI (e.g., Mansbach et al., 1988; Hijzen et al., 1991), and LI (e.g., Solomon et al., 1981; Weiner et al., 1984, 1988; Warburton et al., 1994). However, amphetamine did not induce negative-symptom-like behavior of reduced social interaction (Sams-Dodd, 1998). Moreover, administration of amphetamine has no effect on, nor tends to facilitate rather than impair, learning and memory tasks assessed by the Morris water maze (Brown et al., 2000) and radial-maze (Eckerman et al., 1980; Packard and White, 1989; Ennaceur, 1994), which may be attributed in part to the dopaminergic function and dopamine synthesis in the prefrontal cortex (Brown et al., 2000). The animal models of schizophrenia produced by dopamine agonists are therefore limited by the lack of behaviors analogous to negative and cognitive symptoms of schizophrenia. The animal models by dopamine agonists are surely useful for screening drugs with a predicted action of antagonizing dopamine receptors; however, current research on new drug discovery is shifting towards searching for drugs effective on negative symptoms and conginitive deficits. Models based on direct manipulations of the dopamine system, *per se*, may have exhausted their heuristic potential (Lipska and Weinberger, 2000), although the behavioral sensitization model (described above) using dopamine agonists has become a standard procedure to assess the validity of the potential animal model of schizophrenia.

7.3.2 NMDA Receptor Antagonist

Numerous studies have demonstrated that noncompetitive NMDA receptor antagonists such as PCP, ketamine, and MK801 produce schizophrenia-like symptoms or behaviors. PCP induces a psychotomimetic state that closely resembles schizophrenia in humans. As opposed to amphetamine-induced psychosis, PCP-induced psychosis incorporates negative symptoms (e.g., emotional withdrawal and motor retardation) and cognitive deficits in addition to positive symptoms (e.g., delusions, hallucinations, and formal thought disorder) of schizophrenia (Allen and Young, 1978; Javitt and Zukin, 1991). In rodents, PCP induces disruption in PPI (Mansbach and Geyer, 1989) and this disruption was blocked by atypical antipsychotics clozapine (Bakshi et al., 1994) and olanzapine (Bakshi and Geyer, 1995), but not by a typical antipsychotic haloperidol (Keith et al., 1991). PCP also induced stereotyped behavior and social isolation in rats, and these abnormalities were antagonized by chronic clozapine, but not haloperidol (Sams-Dodd, 1996). As described above, immobility in forced swimming was increased by PCP and this effect was not reversed by antidepressants (Noda et al., 1995, 1997). Acute administration of PCP induces deficits in learning and memory in rodents assessed by the Morris water maze (Wass et al., 2006), radial-maze (Kesner et al., 1983; Danysz et al., 1988; Zhang et al. 2005), and T-maze (Handelmann et al., 1987) paradigms. Although subchronic administration of PCP was reported to produce impaired performance of a spatial working memory task of T-maze (Jentsch et al., 1997), some other studies failed to demonstrate that repeated administration of PCP to rats or mice produces enduring memory impairment assessed with the radial maze (Li et al., 2003; Marquis et al., 2003). Therefore, the cognitive effects of subchronic PCP administration in rats are still controversial in the literature, or it may depend on task parameters (Marquis et al., 2007).

The dissociative anesthetic, ketamine, is also known to cause schizophrenia-like symptoms in humans (Adler et al., 1998, 1999; Lahti et al., 2001). In rodents, ketamine induces disruption in PPI, working memory deficits, and augmented locomotion (e.g., Mansbach and Geyer, 1989; Jones et al., 1990; Verma and Moghaddam, 1996; Imre et al., 2006). MK-801 is extensively used for animal studies and has been shown to induce a wide range of schizophrenia-like behaviors including hyperlocomotion, disruption of PPI, impaired performance in learning and memory tasks, and decreased social behavior (e.g., Mansbach and Geyer, 1989; Heale and Harley, 1990; Jones et al., 1990; Verma and Moghaddam, 1996; Murray and Ridley, 1997; Bakshi and Geyer, 1998; Jacobs et al., 2000; Myhrer, 2003; Zhang et al., 2005).

Administration of NMDA receptor antagonists in pre- or perinatal animals also induces a post-pubertal emergence of motor and sensory gating abnormalities related to schizophrenia, and impaired acquisition of delayed spatial learning tasks (Deutsch et al., 1998; Wang et al., 2001; Andersen et al., 2004). These models may mimic the relevant neurodevelopmental aspects of at least some forms of schizophrenia.

The NMDA-antagonist model of schizophrenia has recently been successfully utilized to develop a novel class of antipsychotic drugs. An agonist of Group II metabotropic glutamate receptors (mGluR2/3) was found to attenuate the disruptive effects of PCP on working memory, stereotypy, locomotion, and cortical glutamate efflux (Moghaddam and Adams, 1998). This behavioral reversal occurred in spite of sustained dopamine hyperactivity, suggesting that the mGluR2/3 antagonist may present a nondopaminergic therapeutic strategy for treatment of psychiatric disorders. On the basis of this observation, Patil et al. (2007) went on to perform a randomized, double-blind, and placebo-controlled trial of LY404039, a selective agonist for mGluR2/3 in patients with schizophrenia and controls; they found that the drug was safe, well tolerated, and effective in treating both positive and negative symptoms of schizophrenia compared to placebo.

7.4 Neurodevelopmental Animal Models

Many epidemiological studies suggest that insults early in life such as fetal hypoxia and prenatal underdevelopment are risk factors for schizophrenia (e.g., Kunugi et al., 1996, 2001; McNeil and Cantor-Graae, 2000; Cannon and Rosso, 2002), which supports the reformulation of schizophrenia as a neurodevelopmental disorder (Murray and Lewis, 1987; Weinberger, 1987). Neurodevelopmental animal models of schizophrenia are based on experimentally induced disruption of brain development during pre- or perinatal period that results in altered brain neurochemistry and aberrant schizophrenia-like behaviors. Two models have been studied most extensively: neonatal hippocampal lesion and Caesarean section (with anoxia).

7.4.1 Neonatal Hippocampal Lesion

The hippocampus is often considered to play an important role in the pathophysiology of schizophrenia. The major findings in schizophrenic patients are decreased volumes, hypometabolism, and cytoarchitectural abnormalities, which are more robust on the left hippocampus, as well as verbal memory impairment (reviewed by Gothelf et al., 2000). Lipska et al. (1993) produced ibotenic acid lesions of the ventral hippocampal (VH) formation in rats on the seventh day after birth. In this animal model, a number of abnormal behaviors do not become evident before adolescence or early adulthood, which is consistent with the fact that schizophrenia develops in adolescence or adulthood. When tested as juveniles (postnatal day 35), rats with the neonatal VH lesions are less social than controls (Lipska et al., 1993; Sams-Dodd et al., 1997), but otherwise they behave normally in motor tests involving exposure to stress and dopamine agonists. In adolescence and adulthood (postnatal day 56 or older), however, lesioned animals display markedly changed behaviors such as motor hyper-responsiveness to stress, stimulants, and enhanced stereotypes that could be linked primarily to increased mesolimbic/nigrostriatal dopamine transmission. They also show enhanced sensitivity to NMDA receptor antagonists (MK-801 and PCP), deficits in PPI and LI, impaired social behaviors, and working memory problems (Lipska and Weinberger, 1993). Antipsychotic drugs normalize some impaired behaviors in the VH lesion rats (Lipska and Weinberger, 1993; Sams-Dodd et al., 1997). Therefore, this animal model has face, constructive, and predictive validity of schizophrenia.

7.4.2 *Cesarean Section and Perinatal Anoxia*

In a series of studies, Boksa and her colleagues have characterized rats, mice, and guinea pigs born by Cesarean section alone or Cesarean section with additional anoxia as a possible animal model of schizophrenia (for review, see, Boksa and El-Khodor, 2003; Boksa, 2004). These animals showed heightened behavioral responses to amphetamine and disrupted PPI relative to vaginally born control animals (El-Khodor and Boksa, 1998; Vaillancourt and Boksa, 2000). Rats with transient anoxia added to Cesarean section tended to be behaviorally hyper-responsive to repeated stress in adulthood as compared to Ceasarean section alone or vaginally born animals (El-Khodor and Boksa, 2000). At the molecular level, animals with Caesarean section showed enhanced amphetamine-induced dopamine release from the nucleus accumbens in response to repeated stress (Brake et al., 1997). Furthermore, the birth insults interacted with stress in adulthood to alter components of dopaminergic neurotransmission such as dopamine receptors and the dopamine transporter (El-Khodor and Boksa, 2001, 2002). Such interactions between early insults and stress in adulthood in the etiology of dopamine-related behaviors and molecular changes might provide clues to elucidate the pathogenesis of schizophrenia.

7.5 Transgenic Animal Models

Recent advances in molecular genetic studies as well as pharmacological findings have provided a number of candidate genes for schizophrenia. Behavioral analysis on genetically engineered mice is a powerful tool to examine the possible mechanisms of such genes giving susceptibility to certain phenotypes in schizophrenia (Takao et al., 2007). Some examples are described below.

7.5.1 *NMDAR1 Knock Down Mice*

The glutamate dysfunction hypothesis points to the possibility that genetically engineered mice with altered NMDA function could be an animal model of schizophrenia. For example, a mouse line that expresses low levels of the NMDA R1 subunit (NR1) of the NMDA receptor was generated (Mohn et al., 1999). These mutant mice show increased locomotor activity, increased acoustic startle reactivity, and deficits in PPI of acoustic startle (Duncan et al., 2004, 2006b; Fradley et al., 2005). Increased locomotor activity was inhibited by olanzapine (atypical antipsychotic) more than haloperidol (typical antipsychotic) (Duncan et al., 2006a). Both drugs increased PPI. These results suggest that the NR1-deficient mouse has face, constructive and predictive validity of animal model of schizophrenia, which has its origin predominantly in glutamate dysfunction.

7.5.2 *HOMER1 Mutant Mice*

Homer proteins are a group of proteins that regulate group I metabotropic glutamate receptor function. Norton et al. (2003) reported an association between schizophrenia and an intronic single nucleotide polymorphism of the *HOMER1* gene. Interestingly, *HOMER1*-knock out mice showed a phenotype analogous to schizophrenia, including deficits in radial-arm maze performance, impaired prepulse inhibition, increased anxiety in a novel objects test, enhanced reactivity to novel environments, decreased instrumental response for sucrose, and enhanced MK-801- and methamphetamine-stimulated motor behavior (Szumlinski et al., 2005). No-net-flux in vivo microdialysis revealed a decrease in extracellular glutamate content in the nucleus accumbens and an increase in the prefrontal cortex (Szumlinski et al., 2005). These observations might provide clues to the mechanisms of glutamate dysfunction theory of schizophrenia.

7.5.3 *DISC1 Mutant Mouse*

DISC1 (Disrupted-In-Schizophrenia 1) was identified from a breakpoint of a balanced translocation (1;11)(q42.1;q14.3) that segregates with schizophrenia and related psychiatric disorders in a large Scottish family (Millar et al., 2000). Subsequent studies have suggested that genetic variations of *DISC1* are associated with mood disorders (bipolar disorder and depression) as well (Hodgkinson et al.,

2004; Hashimoto et al., 2006), indicating that *DISC1* is a risk gene for major mental illnesses. Three mouse *DISC1* mutants have been identified and characterized: two missense variants created by the ENU chemical mutagenesis program (Coghill et al., 2002) and one naturally occurring, out-of-frame deletion close to the *C*-terminus of the protein (Clapcote and Roder, 2006). Mice with mutation Q31L showed depressive-like behavior with deficits in the forced swimming and other measures, which were reversed by the antidepressant bupropion, while L100P mutant mice exhibited schizophrenic-like behavior, with profound deficits in PPI and LI, which were reversed by antipsychotic treatment (Clapcote et al., 2007). These studies therefore demonstrate that *DISC1* missense mutations in mice give rise to phenotypes related to depression and schizophrenia, supporting the role of *DISC1* in major mental illnesses. Furthermore, mice expressing a dominant-negative form of *DISC1* under the CaMKII promoter was generated and found to display several behavioral abnormalities, including hyperactivity, disturbance in PPI, olfactory-associated behavior, and an anhedonia/depression-like deficit (Hikida et al., 2007).

7.5.4 NPAS3 Transcription Factors Knock Out Mice

This is another example of a gene identified from chromosomal translocation. The neuronal PAS domain protein 3 (*NPAS3*) gene encoding a brain-enriched transcription factor was recently found to be disrupted by translocation t(9;14)(q34;q13) carried by a mother and her daughter, both of whom were affected with schizophrenia (Kamnasaran et al., 2003). The mice having either the double-null (Npas1–/– Npas3–/–) or *NPAS3*-null over *NPAS1*-heterozygous (Npas1 + /– Npas3–/–) genotype exhibit behavioral and neuroanatomical abnormalities related to schizophrenia, including impaired social recognition, increased open-field locomotors activity, stereotypic darting behavior, reduced PPI, and decreased brain levels of protein (Erbel-Sieler et al., 2004).

7.5.5 Calcineurin Knock Out Mice

Calcineurin (protein phosphatase 2B) is a member of the serine/threonine protein phosphatase family and the only known phosphatase that can be activated by Ca^{2+} and calmodulin. Calcineurin is positioned downstream of dopaminergic signaling (Greengard et al., 2001) and is also involved in NMDA receptor-mediated synaptic plasticity (Zeng et al., 2001), suggesting its possible role in schizophrenia. In line with this, genetic variations of *PPP3CC*, a gene encoding the calcineurin gamma subunit, were shown to be associated with schizophrenia, supporting the possibility that alterations in calcineurin signaling contribute to schizophrenia (Gerber et al., 2003). Forebrain-specific calcineurin knockout mice were impaired in hippocampus-dependent working and episodic-like memory tasks (Zeng et al., 2001). Subsequent comprehensive behavioral analysis of this mouse revealed increased locomotor activity, enhanced MK-801-induced hyperlocomotion, decreased social interaction, and deficits in PPI and LI (Miyakawa et al., 2003), a phenotype remarkably analogous to schizophrenia.

7.5.6 *NRG1 Mutant Mice*

Neuregulin 1 (*NRG1*), located on chromosome 8p21–12, is a possible candidate gene for schizophrenia because of its critical involvement in neurodevelopment, regulation of glutamate and other neurotransmitter receptor expression, and synaptic plasticity. In line with this, *NRG1* has been one of the most promising genes giving susceptibility to schizophrenia as of the first report of its strong linkage and association with the disease by Stefansson et al. (2002). They also demonstrated that heterozygous mutant mice for *NRG1* and its receptor *ErbB4* have hyperactivity in the novel open-field test, impaired PPI, and fewer functional NMDA receptors than wild-type mice. Further, behavioral abnormalities were partially reversed by an atypical antipsychotic, clozapine. A more recent behavioral phenotyping on the heterozygous *NRG1* mutant mice found a locomotion- and exploration-related hyperactive phenotype and increased sensitivity to the stimulating effect of environmental enrichment on locomotion and exploration (Karl et al., 2007). In the social interaction test, aggressive following was increased in *NRG1* mutants (O'Tuathaigh et al., 2008). These results suggest that *NRG1* mutants may be related to impairment in response to novel environment or social novelty.

7.6 Concluding Remarks

As mentioned earlier, there has so far been no single ideal animal model for schizophrenia, which accords with the fact that the etiology of schizophrenia is still largely unknown and with the possibility that the illness is etiologically heterogeneous. However, each model contains a specific physiological marker, endophenotype, or molecular marker possibly involved in the pathogenesis and pathophysiology of schizophrenia, which provides valuable tools to elucidate molecular mechanisms of the marker or phenotype. Classical pharmacological animal models of schizophrenia, those with NMDA receptor antagonists in particular, still play an important role in the development of new drugs. Indeed, a new class of antipsychotics, which seems to be promising, has recently been developed using the glutamate dysfunction model (Patil et al., 2007). Neurodevelopmental animal models of schizophrenia are important to elucidate the mechanisms underlying the link between early insults, the most well-established risk factor for schizophrenia, and adult onset of schizophrenia at a molecular level. With the advent of rapid progression of techniques in molecular genetics and system biology such as transcriptome and proteome, many molecules have been emerging as a potential key molecule and/or a biomarker of schizophrenia. To analyze their function on behavior, use of genetically engineered mice or other species is now an essential and powerful strategy, although methods of behavioral tests as well as their analyses and interpretations need further refinement in the future.

References

Adler CM, Goldberg TE, Malhotra AK, Pickar D, Breier A (1998) Effects of ketamine on thought disorder, working memory, and semantic memory in healthy volunteers. Biol Psychiatr 43:811–816.

Adler CM, Malhotra AK, Elman I, Goldberg T, Egan M, Pickar D, Breier A (1999) Comparison of ketamine-induced thought disorder in healthy volunteers and thought disorder in schizophrenia. Am J Psychiatr156:1646–1649.

Allen RM, Young SJ (1978) Phencyclidine-induced psychosis. Am J Psychiatr 135:1081–1084.

American Psychiatric Association (1994) Diagnostic and Statistical Manual of Mental Disorders, 4th ed, American Psychiatric Association, Washington, D.C.

Andersen JD, Pouzet B (2004). Spatial memory deficits induced by perinatal treatment of rats with PCP and reversal effect of D-serine. Neuropsychopharmacology 29(6):1080–1090.

Bakshi VP, Geyer MA (1995) Antagonism of phencyclidine-induced deficits in prepulse inhibition by the putative atypical antipsychotic olanzapine. Psychopharmacology122:198–201.

Bakshi VP, Geyer MA (1998) Multiple limbic regions mediate the disruption of prepulse inhibition produced in rats by the noncompetitive NMDA antagonist dizocilpine. J Neurosci 18:8394–8401.

Bakshi VP, Swerdlow NR, Geyer MA (1994) Clozapine antagonizes phencyclidine-induced deficits in sensorimotor gating of the startle response. J Pharmacol Exp Ther 271:787–794.

Boksa P (2004) Animal models of obstetric complications in relation to schizophrenia. Brain Res Brain Res Rev 45:1–17.

Boksa P, El-Khodor BF (2003) Birth insult interacts with stress at adulthood to alter dopaminergic function in animal models: possible implications for schizophrenia and other disorders. Neurosci Biobehav Rev 27:91–101.

Braff DL (1993) Information processing and attention dysfunctions in schizophrenia. Schizophr Bull 19:233–259.

Braff DL, Geyer MA, Swerdlow NR (2001) Human studies of prepulse inhibition of startle: normal subjects, patient groups, and pharmacological studies. Psychopharmacology (Berl). 156: 234–258.

Braff D, Stone C, Callaway E, Geyer M, Glick I, Bali L (1978) Prestimulus effects on human startle reflex in normals and schizophrenics. Psychophysiology 15:339–343.

Brake WG, Noel MB, Boksa P, Gratton A (1997) Influence of perinatal factors on the nucleus accumbens dopamine response to repeated stress during adulthood: an electrochemical study in the rat. Neuroscience 77:1067–1076.

Brown RW, Bardo MT, Mace DD, Phillips SB, Kraemer PJ (2000) D-amphetamine facilitation of morris water task performance is blocked by eticlopride and correlated with increased dopamine synthesis in the prefrontal cortex. Behav Brain Res 114:135–143.

Cannon TD, Rosso IM (2002) Levels of analysis in etiological research on schizophrenia. Dev Psychopathol 14: 653–66; Review.

Clapcote SJ, Roder JC (2006) Deletion polymorphism of Disc1 is common to all 129 mouse substrains: implications for gene-targeting studies of brain function. Genetics; 173:2407–2410.

Clapcote SJ, Lipina TV, Millar JK, Mackie S, Christie S, Ogawa F, Lerch JP, Trimble K, Uchiyama M, Sakuraba Y, Kaneda H, Shiroishi T, Houslay MD, Henkelman RM, Sled JG, Gondo Y,

Porteous DJ, Roder JC (2007) Behavioral phenotypes of Disc1 missense mutations in mice. Neuron 54:387–402.

Coghill EL, Hugill A, Parkinson N, Davison C, Glenister P, Clements S, Hunter J, Cox RD, Brown SD (2002) A gene-driven approach to the identification of ENU mutants in the mouse. Nat Genet 30:255–256.

Creese I, Iversen SD (1973) Blockage of amphetamine induced motor stimulation and stereotypy in the adult rat following neonatal treatment with 6-hydroxydopamine. Brain Res 55:369–382.

Creese I, Burt DR, Snyder SH (1976) Dopamine receptor binding predicts clinical and pharmacological potencies of antischizophrenic drugs. Science 192:481–483.

Danysz W, Wroblewski JT, Costa E (1988) Learning impairment in rats by N-methyl-D-aspartate receptor antagonists. Neuropharmacology 27:653–656.

Deutsch SI, Mastropaolo J, Rosse RB (1998) Neurodevelopmental consequences of early exposure to phencyclidine and related drugs. Clin Neuropharmacol 21:320–332.

Dunn LA, Atwater GE, Kilts CD (1993) Effects of antipsychotic drugs on latent inhibition: sensitivity and specificity of an animal behavioral model of clinical drug action. Psychopharmacology (Berl) 112:315–323.

Duncan GE, Moy SS, Perez A, Eddy DM, Zinzow WM, Lieberman JA, Snouwaert JN, Koller BH (2004) Deficits in sensorimotor gating and tests of social behavior in a genetic model of reduced NMDA receptor function. Behav Brain Res 153:507–519.

Duncan GE, Moy SS, Lieberman JA, Koller BH (2006a) Typical and atypical antipsychotic drug effects on locomotor hyperactivity and deficits in sensorimotor gating in a genetic model of NMDA receptor hypofunction. Pharmacol Biochem Behav 85:481–491.

Duncan GE, Moy SS, Lieberman JA, Koller BH (2006b) Effects of haloperidol, clozapine, and quetiapine on sensorimotor gating in a genetic model of reduced NMDA receptor function. Psychopharmacology (Berl) 184:190–200.

Eckerman DA, Gordon WA, Edwards JD, MacPhail RC, Gage MI (1980) Effects of scopolamine, pentobarbital, and amphetamine on radial arm maze performance in the rat. Pharmacol Biochem Behav12:595–602.

El-Khodor BF, Boksa P (1998) Birth insult increases amphetamine-induced behavioral responses in the adult rat. Neuroscience 87:893–904.

El-Khodor BF, Boksa P (2000) Transient birth hypoxia increases behavioral responses to repeated stress in the adult rat. Behav Brain Res 107:171–175.

El-Khodor B, Boksa P (2001) Caesarean section birth produces long term changes in dopamine D1 receptors and in stress-induced regulation of D3 and D4 receptors in the rat brain. Neuropsychopharmacology 25:423–439.

El-Khodor BF, Boksa P (2002) Birth insult and stress interact to alter dopamine transporter binding in rat brain. Neuroreport 13:201–206.

Ellenbroek BA, Cools AR (2002) Apomorphine susceptibility and animal models for psychopathology: genes and environment, Behav Genet 32:349–361.

Ennaceur A (1994) Effects of amphetamine and medial septal lesions on acquisition and retention of radial maze learning in rats. Brain Res 636:277–285.

Erbel-Sieler C, Dudley C, Zhou Y, Wu X, Estill SJ, Han T, Diaz-Arrastia R, Brunskill EW, Potter SS, McKnight SL (2004) Behavioral and regulatory abnormalities in mice deficient in the NPAS1 and NPAS3 transcription factors. Proc Natl Acad Sci U S A 101:13648–13653.

Feldon J, Weiner I (1991) The latent inhibition model of schizophrenic attention disorder. Haloperidol and sulpiride enhance rats' ability to ignore irrelevant stimuli. Biol Psychiatr 29:635–646.

Fioravanti M, Carlone O, Vitale B, Cinti ME, Clare L (2005) A meta-analysis of cognitive deficits in adults with a diagnosis of schizophrenia. Neuropsychol Rev 15:73–95.

Fradley RL, O'Meara GF, Newman RJ, Andrieux A, Job D, Reynolds DS (2005) STOP knockout and NMDA NR1 hypomorphic mice exhibit deficits in sensorimotor gating. Behav Brain Research163:257–264.

Gerber DJ, Hall D, Miyakawa T, Demars S, Gogos JA, Karayiorgou M, Tonegawa S (2003) Evidence for association of schizophrenia with genetic variation in the 8p21.3 gene, PPP3CC, encoding the calcineurin gamma subunit. Proc Natl Acad Sci U S A 100: 8987–8992.

Geyer MA, Krebs-Thomson K, Braff DL, Swerdlow NR (2001) Pharmacological studies of prepulse inhibition models of sensorimotor gating deficits in schizophrenia: a decade in review. Psychopharmacology (Berl) 156:117–154.

Geyer MA, McIlwain KL, Paylor R (2002) Mouse genetic models for prepulse inhibition: an early review. Mol Psychiatr 7:1039–53.

Gothelf D, Soreni N, Nachman RP, Tyano S, Hiss Y, Reiner O, Weizman A (2000) Evidence for the involvement of the hippocampus in the pathophysiology of schizophrenia. Eur Neuropsychopharmacol 10:389–395.

Graham FK (1975) The more or less startling effects of weak prestimulation. Psychophysiology 12:238–248.

Gray NS, Hemsley DR, Gray JA (1992). Abolition of latent inhibition in acute, but not chronic, schizophrenics. Neurol Psychiatr Brain Res 1:83–89.

Greengard P (2001) The neurobiology of dopamine signaling. Biosci Rep 21:247–269.

Green MF (1996) What are the functional consequences of neurocognitive deficits in schizophrenia? Am J Psychiatr 153:321–330.

Handelmann GE, Contreras PC, O'Donohue TL (1987) Selective memory impairment by phencyclidine in rats. Eur J Pharmacol 140:69–73.

Hashimoto R, Numakawa T, Ohnishi T, Kumamaru E, Yagasaki Y, Ishimoto T, Mori T, Nemoto K, Adachi N, Izumi A, Chiba S, Noguchi H, Suzuki T, Iwata N, Ozaki N, Taguchi T, Kamiya A, Kosuga A, Tatsumi M, Kamijima K, Weinberger DR, Sawa A, Kunugi H (2006) Impact of the DISC1 Ser704Cys polymorphism on risk for major depression, brain morphology and ERK signaling. Hum Mol Genet 15:3024–3033.

Heale V, Harley C (1990) MK-801 and AP5 impair acquisition, but not retention, of the Morris milk maze. Pharmacol Biochem Behav 36:145–149.

Hikida T, Jaaro-Peled H, Seshadri S, Oishi K, Hookway C, Kong S, Wu D, Xue R, Andradé M, Tankou S, Mori S, Gallagher M, Ishizuka K, Pletnikov M, Kida S, Sawa A (2007) Dominant-negative DISC1 transgenic mice display schizophrenia-associated phenotypes detected by measures translatable to humans. Proc Natl Acad Sci U S A 104:14501–14506.

Hijzen TH, Broersen LM, Slangen JL (1991) Effects of subchronic d-amphetamine on prepulse and gap inhibition of the acoustic startle reflex in rats. Biol Psychiatr 1119–1128.

Hoffman HS, Searle JL (1968) Acoustic and temporal factors in the evocation of startle. J Acoust Soc Am 43:269–282.

Hodgkinson CA, Goldman D, Jaeger J, Persaud S, Kane JM, Lipsky RH, Malhotra AK (2004) Disrupted in schizophrenia 1 (DISC1): association with schizophrenia, schizoaffective disorder, and bipolar disorder. Am J Hum Genet 75:862–872.

Hori H, Noguchi H, Hashimoto R, Nakabayashi T, Omori M, Takahashi S, Tsukue R, Anami K, Hirabayashi N, Harada S, Saitoh O, Iwase M, Kajimoto O, Takeda M, Okabe S, Kunugi H (2006) Antipsychoticmedication and cognitive function in schizophrenia. Schizophr Res 86:138–146.

Imre G, Fokkema DS, Den Boer JA, Ter Horst GJ (2006) Dose-response characteristics of ketamine effect on locomotion, cognitive function and central neuronal activity. Brain Res Bull 69:338–345.

Jacobs PS, Taylor BM, Bardgett ME (2000) Maturation of locomotor and Fos responses to the NMDA antagonists, PCP and MK-801. Brain Res Dev Brain Res 122:91–95.

Javitt DC, Zukin SR (1991) Recent advances in the phencyclidine model of schizophrenia. Am J Psychiatr148:1301–1308.

Jentsch JD, Tran A, Le D, Youngren KD, Roth RH (1997) Subchronic phencyclidine administration reduces mesoprefrontal dopamine utilization and impairs prefrontal corticaldependent cognition in the rat. Neuropsychopharmacology 17:92–99.

Jones KW, Bauerle LM, DeNoble VJ (1990) Differential effects of sigma and phencyclidine receptor ligands on learning. Eur J Pharmacol 179: 97–102.

Kalivas PW, Hooks MS, Sorg B (1993) The pharmacology and neural circuitry of sensitization to psychostimulants. Behav Pharmacol 4:315–334.

Kamnasaran D, Muir WJ, Ferguson-Smith MA, Cox DW (2003) Disruption of the neuronal PAS3 gene in a family affected with schizophrenia. J Med Genet 40:325–332.

Karl T, Duffy L, Scimone A, Harvey RP, Schofield PR (2007) Altered motor activity, exploration and anxiety in heterozygous neuregulin 1 mutant mice: implications for understanding schizophrenia. Genes Brain Behav 6:677–687.

Keefe RS, Silva SG, Perkins DO, Lieberman JA (1999) The effects of atypical antipsychotic drugs on neurocognitive impairment in schizophrenia: a review and meta-analysis. Schizophr Bull 25:201–222.

Keefe RS, Eesley CE, Poe MP (2005) Defining a cognitive function decrement in schizophrenia. Biol Psychiatr 57:688–691.

Keith VA, Mansbach RS, Geyer MA (1991) Failure of haloperidol to block the effects of phencyclidine and dizocilpine on prepulse inhibition of startle. Biol Psychiatr 30:557–566.

Kesner RP, Hardy JD, Novak JM (1983) Phencyclidine and behavior: II. Active avoidance learning and radial arm maze performance. Pharmacol Biochem Behav 18:351–356.

Kumari V, Cotter PA, Mulligan OF, Checkley SA, Gray NS, Hemsley DR (1999) Thornton JC, Corr PJ, Toone BK, Gray JA: Effects of d-amphetamine and haloperidol on latent inhibition in healthy male volunteers. J Psychopharmacol 13:398–405.

Kunugi H, Takei N, Murray RM, Saito K, Nanko S (1996) Small head circumference at birth in schizophrenia. Schizophr Res 20:165–170.

Kunugi H, Nanko S, Murray RM (2001) Obstetric complications and schizophrenia: prenatal underdevelopment and subsequent neurodevelopmental impairment. Br J Psychiatr Suppl 40:25–9.

Kunugi H, Tanaka M, Hori H, Hashimoto R, Saitoh O, Hironaka N (2007) Prepulse inhibition of acoustic startle in Japanese patients with chronic schizophrenia. Neurosci Res 59:23–28.

Lahti AC, Weiler MA, Tamara Michaelidis BA, Parwani A, Tamminga CA (2001) Effects of ketamine in normal and schizophrenic volunteers. Neuropsychopharmacology 25:455–467.

Li Z, Kim CH, Ichikawa J, Meltzer HY (2003) Effect of repeated administration of phencyclidine on spatial performance in an eight-arm radial maze with delay in rats and mice. Pharmacol Biochem Behav 75:335–340.

Lipska BK, Weinberger DR (1993) Delayed effects of neonatal hippocampal damage on haloperidol-induced catalepsy and apomorphine-induced stereotypic behaviors in the rat. Brain Res Dev Brain Res 75:213–222.

Lipska BK, Weinberger DR (2000) To model a psychiatric disorder in animals: schizophrenia as a reality test. Neuropsychopharmacology 23:223–239.

Lipska BK, Jaskiw GE, Weinberger DR (1993) Postpubertal emergence of hyperresponsiveness to stress and to amphetamine after neonatal excitotoxic hippocampal damage: a potential animal model of schizophrenia. Neuropsychopharmacology 9:67–75.

Lubow RE (1997) Latent inhibition as a measure of learned inattention: some problems and solutions. Behav Brain Res 88:75–83; Review.

Lubow RE, Gewirtz JC (1995) Latent inhibition in humans: data, theory, and implications for schizophrenia. Psychol Bull 117:87–103; Review.

Lubow RE, Moore (1959) Latent inhibition: The effect of non-reinforced preexposure to the conditioned stimulus. J Comp Physiol Psychol 52:415–419.

Mansbach RS, Geyer MA (1989) Effects of phencyclidine and phencyclidine biologs on sensorimotor gating in the rat. Neuropsychopharmacology 2: 299–308.

Mansbach RS, Geyer MA, Braff DL (1988) Dopaminergic stimulation disrupts sensorimotor gating in the rat. Psychopharmacology (Berl) 94:507–514.

Marquis JP, Goulet S, Doré FY (2003) Schizophrenia-like syndrome inducing agent phencyclidine failed to impair memory for temporal order in rats. Neurobiol Learn Mem. 80:158–167.

Marquis JP, Audet MC, Doré FY, Goulet S (2007) Delayed alternation performance following subchronic phencyclidine administration in rats depends on task parameters. Prog Neuropsychopharmacol Biol Psychiatr 31:1108–1112.

McNeil TF, Cantor-Graae E (2000) Neuromotor markers of risk for schizophrenia. Aust N Z J Psychiatry 34:86–90; Review.

Millar JK, Wilson-Annan JC, Anderson S, Christie S, Taylor MS, Semple CA, Devon RS, Clair DM, Muir WJ, Blackwood DH, Porteous DJ (2000) Disruption of two novel genes by a translocation co-segregating with schizophrenia. Hum Mol Genet 9:1415–1423.

Miyakawa T, Leiter LM, Gerber DJ, Gainetdinov RR, Sotnikova TD, Zeng H, Caron MG, Tonegawa S (2003) Conditional calcineurin knockout mice exhibit multiple abnormal behaviors related to schizophrenia. Proc Natl Acad Sci U S A 100:8987–8992.
Mohn AR, Gainetdinov RR, Caron MG, Koller BH (1999) Mice with reduced NMDA receptor expression display behaviors related to schizophrenia. Cell 98:427–436.
Moghaddam B, Adams BW (1998) Reversal of phencyclidine effects by a group II metabotropic glutamate receptor agonist in rats. Science 281:1349–1352.
Morrens M, Hulstijn W, Lewi PJ, De Hert M, Sabbe BG (2006) Stereotypy in schizophrenia. Schizophr Res 84:397–404.
Morris R (1984) Developments of a water-maze procedure for studying spatial learning in the rat. J Neurosci Meth 11:47–60.
Murray RM, Lewis SW (1987) Is schizophrenia a neurodevelopmental disorder? Br Med J 295:681–682.
Murray TK, Ridley RM (1997) The effect of dizocilpine (MK-801) on conditional discrimination learning in the rat. Behav Pharmacol 8:383–388.
Myhrer T (2003) Neurotransmitter systems involved in learning and memory in the rat: a meta-analysis based on studies of four behavioral tasks. Brain Res Brain Res Rev 41: 268–287; Review.
Myin-Germeys I, Delespaul P, van Os J (2005) Behavioural sensitization to daily life stress in psychosis. Psychol Med 35:733–741.
Noda Y, Yamada K, Furukawa H, Nabeshima T (1995) Enhancement of immobility in a forced swimming test by subacute or repeated treatment with phencyclidine: a new model of schizophrenia. Br J Pharmacol 116:2531–2537.
Noda Y, Mamiya T, Furukawa H, Nabeshima T (1997) Effects of antidepressants on phencyclidine-induced enhancement of immobility in a forced swimming test in mice. Eur J Pharmacol 324:135–140.
Norton N, Williams HJ, Williams NM, Spurlock G, Zammit S, Jones G, Jones S, Owen R, O'Donovan MC, Owen MJ (2003) Mutation screening of the Homer gene family and association analysis in schizophrenia. Am J Med Genet B Neuropsychiatr Genet 120:18–21.
O'Tuathaigh CM, O'Connor AM, O'Sullivan GJ, Lai D, Harvey R, Croke DT, Waddington JL (2008) Disruption to social dyadic interactions but not emotional/anxiety-related behaviour in mice with heterozygous 'knockout' of the schizophrenia risk gene neuregulin-1. Prog Neuropsychopharmacol Biol Psychiatr 32:462–466.
Packard MG, White NM (1989) Memory facilitation produced by dopamine agonists: role of receptor subtype and mnemonic requirements. Pharmacol Biochem Behav 33:511–518.
Patil ST, Zhang L, Martenyi F, Lowe SL, Jackson KA, Andreev BV, Avedisova AS, Bardenstein LM, Gurovich IY, Morozova MA, Mosolov SN, Neznanov NG, Reznik AM, Smulevich AB, Tochilov VA, Johnson BG, Monn JA, Schoepp DD (2007) Activation of mGlu2/3 receptors as a new approach to treat schizophrenia: a randomized Phase 2 clinical trial. Nat Med 13:1102–1107.
Picada J.N., Henriques J.A. and Roesler R (2003) An oxidized form of apomorphine fails to induce stereotypy. Schizophr. Res 63:199–200.
Pierce RC, Kalivas PW (1997) A circuitry model of the expression of behavioral sensitization to amphetamine-like psychostimulants. Brain Res Brain Res Rev 25:192–216.
Porsolt RD, Bertin A, Jalfre M (1978) "Behavioural despair" in rats and mice: strain differences and the effects of imipramine. Eur J Pharmacol 51:291–294.
Powell CM, Miyakawa T (2006) Schizophrenia-Relevant Behavioral Testing in Rodent Models: A Uniquely Human Disorder? Biol Psychiatr 59:1198–1207.
Randrup A, Munkvad I (1974) Pharmacology and physiology of stereotyped behavior. J Psychiatr Res 11:1–10.
Ridley RM (1994) The psychology of perserverative and stereotyped behaviour. Prog Neurobiol 44: 221–231.
Robinson TE, Becker JB (1986) Enduring changes in brain and behavior produced by chronic amphetamine administration: a review and evaluation of animal models of amphetamine psychosis. Brain Res 396:157–198.

Sams-Dodd F (1996) Phencyclidine-induced stereotyped behaviour and social isolation in rats: a possible animal model of schizophrenia. Behav Pharmacol 7:3–23.

Sams-Dodd F (1998a) Effects of continuous D-amphetamine and phencyclidine administration on social behaviour, stereotyped behaviour, and locomotor activity in rats. Neuropsychopharmacology 19:18–25.

Sams-Dodd F, Lipska BK, Weinberger DR (1997) Neonatal lesions of the rat ventral hippocampus result in hyperlocomotion and deficits in social behaviour in adulthood. Psychopharmacology (Berl) 132:303–310.

Sato M, Chen C-C, Akiyama K, Otsuki S (1983) Acute exacerbation of paranoid psychotic state after long-term abstinence in patients with previous methamphetamine psychosis. Biol Psychiatr 18:429–440.

Solomon PR, Crider A, Winkelman JW, Turi A, Kamer RM, Kaplan LJ (1981) Disrupted latent inhibition in the rat with chronic amphetamine or haloperidol-induced supersensitivity: relationship to schizophrenic attention disorder. Biol Psychiatr 16:519–537.

Stefansson H, Sigurdsson E, Steinthorsdottir V, Bjornsdottir S, Sigmundsson T, Ghosh S, Brynjolfsson J, Gunnarsdottir S, Ivarsson O, Chou TT, Hjaltason O, Birgisdottir B, Jonsson H, Gudnadottir VG, Gudmundsdottir E, Bjornsson A, Ingvarsson B, Ingason A, Sigfusson S, Hardardottir H, Harvey RP, Lai D, Zhou M, Brunner D, Mutel V, Gonzalo A, Lemke G, Sainz J, Johannesson G, Andresson T, Gudbjartsson D, Manolescu A, Frigge ML, Gurney ME, Kong A, Gulcher JR, Petursson H, Stefansson K (2002) Neuregulin 1 and susceptibility to schizophrenia. Am J Hum Genet 71:877–892.

Swerdlow NR, Geyer MA (1998) Using an animal model of deficient sensorimotor gating to study the pathophysiology and new treatments of schizophrenia. Schizophr Bull 24:285–301.

Swerdlow NR, Braff DL, Taaid N, Geyer MA (1994) Assessing the validity of an animal model of deficient sensorimotor gating in schizophrenic patients. Arch Gen Psychiatr 51:139–154.

Szumlinski KK, Lominac KD, Kleschen MJ, Oleson EB, Dehoff MH, Schwarz MK, Seeburg PH, Worley PF, Kalivas PW (2005) Behavioral and neurochemical phenotyping of Homer1 mutant mice: possible relevance to schizophrenia. Genes Brain Behav 4:273–288.

Takao K, Yamasaki N, Miyakawa T (2007) Impact of brain-behavior phenotypying of genetically-engineered mice on research of neuropsychiatric disorders. Neurosci Res 58:124–132.

Thornton AE, Van Snellenberg JX, Sepehry AA, Honer W (2006) The impact of atypical antipsychotic medications on long-term memory dysfunction in schizophrenia spectrum disorder: a quantitative review. J. Psychopharmacol 20:335–346.

Ujike H (2002) Stimulant-induced psychosis and schizophrenia: the role of sensitization. Curr Psychiatr Rep 4:177–184.

Vaillancourt C, Boksa P (2000) Birth insult alters dopamine-mediated behavior in a precocial species, the guinea pig. Implications for schizophrenia. Neuropsychopharmacology 23:654–666.

Verma A, Moghaddam B (1996) NMDA receptor antagonists impair prefrontal cortex function as assessed via spatial delayed alternation performance in rats: modulation by dopamine. J Neurosci 16:373–379.

Wang C, McInnis J, Ross-Sanchez M, Shinnick-Gallagher P, Wiley JL, Johnson KM (2001) Long-term behavioral and neurodegenerative effects of perinatal phencyclidine administration: implications for schizophrenia. Neuroscience 107:535–550.

Warburton EC, Joseph MH, Feldon J, Weiner I, Gray JA (1994) Antagonism of amphetamine-induced disruption of latent inhibition in rats by haloperidol and ondansetron: implications for a possible antipsychotic action of ondansetron. Psychopharmacology (Berl) 114: 657–664.

Wass C, Archer T, Pålsson E, Fejgin K, Klamer D, Engel JA, Svensson L (2006) Effects of phencyclidine on spatial learning and memory: nitric oxide-dependent mechanisms. Behav Brain Res 171:147–153.

Weinberger DR (1987) Implications of normal brain development for the pathogenesis of schizophrenia. Arch Gen Psychiatr 44:660–669.

Weiner I, Lubow RE, Feldon J (1984) Abolition of the expression but not the acquisition of latent inhibition by chronic amphetamine in rats. Psychopharmacology (Berl) 83:194–199.

Weiner I, Lubow RE, Feldon J (1988) Disruption of latent inhibition by acute administration of low doses of amphetamine. Pharmacol Biochem Behav 30:871–878.

Weiner I, Shadach E, Barkai R, Feldon J (1997) Haloperidol- and clozapine-induced enhancement of latent inhibition with extended conditioning: implications for the mechanism of action of neuroleptic drugs. Neuropsychopharmacology 16:42–50.

Wolff MC, Leander JD (2003) Comparison of the effects of antipsychotics on a delayed radial maze task in the rat. Psychopharmacology (Berl) 168:410–416.

Yuii K, Suzuki M, Kurachi M (2007) Stress sensitization in schizophrenia. Ann N Y Acad Sci 1113:276–290.

Zeng H, Chattarji S, Barbarosie M, Rondi-Reig L, Philpot BD, Miyakawa T, Bear MF, Tonegawa S (2001) Forebrain-specific calcineurin knockout selectively impairs bidirectional synaptic plasticity and working/episodic-like memory. Cell 107:617–629.

Zhang HT, Huang Y, Suvarna NU, Deng C, Crissman AM, Hopper AT, De Vivo M, Rose GM, O'Donnell JM (2005) Effects of the novel PDE4 inhibitors MEM1018 and MEM1091 on memory in the radial-arm maze and inhibitory avoidance tests in rats. Psychopharmacology (Berl) 179:613–619.

Chapter 8
Synaptoproteomics of Existing and New Animal Models of Depression

Alessandra Mallei*, Roberto Giambelli*, Aram El Khoury, Susanne H. M. Gruber, Laura Musazzi, Valentina S. Barbiero, Daniela Tardito, Barbara Vollmayr, Peter Gass, Aleksander A. Mathé, Giorgio Racagni, and Maurizio Popoli (✉)

Abstract Depression is a severe and life-threatening psychiatric illness whose pathogenesis is still essentially unknown. Proteomic analysis of synaptic terminals (synaptoproteomics) in animal models of depression is a powerful approach to gain insight into the molecular mechanisms underlying vulnerability to mood disorders and the long-term action of drug treatments. Here, we employed two different animal models of depression, the Learned Helplessness rats (a classical behavioral model of depression) and a new model of depression with gene–environment interaction (Flinders Sensitive Line rats subjected to early life stress). Both animal models were treated with the antidepressant escitalopram. Analysis of their synaptoproteomic profile revealed a number of protein spots differently regulated by basic vulnerability and/or early life stress. Using this approach, we obtained information regarding biomarkers that may represent predictors of pathology or response/resistance to drug treatment, as well as potential targets for novel pharmacological and therapeutic strategies.

Abbreviations 2D: Two-dimensional; 5-HT_{1A}: Serotonin 1A receptor; BCA: Bicinchoninic Acid; CHAPS: 3-[(3-Cholamidopropyl)dimethylammonio]-1-propanesulfonate; DA: Dopamine; DAVID: Database for Annotation, Visualization and Integrated Discovery; ESC: Escitalopram; FRL: Flinders Sensitive Line; FSL: Flinders Resistant Line; G × E: Gene–Environment; GO: Gene Ontology; HC: Hippocampus; HPA: Hypothalamic–Pituitary–Adrenal; IBS: Inflammatory Bowel Syndrome; IEF: Isoelectric Focusing; IPA: Ingenuity Pathways Analysis;

M. Popoli
Center of Neuropharmacology, Department of Pharmacological Sciences, University of Milano, Via Balzaretti, 9, 20133, Milano, Italy
maurizio.popoli@unimi.it

* These authors equally contributed.

C.W. Turck (ed.) *Biomarkers for Psychiatric Disorders*,
doi: 10.1007/978-0-387-79251-4_8,

IPG: Immobilized pH Gradient; LH: Learned Helplessness; mRNA: messenger RNA; MS: Maternal separation; nLH: Non-Learned Helplessness; NMDA: *N*-methyl-D-aspartate; P/FC: Prefrontal/frontal cortex; pI: Isoelectric point; PND: Post natal day; REM: Random eye movement; SDS: Sodium dodecyl sulfate; SDS-PAGE: Sodium dodecyl sulfate polyacrylamide gel electrophoresis; Tris: Tris(hydroxymethyl)aminomethane; Veh: Vehicle

8.1 Introduction

The available prescription drugs for the treatment of depression alleviate symptoms in only about 60–70% of patients and many patients experience adverse effects that cause discontinuation leading to cessation of the treatment (Keller and Boland, 1998). Moreover, there is a time lag of 3–4 weeks before the onset of the therapeutic effects can be appreciated. Indeed, individuals respond differently to antidepressant drugs, and often the effects are unpredictable (efficacy, onset of action, adverse effects, placebo effects). Therefore, to optimize response to antidepressant treatments and to minimize adverse effects, it is necessary to identify biomarkers of individual susceptibility and response to drugs. Proteins are the actual effectors of biological functions, and since mRNA and protein expression levels are often not related (Anderson and Seilhamer, 1997; Paulson et al., 2003), information provided by transcriptional studies cannot entirely account for functional changes. Therefore, global analysis of protein expression is a powerful approach to gain insight into the molecular mechanisms underlying vulnerability to psychiatric disorders and the long-term action of drug treatments (Rohlff and Hollis, 2003; Fountoulakis, 2004; Kramer and Cohen, 2004, Vercauteren et al., 2004).

Animal models are important tools for the study of mood disorders because they allow the employment of behavioral and biochemical research methods that cannot be used, for evident ethical reasons, in humans. Therefore, scientists have developed several animal models to evaluate the validity and consistency of diverse experimental approaches to the study of psychiatric diseases. An appropriate model for human mood disorders should accomplish the main criteria of *construct validity* (how closely the model is consistent with the theoretical rationale), *face validity* (how closely the model reproduces symptoms of the pathology) and *predictive validity* (how well the model responds to pharmacological treatment) (Hitzemann, 2000; Urani et al., 2005). These criteria have been used in particular to validate the most important animal models for major depression. In this work we have applied global analysis of protein expression to the study of two animal models. The models employed are a classical behavioral animal model of depression (Learned Helplessness; LH) and a new animal model of depression with gene–environment interaction (Flinders Sensitive Line rats subjected to early life stress).

8.2 The Learned Helplessness Rat Model of Depression

Clinical studies indicate that a broad range of genetic and environmental influences, through combined action in vulnerable individuals, are involved in the pathophysiology of depression (Matthews et al., 2005). Therefore it is reasonable that an external stress may result in changes in behavior in a laboratory animal. For these reasons, one of the most significant and validated animal models of depression applies an uncontrollable stress, which in a number of animals leads to a depressive-like behavioral phenotype (Vollmayr and Henn, 2003). The LH protocol was initially elaborated by Seligman and Overmier (Overmier and Seligman, 1967) and induces the behavioral condition through the application of acute uncontrollable stress. Usually the aversive stimuli are administered by foot shock, but tail shock and loud acoustic sound have also been used. Originally the procedure was carried out with dogs but the paradigm was subsequently investigated and completely developed in rats. The procedure we used consists of a 40-min session of inescapable shock (0.8 mA; single shocks randomized by computer with inter-shock time lasting from 5 to 15 s) and the total shock duration is 20 min (Vollmayr and Henn, 2001). Twenty-four hours after the application of inescapable stress, the animals are tested in a conditioning chamber. The same current is applied with 15 shocks lasting 60 s each with 24 s time interval. The current is accompanied by a light pointed to a lever in order to facilitate its detection and the animals can stop the shock by pressing the lever. Thus, a computer records the time necessary for each animal to terminate each trial, in order to classify the animal behavior. More than 10 failures identify the animals as "learned helpless", whereas with less than five failures they are considered "non-learned helpless" (nLH) (Table 8.1). Generally with Sprague-Dawley rats we found about 20% of LH in such test runs (Vollmayr and Henn, 2001). The LH rats show significant face, construct and predictive validity: they have weight loss, sleep changes, decreased libido, anhedonia (monitored using sucrose preference test) (Willner, 1990; Willner, 1995), and hypothalamic–pituitary–adrenal (HPA) axis alterations. Moreover, various antidepressants are able to reverse helpless behavior: tricyclics, norepinephrine reuptake inhibitors, and serotonin reuptake inhibitors (Vollmayr and Henn, 2001). Therefore, LH animals display the most important behavioral and symptomatological changes associated with

Table 8.1 Parameters for inescapable shock and learned helplessness test, adapted from Wollmayr and Henn 2001

Inescapable shock	Test (after 24h)	Score
Forty-min session of 0.8 mA; Total shock duration 20 min	Fifteen shocks of 0.8 mA lasting 60 s each with an intertrial gap of 24 s	More than 10 "failures" (animal does not press the lever during the 60 s of the shock) classified the rat as LH
	Animals can stop a trial by pressing a lever	Less tan 5 'failures' is classified nLH

depression; the rats maintain helpless behavior for about 2 weeks after acquisition (Henn et al., 2002; Vollmayr et al., 2003).

8.3 The Flinders Sensitive Line (FSL) and the Flinders Resistant Line (FRL) Rat Model of Depression

The Flinders Lines of rats were established by selective breeding of Sprague-Dawley rats for their sensitivity to cholinergic agents in behavioral paradigms. The breeding program developed two separate lines, FSL and FRL, distinct for their sensitivity to the effects of the anticholinesterase agent diisopropyl fluorophosphate (Overstreet and Russell, 1982).

The FSL rats are a well-validated animal model of depression carrying genetic vulnerability associated with distinct features of the pathology and responsiveness to antidepressant drugs. The FSL rats fulfill the major validation criteria of construct, face, and predictive validity for a good animal model (Yadid et al., 2000; Overstreet et al., 2005). Indeed, it has been shown that a subset of depressed individuals display supersensitivity to cholinergic agonists (Janowsky et al., 1980, 1994); therefore, the FSL model is consistent with the cholinergic model of depression (good construct validity). Moreover, this model has many behavioral similarities with depressed individuals, e.g., reduced general activity, decreased appetite, decreased sexual activity, reduced latency of REM sleep and increased length of REM sleep episodes, anhedonia (after chronic mild stress) (Pucilowski et al., 1993), and serotonergic abnormalities (good face validity) (see Table 8.2). In addition, FSL rats show an increased immobility in the Porsolt swim test (a widely

Table 8.2 Behavioral and biochemical features of FSL rats compared with human pathology, adapted from Overstreet et al. 2005

Depressed individuals	FSL rats
Psycomotor retardation	Reduced bar pressing for reward
Anhedonia	Reduced saccharin intake[a]
Reduced appetite; loss of weight	Lower body weight; reduced appetite
Elevated REM sleep	Elevated REM sleep
Reduced REM sleep latency	Reduced REM sleep latency
Reduced T-cell activity	Reduced T-cell activity
Higher incidence of IBS	Greater gut sensitivity to antigen
Higher incidence of asthma	Greater airway sensitivity to antigen
Anxiety of some types	Anxiety in some tasks
Increased cholinergic sensitivity	Increased cholinergic sensitivity
Altered 5-HT_{1A}	Altered 5-HT_{1A}
Reduced Neuropeptide Y	Reduced Neuropeptide Y

IBS inflammatory bowel syndrome; *DA* dopamine; $5\text{-}HT_{1A}$ serotonin 1A receptor
[a]After chronic mild stress

used test for behavioral despair), which is reversible by chronic but not acute treatment with antidepressant drugs (good predictive validity).

8.4 Gene–Environment Interaction: Developing a New Animal Model of Depression

The pathogenesis of depression is still essentially unknown. Genetic epidemiological studies found no evidence of the classic Mendelian inheritance in humans but showed that interaction between multiple genes of modest effect and environmental factors (G×E) confers vulnerability to the disease (Lesch, 2004). Environmental risk factors include maternal stress during pregnancy, deprivation of parental care during childhood, childhood neglect or childhood abuse, parental loss, substance abuse, toxic exposure, and head injury (Caspi and Moffitt, 2006). Indeed, the experience of stressful events in childhood (*early life stress*) was found to increase the risk for the development of mood disorders in adult life (Heim and Nemeroff, 2001; Caspi et al., 2003).

We employed an innovative experimental design to reproduce in an animal model the interaction between environmental adverse events and genetic susceptibility. To reproduce early life stress events, the FSL rats and their control FRL rats were subjected to a standard maternal separation protocol from postnatal day 2 (PND-2) to PND-14 (Plotsky and Meaney, 1993). Indeed, maternal separation has been shown to induce HPA axis alteration such as increase in both adrenocorticotropic hormone and corticosterone plasma levels (Vazquez et al., 2000). Moreover, rats subjected to maternal separation exhibit behavioral abnormalities such as decrease in saccharin intake (anhedonia), increased startle response, and reduced exploration in a novel open field (Caldji et al., 2000; Ladd et al., 2000).

We began recently to characterize this new animal model of depression behaviorally and biochemically. It has been shown that FSL rats subjected to maternal separation have lower body weight and higher immobility time in the Porsolt swim test than FRL maternally separated rats. Indeed, lower body weight and lower motor activity are two features that can be found in depressed individuals. Moreover, treatment with escitalopram (ESC) was found to increase swim duration in FSL and FSL subjected to maternal separation but not in FRL (El Khoury et al., 2006). A strong inhibition of long-term potentiation induction (a cellular model of synaptic plasticity) and marked reduction of *N*-methyl-D-aspartate (NMDA) receptor synaptic expression were found in the hippocampus of the FSL rats. Early life stress exacerbated depressive behavioral phenotype and enhanced synaptic plasticity only in FSL rats. This was accompanied by marked increase of synaptic NMDA receptors, abnormal regulation of presynaptic mechanisms, altered response of synaptic extracellular signal-regulated mitogen-activated protein kinases, and reduced behavioral response to antidepressant treatment (Musazzi et al., submitted). These data suggest that early gene–environment interactions cause lifelong synaptic changes in FSL by inducing maladaptive plasticity, affecting both the course of depressive-like behavior and response to drugs.

8.5 Optimization of Synaptoproteomic Analysis

Two-dimensional (2D) gel electrophoresis coupled with mass spectrometry is a powerful technique that allows the separation and identification of thousands of proteins in a single experiment. 2D electrophoresis consists of two sequential separation steps: the first dimension resolves proteins on the basis of their isoelectric point (isoelectric focusing, IEF) and the second dimension separates the focused proteins on the basis of their molecular mass by sodium dodecyl sulfate polyacrylamide gel electrophoresis (SDS-PAGE). In the last decade, both the development of specific software able to simultaneously analyze thousands of proteins in many gels and the improvement of mass spectrometer sensitivity, boosted the proteomic technology potentials. Nevertheless, the number of proteins expressed in a particular cell or tissue has been estimated to be more than 100,000 (including post-translational variants), making the separation and identification of all the proteins in the sample problematic. Moreover, the presence of highly expressed proteins such as cytoskeletal or housekeeping proteins could mask the detection of proteins expressed at a lower level. A strategy to reduce the complexity of the proteome and to enrich for less abundant proteins is to prefractionate the sample examined (Fountoulakis, 2004; Stasyk and Huber, 2004; Vercauteren et al., 2004).

Synaptosomes are isolated nerve endings obtained by differential centrifugation of brain tissue. They contain the presynaptic terminal, including mitochondria and synaptic vesicles, and part of the postsynaptic site, including the postsynaptic density. Therefore, this subcellular fraction is greatly enriched in proteins involved in synaptic functions. Furthermore, the main biological targets of antidepressant drugs are located at synaptic sites. A few proteomics studies have been carried out on subcellular synaptic fractions (Satoh et al., 2002; Li et al., 2004; Liberatori et al., 2004; Boyd-Kimball et al., 2005; Witzmann et al., 2005; Mello et al., 2007), but no one has thus far attempted the characterization of the synaptic proteome in animal models of depression.

Here we give an overview of the 2D electrophoresis protocol that we are currently using in our laboratory to perform synaptoproteomic studies.

8.5.1 Sample Preparation

Synaptosomes were purified by discontinuous Percoll gradient procedure according to Dunkley and colleagues (Dunkley et al., 1986) with minor modifications (Bonanno et al., 2005). This technique has several advantages over the Ficoll or sucrose gradient methods, i.e., isotonicity can be maintained throughout the isolation; synaptosomes are free of myelin and extrasynaptosomial mitochondria and are functional, viable, and characterized as protein markers (Dunkley et al., 1986; Raiteri and Raiteri, 2000, Barbiero et al., 2007). Briefly, pooled hippocampi (HC) and prefrontal/frontal cortex (P/FC) were homogenized in 10 vol of 0.32 M sucrose, and buffered at pH 7.4 with Tris, using a glass/teflon tissue grinder (clearance, 0.25 mm). The homogenate was centrifuged at 1000*g* to remove nuclei and debris, and the supernatant was gently stratified on a discontinuous Percoll gradient (2, 6, 10, and 20% v/v in Tris-buffered sucrose) and centrifuged at 33,500*g* for 5 min. The layer between 10 and 20% Percoll (synaptosomal fraction) was collected and washed by centrifugation and the resulting pellet was stored at −80°C.

Transmembrane proteins or membrane-associated proteins present in the synaptosomal fraction are difficult to solubilize. Therefore, to ensure optimal protein solubilization and denaturation, and to minimize horizontal streaking due to protein–protein interaction, synaptosome pellets were dissolved in an IEF buffer containing 7 M urea, 2 M thiourea, 40 mM Tris, 3 mM tributylphosphine, 2% CHAPS, and 1% carrier ampholytes plus protease inhibitors (Carboni et al., 2002). Indeed, in order to obtain solubilization of cytosolic and membrane proteins in a single step, we decided to use CHAPS, a surfactant that performed better than others employed in different buffers. Protein content was measured by BCA assay (Pierce Chemical) after dialysis in 1% SDS in water. Salts, small ionic molecules, or other contaminants present in the synaptosomes can potentially interfere with the first-dimension focusing. If necessary, the sample can be subjected to an additional desalting step prior to IEF by means of protein precipitation (2D Clean-Up Kit, GE Healthcare), following the manufacturer's instructions.

8.5.2 Isoelectric Focusing

Immobilized pH gradient (IPG) strips are made by acrylamide derivatives containing either a free carboxylic acid or a tertiary amino group that is copolymerized with acrylamide and bis-acrylamide and immobilized on a plastic sheet. Therefore, the pH gradient is stable and allows better reproducibility between experiments. Since 70% of brain proteins have an isoelectric point (pI) between 3 and 10, with most having a pI range of 4–7, non-linear-gradient pH 3–10 IPG strips were chosen, in contrast to some other groups, to allow better resolution of proteins with pI values between pH 4 and 7. Moreover, we chose to employ 7 cm IPG strips, because this shorter length allows the loading of smaller amounts of synaptosomes without compromising protein resolution. Therefore, a 7-cm nonlinear IPG strip (Bio-Rad) of pH 3–10 was rehydrated for 16 h with 125 μL of IEF buffer containing 115 μg of synaptosomes, 10 mM iodoacetamide as alkylating agent, and a trace of bromophenol blue. IEF was performed at 15°C at a maximum of 4000 V for a total of 28,000 Vh using a Protean IEF Cell (Bio-Rad). The Protean IEF Cell allows the isoelectric focusing of up to 12 strips at the same time. The voltage was increased stepwise from 100 to 4000 V (see Table 8.3 for running conditions).

Table 8.3 IEF voltage settings

Step	Voltage (V)	Time	Volt-hours	Current[a](μA)	Ramp
1[b]	50	2h		50	Slow
2	100	15min		50	Rapid
3	250	1h		50	Rapid
4	1000	10h		50	Rapid
5	4000		28,000Vh	50	Rapid

[a] Current limit for each IEF strip
[b] Optional step to improve sample desalting

8.5.3 SDS-PAGE and Gel Staining

Prior to the second dimension, the IPG strips were equilibrated for 25 min in a solution containing 6 M urea; 2% SDS; 375 mM Tris pH 8.8; and 4 mM tributyl-phosphine. A multiple mini vertical slab gel system (Bio-Rad's Mini-PROTEAN 3 Dodeca Cell) was chosen to perform the SDS-PAGE because it allows the analysis of small amounts of samples while ensuring protein spot resolution. An 8–18%T gradient polyacrylamide gel was used in order to obtain better resolution of proteins ranging 10–250 kD. After the equilibration step, the IPG strips were placed on top of the gradient gels and sealed with 0.5% agarose in running buffer (192 mM glycine, 15 mM Tris, 0.1% SDS plus bromophenol blue). Gels were run at a constant temperature of 15°C at 5 mA per gel for 1 h and 10 mA per gel until the bromophenol blue front reached the bottom of the gel.

Gels with synaptosomal proteins resolved by 2D electrophoresis were then fixed in 10% ethanol and 7% acetic acid for 4 h. SYPRO Ruby fluorescent dye was used to stain proteins in the gels. Indeed, in contrast to the other staining methods, SYPRO Ruby gives very little background is very sensitive (1 ng of protein), and the staining is linear over 3 orders of magnitude. The result is a significant improvement in terms of resolution and number of spots of the maps. The gels were stained for 20 h with SYPRO Ruby and destained for 1 h in a 10% ethanol and 7% acetic acid solution. Afterwards, gels were stored in distilled water until protein spots were excised from the gel.

8.5.4 Two-dimensional Image Analysis

Gel images were acquired by using the Quantity One software and the VersaDoc imaging system (Bio-Rad). Spot detection and map comparison (matchset analysis) were carried out with the PDQuest 7.3 software (Bio-Rad). In brief, the PDQuest software is able to automatically detect protein spots in the acquired 2D maps.

The software creates a set of three images: the original gel image (Figs. 8.1 and 8.2), the filtered image, and the Gaussian image (a synthetic image in which each spot has been remodeled to fit a Gaussian curve, thus representing an ideal spot).

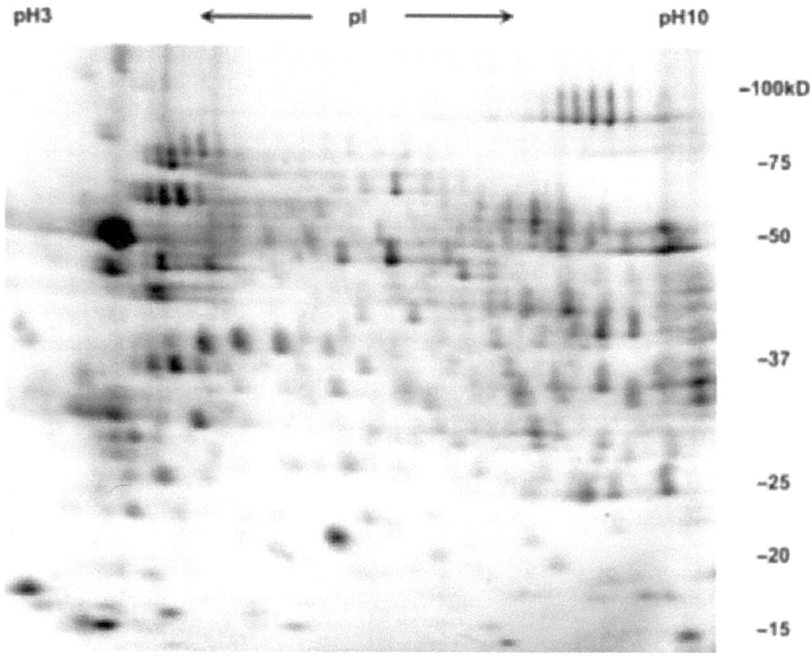

Fig. 8.1 Example of 2D electrophoresis map of synaptosomes obtained from P/FC of FSL/FRL rats

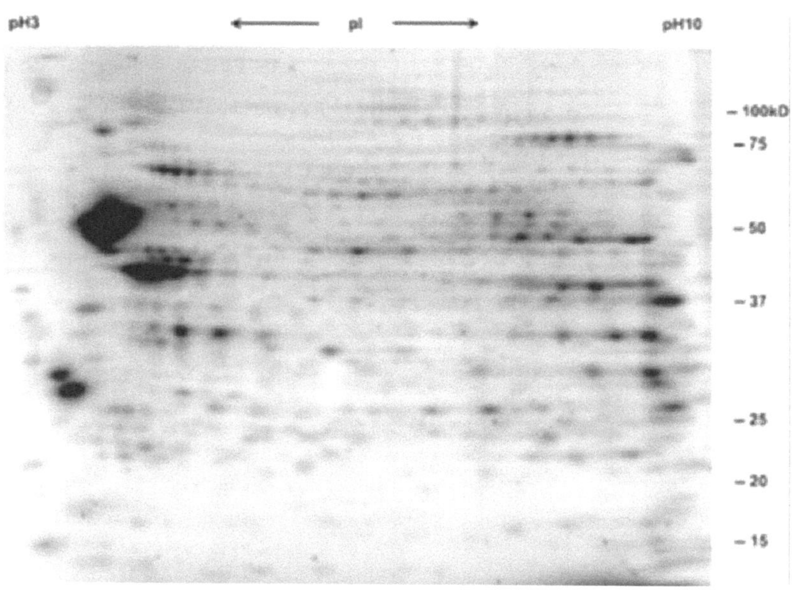

Fig. 8.2 Example of 2D electrophoresis map of synaptosomes obtained from hippocampus of LH – nLH rats

These images were then landmarked and matched to allow the comparison of protein expression patterns across the experimental samples (Figs. 8.3 and 8.4). Tables8.4 and 8.5 summarize the analysis of spot-matching between 2D gels from different experimental groups. Statistical analysis (*t*-test) was carried out by PDQuest

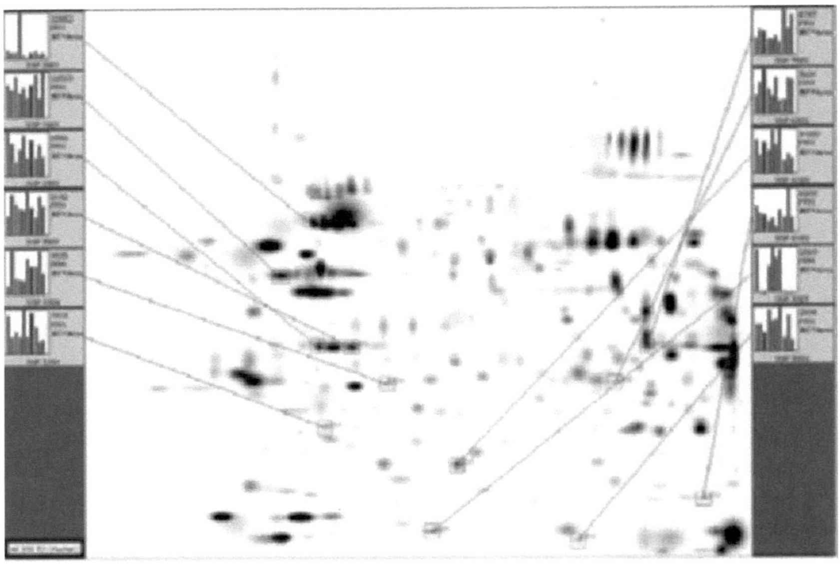

Fig. 8.3 Example of hippocampus master image from synaptosomes of LH–nLH rats. The master is a synthetic image containing the spot data from all the 2D maps in the analysis set. Left and right side bar graphs represent expression of the highlighted proteins in the various comparisons (*See Color Plates*)

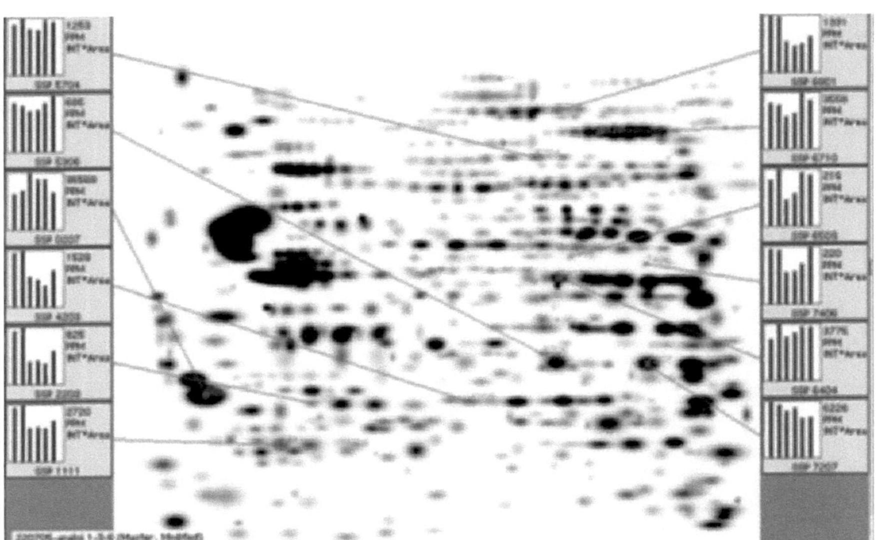

Fig. 8.4 Example of prefrontal/frontal cortex master image of synaptosomes from FSL/FRL. Left and right side bar graphs represent expression of the highlighted proteins in the various comparisons (*See Color Plates*)

Table 8.4 Matching summary of LH–nLH HC synaptosome maps

Gel Name	Spots	Matched[b]	Match Rate 1[c] (%)	Match Rate 2[d] (%)	Corr. Coeff.[e]
LH+veh II	397	316	79	86	0.482
LH+ESC I[a]	283	283	100	77	1.000
nLH+ESC I	291	248	85	67	0.272
nLH+veh I	354	287	81	78	0.538
LH+veh I	404	315	77	86	0.453
nLH+veh II	318	256	80	69	0.546
LH+veh III	353	285	80	77	0.437
nLH+ESC III	256	212	82	57	0.414
LH+ESC II	216	194	89	53	0.632
nLH+ESC II	246	212	86	57	0.376

[a] Master reference map
[b] Number of spots matched to master
[c] Percentage of matched spots relative to the total number of spots on the gel
[d] Percentage of matched spots relative to the total number of spots on the master
[e] Calculated by PDQuest software

Table 8.5 Matching summary of FSL synaptosome maps

Gel Name	Spots	Matched[b]	Match rate 1[c] (%)	Match rate 2[d] (%)	Corr. Coeff.[e]
FSL I	521	415	79	70	0.891
FSL II[a]	588	588	100	99	1
FSL MS I	450	382	84	64	0.9
FSL MS II	476	418	87	70	0.907
FSL MS ESC I	451	392	86	66	0.902
FSL MS ESC II	509	424	83	71	0.894

[a] Master reference map
[b] Number of spots matched to master
[c] Percentage of matched spots relative to the total number of spots on the gel
[d] Percentage of matched spots relative to the total number of spots on the master
[e] Calculated by PDQuest software

software to compare replicate groups and identify sets of proteins that show statistically significant difference with a confidence level of 0.05.

8.6 Experimental Design and Results in the LH-nLH Animal Model

In the study on LH-nLH animal model of depression, we investigated, by using synaptoproteomics, the basal synaptic difference between LH and nLH rats and the changes in the proteome induced by antidepressant treatment in the two animal groups. Both LH and nLH rats were also subchronically treated for 6 days with the selective serotonin reuptake inhibitor ESC (25 mg kg^{-1} day^{-1}). Thus four different experimental groups were prepared:

Table 8.6 Comparisons of experimental groups in LH/nLH HC

Comparisons	Spots up-or downregulated
LH + veh vs. nLH + veh	11
LH + ESC vs. LH + veh	17
nLH + ESC vs. nLH + veh	10

1. LH+v ehicle
2. nLH+v ehicle
3. LH+ESC
4. nLH+ESC

On the last day of drug administration the animals were sacrificed and HC and P/FC were dissected. After preparation of synaptosomes from these brain regions, the "synaptic map" from each group was obtained using the experimental procedures described above (Fig. 8.1). Three comparisons were then performed:

1. (LH + vehicle) vs. (nLH + vehicle)
2. (LH + ESC) vs. (LH + vehicle)
3. (nLH + ESC) vs. (nLH + vehicle)

Maps of both P/FC and HC showed a considerable number of protein spots whose expression levels were differently regulated in LH and nLH groups and after antidepressant treatment (Table8.6).

As an example, with regard to HC synaptosomes (Figs. 8.1 and 8.3) we found 11 proteins differently expressed between basal LH and nLH groups. In detail, nine spots were upregulated and two were downregulated in LH compared to nLH. Following antidepressant treatment, the expression levels of 17 synaptic proteins were modified in LH rats (eight upregulated and nine downregulated). Moreover, 10 proteins were modified by drug administration in nLH rats (six upregulated and only two downregulated by ESC) (Table8.6).

Interestingly, we observed that three proteins that were differently regulated between basal LH and nLH were also modulated by the pharmacological treatment. Similar results were found in the P/FC area.

8.7 Experimental Design and Results in FSL/FRL Animal Model

To mimic early life adverse events, FSL/FRL rats were subjected to a standard maternal separation protocol according to Plotsky and Meaney (1993). Briefly, FSL/FRL pups and dam were removed from the home cage, and the pups were placed in two distinct temperature-controlled cage (30°C) incubators in separate rooms so that no (ultrasonic) communication was possible between the dam and pups. The separation procedure lasted from PND-2 to PND-14 for 3 h daily. At the end of the separation period, both pups and dam were put back in the home cage. FSL/FRL pups were then left undisturbed with their mother until PND-23 when they were weaned.

Treatment with the selective serotonin reuptake inhibitor ESC (25 mg kg^{-1} day^{-1}) was carried out from PND-43 to PND-73 by using drug-containing dietary chow. A week before the end of the treatment, rats were subjected to the Porsolt test to check for drug efficacy (El Khoury et al., 2006). At PND-73, P/FC and HC regions were removed and synaptosomes prepared. The experimental groups were as follows:

1. FSL rats treated with vehicle (FSL veh)
2. FRL rats treated with vehicle (FRL veh)
3. FSL rats subjected to ESC treatment (FSL ESC)
4. FSL rats subjected to maternal separation (FSL MS)
5. FRL rats subjected to maternal separation (FRL MS)
6. FSL rats subjected to maternal separation and ESC treatment (FSL MS ESC)

Proteomic analysis was performed for selected comparisons (Tables 8.7–8.8) and high-quality protein maps were generated (Fig. 8.2 and 8.4).

8.7.1 Results of Synaptoproteomics of Prefrontal/Frontal Cortex

Statistical analysis of 2D maps from P/FC synaptosomes revealed 37 proteins differently regulated in basal FSL vs. FRL rats. Stress of maternal separation significantly dysregulated 48 proteins in FSL and 24 proteins in FRL P/FC synaptosomes. In basal FSL the chronic ESC treatment differently regulated 33 protein spots. In FSL subjected to the stress of maternal separation, chronic ESC treatment modulated seven spots (Table8.7).

Interestingly, in FSL twice as many proteins as in FRL were dysregulated by maternal separation, thereby suggesting an increased basal vulnerability of FSL rats to early life stress. In addition, three proteins downregulated by maternal separation in FSL were upregulated by the subsequent treatment with ESC. Indeed, these three proteins, regulated in opposite directions by maternal separation and antidepressant treatment, may represent biomarkers of both vulnerability to stress and response to drug. Therefore, these three proteins (or respective pathways involved) could represent putative targets for the development of novel drugs with antidepressant

Table 8.7 Comparisons of experimental groups in FSL/FRL PFC

Comparisons	Spots up- or downregulated
FSL + Veh vs. FRL + Veh	37
FSL + ESC vs. FSL + Veh	33
FRL + MS vs. FRL + Veh	24
FSL + MS vs. FSL + Veh	48
FSL + MS + ESC vs. FSL + MS	7

Table 8.8 Comparisons of experimental groups in FSL/FRL HC

Comparisons	Spots up- or down-regulated
FSL + Veh vs. FRL + Veh	15
FSL + ESC vs. FSL + Veh	13
FRL + MS vs. FRL + Veh	17
FSL + MS vs. FSL + Veh	14
FSL + MS + ESC vs. FSL + MS	12

action. Protein spots differently regulated in FSL vs. FRL by either treatment or maternal separation were excised from gels and submitted to identification by mass spectrometry.

8.7.2 Results of Synaptoproteomics of Hippocampus

HC synaptosomes were prepared from the same animal groups. 2D maps were generated and analyzed as before. Statistical analysis revealed 15 protein spots differently regulated in FSL vs. FRL rats. ESC regulated 13 spots in basal FSL. Early life stress regulated 17 spots in FRL and 14 in FSL; subsequent pharmacological treatment with ESC regulated in the FSL rats maternally separated, 12 protein spots (Table8.8).

Interestingly, fewer protein spots were significantly up- or downregulated in the different experimental groups: fewer in HC compared to P/FC synaptosomes. Contrary to P/FC, stress of maternal separation differently regulated in HC almost the same number of proteins in FRL and FSL. However, similar to the results obtained in P/FC, maternal separation regulated different sets of proteins in the two rat lines.

8.8 In silico Analysis of Protein Pathways

The results obtained by synaptoproteomic studies in both animal models will be further analyzed using available bioinformatics platforms. These "in silico" tools allow investigation of gene ontology (GO) terms and help to derive more information about potential cellular pathways that may be modified by stress susceptibility or altered by antidepressant treatments. Several software are available online, among them FatiGO (www.babelomics.org), Ingenuity Pathways Analysis (IPA) (www.ingenuity.com), and DAVID (Database for Annotation, Visualization and Integrated Discovery) (http://david.abcc.ncifcrf.gov/home.jsp).

The main aim of these pathway analysis studies will be a functional classification of the proteins according to GO terms, disease terms, and signaling pathways, in order to understand the biological processes operating behind them using different procedures to scan an apparently disjointed list of proteins.

FatiGO is an application used to find GO terms that are over- and under-represented in a set of proteins and therefore characterizes the biological processes responsible for their differential expressions (Al-Shahrour et al., 2004; Al-Shahrour et al., 2005).

IPA software system is one of the most complete tools that allow obtaining potential pathways theoretically modified by experimental procedures by analyzing protein changes in the experiments. The program is able to generate networks of proteins that are ranked with a score to facilitate data interpretation. The score allows estimating the level of the result obtained, to plan a following trial. Recently, this application was used by Wishart et al. to elaborate the results obtained by 2D-electrophoresis analysis of synaptic proteins isolated from mouse striatum (Wishart et al., 2007).

DAVID is an in silico discovery system developed to facilitate the analysis of large lists of proteins to highlight the potential correlations concerning functional classification, biochemical pathway maps, and conserved protein domains. This web-accessible application operates similar to FatiGO, but additionally allows the simultaneous view of different annotations regarding biological processes, molecular function, and cellular components (Dennis et al., 2003).

8.9 Summary and Perspectives

It is becoming apparent that proteomic analysis of synaptic terminal fractions, with its powerful investigative potential of the neurotransmitter release machinery and synaptic function, performed on animal models with different combinations of genetic background and environmental manipulation, may represent an unprecedented opportunity for research on biomarkers in psychiatric disorders. Recently, we observed an increased interest in the field of biomarker research, owing to their potential use in early diagnosis and prognosis of disease, and in the development of new drug targets. Moreover, biomarkers could provide clinicians with the means to tailor a pharmacological therapy to an individual patient on the basis of the particular endophenotype or vulnerability to adverse effects. Therefore, validated biomarkers could greatly reduce both the individual suffering and the enormous costs of the affective disorders to society. In particular, biomarkers associated with mood disorders are needed, as diagnoses are mainly based on phenotypical categorizing signs and symptoms. A biomarker may be a genetic trait or a biochemical change, such as the expression of proteins, and their study could be accomplished by using different methods or strategies. In this chapter, we have described one of these strategies, consisting of synaptoproteomics applied to different animal models of depression. LH and FSL are, as described above, two well-validated models that replicate symptoms of the pathology. Both animal models were treated with the same antidepressant and their synaptoproteomic profile was characterized to identify molecular correlates of pathology and response to drug and the resulting altered neurobiological pathways. Using this approach, we plan to obtain at the same time information concerning biomarkers that may be used as predictors of pathology and new targets of drug action, available for detection of treatment efficacy as well as for development of novel pharmacological and therapeutic strategies. Moreover, this approach, in combination with functional proteomics and trascriptomics, could provide a powerful strategy for the identification of the complex pathophysiological mechanisms in mood disorders. Indeed, proteome and transcriptome analyses could suggest gene and protein clusters corresponding to the pathological phenotype ("trait biomarkers") and clusters corresponding to modification of phenotype by treatment ("state biomarkers").

Acknowledgments This work was funded by a European Union (6th Framework Program) grant for project GENDEP (contract no. LSHB-CT-2003-503428) and by the Swedish Medical Research Council grant 10414 (to AAM).

References

Al-Shahrour F, Diaz-Uriarte R, Dopazo J (2004) FatiGO: a web tool for finding significant associations of Gene Ontology terms with groups of genes. Bioinformatics 20:578–580

Al-Shahrour F, Minguez P, Vaquerizas JM, Conde L, Dopazo J (2005) BABELOMICS: a suit of web tools for functional annotation and analysis of groups of genes in high-throughput experiments. Nucleic Acids Res 33:W460–W464

Anderson L, Seilhamer J (1997) A comparison of selected mRNA and protein abundances in human liver. Electrophoresis 18:533–537

Barbiero VS, Giambelli R, Musazzi L, Tiraboschi E, Tardito D, Perez J, Drago F, Racagni G, Popoli M (2007) Chronic antidepressants induce redistribution and differential activation of alphaCaM Kinase II between presynaptic compartments. Neuropsychopharmacology advance online publication, 14 March 2007; doi:10.1038/sj.npp.1301378

Bonanno G, Giambelli R, Raiteri L, Tiraboschi E, Zappettini S, Musazzi L, Raiteri M, Racagni G, Popoli M (2005) Chronic antidepressants reduce depolarization-evoked glutamate release and protein interactions favoring formation of SNARE complex in hippocampus. J Neurosci 25:3270–3279

Boyd-Kimball D, Castegna A, Sultana R, Poon HF, Petroze R, Lynn BC, Klein JB, Butterfield DA (2005) Proteomic identification of proteins oxidized by Abeta(1–42) in synaptosomes: implications for Alzheimer's disease. Brain Res 1044:206–215

Caldji C, Francis D, Sharma S, Plotsky PM, Meaney MJ (2000) The effects of early rearing environment on the devepopment of GABAA and central benzodiazepine receptor levels and novelty-induced fearfulness in the rat. Neuropsychopharmacology 22:219–229

Caspi A, Moffitt TE (2006) Gene-environment interactions in psychiatry: joining forces with neuroscience. Nat Rev Neurosci 7:583–590

Caspi A, Sugden K, Moffitt TE, Taylor A, Craig IW, Harrington H, McClay J, Mill J, Martin J, Braithwaite A, Poulton R (2003) Influence of life stress on depression: moderation by a polymorphism in the 5-HTT gene. Science 301:386–389

Carboni L, Piubelli C, Righetti PG, Jansson B, Domenici E (2002) Proteomic analysis of rat brain tissue: comparison of protocols for two-dimensional gel electrophoresis analysis based on different solubilizing agents. Electrophoresis 23:4132–4141

Dennis G Jr, Sherman BT, Hosack DA, Yang J, Gao W, Lane HC, Lempicki RA (2003) DAVID: Database for Annotation, Visualization, and Integrated Discovery. Genome Biol 4:P3

Dunkley PR, Jarvie PE, Heath JW, Kidd GJ, Rostas JA (1986) A rapid method for isolation of synaptosomes on Percoll gradients. Brain Res 372:115–129

El Khoury A, Gruber SHM, M rk A, Mathé AA (2006) Adult life behavioral consequences of early maternal separation are alleviated by escitalopram treatment in a rat model of depression. Prog Neuropsychopharmacol Biol Psychiatr 30:535–540
Fountoulakis M (2004) Application of proteomiics technologies in the investigation of the brain. Mass Spectrom Rev 23:231–258
Heim C, Nemeroff CB (2001) The role of childhood trauma in the neurobiology of mood and anxiety disorders: preclinical and clinical studies. Biol Psychiatr 49:1023–1039
Henn FA, Edwards E, Anderson D, Vollmayr B (2002) Psychotherapy and antidepressant treatment of depression: evidence for similar neurobiological mechanism. World Psychiatr 1:115–117
Hitzemann R (2000) Animal models of psychiatric disorders and their relevance to alcoholism. Alcohol Res Healt 3:149–158
Janowsky DS, Risch SC, Parker D, Huey L, Judd L (1980) Increased vulnerability to cholinergic stimulation in affect disorder patients. Psychopharmacol Bull 16:29–31
Janowsky DS, Overstreet DH, Nurnberger JI Jr (1994) Is cholinergic sensitivity a genetic marker for the affective disorders? Am J Med Genet 54:335–344
Keller MB, Boland RJ (1998) Implications of failing to achieve successful long-term maintenance treatment of recurrent unipolar major depression. Biol Psychiatr 44:348–360
Kramer R, Cohen D (2004) Functional genomics to new drug targets. Nat Rev Drug Discov 3:965–72
Ladd CO, Huot RL, Thrivikraman KV, Nemeroff CB, Meaney MJ, Plotsky PM (2000) Long-term behavioural and neuroendocrine adaptations to adverse early experience. Prog Brain Res 122:81–103
Lesch KP (2004) Gene-environment interaction and the genetics of depression. J Psychiatr Neurosci 29:174–84
Li KW, Hornshaw MP, Van Der Schors RC, Watson R, Tate S, Casetta B, Jimenez CR, Gouwenberg Y, Gundelfinger ED, Smalla KH, Smit AB (2004) Proteomics analysis of rat brain postsynaptic density. Implications of the diverse protein functional groups for the integration of synaptic physiology. J Biol Chem 279:987–1002
Liberatori S, Canas B, Tani C, Bini L, Buonocore G, Godovac-Zimmermann J, Mishra OP, Delivoria-Papadopoulos M, Bracci R, Pallini V (2004) Proteomic approach to the identification of voltage-dependent anion channel protein isoforms in guinea pig brain synaptosomes. Proteomics 4:1335–1340
Matthews K, Christmas D, Swan J, Sorrel E (2005) Animal models of depression: navigating through the clinical fog. Neurosci Biobehav Rev 29:503–513
Mello CF, Sultana R, Piroddi M, Cai J, Pierce WM, Klein JB, Butterfield DA (2007) Acrolein induces selective protein carbonylation in synaptosomes. Neuroscience 147:674–679
Musazzi L, Ryan B, Barbiero VS, Mallei A, Giambelli R, Tardito D, Gruber SHM, El Khoury A, Andersson W, Anwyl R, Racagni G, Mathé AA, Rowan MJ, Popoli M Early-life stress enhances NMDA receptor-dependent synaptic plasticity in a gene-environment rat model of depression. Submitted 2008
Overmier JB, Seligman ME (1967) Effects of inescapable shock upon subsequent escape and avoidance responding. J Comp Physiol Psychol 63:28–33
Overstreet DH, Russell RW (1982) Selective breeding for diisopropyl fuorophosphate-sensitivity: behavioural effects of cholinergic agonists and antagonists. Psychopharmacology (Berl.) 78:150–155
Overstreet DH, Friedman E, Mathé AA, Yadid G (2005) The Flinders Sensitive Line rat: a selectively bred putative animal model of depression. Neurosci Biobehav Rev 29:739–759
Paulson L, Martin P, Persson A, Nilsson CL, Ljung E, Westman-Brinkmalm A, Eriksson PS, Blennow K, Davidsson P (2003) Comparative genome- and proteome analysis of cerebral cortex from MK-801-treated rats. J Neurosci Res 71:526–533
Plotsky PM, Meaney MJ (1993) Early, postnatal experience alters hypothalamic corticotropin-releasing factor (CRF) mRNA, median eminence CRF content and stress-induced release in adult rats. Brain Res Mol Brain Res 18:195–200

Pucilowski O, Overstreet DH, Rezvani AH, Janowsky DS (1993) Chronic mild stress-induced anhedonia: greater effect in a genetic rat model of depression. Physiol Behav 54:1215–1220

Raiteri L, Raiteri M (2000) Synaptosomes still viable after 25 years of superfusion. Neurochem Res 25:1265–1274

Rohlff C, Hollis K (2003) Modern proteomic strategies in the study of complex neuropsychiatric disorders. Biol Psychiatr 53:847–853

Stasyk T, Huber LA (2004) Zooming in: fractionation strategies in proteomics. Proteomics 4:3704–3716

Satoh K, Takeuchi M, Oda Y, Deguchi-Tawarada M, Sakamoto Y, Matsubara K, Nagasu T, Takai Y (2002) Identification of activity-regulated proteins in the postsynaptic density fraction. Genes Cells 7:187–197

Urani A, Chourbaji S, Gass P (2005) Mutant mouse models of depression: candidate genes and current mouse lines. Neurosci Biobehav Rev 29:805–828

Vazquez DM, Lopez JF, Van Hoers H, Watson SJ, Levine S (2000) Maternal deprivation regulates serotonin 1A and 2A receptors in the infant rat. Brain Res 855:76–82

Vercauteren FG, Bergeron JJ, Vandesande F, Arckens L, Quirion R (2004) Proteomic approaches in brain research and neuropharmacology. Eur J Pharmacol 500:385–398

Vollmayr B, Henn FA (2001) Learned helplessness in the rat: improvements in validity and reliability. Brain Res Protoc 8:1–7

Vollmayr B, Henn FA (2003) Stress models of depression. Clin Neurosci Res 3:245–251

Vollmayr B, Simonis C, Weber S, Gass P, Henn FA (2003) Reduced cell proliferation in the dentate gyrus is not correlated with the development of learned helplessness. Biol Psychiatry 54:1035–1040

Willner P (1990) Animal models of depression: an overview. Pharmacol Ther 45:425–455

Willner P (1995) Animal models of depression: validity and applications. Adv Biochem Psychopharmacol 49:19–41

Wishart TM, Paterson JM, Short DM, Meredith S, Robertson KA, Sutherland C, Cousin MA, Dutia MB, Gillingwater TH (2007) Differential proteomic analysis of synaptic proteins identifies potential cellular targets and protein mediators of synaptic neuroprotection conferred by the slow Wallerian degeneration (Wld) gene. Mol Cell Proteomics 8:1318–1330

Witzmann FA, Arnold RJ, Bai F, Hrncirova P, Kimpel MW, Mechref YS, McBride WJ, Novotny MV, Pedrick NM, Ringham HN, Simon JR (2005) A proteomic survey of rat cerebral cortical synaptosomes. Proteomics 5:2177–2201

Yadid G, Nakash R, Deri I, Tamar G, Kinor N, Gispan I, Zangen A (2000) Elucidation of the neurobiology of depression: insights from a novel genetic animal model. Prog Neurobiol 62:353–378

Chapter 9
Animal Models for Anxiety Disorders

Su Guo

Abstract Anxiety disorders are characterized by overwhelming anxiety or fear and are chronic and relentless if left untreated. Current available treatments for anxiety disorders are inadequate and some have severe side effects, thus warranting a better understanding of the etiology and mechanisms underlying anxiety and the development of anxiety disorders. In this chapter, the use of animal models to identify molecular and cellular circuitry that regulate fear or anxiety, and the influence of environment on the development of fear or anxiety, are discussed.

Abbreviations ACTH: Adrenocorticotropin; CORT: Corticosterone; CRH: Corticotrophin-releasing hormoneCS: Conditioned stimulus; 5-HT1A: Serotonin 1A receptor; LTP: Long-term potentiation; MAO-A: Monoamine oxidase A; PTSD: Post-traumatic stress disorder; SSRI: Serotonin reuptake inhibitors; US: Un-conditioned stimulus

9.1 Introduction

Anxiety disorders affect approximately 19 million American adults, and are characterized by overwhelming anxiety and fear. These disorders are chronic, relentless, and can grow progressively worse if not treated. In this chapter, I will first give an overview of various anxiety disorders that are observed in humans. Then I will discuss the research on the underlying mechanisms, which looks into the pharmacology, neural circuitry, genetics, and environmental effects, using various paradigms of fear/anxiety in animal models. Since a majority of currently available paradigms have employed rodent models, they will be discussed most extensively. Other animal

S. Guo
Department of Biopharmaceutical Sciences, Programs in Human Genetics and Biological Sciences, University of California, San Francisco, CA, 94143–2811, USA
su.guo@ucsf.edu

C.W. Turck (ed.) *Biomarkers for Psychiatric Disorders*,
doi: 10.1007/978-0-387-79251-4_9, © Springer Science+Business Media, LLC 2008

models including primates and zebrafish will also be described, wherever data are available. Finally, the impacts of the studies in animal models on future improvement of diagnoses, treatments, and prevention of human anxiety disorders will be discussed.

9.2 Overview of Human Anxiety Disorders

While the relatively mild and brief episodes of fear or anxiety are often protective and beneficial, uncontrollable and persistent emotional and physiological symptoms, as manifested during anxiety disorders, are detrimental to the health and well-being of an individual. These symptoms further pose huge economic burdens on the family and society. Six forms of anxiety disorders have been described in humans: (1) Panic disorder is characterized by unpredictable, rapid-fire attacks of intense anxiety. (2) Generalized anxiety disorder is marked by exaggerated worry and tension, even though there is little or nothing to provoke it. (3) Social phobia, also called social anxiety disorder, involves overwhelming anxiety and excessive self-consciousness in daily social situations. (4) Specific phobias are intense fears of some specific objects that in reality pose little or no danger. (5) Obsessive-compulsive disorder involves anxious thoughts and repetitive rituals, which, when performed, could help reduce anxiety. (6) Post-traumatic stress disorder (PTSD) is characterized by persistent frightening thoughts and memories of a terrifying event.

Currently, two types of treatment are available for anxiety disorders: medication and specific types of psychotherapy. Although these treatments are effective in some people, many patients are left with residual symptoms or experience side effects that limit the use of these currently available mediations. Therefore, improved understanding of the molecular and neural basis of anxiety and anxiety disorders could potentially lead to improved treatments.

9.3 Animal Models of Anxiety Disorders

In order to screen for therapeutic compounds and to understand the molecular and neural basis of anxiety and anxiety disorders, animal models have been extensively employed. Indeed, the ability to respond to a threat or danger is universal in the animal kingdom. Since adequate responses are essential for the survival of the individual and species, these responses have been evolutionarily selected. Animals respond to the potential presence of a threat with characteristic autonomic and behavioral responses that are quantifiable. Through the analysis of innate or learned anxiety-like behaviors in animals at the pharmacological, molecular, genetic, and neural circuit levels, mechanisms that regulate anxiety-like behaviors have begun to emerge.

In the following paragraphs, paradigms that assess either innate anxiety- or learned anxiety- like behaviors in animals will be discussed. Their potential relationship to specific forms of human anxiety disorders is suggested. However, like animal models of any human diseases, the quality of an animal model is typically assessed by three validity criteria: Face validity describes how closely the animal model mimics the key symptoms of the human disorder; predictive validity refers to the success in the animal model of a pharmacological treatment, which is shown to have therapeutic effect in humans; and construct validity states that the cellular and molecular processes in the animal model and the human disorder are analogous (Willner, 1984). In the case of animal models of anxiety disorders, wherever the data are available, these criteria of validation will be discussed.

In addition to modeling human anxiety disorders for the purpose of pharmaceutical drug screening and testing, animal models of anxiety and anxiety disorders provide an opportunity to identify the environmental and genetic factors that control the neuronal circuitry, which in turn underlies the regulation of fear or anxiety.

9.3.1 Models for Generalized Anxiety and Panic Disorders

9.3.1.1 Assessing Innate Anxiety-Like Behaviors

Test of Exploratory Behavior

Many of the currently used models use animals' exploratory behavior as an index of anxiety. It has been proposed that the behavior of animals exposed to a novel situation results from a competition between an exploratory tendency (motivated by curiosity or boredom) and a withdrawal tendency (motivated by fear or anxiety) (Montgomery, 1955). Specific behavioral paradigms that employ animals' exploratory behavior to assess anxiety include the following: (1) The open field assay, which is popular because it requires minimal apparatus and is quickly performed (Walsh and Cummins, 1976). In this assay, locomotion and defecation are often used as indices for measurement. However, the use of locomotion as an index of anxiety is complicated by the fact that mechanisms that change locomotion do not always involve changes in anxiety. Although defecation can be used as an indication of the activation of the autonomic nervous system, food intake or drugs that alter gut motility, rather than anxiety, could affect defecation. Therefore, a better modification of the open field assay, namely "novelty-suppressed feeding," which assesses the animals' approach to the consumption of food located in the center of the open field, appears to be a more sensitive measure of individual differences in emotionality from a psycho-behavioral perspective (Britton and Brittone, 1981). (2) The plus-maze test, which was initially developed by Montgomery (Montgomery, 1955). Pellow and co-workers performed studies to validate it as an animal model of anxiety in rats (Pellow et al., 1985), and Lister validated the test in mice (Lister, 1987).

This test has proved very popular in recent years, in part because it is both rapid and appears to be sensitive to the effects of both anxiolytic (benzodiazepine, barbiturates, and ethanol) and anxiogenic (benzodiazepine receptor inverse agonists, caffeine, and yohimbine) agents. The elevated plus-maze test also has good face validity, in that the reluctance of animals to explore the open arms of the maze probably results from a combination of their aversion to open space and the elevation of the maze. 3) Light-dark transitions, which was initially developed by Crawley and colleagues (Crawley and Goodwin, 1980). This test also possesses a certain degree of face validity, in that light serves as an anxiogenic stimulus, and there is an apparent conflict between the desire to explore and the desire to avoid the brightly lit part of the apparatus. A number of benzodiazepines and meprobamate produce anxiolytic profiles; however, pentobarbital fails to alter the number of transitions at doses that did not increase activity (Crawley and Goodwin, 1980).

Tests of Social Behavior

In addition to testing exploratory behaviors as a way of measuring innate anxiety, assessment of social behaviors, in which aggression is infrequently observed, has also been used to measure innate anxiety. The social interaction test, developed by File and colleagues, uses novelty and high lighting conditions as anxiogenic stimuli (File, 1988). The time spent by pairs of male rats in active social interaction is measured in a test arena under high or low lighting conditions, and/or in a familiar or unfamiliar testing apparatus. Locomotor activity is also measured, and this assists in determining whether any change in social interaction is due to a general stimulant or sedative effects. As in the plus-maze test, when both the time spent in social interaction and locomotor activity are affected by a treatment or a genetic condition, analyses of covariance are performed with locomotor activity as a covariate. Attributing the reduction in social interaction to increased anxiety in this test appears to have some face validity. Ethological studies suggest that rats find high light and novel environment aversive. Moreover, plasma corticosterone levels are higher in rats that are subject to high light conditions or in an unfamiliar environment (File and Peet, 1980). An anxiolytic profile of benzodiazepine is observed following chronic treatment after tolerance has developed to the drug's sedative effects. The benzodiazepine increases the amount of time of interaction under conditions of high lighting or in unfamiliar arena. Other anxiolytics, such as phenobarbitone, have been found to increase social interaction across all test conditions (File and Peet, 1980). Anxiogenic agents appear to reduce the time spent in social interaction without altering locomotor activity. A test of social interaction has been adopted in mice, but is met with a challenge, in that pairs of mice have a tendency to be much more aggressive during the test than pairs of male rats (Lister and Hilakivi, 1988). A disadvantage of the social interaction test is that the test involves the observation of pairs of animals, thereby making it difficult to examine individual difference in behavior with this test, as the behavior of one animal is critically dependent on the behavior of its test partner. Nevertheless, this test may shed light on social phobia, which is a type of anxiety disorder observed in humans.

The ultrasonic vocalization emitted by rat pups following separation from their mother has also been used as an animal model of anxiety (Gardner, 1985). The method has some face validity since the ultrasonic cries are considered to be "distress" calls (Noirot, 1972). One needs to keep in mind, however, that ultrasonic calls are also emitted in response to cold and to tactile stimulation resulting, for example, from retrieval by the mother, or contact with unfamiliar surfaces (Okon, 1972). Therefore, it is important to examine whether any change in ultrasonic calls is secondary to some alteration in body temperature.

Defensive behaviors of rats have also been examined in a visible burrow system, following the presentation of a cat (Blanchard and Blanchard, 1989). The defensive behaviors are examined in great detail as a function of time after the removal of the cat. Withdrawal, immobility, increased risk assessment, and a suppression of nondefensive behaviors such as eating, drinking, grooming, and mounting are quantified. This paradigm has a high degree of face validity as a method to assess fear and anxiety.

Innate Anxiety-Like Behavior in Animals Other Than Rodents

Observer-rating methods have been used to assess anxiety in primates. The observation of struggling in a restraining chair, behaviors such as head- and body- turning, periods of immobility, scratching, and vocalization may be behavioral manifestations of anxiety (Ninan, 1982). In addition, behavioral as well as physiological assessment of the hypothalamic-pituitary-adrenal (HPA) axis has been carried out in monkeys separated from their social groups (Mendoza et al., 1978). However, to assess a direct relationship between the model and anxiety, these behaviors should be selectively elicited by other anxiety-provoking stimuli, and not observed in nonanxious animals. There is often an assumption that behavioral studies in non-human primates do not require the same degree of validation as those using other species. While the phylogenetic closeness to humans makes studies in primates extremely valuable, such studies should be as rigorously controlled as studies in other species.

In recent years, the zebrafish *Danio rerio* has emerged as a vertebrate model organism suitable for unbiased genetic analysis (Driever et al., 1996; Haffter et al., 1996) and large-scale small molecule screening (Zon and Peterson, 2005). As a model for anxiety, zebrafish is still at its infant stage (Guo, 2004). Despite that, several behavioral paradigms have been exploited, which include light-dark transition (Guo, 2004), behavioral responses to alarm substance (Speedie and Gerlai, 2007), anti-predator behavior (Bass and Gerlai, 2008), and novelty-elicited diving response (Levin et al., 2007). Rigorous pharmacological validations of these behavioral paradigms remain to be carried out.

9.3.1.2 Assessing Learned Anxiety-Like Behaviors

When a neutral stimulus such as a tone (called conditioned stimulus) is paired with an aversive stimulus such as a foot shock, the tone can now elicit a constellation of behaviors that are typically used to define a state of fear in animals. Since these behavioral responses are similar in many respects to the constellation of behaviors that are used to diagnose generalized anxiety in humans, assessing this type of conditioned fear yields insights into human anxiety.

The Conditioned Emotional Response

In the prototypic experiment, food-deprived rats are first trained to press a lever for food using intermittent reinforcement. After giving a sufficient number of training sessions to establish stable rates of bar pressing, a neutral stimulus such as a light or a tone, is paired with a shock. After a small number of pairings, the neutral stimulus alone produces a suppression of the lever-pressing behavior (Estes and Skinner, 1941). Although such conditioned suppression is generally viewed as a reflection of a central state of fear, it is clear that a suppression of lever pressing alone is not sufficient to infer a central state of fear. Salient or novel stimuli by themselves can suppress bar pressing. Thus, in addition to suppression of bar pressing, assessment of other phenotypes, which are often observed in conditioned suppression, such as defecation, urination, fast and shallow breathing, and freezing, will further strengthen the link to fear or anxiety. Moreover, drugs that are known to reduce fear or anxiety should also attenuate the conditioned emotional response. Indeed, many drugs have been tested using this paradigm. In general, there has been success in differentiating between drugs that are anxiolytic and non-anxiolytic in humans. For example, benzodiazepine, barbiturates, and opiates generally attenuate the conditioned emotional response, whereas chlorpromazine and amphetamine do not. However, exceptions to this generalization can be found for every drug class, and this assay is in general viewed as too variable to be used for screening novel anxiolytics.

A simpler paradigm than the conditioned suppression of bar pressing is the following: a rat is given a tone (CS) followed by an electric shock (US). After a few tone-shock pairings, defensive responses including freezing, changes in heart rate, blood pressure, hormone release, and pain sensitivity will be elicited by the tone. This simple paradigm of fear conditioning has been used to dissect the neural circuits involved (see later sections).

The Fear Potentiated Startle

The amplitude of the acoustic startle reflex in the rat can be augmented by the presence of a cue (e.g., a light) that has been previously paired with a shock (Brown et al., 1951). As it is common sense that people startle more when they are afraid, this "fear-potentiated startle" has good face validity as a model for anxiety. Indeed, it has been replicated using either an auditory or a visual conditioned stimulus, when startle has been elicited by either a loud sound or an airpuff (Davis, 1986).

Fear-potentiated startle offers a number of advantages for analyzing how drugs affect fear or anxiety. Perhaps the most important advantage is that it is defined as a within-subject difference in startle amplitude in the presence or the absence of the conditioned stimulus, thus making it a sensitive measure, by reducing problems caused by between-subject variability in startle. Indeed, a variety of drugs that reduce fear or anxiety in humans decrease potentiated startle in rats. Drugs like clonidine, morphine, diazepam and buspirone, which differ considerably in their mechanism of action, all block potentiated startle. In most cases, these treatments do not depress startle levels on the noise-alone trials, although drugs like clonidine do have marked depressant effects on both types of trials. In addition, drugs like yohimbine and piperoxane, which are anxiogenic in humans, also increase potentiated startle in rats.

9.3.2 Models for Other Anxiety Disorders

9.3.2.1 Animal Models for Phobic Disorders

A phobia is a fear of a situation that is out of proportion to its danger, can neither be explained or reasoned away, is largely beyond voluntary control, and leads to avoidance of the feared situation. It appears that phobias develop more readily to some stimuli (such as spiders and snakes) than to other dangerous objects. In addition, phobias can be transmitted by observation in animals. For example, Rhesus monkeys raised in captivity do not have a fear of snakes, but rapidly acquire such a fear by watching an adult monkey behaving fearfully in the presence of a snake. Interestingly, the fear of a flower cannot be transmitted in a similar manner (Cook et al., 1985). Some phobias in animals appear to be innate. For example, specific fears of poisonous coral snakes are found in some snake-eating birds that have never seen such a snake before (Smith, 1975, 1977).

Although there has been much research on animal phobias from the ethological and psychological points of view, there is very little pharmacological data from animal studies that can be considered directly relevant to the treatment of human phobic disorders. In humans, behavioral therapy, i.e., exposing patients to their phobic stimuli remains the most efficacious treatment for phobic disorders. Future pharmacological research using animal models will be used in conjunction with the behavioral therapies to benefit patients.

9.3.2.2 Animal Models for Obsessive Compulsive Disorder (OCD)

OCD is an anxiety disorder that afflicts 2–3% of the population with recurrent intrusive thoughts and ritualized actions, causing significant impairment of normal lives. Serotonin reuptake inhibitors (SSRI) are the only effective pharmacological treatment for OCD.

Based on behavioral similarity, naturally occurring repetitive or stereotypic behaviors observed in animals have been examined. These include tail chasing, fur chewing, and weaving, innate motor behaviors that occur during conflict, frustration or stress (displacement behaviors) such as grooming, cleaning and pecking. The effect of SSRIs has been tested in some models (Altemus et al., 1996; Nurnberg et al., 1997). Although some of these models have good predictive validity in addition to face validity, many have not been used since the original publications. To date, three behavioral models are in use: The barbering (Garner et al., 2004), marble burying (Treit, 1985), and signal attenuation models (Joel and Avisar, 2001). For instance, barbering is a common behavior in laboratory mice, where barbers pluck hair from their companions. Because not all mice perform such behavior, it has been viewed as an abnormal form of behavior, which shows similarity to compulsive hair plucking (trichotillomania) in humans (Garner et al., 2004). Although barbering seems to have strong face validity as a model of trichotillomania, it currently lacks pharmacological validation. Barbering has an important advantage over other models, in that it develops spontaneously, and thus may provide insights into the etiology of trichotillomania.

9.4 Neural Circuitry Underlying Anxiety-Like Behaviors in Animal Models

The neural circuitry underlying learned anxiety-like behavior has been studied in the context of the fear-conditioning paradigm. Conditioned fear is mediated by the transmission of information about the CS and US to the amygdala, which then sends output projections to the behavioral, autonomic, and endocrine response control systems located in the brainstem (Davis, 1992; Fanselow, 1994; Kapp et al., 1992; LeDoux, 1992). Pathway tracing and lesion studies suggest that lateral amygdala (LA) is the sensory gateway to the amygdala, and therefore the first possible site in the amygdala where cells processing the CS might be modified by association with the US in fear conditioning. With the basic elements of the circuitry understood from lesion studies, fear plasticity in the amygdala has been studied in the following ways: First, single-unit recording; second, LTP; third, drugs that block LTP are infused into amygdala areas where LTP is believed to occur, and effects on the acquisition of conditioned fear behavior assessed.

The neural circuitry underlying innate anxiety-like behaviors is also thought to involve the amygdala as an important structure: threatening stimuli perceived by the sensory system are transduced to the sensory thalamus and sensory cortex. The lateral nucleus of the amygdala receives input signals from the thalamus, the hippocampus, and the cortex, and projects to the central nucleus of the amygdala, which in turn extensively projects to the hypothalamus, midbrain, and brain stem areas. Indeed, lesioning of the amygdala in rats completely abolishes rats' innate fear defense response toward a cat (Blanchard and Blanchard, 1972; Misslin and Ropartz, 1981).

9.5 Molecular Genetic Understanding of Anxiety in Animal Models

The availability of gene targeting and transgenic technology in mice has led to the finding of many knockout and transgenic animals that exhibit abnormal anxiety-like behaviors. Some of the identified genes fit into the neural systems that have been implicated in regulating anxiety, based on the study of neural circuitry, whereas others seem to be required for general neural functions, implying that the overall well-being of the nervous system is critical to maintain a normal level of anxiety.

Enhancement of the GABA neurotransmitter system function is one of the effective therapeutic treatments for anxiety disorders. Targeted inactivation of GABA synthesis enzymes or certain GABA receptors indeed lead to increased anxiety-like behaviors (Kash et al., 1999; Löw et al., 2000).

Deregulation of the serotoninergic system has also been strongly implicated in anxiety and depression. Targeted inactivation of the 5-HT_{1A} receptor gene leads to increased anxiety-like behavior in mice (Heisler et al., 1998; Parks et al., 1998; Ramboz et al., 1998). Interestingly, through a tissue specific conditional rescue strategy, it is shown that the expression of 5-HT1A in the hippocampus and cortex could rescue the increased anxiety phenotype in the 5-HT1A mutant. Treating the "rescued" mice with doxycycline to shut off 5-HT1A expression during development or during adulthood shows that receptor expression before day 21 is crucial for normal anxiety-like behavior (Gross et al., 2002).

Like SSRIs, noradrenaline reuptake inhibitors are also effective in treating depression. In addition, inhibitors for the monoamine oxidase A (MAO-A), which block the metabolism of noradrenaline, serotonin, and dopamine, are used in the treatment of anxiety disorders. Mutant mice lacking the MAO-A gene have a reduced anxiety-like phenotype (Cases et al., 1995). However, additional studies suggest that a more general alteration in sensory ability or selective alteration in emotional behavior might have contributed to the observed differences (Kim et al., 1997).

Corticotrophin-releasing hormone (CRH) is the primary regulator of HPA axis. Stress induces the release of CRH from the hypothalamus, which causes the release of adrenocorticotropin (ACTH) from the pituitary into the general circulation. ACTH then induces the release of glucocorticocoid stress hormones from the adrenal glands (Miller and O'Callaghan, 2002). CRH activity is mediated by two main receptors, CRH-R1 and CRH-R2 (Eckart et al., 2002). Both receptors also bind urocortin, a neuropeptide related to CRH (Vaughan et al., 1995). Deletion of the CRH gene in mice significantly decreased basal corticosterone (CORT) levels and markedly blunted the stress-induced increase of CORT levels in the circulation. However, at the behavioral level, the CRH mutant mice have a normal anxiety-like phenotype (Dunn and Swiergiel, 1999), suggesting that either CRH is not involved in regulating anxiety-like behavior, or there is a genetic redundancy in the regulation of anxiety-like behavior. Since the CRH transgenic mice have a heightened level of anxiety (Stenzel-poore et al., 1994), this finding suggests that redundancy is likely to be the explanation for the lack of behavioral phenotype in CRH mutant mice. CRH-R1 knockout mice show reduced levels of anxiety (Smith et al., 1998; Timpl et al., 1998). However, the story with CRH-R2 knockout mice is more complicated, as results differ among the three groups that have independently inactivated CRH-R2 (Bale et al., 2000; Coste et al., 2000; Kishimoto et al., 2000). CRH-R2 knockout mice are reported to have either no difference in anxiety-like behaviors or have a heightened anxiety-like phenotype. Discrepancy regarding the role of CRH-R2 could be either due to differences in the procedures, or in the genetic background. Mice deficient in both CRH receptors have an impaired HPA response to stress and a sexual dichotomy in anxiety-related behaviors (Bale et al., 2002; Preil et al., 2001).

9.6 The Role of Environment

In addition to many genetic factors involved in regulating anxiety as described above, human experience suggests that the environment can have a strong influence on an individual's susceptibility to anxiety disorders. Studies using animal models show that maternal behaviors can have an influence on the development of strain-specific difference in anxiety-like behavior in mice (Francis et al., 2003). Thus, future studies of how environmental factors interact with genetic components to affect neural circuitry functions will provide further insights into anxiety and anxiety disorders.

9.7 Summary and Perspectives

Various anxiety disorders manifest in humans. For most of them, effective and safe treatments are not yet available. Studies of fear or anxiety in animal models have led to various behavioral assays, which have been or can be used for pharmaceutical screening of small molecule drugs that may prove useful in the treatment of human anxiety disorders. As discussed in this chapter, the validity of various animal models needs to be carefully evaluated before using them to screen for drugs that are targeted to specific human anxiety disorders. Secondly, the assays should have a high enough throughput so that a large number of diverse chemicals can be screened with relative ease. Because of the complexity and diversity of the chemical space, large-scale screens are necessary to identify novel small molecules to be further evaluated down the pharmaceutical pipeline. In this regard, the newly emerged model organism, the zebrafish *Danio rerio* may prove advantageous for the first round of large-scale chemical screening.

In addition to serving as a platform for drug screening, animal models of anxiety are critical to our fundamental understanding of the molecular and neural mechanisms underlying the regulation of fear or anxiety. Such improvement in our fundamental understanding is important for rationalized drug discovery. Regarding the neural circuitry of fear or anxiety, we have learned that the amygdala is a central component of the circuitry. In the future, it would be important to determine how the "emotional" value of certain sensory stimuli is encoded by the sensory system and conveyed to the amygdala, and how the amygdala further regulates behavioral output. Although gene-targeting studies in mice have led to a vast array of genes, which is involved in anxiety-like behaviors, our knowledge of how genes and environment regulate the development and function of the anxiety circuitry remains rudimentary. Future work using animal models, building upon the current foundation, promises to further uncover the molecular and neural basis of anxiety and anxiety disorders, and lead to effective treatments of these disorders.

References

Altemus, M., Glowa, J. R., Galliven, E., Leong, Y. M., and Murphy, D. L. (1996). Effects of serotonergic agents on food-restriction-induced hyperactivity. Pharm Biochem Behav *53*, 123–131.

Bale, T. L., Contarino, A., Smith, G. W., Chan, R., Gold, L. H., Sawchenko, P. E., Koob, G. F., Vale, W. W., and Lee, K. F. (2000). Mice deficient for corticotropin-releasing hormone receptor-2 display anxiety-like behaviour and are hypersensitive to stress. Nat Genet *24*, 410–414.

Bale, T. L., Picetti, R., Contarino, A., Koob, G. F., Vale, W. W., and Lee, K. F. (2002). Mice deficient for both corticotropin-releasing factor receptor 1 (CRFR1) and CRFR2 have an impaired stress response and display sexually dichotomous anxiety-like behavior. J Neurosci *22*, 193–199.

Bass, S. L., and Gerlai, R. (2008). Zebrafish (Danio rerio) responds differentially to stimulus fish: The effects of sympatric and allopatric predators and harmless fish. Behav Brain Res *186*, 107–117.

Blanchard, D. C., and Blanchard, D. C. (1972). Innate and conditioned reactions to threat in rats with amygdaloid lesions. J Comp Physiol Psychol *81*, 281–290.

Blanchard, R. J., and Blanchard, D. C. (1989). Antipredator defensive behaviors in a visible burrow system. J Comp Psychol *103*, 70–82.

Britton, D. R., and Brittone, K. T. (1981). A sensitive open field measure of anxiolytic drug activity. Pharm Biochem Behav *15*, 577–582.

Brown, J. S., Kalish, H. I., and Farber, I. E. (1951). Conditioned fear as revealed by magnitude of startle response to an auditory stimulus. J Exp Psychol *41*, 317–328.

Cases, O., Seif, I., Grimsby, J., Gaspar, P., Chen, K., Pournin, S., Muller, U., Aguet, M., Babinet, C., and Shih, J. C. (1995). Aggressive behavior and altered amounts of brain serotonin and norepinerphrine in mice lacking MAOA. Science *268*, 1763–1766.

Cook, M., Mineka, S., Wolkenstein, B., and Laitsch, K. (1985). Observational conditioning of snake fear in unrelated rhesus monkeys. J Abnorm Psychol *94*, 591–610.

Coste, S. C., Kesterson, R. A., Heldwein, K. A., Stevens, S. L., Heard, A. D., Hollis, J. H., Murray, S. E., Hill, J. K., Pantely, G. A., Hohimer, A. R., . (2000). Abnormal adaptations to stress and impaired cardiovascular function in mice lacking corticotropin-releasing hormone receptor-2. Nat Genet *24*, 403–409.

Crawley, J., and Goodwin, F. K. (1980). Preliminary report of a simple animal behavior model for the anxiolytic effects of benzodiazepines. Pharmacol Biochem Behav *13*, 167–170.

Davis, M. (1986). Pharmacological and anatomical analysis of fear conditioning using the fear-potentiated startle paradigm. Behav Neurosci *100*, 814–824.

Davis, M. (1992). The role of the amygdala in conditioned fear. In The Amygdala: neurobiological aspects of emotion, memory, and mental dysfunction. (New York, Wiley-Liss).

Driever, W., Solnica-Krezel, L., Schier, A. F., Neuhauss, S. C. F., Malicki, J., Stemple, D. L., Stainier, D. Y. R., Zwartkruis, F., Abdelilah, S., Rangini, Z., . (1996). A genetic screen for mutations affecting embryogenesis in zebrafish. Development *123*, 37–46.

Dunn, A. J., and Swiergiel, A. H. (1999). Behaivoral responses to stress are intact in CRF-deficient mice. Brain Res *845*, 14–20.

Eckart, K., Radulovic, J., Radulovic, M., Jahn, O., Blank, T., (2002). Pharmacology and biology of corticotropin-releasing factor (CRF) receptors. Recept Channel *8*, 163–177.

Estes, W. K., and Skinner, B. F. (1941). Some quantitative properties of anxiety. J Exp Psychol *29*, 390–400.

Fanselow, M. S. (1994). Neural organization of the defensive behavior system responsible for fear. Psychon Bull Rev *1*, 429–438.

File, S. E. (1988). How good is social interaction as a test of anxiety? In Selected models of anxiety, depression, and psychosis., P. Simon, P. Soubrie, and D. Wildlochewr, eds. (Karger, Basel), pp. 151–166.

File, S. E., and Peet, L. A. (1980). The sensitivity of the rat corticosterone response to environmental manipulations and to chronic chlordiazepoxide. Physiol Behav *25*, 753–758.

Francis, D. D., Szegda, K., Campbell, G., Martin, W. D., and Insel, T. R. (2003). Epigenetic sources of behavioral differences in mice. Nat Neurosci *6*, 445–446.

Gardner, C. R. (1985). Distress vocalization in rat pups. A simple screeninng method for anxiolytic drugs. J Pharmac Meth *134*, 275–283.

Garner, J. P., Dufour, B., Gregg, L. E., Weisker, S. M., and Mench, J. A. (2004). Social and husbandry factors affecting the prevalence and severity of barbering ("Whisker trimming") by laboratory mice. Appl Anim Behav Sci *89*, 263–282.

Gross, C., Zhuang, X., Stark, K., Ramboz, S., Oosting, R., Kirby, L., Santarelli, L., Beck, S., and Hen, R. (2002). Serotonin 1A receptor acts during development to establish normal anxiety-like behavior in the adult. Nature *416*, 396–400.

Guo, S. (2004). Linking genes to brain, behavior, and neurological diseases: what can we learn from zebrafish? Gene Brain Behav *3*, 63–74.

Haffter, P., Granato, M., Brand, M., Mullins, M. C., Hammerschmidt, M., Kane, D. A., Odenthal, J., Van Eeden, F. J. M., Jiang, Y. J., Heisenberg, C. P., . (1996). The identification of genes with unique and essential function in the development of the zebrafish, *Danio rerio*. Development *123*, 1–36.

Heisler, L. K., Chu, H. M., Brennan, T. J., Danao, J. A., Bajwa, P., Parsons, L. H., and Tecott, L. H. (1998). Elevated anxiety and anti-depressant-like responses in serotonin 5-HT1A receptor mutant mice. Proc Natl Acad Sci *95*, 15049–15054.

Joel, D., and Avisar, A. (2001). Excessive lever pressing following post-training signal attenuation in rats: a possible animal model of obsessive compulsive disorder? Behav Brain Res *123*, 77–87.

Kapp, B. S., Whalen, P. J., Supple, W. F., and Pascoe, J. P. (1992). Amygdaloid contributions to conditioned arousal and sensory information processing. In The Amygdala: neurobiological aspects of emotion, memory, and mental dysfunction. (New York, Wiley-Liss).

Kash, S. F., Tecott, L. H., Hodge, C., and Baekkeskov, S. (1999). Increased anxiety and altered responses to anxiolytics in mice deficient in the 65-Kda isoform of glutamic acid decarboxylase. Proc Natl Acad Sci *96*, 1698–1703.

Kim, J. J., Shih, J. C., Chen, K., Chen, L., Bao, S., Maren, S., Anagnostaras, S. G., Fanselow, M. S., De Maeyer, E., Seif, I., and Thompson, R. F. (1997). Selective enhancement of emotional, but not, motor, learning in monoamine oxidase A-deficient mice. Proc Natl Acad Sci *94*, 5929–5933.

Kishimoto, T., Radulovic, J., Radulovic, M., Lin, C. R., Schrick, C., Hooshmand, F., Hermanson, O., Rosenfeld, M. G., and Spiess, J. (2000). Deletion of crhr2 reveals an anxiolytic role for corticotropin-releasing hormone receptor-2. Nat Genet *24*, 415–419.

LeDoux, J. E. (1992). Emotion and the amygdala. In The Amygdala: neurobiological aspects of emotion, memory, and mental dysfunction. (New York, Wiley-Liss).

Levin, E. D., Bencan, Z., and Cerutti, D. T. (2007). Anxiolytic effects of nicotine in zebrafish. Physiol Behav *90*, 54–58.

Lister, R. G. (1987). The use of a plus-maze to measure anxiety in the mouse. Psychopharmacology *92*, 180–185.

Lister, R. G., and Hilakivi, L. A. (1988). The effects of novelty, isolation, light, and ethanol on the social behavior of mice. Psychopharmacology *96*, 181–187.

Lüw, K., Crestani, F., Keist, R., Benke, D., Brönig, I., Benson, J. A., Fritschy, J. M., Rölicke, T., Bluethmann, H., Mühler, H., and Rudolph, U. (2000). Molecular and neuronal substrate for the selective attenuation of anxiety. Science *290*, 131–134.

Mendoza, S. P., Smotherman, W. P., Miner, M., Kaplan, J., and Leinve, S. (1978). Pituitary-adrenal response to separation in mother and infant squirrel monkeys. Dev Psychobiol *11*, 169–175.

Miller, D. B., and O'Callaghan, J. P. (2002). Neuroendocrine aspects of the response to stress. Metabolism *51*, 5–10.

Misslin, R., and Ropartz, P. (1981). Effects of amygdala lesions on the responses to novelty in mice. Behav Process *6*, 329–336.

Montgomery, K. C. (1955). The relation between fear induced by novel stimulation and exploratory behavior. J Comp Physiol Psychol *48*, 254–260.

Ninan, P. T. (1982). Benzodiazepine receptor-mediated experimental "anxiety" in primates. Science *218*, 1332–1334.

Noirot, E. (1972). Ultrasounds and maternal behavior in small rodents. Dev Psychobiol *5*, 371–387.

Nurnberg, H. G., Keith, S. J., and Paxton, D. M. (1997). Consideration of the relevance of ethological animal models for human repetitive behavioral spectrum disorders. Biol Psychiatr *41*, 226–229.

Okon, E. E. (1972). Factors affecting ultrasound production in infant rodents. J Zool Lond *168*, 139–148.

Parks, C. L., Robinson, P. S., Sibille, E., Shenk, T., and Toth, M. (1998). Increased anxiety of mice lacking the serotonin 1A receptor. Proc Nätl Acad Sci *95*, 10734–10739.

Pellow, S., Chopin, P., File, S. E., and Briley, M. (1985). Validation of open/closed arm entries in an elevated plus-maze as a measure of anxiety in the rat. J Neurosci Meth *14*, 149–167.

Preil, J., Möller, M. B., Gesing, A., Reul, J. M., Sillaber, I., van Gaalen, M. M., Landgrebe, J., Holsboer, F., Stenzel-Poore, M., and Wurst, W. (2001). Regulation of the hypothalamic-pituitary-adrenocortical system in mice deficient for CRH receptors 1 and 2. Endocrinology *142*, 4946–4955.

Ramboz, S., Oosting, R., Amara, D. A., Kung, H. F., Blier, P., Mendelsohn, M., Mann, J. J., Brunner, D., and Hen, R. (1998). Serotonin 1A receptor knockout: an animal model of anxiety-related disorder. Proc Natl Acad Sci *95*, 14476–14481.

Smith, S. M. (1975). Innate recognition of coral snake pattern by a possible avian predator. Science *187*, 759–760.

Smith, S. M. (1977). Coral snake recognition and stimulus generalization by naive great kiskadees. Nature *265*, 535–536.

Smith, G. W., Aubry, J. M., Dellu, F., Contarino, A., Bilezikjian, L. M., (1998). Corticotropin releasing factor receptor 1-deficient mice display decreased anxiety, impaired stress response, and aberrant neuroendocrine development. Neuron *20*, 1093–1102.

Speedie, N., and Gerlai, R. (2007). Alarm substance induced behavioral responses in zebrafish (Danio rerio). Behav Brain Res *188*(1):168–177.

Stenzel-poore, M. P., Heinrichs, S. C., Rivest, S., Koob, G. F., and Vale, W. W. (1994). overproduction of corticotropin-releasing factor in transgenic mice: a genetic model of anxiogenic behavior. J Neurosci *14*, 2579–2584.

Timpl, P., Spanagel, R., Sillaber, I., Kresse, A., Reul, J. M., and al., e. (1998). Impaired stress response and reduced anxiety in mice lacking functional corticotropin-releasing hormone receptor 1. Nat Genet *19*, 162–166.

Treit, D. (1985). The inhibitory effect of diazepam on defensive burying: anxiolytic vs. analgesic effects. Pharm Biochem Behav *22*, 47–52.

Vaughan, J., Donaldson, C., Bittencourt, J., Perrin, M. H., Lewis, K., and al., e. (1995). Urocortin, a mammalian neuropeptide related to fish urotensin I and to corticotropin-releasing factor. Nature *378*, 287–292.

Walsh, R. N., and Cummins, R. A. (1976). The open field test: a critical review. Psychol Bull *83*, 482–504.

Willner, P. (1984). The validity of animal models of depression. Psychopharmacology *83*, 1–16.

Zon, L. I., and Peterson, R. T. (2005). In vivo drug discovery in the zebrafish. Nat Rev Drug Discovery *4*, 35–44.

Chapter 10
Animal Models of Affective Behaviors and Drug Addiction

Eva Drews*, Astrid Becker*, Andras Bilkei-Gorzo, and Andreas Zimmer(✉)

Abstract Psychiatric disorders have a strong heritability, and like most common disorders many gene variants, each with a small effect, contribute to the genetic risk. One of the challenges in complex genetics is to determine how different alleles interact with each other and with the environment in the context of disease etiology. Here we discuss mouse models that resemble aspects of human psychiatric disorders syndromes, with particular emphasis on depression and drug addiction. These disorders rank among the top 10 causes of disability and morbidity worldwide and often occur together, thus resulting in a greater severity of health-related consequences. Mice have become the premier research organism, due to their excellent genetics. We have a large and ever-increasing number of mouse strains with defined genetic backgrounds, and the possibilities of precise genetic manipulations seem almost unlimited. Furthermore, environmental factors that are thought to influence disease outcome can be controlled, thus enabling researchers to study interactions between genes and the environment.

Abbreviations CMS: Chronic mild stress; Cnr1: Cannabinoid receptor 1; CPP: Conditioned place preference; DMS: Diagnostic and Statistical Manual of Mental Disorders; DNA: Deoxyribonucleic acid; F1: First filial generation; FST: Forced swim test; kHz: Kilohertz; OB: Olfactory bulbectomy; PR: Progressive ratio; QTL: Quantitative trait loci; RI: Recombinant inbred; SSR: Simple sequence repeats; SSRIs: Selective serotonin reuptake inhibitors; STR: Short tandem repeats; TST: Tail suspension test; USV: Ultrasonic vocalisation; VNTR: Variable number tandem repeats

A. Zimmer
Institute of Molecular Psychiatry, Life & Brain Center, Sigmund-Freud-Straße 25, 53127, Bonn, Germany
a.zimmer@uni-bonn.de

*These authors contributed equally to this work.

C.W. Turck (ed.) *Biomarkers for Psychiatric Disorders*,
doi: 10.1007/978-0-387-79251-4_10,

10.1 Introduction

Mice have become the premier vertebrate research organisms because they are readily amenable to a wide spectrum of genetic manipulations (Dierich and Kieffer, 2004; Ristevski, 2005; Brault et al., 2006; Abuin et al., 2007; Gaveriaux-Ruff and Kieffer, 2007; Schmidt-Supprian and Rajewsky, 2007). We can introduce new genes, delete or replace genes, or introduce tissue-specific mutations. Using dual systems that consist of a gene locus that can be activated or changed and specifically engineered transcription factors or DNA recombinases, we can induce gene alterations by administering drugs such as tamoxifen or tetracycline. Genetic manipulations of the murine genome can be restricted to one specific gene, or it can encompass larger chromosomal regions and thus affect many genes. Indeed it is possible to exchange or delete large chromosomal regions. In addition, we can generate allelic series of point mutations with sophisticated methods for chemical mutagenesis and mutation screening. Genetic modifications of different kinds are already available for at least one quarter of all murine genes, and soon we will probably have a mouse mutant for every gene. We can also induce genetic manipulations in specific organs of adult mice with viral vectors (Aronoff and Petersen, 2006; Pfeifer, 2006). And we can do even more than that. We can label cells, organelles or proteins with fluorescent markers (Ristevski, 2005). We can visualize gene activity with reporter genes. We can manipulate ion channels in such a manner that they are activated by light of specific wavelength, thus turning neurons on and off with diodes of different colors (Zhang et al., 2007). In other words, no other mammalian organism comes even close to the diversity of genetic models that we now have available for mice. Such genetically modified mice have already contributed enormously to our understanding of the biological basis of psychiatric disorders (Takao et al., 2007). Although they may not have as highly developed cognitive and social functions as rats or monkeys, mice are used to investigate specific signaling pathways or specific neuronal circuits in normal brain physiology and in pathological mechanisms.

Psychiatric disorders have a strong heritability and, like most other common disorders, they are complex (Meyer-Lindenberg and Weinberger, 2006). This means that these disorders are caused by a combination of different common gene variants, whereas a single gene only contributes a small percentage to the overall genetic risk load. Thus they can be contrasted to the less frequent Mendelian disorders, where one gene mutation causes a disease phenotype. Genes involved in Mendelian disorders are readily identifiable, provided researchers have access to samples (e.g., families with affected and unaffected individuals) that are sufficiently large or informative enough for genetic mapping. In contrast, it has been a rather daunting task to identify the underlying gene variants in complex disorders. The most successful studies so far have used candidate gene approaches where the biological evidence already suggested a possible involvement in the disease process. Only with the advent of chip-based technologies that permit the detection of hundreds of thousands of polymorphisms in each individual, has it been possible to realize systematic whole-genome association studies (Fan et al., 2006; Eberle et al., 2007; Ropers, 2007). Nevertheless, these studies are still daunting and cost-intensive, because they ideally require very large sample sizes with more than 10,000 cases and controls. Currently, we know very little about complex genetics and how common gene variants contribute to disease etiology. Common gene variants often do not affect the protein-coding capacity of a gene locus and thus are thought to change gene regulatory mechanisms. Indeed, disease-associated alleles are often unknown, because association studies only point out specific polymorphisms or haplotypes that are over- or underrepresented in the case vs. control sample. It then remains to be determined which of the observed polymorphisms is responsible for the increased genetic risk – a task that has not often been met with success so far. These small and subtle gene effects are difficult to address with genetically manipulated mouse strains. Although it is possible to generate "humanized" mice that carry gene regulatory alterations in disease-associated alleles, it remains to be shown how informative such mouse models will be (Shultz et al., 2007). Nevertheless, mouse geneticists have used "classical" genetics to breed a large number of diverse inbred mouse strains with interesting phenotype characteristics that may be relevant for human psychiatric disorders. Such mouse strains resemble the human situation and can be exploited to study specific questions related to complex phenotypes. The importance of this resource has been recognized and currently a large effort to create over 1,000 such diverse inbred strains is being funded by the Collaborative Cross project.

In complex genetics, one of the most challenging tasks is to determine how different genes interact with each other and with the environment in the context of disease etiology. One possibility of gene-gene interactions is the simple additive effects and a more or less linear relationship between genetic risk load and disease probability. However, evidence from chromosome substitution strains indicates that this is not the case. Rather it seems that most gene-gene interactions are synergistic and certain gene combinations carry a risk load that is far greater than the sum of the individual effects. Other gene combinations have the tendency to cancel each other (Nadeau et al., personal communication). In addition, environmental factors can lead to the activation of transcription factors by intracellular signaling cascades and concomitant expression changes. These events usually are of rather short duration and typically last for no more than days or weeks, but they may also result in long-term alterations through epigenetic modifications of the chromatin structure. Thus, events during early childhood may affect the gene regulatory processes of the adult organism. Mice offer the great advantage that environmental factors can be easily controlled and varied. Numerous studies have already shown that such environmental variables have a profound influence on the expression of behavioral phenotypes (and others) (Tsankova et al., 2007). Elucidation of these complex effects probably requires systems approaches including expression profiling, proteomics, epigenomics, etc. (Sieberts and Schadt, 2007). The evaluation of animal behaviors is also important for psychiatric phenotypes.

The term "behavior" generally refers to the way in which an organism acts under specific conditions or in relation to specific circumstances. Behavior represents the final integrated result of genetic, biochemical, physiological and neuronal processes. Thus, behavioral responses may be highly sensitive indicators for environmental or physiological changes. However, they are rarely informative about specific pathways that contribute to the behavioral change. In other words, behavioral testings often tell us that something is different with an animal, but not why. Behavioral testings are commonly used in functional neuroscience studies, drug development, or environmental monitoring, e.g., in exploring environmental toxicants (Wurbel, 2002; Wahlsten et al., 2003; McArthur and Borsini, 2006). A behavioral test always has one or several dependent variables that correspond to the behavioral readout. Animal models also have independent variables that induce specific disease-related conditions, such as genetic manipulations, surgeries, exposure to specific environmental factors, or drug administration. There is an ongoing debate about the validity of animal models for human psychiatric disorders that extend to virtually every aspect of animal experimentation including the conceptual, theoretical framework, the suitability of the species, or the usefulness of pharmacological vs. genetic interventions. This debate is often fuelled by the fact that the behavioral responses in such studies are used as an indicator for emotional or cognitive aspects that cannot be directly addressed in experimental animals.

An animal model that resembles a human clinical condition should meet three validity criteria (Sarter and Bruno, 2002; Willner and Mitchell, 2002):

- Construct validity refers to the degree of overlap between the cause or rationale of the human condition and the animal model.
- Face validity reflects the extent to which the animal behavioral responses reflect human symptoms.
- Predictive validity indicates the value of an animal model to predict the efficacy of pharmacological treatments in humans.

In this chapter, we address the strengths and weaknesses of behavioral paradigms by describing experiments related to affective disorders and drug addiction. Drug addiction and depression rank among the top 10 causes of worldwide disability and morbidity in human populations (Murray and Lopez, 1996). These disorders often occur together, thus resulting in a greater severity of health-related consequences such as an increased number of suicides. The reason for the high degree of comorbidity is poorly understood. Many investigators favor the view that one is a cause or consequence of the other, but there is also an increasing body of evidence suggesting that both disorders share some etiologies and pathomechanisms (Paterson and Markou, 2007). Thus, drug addiction and depression are modulated by partially overlapping neuronal circuits (e.g., reward system), neurotransmitter systems (e.g., serotonin, endocannabinoids), hereditary factors (e.g., Cnr1 polymorphisms), and environmental influences (e.g., stress). Many of the diagnostic criteria for depression and drug addiction involve subjective emotional symptoms that cannot be directed in mice. It is thus difficult, if not impossible, to device a behavioral paradigm that meets all validity criteria. Indeed, it is highly questionable if mice can experience emotions that resemble sadness in humans. In practice it is sometimes difficult to evaluate, based on behavioral responses alone, whether symptoms such as reduced locomotor activity or anhedonia have a neuronal cause, or whether a mouse is simply sick. Nevertheless, there are a number of animal models that recapitulate certain aspects or symptoms of depression. This chapter is by no means intended to provide the uninitiated reader with precise experimental protocols. We are rather using these paradigms to demonstrate some general issue that should be considered for the assessment and interpretation of behavioral phenotypes.

10.2 Behavioral Despair Tests

Behavioral despair paradigms are based on the observation that rodents placed in an uncomfortable and inescapable situation initially show escape behaviors followed by periods of inactivity, which was assumed to indicate that the animals had given up hope of escaping (behavioral despair) (Katz, 1982). Antidepressants increase the duration of escape behaviors and, conversely, decrease the duration of immobility (Steru et al., 1987). Immobility in this test is not an indication of hypomobility; it rather reflects the animals' reluctance in continuing the escape attempts. It might represent a correlative to the observation that patients suffering from depression also show deficits in psychomotor tests that require a sustained effort. These paradigms have a high degree of predictive validity, because they are sensitive for all major classes of antidepressant treatments: Monoamine oxidase inhibitors, tricyclic compounds, and selective serotonin reuptake inhibitors (SSRIs), as well as electroconvulsive shocks show efficacy in this test (Borsini and Meli, 1988). The test is relatively simple and reproducible between different laboratories, therefore it is commonly used as a screening test by the pharmaceutical industry. It should be noted that acute drug treatments are effective in these assays, although antidepressant effects in humans are only observed after repeated or prolonged use.

In the forced swim test (FST, Porsolt's test), developed by Porsolt and co-workers, a mouse is placed in a water-filled (23–25°C) cylinder from which it cannot escape (Porsolt et al., 1977a,b). The depth of the water has to be sufficient (at least 10 cm) so that the animal is unable to make contact with the bottom, not even with its tail. Most mice can easily float on the surface and require only slight movements to keep the head above the water. Typically, mice initially show a period of escape-orientated movements in the FST, followed by a period of predominant immobile postures. The duration of this test is usually 5–6 min of which only the immobility time during the last 4 min is scored (Lucki et al., 2001).

Steru and colleagues developed another behavioral despair paradigm called the tail suspension test (TST) (Steru et al., 1987). Here, mice are suspended by their tails for 5–6 min and their immobility time is scored. Automatic scoring devices have been developed for these tests which allow high throughput screens and the evaluation of additional Parameters such as force, power and energy of movements. FST and TST are conceptually very similar, but influenced by different confounding variables. Thus, a disruption of thermoregulatory processes can affect FST responses, while TST is more sensitive to effects on motor functions. Indeed, some mouse strains, including the commonly used C57BL/6J mice, are known to climb up on their tails and cannot be used in the TST assay (Dalvi and Lucki, 1999). Both tests therefore provide complementary or converging information.

10.2.1 Distress Vocalization

When newborn rat and mouse pups lose contact with their littermates or mother, they emit calls in ultrasonic frequencies (30–90 kHz) that induce a searching behavior in the mother. In addition to the absence of olfactory and tactile cues, cold also elicits the calling behavior (Branchi et al., 1998). The ultrasonic vocalization test (USV) in rodent pups is an ethologically valid model for the measurement of emotionality because it correlates with stress reactivity. Antidepressants (de Paulis, 2007) and anxiolytics reduce the duration and number of calls, whereas anxiogenic agents induce changes in the opposite direction. This procedure is frequently used to test altered emotionality in genetically modified mice (Winslow et al., 2000; Santarelli et al., 2001; Scattoni et al., 2007). In mice, the calling reaches a peak at postnatal days 6–8 and disappears during the second week after birth (Branchi et al., 1998). Therefore, it is critically important to precisely match the age of control and test animals. The experimental design of the USV test is quite simple: The pups are placed on a cold (16–20°C) glass jar in a sound-isolated chamber containing an ultrasonic detector. The test typically lasts for 10 min and requires specialized software for the analysis of the calls.

10.2.2 Learned Helplessness

This model was originally developed for dogs and later adopted for rats and mice (Overmier and Seligman, 1967; Seligman and Beagley, 1975). It involves the exposure of mice to inescapable mild footshocks. Such animals may show escape deficiencies when re-exposed to a similar situation with an escape opportunity. It has been demonstrated that a broad spectrum of antidepressants increases the frequency of escape attempts, although the efficacy of different compounds varies substantially between different strains of mice. One major caveat of this model is seen in the fact that only a certain percentage of mice (between 10 and 80% depending on the mouse strain) show the learned helplessness behavior (Cryan and Mombereau, 2004). The 129 mouse strain, from which most embryonic stem cells used for gene targeting are derived, unfortunately seems to perform particularly poorly in this test. Indeed, even non-shocked 129 control mice show a poor escape performance. Thus, it is important to ensure that mice with targeted gene deletions have been crossed to a different strain and contain no residual of the 129 genetic background that may affect the test results. The model is also sensitive to gender effects. Male mice normally show a stronger learned helplessness behavior than females (Caldarone et al., 2000). This model is also influenced by any alterations in pain sensitivity. Thus, genetic or drug effects on nociception that may result in a reduced learned helplessness should be excluded.

10.2.3 Chronic Mild Stress

A number of animal models for depression are based on the exposure of animals to repeated stressors. Many investigators feel that these models offer a high degree of face validity, because episodes of major depression often occur in the context of stress. Katz and collaborators developed a rat model in which animals are exposed to a series of intense stressors (Katz, 1982), which was modified for mice by exposing animals to various relatively mild and unpredictable stressors (Willner et al., 1987). In this chronic mild stress (CMS) model mice are exposed to environmental stressors like tilted cage, food or water deprivation, paired caging, overnight illumination or wet bedding during a prolonged period of usually more than 2 weeks. This chronic mild stress induces long-lasting behavioral and neurochemical changes, which parallel aspects of depressive disorders in humans (Willner, 2005). Behavioral changes in mice are seen in a reduced sensitivity to rewards, termed anhedonia. Anhedonia is, according to Nelson (Nelson and Charney, 1981), a key symptom of the human depressive state and concerning DMS IV (Association, 1994) a core feature of the melancholic subtype. It also has good predictive validity, as anhedonia can be reversed by chronic treatment with several classes of antidepressants (Willner, 1995; Papp et al., 1996). However, this model has a number of disadvantages that limit its widespread use. The CMS model has a very poor reliability across different laboratories and often cannot be reproduced, even in laboratories where it had been successfully established before.

10.3 Bulbectomy

In this model, the olfactory bulbs of adult mice are bilaterally removed via a microsurgical procedure (olfactory bulbectomy, OB). About 2 weeks after surgery, mice show aspects of behavioral alteration, such as hyperactivity in the open field, disturbed eating patterns, anhedonia or memory deficits. The most commonly evaluated behavioral change in this model is hypermotility in the open field (Zueger et al., 2005). These changes are thought to originate from a complex pattern of neurochemical, neuroendocrine and neuroimmune alterations representing compensatory-neuronal-reorganization and alterations in synaptic connections (Jarosik et al., 2007). These changes in synaptic plasticity are noticed in various subcoritcal limbic regions such as the amygdala and hippocampus and are qualitatively similar to those observed in depressed patients (Cryan and Mombereau, 2004; Song and Leonard, 2005). Hence, like the CMS model, the OB model possesses high face validity. Most changes occurring after OB can be attenuated or reversed by prolonged antidepressant treatment (Otmakhova et al., 1992; Kelly et al., 1997; Song and Leonard, 2005).

10.4 Depression and Drug Addiction

Depression is a common comorbidity with drug addiction (Robbins, 1974) and also occurs frequently during abstinence. Both withdrawal from substance abuse and depression have been linked to changes of several neurotransmitter pathways including serotonin, dopamine and glutamate (Palomo et al., 2007). Drug addiction is a complex brain disorder characterized by an inability to control drug use, despite severe adverse effects. The addiction process can be divided into different stages including controlled use, compulsive use, abstinence and relapse. Chronic substance use/abuse is associated with physiological adaptations, such as the development of drug tolerance. The mechanisms that trigger the transition from controlled use to addiction are still largely unknown. Most addiction processes can be studied in mice, as they readily self-administer almost all drugs abused by humans, show adaptive changes after prolonged drug exposure and withdrawal symptoms as well as drug-craving behavior. Thus, animal models have been developed to assess almost all aspects of drug addiction including acute effects (e.g., thermoregulation, motility), tolerance, somatic drug dependence, drug preference and reward and relapse to drug-seeking behavior.

10.5 Drug Self-Administration/Drug Preference

Models that evaluate whether an animal finds a drug rewarding include self-administration and place conditioning paradigms. For substances like ethanol and nicotine, the two-bottle choice paradigm is often the method of choice to evaluate drug preference via oral self-administration (Fig. 10.1). In such experiments the animals have access to a drug-solution and water, which may contain flavors to match the taste of the drug solution. To exclude a positional bias, the position of the bottles is regularly alternated and the proportion of drug consumption relative to total fluid intake is calculated as a preference ratio. Liquid consumption is mostly evaluated by weighing the bottles, although some companies are developing automated devices. Albeit this paradigm is mostly used for alcohol and nicotine, it has also been used with other drugs such as cocaine, barbiturates and benzodiazepines (Meisch et al., 1992; Stewart et al., 1994; Jentsch et al., 1998). It can be argued whether this model has good construct validity. However, the motivation for drug use is certainly different between animals and humans, as social factors (e.g., peer pressure) contribute probably at least as much as the reinforcing properties of the drug during the early stages of drug consumption.

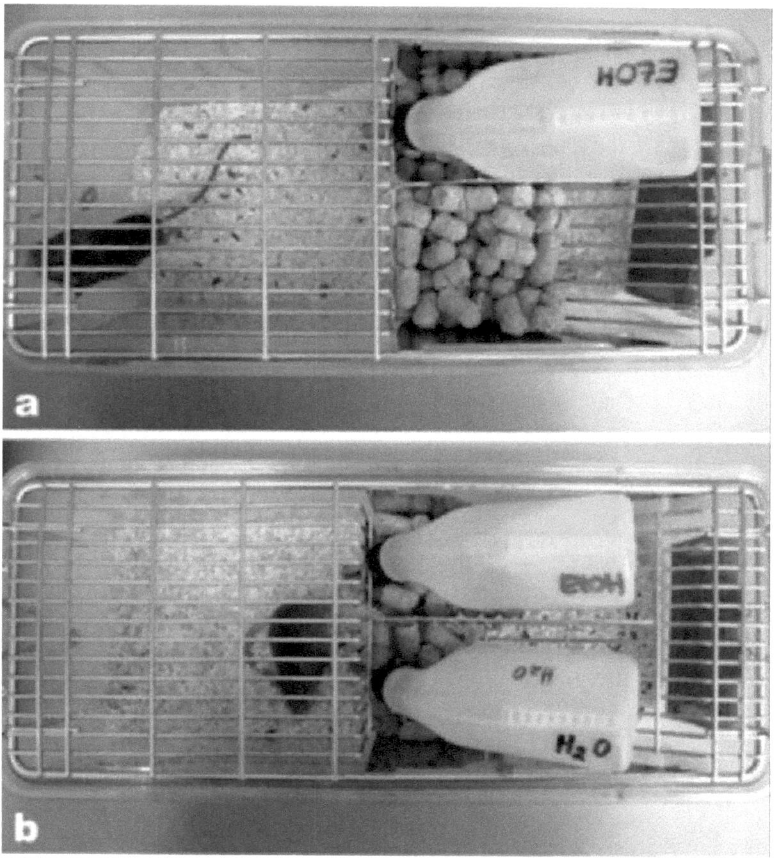

Fig. 10.1 *Forced drinking* (**a**) and *Two bottle-choice paradigm* (**b**). In the *forced drinking paradigm* the animals are forced to consume distinct amounts of drug-solution over an extended period of time. Contrastingly, in the *two bottle-choice paradigm* the animals have the choice between a drug-solution or tap water to evaluate drug preference (*See Color Plates*)

Operant self-administration paradigms were originally developed by Skinner in the 1930s. It involves an apparatus (Skinner box) equipped with one or more operandi (typically a lever or a sensor) that can register behavioral actions and activate the delivery of a reinforcer. The paradigm is based on the concept that behaviors are driven by their outcome. Thus, delivery of a positive reinforcer will strengthen the behavioral response.

Some drugs of abuse have aversive properties that may limit the drug intake and thus the transition to drug addiction. Different protocols were established to minimize these aversive effects. For ethanol, the saccharin fading protocol is a widely used model. Here, the concentration of ethanol in a mixed ethanol/saccharin solution is slowly increased until ethanol is presented alone (Hansson et al., 2007). Different schedules of reinforcement exist to evaluate self-administration behavior in animals. In the fixed ratio schedule of reinforcement, the reward (food, drugs) is delivered after a fixed number of responses (lever presses) have been made.

In operant behavioral tasks like the progressive ratio (PR) test, the instrumental demand for a constant reward increases progressively until responding ceases and reinforcers are no longer obtained. This "break point" is an operational measure for a shift in motivation, where the rewarding value is lower than the effort the animal is willing to expend to obtain it. The PR test is a complex task, where animals are forced to adjust their instrumental behavior from continuous reinforcement to an incremental schedule of responding. This task implies switching from a fixed ratio to an increasing schedule and the dynamic adjustment of behavior according to a cost-benefit analysis. Therefore, it is useful for testing brain mechanisms underlying the translation of motivation and changing values of reinforcement into appropriate behavior (Drews et al., 2005). In addition to oral intake, reinforcers are also administered directly into brain regions of interest.

Replacement of the drug-reinforcer with a neutral substance results in a gradual extinction of the operant behavior. These behaviors can, however, often be reinstated by factors that trigger relapse in humans, such as drug-associated cues, priming with a small amount of the drug, or stress-exposure. This experimental design thus permits the analysis of mechanisms that lead to the relapse of drug-seeking behaviors after a period of abstinence.

Probably the most common model to study the motivational effects of drugs of abuse in experimental animals is the conditioned place preference (CPP) paradigm. Genetically modified animals (knock-out and transgenic animals) are used in the CPP to analyze gene functions in drug reward. Furthermore, drug tolerance and sensitization, as well as extinction and reinstatement procedures, can be examined by means of the CPP paradigm (Tzschentke, 2007).

The CPP paradigm uses a classical conditioning procedure for which a variety of protocols have been adapted to meet specific conceptual requirements. In a simple version, animals are exposed to two or more distinct neutral environments such as two boxes or compartments provided with different floors, marked walls, colors, texture or lighting (Fig. 10.2). First, a potential place bias is determined in an initial session where animals can freely explore the entire apparatus. During the subsequent conditioning phase, animals are alternately restricted to one or the other compartment, where they receive either a drug or a control solution. The environments are thus paired with distinct drug or non-drug states. In a final test session, animals have again the opportunity to move freely in the entire apparatus. If the drug is reinforcing, animals will now spend more time in the drug-associated compartment when compared to the initial session. Conversely, if the drug is aversive, the animal will spend more time in the non-drug associated compartment. The CPP paradigm thus provides indirect information about the positive and negative reinforcing effects of a drug (Carboni and Vacca, 2003).

Fig. 10.2 *Conditioned place preference paradigm*. Animals are exposed to two or more distinct neutral environments that are differentiated from each other. After several conditioning sessions, this paradigm provides information about positive and negative reinforcing effects of a drug (*See Color Plates*)

10.6 Drug Tolerance and Sensitization

Tolerance is defined as a state of progressively decreasing drug response so that increasing drug dosages are needed to produce the desired effect. Two ways of tolerance have been examined, viz., acute and chronic tolerance. Acute tolerance is measured after one or two drug exposures within the same testing session, whereas chronic tolerance is measured after repeated drug exposure (Bennett et al., 2006).

Depending on the drug, there are several ways to determine tolerance in animal models. Measurement of drug-induced hypothermia for example is a commonly used method to determine chronic tolerance to ethanol in rodents. Therefore, drug–naïve mice receive an intraperitoneal injection of an ethanol solution that leads to a measurable loss of body temperature. Most mice will develop a tolerance to the hypothermic effects of ethanol after chronic ethanol administration over a period of several weeks. Thus, the ethanol-induced hypothermia is reduced after chronic exposure. Because of its antinociceptive properties, tolerance to morphine is usually evaluated by the reduction or loss of analgesia after chronic morphine treatment.

On the other hand, some drug effects increase after repeated drug exposure. This sensitization phenomenon is typically observed as an increased locomotor activation after repeated psychostimulant administration. Conceptually, it is sometimes considered an example of non-associative learning in which the gradual strengthening of a response follows repeated administrations of a stimulus (e.g., drugs) (Bell et al., 1995). Sensitization has also been implicated in the vulnerability to relapse. Despite researchers having already found several factors to be associated with the development of sensitization, there is no complete understanding of when tolerance or sensitization will occur after chronic drug use. Factors that are involved in drug sensitization processes include genetic variations of the subjects (Phillips, 1997), the schedule of drug administration (Robinson and Becker, 1986), and the novelty of the environment (Robinson and Berridge, 2000). Some studies have showed that animals more likely become sensitized to a drug if it does not inhibit their goal-directed behavior, e.g., food intake (Wolgin and Hertz, 1995; Pinkston and Branch, 2003). On the other hand, development of drug tolerance often appears if drug-related behaviors (locomotor activity, anxiety) interfere with such goal-directed behaviors (Wolgin and Hertz, 1995).

10.7 Drug Dependence and Withdrawal

Withdrawal follows cessation of chronic drug consumption or treatment in dependent individuals. It produces several symptoms that may provide markers for the study of neurobiological mechanisms of dependence (Bennett et al., 2006). Existing standard rating scales help to easily quantify many of the physical signs occurring after withdrawal of several substances (opiates, nicotine and alcohol).

Mice will become physically dependent on ethanol if they are forced to consume significant amounts over an extended period of time (forced drinking paradigm, Fig. 10.1). When ethanol is removed, such mice will display withdrawal symptoms that may start within hours and last for several days. These symptoms typically include handling-induced convulsions, excitation and increased anxiety-like behaviors (Mayo-Smith, 1997; Watson and Little, 1999). Animal models for these behaviors possess good face validity for human withdrawal, also including tremors, excitation, seizures and anxiety (Kosten and O'Connor, 2003). Ethanol withdrawal – induced convulsions or seizures are easy to measure via turning and lifting the animals and are scored by severity (Watson et al., 1994).

To evaluate physical withdrawal in rodents dependent on opiates, withdrawal symptoms can be induced by the administration of opioid antagonists such as naloxone. After injection of the antagonist, the animals exhibit typical behaviors like wet-dog shakes, tremors, chews, teeth-chattering, diarrhea, etc., that, again, can be quantified by scores. The same approach is often used to determine symptoms of nicotine withdrawal. After administration of a nicotinic antagonist (e.g., mecamylamine), the animals display body shakes, chews, tremors, yawns and scratches. The frequency and time of occurrence is recorded and quantified.

In addition to the physical signs, there are the motivational aspects of drug withdrawal. Measuring these aspects may help to understand the counteradaptive mechanisms that may drive addiction. Animal models for motivational aspects of drug withdrawal include operant schedules (Gellert and Sparber, 1977), place aversion (Stinus et al., 2000), intracranial self-stimulation (Schulteis et al., 1995) and drug discrimination studies (France and Woods, 1989).

10.8 Relapse

Drug addiction is a chronic disorder and relapse to drug addiction can occur even after prolonged periods of abstinence. The most common reasons why humans revert to addictive behaviors include exposure to drug associated cues, drug priming and stress. Conditioned cues acquire reinforcing properties on their own, as they signal the availability of a reinforcer and thus predict reward. Generally it is believed that the same neuronal mechanisms involved in drug reward are also involved in relapse. There are two main theories about how reward pathways mediate relapse to drug-seeking. One postulates that drug reward pathways overlap with relapse pathways (Self and Nestler, 1995; Robinson and Berridge, 2000). The other suggests that an opponent process produces a hypofunctional state of the reward circuitry, which in turn leads to dysphoria or anxiety during withdrawal (Koob et al., 1997). Following this theory, drug-seeking is the attempt of alleviating the aversive feeling of withdrawal.

To assess reinstatement of drug-seeking behavior in mice, the subjects first have to acquire an operant self-administration behavior, followed by an extinction phase. For reinstatement, animals are exposed to conditioned cues (typically a light, smell, or tone), to a small dose of the drug, or to an external stressor (de Wit and Stewart, 1981). Reinstatement is indicated by a preferential activation of the reward-associated operandum, even when it is not rewarded by a drug delivery.

10.9 Analytical Tools

Drug addiction-related behavioral phenotypes in a genetically heterogeneous mouse population are influenced by multiple gene loci, each of which typically contributes to a small effect. Thus, the behaviors vary by degrees and constitute a quantitative trait. Genetic loci contributing to these quantitative traits (quantitative trait loci, QTL) are identified by quantitative genetic approaches (QTL analysis).

Two common strategies are used in mouse QTL studies (Belknap et al., 2001). The fist strategy is based on the breeding of a suitable genetically diverse population of mice, followed by phenotype and genotype analyses of the individuals. Typically, two or more inbred strains of mice are crossed for at least two generations. The progenitor mouse strains should have sufficient variation for the traits of interest and they should be genetically diverse enough to enable genetic mapping (Flint, 2003). The sample size required for the identification of QTL depends largely on the size of the effect that a QTL contributes to phenotypes on interest. Inferences about QTLs can be made if one or more genetic markers are overrepresented (or underrepresented) in individuals scoring in the low or high range of the phenotype spectrum. Genotyping is often done with the use of microsatellite markers, which contain mono, di, tri, or tetra tandem repeats flanked by specific sequences. Such simple sequence repeats (SSR), short tandem repeats (STR), or variable number tandem repeats (VNTR) have been characterized for most commonly used inbred mouse strains.

The second strategy involves the phenotyping of recombinant inbred (RI) strains, which have a unique combination of well-circumscribed genome contributions from two or more progenitor inbred strains. Traditionally, RI strains have been derived by consecutive brother x sister matings (>20 generations), starting from an F1 generation of an intercross of two inbred strains. More advanced RI strains, as those generated in the Collaborative Cross project, are derived from multiple inbred strains.

10.10 Summary and Perspective

In this chapter, we have summarized some mouse models for the analysis of behaviors related to depression and drug addiction, two important complex brain disorders that often occur together. Both disorders are characterized by a spectrum of symptoms, including emotions that may be uniquely human. They are also influenced by a number of social and environmental factors that are irrelevant for mice. Thus, we cannot build rodent models for addiction and depression that reflect all facets of the human condition. Nevertheless, we have a number of mouse models with excellent validity for specific aspects or symptoms of these disorders, which we can utilize to study the neuronal circuits and cellular-molecular mechanisms involved (Arguello and Gogos, 2006). We believe that these findings will result in novel mechanism-based pharmacotherapies for psychiatric disorders.

References

Abuin A, Hansen GM, Zambrowicz B (2007) Gene trap mutagenesis. Handb Exp Pharmacol: 129–147.

Arguello PA, Gogos JA (2006) Modeling madness in mice: one piece at a time. Neuron 52:179–196.

Aronoff R, Petersen CC (2006) Controlled and localized genetic manipulation in the brain. J Cell Mol Med 10:333–352.
Association AP (1994) DSM IV – Diagnostic and statistical manual of psychiatric disorders, 4th Edition. Washington, DC: American Psychiatric Association.
Belknap JK, Hitzemann R, Crabbe JC, Phillips TJ, Buck KJ, Williams RW (2001) QTL analysis and genomewide mutagenesis in mice: complementary genetic approaches to the dissection of complex traits. Behav Genet 31:5–15.
Bell IR, Hardin EE, Baldwin CM, Schwartz GE (1995) Increased limbic system symptomatology and sensitizability of young adults with chemical and noise sensitivities. Environ Res 70:84–97.
Bennett B, Downing C, Parker C, Johnson TE (2006) Mouse genetic models in alcohol research. Trends Genet 22:367–374.
Borsini F, Meli A (1988) Is the forced swimming test a suitable model for revealing antidepressant activity? Psychopharmacology (Berl) 94:147–160.
Branchi I, Santucci D, Vitale A, Alleva E (1998) Ultrasonic vocalizations by infant laboratory mice: a preliminary spectrographic characterization under different conditions. Dev Psychobiol 33:249–256.
Brault V, Pereira P, Duchon A, Herault Y (2006) Modeling chromosomes in mouse to explore the function of genes, genomic disorders, and chromosomal organization. PLoS Genet 2:e86.
Caldarone BJ, George TP, Zachariou V, Picciotto MR (2000) Gender differences in learned helplessness behavior are influenced by genetic background. Pharmacol Biochem Behav 66:811–817.
Carboni E, Vacca C (2003) Conditioned place preference. A simple method for investigating reinforcing properties in laboratory animals. Methods Mol Med 79:481–498.
Cryan JF, Mombereau C (2004) In search of a depressed mouse: utility of models for studying depression-related behavior in genetically modified mice. Mol Psychiatr 9:326–357.
Dalvi A, Lucki I (1999) Murine models of depression. Psychopharmacology (Berl) 147:14–16.
de Paulis T (2007) Drug evaluation: Vilazodone--a combined SSRI and 5-HT1A partial agonist for the treatment of depression. IDrugs 10:193–201.
de Wit H, Stewart J (1981) Reinstatement of cocaine-reinforced responding in the rat. Psychopharmacology (Berl) 75:134–143.
Dierich A, Kieffer BL (2004) Knockout mouse models in pain research. Methods Mol Med 99:269–299.
Drews E, Schneider M, Koch M (2005) Effects of the cannabinoid receptor agonist win 55,212–2 on operant behavior and locomotor activity in rats. Pharmacol Biochem Behav 80:145–150.
Eberle MA, Ng PC, Kuhn K, Zhou L, Peiffer DA, Galver L, Viaud-Martinez KA, Lawley CT, Gunderson KL, Shen R, Murray SS (2007) Power to detect risk alleles using genome-wide tag SNP panels. PLoS Genet 3:1827–1837.
Fan JB, Chee MS, Gunderson KL (2006) Highly Parallel genomic assays. Nat Rev Genet 7:632–644.
Flint J (2003) Analysis of quantitative trait loci that influence animal behavior. J Neurobiol 54:46–77.
France CP, Woods JH (1989) Discriminative stimulus effects of naltrexone in morphine-treated rhesus monkeys. J Pharmacol Exp Ther 250:937–943.
Gaveriaux-Ruff C, Kieffer BL (2007) Conditional gene targeting in the mouse nervous system: Insights into brain function and diseases. Pharmacol Ther 113:619–634.
Gellert VF, Sparber SB (1977) A comparison of the effects of naloxone upon body weight loss and suppression of fixed-ratio operant behavior in morphine-dependent rats. J Pharmacol Exp Ther 201:44–54.
Hansson AC, Bermudez-Silva FJ, Malinen H, Hyytia P, Sanchez-Vera I, Rimondini R, Rodriguez de Fonseca F, Kunos G, Sommer WH, Heilig M (2007) Genetic impairment of frontocortical endocannabinoid degradation and high alcohol preference. Neuropsychopharmacology 32:117–126.

Jarosik J, Legutko B, Unsicker K, von Bohlen und Halbach O (2007) Antidepressant-mediated reversal of abnormal behavior and neurodegeneration in mice following olfactory bulbectomy. Exp Neurol 204:20–28.

Jentsch JD, Henry PJ, Mason PA, Merritt JH, Ziriax JM (1998) Establishing orally self-administered cocaine as a reinforcer in rats using home-cage pre-exposure. Prog Neuropsychopharmacol Biol Psychiatr 22:229–239.

Katz RJ (1982) Animal model of depression: pharmacological sensitivity of a hedonic deficit. Pharmacol Biochem Behav 16:965–968.

Kelly JP, Wrynn AS, Leonard BE (1997) The olfactory bulbectomized rat as a model of depression: an update. Pharmacol Ther 74:299–316.

Koob GF, Caine SB, Parsons L, Markou A, Weiss F (1997) Opponent process model and psychostimulant addiction. Pharmacol Biochem Behav 57:513–521.

Kosten TR, O'Connor PG (2003) Management of drug and alcohol withdrawal. N Engl J Med 348:1786–1795.

Lucki I, Dalvi A, Mayorga AJ (2001) Sensitivity to the effects of pharmacologically selective antidepressants in different strains of mice. Psychopharmacology (Berl) 155:315–322.

Mayo-Smith MF (1997) Pharmacological management of alcohol withdrawal. A meta-analysis and evidence-based practice guideline. American Society of Addiction Medicine Working Group on Pharmacological Management of Alcohol Withdrawal. Jama 278:144–151.

McArthur R, Borsini F (2006) Animal models of depression in drug discovery: a historical perspective. Pharmacol Biochem Behav 84:436–452.

Meisch RA, Lemaire GA, Cutrell EB (1992) Oral self-administration of pentobarbital by rhesus monkeys: relative reinforcing effects under concurrent signalled differential-reinforcement-of-low-rates schedules. Drug Alcohol Depend 30:215–225.

Meyer-Lindenberg A, Weinberger DR (2006) Intermediate phenotypes and genetic mechanisms of psychiatric disorders. Nat Rev Neurosci 7:818–827.

Murray CJL, Lopez AD (1996) Evidence-based health policy – Lessons from the global burden of disease study. Science 274:740–743.

Nelson JC, Charney DS (1981) The symptoms of major depressive illness. Am J Psychiatr 138:1–13.

Otmakhova NA, Gurevich EV, Katkov YA, Nesterova IV, Bobkova NV (1992) Dissociation of multiple behavioral effects between olfactory bulbectomized C57Bl/6J and DBA/2J mice. Physiol Behav 52:441–448.

Overmier JB, Seligman ME (1967) Effects of inescapable shock upon subsequent escape and avoidance responding. J Comp Physiol Psychol 63:28–33.

Palomo T, Archer T, Kostrzewa RM, Beninger RJ (2007) Comorbidity of substance abuse with other psychiatric disorders. Neurotox Res 12:17–27.

Papp M, Moryl E, Willner P (1996) Pharmacological validation of the chronic mild stress model of depression. Eur J Pharmacol 296:129–136.

Paterson NE, Markou A (2007) Animal models and treatments for addiction and depression co-morbidity. Neurotox Res 11:1–32.

Pfeifer A (2006) Lentiviral transgenesis – a versatile tool for basic research and gene therapy. Curr Gene Ther 6:535–542.

Phillips TJ (1997) Behavior genetics of drug sensitization. Crit Rev Neurobiol 11:21–33.

Pinkston JW, Branch MN (2003) Sensitization to cocaine in pigeons: interaction with an operant contingency. Exp Clin Psychopharmacol 11:102–109.

Porsolt RD, Bertin A, Jalfre M (1977a) Behavioral despair in mice: a primary screening test for antidepressants. Arch Int Pharmacodyn Ther 229:327–336.

Porsolt RD, Le Pichon M, Jalfre M (1977b) Depression: a new animal model sensitive to antidepressant treatments. Nature 266:730–732.

Ristevski S (2005) Making better transgenic models: conditional, temporal, and spatial approaches. Mol Biotechnol 29:153–163.

Robbins PR (1974) Depression and drug addiction. Psychiatr Q 48:374–386.

Robinson TE, Becker JB (1986) Enduring changes in brain and behavior produced by chronic amphetamine administration: a review and evaluation of animal models of amphetamine psychosis. Brain Res 396:157–198.

Robinson TE, Berridge KC (2000) The psychology and neurobiology of addiction: an incentive-sensitization view. Addiction 95 Suppl. 2:S91–S117.

Ropers HH (2007) New perspectives for the elucidation of genetic disorders. Am J Hum Genet 81:199–207.

Santarelli L, Gobbi G, Debs PC, Sibille ET, Blier P, Hen R, Heath MJ (2001) Genetic and pharmacological disruption of neurokinin 1 receptor function decreases anxiety-related behaviors and increases serotonergic function. Proc Natl Acad Sci USA 98:1912–1917.

Sarter M, Bruno J (2002) Animal models in biological psychiatry. In: Textbook of Biological Psychiatry (D'Haenen H, Den Boer H, Willner P, eds). Chichester: Wiley.

Scattoni ML, McFarlane HG, Zhodzishsky V, Caldwell HK, Young WS, Ricceri L, Crawley JN (2007) Reduced ultrasonic vocalizations in vasopressin 1b knockout mice. Behav Brain Res. 187:374–378.

Schmidt-Supprian M, Rajewsky K (2007) Vagaries of conditional gene targeting. Nat Immunol 8:665–668.

Schulteis G, Markou A, Cole M, Koob GF (1995) Decreased brain reward produced by ethanol withdrawal. Proc Natl Acad Sci USA 92:5880–5884.

Self DW, Nestler EJ (1995) Molecular mechanisms of drug reinforcement and addiction. Annu Rev Neurosci 18:463–495.

Seligman ME, Beagley G (1975) Learned helplessness in the rat. J Comp Physiol Psychol 88:534–541.

Shultz LD, Ishikawa F, Greiner DL (2007) Humanized mice in translational biomedical research. Nat Rev Immunol 7:118–130.

Sieberts SK, Schadt EE (2007) Moving toward a system genetics view of disease. Mamm Genome 18:389–401.

Song C, Leonard BE (2005) The olfactory bulbectomised rat as a model of depression. Neurosci Biobehav Rev 29:627–647.

Steru L, Chermat R, Thierry B, Mico JA, Lenegre A, Steru M, Simon P, Porsolt RD (1987) The automated Tail Suspension Test: a computerized device which differentiates psychotropic drugs. Prog Neuropsychopharmacol Biol Psychiatr 11:659–671.

Stewart BS, Lemaire GA, Roache JD, Meisch RA (1994) Establishing benzodiazepines as oral reinforcers: midazolam and diazepam self-administration in rhesus monkeys. J Pharmacol Exp Ther 271:200–211.

Stinus L, Caille S, Koob GF (2000) Opiate withdrawal-induced place aversion lasts for up to 16 weeks. Psychopharmacology (Berl) 149:115–120.

Takao K, Yamasaki N, Miyakawa T (2007) Impact of brain-behavior phenotypying of genetically-engineered mice on research of neuropsychiatric disorders. Neurosci Res 58:124–132.

Tsankova N, Renthal W, Kumar A, Nestler EJ (2007) Epigenetic regulation in psychiatric disorders. Nat Rev Neurosci 8:355–367.

Tzschentke TM (2007) Measuring reward with the conditioned place preference (CPP) Paradigm: update of the last decade. Addict Biol 12:227–462.

Wahlsten D, Rustay NR, Metten P, Crabbe JC (2003) In search of a better mouse test. Trends Neurosci 26:132–136.

Watson WP, Little HJ (1999) Correlation between increases in dihydropyridine binding in vivo and behavioural signs of ethanol withdrawal in mice. Alcohol Alcohol 34:35–42.

Watson WP, Misra A, Cross AJ, Green AR, Little HJ (1994) The differential effects of felodipine and nitrendipine on cerebral dihydropyridine binding ex vivo and the ethanol withdrawal syndrome in mice. Br J Pharmacol 112:1017–1024.

Willner P (1995) Animal-models of depression – validity and applications. Depression and Mania 49:19–41.

Willner P (2005) Chronic mild stress (CMS) revisited: consistency and behavioural-neurobiological concordance in the effects of CMS. Neuropsychobiology 52:90–110.

Willner P, Mitchell PJ (2002) The validity of animal models of predisposition to depression. Behav Pharmacol 13:169–188.

Willner P, Towell A, Sampson D, Sophokleous S, Muscat R (1987) Reduction of sucrose preference by chronic unpredictable mild stress, and its restoration by a tricyclic antidepressant. Psychopharmacology (Berl) 93:358–364.

Winslow JT, Hearn EF, Ferguson J, Young LJ, Matzuk MM, Insel TR (2000) Infant vocalization, adult aggression, and fear behavior of an oxytocin null mutant mouse. Horm Behav 37:145–155.

Wolgin DL, Hertz JM (1995) Effects of acute and chronic cocaine on milk intake, body weight, and activity in bottle- and cannula-fed rats. Behav Pharmacol 6:746–753.

Wurbel H (2002) Behavioral phenotyping enhanced--beyond (environmental) standardization. Genes Brain Behav 1:3–8.

Zhang F, Aravanis AM, Adamantidis A, de Lecea L, Deisseroth K (2007) Circuit-breakers: optical technologies for probing neural signals and systems. Nat Rev Neurosci 8:577–581.

Zueger M, Urani A, Chourbaji S, Zacher C, Roche M, Harkin A, Gass P (2005) Olfactory bulbectomy in mice induces alterations in exploratory behavior. Neurosci Lett 374:142–146.

Chapter 11
Neuroimaging Biomarkers in Schizophrenia

Heike Tost and Andreas Meyer-Lindenberg (✉)

Abstract Schizophrenia is a highly heritable and devastating mental disorder with a complex genetic architecture. The combination of multimodal neuroimaging with genetic mapping techniques and the subsequent assessment of genomic variation ("imaging genetics") has recently proved to be exceptionally powerful in advancing our understanding of the neural underpinnings of normal cognition and mental disease states. This upcoming research field has opened doors to a previously inaccessible level of both biological characterization and the validation of schizophrenia risk gene effects. This chapter reviews the current scientific knowledge on neuroimaging biomarkers in schizophrenia and illustrates the major conceptual change in the way in which biological intermediate phenotypes are viewed and pursued in psychiatry today.

Abbreviations ACG: Anterior cingulate gyrus; BOLD: Blood oxygen level dependent; CD–CV: Common disease–common variant; COMT: Catechol-O-methyltransferase; CPT: Continuous performance test; CSF: Cerebrospinal fluid; CSTC: Cortico-striatal-thalamo-cortical; DA: Dopamine; DARPP32: Dopamine and cyclic AMP regulated phosphoprotein; DISC1: Disrupted-in-schizophrenia 1; DLPFC: Dorsolateral prefrontal cortex; DNA: Deoxyribonucleic acid; DTI: Diffusion tensor imaging; DTNBP1: Dysbindin 1; EPS: Extrapyramidal side effects; FEF: Frontal eye fields; fMRI: Functional magnetic resonance imaging; GM: Gray matter; GRM-3: Metabotropic glutamate receptor 3; HLP: Haloperidol; 1H-MRSI: Proton magnetic resonance spectroscopic imaging; 5-HTT: Serotonin transporter promoter; MRI: Magnetic resonance imaging; mRNA: Messenger ribonucleic acid; NAA: N-acetyl-aspartate; NRG1: Neuroregulin 1; PET: Positron emission tomography; PFC: Prefrontal cortex; PPC: Posterior-parietal cortex; PPI: Prepulse inhibition; PPP1R1B: Protein phosphatase 1, regulatory (inhibitor) subunit 1B; SMA: Supplementary motor area; STG: Superior temporal gyrus; VBM: Voxel-based morphometry; V5/hMT: Motion-sensitive visual Processing area; VLPFC: Ventrolateral prefrontal cortex; WM: White matter

---

A. Meyer-Lindenberg
Clinical Brain Disorders Branch, Genes, Cognition, and Psychosis Program, National Institute of Mental Health, National Institutes of Health 9000, Rockville Pike, Bethesda, MD 20892, USA

C.W. Turck (ed.) *Biomarkers for Psychiatric Disorders*,
doi: 10.1007/978-0-387-79251-4_11,

11.1 Introduction

Schizophrenia is a psychiatric disorder imposing significant emotional and financial burdens on the affected individuals, their families, and the social health care system. Although the presumed neurobiological origin was already an integral part of the first formal description of the disorder as "dementia praecox" in 1919 (Kraepelin, 1919), the elucidation of the underlying neurobiological substrates was limited for a long time by the restricted methodological spectrum available to address such questions. Therefore, until the beginning of the 1980s, biomarker research in schizophrenia was characterized primarily by neurobiochemical and postmortem research, complemented by limited insights on the brain systems level, often acquired through crude, painful and potentially dangerous procedures like ventriculography or pneumoencephalography (used to demonstrate ventricular enlargement in affected patients).

Beginning in the 1990s, neuroimaging emerged as a pivotal tool for the noninvasive investigation of mediating neural mechanisms in schizophrenia. Techniques such as magnetic resonance imaging (MRI) and positron emission tomography (PET) revolutionized our understanding of the neural underpinnings of normal cognition and mental disease states by characterizing underlying brain functions and dysfunctions at the neural systems level. Recent methodological developments enabled the successful integration of neuroimaging in modern translational research, allowing for a multimodal research strategy that bridges the gap between genetic and molecular mechanisms and clinical phenomena. The combination of multimodal neuroimaging with genetic mapping techniques and the subsequent assessment of genomic variation ("imaging genetics") have recently proved to be exceptionally powerful in advancing our understanding of both normal cognition and the pathophysiological basis of neuropsychiatric disorders (Meyer-Lindenberg and Weinberger, 2006). This upcoming research field has opened the doors to a previously inaccessible level of both biological characterization and the validation of gene effects of schizophrenia risk and has introduced a major conceptual change in the way in which biological intermediate phenotypes are viewed and pursued in psychiatry today.

During the last few decades, the ongoing effort to elucidate the biological foundation of schizophrenia has led modern psychiatry through recurring phases of enthusiasm and frustration. Although, as reviewed below, major progress has been made with respect to the delineation of schizophrenia biomarkers, there is currently no physical or technical test that can diagnose the disorder with certainty. This lack of a gold standard is a major obstacle in defining valid and reliable biomarkers. What can we demand of a useful biomarker in this situation? It should at least be systematically related to schizophrenia and/or relevant clinical parameters such as severity of symptoms or treatment response. Ideally, it should also be specific to schizophrenia, i.e., it should capture a process at the core of "the disorder." However, at the present state of psychiatric research, this latter goal may prove to be elusive, as both the diagnosis and differential diagnosis of schizophrenia are based on the type and duration of clinical symptoms as specified in current diagnostic manuals, i.e., they follow a procedure based on behavioral observation and introspection that is largely identical to the basic diagnostic methods applied 100 years ago.

On the one hand, the DSM-IV and ICD 10 are essential and successful guidelines for disease management in current clinical practice. At the same time, however, the implemented phenotypic level of description inevitably limits the biological validity of these manuals. If the experience from other complex disorders like diabetes or spinocerebellar ataxia is any guide, then it is unlikely that even the most sophisticated behavioral classifications will ever define a biologically homogenous illness. Thus, a vicious cycle develops in which a biologically heterogeneous group of conditions are given the label of "schizophrenia," adding noise and reducing the power of any attempt to define the underlying biology of the disorder. This lack of precision ultimately limits the amount of physiological understanding to such a degree that no viable alternative disease definition can be supported. This issue is far from being merely academic, as it is precisely this lack of biological understanding that limits the discovery of new treatment options. New interventions are badly needed for this devastating disorder, where currently available therapies, despite being evidence-based and highly effective for the majority of patients, are still not curative. This chapter reviews the current scientific knowledge on neuroimaging biomarkers in schizophrenia. In doing so, the chapter attempts to give consideration to both the legitimate enthusiasm for a vibrant and successful research field as well as its remaining limitations, particularly the lack of innovative drug development necessary for individualized, science-based patient care.

11.2 Schizophrenia: A Genetically Complex Brain Disorder

11.2.1 Clinical Characteristics

Schizophrenia is a chronic, severe, and disabling brain disorder affecting approximately 1% of the population worldwide (Sullivan et al., 2000). During the course of the illness, a wide array of cognitive, perceptual, emotional, and behavioral manifestations may arise that significantly interfere with an individual's occupational and social capabilities. Core "positive" symptoms include psychotic manifestations, such as hearing voices in the absence of auditory input (auditory hallucinations), assigning unusual significance or meaning to normal events or holding false personal beliefs in spite of invalidating evidence (delusions), and disorganized, bizarre, or catatonic behaviors. In contrast, "negative" symptoms are usually less accessible to the observer, more persistent, and not always easily differentiated from antipsychotic side effects. They manifest as diminution or loss of normal functions, usually cognitive deficits, affective flattening, anhedonia, or avolition. Most patients experience their first symptoms in early adolescence, followed by symptoms that are often chronic, but sometimes benign or of variable course. Current evidence supports the idea that the overt manifestation of the disease is preceded by a lasting prodromal phase, in which psychosocial adaptation is often impaired and nonspecific cognitive, volitional, or affective symptoms may be apparent. While the rapid manifestation of symptoms and presence of psychosocial stressors is indicative of a positive outcome, other clinical characteristics like gradual onset of symptoms, positive family history of psychosis, predominance of negative symptoms, and presence of neuroradiological abnormalities suggest a poor prognosis.

11.2.2 Susceptibility Genes

The precise neurobiological foundation of schizophrenia is unknown. Current evidence-based etiological models point towards the key importance of interactions between predisposing vulnerability and environmental factors. Available results from twin and family studies strongly suggest that schizophrenia is a predominately genetic disorder where the proportion of trait variance explained by genetic factors (heritability estimates) is up to 80% (Sullivan et al., 2003). During the last several decades, a multitude of studies have attempted to characterize the predisposing risk genes leading to the development of the disorder. Currently available linkage data disprove the notion of a single causative gene and a simple pattern of inheritance. Like diabetes or cancer, schizophrenia is conceptualized nowadays as a complex genetic disorder that arises from the interaction of multiple risk alleles, each accounting for only a small increment in risk for the development of the disorder (Weinberger, 2005). It is believed that relatively few (i.e., 10–20) yet frequent genetic variants of low attributable risk combine to precipitate the disorder. This theory is known as the common disease–common variant (often abbreviated CD-CV) hypothesis, and it is a major motivation for ongoing genome wide association studies in schizophrenia. Nonetheless, it is also possible that such common variants are not found at a statistically unambiguous level of proof, implying that the alternative to CD–CV, namely the presence of multiple rare mutations of high attributable risk, is true. In the latter case, much larger sample sizes and methodological advances, for example in deep sequencing, will be necessary to identify the genetic culprits. Regardless, the evidence for a strong genetic component to disease development is incontrovertible.

Many factors contribute to the small size of the effects of genes in psychiatry. One such factor is that few variants involve changes in protein structure or function (i.e., they are not "functional" in the traditional sense of a coding variant). More often, however, aspects of gene regulation, which have relatively subtle biological effects, are implicated. The complex genetic architecture of psychiatric disorders implies that no particular constellation of genes or environmental conditions will be characteristic of most ill individuals (Gottesman and Shields, 1967). It is important to note that schizophrenia susceptibility genes do not encode for the psychopathological phenomena themselves, but interfere with the regular functioning of underlying neurotransmitter systems that mediate the emergence of such complex behavioral phenotypes. This suggests that the more behavioral a given phenotype is, the less likely it will be directly predicted by a genotype.

During the last decade, the availability of modern genetic research techniques like linkage analysis and positional cloning resulted in evidence associating schizophrenia with a number of genetic loci. Although still a major research focus in the field, the initial "linkage enthusiasm" was dampened by a myriad of weak results and nonreplications. These are most likely explained by the fact that psychiatric disorders are genetically complex and do not necessarily meet the criteria essential for gaining reliable results with the applied methods (e.g., valid model of inheritance, genetic homogeneity, moderate to major gene effects) (Riley and McGuffin, 2000). In recent years this has led to a reorientation of the field towards a detailed characterization of the schizophrenia risk gene effects in healthy subjects on the molecular, cellular, and systemic level ((Meyer-Lindenberg and Weinberger, 2006), see the chapter on intermediate phenotypes for further discussion). This is furthered by the identification of candidate risk genes at linkage loci, such as catechol-O-methyltransferase (COMT, chromosome 22q11), metabotropic glutamate receptor 3 (GRM-3, chromosome 7q21–22), neuroregulin 1 (NRG 1, 8p12–21), dysbindin 1 (DTNBP1, chromosome 6p22), and disrupted-in-schizophrenia 1 (DISC1, chromosome 1q42) (see (Harrison and Weinberger, 2005) for a comprehensive review of genetic findings). Not all identified risk gene variants are clearly functional. However, the impact of known gene products on the regulation of dopaminergic or glutamatergic neurotransmission and plasticity processes is now increasingly accepted. Further candidates not supported by linkage have been studied, and the incoming genome wide association data will identify further variants at a probably much more stringent level of statistical evidence that can be utilized for a translational approach.

11.2.3 Neurodevelopmental Disease Models

Neurodevelopmental models of schizophrenia have refined the diathesis-stress theory by proposing that the combination of genetic and epigenetic factors leads to a perturbation of regular prefrontal cortex network development that results in a predisposition for the development of psychotic symptoms (Lewis and Levitt, 2002). The observation that the onset of the disorder occurs during early adolescence has driven considerable research toward the elucidation of neural events that occur during this period of ontogenesis, especially normal maturational changes, that may trigger psychopathology in genetically predisposed individuals. One influential hypothesis assumes that schizophrenia emerges from intrauterine disturbances of temporolimbic-prefrontal network formation that manifest clinically after adolescence (Weinberger, 1987). According to this hypothesis, a disturbed neural interaction leads to an impairment of prefrontal lobe functioning that manifests as negative symptoms (e.g., blunted speech, lack of drive) and cognitive deficits, especially in the executive domain (e.g., working memory, selective attention). Due to a deficiency in the prefrontal control exerted on phylogenetically older brain areas, subcortical dopamine release in the basal ganglia is thought to become disinhibited; this phenomenon is linked, possibly because of the relevance of dopamine for the stabilization of cortical neural assemblies, to the emergence of positive symptoms like hallucinations and delusions (Meyer-Lindenberg et al., 2002). On the cellular level, the theory of a neurodevelopmental perturbation in early life is supported by studies reporting subtle cytoarchitectonic aberrations of the entorhinal and prefrontal cortices in schizophrenia (Arnold et al., 1997; Bertram et al., 2007). Further evidence arises from work indicating a role for both candidate susceptibility genes (Brandon et al., 2004; Sei et al., 2007) and known epigenetic risk factors for schizophrenia (e.g., obstetric complications) (Meyer et al., 2007) for neuronal migration and central nervous system development.

11.2.4 Dopaminergic Dysfunction

Associated abnormalities of dopamine (DA) neurotransmission are suggested by many factors, including the fact that antipsychotic agents block dopamine receptors – a cornerstone of the so-called "dopamine hypothesis" of schizophrenia (Baumeister and Francis, 2002). The dopaminergic system originating from the midbrain affects a wide range of motivated behaviors, e.g., prefrontal cognitive functions (Williams and Goldman-Rakic, 1995) and reward-related learning (Schultz, 1998). On the functional level, it has been proposed that prefrontal-striatal interactions, which are modulated by DA, provide a "filter" for information that is competing for cortical processing, a mechanism that could underlie the prefrontal cognitive symptoms in schizophrenia (Pantelis et al., 1997) as well as potential schizophrenia intermediate phenotypes such as abnormal prepulse inhibition of startle (Swerdlow et al., 2001). Furthermore, the clinical potencies of antipsychotic drugs are directly related to their affinities for the dopamine D_2 receptor, suggesting that the clinical phenomena against which these drugs are primarily effective ("positive symptoms") arise from a disturbed D_2 receptor-dependent hyperactive dopaminergic signaling state on the network level (Seeman et al., 2005). Two main dopaminergic systems impact on prefrontal neural circuits; while mesocortical DA projections arise from the ventral tegmental area and target the dorsolateral prefrontal, medial frontal, and anterior cingulate cortex (Tzschentke, 2001), nigrostriatal projections exert their influence on the prefrontal cortex (PFC) via somatotopically organized feedback loops involving the basal ganglia and the thalamus (Gerfen, 2000). Dopaminergic dysfunctions in schizophrenia have been repeatedly observed in both systems (Grace, 1993; Meyer-Lindenberg et al., 2002; Sjoholm et al., 2004; Weinberger, 1987). While a simple model of increased dopaminergic tone throughout the brain has been disproved, disinhibited nigrostriatal dopaminergic neurotransmission is a consistently reported abnormality that has been linked to deficiencies in PFC functioning (McGowan et al., 2004; Meyer-Lindenberg et al., 2002).

Single-cell recordings in animal studies demonstrated that mesocortical dopamine, especially through the D_1 receptor tone that regulates the excitatory synaptic input to prefrontal pyramidal neurons, plays a decisive role in the modulation of working memory functions known to be disturbed in schizophrenia (Fuster, 1990; Goldman-Rakic, 1995; Williams and Goldman-Rakic, 1995). A large body of work has established an "inverted u" shaped relationship between working-memory related activation of PFC neurons and dopaminergic, especially D_1 receptor, stimulation (Williams and Goldman-Rakic, 1995), with dopaminergic tone being essential for optimizing signal to noise ratio, or "tuning" of prefrontal networks (Durstewitz et al., 2000). According to this, a balanced D_1 dopaminergic tone augments the robustness of network representations by making them less susceptible to interfering stimuli or background neural "noise." States of overly excessive or lacking D_1 dopaminergic drive, in contrast, seem to switch the system away from robustness and efficacy, i.e., they facilitate the emergence of prefrontal cognitive deficits and of psychotic symptoms. Taken together, convergent genetic, molecular, cellular, neurophysiological, and neuroimaging evidence suggests that the clinical phenomena associated with schizophrenia arise from a genetically complex disorder of brain development associated with downstream disturbances in dopaminergic neurotransmission and functional impairments of PFC network efficacy.

11.3 Neurostructural Biomarkers

In the last several decades, the highly-replicated finding of ventricular enlargement in schizophrenia stimulated a large number of postmortem studies examining the question of whether or not a primarily neurodegenerative process is involved in the development of the disorder (Henn and Braus, 1999; Nasrallah, 1982; Stevens, 1982). On the cortical level, these studies have provided evidence for a variety of subtle histopathological changes, e.g., aberrantly located neurons in the enthorhinal cortex (Arnold et al., 1997; Jakob and Beckmann, 1986), cortical pyramidal neurons with smaller perikarya (Benes et al., 1991a,b Rajkowska et al., 1998), diminished dendritic arborization (Garey et al., 1998; Glantz and Lewis, 2000), and reduced GABAergic interneuron density (Lewis, 2000; Reynolds et al., 2002). On the subcortical level, these findings were complemented by work suggesting alterations in volume, density, or total number of neurons in the caudate nucleus, putamen, nucleus accumbens, thalamus, hippocampus, and amygdala (Kreczmanski et al., 2007; Walker et al., 2002; Zaidel et al., 1997). Not all of these studies, however, have produced consistent results, and no clear pattern of histological alterations has yet been established. This might be explained, at least in part, by the inherent limitations of postmortem analyses (e.g., postmortem changes due to tissue decomposition, fixation effects), differences in sample characteristics (e.g., with respect to medication status), and the histological methods used. Interestingly, certain morphological signs typically observed in neurodegenerative disorders like Parkinson's disease or Huntington's disease are usually missing in schizophrenia, notably the associated increase in the absolute number of glial cells (i.e., reactive gliosis) (Benes, 1999; Longson et al., 1996). This led to the conclusion that neither the age of onset characteristics nor the overall histopathological picture is consistent with the pattern of a primarily neurodegenerative disorder. Instead, a neurodevelopmental perturbation in early life seems to affect the migration, survival, and connectivity of neurons in corticostriatal and corticolimbic circuitries, thereby promoting the manifestation of overt clinical symptoms in early adulthood (Benes, 1993; Lewis and Levitt, 2002).

Beginning in the mid-1980s, the availability of MRI has revolutionized the field in allowing the noninvasive examination of structural biomarkers in the living brain. From the beginning, the manual delineation of a priori defined regions of interest has been a popular method for obtaining volumetric measures in psychiatric research. With an 81% rate of positive findings, bilateral decreases in hippocampal gray matter volume has been among the most consistently reported volumetric results in schizophrenia (Henn and Braus, 1999). According to a recent thorough meta-analysis, the mean patient hippocampal volume is decreased to 94% (left) and 95% (right) of the volume of normal controls (Wright et al., 2000). Basal ganglia structures, in contrast, have been repeatedly shown to exhibit an overall volumetric increase in affected patients (caudate, putamen, and pallidum) (Goldman et al., accepted for publication; Henn and Braus, 1999; Potvin et al., 2007; Wright et al., 2000). This increase is most likely explained by long-term medication effects associated with typical neuroleptics, as the finding has not been observed in antipsychotic-naïve patients (Chua et al., 2007; Keshavan et al., 1998) or atypically medicated populations (Gur et al., 1998), and has proven to reverse when the medication regimen is switched from conventional to atypical drugs (Lang et al., 2004; Scheepers et al., 2001) (see the chapter on antipsychotic drug effects for further discussion). The question of a disorder-intrinsic pathology of basal ganglia structures in schizophrenia is still a matter of debate (Crespo-Facorro et al., 2007). Recent evidence suggests that significant gray matter volume decrease might indeed be evident at disease onset, but might later be masked in studies performed in medicated populations (Glenthoj et al., 2007). This assertion is further supported by evidence that the basal ganglia volume increase in schizophrenia, while pronounced, is not heritable (Goldman et al., accepted for publication).

In recent years, the broad availability of clinical MRI scanners and the implementation of automated processing techniques facilitated the large-scale analysis of structural imaging datasets. Voxel-based morphometry (VBM) is an efficient whole-brain approach that has been used extensively to identify associated brain structural abnormalities in schizophrenia samples. Among others, increased ventricular CSF, smaller mean cerebral volume, and widespread GM reductions in the frontal, parietal, and temporal cortices and cerebellum have been reported (Hulshoff Pol et al., 2006; Neckelmann et al., 2006; Whitford et al., 2006). Probably due to the heterogeneity of examined patient groups with respect to age at onset, medications status, length of antipsychotic treatment, clinical symptoms, and illness severity, no single region has been consistently reported to be abnormal in schizophrenia to date. A recent comprehensive meta-analysis of VBM studies identifies gray matter deficits in the left superior temporal gyrus and the left medial temporal lobe as most consistently reported findings, and points out the impact of the choice of image preprocessing variables on study outcomes (e.g., smoothing kernel, normalization type) (Honea et al., 2005). On the subcortical level, the availability of fully-automated whole brain segmentation techniques allows for the analysis of structural datasets in large subject cohorts. In one of our own morphometric studies (Goldman et al., accepted for publication) conducted in 221 healthy controls, 169 patients with schizophrenia, and 183 unaffected siblings, the patient group showed a bilateral decrease in hippocampal and cortical gray matter volume and increases in bilateral dorsal striatum compared to the healthy controls. Moreover, evidence for the heritability of reduced gray matter volume as a schizophrenia-associated biological trait was obtained for the hippocampus and cerebral cortex.

Though most likely of neurodevelopmental origin, gray matter volume decreases in schizophrenia have shown a progressive development over time in several studies that are related to illness course and treatment outcome, as evidenced by longitudinal studies performed in first-episode (Cahn et al., 2002; DeLisi et al., 1997; Gur et al., 1998; Lieberman et al., 2001), chronic (Davis et al., 1998; Mathalon et al., 2001), and childhood-onset (Rapoport et al., 1999) patients. Given the postmortem data reviewed above, the neurobiological substrate of these observed changes in MR signal is not clear (Weinberger and McClure, 2002). Recent diffusion tensor imaging (DTI) studies have suggested that the underlying disease-process is not restricted to the gray matter partition, but also involves microstructural alterations of connecting white matter fiber tracts. Findings of disturbed integrity of axonal projections in the anterior cingulate bundle (108, Manoach et al., 2007; Sun et al., 2003), prefronto-limbic projections, and the frontal-thalamic circuitry (Rose et al., 2006) have been reported, among others. Although the total number of DTI studies is still limited, the evidence for an additional disturbance of white matter tract integrity in schizophrenia is quite convincing (Kubicki et al., 2007).

11.4 Functional Neuroimaging Markers

Since its first application in the early 1990s, functional magnetic resonance imaging (fMRI) has emerged as the leading neuroimaging research technique for the examination of brain functional alterations in mental disorders. Meanwhile, the methodological spectrum of the field has developed enormously. Clinical scanners with ultra-fast gradient systems, multichannel array coils, and advanced data acquisition schemes are broadly available today, allowing for whole brain data collection of highly resolved slices within a few seconds. Over the last decade, experimental approaches were successively refined from simple paradigms with blockwise stimulation to rapid event-related task designs and elaborated methods for connectivity analyses of neural network interactions have been successfully established (Meyer-Lindenberg et al., 2005; Meyer-Lindenberg et al., 2005; Stein et al., in press). The popularity of this approach in neuropsychiatry is mirrored by the vast amount of fMRI studies published on schizophrenia research in recent years. The present chapter reviews major findings in this still-expanding research field, focusing on selected neurocognitive domains.

11.4.1 Visual Perception

A number of neuropsychological studies assessing the visual system found impairments in early visual processing in schizophrenia (Butler and Javitt, 2005). Visual backward masking experiments, where patients identify significantly fewer target stimuli than normal controls when the target is briefly obscured by a second visual stimulus (Braff and Saccuzzo, 1981, 1985; Keri et al., 2000; Moritz et al., 2001), provide classic examples of impairments in schizophrenia. The visual system is divided into magnocellular and parvocellular pathways which project to dorsal and ventral visual areas of the cortex, respectively. The magnocellular network or "dorsal stream" covers cortical areas specialized for the handling of motion and depth cues, e.g., the motion-sensitive field V5 (hMT), posterior-parietal cortex (PPC), frontal eye fields (FEF), and the dorsolateral prefrontal cortex (DLPFC) (Ungerleider et al., 1998; Ungerleider and Mishkin, 1982). Electrophysiological and behavioral investigations have suggested preferential involvement of the dorsal visual stream in schizophrenia, as indicated by pronounced deficits during the processing of higher spatial frequencies and moving stimuli (Cadenhead et al., 1998; O'Donnell et al., 1996; Schwartz et al., 1999). A circumscribed processing deficit of motion signals in area V5 has been hypothesized because of observed impairments in target velocity discrimination and associated eye tracking dysfunctions (Chen et al., 1999a,b). Similar deficits are evident in clinically-unaffected relatives of schizophrenia patients, suggesting that these symptoms resemble trait markers of disease vulnerability, i.e., a potential target for diagnosis, treatment, and prevention (Chen et al., 2006).

The precise location of the presumed dorsal stream dysfunction, however, is still a matter of debate. Apart from a "bottom up" processing deficit in V5, thalamic filter dysfunctions and "top down" control deficits of the PFC over lower visual cortices have been hypothesized (Keri et al., 2000; Levin, 1984a,b). fMRI studies have been aimed at delineating the underlying neural correlates of these observed perceptual processing deficits. One of our own studies (Braus et. al. 2002, (Braus et al., 2002)) examined the cortical response to passive visuo-acoustic stimulation without higher order cognitive demands in neuroleptic-naive schizophrenia patients. Compared to healthy controls, significant decreases in activation in the lateral geniculate nucleus of the thalamus as well as higher order areas of the dorsal processing stream (PPC, FEF, and DLPFC) were evidenced in the patient group. With respect to the hypothesis of a circumscribed functional deficiency of V5 in schizophrenia, the fMRI literature is conflicting and suggests, if at all, a task-dependent deficit. Recent work by Lencer and coworkers (2005) reported a focal decrease of the hemodynamic response of V5 in schizophrenia patients during the performance of smooth-pursuit eye movements (Lencer et al., 2005). In contrast, our own follow-up studies pointed to preserved V5 functioning during passive perception and velocity discrimination of moving visual stimuli, but also revealed a significantly decreased activation of higher order control areas of the dorsal stream (PPC, DLPFC) (Tost et al., 2005).

Within the parvocellular "ventral" processing stream, recent work by Yoon and coworkers (2006) indicated that the spatial extent and magnitude of activity in the fusiform face area seemed to be preserved during the perceptual processing of faces in schizophrenia (Yoon et al., 2006). Structural studies of occipital cortex and the optic radiations have also provided support for an anatomical basis for an early visual processing deficit in connected brain areas upstream of the extrastriate cortex. A recent postmortem study by Dorph-Petersen (2007) reported markedly reduced total neuron number (25%) and volume (23%) in the primary visual cortex of patients relative to healthy controls, suggesting altered cortical parcellation of lower visual areas (Dorph-Petersen et al., 2007). In addition, recent DTI work indicating a decrease in fractional anisotropy of the optic radiations in schizophrenia (Butler et al., 2006) points to microstructural alterations of the feeding thalamo-cortical projections.

11.4.2 Auditory Perception

Auditory hallucinations, i.e., the perception of inner voices in the absence of sensory input, are cardinal symptoms of schizophrenia. Neuropsychological theories of the early 1990s assumed that these hallucinations originate from a dysfunctional processing of silent inner articulations (David and Cutting, 1994), a notion that has been examined using neuroimaging methods on the biological level during the last 15 years. Structural and functional abnormalities of the superior temporal gyrus (STG) and associated language areas of the temporal and frontal lobe have emerged as some of the most prominent biomarkers of schizophrenia neuroimaging research. Early fMRI studies provided a plausible explanation for the fact that patients accept the internally generated voices as "real" by demonstrating a pronounced enhancement of activity in auditory and speech processing cortices during hallucinatory experiences (Dierks et al., 1999; Ffytche et al., 1998; McGuire et al., 1993; Silbersweig et al., 1995). Other fMRI studies pointed to a diminished temporal lobe response to external speech during hallucinatory experiences, a phenomenon usually explained by the competition of physiological and pathological processes for limited neural processing capacity (David et al., 1996; Woodruff et al., 1997).

Effective connectivity analyses point to impaired information integration between left superior temporal and anterior cingulate cortices in patients with auditory hallucinations, as evidenced by a study of Mechelli (2007), where the difference between the perception of own vs. unfamiliar voices was examined (Mechelli et al., 2007). Recent work by Weinstein et al. (2006) extends the view on superior temporal lobe dysfunctions in schizophrenia further by showing that the functional anomalies during receptive language processing predict the severity of the patient's thought disorder, a clinical symptom manifesting as irregularities in speech (Weinstein et al., 2006). On the structural level, a significantly reduced leftward asymmetry of the planum temporale is a well-known feature of schizophrenia (the planum temporale is a triangular area on the STG that overlaps with Wernicke's area; cytoarchitectonically, it resembles secondary auditory cortex) (Kwon et al., 1999; Shapleske et al., 1999, Sumich et al., 2005]). Mounting scientific evidence points to a brain structural correlate of the severity of hallucinatory symptoms in schizophrenia, as evidenced by the associated gray matter volume deficits and microstructural changes of white matter projections of the STG (Gaser et al., 2004; Hubl et al., 2004; Neckelmann et al., 2006).

11.4.3 Motor Functioning

Psychomotor disturbances are common manifestations in schizophrenia ranging from subtle neurological soft signs (e.g., coordination deficits) to complex disorders of behavioral control (stereotyped behaviors, catatonic symptoms) (Schroeder et al., 1991; Vrtunski et al., 1986). The examination of motor functioning in schizophrenia was very popular in early psychiatric neuroimaging research and usually involved simple block-design experiments alternating between the performance of repetitive motor activities (e.g., sequential finger opposition) and rest conditions. Major findings of these early works (which have not always consistently replicated in subsequent studies (Braus et al., 1999; Buckley et al., 1997; Mattay et al., 1997; Schröder et al., 1995, 1999 Wenz et al., 1994)) include the observation of hypoactivation in primary motor, sensorimotor and supplementary motor cortices (SMA), Moreover, a disrupted functional asymmetry (i.e., reduced laterality) of motor cortices during motor task performance has been reported repeatedly (Bertolino et al., 2004; Mattay et al., 1997; Rogowska et al., 2004). Within the highly lateralized motor circuitry this observation points to a deficient ipsilateral inhibition or poor focal recruitment of cortical motor areas in schizophrenia. In line with the neurodevelopmental disease model, an underlying "interhemispheric disconnectivity" arising from the functional breakdown of transcallosal glutamergic projections has been suggested (Mattay et al., 1997) and encouraged by corresponding electrohysiological findings (Hoppner et al., 2001).

Although the neurofunctional basis of motor symptoms in schizophrenia is still barely understood, corresponding fMRI studies have been rare in recent years. Among other explanations, the pronounced impact of medication confounds on motor loop functioning (Tost et al., 2006) and the comparative lack of fMRI studies in neuroleptic-naïve patient cohorts might account for this situation. Sparse available evidence indicates the capacity of atypical antipsychotics to restore the disturbed lateralization and connectivity patterns observable during motor task performance (Bertolino et al., 2004; Stephan et al., 2001). A recent study by Vink (2006) points to a functional anomaly of the striatum that is suggestive of an endophenotypic trait marker. A lack of anticipation-related increases in striatal activation during motor inhibition was observable in both schizophrenia patients and their unaffected relatives, but not in healthy controls (Vink et al., 2006). On the structural level, there is accumulating experimental support for an association between reported motor findings and gray matter volume decreases in higher order modules of the motor loop (e.g., SMA) and microstructural anomalies of interconnecting projections (Exner et al., 2006; Goghari et al., 2005; Matsui et al., 2002).

11.4.4 Cognition

Schizophrenia patients exhibit a wide array of impairments in higher order cognitive functions such as memory, learning, and attention (Joyce and Roiser, 2007). Neuropsychological studies in schizophrenia have demonstrated preferential involvement of cognitive domains known to be dependent on dorsolateral and medial PFC efficiency, especially working memory, cognitive flexibility, selective attention, and response inhibition. Within this spectrum, the neurobiological basis of working memory dysfunctions has been most extensively examined. Unlike short-term memory, working memory paradigms challenge the active storage and online manipulation of information during task performance, often assessed in the form of so-called "*n*-back" tasks. Here, participants are required to constantly monitor a sequence of stimulus presentations and react to items that match the one presented *n* stimuli before. The popularity of this paradigm is explained by its obvious experimental advantages; while the working memory load is manipulated systematically (parametric increase of the factor *n* to 2-back, 3-back, etc.), the stimulus and response conditions are kept constant. Previous experimental work in animals and humans has convincingly emphasized the importance of the mesocortical "dopaminergic tone," or D_1 receptor occupancy levels of the PFC for *n*-back task performance (Callicott et al., 1999; Egan et al., 2001; Goldman-Rakic, 1995; Mattay et al., 2003; Williams and Goldman-Rakic, 1995).

During the last decade, a number of neuroimaging studies have been performed to unravel the neural basis of working memory deficits in schizophrenia. Most studies have reported disorder-related functional anomalies of the dorsolateral prefrontal cortex (DLPFC) during task performance (Brodmann areas 46, 9) and disturbances in the functional coupling of DLPFC and hippocampus (Meyer-Lindenberg et al., 2001; Meyer-Lindenberg et al., 2005). The exact nature of this dysfunction, however, is still a matter of debate, as DLPFC hypofunctions (Andreasen et al., 1997; Barch et al., 2001; Paulman et al., 1990; Weinberger et al., 1986), increased activations (Callicott et al., 2000), unchanged activations (Manoach et al., 1999), and combinations of hyper- and hypoactive states (Callicott et al., 2003) have been observed. Among others, differences in the clinical characteristics of study populations, influence of medication, performance effects, and task load differences have been discussed in this context (Tan et al., 2005). These discrepant findings have motivated leading researchers in the field to question the traditional theory of a pure "hypofrontality" in schizophrenia and to strive for more complex pathophysiological models of prefrontal cortical dysfunction (Callicott et al., 2000 2003; Manoach, 2003). A recent study performed in first-episode schizophrenia patients points to a compensatory recruitment of interconnected (but hierarchically inferior) parts of the ventrolateral prefrontal cortex (VLPFC, Brodmann areas 44, 45, 47) (Tan et al., 2005). This observation has led to the conclusion that the functional specialization and hierarchical organization of the prefrontal cortex is compromised in schizophrenia, resulting in reduced functional efficacy of the DLPFC and, subsequently, a compensatory recruitment of less specialized areas in the VLPFC (Tan et al., 2006).

Attentional deficits, likely at the root of the "peculiar association disturbances" (Bleuler, 1911) experienced as thought disorder, incoherence, and disorganized speech by the observer, have been viewed as a core element of schizophrenia since the first clinical descriptions of the condition. In schizophrenia research, the most popular experimental setup for the assessment of attentional dysfunctions is the so-called "continuous performance test" (CPT), a terminological label covering a variety of nonstandardized task designs. Apart from the common task requirements for challenging selective attention (e.g., selective responses to certain targets, inhibition of inadequate reactions to nontargets), the individual configuration of the CPT paradigm defines the cognitive domains and brain areas tapped. Popular task setups range from degraded CPT (additional perceptual load), to contingent CPT (additional working memory load, e.g., CPT-AX, CPT-IP, and CPT-double-T), and CPT paradigms employing distractor stimuli (additional cognitive interference load). The considerable experimental heterogeneity must be taken into account when interpreting available neuroimaging findings of the field.

Thus far, most functional imaging studies have used variations of the contingent CPT to explore the neurobiological basis of attentional dysfunctions in schizophrenia. Likely due to the moderate working memory load of these tasks, DLPFC (MacDonald and Carter, 2003; Perlstein et al., 2003; Volz et al., 1999) and VLPFC (Eyler et al., 2004) abnormalities are among the most replicated findings. The observed deficits in dorsolateral PFC activation are also present in first-episode neuroleptic-naive patients (Barch et al., 2001), suggesting that they are unlikely explained solely by a confounded medication effect. Furthermore, increased recruitment of postcentral, medial temporal, and cerebellar areas, as well as attenuated interregional connectivity of the medial frontal gyrus, anterior cingulate gyrus (ACG), and cerebellum have been reported (Eyler et al., 2004; Honey et al., 2005). A combined fMRI and VBM study by Salgado-Pineda (2004) suggests that an underlying structural impairment of ACG and inferior parietal cortex might contribute to the observed functional and behavioral deficits (Salgado-Pineda et al., 2004). Paradigms with a high load on interference management consistently detected the ACG as "dysfunctional bottleneck" in schizophrenia. Reported anomalies include regional hypoperfusion (Carter et al., 1997), absence or abnormal localization of dorsal ACG activation (Heckers et al., 2004), as well as morphological and microstructural anomalies of the ACG and cingulum fiber bundle [108, Sun et al., 2003; Yücel et al., 2002] that coincide with the observed functional anomalies (Artiges et al., 2006).

11.5 Intermediate Phenotypes: Concept and Applications

Although genes are known to contribute substantially to the development of many psychiatric disorders, their mechanisms of action have long appeared elusive. As discussed previously, the genetic architecture of psychiatric risk is complex and is dominated by multiple interacting genetic and environmental factors, with individual risk gene variants only explaining a small amount of the total phenotypic variance. In recent years, the development of imaging genetics – a strategy for mapping neural structure and activity as a function of genotype in living humans – has inspired a conceptual transformation in psychiatry (Meyer-Lindenberg and Weinberger, 2006). The following section introduces the intermediate phenotype concept in neuroimaging research and illustrates the power of this approach for characterizing the mediating neural mechanisms that bridge the gap from DNA sequence to pathological behavior.

11.5.1 Intermediate Phenotypes in Neuropsychiatry

The path from genes over neural system interactions to normal behavioral and psychiatric phenotypes is complex. Multiple genetic risk variants interact with each other and the environment, and impact, in differing proportions, on multiple neural systems that are linked to a variety of different cognitive and behavioral domains (see Fig. 11.1). Prior research in the field clearly pointed out that there is no one-to-one mapping between genes and neural system mechanisms, or between neural mechanisms and behavior. Thus, it is not surprising that the classification of disorder entities based on the overt phenotypes has thus far only proved to be of limited value in the genetic dissection of psychiatric diseases. In an effort to overcome this limitation, two major routes have been pursued to identify genetic variants linked to psychiatric disorders (see Fig. 11.2).

The gene discovery approach uses behavioral or neural systems phenotypes to reduce phenotypic heterogeneity and increase the power of genetic association studies. This strategy is intimately connected with the so-called "endophenotype concept," a term that was adopted from evolutionary biology by Gottesman and Shields (Gottesman and Shields, 1967) and later translated into the modern psychiatric context by Gottesman and Gould (Gottesman and Gould, 2003). The term "endophenotype" (from the Greek word *endos* for interior, within) describes a biological trait that is expected to be closer to the genetic substrate, i.e., a trait marker that is directly associated with underlying vulnerability yet distinct from the disease phenotype itself. The manifestation may be neurophysiological, biochemical, endocrinological, neuroanatomical, or neuropsychological in nature and is thought to assist in both "the identification of aberrant genes in the hypothesized polygenic systems conferring vulnerabilities to disorders" and the decomposition of psychiatric diagnosis into biologically-valid disease entities (Gottesman and Gould, 2003). One example is the deficient inhibition of the P50 auditory-evoked response to auditory stimulation evident in schizophrenia patients and their unaffected relatives that led to an association

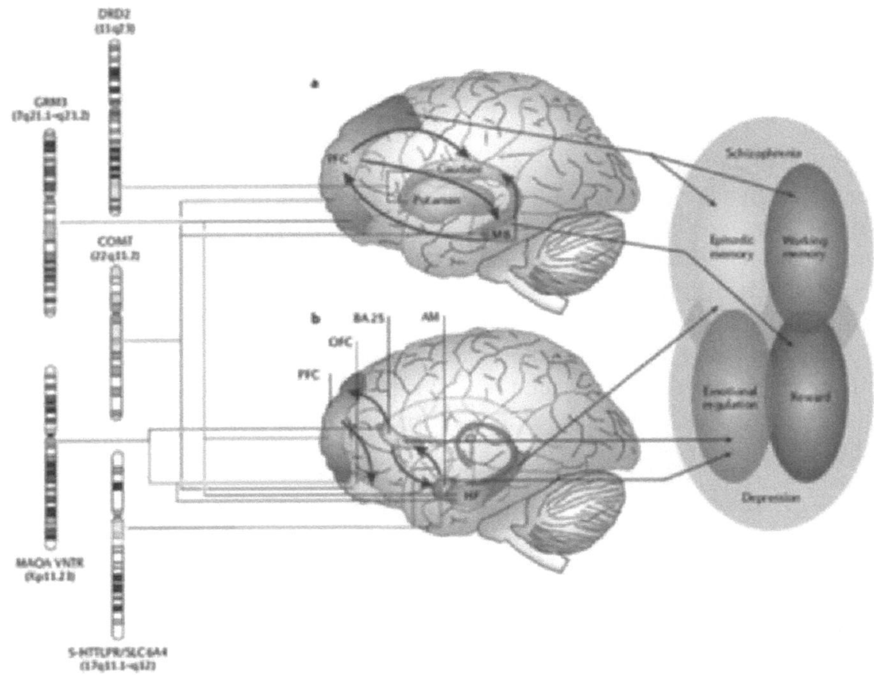

Fig. 11.1 The complex path from genes to behavioral and disease phenotype: mediation through brain circuitry. Multiple genetic risk variants affect, through interaction with each other and the environment, multiple neural systems linked to several neuropsychological and behavioral domains that are impaired, in differing proportions, in psychiatric diseases. As examples, the following genetic variants are depicted (chromosomal variation in parentheses): GRM3 single nucleotide polymorphism 4 (7q21.1–q21.2) [Egan et al., 2004], dopamine receptor D2 (DRD2) Taq 1a (11q23) [Cohen et al., 2005], catechol-O-methyltransferase (COMT) Val66Met (22q11.2) [Egan et al., 2001, Meyer-Lindenberg et al., 2006], serotonin transporter length polymorphism (5-HTTLPR/SLC6A4) (17q11.1–q12) [Hariri et al., 2002, Pezawas et al., 2005] and monoamine oxidase A variable number tandem repeat (MAOA VNTR) (Xp11.23) [Meyer-Lindenberg et al., 2006]. These are shown to affect a circuit that links the prefrontal cortex (PFC) with the midbrain (MB) and striatum (caudate and putamen) (**a**), which is relevant for schizophrenia, and a circuit that connects the amygdala (AM) with regulatory cortical and limbic areas (**b**), which is implicated in depression and anxiety. These circuits, in turn, are shown to mediate risk for schizophrenia and depression and various neuropsychological functions. BA 25, Brodmann's area 25; HF, hippocampal formation; OFC, orbitofrontal cortex. (reprinted with permission from [Meyer-Lindenberg and Weinberger, 2006]) (*See Color Plates*)

of the disorder with the alpha 7-nicotinic cholinergic receptor gene (Freedman et al., 1997). Another study (Gasperoni et al., 2003) used working memory performance as biological trait marker to identify a significant association with a region of the genome that have been previously suggested in traditional linkage studies of schizophrenia (Ekelund et al., 2000; He et al., 1996; St Clair et al., 1990). The term "endophenotype" is nonetheless somewhat misleading, because there is nothing "hidden" about these phenomena in the context of neurobiology and neuroimaging. We thus prefer the term "intermediate phenotype" both because it implies a biological trait that is part of a predictable path from gene to behavior and because the phenotypes and mechanisms are not secondary, but rather primary.

To actually further disorder-related gene discovery, the biological trait markers have to meet certain criteria (Almasy and Blangero, 2001; Gottesman and Gould, 2003; Waldman, 2005; Weinberger, 1999). These markers must be sufficiently heritable, possess good metric properties, be stable over time, show increased cosegregation with the disorder, and should be expressed in unaffected relatives of patients, among other criteria. Published work has suggested that few, if any, of the phenomena discussed as "endophenotypes" in the current psychiatric literature actually meet these standards. This facet of the intermediate phenotype concept therefore closely meshes with criteria that biomarkers should fulfill. However, the recent success of imaging genetics indicates that a greater explanatory power is achieved when genetic variants are used as tools for the discovery and characterization of intermediate phenotypes, i.e., associated functional variations of the level of the mediating neural substrate.

The neural mechanism approach uses genes known to be associated with psychiatric disorders or behavioral traits to uncover the underlying neural mechanisms that mediate these complex phenotypes. Many risk gene associations to brain-based phenotypes are observed even in healthy individuals. One assumption of this strategy is that the penetrance of gene effects is greater at the neurobiological level than at the level of complex behavior, and that these gene effects are traceable at the brain level in carriers of risk alleles even if the phenotype shows no clinical signs of the disorder. The first evidence for greater penetrance of schizophrenia risk genes on the level of more simple and biologically based phenotypes arose in the early 1990s, when the neuropsychological evaluation of monozygotic twins discordant for schizophrenia evidenced a similar profile of prefrontal cognitive deficits (Goldberg et al., 1990). In the imaging genetics field, successful examples for this approach are the use of the catechol-O-methyltransferase (COMT) $Val^{158}Met$ polymorphism to characterize prefrontal function and prefrontal–midbrain interactions linked to the risk for schizophrenia (Egan et al., 2001; Meyer-Lindenberg et al., 2005), and the modulation of amygdala-cingulate circuitry function by the serotonin transporter (5-HTT) promoter polymorphism which has been linked to the risk for anxiety and depression (Hariri et al., 2002; Pezawas et al., 2005). In this context, imaging genetics has proven a powerful tool for elucidating the neural mechanisms that translate genetic effects into complex phenotypes.

11.5.2 Characterization of Schizophrenia Risk Gene Effects

As outlined previously, disturbances in dopaminergic transmission and related dysfunctions of striatal-prefrontal and DLPFC-hippocampal circuitries are well-known features of schizophrenia. Based on the intermediate phenotype strategy, related

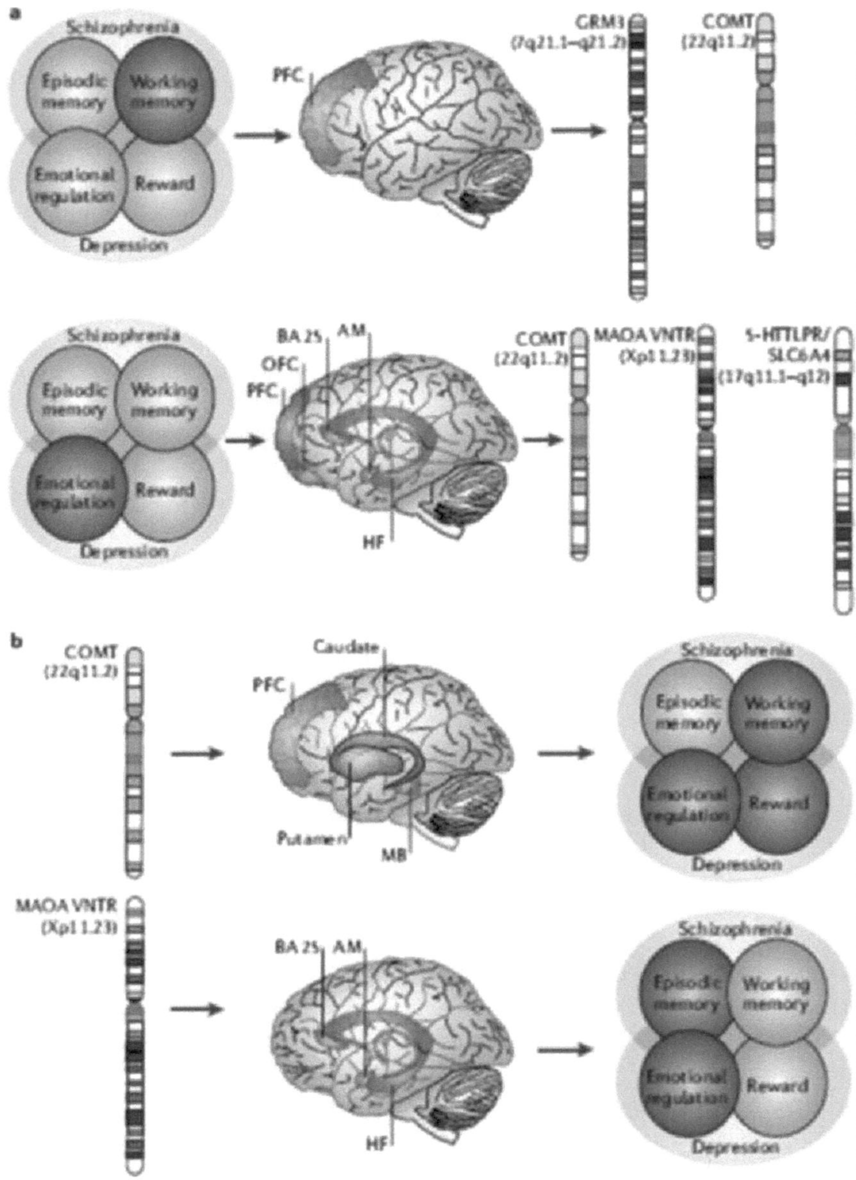

Fig. 11.2 Intermediate phenotypes as tools for gene discovery versus neural mechanism characterization. Examples of two alternative approaches to the identification of genetic variants linked to psychiatric disorders are illustrated, with the relevant genes, neural systems and behavioral phenotypes highlighted in red, and arrows indicating the direction of research inference. (**a**) In the gene discovery approach, behavioral or neural systems phenotypes are used to reduce genetic complexity and increase penetrance to identify genes implicated in psychiatric disorders. In the figure, prefrontal cortex (PFC) dysfunction has been linked to catechol-*O*-methyltransferase (COMT) and GRM3 genetic variation [Egan et al., 2004, Meyer-Lindenberg et al., 2006], and emotional regulation has been linked to variation in COMT, monoamine oxidase A (MAOA) and

dysfunctions are expected in genetically predisposed healthy individuals if the evidenced dysfunctions are actually an expression of the genetic risk for schizophrenia rather than the clinical phenotype itself. Results from several studies of healthy relatives of patients with schizophrenia support this idea. The best evidence likely arises from intermediate phenotype studies on catechol-O-methyltransferase (COMT), a major enzyme degrading cortical dopamine. Since dopamine transporters are scarce in prefrontal cortex (Lewis et al., 2001), COMT is a critical determinant of dopamine flux in the PFC (Tunbridge et al., 2006). A common val$^{(108/158)}$met substitution affects the stability of the COMT protein, leading to a significant decrease in enzyme activity in brain and lymphocytes (Chen et al., 2004). The COMT gene is located at 22q11.2, a region implicated in schizophrenia by linkage (Owen et al., 2004) and by the 22q11.2 syndrome, a hemideletion associated with strongly increased risk of schizophrenia-like illness (Murphy, 2002). Studies from our own laboratory have demonstrated that this coding variant impacts prefrontal cortex activation during working memory performance (Egan et al., 2001) and also modulates the PFC-dependent neuropsychological performance (Goldberg et al., 2003) and cortical response to amphetamine in healthy subjects (Mattay et al., 2003). The latter finding suggests that the COMT genotype places individuals at predictable points along the putative "inverted u-shaped" curve that links prefrontal dopamine stimulation, neuronal activity, and PFC efficiency. Homozygotes for the val-encoding allele (maximal COMT activity, less synaptic dopamine) are thought to be positioned to the left of met allele carriers (less COMT activity, more synaptic dopamine), which seem to be optimally located near the peak of that curve (see Fig. 11.3). Additional evidence from one of our own PET studies (Meyer-Lindenberg et al., 2005) indicates that COMT genotype impacts on prefrontal regulation of midbrain dopamine synthesis in a genotype-dependent directionality consistent with the inverted-u shaped model (Akil et al., 2003). These findings validate and extend the concept of cortical inefficiency as a key endomechanism that contributes to risk for schizophrenia. The genetic risk associated with this common

Fig. 11.2 (continued) the serotonin transporter length polymorphism (5-HTTLPR/SLC6A4) [Heinz et al., 2005, Meyer-Lindenberg et al., 2006, Pezawas et al., 2005], and so could have been hypothetically employed as a phenotype to identify these genes. (**b**) In the neural mechanism approach, genes known to be associated with psychiatric disorders or behavioral traits are used to discover neural mechanisms mediating their complex emergent phenotypic associations, implicating these mechanisms in the psychiatric disorders to which they have been linked. Examples include the use of the COMT Val158Met polymorphism to characterize prefrontal function and prefrontal-midbrain interactions linked to risk for schizophrenia [Egan et al., 2001, Meyer-Lindenberg et al., 2005], and the delineation of cingulate circuitry regulating amygdala (AM) function mediating risk for anxiety and depression through an investigation of the MAOA variable number tandem repeat (VNTR) [Hariri et al., 2002, Pezawas et al., 2005]. BA 25, Brodmann's area 25; HF, hippocampal formation; MB, midbrain; OFC, orbitofrontal cortex (reprinted with permission from [Meyer-Lindenberg and Weinberger, 2006]) (*See Color Plates*)

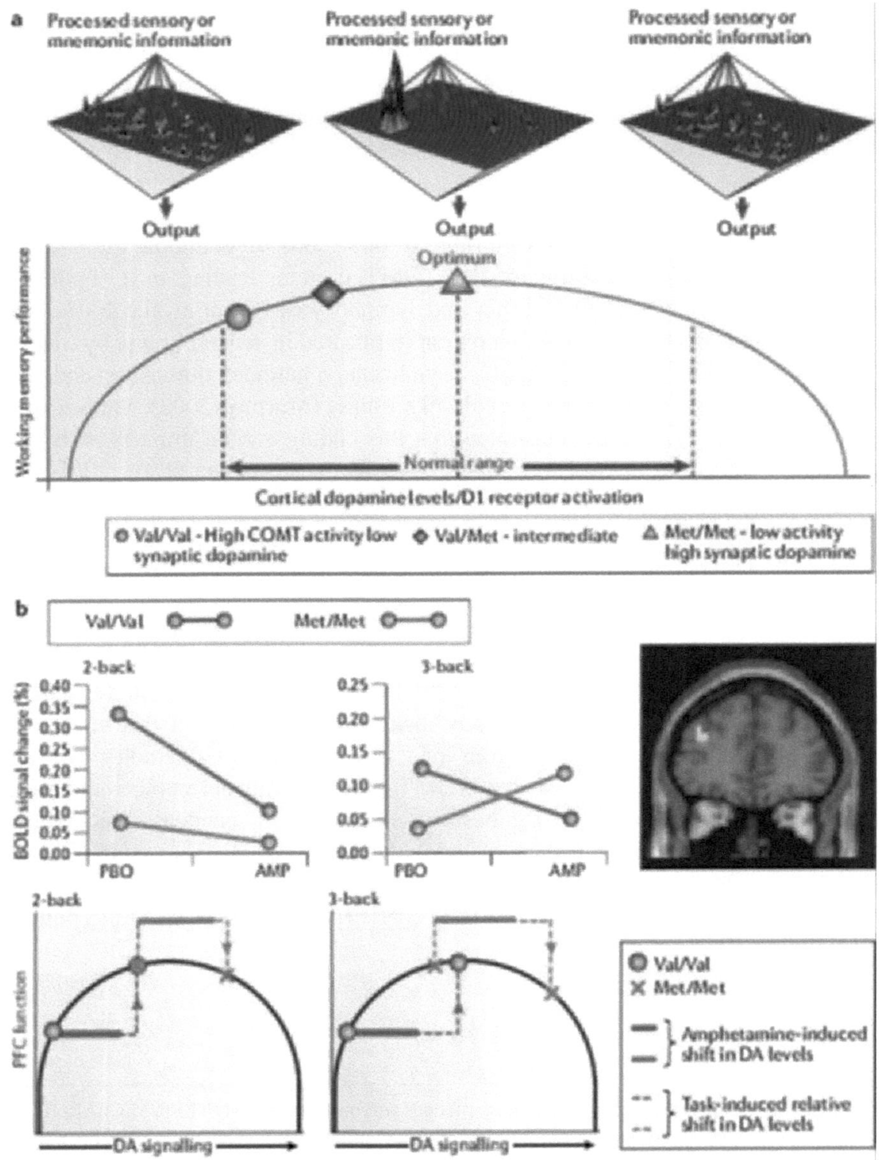

Fig. 11.3 Effects of COMT Val(108/158)Met on prefrontal cortex activity linked to extracellular dopamineConvergent evidence has established an inverted u-shaped relationship between working-memory-related activation of dorsolateral prefrontal cortex (PFC) neurons and dopaminergic tone (especially D1 receptor stimulation) [Williams and Goldman-Rakic, 1995], which has been shown to be essential for optimizing signal-to-noise ratio, or tuning of the neuronal response. This figure demonstrates the use of genetic variation in catechol-O-methyltransferase (COMT) to mechanistically characterize neural circuits relevant for schizophrenia that are dependent on this relationship.

variant is probably mediated by a reduced signal-to-noise ratio in the PFC, an idea supported by the finding that working-memory-related and working-memory-unrelated activity in the PFC are inversely coupled to midbrain dopamine synthesis and directionally dependent on COMT genotype (Meyer-Lindenberg et al., 2005).

Other imaging genetic approaches have focused on the neural systems level characterization of schizophrenia risk genes involved in the modulation of glutamatergic neurotransmission and glutamate-dopamine interactions. Glutamate is the most important excitatory neurotransmitter in the cortex, and the excitability of glutamate neurons is regulated in part by dopamine (Simons and Spiers, 2003). GRM3, which encodes a metabotropic glutamate receptor responsible for modulating synaptic glutamate, is a candidate gene linked to schizophrenia (Egan et al., 2004). A single nucleotide polymorphism (SNP4 A allele) has been associated with poorer performance on several cognitive tests, lower prefrontal NAA levels, and disturbed DLPFC-hippocampal activation during episodic memory performance and encoding (de Quervain and Papassotiropoulos, 2006; Egan et al., 2001, 2004 2004). This implicates a specific molecular pathway by which genetic variation in GRM3 alters glutamatergic neurotransmission, prefrontal and hippocampal physiology and cognition, thereby increasing the risk for schizophrenia. Another important schizophrenia risk gene is Disrupted-in-schizophrenia 1 (DISC1), a candidate gene predominantly expressed within the hippocampus (Callicott et al., 2005). It has been shown that a common nonconservative single nucleotide polymorphism (Ser704Cys; rs821616) in this gene predicts reduced hippocampal gray matter volume and abnormal activation of the hippocampus during working memory performance in healthy subjects, thereby mirroring findings in patients with schizophrenia (Meyer-Lindenberg et al., 2005) and their siblings (Callicott et al., 2003). Hippocampal volume, together with DLPFC gray matter volume and memory performance, was also found to be affected by a DISC1 haplotype overtransmitted to patients with schizophrenia in a twin cohort (Cannon et al., 2005).

Recent work (Meyer-Lindenberg et al., 2007) of our laboratory has focused on DARPP32, which is expressed in regions receiving dopaminergic innervation, especially the neostriatum (caudate and putamen) (Svenningsson et al., 2004). This protein,

Fig. 11.3 (continued) (**a**) An inverted u-shaped curve [Goldman-Rakic et al., 2000] links extracellular dopamine to prefrontal signal-to-noise ratio (*top*) and working memory performance (*bottom*): homozygotes for the Val-encoding allele are positioned at the left (suboptimally low dopamine due to high COMT activity) and Met homozygotes are near the optimum (higher dopamine due to reduced COMT activity). Heterozygotes are intermediate. (**b**) Increasing synaptic dopamine (DA) by the administration of amphetamine (AMP) dissociates the functional states of Val and Met homozygotes [Mattay et al., 2003]. At medium working memory load level (2-back task, where subjects need to remember a stimulus presented 2 trials back), Val homozygotes, on the left of the optimum on the inverted u curve, profit from increased dopamine, whereas Met-homozygotes, near the optimum, show little change (*left*). At high load level (3-back task), dopamine increase by drug intervention now pushes Met homozygotes into the suboptimally high range of dopamine stimulation, leading to reduced prefrontal efficiency (right; localization of activity in the prefrontal cortex is shown on the far right). PBO, placebo (reprinted with permission from [Meyer-Lindenberg and Weinberger, 2006]) (*See Color Plates*)

a phosphatase encoded by PPP1R1B, acts as a central molecular switch in dopaminoceptive neurons for the integration of dopamine and glutamate signals. DARPP32 is a key node in a final common pathway of psychotomimetics in both the frontal cortex and striatum, making it an attractive candidate gene for schizophrenia. Imaging genetics showed a pronounced and convergent effect on the structure and activation of the striatum, as well as on prefrontal–striatal interactions, indicating that PPP1R1B might contribute to risk for schizophrenia by causing disturbed gating (Swerdlow et al., 2001) subsequent to impaired fronto-striatal function.

11.5.3 Methodological Challenges

Compared to other neuroimaging disciplines, the imaging genetics approach bears inherent methodological difficulties that introduce new challenges to the neuroimaging field. Due to the small effect size of individual psychiatric risk genes, and the functional significance of complex gene-by-gene interactions, large-scale imaging datasets are necessary to identify valid gene effects on the neural system level. This standard that has not been fulfilled by most of the studies reporting significant effects to date which questions the robustness of published results. At the same time, a bias against the publication of negative results is very likely. Further research, ideally with contrasting genes of no known or likely function with known functional variants, is necessary to achieve a principled assessment of expected false positive rates that can guide recommendations for statistical inference in this field.

On the experimental side, the success of the intermediate phenotype approach in functional imaging research is crucially dependent on the suitability of the employed paradigms. Efforts to establish genetic effects on the biological level across participants will only be successful if a given experiment satisfies both, the induction of reliable and robust brain responses and the mapping of neural systems that are actually related to the disease under study. Imaging data analysis in the context of genetically complex disorders is challenged by the requirement to develop statistical models that account for multiple interacting genetic variants in a gene, between separate genes, and gene-by-environment interactions. Recent work from our laboratory shows that statistical methods adapted from haplotype regression are feasible and powerful tools to map the nonlinear interaction of multiple functional variants in the COMT gene on working memory-related prefrontal functions (Meyer-Lindenberg et al., 2006).

In the face of whole-genome scans of hundreds of thousands of genetic variants, the question of which polymorphisms should be selected for the characterization of genetic neural mechanisms introduces a new kind of multiple comparison problem (induced by the number of studied genetic variants) that must be addressed in future research. The situation is aggravated by the fact that many variants that are statistically associated with psychiatric disease are intronic and of no known functional consequence. In the absence of reliable information on the heritability and reliability of most imaging phenotypes currently in use, a statistically significant result in neuroimaging by itself is not sufficient to establish the functionality of a given polymorphism. The tight integration of imaging genetics in translational approaches characterizing intermediate phenotypes will be crucial in future psychiatric research, i.e., by demonstrating compatible genetic effects on mRNA expression, protein levels, or cellular physiology.

11.6 Imaging of Antipsychotic Drug Effects

Modern disease management in schizophrenia can be dated back to the early 1950s where the discovery of the antipsychotic properties of chlorpromazine set the ball rolling to the greatest psychiatric revolutions of the twentieth century (Lopez-Munoz et al., 2005). Chlorpromazine, and later other neuroleptic substances like haloperidol or fluphenazine, have successfully replaced ineffective and crude interventions like insulin shock therapy and lobotomy, allowing for a huge wave of patient discharges from psychiatric hospitals (Granger and Albu, 2005). These "first generation" antipsychotics proved to be efficient in relieving positive symptoms, although they also bear the potential for unwanted and bothersome treatment outcomes like extrapyramidal side effects (EPS) or tardive dyskinesia. In the last fifteen years, the therapeutic options for schizophrenia have substantially been expanded through the development of novel antipsychotic substances with more beneficial effects (e.g., clozapine, amisulpride, olanzapine, aripiprazole). These so-called "atypical" antipsychotics are a heterogenous group of drugs united by the absence of EPS, and a superior impact on persistent negative symptoms and cognitive deficits (Meltzer et al., 1994, Meltzer and McGurk, 1999) (see (Miyamoto et al., 2005) for a comprehensive review of the psychopharmacology of first- and second generation antipsychotics).

At the turn of the last century, noninvasive neuroimaging techniques evolved into the major tools for the examination of treatment-related alterations of brain function, structure, and metabolism (Arango et al., 2003; Bertolino et al., 2001; Braus et al., 2001; Gur et al., 1998; Heitmiller et al., 2004; Lang et al., 2004; Scheepers et al., 2001). fMRI studies in neuroleptic-naïve and drug-free patient cohorts have repeatedly reported favorable effects of atypical antipsychotics on disturbed functional imaging patterns in schizophrenia (Lund et al., 2002; Ramsey et al., 2002; Snitz et al., 2005). In the motor domain, Bertolino and coworkers (2004) reported a normalization of sensorimotor hypoactivations under olanzapine (Bertolino et al., 2004). With the same agent, Stephan et al. (2001) evidenced a restoration of functional connectivity patterns in the cortico-thalamic-cerebellar-cortical motor circuitry (Stephan et al., 2001). The "restoring" functional effects of atypical antipsychotics were not restricted to the motor domain, but also observable during verbal fluency (Jones et al., 2004), passive auditory stimulation (Jones et al., 2004), conflict detection (Snitz et al., 2005), and working memory performance (Meisenzahl et al., 2006). Between-class differences of antipsychotic agents have mainly been examined with cross-sectional designs. In a recent fMRI study on sensorimotor gating, Kumari (2007) (Kumari et al., 2007) examined prepulse inhibition (PPI) of the eye-blink startle response in healthy controls and schizophrenia patients on stable doses of typical versus atypical antipsychotics. The study confirms older findings in the field (Braus et al., 1999; Honey et al., 1999) by suggesting that atypical antipsychotics, but not first generation neuroleptics, ameliorate PPI deficits in schizophrenia and restore associated dysfunctional brain patterns. In summary, although atypical and typical agents have not been systematically compared and the group of atypical antipsychotics is likely heterogeneous in neurochemistry, clinical efficacy and systems-level effects, current neuroimaging evidence supports the notion of a superior effect of second-generation antipsychotics over classical neuroleptics on some brain functional measures in schizophrenia.

On the structural level, several proton magnetic resonance spectroscopic imaging (1H-MRSI) studies focusing on N-acetylaspartate (NAA) concentrations in the ACG and PFC support the notion that atypical antipsychotics selectively improve neuronal integrity in schizophrenia (Bertolino et al., 2001; Braus et al., 2001; Ende et al., 2000; Szulc et al., 2005). Lieberman et al. (2005) addressed the issue of differential treatment effects of conventional versus atypical drugs on brain volume in a thorough multicenter study in 161 patients with first-episode psychosis (Lieberman et al., 2005). Compared to the baseline scan, the patients under haloperidol (HLP) treatment showed a significant reduction in absolute gray matter volume over time, while the brain volume of the olanzapine and the healthy control groups remained constant. Although the precise neurobiological basis of the differential treatment effect is still unknown and a matter under investigation, this finding underscores the favorable pharmacological profile of atypical antipsychotics in schizophrenia. Possible explanations include a "neuroprotective" effect of atypical drugs on the progressive brain morphometric changes associated with the disorder (possibly mediated by N-methyl-D-aspartate receptor antagonism (Duncan et al., 2000), increased expression of trophic factors (Bai et al., 2003; Fumagalli et al., 2003), or stimulation of neurogenesis (Halim et al., 2004; Wakade et al., 2002)). Alternately, a toxic effect of HLP might be effective on the cellular level that accelerates the disease-inherent pathomorphological changes (possibly mediated by the induction of oxidative stress and excitatory neurotoxicity (Goff et al., 1995; Post et al., 1998; Wright et al., 1998)). Future clinical studies are necessary to confirm the potency of different antipsychotic treatment options to alter the biological course of the disorder.

Only a few neuroimaging studies thus far have examined the psychopharmacological effects of antipsychotic drugs in healthy volunteers. In a double-blind cross-over design, Abler (2007) examined the influence of a single dose of olanzapine (5 mg) on reward-related brain activations with a delayed incentive paradigm. Compared to the placebo condition, the authors observed an attenuated BOLD-response in the ventral striatum, anterior cingulate, and inferior frontal cortex, as well as a diminished acceleration of reward-related reaction times (Abler et al., 2007). Own prior work on haloperidol challenge effects in the visual (Brassen et al., 2003) and motor system (Tost et al., 2006) confirm a primarily dampening effect of antipsychotic drugs on task-related functional and behavioral measures. On the systems level, we prospectively examined the influence of HLP on cortico-striatal-thalamo-cortical (CSTC) motor loop functioning with a sequential finger opposition task. Contrary to the predictions of the nigrostriatal CSTC network model (Alexander et al., 1986, 1990) a significant drug-induced BOLD-signal decrease in the striatum and a lateralized activation loss of ipsilateral higher order motor cortices and contralateral cerebellum was observed. Neuropsychological implications of Seamans & Durstewitz's (2001) two-state model of prefrontal functioning (Durstewitz et al., 2000a,b; Seamans et al., 2001) were evaluated with a test battery of potentially D_1- (e.g., working memory) or D_2-sensitive cognitive measures (e.g., motor speed, attentional set shifting). The evidenced selectivity of HLP for the D_2-sensitive tasks supports the hypothesis that mesocortical D_1 and D_2 receptors exert a differential effect on PFC functions. Prospective evaluations confirmed the normalization tendency of the observed functional and behavioral effects with declining drug levels. These results emphasize the necessity of controlling for acute medication effects in psychiatric populations and confirm the potency of noninvasive neuroimaging methods for probing the predictive accuracy of neurophysiological models under drug challenge conditions (Tost et al., 2006).

11.7 Summary and Perspectives

Schizophrenia is a highly heritable and devastating mental disorder with a complex genetic architecture. Current evidence suggests that the path to psychopathology is laid by the interaction of multiple risk alleles, each accounting for only a small increment in risk for the development of the disorder. Neurodevelopmental disease models propose that the combination of genetic and environmental factors leads to a perturbation of regular temporolimbic-prefrontal network formation that manifest clinically after adolescence. On the neural systems level, downstream effects include disturbances in dopaminergic neurotransmission and impairments of PFC efficacy. The availability of modern neuroimaging techniques has recently revolutionized our understanding of the mediating neural mechanisms. Numerous cognitive, functional, morphological and metabolic anomalies in the brain have been described in the last 15 years, suggesting a heterogeneous disease entity rather than a circumscribed pathology. Core pathophysiological findings in patients include working memory deficits, associated activation anomalies and functional dyscoupling of the hippocampus and PFC, as well as gray matter volume reductions of the PFC and the medial temporal lobe. Similar findings have been reported for healthy subjects genetically at risk for schizophrenia, suggesting that these anomalies reflect valid intermediate phenotypes of the underlying disease vulnerability.

According to the Surgeon General's Report on Mental Health (available at http://www.surgeongeneral.gov/), four out of ten severe causes of disabilities in developed nations are related to mental disorders – schizophrenia being one of them. Mental illness not only imposes a tremendous emotional burden on the affected individuals and their families but also presents immense treatment expenses for society, costing an estimated $70 billion per year in the USA alone. This pressing issue makes it imperative that all available resources are concentrated towards the improvement of current prevention, diagnosis and treatment options. Future basic research in neuroscience will face a growing challenge to bridge the gap between theory and treatment through applying focus to questions that will likely to lead to future applications.

Within a translational framework, imaging genetics proved to be a promising tool for the delineation and biological validation of core pathophysiological processes in schizophrenia. The characterization of neural mechanisms associated with the genetic risk for the disorder is one component on the path to identifying the converging molecular pathways and their neuronal and systems-level targets, that ultimately lead to the manifestation of psychopathological phenomena – and their therapy. Establishment of viable biomarkers and the construction of valid experimental models in animals based on this pathophysiological understanding are critical steps in this endeavor. Future research in this field is not only expected to change our view of the taxonomy and pathophysiology of psychiatric disease, but also to point the way to new treatment targets and more principled clinical management.

Acknowledgments The authors thank Shabnam Hakimi for helpful comments on the manuscript.

References

Abler B, Erk S, Walter H (2007) Human reward system activation is modulated by a single dose of olanzapine in healthy subjects in an event-related, double-blind, placebo-controlled fMRI study. Psychopharmacology (Berl) 191(3): 823–833.

Akil M, Kolachana BS, Rothmond DA, Hyde TM, Weinberger DR, Kleinman JE (2003) Catechol-O-methyltransferase genotype and dopamine regulation in the human brain. J Neurosci 23(6): 2008–2013.

Alexander GE, DeLong MR, Strick PL (1986) Parallel organization of functionally segregated circuits linking basal ganglia and cortex. Annu Rev Neurosci 9: 357–381.

Alexander GE, Crutcher MD, DeLong MR (1990) Basal ganglia-thalamocortical circuits: parallel substrates for motor, oculomotor, "prefrontal" and "limbic" functions. Prog Brain Res 85: 119–146.

Almasy L and Blangero J (2001) Endophenotypes as quantitative risk factors for psychiatric disease: rationale and study design. Am J Med Genet 105(1): 42–44.

Andreasen NC, O'Leary DS, Flaum M, Nopoulos P, Watkins GL, Boles Ponto LL, Hichwa RD (1997) Hypofrontality in schizophrenia: distributed dysfunctional circuits in neuroleptic-naive patients. Lancet 349(9067): 1730–1734.

Arango C, Breier A, McMahon R, Carpenter WT, Buchanan RW (2003) The relationship of clozapine and haloperidol treatment response to prefrontal, hippocampal and caudate brain volumes. Am J Psychiatr 160(8): 1421–1427.

Arnold SE, Ruscheinsky DD, Han LY (1997) Further evidence of abnormal cytoarchitecture of the entorhinal cortex in schizophrenia using spatial point pattern analyses. Biol Psychiatr 42(8): 639–647.

Artiges E, Martelli C, Naccache L, Bartres-Faz D, Leprovost JB, Viard A, Paillere-Martinot ML, Dehaene S, Martinot JL (2006) Paracingulate sulcus morphology and fMRI activation detection in schizophrenia patients. Schizophr Res 82(2–3): 143–151.

Bai O, Chlan-Fourney J, Bowen R, Keegan D, Li XM (2003) Expression of brain-derived neurotrophic factor mRNA in rat hippocampus after treatment with antipsychotic drugs. J Neurosci Res 71(1): 127–131.

Barch DM, Carter CS, Braver TS, Sabb FW, MacDonald A, 3rd, Noll DC, Cohen JD (2001) Selective deficits in prefrontal cortex function in medication-naive patients with schizophrenia. Arch Gen Psychiatr 58(3): 280–288.

Baumeister AA and Francis JL (2002) Historical development of the dopamine hypothesis of schizophrenia. J Hist Neurosci 11(3): 265–277.

Benes FM (1993) The relationship between structural brain imaging and histopathologic findings in schizophrenia research. Harv Rev Psychiatr 1(2): 100–109.

Benes FM (1999) Evidence for altered trisynaptic circuitry in schizophrenic hippocampus. Biol Psychiatr 46(5): 589–599.

Benes FM, McSparren J, Bird ED, SanGiovanni JP, Vincent SL (1991a) Deficits in small interneurons in prefrontal and cingulate cortices of schizophrenic and schizoaffective patients. Arch Gen Psychiatr 48(11): 996–1001.

Benes FM, Sorensen I, Bird ED (1991b) Reduced neuronal size in posterior hippocampus of schizophrenic patients. Schizophr Bull 17(4): 597–608.

Bertolino A, Callicott JH, Mattay VS, Weidenhammer KM, Rakow R, Egan MF, Weinberger DR (2001) The effect of treatment with antipsychotic drugs on brain N-acetylaspartate measures in patients with schizophrenia. Biol Psychiatr 49(1): 39–46.

Bertolino A, Blasi G, Caforio G, Latorre V, De Candia M, Rubino V, Callicott JH, Mattay VS, Bellomo A, Scarabino T, Weinberger DR, Nardini M (2004) Functional lateralization of the sensorimotor cortex in patients with schizophrenia: effects of treatment with olanzapine. Biol Psychiatr 56(3): 190–197.

Bertram I, Bernstein HG, Lendeckel U, Bukowska A, Dobrowolny H, Keilhoff G, Kanakis D, Mawrin C, Bielau H, Falkai P, Bogerts B (2007) Immunohistochemical evidence for impaired neuregulin-1 signaling in the prefrontal cortex in schizophrenia and in unipolar depression. Ann NY Acad Sci 1096: 147–156.

Bleuler E, Dementia praecox or the group of schizophrenias (1911). 1950, International Universities Press: New York.

Braff DL and Saccuzzo DP (1981) Information processing dysfunction in paranoid schizophrenia: a two- factor deficit. Am J Psychiatr 138(8): 1051–1056.

Braff DL and Saccuzzo DP (1985) The time course of information-processing deficits in schizophrenia. Am J Psychiatr 142(2): 170–174.

Brandon NJ, Handford EJ, Schurov I, Rain JC, Pelling M, Duran-Jimeniz B, Camargo LM, Oliver KR, Beher D, Shearman MS, Whiting PJ (2004) Disrupted in Schizophrenia 1 and Nudel form a neurodevelopmentally regulated protein complex: implications for schizophrenia and other major neurological disorders. Mol Cell Neurosci 25(1): 42–55.

Brassen S, Tost H, Hoehn F, Weber-Fahr W, Klein S, Braus DF (2003) Haloperidol challenge in healthy male humans: a functional magnetic resonance imaging study. Neurosci Lett 340(3): 193–196.

Braus DF, Ende G, Weber-Fahr W, Sartorius A, Krier A, Hubrich-Ungureanu P, Ruf M, Stuck S, Henn FA (1999) Antipsychotic drug effects on motor activation measured by functional magnetic resonance imaging in schizophrenic patients. Schizophr Res 39(1): 19–29.

Braus DF, Ende G, Weber-Fahr W, Demirakca T, Henn FA (2001) Favorable effect on neuronal viability in the anterior cingulate gyrus due to long-term treatment with atypical antipsychotics: an MRSI study. Pharmacopsychiatry 34(6): 251–253.

Braus DF, Weber-Fahr W, Tost H, Ruf M, Henn FA (2002) Sensory information processing in neuroleptic-naive first-episode schizophrenic patients: a functional magnetic resonance imaging study. Arch Gen Psychiatr 59(8): 696–701.

Buckley PF, Friedman L, Wu D, Lai S, Meltzer HY, Haacke EM, Miller D, Lewin JS (1997) Functional magnetic resonance imaging in schizophrenia: initial methodology and evaluation of the motor cortex. Psychiatr Res Neuroimaging 74: 13–23.

Butler PD and Javitt DC (2005) Early-stage visual processing deficits in schizophrenia. Curr Opin Psychiatr 18(2): 151–157.

Butler PD, Hoptman MJ, Nierenberg J, Foxe JJ, Javitt DC, Lim KO (2006) Visual white matter integrity in schizophrenia. Am J Psychiatr 163(11): 2011–2013.

Cadenhead KS, Serper Y, Braff DL (1998) Transient versus sustained visual channels in the visual backward masking deficits of schizophrenia patients. Biol Psychiatr 43(2): 132–138.

Cahn W, Hulshoff Pol HE, Lems EB, van Haren NE, Schnack HG, van der Linden JA, Schothorst PF, van Engeland H, Kahn RS (2002) Brain volume changes in first-episode schizophrenia: a 1-year follow-up study. Arch Gen Psychiatr 59(11): 1002–1010.

Callicott JH, Mattay VS, Bertolino A, Finn K, Coppola R, Frank JA, Goldberg TE, Weinberger DR (1999) Physiological characteristics of capacity constraints in working memory as revealed by functional MRI. Cereb Cortex 9(1): 20–26.

Callicott JH, Bertolino A, Mattay VS, Langheim FJ, Duyn J, Coppola R, Goldberg TE, Weinberger DR (2000) Physiological dysfunction of the dorsolateral prefrontal cortex in schizophrenia revisited. Cereb Cortex 10(11): 1078–1092.

Callicott JH, Egan MF, Mattay VS, Bertolino A, Bone AD, Verchinksi B, Weinberger DR (2003) Abnormal fMRI response of the dorsolateral prefrontal cortex in cognitively intact siblings of patients with schizophrenia. Am J Psychiatr 160(4): 709–719.

Callicott JH, Mattay VS, Verchinski BA, Marenco S, Egan MF, Weinberger DR (2003) Complexity of prefrontal cortical dysfunction in schizophrenia: more than up or down. Am J Psychiatr 160(12): 2209–2215.

Callicott JH, Straub RE, Pezawas L, Egan MF, Mattay VS, Hariri AR, Verchinski BA, Meyer-Lindenberg A, Balkissoon R, Kolachana B, Goldberg TE, Weinberger DR (2005) Variation in DISC1 affects hippocampal structure and function and increases risk for schizophrenia. Proc Natl Acad Sci USA 102(24): 8627–8632.

Cannon TD, Hennah W, van Erp TG, Thompson PM, Lonnqvist J, Huttunen M, Gasperoni T, Tuulio-Henriksson A, Pirkola T, Toga AW, Kaprio J, Mazziotta J, Peltonen L (2005) Association of DISC1/TRAX haplotypes with schizophrenia, reduced prefrontal gray matter, and impaired short- and long-term memory. Arch Gen Psychiatr 62(11): 1205–1213.

Carter CS, Mintun M, Nichols T, Cohen JD (1997) Anterior cingulate gyrus dysfunction and selective attention deficits in schizophrenia: [15O]H2O PET study during single-trial Stroop task performance. Am J Psychiatr 154(12): 1670–1675.

Chen Y, Levy DL, Nakayama K, Matthysse S, Palafox G, Holzman PS (1999a) Dependence of impaired eye tracking on deficient velocity discrimination in schizophrenia. Arch Gen Psychiatr 56(2): 155–161.

Chen Y, Palafox GP, Nakayama K, Levy DL, Matthysse S, Holzman PS (1999b) Motion perception in schizophrenia. Arch Gen Psychiatr 56(2): 149–154.

Chen J, Lipska BK, Halim N, Ma QD, Matsumoto M, Melhem S, Kolachana BS, Hyde TM, Herman MM, Apud J, Egan MF, Kleinman JE, Weinberger DR (2004) Functional analysis of genetic variation in catechol-O-methyltransferase (COMT): effects on mRNA, protein, and enzyme activity in postmortem human brain. Am J Hum Genet 75(5): 807–821.

Chen Y, Bidwell LC, Norton D (2006) Trait vs. state markers for schizophrenia: identification and characterization through visual processes. Curr Psychiatr Rev 2(4): 431–438.

Chua SE, Cheung C, Cheung V, Tsang JT, Chen EY, Wong JC, Cheung JP, Yip L, Tai KS, Suckling J, McAlonan GM (2007) Cerebral grey, white matter and csf in never-medicated, first-episode schizophrenia. Schizophr Res 89(1–3): 12–21.

Cohen MX, Young J, Baek JM, Kessler C, Ranganath C (2005) Individual differences in extraversion and dopamine genetics predict neural reward responses. Brain Res Cogn Brain Res 25(3): 851–861.

Crespo-Facorro B, Roiz-Santianez R, Pelayo-Teran JM, Gonzalez-Blanch C, Perez-Iglesias R, Gutierrez A, de Lucas EM, Tordesillas D, Vazquez-Barquero JL (2007) Caudate nucleus volume and its clinical and cognitive correlations in first episode schizophrenia. Schizophr Res 91(1–3): 87–96.

David A and Cutting J, eds. The neuropsychological origin of auditory hallucinations. Neuropsychology of schizophrenia. (1994), Lawrence Erlbaum Associates: Hove, Sussex. 269–313.

David AS, Woodruff PW, Howard R, Mellers JD, Brammer M, Bullmore E, Wright I, Andrew C, Williams SC (1996) Auditory hallucinations inhibit exogenous activation of auditory association cortex. Neuroreport 7(4): 932–936.

Davis KL, Buchsbaum MS, Shihabuddin L, Spiegel-Cohen J, Metzger M, Frecska E, Keefe RS, Powchik P (1998) Ventricular enlargement in poor-outcome schizophrenia. Biol Psychiatr 43(11): 783–793.
de Quervain DJ and Papassotiropoulos A (2006) Identification of a genetic cluster influencing memory performance and hippocampal activity in humans. Proc Natl Acad Sci USA 103(11): 4270–4274.
DeLisi LE, Sakuma M, Tew W, Kushner M, Hoff AL, Grimson R (1997) Schizophrenia as a chronic active brain process: a study of progressive brain structural change subsequent to the onset of schizophrenia. Psychiatr Res 74(3): 129–140.
Dierks T, Linden DE, Jandl M, Formisano E, Goebel R, Lanfermann H, Singer W (1999) Activation of Heschl's gyrus during auditory hallucinations. Neuron 22(3): 615–621.
Dorph-Petersen KA, Pierri JN, Wu Q, Sampson AR, Lewis DA (2007) Primary visual cortex volume and total neuron number are reduced in schizophrenia. J Comp Neurol 501(2): 290–301.
Duncan GE, Miyamoto S, Leipzig JN, Lieberman JA (2000) Comparison of the effects of clozapine, risperidone, and olanzapine on ketamine-induced alterations in regional brain metabolism. J Pharmacol Exp Ther 293(1): 8–14.
Durstewitz D, Seamans JK, Sejnowski TJ (2000a) Dopamine-mediated stabilization of delay-period activity in a network model of prefrontal cortex. J Neurophysiol 83(3): 1733–1750.
Durstewitz D, Seamans JK, Sejnowski TJ (2000b) Neurocomputational models of working memory. Nat Neurosci 3(Suppl): 1184–1191.
Egan MF, Goldberg TE, Kolachana BS, Callicott JH, Mazzanti CM, Straub RE, Goldman D, Weinberger DR (2001) Effect of COMT Val108/158 Met genotype on frontal lobe function and risk for schizophrenia. Proc Natl Acad Sci USA 98(12): 6917–6922.
Egan MF, Straub RE, Goldberg TE, Yakub I, Callicott JH, Hariri AR, Mattay VS, Bertolino A, Hyde TM, Shannon-Weickert C, Akil M, Crook J, Vakkalanka RK, Balkissoon R, Gibbs RA, Kleinman JE, Weinberger DR (2004) Variation in GRM3 affects cognition, prefrontal glutamate, and risk for schizophrenia. Proc Natl Acad Sci USA 101(34): 12604–12609.
Ekelund J, Lichtermann D, Hovatta I, Ellonen P, Suvisaari J, Terwilliger JD, Juvonen H, Varilo T, Arajarvi R, Kokko-Sahin ML, Lonnqvist J, Peltonen L (2000) Genome-wide scan for schizophrenia in the Finnish population: evidence for a locus on chromosome 7q22. Hum Mol Genet 9(7): 1049–1057.
Ende G, Braus DF, Walter S, Weber-Fahr W, Soher B, Maudsley AA, Henn FA (2000) Effects of age, medication, and illness duration on the N-acetyl aspartate signal of the anterior cingulate region in schizophrenia. Schizophr Res 41(3): 389–395.
Exner C, Weniger G, Schmidt-Samoa C, Irle E (2006) Reduced size of the pre-supplementary motor cortex and impaired motor sequence learning in first-episode schizophrenia. Schizophr Res 84(2–3): 386–396.
Eyler LT, Olsen RK, Jeste DV, Brown GG (2004) Abnormal brain response of chronic schizophrenia patients despite normal performance during a visual vigilance task. Psychiatr Res 130(3): 245–257.
Ffytche DH, Howard RJ, Brammer MJ, David A, Woodruff P, Williams S (1998) The anatomy of conscious vision: an fMRI study of visual hallucinations. Nat Neurosci 1(8): 738–742.
Freedman R, Coon H, Myles-Worsley M, Orr-Urtreger A, Olincy A, Davis A, Polymeropoulos M, Holik J, Hopkins J, Hoff M, Rosenthal J, Waldo MC, Reimherr F, Wender P, Yaw J, Young DA, Breese CR, Adams C, Patterson D, Adler LE, Kruglyak L, Leonard S, Byerley W (1997) Linkage of a neurophysiological deficit in schizophrenia to a chromosome 15 locus. Proc Natl Acad Sci USA 94(2): 587–592.
Fumagalli F, Molteni R, Roceri M, Bedogni F, Santero R, Fossati C, Gennarelli M, Racagni G, Riva MA (2003) Effect of antipsychotic drugs on brain-derived neurotrophic factor expression under reduced N-methyl-D-aspartate receptor activity. J Neurosci Res 72(5): 622–628.
Fuster JM (1990) Prefrontal cortex and the bridging of temporal gaps in the perception-action cycle. Ann NY Acad Sci 608: 318–329; discussion 330–316.

Garey LJ, Ong WY, Patel TS, Kanani M, Davis A, Mortimer AM, Barnes TR, Hirsch SR (1998) Reduced dendritic spine density on cerebral cortical pyramidal neurons in schizophrenia. J Neurol Neurosurg Psychiatr 65(4): 446–453.
Gaser C, Nenadic I, Volz HP, Buchel C, Sauer H (2004) Neuroanatomy of 'hearing voices': a frontotemporal brain structural abnormality associated with auditory hallucinations in schizophrenia. Cereb Cortex 14(1): 91–96.
Gasperoni TL, Ekelund J, Huttunen M, Palmer CG, Tuulio-Henriksson A, Lonnqvist J, Kaprio J, Peltonen L, Cannon TD (2003) Genetic linkage and association between chromosome 1q and working memory function in schizophrenia. Am J Med Genet B Neuropsychiatr Genet 116(1): 8–16.
Gerfen CR (2000) Molecular effects of dopamine on striatal-projection pathways. Trends Neurosci 23(Suppl 10): S64–S70.
Glantz LA and Lewis DA (2000) Decreased dendritic spine density on prefrontal cortical pyramidal neurons in schizophrenia. Arch Gen Psychiatr 57(1): 65–73.
Glenthoj A, Glenthoj BY, Mackeprang T, Pagsberg AK, Hemmingsen RP, Jernigan TL, Baare WF (2007) Basal ganglia volumes in drug-naive first-episode schizophrenia patients before and after short-term treatment with either a typical or an atypical antipsychotic drug. Psychiatr Res 154(3): 199–208.
Goff DC, Tsai G, Beal MF, Coyle JT (1995) Tardive dyskinesia and substrates of energy metabolism in CSF. Am J Psychiatr 152(12): 1730–1736.
Goghari VM, Lang DJ, Flynn SW, Mackay AL, Honer WG (2005) Smaller corpus callosum subregions containing motor fibers in schizophrenia. Schizophr Res 73(1): 59–68.
Goldberg TE, Ragland JD, Torrey EF, Gold JM, Bigelow LB, Weinberger DR (1990) Neuropsychological assessment of monozygotic twins discordant for schizophrenia. Arch Gen Psychiatry 47(11): 1066–1072.
Goldberg TE, Egan MF, Gscheidle T, Coppola R, Weickert T, Kolachana BS, Goldman D, Weinberger DR (2003) Executive subprocesses in working memory: relationship to catechol-O-methyltransferase Val158Met genotype and schizophrenia. Arch Gen Psychiatr 60(9): 889–896.
Goldman-Rakic PS (1995) Cellular basis of working memory. Neuron 14(3): 477–485.
Goldman-Rakic PS, Muly EC, 3rd, Williams GV (2000) D(1) receptors in prefrontal cells and circuits. Brain Res Brain Res Rev 31(2–3): 295–301.
Goldman AL, Pezawas L, Mattay VS, Fischl B, Verchinski BA, Zoltick B, Weinberger DR, Meyer-Lindenberg A (2008) Heritability of brain morphology related to schizophrenia: a large-scale automated MRI segmentation study. Biol Psychiatry 63(5): 475–483.
Gottesman II and Gould TD (2003) The endophenotype concept in psychiatry: etymology and strategic intentions. Am J Psychiatr 160(4): 636–645.
Gottesman II and Shields J (1967) A polygenic theory of schizophrenia. Proc Natl Acad Sci USA 58(1): 199–205.
Grace AA (1993) Cortical regulation of subcortical dopamine systems and its possible relevance to schizophrenia. J Neural Transm Gen Sect 91(2–3): 111–134.
Granger B and Albu S (2005) The haloperidol story. Ann Clin Psychiatr 17(3): 137–140.
Gur RE, Cowell P, Turetsky BI, Gallacher F, Cannon T, Bilker W, Gur RC (1998) A follow-up magnetic resonance imaging study of schizophrenia. Relationship of neuroanatomical changes to clinical and neurobehavioral measures. Arch Gen Psychiatr 55(2): 145–152.
Gur RE, Maany V, Mozley PD, Swanson C, Bilker W, Gur RC (1998) Subcortical MRI volumes in neuroleptic-naive and treated patients with schizophrenia. Am J Psychiatr 155(12): 1711–1717.
Halim ND, Weickert CS, McClintock BW, Weinberger DR, Lipska BK (2004) Effects of chronic haloperidol and clozapine treatment on neurogenesis in the adult rat hippocampus. Neuropsychopharmacology 29(6): 1063–1069.
Hariri AR, Mattay VS, Tessitore A, Kolachana B, Fera F, Goldman D, Egan MF, Weinberger DR (2002) Serotonin transporter genetic variation and the response of the human amygdala. Science 297(5580): 400–403.
Harrison PJ and Weinberger DR (2005) Schizophrenia genes, gene expression, and neuropathology: on the matter of their convergence. Mol Psychiatry 10(1): 40–68; image 45.

He L, Carothers A, Blackwood DH, Teague P, Maclean AW, Brown J, Wright AF, Muir WJ, Porteous DJ, St Clair DM (1996) Recombination patterns around the breakpoints of a balanced 1;11 autosomal translocation associated with major mental illness. Psychiatr Genet 6(4): 201–208.

Heckers S, Weiss AP, Deckersbach T, Goff DC, Morecraft RJ, Bush G (2004) Anterior cingulate cortex activation during cognitive interference in schizophrenia. Am J Psychiatr 161(4): 707–715.

Heinz A, Braus DF, Smolka MN, Wrase J, Puls I, Hermann D, Klein S, Grusser SM, Flor H, Schumann G, Mann K, Buchel C (2005) Amygdala-prefrontal coupling depends on a genetic variation of the serotonin transporter. Nat Neurosci 8(1): 20–21.

Heitmiller DR, Nopoulos PC, Andreasen NC (2004) Changes in caudate volume after exposure to atypical neuroleptics in patients with schizophrenia may be sex-dependent. Schizophr Res 66(2–3): 137–142.

Henn FA and Braus DF (1999) Structural neuroimaging in schizophrenia. An integrative view of neuromorphology. Eur Arch Psychiatry Clin Neurosci 249(Suppl 4): 48–56.

Honea R, Crow TJ, Passingham D, Mackay CE (2005) Regional deficits in brain volume in schizophrenia: a meta-analysis of voxel-based morphometry studies. Am J Psychiatr 162(12): 2233–2245.

Honey GD, Bullmore ET, Soni W, Varatheesan M, Williams SC, Sharma T (1999) Differences in frontal cortical activation by a working memory task after substitution of risperidone for typical antipsychotic drugs in patients with schizophrenia. Proc Natl Acad Sci USA 96(23): 13432–13437.

Honey GD, Pomarol-Clotet E, Corlett PR, Honey RA, McKenna PJ, Bullmore ET, Fletcher PC (2005) Functional dysconnectivity in schizophrenia associated with attentional modulation of motor function. Brain 128(Pt 11): 2597–2611.

Hoppner J, Kunesch E, Grossmann A, Tolzin CJ, Schulz M, Schlafke D, Ernst K (2001) Dysfunction of transcallosally mediated motor inhibition and callosal morphology in patients with schizophrenia. Acta Psychiatr Scand 104(3): 227–235.

Hubl D, Koenig T, Strik W, Federspiel A, Kreis R, Boesch C, Maier SE, Schroth G, Lovblad K, Dierks T (2004) Pathways that make voices: white matter changes in auditory hallucinations. Arch Gen Psychiatr 61(7): 658–668.

Hulshoff Pol HE, Schnack HG, Mandl RC, Brans RG, van Haren NE, Baare WF, van Oel CJ, Collins DL, Evans AC, Kahn RS (2006) Gray and white matter density changes in monozygotic and same-sex dizygotic twins discordant for schizophrenia using voxel-based morphometry. Neuroimage 31(2): 482–488.

Jakob H and Beckmann H (1986) Prenatal developmental disturbances in the limbic allocortex in schizophrenics. J Neural Transm 65(3–4): 303–326.

Jones HM, Brammer MJ, O'Toole M, Taylor T, Ohlsen RI, Brown RG, Purvis R, Williams S, Pilowsky LS (2004) Cortical effects of quetiapine in first-episode schizophrenia: a preliminary functional magnetic resonance imaging study. Biol Psychiatr 56(12): 938–942.

Joyce EM and Roiser JP (2007) Cognitive heterogeneity in schizophrenia. Curr Opin Psychiatr 20(3): 268–272.

Keri S, Antal A, Szekeres G, Benedek G, Janka Z (2000) Visual information processing in patients with schizophrenia: evidence for the impairment of central mechanisms. Neurosci Lett 293(1): 69–71.

Keshavan MS, Rosenberg D, Sweeney JA, Pettegrew JW (1998) Decreased caudate volume in neuroleptic-naive psychotic patients. Am J Psychiatr 155(6): 774–778.

Kraepelin E, Dementia praecox and paraphrenia. (1919), Edinburgh: Livingstone.

Kreczmanski P, Heinsen H, Mantua V, Woltersdorf F, Masson T, Ulfig N, Schmidt-Kastner R, Korr H, Steinbusch HW, Hof PR, Schmitz C (2007) Volume, neuron density and total neuron number in five subcortical regions in schizophrenia. Brain 130(Pt 3): 678–692.

Kubicki M, Westin CF, Nestor PG, Wible CG, Frumin M, Maier SE, Kikinis R, Jolesz FA, McCarley RW, Shenton ME (2003) Cingulate fasciculus integrity disruption in schizophrenia: a magnetic resonance diffusion tensor imaging study. Biol Psychiatr 54(11): 1171–1180.

Kubicki M, McCarley R, Westin CF, Park HJ, Maier S, Kikinis R, Jolesz FA, Shenton ME (2007) A review of diffusion tensor imaging studies in schizophrenia. J Psychiatr Res 41(1–2): 15–30.
Kumari V, Antonova E, Geyer MA, Ffytche D, Williams SC, Sharma T (2007) A fMRI investigation of startle gating deficits in schizophrenia patients treated with typical or atypical antipsychotics. Int J Neuropsychopharmacol 10(4): 463–477.
Kwon JS, McCarley RW, Hirayasu Y, Anderson JE, Fischer IA, Kikinis R, Jolesz FA, Shenton ME (1999) Left planum temporale volume reduction in schizophrenia. Arch Gen Psychiatr 56(2): 142–148.
Lang DJ, Kopala LC, Vandorpe RA, Rui Q, Smith GN, Goghari VM, Lapointe JS, Honer WG (2004) Reduced basal ganglia volumes after switching to olanzapine in chronically treated patients with schizophrenia. Am J Psychiatr 161(10): 1829–1836.
Lencer R, Nagel M, Sprenger A, Heide W, Binkofski F (2005) Reduced neuronal activity in the V5 complex underlies smooth-pursuit deficit in schizophrenia: evidence from an fMRI study. Neuroimage 24(4): 1256–1259.
Levin S (1984a) Frontal lobe dysfunctions in schizophrenia--I. Eye movement impairments. J Psychiatr Res 18(1): 27–55.
Levin S (1984b) Frontal lobe dysfunctions in schizophrenia--II. Impairments of psychological and brain functions. J Psychiatr Res 18(1): 57–72.
Lewis DA (2000) GABAergic local circuit neurons and prefrontal cortical dysfunction in schizophrenia. Brain Res Brain Res Rev 31(2–3): 270–276.
Lewis DA and Levitt P (2002) Schizophrenia as a disorder of neurodevelopment. Annu Rev Neurosci 25: 409–432.
Lewis DA, Melchitzky DS, Sesack SR, Whitehead RE, Auh S, Sampson A (2001) Dopamine transporter immunoreactivity in monkey cerebral cortex: regional, laminar, and ultrastructural localization. J Comp Neurol 432(1): 119–136.
Lieberman J, Chakos M, Wu H, Alvir J, Hoffman E, Robinson D, Bilder R (2001) Longitudinal study of brain morphology in first episode schizophrenia. Biol Psychiatr 49(6): 487–499.
Lieberman JA, Tollefson GD, Charles C, Zipursky R, Sharma T, Kahn RS, Keefe RS, Green AI, Gur RE, McEvoy J, Perkins D, Hamer RM, Gu H, Tohen M (2005) Antipsychotic drug effects on brain morphology in first-episode psychosis. Arch Gen Psychiatr 62(4): 361–370.
Longson D, Deakin JF, Benes FM (1996) Increased density of entorhinal glutamate-immunoreactive vertical fibers in schizophrenia. J Neural Transm 103(4): 503–507.
Lopez-Munoz F, Alamo C, Cuenca E, Shen WW, Clervoy P, Rubio G (2005) History of the discovery and clinical introduction of chlorpromazine. Ann Clin Psychiatr 17(3): 113–135.
Lund A, Kroken R, Thomsen T, Hugdahl K, Smievoll AI, Barndon R, Iversen J, Landro NI, Sundet K, Rund BR, Ersland L, Lundervold A, Asbjornsen A (2002) "Normalization" of brain activation in schizophrenia. An fMRI study. Schizophr Res 58(2–3): 333–335.
MacDonald AW, 3rd and Carter CS (2003) Event-related FMRI study of context processing in dorsolateral prefrontal cortex of patients with schizophrenia. J Abnorm Psychol 112(4): 689–697.
Manoach DS (2003) Prefrontal cortex dysfunction during working memory performance in schizophrenia: reconciling discrepant findings. Schizophr Res 60(2–3): 285–298.
Manoach DS, Press DZ, Thangaraj V, Searl MM, Goff DC, Halpern E, Saper CB, Warach S (1999) Schizophrenic subjects activate dorsolateral prefrontal cortex during a working memory task, as measured by fMRI. Biol Psychiatr 45(9): 1128–1137.
Manoach DS, Ketwaroo GA, Polli FE, Thakkar KN, Barton JJ, Goff DC, Fischl B, Vangel M, Tuch DS (2007) Reduced microstructural integrity of the white matter underlying anterior cingulate cortex is associated with increased saccadic latency in schizophrenia. Neuroimage 37(2): 599–610.
Mathalon DH, Sullivan EV, Lim KO, Pfefferbaum A (2001) Progressive brain volume changes and the clinical course of schizophrenia in men: a longitudinal magnetic resonance imaging study. Arch Gen Psychiatr 58(2): 148–157.
Matsui M, Yoneyama E, Sumiyoshi T, Noguchi K, Nohara S, Suzuki M, Kawasaki Y, Seto H, Kurachi M (2002) Lack of self-control as assessed by a personality inventory is related to reduced volume of supplementary motor area. Psychiatr Res 116(1–2): 53–61.

Mattay VS, Callicott JH, Bertolino A, Santha AK, Tallent KA, Goldberg TE, Frank JA, Weinberger DR (1997) Abnormal functional lateralization of the sensorimotor cortex in patients with schizophrenia. Neuroreport 8(13): 2977–2984.

Mattay VS, Goldberg TE, Fera F, Hariri AR, Tessitore A, Egan MF, Kolachana B, Callicott JH, Weinberger DR (2003) Catechol O-methyltransferase val158-met genotype and individual variation in the brain response to amphetamine. Proc Natl Acad Sci USA 100(10): 6186–6191.

McGowan S, Lawrence AD, Sales T, Quested D, Grasby P (2004) Presynaptic dopaminergic dysfunction in schizophrenia: a positron emission tomographic [18F]fluorodopa study. Arch Gen Psychiatr 61(2): 134–142.

McGuire PK, Shah GM, Murray RM (1993) Increased blood flow in Broca's area during auditory hallucinations in schizophrenia. Lancet 342(8873): 703–706.

Mechelli A, Allen P, Amaro E, Jr., Fu CH, Williams SC, Brammer MJ, Johns LC, McGuire PK (2007) Misattribution of speech and impaired connectivity in patients with auditory verbal hallucinations. Hum Brain Mapp 28(11): 1213–22.

Meisenzahl EM, Scheuerecker J, Zipse M, Ufer S, Wiesmann M, Frodl T, Koutsouleris N, Zetzsche T, Schmitt G, Riedel M, Spellmann I, Dehning S, Linn J, Bruckmann H, Moller HJ (2006) Effects of treatment with the atypical neuroleptic quetiapine on working memory function: a functional MRI follow-up investigation. Eur Arch Psychiatr Clin Neurosci 256(8): 522–531.

Meltzer HY and McGurk SR (1999) The effects of clozapine, risperidone, and olanzapine on cognitive function in schizophrenia. Schizophr Bull 25(2): 233–255.

Meltzer HY, Lee MA, Ranjan R (1994) Recent advances in the pharmacotherapy of schizophrenia. Acta Psychiatr Scand Suppl 384: 95–101.

Meyer-Lindenberg A and Weinberger DR (2006) Intermediate phenotypes and genetic mechanisms of psychiatric disorders. Nat Rev Neurosci 7(10): 818–827.

Meyer-Lindenberg A, Poline JB, Kohn PD, Holt JL, Egan MF, Weinberger DR, Berman KF (2001) Evidence for abnormal cortical functional connectivity during working memory in schizophrenia. Am J Psychiatr 158(11): 1809–1817.

Meyer-Lindenberg A, Miletich RS, Kohn PD, Esposito G, Carson RE, Quarantelli M, Weinberger DR, Berman KF (2002) Reduced prefrontal activity predicts exaggerated striatal dopaminergic function in schizophrenia. Nat Neurosci 5(3): 267–271.

Meyer-Lindenberg A, Hariri AR, Munoz KE, Mervis CB, Mattay VS, Morris CA, Berman KF (2005) Neural correlates of genetically abnormal social cognition in Williams syndrome. Nat Neurosci 8(8): 991–993.

Meyer-Lindenberg A, Kohn PD, Kolachana B, Kippenhan S, McInerney-Leo A, Nussbaum R, Weinberger DR, Berman KF (2005) Midbrain dopamine and prefrontal function in humans: interaction and modulation by COMT genotype. Nat Neurosci 8(5): 594–596.

Meyer-Lindenberg AS, Olsen RK, Kohn PD, Brown T, Egan MF, Weinberger DR, Berman KF (2005) Regionally specific disturbance of dorsolateral prefrontal-hippocampal functional connectivity in schizophrenia. Arch Gen Psychiatr 62(4): 379–386.

Meyer-Lindenberg A, Buckholtz JW, Kolachana B A RH, Pezawas L, Blasi G, Wabnitz A, Honea R, Verchinski B, Callicott JH, Egan M, Mattay V, Weinberger DR (2006) Neural mechanisms of genetic risk for impulsivity and violence in humans. Proc Natl Acad Sci USA 103(16): 6269–6274.

Meyer-Lindenberg A, Nichols T, Callicott JH, Ding J, Kolachana B, Buckholtz J, Mattay VS, Egan M, Weinberger DR (2006) Impact of complex genetic variation in COMT on human brain function. Mol Psychiatr 11(9): 867–877, 797.

Meyer-Lindenberg A, Straub RE, Lipska BK, Verchinski BA, Goldberg T, Callicott JH, Egan MF, Huffaker SS, Mattay VS, Kolachana B, Kleinman JE, Weinberger DR (2007) Genetic evidence implicating DARPP-32 in human frontostriatal structure, function, and cognition. J Clin Invest 117(3): 672–682.

Meyer U, Yee BK, Feldon J (2007) The neurodevelopmental impact of prenatal infections at different times of pregnancy: the earlier the worse? Neuroscientist 13(3): 241–256.

Miyamoto S, Duncan GE, Marx CE, Lieberman JA (2005) Treatments for schizophrenia: a critical review of pharmacology and mechanisms of action of antipsychotic drugs. Mol Psychiatr 10(1): 79–104.

Moritz S, Ruff C, Wilke U, Andresen B, Krausz M, Naber D (2001) Negative priming in schizophrenia: effects of masking and prime presentation time. Schizophr Res 48(2–3): 291–299.

Murphy KC (2002) Schizophrenia and velo-cardio-facial syndrome. Lancet 359(9304): 426–430.

Nasrallah HA (1982) Neuropathology of the corpus callosum in schizophrenia. Br J Psychiatr 141: 99–100.

Neckelmann G, Specht K, Lund A, Ersland L, Smievoll AI, Neckelmann D, Hugdahl K (2006) Mr morphometry analysis of grey matter volume reduction in schizophrenia: association with hallucinations. Int J Neurosci 116(1): 9–23.

O'Donnell BF, Swearer JM, Smith LT, Nestor PG, Shenton ME, McCarley RW (1996) Selective deficits in visual perception and recognition in schizophrenia. Am J Psychiatr 153(5): 687–692.

Owen MJ, Williams NM, O'Donovan MC (2004) The molecular genetics of schizophrenia: new findings promise new insights. Mol Psychiatr 9(1): 14–27.

Pantelis C, Barnes TR, Nelson HE, Tanner S, Weatherley L, Owen AM, Robbins TW (1997) Frontal-striatal cognitive deficits in patients with chronic schizophrenia. Brain 120 (Pt 10): 1823–1843.

Paulman RG, Devous MD, Sr., Gregory RR, Herman JH, Jennings L, Bonte FJ, Nasrallah HA, Raese JD (1990) Hypofrontality and cognitive impairment in schizophrenia: dynamic single-photon tomography and neuropsychological assessment of schizophrenic brain function. Biol Psychiatr 27(4): 377–399.

Perlstein WM, Dixit NK, Carter CS, Noll DC, Cohen JD (2003) Prefrontal cortex dysfunction mediates deficits in working memory and prepotent responding in schizophrenia. Biol Psychiatr 53(1): 25–38.

Pezawas L, Meyer-Lindenberg A, Drabant EM, Verchinski BA, Munoz KE, Kolachana BS, Egan MF, Mattay VS, Hariri AR, Weinberger DR (2005) 5-HTTLPR polymorphism impacts human cingulate-amygdala interactions: a genetic susceptibility mechanism for depression. Nat Neurosci 8(6): 828–834.

Post A, Holsboer F, Behl C (1998) Induction of NF-kappaB activity during haloperidol-induced oxidative toxicity in clonal hippocampal cells: suppression of NF-kappaB and neuroprotection by antioxidants. J Neurosci 18(20): 8236–8246.

Potvin S, Mancini-Marie A, Fahim C, Mensour B, Levesque J, Karama S, Beauregard M, Rompre PP, Stip E (2007) Increased striatal gray matter densities in patients with schizophrenia and substance use disorder: a voxel-based morphometry study. Psychiatry Res 154(3): 275–279.

Rajkowska G, Selemon LD, Goldman-Rakic PS (1998) Neuronal and glial somal size in the prefrontal cortex: a postmortem morphometric study of schizophrenia and Huntington disease. Arch Gen Psychiatr 55(3): 215–224.

Ramsey NF, Koning HA, Welles P, Cahn W, van der Linden JA, Kahn RS (2002) Excessive recruitment of neural systems subserving logical reasoning in schizophrenia. Brain 125(Pt 8): 1793–1807.

Rapoport JL, Giedd JN, Blumenthal J, Hamburger S, Jeffries N, Fernandez T, Nicolson R, Bedwell J, Lenane M, Zijdenbos A, Paus T, Evans A (1999) Progressive cortical change during adolescence in childhood-onset schizophrenia. A longitudinal magnetic resonance imaging study. Arch Gen Psychiatr 56(7): 649–654.

Reynolds GP, Beasley CL, Zhang ZJ (2002) Understanding the neurotransmitter pathology of schizophrenia: selective deficits of subtypes of cortical GABAergic neurons. J Neural Transm 109(5–6): 881–889.

Riley BP and McGuffin P (2000) Linkage and associated studies of schizophrenia. Am J Med Genet 97(1): 23–44.

Rogowska J, Gruber SA, Yurgelun-Todd DA (2004) Functional magnetic resonance imaging in schizophrenia: cortical response to motor stimulation. Psychiatr Res Neuroimaging 130: 227–243.

Rose SE, Chalk JB, Janke AL, Strudwick MW, Windus LC, Hannah DE, McGrath JJ, Pantelis C, Wood SJ, Mowry BJ (2006) Evidence of altered prefrontal-thalamic circuitry in schizophrenia: an optimized diffusion MRI study. Neuroimage 32(1): 16–22.

Salgado-Pineda P, Junque C, Vendrell P, Baeza I, Bargallo N, Falcon C, Bernardo M (2004) Decreased cerebral activation during CPT performance: structural and functional deficits in schizophrenic patients. Neuroimage 21(3): 840–847.

Scheepers FE, de Wied CC, Hulshoff Pol HE, van de Flier W, van der Linden JA, Kahn RS (2001) The effect of clozapine on caudate nucleus volume in schizophrenic patients previously treated with typical antipsychotics. Neuropsychopharmacology 24(1): 47–54.

Schroeder J, Niethammer R, Geider FJ, Reitz C, Binkert M, Jauss M, Sauer H (1991) Neurological soft signs in schizophrenia. Schizophr Res 6(1): 25–30.

Schröder J, Wenz F, Schad LR, Baudendistel K, Knopp MV (1995) Sensorimotor cortex and supplementary motor area changes in schizophrenia. A study with functional magnetic resonance imaging. Br J Psychiatr 167(2): 197–201.

Schröder J, Essig M, Baudendistel K, Jahn T, Gerdsen I, Stockert A, Schad LR, Knopp MV (1999) Motor dysfunction and sensorimotor cortex activation changes in schizophrenia: a study with functional magnetic resonance imaging. Neuroimage 9(1): 81–87.

Schultz W (1998) Predictive reward signal of dopamine neurons. J Neurophysiol 80(1): 1–27.

Schwartz BD, Maron BA, Evans WJ, Winstead DK (1999) High velocity transient visual processing deficits diminish ability of patients with schizophrenia to recognize objects. Neuropsychiatr Neuropsychol Behav Neurol 12(3): 170–177.

Seamans JK, Gorelova N, Durstewitz D, Yang CR (2001) Bidirectional dopamine modulation of GABAergic inhibition in prefrontal cortical pyramidal neurons. J Neurosci 21(10): 3628–3638.

Seeman P, Weinshenker D, Quirion R, Srivastava LK, Bhardwaj SK, Grandy DK, Premont RT, Sotnikova TD, Boksa P, El-Ghundi M, O'Dowd BF, George SR, Perreault ML, Mannisto PT, Robinson S, Palmiter RD, Tallerico T (2005) Dopamine supersensitivity correlates with D2High states, implying many paths to psychosis. Proc Natl Acad Sci USA 102(9): 3513–3518.

Sei Y, Ren-Patterson R, Li Z, Tunbridge EM, Egan MF, Kolachana BS, Weinberger DR (2007) Neuregulin1-induced cell migration is impaired in schizophrenia: association with neuregulin1 and catechol-o-methyltransferase gene polymorphisms. Mol Psychiatry 12(10): 946–957.

Shapleske J, Rossell SL, Woodruff PW, David AS (1999) The planum temporale: a systematic, quantitative review of its structural, functional and clinical significance. Brain Res Brain Res Rev 29(1): 26–49.

Silbersweig DA, Stern E, Frith C, Cahill C, Holmes A, Grootoonk S, Seaward J, McKenna P, Chua SE, Schnorr L, (1995) A functional neuroanatomy of hallucinations in schizophrenia. Nature 378(6553): 176–179.

Simons JS and Spiers HJ (2003) Prefrontal and medial temporal lobe interactions in long-term memory. Nat Rev Neurosci 4(8): 637–648.

Sjoholm H, Bratlid T, Sundsfjord J (2004) 123I-beta-CIT SPECT demonstrates increased presynaptic dopamine transporter binding sites in basal ganglia in vivo in schizophrenia. Psychopharmacology (Berl) 173(1–2): 27–31.

Snitz BE, MacDonald A, 3rd, Cohen JD, Cho RY, Becker T, Carter CS (2005) Lateral and medial hypofrontality in first-episode schizophrenia: functional activity in a medication-naive state and effects of short-term atypical antipsychotic treatment. Am J Psychiatr 162(12): 2322–2329.

St Clair D, Blackwood D, Muir W, Carothers A, Walker M, Spowart G, Gosden C, Evans HJ (1990) Association within a family of a balanced autosomal translocation with major mental illness. Lancet 336(8706): 13–16.

Stein JL, Wiedholz LM, Bassett DS, Weinberger DR, Zink CF, Mattay VS, Meyer-Lindenberg A (2007) A validated network of effective amygdala connectivity. Neuroimage 36(3): 736–745.

Stephan KE, Magnotta VA, White T, Arndt S, Flaum M, O'Leary DS, Andreasen NC (2001) Effects of olanzapine on cerebellar functional connectivity in schizophrenia measured by fMRI during a simple motor task. Psychol Med 31(6): 1065–1078.

Stevens JR (1982) Neuropathology of schizophrenia. Arch Gen Psychiatry 39(10): 1131–1139.

Sullivan PF, Neale MC, Kendler KS (2000) Genetic epidemiology of major depression: review and meta-analysis. Am J Psychiatr 157(10): 1552–1562.

Sullivan PF, Kendler KS, Neale MC (2003) Schizophrenia as a complex trait: evidence from a meta-analysis of twin studies. Arch Gen Psychiatr 60(12): 1187–1192.

Sumich A, Chitnis XA, Fannon DG, O'Ceallaigh S, Doku VC, Faldrowicz A, Sharma T (2005) Unreality symptoms and volumetric measures of Heschl's gyrus and planum temporal in first-episode psychosis. Biol Psychiatr 57(8): 947–950.

Sun Z, Wang F, Cui L, Breeze J, Du X, Wang X, Cong Z, Zhang H, Li B, Hong N, Zhang D (2003) Abnormal anterior cingulum in patients with schizophrenia: a diffusion tensor imaging study. Neuroreport 14(14): 1833–1836.

Svenningsson P, Nishi A, Fisone G, Girault JA, Nairn AC, Greengard P (2004) DARPP-32: an integrator of neurotransmission. Annu Rev Pharmacol Toxicol 44: 269–296.

Swerdlow NR, Geyer MA, Braff DL (2001) Neural circuit regulation of prepulse inhibition of startle in the rat: current knowledge and future challenges. Psychopharmacology (Berl) 156(2–3): 194–215.

Szulc A, Galinska B, Tarasow E, Dzienis W, Kubas B, Konarzewska B, Walecki J, Alathiaki AS, Czernikiewicz A (2005) The effect of risperidone on metabolite measures in the frontal lobe, temporal lobe, and thalamus in schizophrenic patients. A proton magnetic resonance spectroscopy (1H MRS). Pharmacopsychiatry 38(5): 214–219.

Tan HY, Choo WC, Fones CS, Chee MW (2005) fMRI study of maintenance and manipulation processes within working memory in first-episode schizophrenia. Am J Psychiatr 162(10): 1849–1858.

Tan HY, Sust S, Buckholtz JW, Mattay VS, Meyer-Lindenberg A, Egan MF, Weinberger DR, Callicott JH (2006) Dysfunctional prefrontal regional specialization and compensation in schizophrenia. Am J Psychiatr 163(11): 1969–1977.

Tost H, Ende G, Ruf M, Henn FA, Meyer-Lindenberg A (2005) Functional imaging research in schizophrenia. Int Rev Neurobiol 67: 95–118.

Tost H, Meyer-Lindenberg A, Klein S, Schmitt A, Hohn F, Tenckhoff A, Ruf M, Ende G, Rietschel M, Henn FA, Braus DF (2006) D2 antidopaminergic modulation of frontal lobe function in healthy human subjects. Biol Psychiatr 60(11): 1196–1205.

Tunbridge EM, Harrison PJ, Weinberger DR (2006) Catechol-o-Methyltransferase, cognition, and psychosis: val(158)met and beyond. Biol Psychiatry 60(2): 141–151.

Tzschentke TM (2001) Pharmacology and behavioral pharmacology of the mesocortical dopamine system. Prog Neurobiol 63(3): 241–320.

Ungerleider LG, Courtney SM, Haxby JV (1998) A neural system for human visual working memory. Proc Natl Acad Sci USA 95(3): 883–890.

Ungerleider LG and Mishkin M, Two cortical visual systems, in Analysis of visual behavior, Ingle, DJ, Goodale MA, and Mansfield RJ, Editors. (1982), MIT: Cambridge, MA. 549–586.

Vink M, Ramsey NF, Raemaekers M, Kahn RS (2006) Striatal dysfunction in schizophrenia and unaffected relatives. Biol Psychiatr 60(1): 32–39.

Volz H, Gaser C, Hager F, Rzanny R, Ponisch J, Mentzel H, Kaiser WA, Sauer H (1999) Decreased frontal activation in schizophrenics during stimulation with the continuous performance test--a functional magnetic resonance imaging study. Eur Psychiatr 14(1): 17–24.

Vrtunski PB, Simpson DM, Weiss KM, Davis GC (1986) Abnormalities of fine motor control in schizophrenia. Psychiatr Res 18: 275–284.

Wakade CG, Mahadik SP, Waller JL, Chiu FC (2002) Atypical neuroleptics stimulate neurogenesis in adult rat brain. J Neurosci Res 69(1): 72–79.

Waldman ID (2005) Statistical approaches to complex phenotypes: evaluating neuropsychological endophenotypes for attention-deficit/hyperactivity disorder. Biol Psychiatr 57(11): 1347–1356.
Walker MA, Highley JR, Esiri MM, McDonald B, Roberts HC, Evans SP, Crow TJ (2002) Estimated neuronal populations and volumes of the hippocampus and its subfields in schizophrenia. Am J Psychiatr 159(5): 821–828.
Weinberger DR (1987) Implications of normal brain development for the pathogenesis of schizophrenia. Arch Gen Psychiatry 44(7): 660–669.
Weinberger DR (1999) Schizophrenia: new phenes and new genes. Biol Psychiatry 46(1): 3–7.
Weinberger DR (2005) Genetic mechanisms of psychosis: in vivo and postmortem genomics. Clin Ther 27(Suppl A) S8–S15.
Weinberger DR and McClure RK (2002) Neurotoxicity, neuroplasticity, and magnetic resonance imaging morphometry: what is happening in the schizophrenic brain? Arch Gen Psychiatr 59(6): 553–558.
Weinberger DR, Berman KF, Zec RF (1986) Physiologic dysfunction of dorsolateral prefrontal cortex in schizophrenia. I. Regional cerebral blood flow evidence. Arch Gen Psychiatr 43(2): 114–124.
Weinstein S, Werker JF, Vouloumanos A, Woodward TS, Ngan ET (2006) Do you hear what I hear? Neural correlates of thought disorder during listening to speech in schizophrenia. Schizophr Res 86(1–3): 130–137.
Wenz F, Schad LR, Knopp MV, Baudendistel KT, Flomer F, Schroder J, van Kaick G (1994) Functional magnetic resonance imaging at 1.5 T: activation pattern in schizophrenic patients receiving neuroleptic medication. Magn Reson Imaging 12(7): 975–982.
Whitford TJ, Grieve SM, Farrow TF, Gomes L, Brennan J, Harris AW, Gordon E, Williams LM (2006) Progressive grey matter atrophy over the first 2–3 years of illness in first-episode schizophrenia: a tensor-based morphometry study. Neuroimage 32(2): 511–519.
Williams GV and Goldman-Rakic PS (1995) Modulation of memory fields by dopamine D1 receptors in prefrontal cortex. Nature 376(6541): 572–575.
Woodruff PW, Wright IC, Bullmore ET, Brammer M, Howard RJ, Williams SC, Shapleske J, Rossell S, David AS, McGuire PK, Murray RM (1997) Auditory hallucinations and the temporal cortical response to speech in schizophrenia: a functional magnetic resonance imaging study. Am J Psychiatr 154(12): 1676–1682.
Wright AM, Bempong J, Kirby ML, Barlow RL, Bloomquist JR (1998) Effects of haloperidol metabolites on neurotransmitter uptake and release: possible role in neurotoxicity and tardive dyskinesia. Brain Res 788(1–2): 215–222.
Wright IC, Rabe-Hesketh S, Woodruff PW, David AS, Murray RM, Bullmore ET (2000) Meta-analysis of regional brain volumes in schizophrenia. Am J Psychiatr 157(1): 16–25.
Yoon JH, D'Esposito M, Carter CS (2006) Preserved function of the fusiform face area in schizophrenia as revealed by fMRI. Psychiatr Res 148(2–3): 205–216.
Yücel M, Pantelis C, Stuart GW, Wood SJ, Maruff P, Velakoulis D, Pipingas A, Crowe SF, Tochon-Danguy HJ, Egan GF (2002) Anterior cingulate activation during Stroop task performance: a PET to MRI coregistration study of individual patients with schizophrenia. Am J Psychiatr 159(2): 251–254.
Zaidel DW, Esiri MM, Harrison PJ (1997) Size, shape, and orientation of neurons in the left and right hippocampus: investigation of normal asymmetries and alterations in schizophrenia. Am J Psychiatr 154(6): 812–818.

Chapter 12
Sleep EEG Provides Biomarkers in Depression

Mayumi Kimura(✉) and Axel Steiger

Abstract Patients with depression frequently report impaired sleep. Objective sleep is recorded by sleep electroencephalogram (EEG). Characteristic sleep-EEG changes in affective disorders include disinhibition of rapid-eye-movement (REM) sleep (shortened REM latency, prolonged first REM periods, elevated REM density, and a measure of the amount of REMs), impaired sleep continuity and changes of nonREM sleep (decreases of slow wave sleep and stage 2 sleep). Disinhibition of REM sleep is also found in animal models of depression. Elevated REM density characterises an endophenotype in high-risk subjects with a positive family history of affective disorders. Most, but not all, antidepressants suppress REM sleep in patients, healthy volunteers and laboratory animals, whereas nonREM sleep and sleep continuity are either impaired or improved by these drugs. Certain sleep-EEG variables or groups of variables appear to predict the response to antidepressive pharmacotherapy or the long-term course of the illness. Overactivity of the hypothalamo-pituitary-adrenocortical system contributes to the sleep-EEG changes in depression. Sleep-EEG variables appear to provide biomarkers for the course of depression and for the screening of antidepressants.

Abbreviations ACTH: Adrenocorticotropic hormone; CRH: Corticotropin-releasing hormone; DEX-CRH test: Dexamethasone/corticotropin-releasing hormone test; EEG: Electroencephalogram; HAB mice: High anxiety-related behavior mice; HAMD: Hamilton Depression Score; HL mice: Helpless mice; HPA system: Hypothalamo-pituitary adrenocortical system; 5-HT: 5-Hydroxytryptamine (=serotonin); LAB mice: Low anxiety-related behavior mice; NaRIs: Selective noradrenaline reuptake inhibitors; NHL mice: Nonhelpless mice; REM: Rapid eye movement; SOREMPs: Shortened REM latency or sleep onset REM periods; SNRIs: Selective serotonin and noradrenaline reuptake inhibitors; SSRIs: Selective serotonin reuptake inhibitors; SWA: Slow-wave activity

---

M. Kimura
Neurogenetics of Sleep, Max Planck Institute of Psychiatry, Kraepelinstrasse 2 – 10, 80804 Munich, Germany
kimura@mpipsykl.mpg.de

C.W. Turck (ed.) *Biomarkers for Psychiatric Disorders*,
doi: 10.1007/978-0-387-79251-4_12,

12.1 Introduction

Impaired sleep is a frequent symptom of affective disorders. Objective sleep data are recorded by the sleep electroencephalogram (EEG). There are several major reasons why psychiatrists became interested in this method in the 1970s. David Kupfer suggested that the rapid-eye-movement (REM) latency, the interval between sleep onset and REM sleep is an indicator of depression (Kupfer and Foster, 1972). On the other hand it was observed that most antidepressants suppress REM sleep (Chen, 1979). Whereas it was thought previously that a short REM latency would help to distinguish between certain subtypes of depression and that suppression of REM sleep is the way antidepressants act (Vogel et al., 1975), research of the last 30 years has revealed much more complex results.

Mammalian sleep consists of periods of REM and of nonREM sleep. Normal human sleep is composed of the cyclic occurrence of nonREM and REM sleep periods. According to the criteria of Rechtschaffe271

n and Kales (Rechtschaffen and Kales, 1968) visually scored nonREM sleep consists of stages 1–4. During sleep stage 1, a slowing of EEG activity is found characterizing the transition from drowsiness to light sleep. The occurrence of sleep spindles and K-complex wave forms characterizes stage 2. Synchronized slow waves occur during sleep stages 3 and 4, slow wave sleep (SWS). REM sleep is defined by faster EEG activity, rapid horizontal eye movements recorded by electrooculography and skeletal muscle hypotonia recorded by electromyography. In most experimental animals, sleep EEGs can be monitored as in humans (Allison and Van Twyver, 1970). Vigilance states are also classified as either REM or non-REM sleep rather than wakefulness. However, unlike humans the sleep structures in animals are often polyphasic, i.e., sleep-wake patterns are very episodic. Further, animal nonREM sleep is not categorized into several stages, although its depth could be interpreted according to the power of slow-wave activity (SWA) (Borbély, 1982). Both in humans and animals, sleep microarchitecture is investigated by computerized quantitative EEG analysis including amplitudes and power in the standard EEG bands i.e., beta, sigma, alpha, theta and delta.

In a young, normal human subject, sleep stages 1–4 occur shortly after going to bed. After a mean duration of 90 min of the first nonREM period, the first REM period is found, which is relatively short. The major amount of SWS and SWA or significant accumulation of delta power during the night is found in the first non-REM period. The duration of the REM periods increases throughout the night. Accordingly during the first half of the night SWS preponderates, whereas the second half of the night is dominated by stage 2 and REM sleep. Four to five sleep cycles consisting of one episode of nonREM and one of REM sleep occur during one night. In the case of animals, this does not apply. The duration of one nonREM sleep episode does not last long. It lasts for a few seconds to a few minutes and is often interrupted with short periods of wakefulness or REM sleep. However, in rats and mice, as in humans, dominant appearance of nonREM sleep occurs first and then more REM sleep appears later during their rest phase, i.e., the light period.

12.2 Sleep EEG in Patients with Depression

Sleep is impaired in nearly all patients with major depression. More than 80% suffer from insomnia, whereas it is estimated that 15–35% complain of hypersomnia (Hawkins et al., 1985; Armitage, 2007).

Various studies have reported the following sleep-EEG changes in depression (Reynolds and Kupfer, 1987; Benca et al., 1997; Armitage, 2007):

1. *Impaired sleep continuity including prolonged sleep latency, intermittent awakening, early morning awakening*
2. *Disinhibition of REM sleep: shortened REM latency or sleep onset REM periods (SOREMPs) (see below), prolonged first REM period, increased REM density (measure of the frequency of rapid eye movements), particularly during the first REM period*
3. *Changes of nonREM sleep, including decrease of SWS, SWA and stage 2 sleep, shift of SWS and SWA from the first to the second nonREM period in younger patients*

In normal subjects, as well as in patients with depression, sleep EEG is influenced by age and gender. As early as during the third decade of the life span SWS and SWA start to decrease. The menopause is a major turning point in the sleep quality of women, whereas in men sleep quality declines continuously with age. Accordingly an interaction between the effects of age and illness is found in depressed patients. Figure 12.1 shows characteristic hypnograms of young and elderly normal subjects and patients with depression. EEG changes are more distinct in postmenopausal than in premenopausal patients (Antonijevic et al., 1999). In a study by Lauer et al. (Lauer et al., 1991) age-related sleep-EEG changes were obvious in both depressed patients and normal controls. In patients, some of these

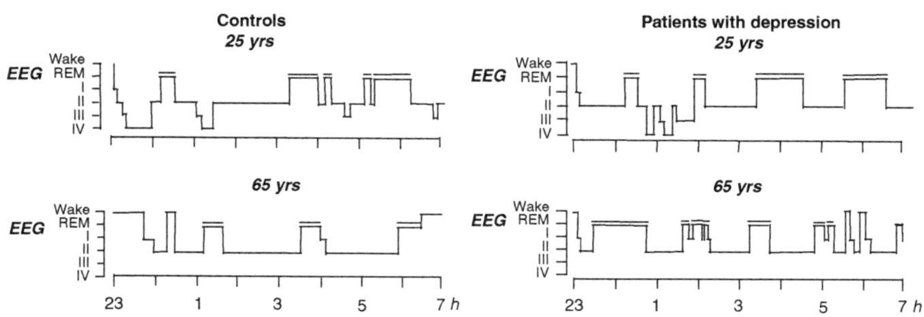

Fig. 12.1 Sleep pattern of young and elderly normal controls and patients with depression. *REM* rapid-eye-movement sleep, *I–IV* stages of non REM sleep
Steiger A: Neuroendocrinology of sleep disorders. In: Textbook of Biological Psychiatry (D'haenen H, den Boer JA, Westenberg H, Willner P, eds.) 2002, Fig. XXIV-3.1. Copyright John Wiley & Sons, Ltd, London. Reproduced with permission

changes occurred early and were more pronounced. REM latency was clearly affected by age, but there were no differences between patients and controls until the middle of the fourth decade. REM density however did not vary with age and was increased in the patients. The amount of SWS decreased with age, but no differences were found between patients and controls at any particular age. The findings of these studies were confirmed by another study by Riemann et al. (Riemann et al., 1991).

In patients with depression, REM latency shows a bimodal distribution (Schulz et al., 1979). SOREMPs (REM latency 0–20 min) were found in 20–30% of the patients, whereas REM latency between 20–40 min was rare. A second peak was found between 40 and 60 min, which is just below the normal range. SOREMPs were frequently observed in older patients and in patients with psychotic depression. In these samples no bimodal distribution of REM latency was found (Ansseau et al., 1984; Kupfer et al., 1986a). Disinhibition of REM sleep was found in depressed patients with insomnia and in patients with hypersomnia as well (Gillin et al., 1984).

Two studies compared sleep EEG between acute depression and stable remission for several weeks. The patients were drugfree at the initial examination, then treated with antidepressants and were drugfree again for several weeks at the second examination. Sleep-EEG variables did not differ between acute depression and recovery (Rush et al., 1986; Steiger et al., 1989). In one of these studies, stage 4 sleep decreased after recovery compared to acute depression (Steiger et al., 1989). These data suggest that persisting sleep-EEG changes in remitted patients may represent a biological scar.

In drugfree depressed patients, who were investigated during several consecutive days, sleep-EEG changes were not stable and changed between days (Schulz et al., 1978). In another study, 12 drugfree depressed patients were investigated for 3 weeks every night. Whereas depression improved, REM latency, sleep continuity and nonREM-sleep variables showed characteristic permanent changes for patients with depression (Coble et al., 1979).

A reduced delta power was found throughout the night in patients with depression (Borbély et al., 1984; Kupfer et al., 1984; Kupfer et al., 1986b).

There is a controversy as to whether the severity of depression and sleep-EEG variables are related. Inverse correlations between REM latency and the Hamilton Depression (HAMD) score were found in some (Kupfer and Foster, 1972; Kupfer, 1976; Spiker et al., 1978) studies, whereas in another report (Feinberg et al., 1982) no relationship between sleep EEG and the severity of depression was shown.

In depressed children and adolescents most studies found no difference of sleep-EEG variables compared to normal controls (Taub et al., 1978; Puig-Antich et al., 1982; Goetz et al., 1987), whereas similar changes as in depressed adults were found in depressed children by Kupfer et al. (Kupfer et al., 1979) and in depressed adolescents by Lahmeyer et al. (Lahmeyer et al., 1983).

The hypothesis that a shortened REM latency is a specific marker of depression was challenged by reports of similar changes in mania (Hudson et al., 1988), schizophrenia (Zarcone,. et al., 1987), schizoaffective disorder (Reich et al., 1975), panic disorder (Uhde et al., 1984), obsessive compulsive disorder (Insel et al., 1982), eating disorders (Katz et al., 1984) and sexual impotence (Schmidt and Nofzinger, 1988). The finding that sleep-EEG changes persist in remitted depressed patients (Rush et al., 1986; Steiger et al., 1989) led to the hypothesis that a history of depression or a comorbidity with depression may explain a shortened REM latency in these disorders. This view is supported by two studies by Lauer and colleagues. They compared normal controls with three groups of patients with (i) major depression, (ii) anorexia and (iii) bulimia. When compared with normal controls, the latter two groups showed no difference whereas REM density was enhanced in the depressed patients. Other variables did not differ between the groups (Lauer et al., 1990). When patients with depression, patients with panic disorder who were never depressed and normal controls were compared, differences were found only during the first sleep cycle. SWS was reduced and the duration and the REM density of the first REM period were increased in the group of depressed patients, whereas the duration of this cycle was reduced in the patients with panic disorder. REM latency was shortened in both groups of patients when compared to controls (Lauer et al., 1992).

Sleep-EEG changes appear to be closely related to the development and to the course of depression. In a long-term investigation an increasing abnormality of REM variables during subsequent episodes was described. SWS however did not show differences between episodes (Kupfer et al., 1991). The comparison of first episodes of depression with recurrent episodes suggested that REM latency and decreased SWS are state independent and do not distinguish between first and recurrent episodes (Thase et al., 1995). Increased phasic REM sleep and lower sleep efficiency however were found in patients with recurrent unipolar depression (Jindal et al., 2002).

Several studies suggest a relationship between sleep-EEG changes before treatment and treatment outcome. An association between shortened REM latency before treatment and favorable response to antidepressant drug therapy was described (Svendsen and Christensen, 1981; Rush et al., 1989; Heiligenstein et al., 1994). A cluster of several disturbed sleep-EEG variables (REM latency, REM density and sleep efficiency) enhanced the predictive value for treatment response to psychotherapy (Thase et al., 1996, 1997). Decreased REM density after psychotherapy was the most robust correlate of remission, whereas REM density before treatment was higher in patients who failed to remit after psychotherapy (Buysse et al., 1999, 2001).

12.3 Sleep EEG in Animal Models of Depression and Anxiety

There are several animal models of depression and anxiety that have been established from selectively bred or genetically vulnerable strains (Henn and Vollmayr, 2005). They are mostly produced from rat or mouse strains, which represent specific behavioral phenotypes and high stress-sensitive markers such as elevated levels of plasma corticosterone. In most of these models, sleep alterations are observed, or animals that show sleep patterns like depressed patients are claimed as the study model of depression. In animal models, however, insomniac patterns are not very evident; neither significantly prolonged wakefulness nor dramatic reduction of SWS is characteristic in them. In contrast, alteration in REM sleep is rather consistent (Adrian et al., 1991; Grønli et al., 2004).

Dugovic and co-workers studied 2 different types of depression models (Dugovic et al., 1999, 2000), and both were characterized by increased REM sleep. In the earlier study with prenatally stressed rats, constantly elevated REM sleep was demonstrated throughout the day (Dugovic et al., 1999), which positively correlated with increased levels of plasma corticosterone. The occurrence of sleep fragmentation during the dark period was also correlated with plasma corticosterone levels, suggesting that more REM sleep and more fragmented sleep tend to occur when the stress level is higher (Bouyer et al., 1998; Marinesco et al., 1999; Tiba et al., 2004). In the later work with Wister-Kyoto rats (Dugovic et al., 2000), enhanced REM sleep was also reported, particularly in the middle of the light period compared with control Wister rats. Although Wister-Kyoto rats expressed persistently lower EEG power spectra across all the frequency bands during REM sleep, they normally responded to total sleep deprivation, e.g., increased delta power, but spent much longer periods in REM sleep during recovery than Wister rats, especially during the first half of the dark period.

A mouse model of depression, with nonhelpless (NHL) vs. helpless (HL) mice, which were developed by selective breeding according to their vulnerability from the tail suspension test result, was investigated by Adrian and co-workers (El Yacoubi et al., 2003). Compared with NHL mice, HL mice that showed 'depression-like' behavior (longer immobility, etc.) and spent less time in wakefulness but more in REM sleep. The enhancement of REM sleep was again clearly shown in this model, as in the others above. The same was seen in Flinder Sensitive Line rats (Shiromani et al., 1988; Overstreet et al., 2005) that are cholinergic-agent sensitive, similar to depressed patients. Therefore, such higher trends towards REM sleep or disinhibition of REM sleep, as seen in the patients, can be a specific marker to claim the animal model of depression.

12.4 Sleep EEG in High Risk Probands for Affective Disorders

In the Munich Vulnerability Study on Affective Disorders, a prospective high-risk design was applied. Healthy probands were investigated, who had a high genetic load for affective disorders because of a positive family history of the disease. The aim was to identify premorbid vulnerability factors. At the index investigation REM density was increased and the amount of SWS during the first nonREM period was reduced in these high risk probands (Lauer et al., 1995). In a follow-up investigation about 4 years later, these findings were stable (Modell et al., 2002). In a subsample of the high risk probands, the cholinomimetic RS86 was given to perform the cholinergic REM sleep-induction test. Whereas at baseline REM latency did not differ between high risk probands and normal controls without a high genetic load for affective disorders, the REM latency was distinctly shortened after cholinergic stimulation in the high risk probands (Schreiber et al., 1992). This observation points to a threshold cholinergic dysfunction in high risk probands. The response pattern of the high risk probands in the cholinergic REM sleep induction test was shown to be predictive for the onset of the first depressive episode (Lauer et al., 2004). Twenty subjects of the initial sample of 82 high risk probands of the Munich Vulnerability Study developed an affective disorder during the follow-up period. The premorbid sleep EEG of these affected high-risk probands showed an increased REM density during the total night and during the first REM period when compared with normal controls without personal or family history of psychiatric disorder (Modell et al., 2005). These findings support the view that increased REM density meets all major requirements for a biological vulnerability marker for affective disorders: (1) it is found in patients with acute depression as well as in the clinically remitted state; (2) it is already present in healthy first-degree relatives of patients; (3) it is stable over time, and (4) it is of predictive value for the onset of the disorder. The authors recommend REM density as a possible endophenotype in family studies (Modell et al., 2005).

12.5 Effects of Antidepressants on Sleep EEG

Many studies report the suppression of REM sleep in patients, normal control subjects and laboratory animals as a common effect of most antidepressants. This effect in humans is shared by tricyclics (Dunleavy et al., 1972; Dunleavy and Oswald, 1973; Passouant et al., 1973) except for trimipramine (Sonntag et al., 1996) and iprindol (Chen, 1979), by irreversible (Cramer and Ohlmeier, 1967; Akindele et al., 1970; Wyatt et al., 1971; Cohen et al., 1982; Landolt et al., 2001) and short-acting reversible (Steiger et al., 1994) monoaminoxidase inhibitors, tetracyclics (Chen, 1979), selective serotonin reuptake inhibitors (SSRIs) (Shipley et al., 1984), selective noradrenaline reuptake inhibitors (NaRIs) (Künzel et al., 2004) and selective serotonin and noradrenaline reuptake inhibitors (SNRIs) (Kluge et al., 2007). The few antidepressants which lack a suppressive effect on REM sleep include buproprione (Nofzinger et al., 1995), the serotonin reuptake enhancer tianeptine (Murck et al., 2003) and the noradrenergic and specific serotoninergic antidepressant mirtazapine (Ruigt et al., 1990; Aslan et al., 2002; Winokur et al., 2003; Schmid et al., 2006). The signs of REM suppression include increase of REM latency, decrease of the time spent in REM sleep and of REM density. Withdrawal of REM-suppressing antidepressants is followed by REM rebounds including decrease of REM latency, increase of time spent in REM sleep and of REM density exceeding baseline values. In normal controls who received various antidepressants for 2 weeks the REM rebound persisted after 1 week (Steiger et al., 1993b).

The potency of various substances to suppress REM sleep differs. The most distinct effects were seen after clomipramine (Steiger, 1988) and the irreversible monoaminoxidase inhibitors phenelzine and tranylcypromine, which are capable of abolishing REM sleep totally. Cessation of these monoaminoxidase inhibitors was followed by REM rebound after a latency of up to 1 week (Akindele et al., 1970). A nearly total REM suppression was also found in healthy men after a single dose of the combined SSRI and serotonin 5-HT_{1A} receptor agonist vilazodone (EMD 68843) (Murck et al., 2001). Besides differences between substances, the dosage also influences the amount of REM-sleep changes (Steiger, 1988) (see Fig. 12.2).

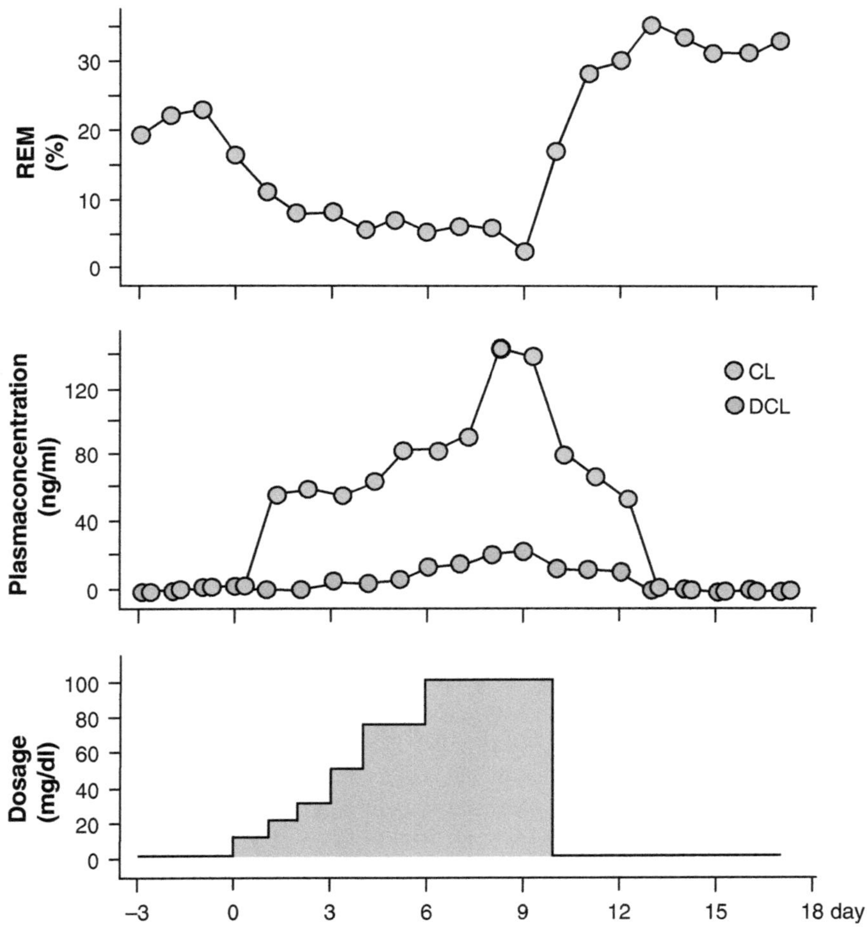

Fig. 12.2 Course of the percentage of REM sleep and the plasma concentration of clomipramine and desmethylclomipramine in a normal subject before, during and after administration of clomipramine. Steiger A: Unterschiede in den Wirkungen von Antidepressiva auf den Schlaf. Psychopharmakotherapie, 3 (1999) 91–95, Fig. 12.1. Copyright Wissenschaftliche Verlagsgesellschaft mbH, Stuttgart. Reproduced with permission (*See Color Plates*)

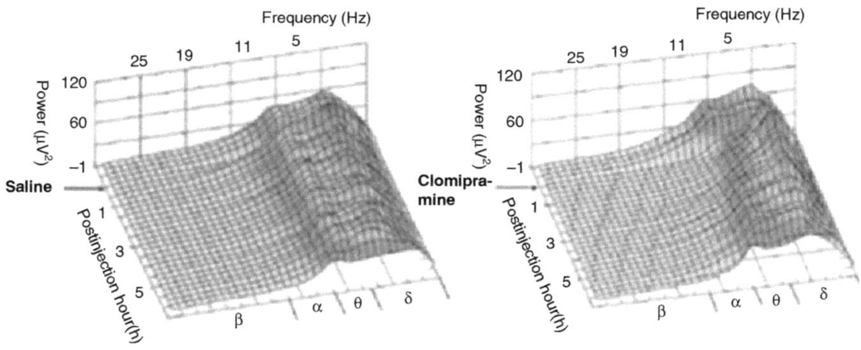

Fig. 12.3 Changes in power spectra of EEGs after saline (*left*) or clomipramine (*right*) injection in the same mice (n = 8). Either vehicle (saline) or clomipramine (30 mg/kg) was given intraperitoneally in the middle of the 12 h-light period, indicated by arrows. Relatively high-intensified power values are illustrated in orange color vs. lower intensity colored in blue. Compared with changes after saline, clomipramine immediately increased delta power but in a much narrower band range, whereas theta power was remarkably diminished for 3 h after injection (*See Color Plates*)

In the case of animal studies, REM-sleep suppression also occurred after both low and high dose administration of clomipramine tested in normal C57BL/6J mice [3 and 30 mg/kg body weight (bw), intraperitoneally (ip)]. The suppression of REM sleep, whereas nonREM sleep was increased, was induced in a dose-dependent manner; the high dose showed more distinct changes in sleep patterns. The latency of the effect appearing in sleep EEG was rather short. Within 15 min after ip injected clomipramine, sharp increases in delta power and broad reduction of theta power were observed (see Fig. 12.3, *right*), in comparison with the power spectra after saline injection (Fig. 12.3, *left*). The reduction of theta power returned to baseline value for a few hours during postinjection; however, the absolute time spent in REM sleep remained significantly lower than for controls for 5 h after injection (Fig. 12.4). According to the results of this animal study, the effects of the antidepressant, clomipramine, on CNS function could be characterized by specific alteration in sleep EEG similarly observed in humans. However, the results of trimipramine, when we tested ip in BL/6J mice with 10 and 30 mg/kg bw, respectively, were not consistent with the human data. In the mice, high dose trimipramine also decreased REM sleep with dramatic suppression of EEG power across all frequency bands (Kimura et al., in press).

In addition to antidepressants, other substances like alcohol, amphetamines and barbiturates also suppress REM sleep. After subchronic administration of these drugs an adaptation is found, whereas REM suppression after antidepressants persists for longer periods (Vogel, 1983). However, a weak adaptation was found for tricyclics (Gillin et al., 1978; Shipley et al., 1985). In contrast, irreversible monoaminoxidase inhibitors prompted a constant suppression for several months (Akindele et al., 1970; Wyatt et al., 1971). However, in a group of three patients, small amounts of REM sleep recurred after 3-6 months of treatment with phenelzine (Landolt and de Boer, 2001). Vogel et al. (Vogel et al., 1975) reported that selective REM-sleep deprivation, but not selective nonREM-sleep deprivation for 3 weeks,

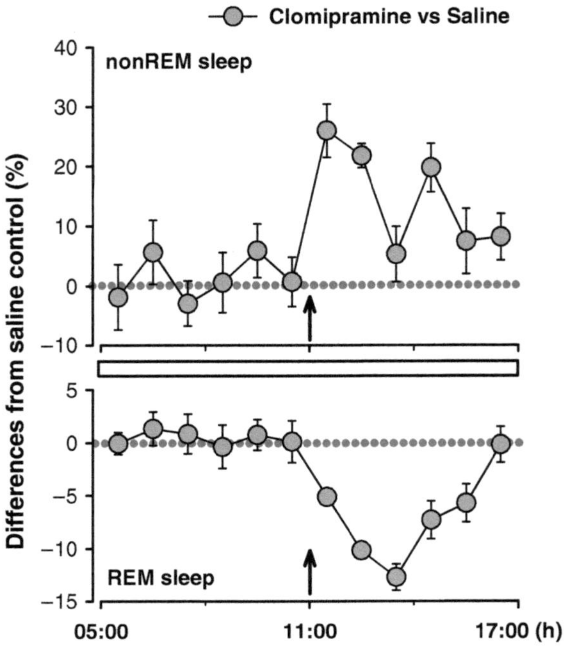

Fig. 12.4 Hourly changes in nonREM and REM sleep before and after clomipramine in mice (indicated with circles, from the same data shown in Fig. 12.3), compared with the saline-treated day (dotted lines). Clomipramine (30 mg/kg) or saline was injected ip in the middle of the 12 h-light period, indicated by arrows. After clomipramine, nonREM sleep increased, but REM sleep was suppressed during the rest of the light period (*See Color Plates*)

had antidepressive effects. This observation and the fact that most antidepressants suppress REM sleep, support the hypothesis that REM suppression is the mechanism of action of antidepressants. However, these findings were not reproduced by another study (Grözinger et al., 2002). Furthermore the hypothesis by Vogel et al. is challenged by the antidepressive effects of drugs like trimipramine and mirtazapine. These drugs even increase REM sleep during the treatment of depression (Wiegand et al., 1986; Sonntag et al., 1996, 2006).

REM suppression appears to be a distinct, though not absolutely required, hint for the antidepressive properties of a compound. This is illustrated by the comparison of the effects of the stereoisomeres of oxaprotiline, R(-)oxaprotiline and S(+)oxaprotiline. S(+)oxaprotiline was superior to R(-)oxaprotiline in its antidepressive effects. S(+)oxaprotiline, but not R(-)oxaprotiline suppressed REM sleep in depressed patients (Steiger et al., 1993a).

In contrast to the similarity of the effects of the vast majority of antidepressants on REM sleep, there are differences in their effects on sleep continuity and on non-REM sleep. Most tricyclics increase SWS (Chen, 1979), whereas clomipramine (Steiger, 1988) and imipramine (Sonntag et al., 1996) decrease SWS. In depressed patients SWS remains unchanged after desipramine (Shipley et al., 1985), nortriptyline (Kupfer et al., 1982) and amitriptyline (Gillin et al., 1978). In response to SSRIs SWS is largely unaffected, whereas sleep continuity is impaired by an increase of intermittent wakefulness (Saletu et al., 1991; Sharpley et al., 1996). In response to the NaRI reboxetine, intermittent wakefulness and stage 2 sleep increased and sleep efficiency decreased (Künzel et al., 2004). Vilazodone treatment abolished REM sleep nearly totally, SWS and SWA increased in the first and third one-thirds of the night, whereas wakefulness was enhanced in the second and third one-thirds of the night (Murck et al., 2001) (see Figs. 12.5–12.7). The SNRI duloxetine increased sleep stage 3 in patients with depression (Kluge et al., 2007). In laboratory animals, the SSRI paroxetine and the NaRI reboxetine showed similar outcomes to the effects on human sleep. Paroxetine is reported to inhibit REM sleep in rats (Gervasoni et al., 2002; Smith et al., 2007; Sánchez et al., 2007). When tested in mice, paroxetine tended to also diminish an appearance of REM sleep and further altered sleep continuity as fragmented nonREM sleep (Dalal et al., in press). Reboxetine consistently suppressed REM sleep in rats and mice and decreased light SWS in rats (Sánchez et al., 2007) but increased nonREM sleep in mice (Kimura et al., in press).

Even at day 2 of treatment of patients with depression with mirtazapine, sleep continuity was improved as total sleep time and sleep efficiency increased and time spent awake decreased. This effect persisted at day 28, when SWS, low delta, theta and alpha activity increased (Schmid et al., 2006) (see Figs. 12.8 and 12.9).

These results suggest that sleep-EEG variables may help to predict the future course of depression. Specific changes in EEG traces would also help in verifying

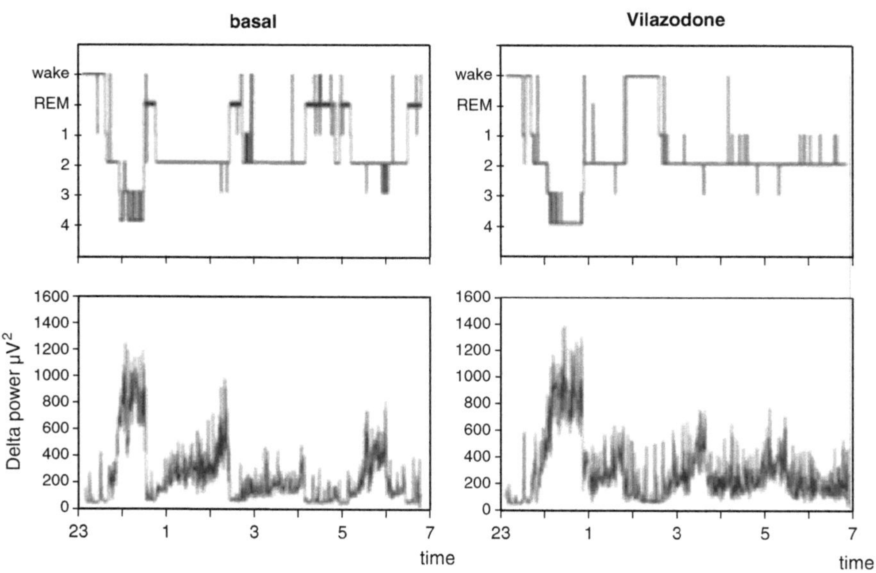

Fig. 12.5 Sleep pattern in one subject after placebo and Vilazodone. The top figures show hypnograms after conventional criteria. The marked rapid eye movement (REM) suppression is clearly visible. The bottom figures show the corresponding time course of the delta power, showing a pronounced increase of the slow wave EEG activity. Springer, Psychopharmacology, 155, 2002, p. 189, Distinct temporal pattern of the effects of the combined serotonin-reuptake inhibitor and 5-HT1A agonist EMD 68843 on the sleep EEG in healthy men, Murck H, Frieboes RM, Antonijevic IA, Steiger A, Fig. 1. With kind permission from Springer Science and Business Media

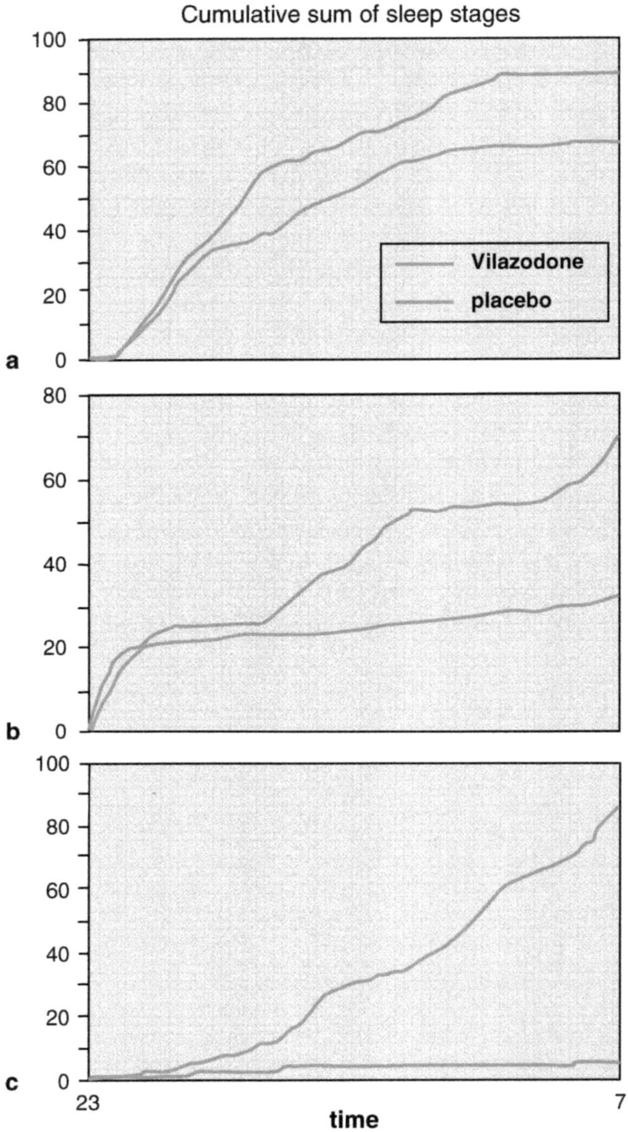

Fig. 12.6 Time course of the accumulation of slow wave sleep (SWS; **a**), wakefulness (**b**), and REM (**c**). Three distinct intervals, which correspond with the thirds of the night, can be distinguished. The increase in SWS after Vilazodone occurs selectively in the first and third one-third, the increase in wakefulness in the second one-third and at the end of the sleep period, whereas REM is suppressed throughout the night. Springer, Psychopharmacology, 155, 2002, p. 189, Distinct temporal pattern of the effects of the combined serotonin-reuptake inhibitor and 5-HT1A agonist EMD 68843 on the sleep EEG in healthy men, Murck H, Frieboes RM, Antonijevic IA, Steiger A, Fig. 2. With kind permission from Springer Science and Business Media (*See Color Plates*)

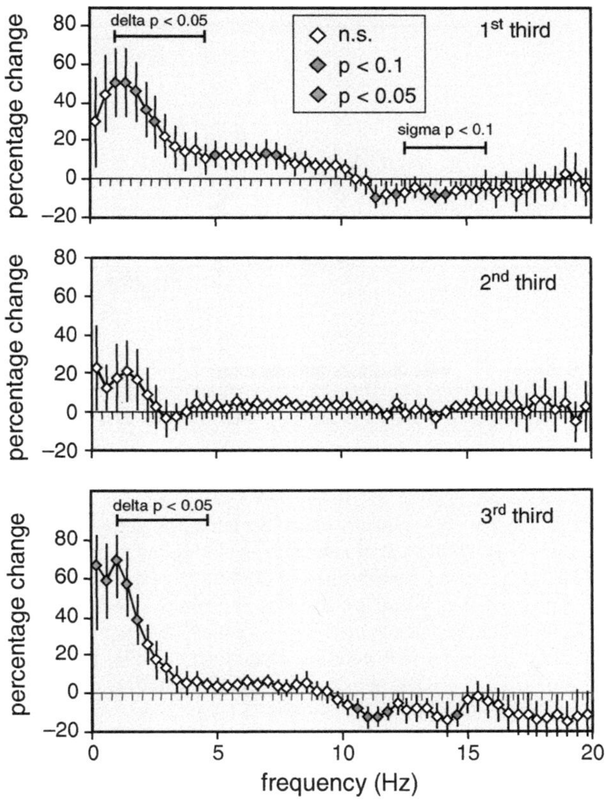

Fig. 12.7 Spectral analysis for nonREM sleep for each one-third of the total night. After Vilazodone delta power is significantly increased in the first and the last one-third, while sigma power shows a trend to decrease during these intervals. Springer, Psychopharmacology, 155, 2002, p. 189, Distinct temporal pattern of the effects of the combined serotonin-reuptake inhibitor and 5-HT1A agonist EMD 68843 on the sleep EEG in healthy men, Murck H, Frieboes, RM, Antonijevic IA, Steiger A, Fig. 3. With kind permission from Springer Science and Business Media (*See Color Plates*)

the efficacy of potential antidepressants at an early stage of treatment, especially in reference to alteration in REM sleep parameters (latency, REM densities, etc.). The increase of REM latency after two days of treatment with amitriptyline correlated with the therapeutic outcome after 4 weeks (Kupfer et al., 1976). A similar result was reported for imipramine (Sonntag et al., 1996), but not for clomipramine (Riemann and Berger, 1990).

12.6 State and Vulnerability Markers During Drug Therapy

A subgroup of 38 depressed patients from a drug trial comparing the efficacy of the serotonin reuptake enhancer tianeptine and the SSRI paroxetine was studied with regard to their effects on sleep EEG, particularly in relation to treatment response.

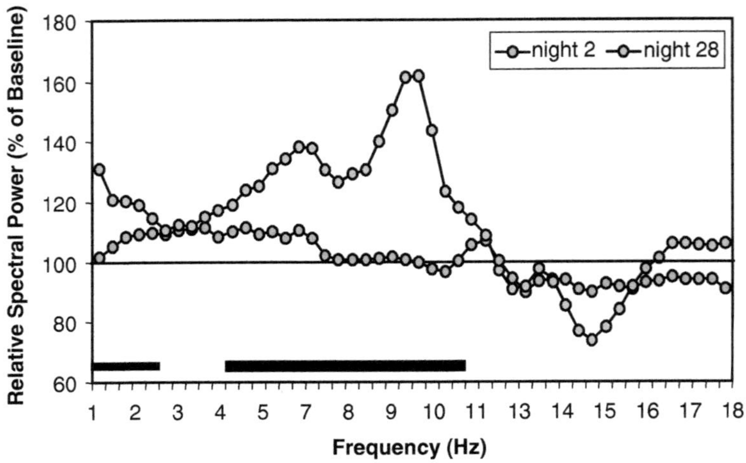

Fig. 12.8 Spectral EEG composition of the largest common amount of nonREM sleep (=155.5 min) 2 days (night 2) and 4 weeks (night 28) after beginning of mirtazapine administration. The values for all frequency bins are plotted as a percentage of baseline night (=100%). The frequency ranges in which the values varied significantly (ANOVA, $p < 0.05$) over all three examination nights are indicated by black bars above the abscissa. Schmid DA, Wichniak A, Uhr M, Ising M, Brunner H, Held K, Weikel J, Sonntag A, Steiger A: Changes of sleep architecture, spectral composition of sleep EEG, the nocturnal secretion of cortisol, ACTH, GH, prolactin, melatonin, ghrelin and leptin and the DEX-CRH-test in depressed patients during treatment with mirtazapine. Neuropsychopharmacology, 31 (2006) 832–844, Fig. 1 (*See Color Plates*)

Sleep EEG was recorded at days 7 and 42 after the start of treatment with either compound. Spectral analysis of the nonREM-sleep EEG revealed a strong decline in the higher sigma frequency range (14–16 Hz) in male treatment responders independent of medication. Nonresponders however did not show marked changes in this frequency range independent of gender. The patients receiving paroxetine showed less REM sleep and more intermittent wakefulness compared to the patients receiving tianeptine. REM density after 1 week of treatment was a predictor of treatment response in the whole sample. Changes in REM density were inversely correlated to the changes in the HAMD Score in the paroxetine, but not in the tianeptine group. These data demonstrate the importance of taking gender into account in the study of the biological effects of drugs. Furthermore the study points to the importance of the highest sigma frequency range and of REM density as markers of treatment response (see Figs. 12.10–12.12) (Murck et al., 2003).

Hatzinger et al. (Hatzinger et al., 2004) investigated sleep-EEG profiles of patients with depression in order to determine whether (1) the retrospective clinical course as reflected by the number of episodes and (2) the prospective long term outcome in the follow-up as reflected by the occurrence of further episodes are associated with certain sleep-EEG variables. Fifteen patients with depression who participated in an earlier controlled treatment study with trimipramine over 6 weeks participated in an exploratory follow-up study. This study showed that the retrospectively assessed long

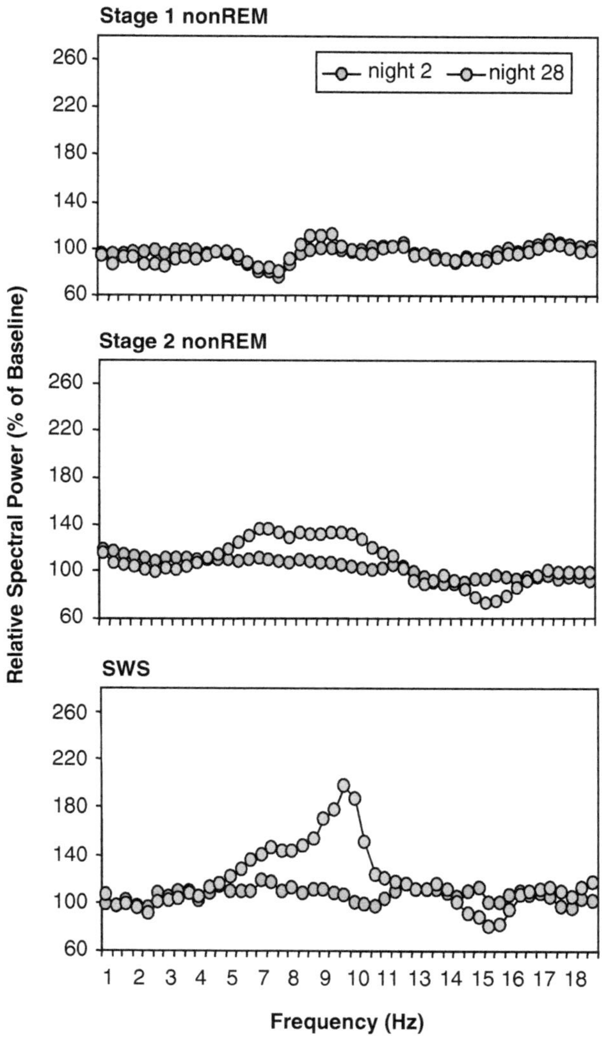

Fig. 12.9 Spectral EEG composition in each of nonREM sleep stages 2 days (night 2) and 4 weeks (night 28) after beginning of mirtazapine administration. The values for all frequency bins are plotted as a percentage of baseline night (= 100%). The increase of alpha activity was apparent only at week 4 and was, in particular, related to SWS. Schmid DA, Wichniak A, Uhr M, Ising M, Brunner H, Held K, Weikel J, Sonntag A, Steiger A: Changes of sleep architecture, spectral composition of sleep EEG, the nocturnal secretion of cortisol, ACTH, GH, prolactin, melatonin, ghrelin and leptin and the DEX-CRH-test in depressed patients during treatment with mirtazapine. Neuropsychopharmacology 31 (2006) 832–844, Fig. 2 (*See Color Plates*)

term course of depression is associated with sleep-EEG variables during the acute episode of depression. The lower the sleep continuity (total sleep time, sleep efficiency index, wake time and number of awakenings) the higher is the number of previous episodes. At the end of drug treatment this strong association disappears,

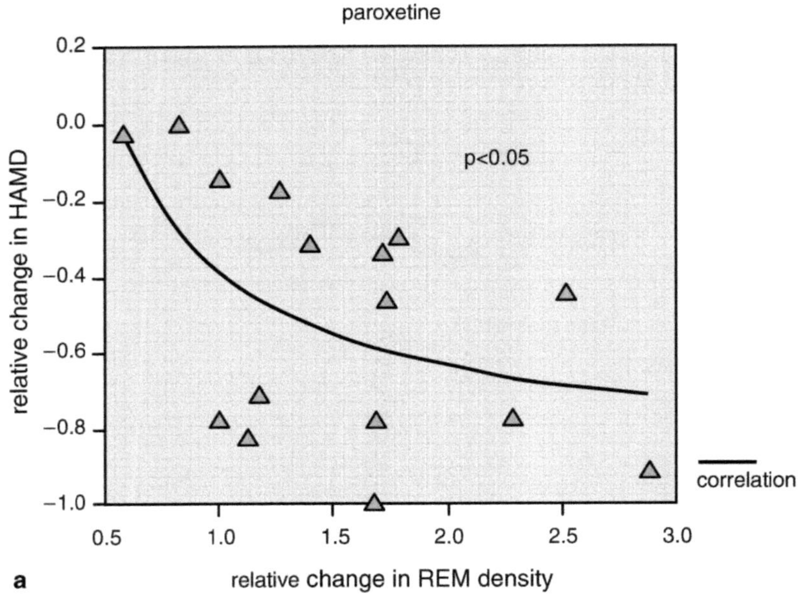

a

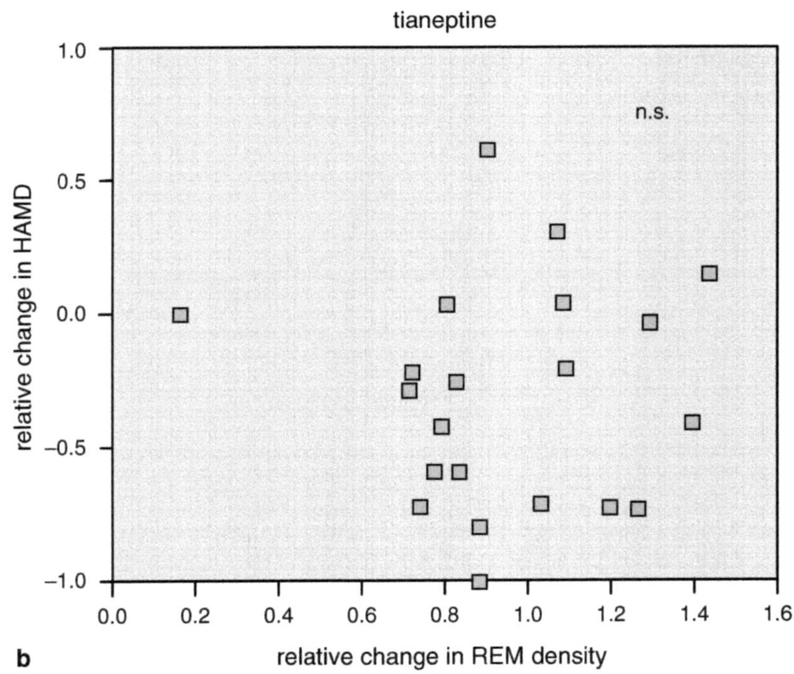

b

Fig. 12.10a Correlation between the relative change in the Hamilton Depression Score (HAMD) (difference HAMD at day 7 – HAMD at day 42) in the course of treatment with paroxetine in relation to the relative change in REM density. REM density increases with the decline in REM. **b** No correlation is found after tianeptine. Murck H, Nickel T, Künzel H, Antonijevic IA, Schill J, Zobel A, Steiger A, Sonntag A, Holsboer F: State markers of depression in sleep EEG: Dependency on drug and gender in patients treated with tianeptine or paroxetine Neuropsychopharmacology 28 (2003) 348–358, Fig. 1a and 1b (*See Color Plates*)

whereas a close association is found between less SWS, particularly during the first third of sleep period, elevated REM density (by trend) and the number of previous episodes. A clear association between the prospective long term course and sleep EEG was shown: decreased SWS and increased REM density at the end of treatment were related to an increased recurrence rate between the treatment study and the follow-up. Furthermore, these sleep-EEG variables were related to the disturbances in hypothalamo-pituitary adrenocortical (HPA) system, as was mirrored by abnormal dexamethasone/corticotropin-releasing hormone (DEX-CRH) test results. These results indicate that patients with an unfavorable long-term course of depression experience increasing aberrant sleep regulation and that these alterations may be of predictive value, not only for treatment response during acute depression but also for recurrences in the long term outcome of depression. Furthermore these predictive sleep-EEG markers of the long-term course of depression are closely related to HPA system activity. The more the sleep-EEG markers were disturbed, the more the HPA system deteriorated. This finding supports the view that HPA overactivity contributes to the sleep EEG changes found in depression.

12.7 The Role of the HPA System in sleep EEG-Changes in Depression

Elevated nocturnal cortisol (Steiger et al., 1989; Linkowski et al., 1987) and ACTH (Linkowski et al., 1987) levels are state markers of acute episodes of depression. Administration of corticotropin-releasing hormone (CRH), the key hormone of the HPA system, prompts shallow sleep in rats (Ehlers et al., 1986) and rabbits (Opp et al., 1989). Similarly, pulsatile injections of CRH to normal young men induce some sleep changes such as decrease of SWS, a blunted growth hormone surge and elevated cortisol levels, which are characteristic of depression (Holsboer et al., 1988). Subchronic administration of the synthetic glucocorticoid receptor agonist methylprednisolone to female patients with multiple sclerosis prompts sleep-EEG changes resembling those found in acute depressive episode (increased REM density, shortened REM latency, shift of SWS and SWA from the first to the second sleep cycle) (Antonijevic and Steiger 2003).

Even in kindergarten children, associations between unfavorable sleep-EEG patterns, increased HPA activity and more difficult behavioral and psychosocial dimensions were observed (Hatzinger et al., 2008). In preschool children sleep EEG was recorded for a night and for baseline HPA-activity assessment saliva samples were collected immediately after awakening. Furthermore saliva samples before, during and after a psychological challenge were used to assess the HPA activity under stress conditions. After cluster analysis children labeled as "poor" sleepers showed significantly increased morning cortisol values as compared to "good" sleepers. Increased cortisol secretion under stress conditions was associated with an increased number of awakenings after sleep onset, and more time spent in sleep stages 1 and 2. Furthermore, psychological difficulties such as impulsivity, overanxiousness, or social inhibition were associated with low sleep efficiency.

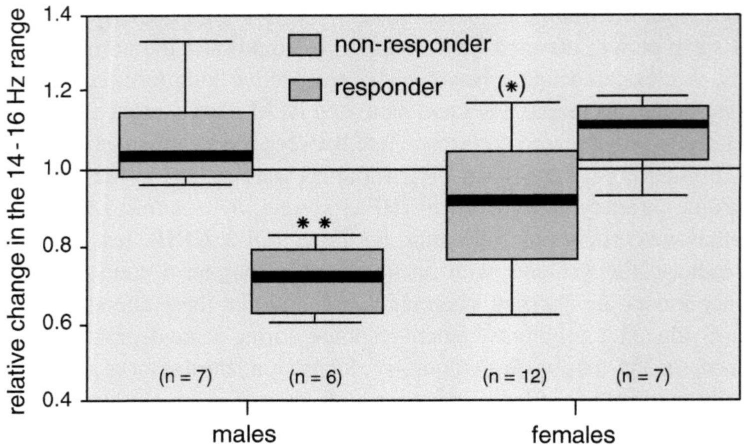

Fig. 12.11 Relative change of EEG power in the 14–16 Hz range divided into gender comparing responders and nonresponders. Male and female subjects behave differentially concerning treatment response (for details, see text). Murck H, Nickel T, Künzel H, Antonijevic IA, Schill J, Zobel A, Steiger A, Sonntag A, Holsboer F: State markers of depression in sleep EEG: Dependency on drug and gender in patients treated with tianeptine or paroxetine. Neuropsychopharmacology 28 (2003) 348–358, Fig. 2 (*See Color Plates*)

The CRH-1-receptor antagonist R121919 was given in two different dose escalations to inpatients with major depression. A random subgroup of 10 patients underwent three sleep-EEG recordings (baseline before treatment, at the end of the first week and at the end of the fourth week of active treatment). SWS increased compared with baseline after 1 and 4 weeks. There was a trend for a decreased number of awakenings and REM density during the same period. Separate evaluation of these changes for both panels showed no significant effect of lower doses, whereas in the higher doses after R121919, REM density decreased and SWS increased significantly between baseline and week 4. Furthermore positive associations between HAMD scores and SWS at the end of active treatment were found. These data suggest that CRH-1-receptor antagonism has a normalizing influence on sleep EEG in depressed patients (Held et al., 2004).

Sleep alterations have also been seen as mentioned earlier in animal models of depression that associate with elevated levels of plasma corticosterone. This implies the positive correlation of disturbed sleep with high HPA activity (Buckley and Schatzberg, 2005). However, it is rather complicated for such disease models to unveil which stage of the HPA system or stress hormones predominantly or simultaneously affect sleep structures. To determine whether the initial modulator of the HPA axis, CRH, is involved in altered sleep, R121919 was tested in stressed rats showing selective low (LAB) or high (HAB) anxiety-related behavior (Lancel et al., 2002). Although in these rats no distinct differences in sleep patterns during baseline were found with relation to their innate anxiety, stress-induced sleep attenuation in HAB rats was normalized with R121919 treatment to the level of

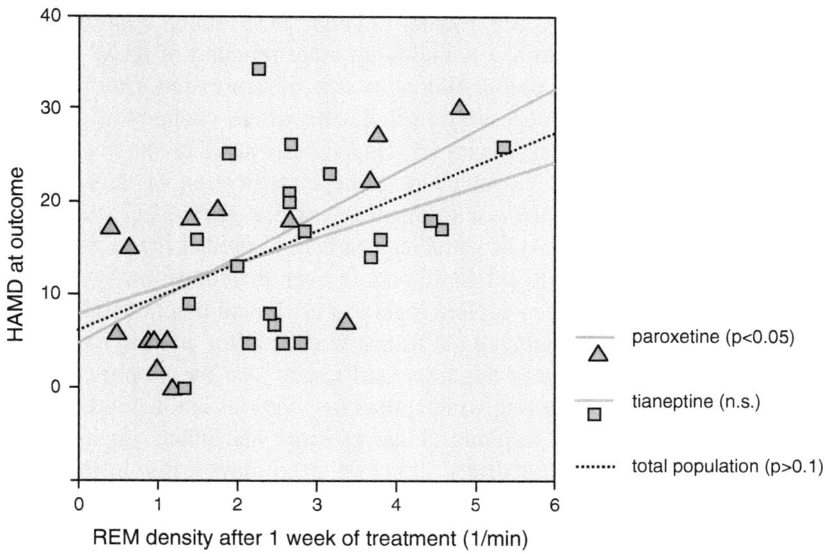

Fig. 12.12 HAMD at day 42 in relation to the REM density at day 7. With paroxetine, the REM density at day 7 is correlated to treatment outcome, whereas with tianeptine no significant correlation existed. Murck H, Nickel T, Künzel H, Antonijevic IA, Schill J, Zobel A, Steiger A, Son ntag A, Holsboer F: State markers of depression in sleep EEG: Dependency on drug and gender in patients treated with tianeptine or paroxetine. Neuropsychopharmacology 28 (2003) 348–358, Fig 3 (*See Color Plates*)

LAB rats that showed weaker stress responses. Further, another CRH-1-receptor antagonist, DMP696, was tested in a conditional transgenic mouse mutant that overexpresses CRH only in the central nervous system but keeps undisturbed HPA activity (Lu et al., in press). This conditional CRH-overexpressing mouse model exhibited elevated REM sleep during baseline and larger REM-sleep rebound after sleep deprivation (Kimura et al., 2006). When DMP696 was given orally 1 h before the termination of 6 h-sleep deprivation, the levels of increased REM sleep during recovery in hetero- and homozygous mice became indistinguishable compared to those shown in control and wildtype mice (Kimura et al., 2007). The results demonstrated that CRH-1-receptor antagonism plausibly cures stress-induced or CRH-derived sleep alteration. Significant effects of CRH-1-receptor antagonists seem to be present only in a pathological situation, which can be probed by the sleep-EEG measurement.

12.8 Perspectives

In summary, sleep-EEG variables appear to provide biomarkers for the course and drug treatment of depression. Elevated REM density was shown to characterize an endophenotype in family studies of depression. REM disinhibition, sleep discontinuity and lower SWS and SWA are robust findings in patients with depression. Particularly, changes of REM sleep, including more rebound of REM sleep after sleep deprivation, are also found in animal models of depression. Careful psychopathological assessment is necessary to delineate which changes of sleep-EEG variables are specific for certain diagnoses. On the other hand, biomarkers including sleep-EEG variables should be included in the classification of depression and other psychiatric disorders. Various studies suggest that sleep-EEG variables like REM latency or certain groups of variables can help to predict the response to the actual treatment with a certain antidepressant or even the course for several years. So far, sleep EEG has not been applied regularly in clinical practice to help physicians to choose the drug promising the fastest remission for an individual patient. Further research should address this issue and search also for simplified methods of sleep-EEG recording, allowing wide-spread use. Various antidepressants modulate sleep EEG in different fashions. Some promote and others impair sleep. It should be clarified whether the drugs' effect on sleep plays a role in their overall clinical effect, e.g., does recovery from depression occur faster after a sleep-improving than after a sleep-impairing substance, and is remission more stable after sleep improving compounds? Most antidepressants suppress REM sleep in humans and animals. However, REM suppression is not the prerequisite for a substance to act as an antidepressant. Therefore another actual research task is to identify a sleep-EEG variable or a cluster of such variables which can be used to screen substances for their capacity to be used as antidepressants.

References

Adrian J, Dugovic C, Martin P (1991) Sleep-wakefulness patterns in the helpless rat. Physiol Behav 49:257–262

Akindele MO, Evans JI, Oswald I (1970) Monoamine oxidase inhibitors, sleep and mood. Electroencephalogr Clin Neurophysiol 29:47–56

Allison T, Van Twyver H (1970) The evolution of sleep. Nat Hist 79:56–65

Ansseau M, Kupfer DJ, Reynolds CF, III, McEachran AB (1984) REM latency distribution in major depression: clinical characteristics associated with sleep onset REM periods. Biol Psychiatr 1651–1666

Antonijevic IA, Steiger A (2003) Depression-like changes of the sleep-EEG during high dose corticosteroid treatment in patients with multiple sclerosis. Psychoneuroendocrinology 28:780–795

Antonijevic IA, Murck H, Frieboes RM, Holsboer F, Steiger A (1999) On the gender differences in sleep-endocrine regulation in young normal humans. Neuroendocrinology 70:280–287

Armitage R (2007) Sleep and circadian rhythms in mood disorders. Acta Psychiatr Scand 115:104–115

Aslan S, Isik E, Cosar B (2002) The effects of mirtazapine on sleep: a placebo controlled, double-blind study in young healthy volunteers. Sleep 25:677–679

Benca RM, Okawa M, Uchiyama M, Ozaki S, Nakajima T, Shibui K, Obermeyer WH (1997) Sleep and mood disorders. Sleep Med Rev 1:45–56

Borbély AA (1982) A two process model of sleep regulation. Hum Neurobiol 1:195–206

Borbély AA, Tobler I, Loepfe M, Kupfer DJ, Ulrich RF, Grochocinski V, Doman J, Matthews G (1984) All-night spectral analysis of the sleep EEG in untreated depressives and normal controls. Psychiatr Res 12:27–33

Bouyer JJ, Vallée M, Deminière JM, Le Moal M, Mayo W (1998) Reaction of sleep-wakefulness cycle to stress in related to differences in hypothalamo-pituitary-adrenal axis reactivity in rat. Brain Res 804:114–124

Buckley TM, Schatzberg AF (2005) On the interactions of the hypothalamic-pituitary-adrenal (HPA) axis and sleep: Normal HPA axis activity and circadian rhythm, exemplary sleep disorders. J Clin Endocrinol Metab 90:3106–3114

Buysse DJ, Tu XM, Cherry CR, Begley AE, Kowalski J, Kupfer DJ, Frank E (1999) Pretreatment REM sleep and subjective sleep quality distinguish depressed psychotherapy remitters and nonremitters. Biol Psychiatr 45:205–213

Buysse DJ, Hall M, Begley A, Cherry CR, Houck PR, Land S, Ombao H, Kupfer DJ, Frank E (2001) Sleep and treatment response in depression: new findings using power spectral analysis. Psychiatr Res 103:51–67

Chen CN (1979) Sleep, depression and antidepressants. Br J Psychiatr 135:385–402

Coble PA, Kupfer DJ, Spiker DG, Neil JF, McPartland RJ (1979) EEG sleep in primary depression. A longitudinal placebo study. J Affect Disord 1:131–138

Cohen RM, Pickar D, Garnett D, Lipper S, Gillin JC, Murphy DL (1982) REM sleep suppression induced by selective monoamine oxidase inhibitors. Psychopharmacology 78:137–140

Cramer H, Ohlmeier D (1967) Ein Fall von Tranylcypromin und Trifluoperazin -(Jatrosom)-Sucht: Psychopathologische, schlafphysiologische und biochemische Untersuchungen. Arch Psychiatr Nervenkr 210:182–197

Dalal MA, Flachskamm C, Uhr M, Kimura (in press) What does paroxetine do with sleep? its effects in mdr 1a/1b(-/-) and FVB wildtype mice. J Sleep Res

Dugovic C, Maccari S, Weibel L, Turek FW, Van Reeth O (1999) High corticosterone levels in prenatally stressed rats predict persistent paradoxical sleep alterations. J Neurosci 19:8656–8664

Dugovic C, Solberg LC, Redei E, Van Reeth O, Turek FW (2000) Sleep in the Wister-Kyoto rat, a putative genetic animal model for depression. Neuroreport 11:627–631

Dunleavy DL, Oswald I (1973) Phenelzine, mood response, and sleep. Arch Gen Psychiatr 28:353–356

Dunleavy DL, Brezinova V, Oswald I, MacLean AW, Tinker M (1972) Changes during weeks in effects of tricyclic drugs on the human sleeping brain. Br J Psychiatr 120:663–672

Ehlers CL, Reed TK, Henriksen SJ (1986) Effects of corticotropin-releasing factor and growth hormone- releasing factor on sleep and activity in rats. Neuroendocrinology 42:467–474

El Yacoubi M, Bouali S, Popa D, Naudon L, Leroux-Nicollet I, Hamon M, Costentin J, Adrian J, Vaugeois J-M (2003) Behavioral, neurochemical, and electrophysiological characterization of a genetic mouse model of depression. Proc Natl Acad Sci USA 100:6227–6623

Feinberg M, Gillin JC, Carroll BJ, Greden JF, Zis AP (1982) EEG studies of sleep in the diagnosis of depression. Biol Psychiatr 17:305–316

Gervasoni D, Panconi E, Henninot V, Boissard R, Barbagli B, Fort P, Luppi PH (2002) Effect of chronic treatment with milnacipran on sleep architecture in rats compared with paroxetine and imipramine. Pharmacol Biochem Behav 73:557–563

Gillin JC, Wyatt RJ, Fram D, Snyder F (1978) The relationship between changes in REM sleep and clinical improvement in depressed patients treated with amitriptyline. Psychopharmacology 59:267–272

Gillin JC, Sitaram N, Wehr T, Duncan W, Post R, Murphy DL, Wyatt RJ, Bunney WE (1984) Sleep and affective illness. In: Post RM, Ballenger, JC (eds) Neurobiology of mood disorders. Williams and Wilins, Baltimore, pp 157–189

Goetz RR, Puig-Antich J, Ryan N, Rabinovich H, Ambrosini PJ, Nelson B, Krawiec V (1987) Electroencephalographic sleep of adolescents with major depression and normal controls. Arch Gen Psychiatr 44:61–68

Grønli J, Murison R, Bjørvatn B, Sørensen E, Portas CM, Ursin R (2004) Chronic mild stress affects sucrose intake and sleep in rats. Behav Brain Res 150:139–147

Grüzinger M, Kogel P, Rüschke J (2002) Effects of REM sleep awakenings and related wakening paradigms on the ultradian sleep cycle and the symptoms in depression. J Psychiatr Res 36:299–308

Hatzinger M, Hemmeter UM, Brand S, Ising M, Holsboer-Trachsler E (2004) Electroencephalographic sleep profiles in treatment course and long-term outcome of major depression: association with DEX/CRH-test response. J Psychiatr Res 38:453–465

Hatzinger M, Brand S, Perren S, Stadelmann S, von Wyl A, von Klitzing K, Holsboer-Trachsler E (2008) Electroencephalographic sleep profiles and hypothalamic-pituitary-adrenocortical (HPA) activity in kindergarten children: Early indication of poor sleep quality associated with increased cortisol secretion. J Psychiatr Res 42:532–543

Hawkins DR, Taub JM, Van de Castle RL (1985) Extended sleep (hypersomnia) in young depressed patients. Am J Psychiatr 142:905–910

Heiligenstein JH, Faries DE, Rush AJ, Andersen JS, Pande AC, Roffwarg HP, Dunner D, Gillin JC, James SP, Lahmeyer H (1994) Latency to rapid eye movement sleep as a predictor of treatment response to fluoxetine and placebo in nonpsychotic depressed outpatients. Psychiatr Res 52:327–339

Held K, Könzel H, Ising M, Schmid DA, Zobel A, Murck H, Holsboer F, Steiger A (2004) Treatment with the CRH1-receptor antagonist R121919 improves sleep EEG in patients with depression. J Psychiatr Res 38:129–136

Henn FA, Vollmayr B (2005) Stress models of depression: Forming genetically vulnerable strains. Neurosci Behav Rev 29:799–804

Holsboer F, von Bardeleben U, Steiger A (1988) Effects of intravenous corticotropin-releasing hormone upon sleep-related growth hormone surge and sleep EEG in man. Neuroendocrinology 48:32–38

Hudson JI, Lipinski JF, Frankenburg FR, Grochocinski VJ, Kupfer DJ (1988) Electroencephalographic sleep in mania. Arch Gen Psychiatr 45:267–273

Insel TR, Gillin C, Moore A, Wallace B.M., Loewenstein RJ, Murphy DL (1982) The sleep of patients with obsessive-compulsive disorder. Arch Gen Psychiatr 39:1372–1377

Jindal RD, Thase ME, Fasiczka AL, Friedman ES, Buysse DJ, Frank E, Kupfer DJ (2002) Electroencephalographic sleep profiles in single-episode and recurrent unipolar forms of major depression: II. Comparison during remission. Biol Psychiatr 51:230–236

Katz JL, Kuperberg A, Pollack CP, Walsh BT, Zumoff B, Weiner H (1984) Is there a relationship between eating disorder and affective disorder? New evidence from sleep recordings. Am J Psychiatr 141:753–759

Kimura M, Müller-Preuss P, Wiesner E, Lu A, Deussing JM (2006) Site-specific CRH overexpression alters sleep in transgenic mice. J Sleep Res 15 (Suppl 1): 101

Kimura M, Müller-Preuss P, Wiesner E, Flachskamm C, Lu A, Deussing JM (2007) REM sleep enhancement in CNS-specific CRH-overexpressing mice. Neurosci Res 58 (Suppl 1):S32

Kimura M, Müller-Preuss P, Wiesner E, Höhne A, Holsboer H (in press) Depression and sleep: A comparative study on EEG activity after different antidepressant treatment in mice. J Sleep Res

Kluge M, Schössler P, Steiger A (2007) Duloxetine increases stage 3 sleep and suppresses rapid eye movement (REM) sleep in patients with major depression. Eur Neuropsychopharmacol 17:527–531

Könzel HE, Murck H, Held K, Ziegenbein M, Steiger A (2004) Reboxetine induces similar sleep EEG changes like SSRI`s in patients with depression. Pharmacopsychiatry 37:193–195

Kupfer DJ (1976) REM latency: A psychobiologic marker for primary depressive disease. Biol Psychiatr 11:159–174

Kupfer DJ, Foster FG (1972) Interval between onset of sleep and rapid-eye-movement sleep as an indicator of depression. Lancet 2:684–686

Kupfer DJ, Foster FG, Reich L, Thompson SK, Weiss B (1976) EEG sleep changes as predictors in depression. Am J Psychiatr 133:622–626

Kupfer DJ, Coble P, Kane J, Petti T, Conners CK (1979) Imipramine and EEG sleep in children with depressive symptoms. Psychopharmacology 60:117–123

Kupfer DJ, Spiker DG, Rossi A, Coble PA, Shaw D, Ulrich R (1982) Nortriptyline and EEG sleep in depressed patients. Biol Psychiatr 17:535–546

Kupfer DJ, Ulrich RF, Coble PA, Jarrett DB, Grochocinski VJ, Doman J, Matthews G, Borbély AA (1984) Application of automated REM and slow wave sleep analysis: II. Testing the assumptions of the two-process model of sleep regulation in normal and depressed subjects. Psychiatr Res 13:335–343

Kupfer DJ, Reynolds CF, III, Grochocinski VJ, Ulrich RF, McEachran A (1986a) Aspects of short REM latency in affective states: a revisit. Psychiatr Res 17:49–59

Kupfer DJ, Reynolds CF3, Ulrich RF, Grochocinski VJ (1986b) Comparison of automated REM and slow-wave sleep analysis in young and middle-aged depressed subjects. Biol Psychiatr 21:189–200

Kupfer DJ, Ehlers CL, Frank E, Grochocinski VJ, McEachran AB (1991) EEG sleep profiles and recurrent depression. Biol Psychiatr 30:641–655

Lahmeyer HW, Poznanski EO, Bellur SN (1983) EEG sleep in depressed adolescents. Am J Psychiatr 140:1150–1153

Lancel M, Müller-Preuss P, Wigger A, Landgraf R, Holsboer F (2002) The CRH1 receptor antagonist R121919 attenuates stress-elicited sleep disturbances in rats, particularly in those with high innate anxiety. J Psychiatr Res 36:197–208

Landolt HP, de Boer LP (2001) Effect of chronic phenelzine treatment on REM sleep: report of three patients. Neuropsychopharmacology 25 (Suppl 5):S63–S67

Landolt HP, Raimo EB, Schnierow BJ, Kelsoe JR, Rapaport MH, Gillin JC (2001) Sleep and sleep electroencephalogram in depressed patients treated with phenelzine. Arch Gen Psychiatr 58:268–276

Lauer CJ, Krieg JC, Riemann D, Zulley J, Berger M (1990) A polysomnographic study in young psychiatric inpatients: major depression, anorexia nervosa, bulimia nervosa. J Affect Disord 18:235–245

Lauer C, Riemann D, Wiegand M, Berger M (1991) From early to late adulthood. Changes in EEG sleep of depressed patients and healthy volunteers. Biol Psychiatr 29:979–993

Lauer CJ, Krieg JC, Garcia-Borreguero D, Özdaglar A, Holsboer F (1992) Panic disorder and major depression: A comparative electroencephalogramic sleep study. Psychiatr Res 44:41–54

Lauer CJ, Schreiber W, Holsboer F, Krieg JC (1995) In quest of identifying vulnerability markers for psychiatric disorders by all-night polysomnography. Arch Gen Psychiatr 52:145–153

Lauer CJ, Modell S, Schreiber W, Krieg JC, Holsboer F (2004) Prediction of the development of a first major depressive episode with a rapid eye movement sleep induction test using the cholinergic agonist RS 86. J Clin Psychopharmacol 24:356–357

Linkowski P, Mendlewicz J, Kerkhofs M, Leclercq R, Golstein J, Brasseur M, Copinschi G, Van Cauter E (1987) 24-hour profiles of adrenocorticotropin, cortisol, and growth hormone in major depressive illness: effect of antidepressant treatment. J Clin Endocrinol Metab 65:141–152

Lu A, Steiner MA, Whittle N, Vogl AM, Walser SM, Ableitner M, Refojo D, Ekker M, Rubenstein JL, Stalla GK, Singewald N, Holsboer F, Wotjak CT, Wurst W, Deussing JM (in press) Conditional mouse mutants highlight mechanisms of corticotropin-releasing hormone effects on stress-coping behavior. Mol Psychiatry

Marinesco S, Bonnet C, Cespuglio R (1999) Influence of stress duration on the sleep rebound induced by immobilization in the rat: a possible role for corticosterone. Neuroscience 92:921–933

Modell S, Ising M, Holsboer F, Lauer CJ (2002) The Munich vulnerability study on affective disorders: Stability of polysomnographic findings over time. Biol Psychiatr 52:430–437

Modell S, Ising M, Holsboer F, Lauer CJ (2005) The Munich vulnerability study on affective disorders: premorbid polysomnographic profile of affected high-risk probands. Biol Psychiatr 58:694–699

Murck H, Frieboes RM, Antonijevic IA, Steiger A (2001) Distinct temporal pattern of the effects of the combined serotonin-reuptake inhibitor and 5-HT1A agonist EMD 68843 on the sleep EEG in healthy men. Psychopharmacology 155:187–192

Murck H, Nickel T, Könzel H, Antonijevic IA, Schill J, Zobel A, Steiger A, Sonntag A, Holsboer F (2003) State markers of depression in sleep EEG: Dependency on drug and gender in patients treated with tianeptine or paroxetine. Neuropsychopharmacology 28:348–358

Nofzinger EA, Reynolds CF3, Thase ME, Frank E, Jennings JR, Fasiczka AL, Sullivan LR, Kupfer DJ (1995) REM sleep enhancement by bupropion in depressed men. Am J Psychiatr 152:274–276

Opp M, Obál F, Jr, Krueger JM (1989) Corticotropin-releasing factor attenuates interleukin 1-induced sleep and fever in rabbits. Am J Physiol 257:R528–R535

Overstreet DH, Friedman E, Mathé AA, Yadid G (2005) The Flinders Sensitive Line rat: a selectively bred putative animal model of depression. Neurosci Biobehav Rev 29:739–759

Passouant P, Cadilhac J, Billiard M, Besset A (1973) La suppression du sommeil paradoxal par la clomipramine. Thérapie 28:379–392

Puig-Antich J, Goetz R, Hanlon C, Davies M, Thompson J, Chambers WJ, Tabrizi MA, Weitzman ED (1982) Sleep architecture and REM sleep measures in prepubertal children with major depression: a controlled study. Arch Gen Psychiatr 39:932–939

Rechtschaffen A, Kales A (1968) A Manual of Standardized Terminology, Techniques and Scoring System for Sleep Stages of Human Subjects. US Department of Health, Education & Welfare, Neurological Information Network, Bethesda, MD

Reich L, Weiss BL, Coble P, McPartland R, Kupfer DJ (1975) Sleep disturbance in schizophrenia. A revisit. Arch Gen Psychiatr 32:51–55

Reynolds CF3, Kupfer DJ (1987) Sleep research in affective illness: state of the art circa 1987. Sleep 10:199–215

Riemann D, Berger M, (1990) The effects of total sleep deprivation and subsequent treatment with clomipramine on depressive symptoms and sleep electroencephalography in patients with a major depressive disorder. Acta Psychiatr Scand 81:24–31

Riemann D, Hohagen F, Lauer C, Berger M (1991) Longterm evolution of sleep in depression. In: Smirne S, Franceschi, M, Ferini-Strambi, L (eds) Sleep and aging. Masson, Paris, pp 195–204

Ruigt GSF, Kemp B, Groenhout CM, Kamphuisen HA (1990) Effect of the antidepressant Org 3770 on human sleep. Eur J Clin Pharmacol 38:551–554

Rush AJ, Erman MK, Giles DE, Schlesser MA, Carpenter G, Vasavada N, Roffwarg HP (1986) Polysomnographic findings in recently drug-free and clinically remitted depressed patients. Arch Gen Psychiatr 43:878–884

Rush AJ, Giles DE, Jarrett RB, Feldman-Koffler F, Debus JR, Weissenburger J, Orsulak PJ, Roffwarg HP (1989) Reduced REM latency predicts response to tricyclic medication in depressed outpatients. Biol Psychiatr 26:61–72

Saletu B, Frey R, Krupka M, Anderer P, Grunberger J, See WR (1991) Sleep laboratory studies on the single-dose effects of serotonin reuptake inhibitors paroxetine and fluoxetine on human sleep and awakening qualities. Sleep 14:439–447

Sánchez C, Brennum LT, Stórustovu S, Kreilgård, Mørk A (2007) Depression and poor sleep: The effect of monoaminergic antidepressants in a pre-clinical model in rats. Pharmacol Biochem Behav 86:468–476

Schmid DA, Wichniak A, Uhr M, Ising M, Brunner H, Held K, Weikel J, Sonntag A, Steiger A (2006) Changes of sleep architecture, spectral composition of sleep EEG, the nocturnal secretion of cortisol, ACTH, GH, prolactin, melatonin, ghrelin and leptin and the DEX-CRH test in depressed patients during treatment with mirtazapine. Neuropsychopharmacology 31:832–844

Schmidt HS, Nofzinger EA, (1988) Short REM latency in impotence without depression. Biol Psychiatr 24:25–32

Schreiber W, Lauer CJ, Krumrey K, Holsboer F, Krieg JC (1992) Cholinergic REM sleep induction test in subjects at high risk for psychiatric disorders. Biol Psychiatr 32:79–90

Schulz H, Lund R, Doerr P (1978) The measurement of change in sleep during depression and remission. Arch Psychiatr Nervenkr 225:233–245

Schulz H, Lund R, Cording C, Dirlich G (1979) Bimodal distribution of REM sleep latencies in depression. Biol Psychiatr 14:595–600

Sharpley AL, Williamson DJ, Attenburrow ME, Pearson G, Sargent P, Cowen PJ (1996) The effects of paroxetine and nefazodone on sleep: a placebo controlled trial. Psychopharmacology 126:50–54

Shipley JE, Kupfer DJ, Dealy RS, Griffin SJ, Coble PA, McEachran AB, Grochocinski VJ (1984) Differential effects of amitriptyline and of zimelidine on the sleep electroencephalogram of depressed patients. Clin Pharmacol Ther 36:251–259
Shipley JE, Kupfer DJ, Griffin SJ, Dealy RS, Coble PA, McEachran AB, Grochocinski VJ, Ulrich R, Perel JM (1985) Comparison of effects of desipramine and amitriptyline on EEG sleep of depressed patients. Psychopharmacology 85:14–22
Shiromani PJ, Overstreet DH, Levy D, Goodrich CA, Campbell SS, Gillin JC (1988) Increased REM sleep in rats selectively bred for cholinergic hyperactivity. Neuropsychopharmacology1:127–133
Smith MI, Piper DC, Duxon MA, Upton N (2007) Effect of SB-243213, a selective 5-HT_{2C} receptor antagonist, on the rat sleep profile: A comparison to paroxetine. Pharmacol Biochem Behav 71:599–605
Sonntag A, Rothe B, Guldner J, Yassouridis A, Holsboer F, Steiger A (1996) Trimipramine and imipramine exert different effects on the sleep EEG and on nocturnal hormone secretion during treatment of major depression. Depression 4:1–13
Spiker DG, Coble P, Cofsky J, Foster FG, Kupfer DJ (1978) EEG sleep and severity of depression. Biol Psychiatr 13:485–488
Steiger A (1988) Effects of clomipramine on sleep-EEG and nocturnal penile tumescence (NPT), a long-term study in a healthy man. J Clin Psychopharmacol 8:349–354
Steiger A, von Bardeleben U, Herth T, Holsboer F (1989) Sleep EEG and nocturnal secretion of cortisol and growth hormone in male patients with endogenous depression before treatment and after recovery. J Affect Disord 16:189–195
Steiger A, Gerken A, Benkert O, Holsboer F (1993a) Differential effects of the enantiomers R(-) and S(+) oxaprotiline on major endogenous depression, the sleep EEG and neuroendocrine secretion: studies on depressed patients and normal controls. Eur Neuropsychopharmacol 3:117–126
Steiger A, von Bardeleben U, Guldner J, Lauer C, Rothe B, Holsboer F (1993b) The sleep EEG and nocturnal hormonal secretion. Studies on changes during the course of depression and on effects of CNS- active drugs. Prog Neuropsychopharmacol Biol Psychiatr 17:125–137
Steiger A, Benkert O, Holsboer F (1994) Effects of long-term treatment with the MAO-A inhibitor moclobemide on sleep EEG and nocturnal hormonal secretion in normal men. Neuropsychobiology 30:101–105
Svendsen K, Christensen PG (1981) Duration of REM sleep latency as predictor of effect of antidepressant therapy. A preliminary report. Acta Psychiatr Scand 64:238–243
Taub JM, Hawkins DR, Van de Castle RL (1978) Electrographic analysis of the sleep cycle in young depressed patients. Biol Psychol 7:203–214
Thase ME, Kupfer DJ, Buysse DJ, Frank E, Simons AD, McEachran AB, Rashid KF, Grochocinski VJ (1995) Electroencephalographic sleep profiles in single-episode and recurrent unipolar forms of major depression: I. Comparison during acute depressive states. Biol Psychiatr 38:506–515
Thase ME, Simons AD, Reynolds CF, III (1996) Abnormal electroencephalographic sleep profiles in major depression: association with response to cognitive behavior therapy. Arch Gen Psychiatr 53:99–108
Thase ME, Kupfer DJ, Fasiczka AJ, Buysse DJ, Simons AD, Frank E (1997) Identifying an abnormal electroencephalographic sleep profile to characterize major depressive disorder. Biol Psychiatr 41:964–973
Tiba PA, Tufik S, Suchecki D (2004) Effects of maternal separation on baseline sleep and cold stress-induced sleep rebound in adult Wister rats. Sleep 27:1146–1153
Uhde TW, Roy-Byrne P, Gillin JC, Mendelson WB, Boulenger JP, Vittone BJ, Post RM (1984) The sleep of patients with panic disorder: a preliminary report. Psychiatr Res 12:251–259
Vogel GW (1983) Evidence for REM sleep deprivation as the mechanism of action of antidepressant drugs. Prog Neuropsychopharmacol Biol Psychiatr 7:343–349
Vogel GW, Thurmond A, Gibbons P, Sloan K, Walker M (1975) REM sleep reduction effects on depression syndromes. Arch Gen Psychiatr 32:765–777

Wiegand M, Berger M, Zulley J, von Zerssen D (1986) The effect of trimipramine on sleep in patients with major depressive disorder. Pharmacopsychiatry 19:198–199

Winokur A, DeMartinis NA, III, McNally DP, Gary EM, Cormier JL, Gary KA (2003) Comparative effects of mirtazapine and fluoxetine on sleep physiology measures in patients with major depression and insomnia. J Clin Psychiat 64:1224–1229

Wyatt RJ, Fram DH, Kupfer DJ, Snyder F (1971) Total prolonged drug-induced REM sleep suppression in anxious depressed patients. Arch Gen Psychiatr 24:145–155

Zarcone VP, Jr., Benson KL, Berger PA (1987) Abnormal rapid eye movement latencies in schizophrenia. Arch Gen Psychiatry 44:45–48

Chapter 13
Strategies to Identify Biomarkers for Depression

Marcelo Paez-Pereda(✉) and Markus Panhuysen

Abstract Usually, disease biomarkers are directly linked to the biological processes that are involved in the causation and progress of the condition. However, in the case of depression and other psychiatric disorders, the mechanisms responsible for the disease are often poorly understood. Factors that contribute to the difficulty of finding biomarkers for psychiatric diseases include the heterogeneity of each psychiatric disease, the complexity of diagnosis, and the difficult accessibility to biological samples of affected tissues. Here, we present possible approaches for the identification of depression biomarkers. The first, a knowledge-based approach, makes use of our current understanding of the underlying disease mechanisms. As an example, the stress response cascade provides a source for numerous candidate biomarkers. Second, reverse-engineering the biological effects and signaling mechanisms of known antidepressants is another promising way to obtain useful biomarkers. Third, biomarkers could be identified by means of unbiased research. The different "omics" technologies have recently yielded promising results in comparing biological parameters of depressive patients vs. control subjects. However, many of these studies only show an increased mean value for a candidate biomarker in depressive patients, while correlative and longitudinal studies are rare. The validity of these approaches remains to be confirmed by further studies in clinical and epidemiological settings using larger numbers of samples.

Abbreviations ACTH: Adrenocorticotropic hormone; BDNF: Brain-derived neurotrophic factor; cAMP: Cyclic adenosine monophosphate; CNS: Central nervous system; Cox-2: Cyclooxigenase 2; CREB: cAMP responsive element binding protein; CRH: Corticotropin-releasing hormone; CRHR: CRH receptor; CRP: C-reactive protein; CSF: Cerebrospinal fluid; EEG: Electroencephalography; EPO: Erythropoietin; FKBP5: FK506-binding immunophilin 5; fMRI: Functional magnetic resonance imaging; GR: Glucocorticoid

M. Paez-Pereda
Affectis Pharmaceuticals, Kraepelinstrasse 2, 80804 Munich, Germany,
paez-pereda@affectis.com

C.W. Turck (ed.) *Biomarkers for Psychiatric Disorders*,
doi: 10.1007/978-0-387-79251-4_13, © Springer Science+Business Media, LLC 2008

receptor; HPA: Hypothalamic–pituitary–adrenal axis; HPT: Hypothalamic–pituitary–thyroid axis; hTPH2: Human tryptophan hydroxylase 2; IL-1: Interleukin-1; IL-6: Interleukin-6; IRS: Inflammatory response system; MR: Mineralocorticoid receptor; MRI: Magnetic resonance imaging; NMDA: *N*-methyl-D-aspartate; NT-proBNP: *N*-terminal pro-B-type natriuretic peptide; P2X7: Purinergic receptor, ligand-gated ion channel 7; PGE2: Prostaglandin E2; SNP: Single nucleotide polymorphism; T3: Triiodothyronin; TNF-α: Tumor necrosis factor alpha; TSH: Thyroid-stimulating hormone.

13.1 Introduction

Disease biomarkers can be defined as observations that either have diagnostic or predictive value in relation to an illness, or allow estimating a patient's susceptibility to a specific treatment. Biomarkers can potentially be found in any biomedical context, and could be classified accordingly as genomic, transcriptomic, proteomic, or metabolomic. In addition, other key types of biomarkers are conceivable: e.g., imaging biomarkers, obtained by the use of techniques such as functional magnetic resonance imaging (fMRI). Especially in the context of polygenic diseases, a valid biomarker may turn out to be a combination of different observations rather than a single one. These kinds of combinatorial biomarkers, also called *signatures*, are to date best established in – but definitely not limited to – the area of gene expression profiling.

Zooming in on psychiatric diseases, the availability and usefulness of different types of biomarkers vary (see Table13.1).

Genomic biomarkers such as small nucleotide polymorphisms (SNPs) are readily available and may turn out to be very informative. Transcriptomic biomarkers can be identified only via comparative expression profiling of given tissues. Owing to its limited availability, expression profiling of the central nervous system (CNS) tissue is not a choice for biomarker identification. However, since the brain–immune–endocrine network is believed to be tightly interdependent, gene expression changes measured in peripheral blood leucocytes could be an alternative. Similar restrictions exist for proteomic biomarkers, with the exception of the cerebrospinal fluid (CSF). CSF is produced by cells in the choroid plexus and contains about 0.3% plasma proteins (Felgenhauer, 1974). Changes of CSF plasma protein and neurotransmitter

Table13. 1 Types of biomarkers and their availability for psychiatric diseases

Type	indicator	availability
Genomic	Small nucleotide polymorphisms	Excellent
Transcriptomic	Change in gene expression, measured via transcript abundance	Very limited (blood)
Proteomic	Change in protein concentration or modification	Limited (CSF, blood)
Metabolomic	Enzyme activity, concentration of metabolites or other molecules	Good (blood plasma, CSF, urine, saliva, feces)

concentration may represent ongoing processes in the brain much better than observations obtained by any other sampling method and could therefore be suitable as a source for depression biomarkers. Unfortunately, CSF can be obtained only by a highly invasive method called *lumbar puncture*, and the process of sample collection itself could – due to its stressful character – cause changes in the patient's CSF composition. Metabolomic biomarkers were already in use in the pregenomic era. They can be obtained, e.g., from blood, CSF, urine, saliva, or feces. Paracrine- or autocrine-acting molecules within the brain, which otherwise would be a rich source for disease biomarkers, are out of reach, if they do not spill over into the CSF or blood. Additionally, changes in the blood, CSF, urine, etc., do not specifically reflect events in brain areas involved in depression.

The use of biomarkers in psychiatry is not an invention of our days, even though in the past, without help of the modern "omics" technologies, attempts to develop diagnostic tools have sometimes gone wrong. As an example, health professionals in 1912 invented a "blood test for madness", an immunoserodiagnostic procedure that was spuriously believed for years to identify patients with dementia praecox (Noll, 2006). Interestingly, state-of-the-art expression-profiling techniques allowed Tsuang and colleagues to identify potential disease signatures for schizophrenia and bipolar disorder in the blood of patients (Tsuang et al., 2005). However, their results have already been put into perspective by a follow-up study (Yao et al., 2007), underscoring how big the challenge of valid biomarker identification still is.

13.2 Identification of Biomarkers Based on Disease Mechanism

One possible approach to search for biomarkers is to first define the altered biological processes that cause the disease and are affected during its course. Once the affected biological processes are identified, single parameters representative of these alterations can be chosen as biomarkers depending on their accessibility, possibility of quantification, sensitivity, and predictive value. Most psychiatric diseases do not have defined single causes; they affect different biological processes simultaneously and they overlap with other conditions resulting in a highly complex diagnosis (Kraemer et al., 2002; Shaw et al., 2007). In particular, in depression, the diagnostic procedures themselves are phenomenological in nature and difficult to quantify. The situation is further complicated by the difficult accessibility of the affected tissue in depressed patients. There is, however, reasonable consensus about some biological processes that are associated with major depression although they might be differently affected in different individuals. These processes involve, for example, the neurotransmitter system, the neuroendocrine axes – and above all the hypothalamus–pituitary–adrenal axis (HPA) that regulates the response to stress – or production of endogenous neurotrophins. Increasing evidence also points towards involvement of the inflammatory response system (IRS) in the etiology of major depression (van and Maes, 1999). In addition, several other processes have been associated with depression, such as several neuropeptide signaling pathways. All these processes could be used as a starting point to identify biomarkers for depression.

13.2.1 The Corticosteroid Hypothesis of Depression

Although depression is not associated to any well-defined lesion in the brain in contrast to neurologic diseases such as Parkinson's or Alzheimer's disease, it is likely that malfunction of the circuits involving the prefrontal and cingulate cortex, amygdala, and hippocampus mediates the diverse symptoms of the disease. Classical studies in depression have found alterations of cognitive parameters besides the most obvious emotional manifestation of the disease (Mathews and MacLeod, 2005). One particular biological process that is involved in the regulation of emotion and cognition is the response to stress. It is not surprising, therefore, that the brain areas involved in the control of the stress response also show alterations of diverse parameters in depression (Drevets, 2003). An abnormal, excessive activation of the stress response measured by the production of hormones of the HPA axis – corticotropin-releasing hormone (CRH), adrenocorticotropic hormone (ACTH), and cortisol – has been consistently observed in depressed patients (Nemeroff, 1988; de Kloet et al., 2005). Some patients with depression have elevated cortisol levels and cortisol is not suppressed by dexamethasone administration as it happens in normal individuals. Moreover, the ACTH response to intravenous CRH is blunted as compared to the increase normally seen in healthy individuals (Ising et al., 2005). These biochemical parameters are among the most consistent observations in patients with depression and can be considered good biomarkers for the disease. Other clinical facts also support the use of HPA alterations as biomarkers of depression; after effective antidepressant treatment in responsive patients, the HPA alterations are normalized (Ising et al., 2005). This normalization of the HPA axis happens before the clinical symptoms show improvement, and individual patients with a normalized HPA axis are less likely to relapse. These observations add to the potential predictive value of the HPA axis alterations as depression biomarkers. More detailed clinical studies will determine the real positive and negative predictive value of HPA alterations as depression biomarkers. Whether HPA axis alterations are a primary cause of depression remains controversial but these alterations certainly play a role in the generation of many symptoms of depression and they also have an impact in the course of the disease. However, only a proportion of depressed patients show elevated cortisol levels and HPA axis alterations and this limits the value of these biochemical parameters as universal markers of depression.

13.2.2 The Hypothalamus–Pituitary–Thyroid Axis in Depression

Patients suffering from hypothyroidism also frequently show depressive symptoms (Loosen, 1986). Furthermore, the antidepressant effect of tricyclic antidepressants is accelerated in women, and patients seemingly unresponsive to treatment with tricyclics can be turned into responders by co-administration of small amounts of thyroid hormone (Prange and Loosen, 1980). Despite the fact that most patients with major depression have a normal thyroid gland function (Fountoulakis et al., 2006),

T3 concentration may have potential as a prognostic biomarker, since its plasma concentration was shown to be inversely related to the time to recurrence of major depression (Joffe and Marriott, 2000). Significant reductions in nocturnal serum TSH and serum T3 concentrations have also been observed in depressed patients (Rubin et al., 1987), further supporting T3 as a candidate disease biomarker.

13.2.3 The Neurotrophin Hypothesis of Depression

One of the consequences of the pathologic response to stress, in particular the elevated cortisol levels, is the damage of hippocampal neurons, which involves the reduction of dendritic spines and branching (Joels et al., 2004; McEwen et al., 2002). In some models of chronic stress and depression, high glucocorticoid levels result in neuronal cell death in the hippocampus. Although still controversial, studies have reported a reduction of the hippocampus size in depressed patients, which could be the result of cell death and changes in neuroplasticity in this brain area (Geuze et al., 2005). Among the molecular events resulting from chronic stress and high glucocorticoid levels in the hippocampus is the reduction of the production of neurotrophins, mainly brain-derived neurotrophic factor (BDNF) (Nestler et al., 2002; Manji et al., 2003). These reduced neurotrophin levels could explain the reduction in the number of dendrites and synapses in this brain area. A reduction of BDNF in the serum of patients with depression has been reported but this reduction is not only associated to depression and has been observed in stressed individuals and in neurological diseases (Nestler et al., 2002). This lack of selectivity precludes the use of BDNF as a biomarker for depression.

One of the responses to brain injury and neuronal apoptosis is the production of erythropoietin, which mainly happens in the brain under hypoxic conditions. Erythropoietin has been repeatedly shown to have potent neuroprotective effects in different models (Brines and Cerami, 2006). Therefore, the release of erythropoietin (EPO) after insult to brain cells could constitute a compensatory mechanism to prevent further damage. EPO can be readily measured in the CSF and has been studied in neurological and psychiatric diseases. Interestingly, depressed patients show elevated levels of EPO in the CSF (Nakamura et al., 1998). Therefore, EPO could be an interesting biomarker for depression. However, it still remains to be shown how reproducible the changes in EPO levels in the CSF are.

13.2.4 Neurogenic Inflammatory Mediators as Depression Biomarkers

On the basis of the scientific evidence supporting an involvement of inflammatory mediators in depression (van and Maes, 1999), many correlative studies have focused on the levels of prostaglandins (PGs) and interleukins (ILs) in depressive patients in comparison to normal subjects. For example, one of the most consistent findings is that IL-6, C-reactive protein (CRP), and TNF-α are increased in the plasma and CSF of a subgroup of depressive patients (Maes, 1995). IL-1 β levels have also been found to be altered in the CSF of depressed patients (Levine et al., 1999). PGE2 and thromboxane B2 are increased in the plasma of depressive patients (Lieb et al., 1983). The production of PGE2 is also increased in the lymphocytes of depressed patients, and the salivary concentration of PGE2 is significantly higher than normal. On the other hand, antidepressants inhibit PGE2 synthesis. These results indicate that PGE2 and other inflammatory mediators found in plasma and CSF could be useful biomarkers for depression. These biomarkers could be used not only to improve the diagnosis of depression but also to follow the course of the disease during clinical trials and in clinical setups. The accessibility to plasma and saliva, in which these inflammatory mediators can be measured, makes them attractive candidate biomarkers that could easily be used routinely in the psychiatric clinic. The fact that these molecules are increased only in subgroups of patients and the heterogeneity of the previous studies make more detailed studies necessary in order to robustly define this subgroup in terms of clinical symptoms and alteration of physiological mechanisms. In particular, the normal variation of cytokine levels such as IL-6 and TNF-α in healthy subjects and in pathological states not related to depression needs further characterization (Pollmacher et al., 2002).

The relevance of these markers in depression could also have an impact in the better overall care for the patient. For example, depression and increased inflammatory biomarkers are proposed to be risk factors for coronary heart disease. However, the mechanistic bases for this covariance in parameters of depression, inflammation, and coronary heart disease are not known (Shimbo et al., 2005). A marker of cardiovascular risk, *N*-terminal pro-B-type natriuretic peptide(NT-proBNP), has also been found elevated in depressed patients (Politi et al., 2007). Another field in which inflammatory biomarkers in combination with depression are relatively frequent is cancer (Jehn et al., 2006). It has been found that patients with cancer-related depression and high inflammatory markers also present cachexia and anorexia (Illman et al., 2005).

Cox-2 is a key molecule activated in macrophages and microglia during the activation of the inflammatory pathway (Yiangou et al., 2006). Therefore, it is expected that Cox-2 and PGE2 could also participate in the pathways that contribute to the symptoms of depression (Muller et al., 2004). In line with this hypothesis, selective Cox-2 inhibitors have been shown to be efficacious in the treatment of psychiatric diseases such as schizophrenia and depression in human patients (Muller et al., 2004, 2006). In these studies, the main therapeutic effects were observed in cognition, which fits with the localization of the neurogenic inflammatory mediators in the brain. Altogether, these genetic data and pathway analyses point to neuroinflammatory mediators as good candidates for biomarkers of depression and schizophrenia.

Recent evidence points to a pivotal role of increased glutamate neurotransmission in the pathogenesis of depression besides the alterations in serotonergic neurotransmission and the HPA axis. For example, magnetic resonance studies have found increased glutamate levels in the cortex in depressed patients (Kugaya and Sanacora, 2005). Consistent with a model of excessive glutamatergic neurotransmission in depression, several antiglutamatergic agents, such as riluzole, lamotrigine, and ketamine, have antidepressant effects (Du et al., 2007; Zarate et al., 2006). Glial cells play a primary role in synaptic glutamate removal. Therefore, alterations in glial cell metabolism may explain the increase in glutamate neurotransmission (Kugaya and Sanacora, 2005). One of the consequences of increased glutamatergic neurotransmission is the overexpression of Cox-2, which results in increased PGE2 levels. In agreement with the increase of these inflammatory mediators, an immune activation including increased production of proinflammatory cytokines has repeatedly been described in patients with major depression (Dunn et al., 2005). These neuroinflammatory mediators are known to alter cognition in rodents and humans (Rafnsson et al., 2007). It is also well known that infection and inflammatory mediators trigger sickness behavior, which bears similarities to some symptoms of depression such as reduced motivation, fatigue, sleep disturbances, etc. (Muller and Schwarz, 2007). Therefore, it is tempting to speculate that neuroinflammatory mediators could play a part in the pathogenesis of depression. This mechanism would consist in a pathologically high glutamatergic neurotransmission triggered by an exacerbated response to stress. This glutamate receptor activation, mainly in the hippocampus and amygdala, would produce plastic changes such as reduction in the number of hippocampal dendrites. At the functional level, the glutamatergic activation would produce cognitive and emotional alterations that would eventually result in the symptoms of depression.

13.2.5 Other Mechanisms Altered in Depression

The cyclic nature of depression is manifested not only by the recurrence of the symptoms in long timescales but also through the diurnal variations in its symptomatology. One consistent finding in a subgroup of depressive patients is the alteration of the sleep–wake rhythm and characteristic changes in sleep architecture that result in reduced sleep quality (Lam, 2006). The cyclic changes in core body temperature also show disturbances (Srinivasan et al., 2006). All these alterations suggest that the circadian time-keeping mechanism is related to the pathophysiology of depression. This hypothesis led to the study of melatonin, a hormone that varies in a diurnal cycle, in depressive patients (Srinivasan et al., 2006). Melatonin can be detected in saliva and plasma, and its main metabolite, 6-sulfatoxymelatonin, can be detected in urine. Melatonin levels show significant alterations in depressive patients during the acute phase of the illness. Not only are the levels of melatonin altered, but also the timing on melatonin secretion during the diurnal cycle (Srinivasan et al., 2006). However, normative values for melatonin levels in normal subjects and depressive patients still need to be defined in larger studies. In particular, in seasonal affective disorder, melatonin level seems to be a good candidate for a biomarker (Lewy et al., 2006). Antidepressant treatment changes melatonin levels and the administration of melatonin could have therapeutic effects in some types of mood disorder (McClung, 2007). This strengthens the link between mood and melatonin levels during the circadian cycle. Furthermore, sleep deprivation, a treatment that is currently employed to treat mood disorders, is thought to act by shifting or resetting the circadian clock. Although the therapeutic effect is not sustained, it is effective in a large number of patients for a very short period.

In the last decades, many proteins that control the circadian rhythm and different parameters of sleep have been identified. Initial studies have unveiled relationships between mutations and markers in these genes and psychiatric disease (McClung, 2007; Bunney and Bunney, 2000).

13.3 Identification of Biomarkers Based on Antidepressants' Mechanism of Action

13.3.1 Effects of Antidepressants on Neurotransmission

Although there is only a limited knowledge of the causative mechanisms of depression, there are effective pharmacological treatments that have been discovered by chance. Tricyclic compounds were the first substances that produced noticeable antidepressant effects. Their mechanism of action is not fully understood, but they have been characterized as inhibitors of the reuptake of serotonin. Through this mechanism, antidepressants increase the levels of serotonin in the synapse, which results in plastic changes in neurotransmission in the long term. Second-generation compounds developed on the basis of the selective inhibition of the serotonin transporter also proved to be efficacious antidepressants in a significant proportion of depressed patients. Monoamine oxidase inhibitors act through an analogous mechanism by elevating monoamine levels. However, the elevation of serotonin and other monoamines happen shortly after the first administration of antidepressants but the improvement of the symptoms of depressions takes several weeks to develop. This indicates that the mechanism of action of antidepressants is complex and probably involves different sequential steps.

On the basis of the mechanism of action of antidepressants, a dysregulation of central serotonin neurotransmission produced by genetic factors has been suspected as a possible cause of major depression. Many studies have been conducted to substantiate this hypothesis. For example, an SNP in human tryptophan hydroxylase-2 (hTPH2), the rate-limiting enzyme of neuronal serotonin synthesis, has been found with a higher frequency in depressed patients as compared to control subjects (Zhang et al., 2005). This functional change in the hTPH2 sequence results in approximately 80% lower serotonin production. This loss-of-function mutation in hTPH2 suggests that defects in brain serotonin synthesis may represent an important risk factor for depression. Associations between hTPH2 polymorphisms and unipolar depression, bipolar disorder, and suicidal behavior have been reported. However, other studies carried out in independent population samples have not found an association between this SNP in hTPH2 and depression (Delorme et al., 2006). This inconsistency in the results across studies prevents the use of this mutation as a diagnostic tool for depression.

13.3.2 Effects of Antidepressants on cAMP Signaling

Among the molecular changes produced by all the major classes of antidepressants, one of the most consistent is the activation of the transcription factor cAMP responsive element binding protein (CREB) in several brain regions including the hippocampus (Carlezon et al., 2005). This clearly demonstrates the involvement of the cAMP pathway in the response to antidepressants. CREB is phosphorylated by protein kinases, which are in turn activated by cAMP or intracellular calcium. These events typically happen after the activation of G-protein coupled receptors at the cell membrane. cAMP levels can also be increased by inhibiting its degradation by phosphodiesterases. The enhancing effect of phosphodiesterase inhibitors on antidepressant effectiveness further proves the important role of the cAMP pathway in antidepressant action. One of the consequences of CREB activation by antidepressants is the increase of BDNF transcription. All these observations suggest the following sequence of events: antidepressants activate the cAMP pathway, which leads to CREB activation that then enhances BDNF production; BDNF, in turn, reverts the neuronal damage caused by the excess of cortisol produced by an exacerbated stress response. Despite a high number of reports offering support to this potential mechanism for antidepressant action, this hypothesis remains controversial. The ability of antidepressants to repair neuronal damage does not seem to depend on BDNF, and mice deficient in CREB function that fail to induce BDNF show, however, a behavioral response to antidepressant treatment. On the other hand, none of the pathways and factors involved in this mechanism activated by antidepressants seems to be specific for depression or for the antidepressant response. Except for the secreted neurotrophins, the factors involved in this mechanism are all intracellular. Considering all these limitations, it is not possible to use them as biomarkers.

13.3.3 Effects of Antidepressants on Neurogenesis

One of the biological processes that have also been implicated in the response to antidepressants is neurogenesis. Antidepressant treatment has been shown to increase adult hippocampal neurogenesis (Dranovsky and Hen, 2006; Duman and Monteggia, 2006). It has also been shown that the ablation of neurogenesis in the hippocampus through radiation can impair the behavioral response to antidepressant treatment

(Santarelli et al., 2003). This points to adult neurogenesis as a mediator of antidepressant effects and it would explain the extended time course of the improvement in the symptoms of depression after treatment. The factors that mediate the increase in neurogenesis produced by an antidepressant remain to be clearly identified. On the other hand, stress reduces neurogenesis and this reduction can be prevented by antidepressant treatment indicating a participation of stress hormones (Dranovsky and Hen, 2006). Moreover, the ability of glucocorticoids to block neurogenesis appears to be dependent on stimulation of the *N*-methyl-D-aspartate (NMDA) receptor (Nacher and McEwen, 2006). Therefore, stress can negatively affect neurogenesis at multiple stages and GR, MR, and NMDA receptors are likely to mediate this effect.

Because of the prominent role of glucocorticoids and the HPA axis in the mechanisms involved in depression and antidepressant response, efforts have been made to find genetic markers in the pathways that mediate glurocorticoid action in the brain. For example, Binder and colleagues demonstrated that a functional polymorphism in FKBP5 is associated with an increased recurrence of depressive episodes and a faster response to antidepressant treatment with modulators of monoamines (Binder et al., 2004). FKBP5 is a glucocorticoid receptor chaperone binding protein and the polymorphism results in increased intercellular FKBP5 protein, which modulates the activity of the glucocorticoid receptor and, therefore, the HPA axis response to stress. Accordingly, the FKBP5 polymorphism is also associated with a higher corticotropin release following suppression of the ACTH with dexamethasone in the depressed patients. These data further emphasize the strong link between the HPA axis and the response to antidepressants that regulate monoamine levels. More recently, it has been shown that a polymorphism in the CRH receptor (*CRHR1*) gene is associated with an increased risk to develop a seasonal pattern and an early age of onset of the first depressive episode. On the other hand, genetic variation in the *CRHR2* gene is associated with a weaker or no response to citalopram treatment (Papiol et al., 2007). The presence of these polymorphisms in the genes that control the HPA axis and their association to specific responses to antidepressant treatment indicate that they could be used to predict efficacy before starting a treatment. Such a biomarker would be extremely useful to better design and analyze clinical trials and eventually in a clinical context.

13.4 Identification of Biomarkers by Unbiased Approaches

13.4.1 Genetic Studies: Focus on P2X7

Unbiased approaches have been used to identify depression biomarkers besides the approaches based on the mechanisms affected by the disease. Barden et al. have used a sample from Saguenay-Lac St-Jean, an isolated population with high incidence of bipolar disorder, owing to a higher degree of homogeneity in the diagnosis and genetic background. A genomewide linkage analysis identified chromosomal regions linked to bipolar disorder (Morissette et al., 1999). In some of the families included in this study, which show high incidence of bipolar disorder, a stronger association was found between the disease and a segment of chromosome 12. Other independent studies in different populations have confirmed the presence in this chromosome of genes that increase the risk to develop mood disorders (Abkevich et al., 2003; Hamet and Tremblay, 2005). Within this chromosomal region, the purinergic receptor P2X7 showed an SNP associated with bipolar disorder. This polymorphism produces an amino acid change in the cytoplasmic domain of *P2X7*. Subsequent studies on the same population and on an independent sample from an open caucasian population found an association between the polymorphism in the *P2X7* gene and an increased risk to develop major depression (Lucae et al., 2006). More recently, the same chromosomal region was found to be associated to anxiety disorders (Erhardt et al., 2007). Further studies in independent and bigger samples are necessary to demonstrate whether this mutation in *P2X7* can be a reliable biomarker for susceptibility to major depression.

P2X7 expression in peripheral tissues, mainly in immune cells, is required for IL-1 processing and secretion (Ferrari et al., 2006). Therefore, P2X7 has been associated with inflammation in many studies. In the brain, P2X7 is probably expressed in glial cells, where it could regulate a neurogenic inflammmatory cascade that might eventually result in changes in neurotransmission (Bennett, 2007).

13.4.2 Disease Signatures

Modern technologies allow analyzing gene expression levels and nucleotide polymorphisms in a massively parallel manner. These high-throughput approaches facilitate the unbiased discovery of previously unknown gene expression or SNP patterns associated with specific phenotypes. However, the discovery of disease signatures for psychiatric diseases is still in its infancy. Most gene expression profiling studies in psychiatric research focus on postmortem brain tissue of patients and controls (Sequeira and Turecki, 2006), which can be informative concerning the underlying disease mechanisms, but represent an unfavorable approach for discovery of clinically relevant biomarkers as predictors for disease or treatment susceptibility or as a diagnostic tool. As mentioned earlier, attempts to identify potential biomarkers or disease signatures for psychiatric disorders using lymphocyte expression profiling have been performed. These studies focus on schizophrenia (Tsuang et al., 2005; Glatt et al., 2005; Vawter et al., 2004; Yao et al., 2007; Zvara et al., 2005), with one study including also bipolar disorder (Tsuang et al., 2005). These studies are complicated by the fact that most patients are on medication that could have strong distorting effects on lymphocytic gene expression. Only one of these studies took this into account and used drug-naive patients (Zvara et al., 2005). To our knowledge, no study has so far been published that tried to identify a depression signature or to establish disease subclasses using lymphocyte expression profiling. Also proteomic approaches to discover clinically relevant biomarkers or disease signatures for depression are extremely rare. One study was performed with pooled CSF proteins from patients with major depression, some of whom had attempted suicide (Brunner et al., 2005). In this study, a 33-kD protein was found to be present in the suicide attempters' group, which was not present in the non-attempters' group; however, owing to the limited material, the protein could not be identified.

Since the SNP genotyping technology has become more high throughput, genomewide association studies of genotypes for polygenic diseases are getting more and more attention. Today, most studies still focus on single-SNP associations. The effects of the identified variants are very weak and explain only a small fraction of the genetic component of a disease. Initial feasibility studies for two and more loci analyses have been performed, with promising results (Marchini et al., 2005). Incorporating epistatic effects into association studies could ultimately lead to the identification of genomic signatures for the disease or the susceptibility to develop it.

13.5 Summary and Perspectives

More than 50 years after the first evidence of the involvement of an imbalance of neurochemicals in major depression, there still is a strong unmet need to diagnose and monitor major depression in a reliable, reproducible, and objective way. Diagnostic methods still rely on phenomenology, which strongly depends on the background and training of the evaluator. The abnormalities in biochemical and biological processes that cause the disease are only hypothetical. The main causes for this situation are the difficult access to affected tissue in the brain of depressed patients and the heterogeneity of the disease cause and manifestation. However, productive efforts have been made on the basis of the HPA axis and monoamine pathways as a source for candidate biomarkers for depression. In particular, in the case of the HPA axis, biochemical and genetic tests have been proposed as biomarkers for depression and for the response to antidepressant treatment. In fact, one of the most robust biochemical tests used in depressive patients is the measure of the combined dexamethasone/CRH test. This test provides an estimation of the activity of the corticosteroid receptors in the brain by measuring plasma samples. The hormonal response in this test correlates to disease presentation and to the response to antidepressant treatment. Genotyping of genes related to this hormonal axis as well as to the monoamine pathway could potentially provide the necessary tools to administer patients a personalized treatment. A combination of these biochemical and genetic tests will be necessary to improve the design and evaluation of clinical trials and help develop better and more specific antidepressants. Other promising candidate biomarkers that deserve further study are the ones based on alternative mechanisms for the pathophysiology of depression such as the markers related to cellular stress (such as hypoxia or injury) or to neuroinflammation. Candidate biomarkers such as EPO, IL-6, CRP, PGE2, and other inflammatory mediators can be measured in plasma and other accessible body fluids. They have already been correlated with depression and with the response to antidepressants. Unbiased approaches based on proteomics and metabolomics are likely to provide further interesting candidate biomarkers in the near future. These unbiased approaches could identify protein products, metabolites and small signaling molecules that are probably altered in depressive patients. Studies using a large number of samples, as also longitudinal and transversal studies, are necessary to reach these goals. Most probably, the population of depressive patients will have to be split into subgroups, which manifest changes in these biomarkers in correlation to discrete symptoms of the disease. Electroencephalography (EEG) and MRI offer new opportunities to define robust parameters to be compared among different patient populations. These methods are likely to improve in the near future to qualify as good biomarkers for diagnosing depression and to monitor therapeutic effects. In particular, quantitative EEG and functional brain imaging techniques are suitable for this purpose (Pien et al., 2005). Changes in specific parameters measured by EEG and MRI probably represent biological alterations that are mechanistically related to the cause of the disease and are less prone to the biases implicit in phenomenological diagnosis of the disease. These changes could be, therefore, considered intermediate phenotypes between the genetic and biochemical cause of the disease and the complex psychological and behavioral manifestation of the symptoms of depression.

References

Abkevich V, Camp NJ, Hensel CH, Neff CD, Russell DL, Hughes DC, Plenk AM, Lowry MR, Richards RL, Carter C, Frech GC, Stone S, Rowe K, Chau CA, Cortado K, Hunt A, Luce K, O'Neil G, Poarch J, Potter J, Poulsen GH, Saxton H, Bernat-Sestak M, Thompson V, Gutin A, Skolnick MH, Shattuck D, and Cannon-Albright L (2003) Predisposition locus for major depression at chromosome 12q22–12q23.2. Am J Hum Genet 73:1271–1281

Bennett MR (2007) Synaptic P2X7 receptor regenerative-loop hypothesis for depression. Aust NZ J Psychiatr 41:563–571

Binder EB, Salyakina D, Lichtner P, Wochnik GM, Ising M, Putz B, Papiol S, Seaman S, Lucae S, Kohli MA, Nickel T, Kunzel HE, Fuchs B, Majer M, Pfennig A, Kern N, Brunner J, Modell S, Baghai T, Deiml T, Zill P, Bondy B, Rupprecht R, Messer T, Kohnlein O, Dabitz H, Bruckl T, Muller N, Pfister H, Lieb R, Mueller JC, Lohmussaar E, Strom TM, Bettecken T, Meitinger T, Uhr M, Rein T, Holsboer F, and Muller-Myhsok B (2004) Polymorphisms in FKBP5 are associated with increased recurrence of depressive episodes and rapid response to antidepressant treatment. Nat Genet 36:1319–1325

Bowden NA, Weidenhofer J, Scott RJ, Schall U, Todd J, Michie PT, and Tooney PA (2006) Preliminary investigation of gene expression profiles in peripheral blood lymphocytes in schizophrenia. Schizophr Res 82:175–183

Brines M and Cerami A (2006) Discovering erythropoietin's extra-hematopoietic functions: biology and clinical promise. Kidney Int 70:246–250

Brunner J, Bronisch T, Uhr M, Ising M, Binder E, Holsboer F, and Turck CW (2005) Proteomic analysis of the CSF in unmedicated patients with major depressive disorder reveals alterations in suicide attempters. Eur Arch Psychiatr Clin Neurosci 255:438–440

Bunney WE and Bunney BG (2000) Molecular clock genes in man and lower animals: possible implications for circadian abnormalities in depression. Neuropsychopharmacology 22:335–345

Carlezon WA, Jr., Duman RS, and Nestler EJ (2005) The many faces of CREB. Trends Neurosci 28:436–445

de Kloet ER, Joels M, and Holsboer F (2005) Stress and the brain: from adaptation to disease. Nat Rev Neurosci 6:463–475

Delorme R, Durand CM, Betancur C, Wagner M, Ruhrmann S, Grabe HJ, Nygren G, Gillberg C, Leboyer M, Bourgeron T, Courtet P, Jollant F, Buresi C, Aubry JM, Baud P, Bondolfi G, Bertschy G, Perroud N, and Malafosse A (2006) No human tryptophan hydroxylase-2 gene R441H mutation in a large cohort of psychiatric patients and control subjects. Biol Psychiatr 60:202–203

Dranovsky A, and Hen R (2006) Hippocampal neurogenesis: regulation by stress and antidepressants. Biol Psychiatr 59:1136–1143

Drevets WC (2003) Neuroimaging abnormalities in the amygdala in mood disorders. Ann NY Acad Sci 985:420–444

Du J, Suzuki K, Wei Y, Wang Y, Blumenthal R, Chen Z, Falke C, Zarate CA, Jr and Manji HK (2007) The anticonvulsants lamotrigine, riluzole, and valproate differentially regulate AMPA receptor membrane localization: relationship to clinical effects in mood disorders. Neuropsychopharmacology 32:793–802

Duman RS, and Monteggia LM (2006) A neurotrophic model for stress-related mood disorders. Biol Psychiatr 59:1116–1127

Dunn AJ, Swiergiel AH, and de BR (2005) Cytokines as mediators of depression: what can we learn from animal studies? Neurosci Biobehav Rev 29:891–909

Erhardt A, Lucae S, Unschuld PG, Ising M, Kern N, Salyakina D, Lieb R, Uhr M, Binder EB, Keck ME, Muller-Myhsok B, and Holsboer F (2007) Association of polymorphisms in P2RX7 and CaMKKb with anxiety disorders. J Affect Disord 101:159–168

Felgenhauer K (15–12–1974) Protein size and cerebrospinal fluid composition. Klin Wochenschr 52:1158–1164

Ferrari D, Pizzirani C, Adinolfi E, Lemoli RM, Curti A, Idzko M, Panther E, and Di VF (2006) The P2X7 receptor: a key player in IL-1 processing and release. J Immunol 176:3877–3883

Fountoulakis KN, Kantartzis S, Siamouli M, Panagiotidis P, Kaprinis S, Iacovides A, and Kaprinis G (2006) Peripheral thyroid dysfunction in depression. World J Biol Psychiatr 7:131–137

Geuze E, Vermetten E, and Bremner JD (2005) MR-based in vivo hippocampal volumetrics: 2. Findings in neuropsychiatric disorders. Mol Psychiatr 10:160–184

Glatt SJ, Everall IP, Kremen WS, Corbeil J, Sasik R, Khanlou N, Han M, Liew CC, and Tsuang MT (2005) Comparative gene expression analysis of blood and brain provides concurrent validation of SELENBP1 up-regulation in schizophrenia. Proc Natl Acad Sci USA 102:15533–15538

Hamet P and Tremblay J (2005) Genetics and genomics of depression. Metabolism 54:10–15

Illman J, Corringham R, Robinson D, Jr., Davis HM, Rossi JF, Cella D, and Trikha M (2005) Are inflammatory cytokines the common link between cancer-associated cachexia and depression? J Support Oncol 3:37–50

Ising M, Kunzel HE, Binder EB, Nickel T, Modell S, and Holsboer F (2005) The combined dexamethasone/CRH test as a potential surrogate marker in depression. Prog Neuropsychopharmacol Biol Psychiatr 29:1085–1093

Jehn CF, Kuehnhardt D, Bartholomae A, Pfeiffer S, Krebs M, Regierer AC, Schmid P, Possinger K, and Flath BC (2006) Biomarkers of depression in cancer patients. Cancer 107:2723–2729

Joels M, Karst H, Alfarez D, Heine VM, Qin Y, van RE, Verkuyl M, Lucassen PJ, and Krugers HJ (2004) Effects of chronic stress on structure and cell function in rat hippocampus and hypothalamus. Stress 7:221–231

Joffe RT and Marriott M (2000) Thyroid hormone levels and recurrence of major depression. Am J Psychiatr 157:1689–1691

Kraemer HC, Schultz SK, and Arndt S (2002) Biomarkers in psychiatry: methodological issues. Am J Geriatr Psychiatr 10:653–659

Kugaya A and Sanacora G (2005) Beyond monoamines: glutamatergic function in mood disorders. CNS Spectrum 10:808–819

Lam RW (2006) Sleep disturbances and depression: a challenge for antidepressants. Int Clin Psychopharmacol 21(Suppl 1):S25–S29

Levine J, Barak Y, Chengappa KN, Rapoport A, Rebey M, and Barak V (1999) Cerebrospinal cytokine levels in patients with acute depression. Neuropsychobiology 40:171–176

Lewy AJ, Lefler BJ, Emens JS, and Bauer VK (2006) The circadian basis of winter depression. Proc Natl Acad Sci USA 103:7414–7419

Lieb J, Karmali R, and Horrobin D (1983) Elevated levels of prostaglandin E2 and thromboxane B2 in depression. Prostaglandins Leukot Med 10:361–367

Loosen PT (1986) Hormones of the hypothalamic-pituitary-thyroid axis: a psychoneuroendocrine perspective. Pharmacopsychiatry 19:401–415

Lucae S, Salyakina D, Barden N, Harvey M, Gagne B, Labbe M, Binder EB, Uhr M, Paez-Pereda M, Sillaber I, Ising M, Bruckl T, Lieb R, Holsboer F, and Muller-Myhsok B (2006) P2RX7, a gene coding for a purinergic ligand-gated ion channel, is associated with major depressive disorder. Hum Mol Genet 15:2438–2445

Maes M (1995) Evidence for an immune response in major depression: a review and hypothesis. Prog Neuropsychopharmacol Biol Psychiatr 19:11–38

Manji HK, Quiroz JA, Sporn J, Payne JL, Denicoff K, Gray A, Zarate CA, Jr and Charney DS (2003) Enhancing neuronal plasticity and cellular resilience to develop novel, improved therapeutics for difficult-to-treat depression. Biol Psychiatr 53:707–742

Marchini J, Donnelly P, and Cardon LR (2005) Genome-wide strategies for detecting multiple loci that influence complex diseases. Nat Genet 37:413–417

Mathews A and MacLeod C (2005) Cognitive vulnerability to emotional disorders. Annu Rev Clin Psychol 1:167–195

McClung CA (2007) Circadian genes, rhythms and the biology of mood disorders. Pharmacol Ther 114:222–232

McEwen BS, Magarinos AM, and Reagan LP (2002) Structural plasticity and tianeptine: cellular and molecular targets. Eur Psychiatr 17(Suppl 3):318–330

Morissette J, Villeneuve A, Bordeleau L, Rochette D, Laberge C, Gagne B, Laprise C, Bouchard G, Plante M, Gobeil L, Shink E, Weissenbach J, and Barden N (1999) Genome-wide search for linkage of bipolar affective disorders in a very large pedigree derived from a homogeneous population in quebec points to a locus of major effect on chromosome 12q23-q24. Am J Med Genet 88:567–587

Muller N and Schwarz MJ (2007) The immune-mediated alteration of serotonin and glutamate: towards an integrated view of depression. Mol Psychiatr 12:988–1000

Muller N, Riedel M, and Schwarz MJ (2004a) Psychotropic effects of COX-2 inhibitors--a possible new approach for the treatment of psychiatric disorders. Pharmacopsychiatry 37:266–269

Muller N, Strassnig M, Schwarz MJ, Ulmschneider M, and Riedel M (2004b) COX-2 inhibitors as adjunctive therapy in schizophrenia. Expert Opin Investig Drugs 13:1033–1044

Muller N, Schwarz MJ, Dehning S, Douhe A, Cerovecki A, Goldstein-Muller B, Spellmann I, Hetzel G, Maino K, Kleindienst N, Moller HJ, Arolt V, and Riedel M (2006) The cyclooxygenase-2 inhibitor celecoxib has therapeutic effects in major depression: results of a double-blind, randomized, placebo controlled, add-on pilot study to reboxetine. Mol Psychiatr 11:680–684

Nacher J and McEwen BS (2006) The role of N-methyl-D-asparate receptors in neurogenesis. Hippocampus 16:267–270

Nakamura T, Ebihara I, Shimada N, and Koide H (1998) Elevated levels of erythropoietin in cerebrospinal fluid of depressed patients. Am J Med Sci 315:199–201

Nemeroff CB (1988) The role of corticotropin-releasing factor in the pathogenesis of major depression. Pharmacopsychiatry 21:76–82

Nestler EJ, Barrot M, DiLeone RJ, Eisch AJ, Gold SJ, and Monteggia LM (28–3–2002) Neurobiology of depression. Neuron 34:13–25

Noll R (2006) The blood of the insane. Hist Psychiatr 17:395–418
Papiol S, Arias B, Gasto C, Gutierrez B, Catalan R, and Fananas L (2007) Genetic variability at HPA axis in major depression and clinical response to antidepressant treatment. J Affect Dissord 104:83–90
Pien HH, Fischman AJ, Thrall JH, and Sorensen AG (15–2–2005) Using imaging biomarkers to accelerate drug development and clinical trials. Drug Discov Today 10:259–266
Politi P, Minoretti P, Piaggi N, Brondino N, and Emanuele E (2007) Elevated plasma N-terminal ProBNP levels in unmedicated patients with major depressive disorder. Neurosci Lett 417:322–325
Pollmacher T, Haack M, Schuld A, Reichenberg A, and Yirmiya R (2002) Low levels of circulating inflammatory cytokines–do they affect human brain functions? Brain Behav Immun 16:525–532
Prange AJ, Jr and Loosen PT (1980) Some endocrine aspects of affective disorders. J Clin Psychiatr 41:29–34
Rafnsson SB, Deary IJ, Smith FB, Whiteman MC, Rumley A, Lowe GD, and Fowkes FG (2007) Cognitive decline and markers of inflammation and hemostasis: the Edinburgh Artery Study. J Am Geriatr Soc 55:700–707
Rubin RT, Poland RE, Lesser IM, and Martin DJ (1987) Neuroendocrine aspects of primary endogenous depression–IV. Pituitary-thyroid axis activity in patients and matched control subjects. Psychoneuroendocrinology 12:333–347
Santarelli L, Saxe M, Gross C, Surget A, Battaglia F, Dulawa S, Weisstaub N, Lee J, Duman R, Arancio O, Belzung C, and Hen R (2003) Requirement of hippocampal neurogenesis for the behavioral effects of antidepressants. Science 301:805–809
Sequeira A and Turecki G (2006) Genome wide gene expression studies in mood disorders. OMICS 10:444–454
Shaw LM, Korecka M, Clark CM, Lee VM, and Trojanowski JQ (2007) Biomarkers of neurodegeneration for diagnosis and monitoring therapeutics. Nat Rev Drug Discov 6:295–303
Shimbo D, Chaplin W, Crossman D, Haas D, and Davidson KW (2005) Role of depression and inflammation in incident coronary heart disease events. Am J Cardiol 96:1016–1021
Srinivasan V, Smits M, Spence W, Lowe AD, Kayumov L, Pandi-Perumal SR, Parry B, and Cardinali DP (2006) Melatonin in mood disorders. World J Biol Psychiatr 7:138–151
Tsuang MT, Nossova N, Yager T, Tsuang MM, Guo SC, Shyu KG, Glatt SJ, and Liew CC (2005) Assessing the validity of blood-based gene expression profiles for the classification of schizophrenia and bipolar disorder: a preliminary report. Am J Med Genet B Neuropsychiatr Genet 133:1–5
van WD and Maes M (1999) Activation of the inflammatory response system: a new look at the etiopathogenesis of major depression. Neuro Endocrinol Lett 20:11–17
Vawter MP, Ferran E, Galke B, Cooper K, Bunney WE, and Byerley W (2004) Microarray screening of lymphocyte gene expression differences in a multiplex schizophrenia pedigree. Schizophr Res 67:41–52
Yao Y, Schroder J, and Karlsson H (2007) Verification of proposed peripheral biomarkers in mononuclear cells of individuals with schizophrenia. J Psychiatr Res 42:639–643
Yiangou Y, Facer P, Durrenberger P, Chessell IP, Naylor A, Bountra C, Banati RR, and Anand P (2006) COX-2, CB2 and P2X7-immunoreactivities are increased in activated microglial cells/macrophages of multiple sclerosis and amyotrophic lateral sclerosis spinal cord. BMC Neurol 6:12
Zarate CA, Jr., Singh JB, Carlson PJ, Brutsche NE, Ameli R, Luckenbaugh DA, Charney DS, and Manji HK (2006) A randomized trial of an N-methyl-D-aspartate antagonist in treatment-resistant major depression. Arch Gen Psychiatr 63:856–864
Zhang X, Gainetdinov RR, Beaulieu JM, Sotnikova TD, Burch LH, Williams RB, Schwartz DA, Krishnan KR, and Caron MG (2005) Loss-of-function mutation in tryptophan hydroxylase-2 identified in unipolar major depression. Neuron 45:11–16
Zvara A, Szekeres G, Janka Z, Kelemen JZ, Cimmer C, Santha M, and Puskas LG (2005) Overexpression of dopamine D2 receptor and inwardly rectifying potassium channel genes in drug-naive schizophrenic peripheral blood lymphocytes as potential diagnostic markers. Dis Markers 21:61–69

Chapter 14
Pharmacogenetics of Antidepressant Response

Alessandro Serretti(✉), Antonio Drago, and Michael N. Liebman

Abstract Nowadays, the efficacy of antidepressant treatment and the profile of side effects are not predictable: sociodemographic and clinical variables have not provided clinicians with consistent support. Emerging evidence reports promising results from the genetic approach: pharmacogenetics identifies which genetic variations are associated with drug treatment outcome. Both pharmacodynamic and pharmacokinetic key genes have been investigated, and even though consistent conclusions are yet to come, several lines of evidence have led researchers to assume a promising future of effective pharmacogenetic-based pretreatment assessment. The most important antidepressant pharmacogenetic results in this field are reviewed in this chapter and future research lines are proposed.

Abbreviations 5-HT1A: Serotonin receptor 1 A; 5-HT2A: Serotonin receptor 2A; 5-HT6: Serotonin receptor 6; ABCB1: ATP-binding cassette subfamily B member 1; Ach: Acetylcholine; AD: Antidepressant; B1AR: Beta1 adrenoceptor; B2AR: Beta2 adrenoceptor; CLOCK: Circadian Locomotor Output Cycles Kaput; COMT: Catechol-*O*-methyl transferase; CYP: Cytochrome; CRH: Corticotropin releasing hormone; D2: Dopamine (receptor) type 2; D3: Dopamine (receptor) type 3; D4: Dopamine (receptor) type 4; DRD2: (gene for the) dopamine receptor D 2; DRD3: (gene for the) dopamine receptor D 3; DRD4: (gene for the) dopamine receptor D 4; IL: Interleukin; LSD: Lysergic acid diethylamide; MAO-A: Mono amino oxidase A; MB-COMT: Membrane-bound COMT; MDD: Major depressive disorder; NET: Norepinephrine transporter; NO: Nitric oxide synthase; OCD: Obsessive compulsive disorder; SNP: Single nucleotide polymorphism; SSRIs: Selective Serotonin Reuptake Inhibitors; S-COMT: Soluble form of COMT; TCA: Tricyclic antidepressants; U.S. FDA: Food and Drug Administration (United States); VNTR: Variable number tandem repeat

---

A. Serretti
Institute of Psychiatry, University of Bologna, Bologna, Italy
alessandro.serretti@unibo.it

C.W. Turck (ed.) *Biomarkers for Psychiatric Disorders*,
doi: 10.1007/978-0-387-79251-4_14, © Springer Science+Business Media, LLC 2008

14.1 Introduction

At present, clinical phenotypes of depression do not allow the selection of a specific treatment for a specific patient or clinical condition, and when an antidepressant treatment is started, the antidepressant effect does not always lead to complete remission. In fact, even though different classes of antidepressant drugs (AD) have been used to treat depressive syndromes, the treatment efficacy is often incomplete (60–70% of patients do not experience remission, 30–40% do not show significant response) (Moncrieff and Kirsch, 2005) regardless of the initial choice of standard psychiatric medication (Entsuah et al., 2001; Bauer, 2002). Moreover, antidepressant response is usually associated with a 2 to 4 week lag before improvement; even though melioration can occur during the first 2 weeks of treatment, this is difficult to distinguish from placebo effect (Mitchell, 2006). Accordingly, clinical guidelines still recommend waiting for at least 4–6 weeks before switching to another AD when an antidepressant response is not achieved. This exposes patients to an ineffective therapy period, higher risk of worsening clinical conditions, higher risk of premature discontinuation (Masand, 2003), and, worse, a feeling of hopelessness possibly leading to high suicidal risk. As a result, patients run a high risk of recurrence of major depressive episodes or worsening of the mental suffering, and have to stay in hospital, on average, for longer periods involving higher costs. Ultimately, the side-effect profiles are not predictable, and even though there is wide interindividual variability, side effects are so common (40–90% (Cramer and Rosenheck, 1998)) that the clinical choice of a specific drug is partially determined by the probability of occurrence of unwanted effects, given the general knowledge of the properties of drugs. Therefore, to reduce the suffering of patients and minimize costs, it would be desirable to know in advance whether a drug is likely to be effective and tolerable (clinical and anamnestic variants have not been found to be helpful in this direction (Nierenberg, 2003)), and whether the genetically determined investigation of pharmacological responses could hold more opportunities (Perlis, 2007; O'Reilly et al., 1994; Serretti, 1998). The study of how an individual's genetic inheritance affects the body's response to drugs is investigated by pharmacogenetics: it examines how the genetic makeup affects antidepressant response by combining traditional pharmaceutical sciences, such as biochemistry, with annotated knowledge of genes, proteins, and single nucleotide polymorphisms (SNPs). The first step in the advance of pharmacogenetics is finding the candidate polymorphisms: there are millions of SNPs in the human genome, and even though they usually cosegregate in small groups, the identification of the key variations remains important but is difficult to achieve. Two strategies are available to find suitable genes: an inductive approach (genomewide analysis) based on wide genome association studies, looking for statistical combinations of genetic variations and clinical phenotypes, independent of pharmacological and physiological studies, and a deductive approach (candidate gene approach) investigating key variations in the chosen genes on the basis of the knowledge of drug therapeutic mechanisms and pathophysiological hypothesis of depression, or evidence-based associations. These lines of research are mutually dependent: induction of new elements gives more details to improve and criticize previous theoretical models of depression, whereas deductive efforts stimulate and address new lines of coherent and focused research.

Generally, pharmacogenetic studies investigate genes related to the two processes involved in the actions of a drug, *pharmacokinetics* and *pharmacodynamics*. Pharmacokinetics describes the way in which a drug is distributed in or cleared from the body and involves absorption of the drug, distribution through hydrophilic and hydrophobic spaces, metabolism, and excretion. Of these processes, genes related to metabolism have received the most attention. Pharmacodynamics examines the drug's interaction with its receptors and transporters and with downstream processes such as second-messenger systems (Perlis, 2007). The most promising results for the pharmacokinetics field have been reported for the genetic variations of CYP 2D6 and P-glycoprotein. The number of interesting pharmacodynamic targets appears to be quite larger, and the most important are tryptophan hydroxylase, catechol-*O*-methyltransferase (COMT), monoamine oxidase (MAO-A), serotonin transporter (SERT), norepinephrine transporter, dopamine transporter, monoamine receptors (5-HT1A, 5-HT2A, 5-HT6, beta1 adrenoceptor, mainly), dopamine receptors, G-protein beta-3 subunit, CRH receptor I, glucorticoid receptor, angiotensin converting enzyme, CLOCK, nitric oxide synthase, and interleukin 1-beta.

There is growing interest in the potential of pharmacogenetics in applications for new drug approvals and review of existing drugs: the U.S. Food and Drug Administration (FDA) has begun to focus on the near-term benefits from such an approach (see for example, http://www.fda.gov/OHRMS/DOCKETS/AC/05/slides/2005–4194S1_Slide-Index.htm((Perlis, 2007).

14.2 Pharmacokinetics

14.2.1 Cytochrome P450

Cytochrome P450 (CYP) describes a class of heme-containing proteins that represent the major enzymes responsible for the oxidation and reduction of numerous endogenous substrates and drugs. Over 50 isoenzymes that catalyze the oxidation of many drugs and chemicals are known so far within this family of proteins, and in humans many isoforms are known and seven are the most frequent: CYP1A, CYP2A6, CYP2B6, CYP2C, CYP2D6, CYP2E1, and CYP3A enzymes. The CYP family of liver enzymes is responsible for the catabolism of at least 30 different classes of drugs. DNA variations in genes coding for these enzymes can influence their ability to metabolize certain drugs, thereby causing drug overdose in patients. CYP2D6 and CYP2C19 have given the most interesting results. CYP2D6 gene is found in position 22q13.1, and it spans 4,382 bases. Its variations have been associated with a long list of medical conditions: arterial hypertension (Zateyshchikov, 2007), leukemia (Eyada, 2007), apnea in children (Eyada, 2007), thyroid cancer (Lemos,

2007), Alzheimer's and Parkinson's diseases (Golab-Janowska, 2007; Dick, 2007; Duric, 2007; Bialecka, 2007; Vilar, 2007), hepatic disease (Olsavsky, 2007), pulmonary disease (Arif, 2007; Yan and Wu, 2007), breast cancer (Marsh and McLeod, 2007), and porphyrias (Lavandera, 2006; and others), and it is implicated in the metabolism of most antidepressant drugs (Lin and Lu, 1998). So far, up to 75 different alleles have been reported for CYP2D6, (descriptions are available at http://www.imm.ki.se/cypalleles), more than 15 of these encoding an inactive or no enzyme at all, while others consist of gene duplications (Bertilsson, 2002). Those gene variants are associated with different drug metabolism rates – individuals are classified as poor (PM), intermediate (IM), extensive (EM) and ultrarapid (UM) metabolizers according to their inherited genetic profile (Nebert and Dieter, 2002; Thuerauf and Lunkenheimer, 2006). The CYP2D6 PM phenotype occurs in as much as 5–10% in the Caucasian population, whereas it is much rarer in Black Americans and in Orientals: in a recent study, the frequency of PMs and UMs was 0.22% and 1.25%, respectively in a Korean sample (Lee, 2006). Up to 3% of PMs in non-white populations have been reported recently (de Leon, 2007). The velocity of the enzyme seems to be influenced by the number of gene copies, as UMs have been shown to have multiple copies of the CYP2D6 gene with a direct influence on plasma drug concentration: standard doses will show inadequate therapeutic antidepressant response (nortriptyline) in subjects with many gene copies of CYP2D6 (Dalen, 1998). Accordingly, side effects will be more frequent in PM subjects. High normal doses of the drug may be required for patients with 2–4 copies (Bertilsson, 2002). Tricyclic antidepressants (TCAs) have small therapeutic "windows" because of their side-effect profile: a dose adjustment based on CYP2D6 variants has been proposed (de Leon, 2007; Kirchheiner, 2001). On the basis of two studies (Balant-Gorgia, 1982; Baumann, 1986), PMs of CYP2D6 ought to receive about 50% of the recommended dose of amitriptyline (Kirchheiner, 2001). As clomipramine elimination via CYP2D6 has been hypothesized to be saturable (Danish University Antidepressant, 1999), dose recommendations are valid only for the medium dose range: PMs of substrates of CYP2D6 may require 60% of the recommended dose and EMs 120% (Thuerauf and Lunkenheimer, 2006). It must be underlined that these variations can be genetically predicted before the treatment starts, even though several other important variants, such as other ongoing drug therapies, age, sex, ethnicity, medical diseases and so on, have a strong impact.

In regard to other classes of antidepressants, CYP2D6 variants have been shown to influence plasma levels of paroxetine and venlafaxine (Ozdemir et al., 1999; Veefkind et al., 2000; van der Weide et al., 2005; Shams, 2006). In contrast, no association between CYP2D6 genotype and tolerability and efficacy of paroxetine in a sample of geriatric patients was reported in another study (Murphy, 2003), which could be due to the flat dose–response curve (Brosen and Naranjo, 2001). Fluoxetine and norfluoxetine are also influenced by the genetic legacy of CYP2D6 (and 2C9) (Scordo, 2005). Fluvoxamine therapeutic effects are associated with the SER receptor and CYP2D6 polymorphisms in Japanese subjects (Suzuki et al., 2006).

The S+ enantiomer of mirtazapine has been reported to be highly influenced by the presence of poor, extensive, or ultrarapid CYP2D6 genotype (Brockmoller et al., 2007; Kirchheiner, 2004).

Also, CYP2C19 shows interesting genetic polymorphisms, and the difference in phenotype between people carrying the different isoforms allowed, making the same distinction between individuals with a normal catalytic function (extensive metabolizers) and a group of individuals with impaired catalytic capacity (poor metabolizers) that has been proposed for the CYP2D6 (Smith, 1998), 1998). CYP2C19 is located in position 10q24.1–q24.3, and it spans 90,636 bases. Three hundred and eleven CYP2C19 SNPs are known so far. The frequency of the CYP2C19 PM phenotype observed in Orientals is about 20%, while in Caucasians it reaches only 2–5% (Smith, 1998a). It was recently shown that citalopram treatment efficacy and side effects are influenced by CYP2C19 in a sample of Chinese patients (Yin, 2006). It has also been reported that the lowest risk of side effects associated with amitriptyline therapy was observed for carriers of two functional CYP2D6 alleles combined with only one functional CYP2C19 allele (0 of 13 (0%) vs. 9 of 11 (81.8%) for the high-risk group) in a sample of fifty Caucasian inpatients with at least medium-grade depressive disorder who received amitriptyline at a fixed dose of 75 mg twice a day (Steimer, 2005).

It must be remembered that many drugs can influence their own and other drugs' catabolic profiles. For example, selective serotonin reuptake inhibitors (SSRIs) display a distinct profile of CPY inhibition; fluvoxamine is a potent CYP1A2 and CYP2C19 inhibitor and a moderate CYP2C9, CYP2D6, and CYP3A4 inhibitor. Fluoxetine and paroxetine are potent CYP2D6 inhibitors, whereas the main metabolite of fluoxetine, norfluoxetine, has a moderate inhibitory effect on CYP3A4. Sertraline is a moderate CYP2D6 inhibitor; only citalopram appears to have little effect on the major CYP isoforms (Rao, 2007). Fluoxetine deserves special attention, as the inhibitory effects on CYP activity can persist for several weeks after fluoxetine discontinuation because of the long half-life of fluoxetine and its metabolite norfluoxetine (Hemeryck and Belpaire, 2002).

This must be taken into consideration when polytherapy treatment is started: for example, the impact of CYP2D6 variants might be greater for combined SSRI + TCA treatments; co-administration of paroxetine and desipramine in EM, which had at least two functional copies of the CYP2D6 gene, was found to result in a 5-fold decrease in desipramine clearance (Brosen et al., 1993).

14.2.2 P-glycoprotein

P-glycoprotein, an ATP-binding transporter protein, is a product of the ABCB1 gene. It is a plasma membrane transporter that exports certain drugs as well as endogenous substances against a concentration gradient in the intestines, kidneys, and testes. It also constitutes an important part of the blood–brain barrier, where it exports its substrates out of the brain back into the circulation. The gene encoding P-glycoprotein – formerly MDR1 (multi drug resistance), now ABCB1 – is localized to chromosome 7 in position 7q21.1 (Fojo, 1987), (Trent and Witkowski, 1987).

So far, several MDR1 SNPs have been identified, and mutations at positions 2,677 and 3,435 were associated with alteration of P-glycoprotein expression and/or function (Eichelbaum et al., 2004); but, generally, the contribution of the investigated ABCB1 genetic variations to antidepressant response appears to be poor (Kirchheiner, 2006a).

Anyway, in a recent study it has been reported that the 3,435 C>T variation was associated with higher fluvoxamine plasma levels in the CT + TT genotype group than in the CC genotype group (Fukui, 2007), whereas the same variation was found to be associated with a higher risk of nortriptyline-induced postural hypotension in patients treated for major depression (Roberts, 2002). Lack of association was reported for the G2677T/a sequence variation and therapeutic response in depressed inpatients treated with amitriptyline (Laika et al., 2006).

14.3 Pharmacodynamics

14.3.1 Monoamine Metabolic Enzymes

14.3.1.1 Tryptophan Hydroxylase

Tryptophan hydroxylase (TPH) catalyzes the rate-limiting step in serotonin (5-HT) biosynthesis. Long-term treatment of rats with SSRIs has been shown to upregulate mRNA and protein levels of TPH (Kim et al., 2002), which supports the involvement of TPH in the pharmacological action of AD. TPH has two isoforms: TPH1 and TPH2. TPH1 is ubiquitous. Its gene, localized to chromosome 11 (11p15.3-p14) (Craig, 1991), is predominantly expressed in peripheral organs such as the gut, pineal gland, spleen, and thymus, and less frequently in the brain. A biallelic SNP on position 218 (TPH1 A218C) is located in a potential GATA transcription factor-binding site and was reported to influence gene transcription: the rarer TPH1*a allele was reported to be associated with decreased serotonin synthesis (Jonsson, 1997). According to the monoaminergic theory of depression, the presence of the TPH1*a allele was found to be associated with suicidal behavior, and a poor response to antidepressant (SSRI) treatment, even though conflicting results can also be found, especially in non-Caucasian samples (Rujescu, 2003; Bellivier et al., 2004; Serretti, 2001b; Yoshida, 2002; Ham, 2005; Hong, 2006).

An independent U.S. study reported no effect of the TPH1 A218C SNP on fluoxetine treatment but a significant effect of three different TPH1 polymorphisms (Peters, 2004).

TPH2 seems to be more selectively expressed in brain areas compared to TPH1 (Zill, 2007; Sakowski, 2006). The TPH2 gene is in position 12q21.1, and its variations have been associated with major depression (Zill, 2004a) and suicidal behavior

(Zhou, 2005). Zhang et al. (2004) found that 5HT levels from cells expressing arginine at position 447 were reduced by approximately 55% compared to cells expressing proline at position 447. Moreover, in individuals with unipolar major depression, Zhang et al. (2005) identified a 1,463G–A transition in the TPH2 gene. This functional SNP in TPH2 replaces the highly conserved Arg441 with His, which resulted in approximately 80% loss of function in serotonin production when TPH2 was expressed in PC12 cells. SNP analysis in a cohort of 87 patients with unipolar major depression revealed that 9 patients carried the mutant allele, whereas among 219 controls, 3 subjects carried this mutation. In addition, this functional SNP was not found in a cohort of 60 patients with bipolar disorder. Positive association results were found in a study by Peters and colleagues in 2004 (Peters, 2004), even though those results were not replicated by another study (Garriock, 2005) with a similar sample for ethnicity and gender distribution.

Other recently investigated polymorphisms (rs11178997 in the promoter of the gene (Scheuch, 2007); rs7305115 and rs4290270 (Lim, 2007); rs4448731 and rs4641527 (de Lara, 2007); rs4570625 (Gutknecht, 2007)) have been associated with a different expression or function of TPH2 or impact on depressive or suicidal history, with possible interesting advantages in pharmacogenetic investigations.

Interestingly, Clark et al. (2005) suggested that glucocorticoid-mediated reduction of TPH2 message might have relevance to the etiology of major depression in cases where elevated glucocorticoids are one hallmark of the disease, since TPH2 mRNA was regulated by glucocorticoids but not by estradiol.

14.3.1.2 Catechol-O-Methyltransferase

COMT is involved in the catabolic pathways of noradrenaline (NE) and dopamine (DA). Moreover, this enzyme can indirectly affect brain serotonin tone, given the reciprocal interactions between DA and serotonin (5-HT). In fact, there is convincing evidence of interactions between serotonergic and dopaminergic systems and it seems that an increase of dopamine concentration in the whole brain could be a limiting factor for the antidepressant-like effect of antidepressants (Arias, 2006; Mossner, 2006). Consistently with this, it has been reported that the antidepressant treatment (venlafaxine) was associated with decreased and increased binding profile for the serotonin and dopamine transporters respectively (Shang, 2007). Similar results were reported for citalopram and paroxetine (Shang, 2007), but not for bupropion, which probably has a noradrenergic and dopaminergic but not serotonergic direct action. The COMT gene has been mapped to chromosome 22 (22q11.1—q11.2) (Grossman et al., 1992). Lachman and collaborators reported a functional polymorphism (SNP) consisting of a transition of guanine to adenine at codon 158 leading to a Val–Met substitution in MB-COMT (and in position 108 in S-COMT) (Lachman, 1996). The presence of the Met allele has been reported to be associated with a lower enzymatic activity (Mannisto and Kaakkola, 2005; Weinshilboum et al., 1999). Consistent with the previously shown assumption, this polymorphism has been associated with higher risk of suicidal behavior and personality traits (Craddock et al., 2006) and a worse response to mirtazapine (but not paroxetine) (Szegedi, 2005) and citalopram (Arias, 2006). In the study performed by Arias and colleagues in 2006, it was reported that homozygosity for the Met allele carried an odds ratio (OR) of 2.21 for nonremission with regard to citalopram treatment: no influence of the investigated COMT variant was found for all the other SSRIs.

14.3.1.3 Monoamine Oxidase A

Monoamine Oxidase A (MAO-A) is a major degrading enzyme in the metabolic pathways of monoamine neurotransmitters (NE, DA, 5-HT). Interestingly, its absence in human beings has been found to be life compatible (Sims, 1989) and associated with a psychiatric-like syndrome characterized by borderline mental retardation and abnormal behavior such as impulsive aggression, attempted rape, and exhibitionism. This syndrome was found to be associated with a punctual nonsense mutation in the MAO-A gene (Brunner, 1993). Moreover, MAO-A genetic variations are supposed to influence the mechanism of action of SSRIs through an interaction with SERT (Maes and Meltzer, 1995). The gene encoding MAO-A is located in position Xp11.23 (Sabol et al., 1998) and is composed of 15 exons and spans 90,659 bp. One hundred and fifty-five variations are known so far. A polymorphism located 1.2 kb upstream of the MAO-A coding sequences (VNTR) was reported to affect the transcription of the MAO-A promoter: alleles with 3.5 or 4 copies of the repeat sequence are transcribed 2–10 times more efficiently than those with 3 or 5 copies of the repeat, suggesting an optimal length for the regulatory region (Sabol et al., 1998).

Bipolar disorder as well as suicidal tendency, personality features, aggressive behavior, alcoholism, and AD response in females have been associated with this and other polymorphisms in the MAO-A gene sequence (Newman, 2005; Manuck, 2000; Preisig et al., 2005; Bondy et al., 2006; Serretti, 2006b; Yu, 2005; Parsian, 1999). No findings of the association with AD response could be found in the literature (Yoshida, 2002; Muller, 2000; Cusin, 2002).

Finally, it has been recently reported that the T941G polymorphism in the MAO-A gene is associated with mirtazapine response in females (Tadic, 2007).

14.3.2 *Monoamine Transporters*

14.3.2.1 Serotonin Transporter

The SERT, SLC6A4, is a target of primary interest in AD pharmacogenetics: it is the principal site of action of many antidepressant drugs (SSRI, TCA) and mediates the behavioral and toxic effects of cocaine and amphetamines. Moreover, SERT knockout mice show increased anxiety and inhibited exploratory locomotion, together with a reduction in aggressive behavior and home cage activity. This effect is further enhanced by desensitization of 5-HT1A and 5-HT1B receptors (Holmes, 2003a,b).

Human SERT is encoded by a single gene (SLC6A4) localized on chromosome 17 (17q11.1-q12) (Ramamoorthy, 1993). The gene spans 31 kb and consists of 14 exons (Lesch, 1994), and within this coding frame, up to 171 polymorphisms are known so far (see http://www.ncbi.nlm.nih.gov/sites/entrez for a complete list). The most promising variations seem to be those located in the promoter zone, especially the 5-HTTLPR. Heils and colleagues described this polymorphism (5-HTTLPR) as a 44-bp insertion/deletion (I/D) involving 2 units in a sequence of 16 repeated elements. The presence of different alleles could affect SERT expression (Heils, 1996): the long (l) 5-HTTLPR allele has twice the SERT expression in the basal state than the short (s) form. It is reasonable to assume that this functional variation influences antidepressant therapy. Moreover, a growing body of evidence links 5-HTTLPR genotypes to a variety of psychiatric disorders with affective symptomatology (e.g., depression, bipolar disorder, anxiety disorders, eating disorders, substance abuse) and pathological behaviors and personality traits related to anxiety, impulsivity, and stress (Serretti, 2006a,b). Since 1998, the 5-HTTLPR polymorphism has been investigated as a marker of AD response in about 20 studies (Binder and Holsboer, 2006; Serretti, 2005b).

A recent meta-analysis confirmed the role of 5-HTTLPR in antidepressant response (Serretti, 2007a), showing that 5-HTTLPR, s/s patients have a selective and slower improvement of depressive "core" and somatic anxiety symptoms (Serretti, 2007a,b). The s/l genotype was recently found to be associated with an OR of 2.37 concerning adverse effects during SSRIs treatment (dermatologic reactions, weight change, and fatigue above all), and the s/s genotype carried an OR of 1.77 (Smits, 2007). These findings are generally well replicated when Caucasian samples were under investigation (Smeraldi, 1998; Zanardi, 2001; Pollock, 2000; Rausch, 2002; Murphy, 2004), whereas studies involving Asian samples usually report conflicting results: short 5-HTTLPR allele associated with better outcome (Kang, 2007; Kim, 2000; Yoshida, 2002), no effect of 5-HTTLPR genotypes (Takahashi, 2002; Yoshida, 2004), or long 5-HTTLPR allele associated with better outcome (Hong, 2006; Yu, 2002).

It must be remembered that compared to Western populations carriage of the l-allele is much less frequent in Asian populations: inconsistent results in Oriental samples could have been influenced by a changed finding, and further study with a larger sample is needed. Interestingly, in 2005, Hu et al. (2005) reported that only the A allele carriers at the A/G SNP within 5-HTTLPR insertion polymorphism yield high mRNA levels, and the L(G) carriers actually behave equivalent to the low-expressing short allele. This could partially explain the inconsistent evidence throughout the studies. A recent study consistently reported that the low-expression allele (S or L(G)) was one of the strongest risk factors associated with adverse effect burden (Hu, 2007).

Another polymorphism influencing SERT expression was identified by Ogilvie and colleagues within intron 2 (STin2) and described as a 17-bp VNTR polymorphism. Moreover, it was reported that STin2 can play a role as a risk factor in depressive disorder (MacKenzie and Quinn, 1999; Ogilvie, 1996; Gutierrez, 1998) and suicide behavior (Gaysina, 2006; Lopez de Lara, 2006), creating a synergistic effect with 5-HTTLPR (Hranilovic, 2004). STin2 also affected antidepressant response in a large Korean sample (Kim, 2000) and it was proposed that a STin2 10/12 genotype might be associated with a poorer antidepressant effect, especially in Asian samples (Smits, 2004); but the finding was not replicated by other studies (Hong, 2006; Ito, 2002). Also, a lack of association with side effects was reported (Smits, 2007).

More recently, an SNP (rs25531) located just upstream of the 5-HTTLPR revealed a significant influence on antidepressant response and, intriguingly, a moderation effect on 5-HTTLPR alleles (Kraft, 2005; Smeraldi, 2006). It has been reported recently that rs25531 is the same SNP that Hu described in 2005, the I/D variation in HTTLPR, and might play a role in anxiety clusters of symptoms in patients with obsessive compulsive disorder (OCD) (Wendland, 2006, 2007).

Finally, in an interesting clinical efficacy oriented study, it has been demonstrated that a genetic assessment on HTTLPR before antidepressant treatment, identifying the risk of non-response subjects, is associated with a better clinical outcome when applied as a pretreatment systematic method (Smits, 2007).

14.3.2.2 Norepinephrine Transporter

The SLC6A2 gene encodes a norepinephrine (noradrenaline) transporter, which is responsible for reuptake of norepinephrine into presynaptic nerve terminals and is a regulator of norepinephrine homeostasis (Kim, 2006). The gene is localized at position 16q12.2 (Bruss, 1993), has got 14 exons, and spans 45,934 bp. Two hundred and sixty-seven genetic variations are known so far. The reuptake of noradrenaline occurs via a specific Na(+)- and Cl(–)-dependent transport system which is the target for TCAs such as desipramine and imipramine. Genetic variations have been proved to be influencing the functions: A369P variant is associated with lack of transport activity, while N292T was found to impede surface expression of NET when coexpressed, while F528C demonstrated increased functionality and, remarkably, exhibited both insensitivity to downregulation by PKC and a decrease in potency for the tricyclic antidepressant desipramine (Hahn et al., 2005). Moreover, it was recently found that the norepinephrine reuptake inhibitor (NRI) response was associated with the G1287A polymorphism: GG genotype was associated with better response (Kim, 2006). Consistently, one study determined whether NET gene variants could affect response to milnacipram (Yoshida, 2004), and significant associations were reported with the T128C (T allele predicting a better response). Anyway, the results are not unequivocal, and replication studies are warranted.

14.3.2.3 Dopamine Transporter

The gene for dopamine transporter (DAT1) – chromosome 5p15,3 (Giros, 1992) spanning 52,635 bp, with 502 known variations – has a 40-bp VNTR polymorphism in exon 15 which affects DAT expression (Fuke, 2001) associated with a faster onset of AD response when the allelic variant associated with enhanced expression (10 repeat variant) is present (Kirchheiner, 2006a,b). This is consistent with the hypothesis that an enhanced dopaminergic tone impairs, at least in part, antidepressant efficacy.

14.3.3 Monoamine Receptors

14.3.3.1 5-HT1A

Serotonin-1A receptors (5-HT1A) are present pre- and post-synaptically in different brain areas. At the level of serotonin cell bodies in the mid-brain dorsal raphe nucleus, 5-HT1A receptor acts as autoreceptor in a short negative feedback loop, and the activation inhibits the firing of serotonin neurons and diminishes the release of this neurotransmitter in the prefrontal cortex. Several AD compounds desensitize raphe 5-HT1A autoreceptors resulting in enhanced 5-HT neurotransmission, and this is thought to be associated with the antidepressant effect of those drugs. Agents that block 5-HT1A autoreceptors (e.g., pindolol) may accelerate the onset of AD action (Perez, 1997). Probably, postsynaptic 5-HT1A receptors mediate different actions: postmortem brains from depressed suicide victims vs. nondepressed individuals display elevated 5-HT1A density in the raphe nuclei (autoreceptors) but not at postsynaptic sites, which may lead to decreased serotonergic activity (Stockmeier, 1998).

Human 5-HT1A gene is located in position 5q11.2-q13 (Kobilka, 1987) and spans 1,269 bp, and 45 genetic variations are known so far. Its transcription is modulated by a common C(-1019)G SNP in its upstream regulatory region. The 5-HT1A G(-1019) allele in 5-HT1A upstream regulatory region fails to bind identified repressors Deaf-1 and Hes5, abolishing Deaf-1 action and impairing Hes5 action. This leads to the upregulation of receptor expression (Lemonde, 2003; Albert and Lemonde, 2004). This mechanism might mediate the association of the G(-1019) allele with depression and suicide (Lemonde, 2003). This allele may also contrast the therapeutic effect of antidepressant drugs by increasing inhibitory 5-HT1A autoreceptors. The hypothesis was confirmed by three independent Western studies (Lemonde, 2004; Serretti, 2004d; Arias, 2005). Similarly, two Asian studies reported a better response to fluoxetine in 5-HT1A(-1019)C/C homozygotes (Hong, 2006; Yu, 2006). A different Gly272Asp polymorphism was explored in Japanese outpatients with major depressive disorder (MDD) treated with fluvoxamine (Suzuki et al., 2004). Asp allele carriers showed a more marked reduction in depressive symptomatology compared to Gly/Gly homozygotes. This finding was not confirmed by subsequent studies (Yu, 2006).

14.3.3.2 Serotonin Receptor 2A (5-HT2A)

The activation of 5-HT2A receptors in medial prefrontal cortex and anterior cingulate cortex is thought to mediate the hallucinogenic properties of LSD, whereas in amygdala the 5-HT2A receptor activation is a component of antidepressant response. A considerable amount of data supports the hypothesis that HTR2A is involved in MDD. Both imaging and clinical studies (Bhagwagar, 2006; Yatham, 1999; Meyer, 2001; Yamauchi, 2005) have reported that drugs with HTR2A agonist properties may have acute euphoriant effects (Newton, 1996). Moreover, paroxetine may exert its effects also through inducing downregulation of 5HT2A receptors (Meyer, 2001; Maj, 1996) and nefazodone exerts its antidepressant effects partially through HTR2A receptor antagonism (Hemrick-Luecke et al., 1994). Also, controversial findings can be found (Hrdina and Vu, 1993; Akin, 2005). Moreover, the 5-HT2A receptors may mediate some of the AD effects seen in experimental animal models of depression (Skrebuhhova et al., 1999).

The 5-HT2A gene is located in position 13q14-q21, consists of three exons separated by two introns, and spans over 20 kb (Chen, 1992; Campbell, 1997). Three have been implicated in AD response.

5-HT2A T(102)C and 5-HT2A G(–1438)A are in linkage disequilibrium (LD) so they can be considered together (Spurlock, 1998). T-allele of 102 T/C SNP is in complete LD with the A-allele at –1438 and C-allele of 102 T/C SNP with G-allele at –1438 (Myers, 2007). Positive association studies can be found in the literature (Murphy, 2003; Suzuki et al., 2006; Choi, 2005; Bishop, 2006; Kato, 2006), and studies reporting no association results can be found as well (Cusin, 2002; Minov, 2001; Sato, 2002; Yoshida, 2003; Anguelova et al., 2003). With regard to the 5HT2A C(1420)T variation, one study showed a marginal association of T/T genotype with a worse response to SSRI treatment (Cusin, 2002).

14.3.3.3 Serotonin Receptor 6 (5-HT6)

This subtype is coded in position 1p36.13 (Kohen, 1996) and is a G-protein coupled receptor which stimulates adenylyl cyclase via G(s) protein coupling together with 5-HT4 and 5-HT7. Animal studies reported a role for this receptor in some behavioral variables (novelty seeking) and instrumental learning (Ballaz et al., 2007; Mitchell et al., 2007). Recently, a discrete involvement of this receptor in the AD mechanism has been reported (Svenningsson, 2007; Wesolowska and Nikiforuk, 2007). The genetic sequence is 14,276 bp long, with three exons and two introns, and 120 genetic variations are known so far. AD response was found to be influenced by a silent polymorphism T(267)C in the first exon (Kohen, 1996; Lee, 2005): C/T genotype carriers were found to display greater efficacy of the AD treatment, even though this was not previously confirmed (Lee, 2005; Wu, 2001). The 267C/T in exon 1 and a trinucleotide repeat polymorphism (GCC(2/3)) in the 5'-upstream region of the gene were found not to be associated with suicidal behavior (Okamura, 2005).

14.3.3.4 Beta1 Adrenoceptor

A number of pharmacologically well-characterized subtypes of adrenergic receptors (BARs) are known, including alpha-1, alpha-2, beta-1, and beta-2. Of these, both

B1AR and B2AR stimulate adenylate cyclase, although they subserve different physiologic functions. A variety of drugs, both agonists and antagonists, selective for either beta-1 or beta-2 receptors, have important applications in clinical medicine. B1AR serves as an important regulator of central nervous system (CNS) mediated behavior and of several neural functions, including mood, memory, neuroendocrine control, and stimulation of autonomic function, and are involved in the mediation of AD effects (Crissman et al., 2001). An influence of genetic variations in the B1AR coding sequence has been suggested by a recent animal study (Crowley, 2006), even though evidence is lacking in this field. Anyway, the role of the B1AR in AD treatment efficacy is consistent with the clinical experience that beta-blocker medications can be associated with side effects such as depression and lethargy (Kirigiti, 2000). Indeed, its molecular action is quite complex: Xu et al. (2003) (Xu, 2003) presented evidence that B2AR and B1AR form heterodymers when coexpressed in cultured cells. Moreover, the B2AR expression affects the internalization and ligand-binding characteristics of B1AR.

The Beta1 adrenergic receptor gene (ADRB1) is located in 10q24-q26 (Yang-Feng, 1990), spans 1,714 bp, and is constituted by a single exon. Eighty polymorphisms are known so far, and a recently identified functional polymorphism in the B1AR G(1165)C leading to the amino acid variation Gly389Arg was associated with an enhanced coupling to the stimulatory G(s) protein and increased adenylyl cyclase activation. Those elements are often observed in affective disorders (Mason, 1999; Zill, 2003), and this has been partially confirmed in a large sample study reporting that even though the G(1165)C variation did not influence the depressive phenotype, a tendency for a relation between C/C homozygosity and a better and even faster response to antidepressant treatment existed in those patients (Zill, 2003).

14.3.3.5 Dopamine Receptors

The dopamine system is highly involved in depressive spectrum symptomatology (Geracitano, 2006): it has been suggested that the pathophysiologic process in melancholic depression involves a decreased dopaminergic neurotransmission owing to hypersensitive inhibitory 5-HT2 hetero receptors located on dopaminergic neurons. Treatment with most antidepressant drugs downregulate these receptors, which allow increased dopaminergic firing and an antidepressant effect. Moreover, because the downregulation of 5-HT2 receptors coincides with the emergence of an antidepressant effect, this would consistently explain the therapeutic time lag (Landen and Thase, 2006). Finally, an interaction between the serotonergic and dopaminergic systems in the nucleus accumbens has been established, as motivation and hedonia have been associated with DA release in the nucleus accumbens (Zangen, 2001).

Many dopamine receptors are known so far: within this group, D2 receptors have been widely investigated. D2 is a G-protein coupled receptor that inhibits adenylyl cyclase activity. A missense mutation in this gene causes myoclonus dystonia, while other mutations have been associated with schizophrenia. Alternative splicing of this gene results in two transcript variants encoding different isoforms (mainly the long and short forms). A third variant has been described, but it has not been determined whether this form is normal or due to aberrant splicing. D2 are important receptors in terms of pharmacodynamic actions, as they are the first target of antipsychotic treatment, the long isoform primarily (Usiello, 2000), especially for a positive cluster of symptoms, but there are also some lines of evidence addressing an important role in antidepressant treatment as results from animal, pharmacological, and genetic investigations (Willner et al., 2005; Dziedzicka-Wasylewska, 2004).

The DRD2 gene is located at position 11q23.2, spans 65,576 bp, and is composed of seven exons. Two hundred and forty SNPs are known so far. A functional polymorphism in DRD2 gene, which causes a structural change from Ser to Cys at codon 311 (DRD2 Ser311Cys) (Itokawa, 1993), showed no significant influence on antidepressant response in some studies (Serretti, 2001a; Benedetti, 2003a; Serretti, 2004a–d).

The DRD4 gene has also been investigated as a possible candidate gene in the pharmacogenetics of antidepressant response. The D4 subtype is a G-protein coupled receptor that inhibits adenylyl cyclase. It is a target for drugs that treat schizophrenia and Parkinson's disease. This gene is coded in position 11p15.5 (Van Tol, 1991) and has considerable homology to DRD2 and DRD3. Mutations in this gene have been associated with various behavioral phenotypes, including autonomic nervous system dysfunction, attention deficit/hyperactivity disorder, and the personality trait of novelty seeking. This gene contains a polymorphic number (2–10 copies) of tandem 48 nt repeats in exon 3 (DRD4 exon3 VNTR). This makes this gene one of the most variable (Van Tol, 1991). Moreover, DRD4 is expressed in limbic areas involved in cognition and emotion, and high novelty seeking was found to be associated with the seven repeat alleles independent of ethnicity, sex, or age (Ebstein, 1996; Benjamin, 1996), even though there are also conflicting findings (Gelernter, 1997; Malhotra and Goldman, 2000). With regard to antidepressant responses, some studies have reported no association with this polymorphism (Serretti, 1999, 2001a,b); however in a recent work, DRD4 exon 3 variants revealed a significant modulation effect on various antidepressant drugs (Garriock, 2006).

14.3.4 Intracellular Signal Transduction Pathways

14.3.4.1 G-Protein Beta-3 Subunit

The G proteins are heterotrimers consisting of alpha, beta, and gamma subunits that dissociate after receptor activation. These proteins convey signals in cells initiated by the activation of many receptors, which are then translated into various intracellular systems through interaction with the diverse effector systems (Wess, 1998). It has been estimated that about 80% of all known hormones, neurotransmitters, and neuromodulators elicit cellular responses through G proteins coupled to a variety of intracellular effectors (Chen, 1999). The high degree of complexity generated by the interactions of G protein–coupled receptors may be one mechanism by which neurons acquire the flexibility for generating the wide range of responses observed in the nervous system (Chen, 1999), this giving the right to assume a possible involvement also in the pharmacogenetics of antidepressant response.

Levine et al. mapped the gene for the subunit beta 3 to 12 pter-p12.3 (Levine, 1990); the entire gene spans 7.5 kb and is composed of 11 exons and 10 introns (Rosskopf, 2000).

A polymorphism (C825T, rs5443) was first identified in exon 10 of the beta 3 subunit (GNB3) of pertussis toxin-sensitive Gi-type proteins (Siffert, 1998). The T allele is associated with the occurrence of a splice variant, called G beta 3 s, resulting from the deletion of nucleotides 498–620 of exon 9. The T-allele-associated splice variant remains biologically active (Siffert, 1998), even though it appears to be less active than the wild form in terms of modulation of ion channels, and in forming heterodymers with other proteins: it has been suggested that pathological conditions in patients carrying the homozygous C825T allele may result from a functional knockout of G beta 3 (Ruiz-Velasco and Ikeda, 2003). GNB3 (825)T variant was found to predict a better AD response in four independent studies (Zill, 2000; Joyce, 2003; Serretti, 2003a,b; Lee, 2004). Hong and colleagues reported a negative result in a study of an Asian sample (Hong, 2006).

14.3.5 HPA Axis and Stress-Hormone System

Dysfunction of the hypothalamic–pituitary–adrenal (HPA) axis is one of the most robust findings in many patients MDD (up to 70%) (Holsboer, 2000). It was reported that the alterations of corticotropin releasing hormone (CRH) function contribute to the pathogenesis of depression: concentrations of CRH in the cerebrospinal fluid (CSF) are elevated (Liu et al., 2002; Nemeroff, 1984). Imaging studies on depressive disorders and suicide behavior, together with animal studies on models of depression and human studies, confirmed the primary role of CRH in the psychiatric field (Nemeroff, 1988; Raadsheer, 1994; Brady, 1992; Gold and Chrousos, 2002; Michelson, 1997). There are two primary receptor subtypes for the CRH in the CNS: corticotropin-releasing hormone receptor1 and 2 (CRHR1and CRHR2). CRHR1 is considered to play a key role in mediating the CRH-elicited effects in depression and anxiety (Van Pett, 2000).

14.3.5.1 CRH Receptor I

The CRH receptor is a potent mediator of endocrine, autonomic, behavioral, and immune responses to stress. CRH, also called corticotropin-releasing factor (CRF), is a 41-amino acid peptide synthesized in the hypothalamus and capable of stimulating the production of adrenocorticotropic hormone (ACTH) and other proopiomelanocortin products of the anterior pituitary. It is the principal neuroregulator of the HPA axis and plays an important role in coordinating the endocrine, autonomic, and behavioral responses to stress and immune challenge. Sakai et al.

(1998) determined that the CRHR1 gene contains at least 14 exons spanning 20 kb of genomic DNA. The CRHR1 isoforms appear to originate from the same gene by alternative splicing. The isoform with the highest CRH affinity and the ability to transduce the most cAMP accumulation in response to CRH binding is encoded by 13 exons and excludes exon 6.

CRHR1 antagonists have consistently demonstrated antidepressant properties in experimental animal and human studies (Seymour et al., 2003; Overstreet and Griebel, 2004; Kehne, 2007). Some evidence points to a relevance of CRHR1 variants and antidepressant response, in particular an association within rs242941 G/G genotype and homozygous GAG haplotype of the three SNPs and fluoxetine therapeutic response (Licinio, 2004; Liu, 2007).

14.3.5.2 Glucorticoid Receptor

Glucocorticoid hormones, like other classes of steroid hormones, exert their cellular action by complexing with a specific cytoplasmic receptor, which in turn translocates to the nucleus and binds to specific sites on chromatin. The glucocorticoid receptor was the first transcription factor to be isolated and studied in detail (Muller and Renkawitz, 1991). The glucocorticoid receptor (GCCR) is crucial for gene expression. It is a 94-kD polypeptide and it is thought to have distinct steroid- and DNA-binding domains.

Francke and colleagues (Francke and Foellmer, 1989) demonstrated by in situ hybridization that the GRL gene is located on 5q31-q32.

A research group at the Max Planck Institute of Psychiatry in Munich, Germany, collected preliminary evidence that a functional polymorphism of the glucorticoid receptor (GR) gene (ER22/23EK) and a series of SNPs within the gene encoding the hsp90 co-chaperone FKBP5 (a part of the mature GR heterocomplex that regulates GR sensitivity) could modulate the onset of response to various classes of antidepressant drugs (Binder, 2004). A more recent study reported that homozygous carriers of the BclI polymorphism and ER22/23EK carriers had an increased risk of developing a major depressive episode, and even though no genetic associations with functional HPA-axis measures in depressed patients have been found, the ER22/23EK carriers showed a significantly faster clinical response to antidepressant therapy as well as a trend toward better cognitive functioning during depression (van Rossum, 2006).

14.3.6 ACE-Substance P System

14.3.6.1 Angiotensin Converting Enzyme

Angiotensin converting enzyme (ACE) is associated with a series of actions influencing blood pressure through the renina–angiotensin cascade, interfering with the secretion of hormones (ACTH, CRH) (Jezova, 1998). It is also expressed in the CNS, where its primary function comprises degradation of neuropeptides including substance P (SP). An influence of the SP over the biological mechanisms that may underline the pathophysiology of depression has been hypothesized, consistently with the finding that the administration of SP agonists was reported to have antidepressant effects, and the SP concentration was found to be diminished after antidepressant treatment (MAO-I) (Kramer, 1998 ; Nutt, 1998 ; Shirayama, 1996).

The presence of a deletion variant (D/) in the ACE gene was found to be associated with higher ACE plasma levels (Rigat, 1990); consistently, it was found to be associated with higher SP levels (Arinami, 1996) and a faster response to antidepressant treatments (Baghai, 2001), including total sleep deprivation (Baghai, 2003), particularly among females (Baghai, 2004). Further, D/D genotype carriers displayed the highest cortisol response in the Dex-CRH test administered at admission (Baghai, 2002).

More recently, another component of the ACE–SP system, the angiotensin II receptor gene (ATI), was included among outcome predictors in major depression (Bondy, 2005).

14.3.7 Endogenous CLOCK System

14.3.7.1 Circadian Locomotor Output Cycles Kaput (CLOCK)

Circadian rhythms of the simplest to the most complex biological systems are ruled by a tiny brain region of the hypothalamic suprachiasmatic nucleus (SCN), which expresses several CLOCK genes in a rhythmic fashion. These rhythms are regulated by positive and negative gene-expression feedback loops that consist of transcriptional and post-translational mechanisms (Reppert and Weaver, 2002). At least two clock-gene products, CLOCK and ARNTL (aryl hydrocarbon receptor nuclear translocator-like), function as transcription factors by binding to E-box enhancers in the promoter regions of other clock genes, including period (Per) genes. In turn, products of the Per genes can regulate the expression of other clock gene transcription factors. CLOCK genes can also influence other non-circadian genes, the so-called clock-controlled genes (CCGs), acting as transcription factors, and this could be related to pharmacological actions: in fact, within the candidate CCGs there are the dopamine and noradrenaline transporters, some dopamine receptors, and the tyrosine hydroxylase (Manev and Uz, 2006). This model is far more complex as psychoactive drugs can influence other promoter zones in the genetic sequence such as cAMP response element (CREB) and activating protein 1 (AP-1) near the E box in the promoter sequence.

Animal experiments reported that the mutant CLOCK mice with evening type behavior had worse performances at spatial learning tests, together with an altered cholinergic tone. Normal serotonergic and dopaminergic tones were also reported (Sei, 2006). Moreover, repeated administration of fluoxetine or cocaine influenced CLOCK gene expression in mice, together with enhanced serotonin *N*-acetyltransferase (Uz, 2005).

The CLOCK gene is in position 4q12 and contains 20 exons (Steeves, 1999).

One polymorphism in the 3' flanking region of the CLOCK gene, a T-to-C substitution at position 3,111 (CLOCK 3111T/C), is known to affect mRNA stability and half-life (Mignone, 2002). In healthy subjects the C allele was associated with significantly higher "eveningness", delay in preferred timing for activity or sleep (Katzenberg, 1998). In mood disorders the same C variant was coupled with higher recurrence rates in bipolar patients (Benedetti, 2003a,b), increased lifetime sleep disturbances (Serretti, 2003a,b), and persistence of insomnia during antidepressant treatment (Serretti, 2005).

14.3.8 Other Relevant Genes

14.3.8.1 Nitric Oxide Synthase

Nitric oxide (NO) is produced from its precursor L-arginine by the enzyme NO synthase (NOS), which includes at least three distinct isoforms – neuronal (NOS1), endothelial, and inducible NOS. The NOS1 gene was mapped to chromosome 12q24.2-q24.31 (Kishimoto, 1992; Xu, 1993) and spans 148,604 bp, and 748 genetic variations of its coding sequence are known so far. Recent studies have implicated NOS in the mechanism that underlies the therapeutic efficacy of antidepressant medication (Wegener, 2003; Suzuki, 2003; Paul, 2001). Animal models have suggested that the aggressive behavior associated with NOS gene ablation is mediated by the serotonergic system. The excessive aggressiveness and impulsiveness of NOS knockout mice is caused by selective decrements in serotonin (5-HT) turnover and deficient 5-HT(1A) and 5-HT(1B) receptor function in brain regions regulating emotion (Chiavegatto, 2001). Even though these lines of evidence suggest a possible role of this gene's variations in the pharmacogenetics of antidepressant therapy, there are not many pharmacogenetic investigations so far involving the NOS coding sequence. Moreover, in 2003 Yu reported no association results within NOS1 C276T polymorphism and antidepressant (fluoxetine) response or risk for experience of a depressive episode (Yu, 2003a).

14.3.8.2 Interleukin 1-Beta

One potential pathway by which depression may impact health is through modulation of immune function. Depressed individuals have been shown to display reductions in measures of cellular immune competence as well as elevated markers of systemic inflammation (Dentino, 1999; Herbert and Cohen, 1993; Kiecolt-Glaser and Glaser, 2002; Zorrilla, 2001). Moreover, there are several lines of evidence suggesting strong influences in both the direction from cytokines to neurotransmitters and from neurotransmitters to cytokines. For example, it has been reported that IL-1 activates brain noradrenergic (Dunn, 1988; Kabiersch, 1988) and serotonergic systems (Gemma, 1997; Linthorst, 1994; Shintani, 1993), reduces acetylcholine release in the hippocampus (Rada, 1991), and potentiates GABA effects (Li et al., 1993; Miller, 1991; Zeise et al., 1992). Moreover, noradrenaline enhances, and ACh inhibits, the release of IL-1β from neurons in hypothalamic explants (Tringali, 1996), while serotonin increases IL-1β mRNA in the hypothalamus (Gemma et al., 2003).

IL-1, produced mainly by blood monocytes, mediates the host reactions of acute phase response. There are some lines of evidence suggesting an altered IL-1 level in mood disorders (Maes, 1993; Anisman, 1999; Anisman et al., 2002; Licinio and Wong, 1999), even though there are also recent negative reports (Yang, 2007).

Webb et al. (Auron, 1985) assigned the IL-1 gene to the chromosome 2q13-q21. IL-1 beta was assigned to the end of 18q (Le Beau and Rowley, 1986). Four SNPs have been reported in the IL-1 beta gene: –31C/T (promoter), –511C/T (promoter), +3954C/T (exon 5) and A/G (intron 4) at position 5810 (Di Giovine et al., 1992; Pociot, 1992; Guasch et al., 1996). The –31 SNP is in strong LD with the –511 SNP (El-Omar, 2000). Homozygosity for the –511T allele of the IL-1beta gene was found to be associated with a trend of lower severity of depressive symptoms and more favorable fluoxetine response (Yu, 2003a,b).

14.4 Conclusions and Perspectives

All observed gene variants do not reach the putative 50% of variance explained by genetic factors in the complex trait of antidepressant response. Pharmacogenetics can aid finding more candidates. New technology (Illumina and Affymetrix) provides researchers with the possibility to investigate up to 500,000 or 600,000 SNPs in a single test, that is, to perform a complete genome SNP investigation. This will help find new associations with clinical features of psychiatric diseases, even though this methodology, together with the DNA microarray (DNA chip) analysis, is at risk of false-positive findings. The development of new experimental designs that combine the methods of linkage analysis, pharmacogenomics, and proteomics could help in this direction, and examples of such sequential approaches have already been published with promising results (Lachman, 1997; Niculescu, 2000).

Moreover, there are other key points to deal with: the question of phenotype is less relevant, as genetic variations have been shown to be more strictly associated with clusters of symptoms (such as "somatic complaints") than with complex phenotypes (such as "depressive disorder") (Yu, 2002; Serretti, 2005a,b; Kato, 2005), and this could be responsible for the inconsistent findings in the literature. Finally, the same genetic variation can be associated with different symptoms. For example, HTTLPR polymorphism has been associated with different characteristics of mood disorders – age of onset (Bellivier, 2002; Nobile, 2004), illness recurrence (Cusin, 2001; Rousseva, 2003), drug response, reactivity to stressful life events (Caspi, 2003), personality traits (Park, 2004) and several psychiatric diagnoses such as alcoholism (Feinn et al., 2005), smoking (Kremer, 2005), psychosomatic disorders (Yeo, 2004), eating disorders (Matsushita, 2004 ; Steiger, 2005), suicide (Courtet, 2001), autism (Bartlett, 2005) and attention deficit hyperactivity disorder (Bobb, 2005).

The most relevant pharmacogenetic studies are presented in Table 14.1.

Future studies will clarify whether such phenotypes are all simultaneously present or at different times in the same individuals.

Table 14.1 Pharmacogenetic Studies

Reference	Gene	Drug	Results
Wilkie, 2007	GNB3, CREB1, BDNF, CREBBP	Various	Better response associated with C allele of GNB3 C825T variation
Fukui, 2007	ABCB1	Fluvoxamine	TT Genotype associated with higher plasma concentration
Pae, 2007	DTNBP1	Various	rs2005976 A Allele associated with worse response to treatment; rs760761 associated
Liu, 2007	CRHR1	Fluoxetine	rs242941 G/G genotype and homozygous GAG haplotype of (rs1876828, rs242939 and rs242941) are associated with fluoxetine therapeutic response
Serretti, 2007	SERTPR	Various	s/s Patients showed a selective and slower improvement of depressive "core" and somatic anxiety symptoms, but not for other issues
Serretti, 2007	SERTPR	Various	Meta-analysis s/s variant is associated with worse treatment response
Ham, 2007	TPH1	Citalopram	The remission rate was worse A/A and A/C genotypes
Kirchheiner, 2007	SLC6A4, DAT1	Various	SLC6A4: no association DAT1: 9/10 genotype OR=1.6; 9/9 OR=6.0 for no or poor response
Kraft, 2007	SLC6A4 (5-HTT)	Citalopram	No association
Kim, 2006	SLC6A2 (NET), SLC6A4 (5-HTT)	Fluoxetine or sertraline	NET: G1287A polymorphism: GG genotype was associated with better response. 5-HTT: s associated with better response.
Choi, 2006	BNDF	Citalopram	M allele (Val66Met) associated with better response to treatment
Shams, 2006	CYP2D6	Venlafaxine	A PM phenotype of CYP2D6 increases the risk of side effects
Yin, 2006	CYP2C19	Citalopram	A PM phenotype influences drug plasma concentrations and side effects risk
Smeraldi, 2006	SERTPR	Fluvoxamine	l variant associated with better and faster response; 16F *l → partial response, 16D *l → better response than 16A *l
McMahon, 2006	HTR2A	Citalopram	Participants homozygous for the rs7997012 A allele ↓ 18% risk of no response to treatment
Laika et al., 2006	ABCB1	Amitriptyline	No association
van Rossum, 2006	GR	Various	ER22/23EK-carriers had faster clinical response to antidepressant
Sugai, 2006	HTR3A and HTR3B	Paroxetine	Tyr/Tyr genotype (Tyr129Ser polymorphism) associated with nausea
Popp, 2006	HTTLPR, HTT-VNTR	Various	HTT-VNTR 2.10/2.10 associated with higher side effects HTTLPR s/s associated with higher side effects

Arias, 2006	COMT	Various	In citalopram treatment, small effect on treatment result
Yu, 2006	HTR1A	Fluoxetine	Female patients with –1019C/C genotype associated with better response
Hong, 2006	HTR1A, SERTPR	Fluoxetine	HTR1A –1019C/C and SERTPR l/l associated with a better response
Suzuki et al., 2006	HTR1A, CYP2D6	Fluvoxamine	CYP2D6 LM with G/G 5-HT2A A-1438G polymorphism 4.242-fold and LM with the A/G genotype had a 4.147-fold higher risk of gastrointestinal side effects
Kato, 2006	5HTTLPR, HTR2A, HTR3A, HTR3B	Fluoxetine and paroxetine	5HTTLPR: L allele associated with better response HTR2A: -1438G/G associated with a good response HTR3A: 178C/C genotype associated with an antidepressant response
Ham, 2005	TPH1	Various	No association
Choi, 2005	HTR2A	Citalopram	GG genotype associated with treatment response
Kraft, 2005	SERTPR	Fluoxetine	rs25531 associated with treatment response
Kato, 2005	5HTTLPR	Fluoxetine and paroxetine	s allele associated with worse response to treatment
Bondy, 2005	ACE	Various	ACE II associated with poorer improvement more than 70% of the AT1 CC homozygotes had a 50% reduction in the HAMD-17 scale within 4 weeks of treatment.
Lee, 2005	HTR6	Various	CT associated with better treatment response than the homozygote CC + TT, especially in the somatic-anxiety subcategory
Szegedi, 2005	COMT	Mirtazapine and paroxetine	COMT polymorphism influences mirtazapine but not paroxetine response
Serretti, 2004	HTR1A	Fluvoxamine	C/C genotype (C(-1019)G) carriers showed a better response to fluvoxamine
Serretti, 2004	5HTTLPR, TPH, MAO-A, CLOCK, PER3	Various	CLOCK TT more relapses in 6 month follow up, no other significant finding
Serretti, 2004	5-HT1A	fluoxetine or nefadozone combined with pindolol or flibanserin alone	G allele associated with worse response
Tsai, 2004	5-HT1B	Fluoxetine	No association
Yoshida, 2004	NET, 5-HTT	Milnacipram	NET T allele (T-182C) associated with a superior antidepressant response NET A/A (G1287A) associated with a slower onset of therapeutic response No influence of 5-HTT

(continued)

Table 14.1 (continued)

Reference	Gene	Drug	Results
Zill, 2004	DTNBP1	Various	No association
Serretti, 2004	SERTPR, TPH,	fluvoxamine or paroxetine	SERT s/s associated with poorer response to treatment TPH no significant result
Grasmader, 2004	CYP 2C9, 2C19 and 2D6	Various	Significant influence of the CYP2D6 genotype, minor influence of the CYP2C19 genotype and no influence of the CYP2C9 genotype on antidepressant plasma concentrations
Lee, 2004	SERTPR	Long term antidepressant treatment	s/s Genotype was associated with poorer outcome
Baghai, 2004	ACE	Various	l allele associated with poorer response
Suzuki et al., 2004	5-HT1A	fluvoxamine	Gly/Gly genotype associated with poorer response
Rau, 2004	CYP2D6	Various	PM associated with more side effects
Peters, 2004	HTR1A, HTR2A, HTR2C, MAOA, SLC6A4, TPH1, and TPH2	Fluoxetine	Significant haplotype associations were found in all but the HTR1A and HTR2C genes
Lee, 2004	GNB3		T allele (C825T) had a severe symptomatology and a better response to antidepressant treatment than patients without the T allele
Durham, 2004	5HTTLPR	Sertraline	S allele associated with slower response
Charlier, 2003	CYP2D6	Fluoxetine or paroxetine	PM had higher plasma concentrations
Arias, 2003	5-HTTLPR	Citalopram	s/s Geneotype associated with non remission
Joyce, 2003	5-HTTLPR, GB3	Fluoxetine or nortriptyline	< 25 years: the T allele GB3 associated with poorer response to nortriptyline, HTTLPR variant was not associated > 25 years: GB3 was not associated with antidepressant response, HTTLPR s/s genotype associated with a poorer response to both fluoxetine and nortriptyline.
De Luca, 2003	MDR1	Various	No association with iatrogenic mania
Perlis, 2003	5-HTTLPR	Fluoxetine	s Allele associated with insomnia or agitation with fluoxetine treatment
Yoshida, 2003	MAOA, HTR2A	Fluvoxamine	MAOA VNTR associated with fluvoxamine induced nausea
Ohara, 2003	CYP2D6	Fluvoxamine	No association with fluvoxamine concentration
Takahashi, 2002	5-HTTLPR (Stin2), TPH	Fluvoxamine	No association with fluvoxamine induced nausea

Murphy, 2004	5HTTLPR	Mirtazapine and paroxetine	s Allele associated with more side effects
Yu, 2003	NOS C276T polymorphism	Fluoxetine	No association
Murphy, 2003	5HT2A T102C CYP 2D6	Paroxetine, mirtazapine	The presence and the severity of paroxetine-induced side effects were strongly associated with C/C genotype.
Serretti, 2003	G-protein beta3 C825T	Fluvoxamine, paroxetine	TT homozygous associated with response ($p = 0.009$)
Zill, 2003	ADRB1 G1165C polymorphism	TCA SSRI, NARI, NSRI	Tendency for association between CC Homozygous and better and faster antidepressant response
Yu, 2003	IL-1beta C-511T polymorphism	Fluoxetine	Homozygous for the -511T allele had a trend of more favorable fluoxetine response
Yu, 2002	SERTPR	Fluoxetine	l/l Genotype shows a better response ($p = 0.013$
Rausch, 2002	SERTPR	Fluoxetine	l/l Subjects show a more rapid response, or perhaps a more frequent placebo response
Ito, 2002	SERT-STin2	Fluvoxamine	No association
Cusin, 2002	MAOA VNTR, 5HT2A T102C	Fluvoxamine, paroxetine	Marginal association between 5HT2A C variants and AD response; No association with MAOA genotypes
Sato, 2002	5HT2A	Fluvoxamine	No association
Yoshida, 2002	MAOA VNTR, TPH A218C	Fluvoxamine	No association in a Japanese sample
Hong et al., 2002.	ACE I/D polymorphism	Venlafaxine, Fluoxetine	No association
Yoshida, 2002	SERTPR	Fluvoxamine	s Variant more frequent in responsive individuals than in non responsive
Zanardi, 2001	SERTPR	Fluvoxamine	l allele subjects were more likely to respond (all sample $p = 0.029$ – without pindolol $p = 0.002$)
Arias, 2001	SERTPR	Citalopram	s/s Genotype was significantly more frequent in no remission group ($p = 0.006$)
Mundo, 2001	SERTPR	SSRI, TCA	Patients with manic or hypomanic episodes induced by AD treatment had an excess of s alleles (p<0001)
Minov, 2001	SERTPR, 5HT2A T102C	Various ADs	Association between C variants and AD response
Serretti, 2001	SERTPR, TPH A218C	Fluvoxamine	A/A genotype was associated with slower response (no pindolol $p = 0.001$)
Serretti, 2001	TPH A218C	Paroxetine	A/A and A/C genotypes were associated with slower response (no pindolol $p = 0.005$)

(continued)

Table 14.1 (continued)

Reference	Gene	Drug	Results
Baghai, 2001	ACE I/D polymorphism	Various ADs	D allele associated with better outcome
Wu, 2001	5HT6 C267T	Venlafaxine, fluoxetine	No association
Serretti, 2001	DRD2 S311C, DRD4 VNTR	Fluvoxamine, paroxetine	No association
Zanardi, 2000	SERTPR	Paroxetine	s allele associated with less favorable and slower response ($p < 0.001$)
Muller, 2000	MAOA VNTR	Moclobemide	No association
Pollock, 2000	SERTPR	Paroxetine	s Allele associated with slower response ($p = 0.028$)
Kim, 2000	SERTPR SERT-STin2	Fluoxetine, paroxetine	s/s genotype showed better response ($p = 0.007$)
Smeraldi, 1998	SERTPR	Fluvoxamine	l Allele subjects were more likely to respond ($p = 0.017$)

References

Akin, D., et al., Signal transduction abnormalities in melancholic depression. Int J Neuropsychopharmacol, 2005. 8(1)5–16.

Albert, P.R. and S. Lemonde, 5-HT1A receptors, gene repression, and depression: guilt by association. Neuroscientist, 2004. 10(6):575–593.

Anguelova, M., C. Benkelfat, and G. Turecki, A systematic review of association studies investigating genes coding for serotonin receptors and the serotonin transporter: II. Suicidal behavior. Mol Psychiatr, 2003. 8(7):646–653.

Anisman, H., et al., Interleukin-1 beta production in dysthymia before and after pharmacotherapy. Biol Psychiatr, 1999. 46(12):1649–1655.

Anisman, H., L. Kokkinidis, and Z. Merali, Further evidence for the depressive effects of cytokines: anhedonia and neurochemical changes. Brain Behav Immun, 2002. 16(5):544–556.

Arias, B., et al. Genetic variability in the promoter region of the serotonin transporter gene is associated with clinical remission of major depression after long term treatment with citalopram. In World Federation of societies of Biological Psychiatry. 2001. Berlin, Germany: The World Journal of Biological Psychiatry.

Arias, B., et al., 5-HTTLPR polymorphism of the serotonin transporter gene predicts non-remission in major depression patients treated with citalopram in a 12-weeks follow up study. J Clin Psychopharmacol, 2003. 23(6):563–567.

Arias, B., et al., Evidence for a combined genetic effect of the 5-HT1A receptor and serotonin transporter genes in the clinical outcome of major depressive patients treated with citalopram. J Psychopharmacol, 2005. 19(2):166–172.

Arias, B., et al., Analysis of COMT gene (Val 158 Met polymorphism) in the clinical response to SSRIs in depressive patients of European origin. J Affect Disord, 2006. 90(2–3):251–256.

Arif, E., et al., Association of CYP2E1 and NAT2 gene polymorphisms with chronic obstructive pulmonary disease. Clin Chim Acta, 2007. 382(1–2):37–42.

Arinami, T., et al., An insertion/deletion polymorphism in the angiotensin converting enzyme gene is associated with both brain substance P contents and affective disorders. Biol Psychiatr, 1996. 40(11):1122–1127.

Auron, P.E., et al., Human and murine interleukin 1 possess sequence and structural similarities. J Mol Cell Immunol, 1985. 2(3):169–177.

Baghai, T., et al., Hypothalamic-pituitary-adrenocortical axis dysregulation in patients with major depression is influenced by the insertion/deletion polymorphism in the angiotensin I-converting enzyme gene. Neurosci Lett, 2002. 328(3):299–303.

Baghai, T., et al., Influence of a functional polymorphism within the angiotensin I-converting enzyme gene on partial sleep deprivation in patients with major depression. Neurosci Lett, 2003. 339(3):223–226.

Baghai, T.C., et al., Possible influence of the insertion/deletion polymorphism in the angiotensin I-converting enzyme gene on therapeutic outcome in affective disorders. Mol Psychiatr, 2001. 6(3):258–259.

Baghai, T.C., et al., The angiotensin I converting enzyme insertion/deletion polymorphism influences therapeutic outcome in major depressed women, but not in men. Neurosci Lett, 2004. 363(1):38–42.

Balant-Gorgia, A.E., et al., Role of oxidation polymorphism on blood and urine concentrations of amitriptyline and its metabolites in man. Arch Psychiatr Nervenkr, 1982. 232(3):215–222.

Ballaz, S.J., H. Akil, and S.J. Watson, Analysis of 5-HT6 and 5-HT7 receptor gene expression in rats showing differences in novelty-seeking behavior. Neuroscience, 2007. 147(2):428–438.

Bartlett, C.W., et al., Three autism candidate genes: a synthesis of human genetic analysis with other disciplines. Int J Dev Neurosci, 2005. 23(2–3):221–234.

Bauer, M., et al., World Federation of Societies of Biological Psychiatry (WFSBP) Guidelines for Biological Treatment of Unipolar Depressive Disorders, Part 1: Acute and continuation treatment of major depressive disorder. World J Biol Psychiatr, 2002. 3(1):5–43.

Baumann, P., et al., Amitriptyline pharmacokinetics and clinical response: I. Free and total plasma amitriptyline and nortriptyline. Int Clin Psychopharmacol, 1986. 1(2):89–101.

Bellivier, F., et al., Serotonin transporter gene polymorphism influences age at onset in patients with bipolar affective disorder. Neurosci Lett, 2002. 334(1):17–20.

Bellivier, F., P. Chaste, and A. Malafosse, Association between the TPH gene A218C polymorphism and suicidal behavior: a meta-analysis. Am J Med Genet B Neuropsychiatr Genet, 2004. 124(1):87–91.

Benedetti, F., et al., Dopamine receptor D2 and D3 gene variants are not associated with the antidepressant effect of total sleep deprivation in bipolar depression. Psychiatr Res, 2003a. 118(3):241–247.

Benedetti, F., et al., Influence of CLOCK gene polymorphism on circadian mood fluctuation and illness recurrence in bipolar depression. Am J Med Genet, 2003b. 123B(1):23–26.

Benjamin, J., et al., Population and familial association between the D4 dopamine receptor gene and measures of Novelty Seeking. Nat Genet, 1996. 12(1):81–84.

Bertilsson, L., et al., Molecular genetics of CYP2D6: clinical relevance with focus on psychotropic drugs. Br J Clin Pharmacol, 2002. 53(2):111–122.

Bhagwagar, Z., et al. Increased 5-HT(2A) receptor binding in euthymic, medication-free patients recovered from depression: a positron emission study with (11)C]MDL 100,907. Am J Psychiatr, 2006. 163(9):1580–1587.

Bialecka, M., et al., Polymorphisms of catechol-0-methyltransferase (COMT), monoamine oxidase B (MAOB), N-acetyltransferase 2 (NAT2) and cytochrome P450 2D6 (CYP2D6) gene in patients with early onset of Parkinson's disease. Parkinsonism Relat Disord, 2007. 13(4):224–229.

Binder, E., et al., Polymorphisms in FKBP5 are associated with increased recurrence of depressive episodes and rapid response to antidepressant treatment. Nat Genet, 2004. 36(12):1319–1325.

Binder, E.B. and F. Holsboer, Pharmacogenomics and antidepressant drugs. Ann Med, 2006. 38(2):82–94.

Bishop, J.R., et al., Serotonin 2A -1438 G/A and G-protein Beta3 subunit C825T polymorphisms in patients with depression and SSRI-associated sexual side-effects. Neuropsychopharmacology, 2006. 31(10):2281–2288.

Bobb, A.J., et al., Molecular genetic studies of ADHD: 1991 to 2004. Am J Med Genet B Neuropsychiatr Genet, 2005. 132(1):109–125.

Bondy, B., et al., Genetic variants in the angiotensin I-converting-enzyme (ACE) and angiotensin II receptor (AT1) gene and clinical outcome in depression. Prog Neuropsychopharmacol Biol Psychiatr, 2005. 29(6):1094–1099.

Bondy, B., A. Buettner, andZill, Genetics of suicide. Mol Psychiatr, 2006. 11(4):336–351.

Brady, L.S., et al., The antidepressants fluoxetine, idazoxan and phenelzine alter corticotropin-releasing hormone and tyrosine hydroxylase mRNA levels in rat brain: therapeutic implications. Brain Res, 1992. 572(1–2):117–125.

Brockmoller, J., I. Meineke, and J. Kirchheiner, Pharmacokinetics of mirtazapine: enantioselective effects of the CYP2D6 ultra rapid metabolizer genotype and correlation with adverse effects. Clin Pharmacol Ther, 2007. 81(5):699–707.

Brosen, K. and C. Naranjo, Review of pharmacokinetic and pharmacodynamic interaction studies with citalopram. Eur Neuropsychopharmacol, 2001. 11(4):275–283.

Brosen, K., J. Hansen, and K. Nielsen, Inhibition by paroxetine of desipramine metabolism in extensive but not in poor metabolizers of sparteine. Eur J Clin Pharmacol, 1993. 44(4):344–355.

Brunner, H.G., et al., Abnormal behavior associated with a point mutation in the structural gene for monoamine oxidase A. Science, 1993. 262(5133):578–580.

Bruss, M., et al., Chromosomal mapping of the human gene for the tricyclic antidepressant-sensitive noradrenaline transporter. Hum Genet, 1993. 91(3):278–280.

Campbell, D., et al., Fine mapping of the human 5-HTR2a gene to chromosome 13q14 and identification of two highly polymorphic linked markers suitable for association studies in psychiatric disorders. Genet Test, 1997. 1(4):297–299.

Caspi, A., et al., Influence of life stress on depression: moderation by a polymorphism in the 5-HTT gene. Science, 2003. 301(5631):386–389.
Charlier, C., et al., Polymorphisms in the CYP 2D6 gene: association with plasma concentrations of fluoxetine and paroxetine. Ther Drug Monit, 2003. 25(6):738–742.
Chen, G., et al., Regulation of signal transduction pathways and gene expression by mood stabilizers and antidepressants. Psychosom Med, 1999. 61(5):599–617.
Chen, K., et al., The human 5-HT2 receptor is encoded by a multiple intron-exon gene. Brain Res Mol Brain Res, 1992. 14(1–2):20–26.
Chiavegatto, S., et al., Brain serotonin dysfunction accounts for aggression in male mice lacking neuronal nitric oxide synthase. Proc Natl Acad Sci U S A, 2001. 98(3):1277–1281.
Choi, M., et al., Serotonin receptor 2A gene polymorphism (-1438A/G) and short-term treatment response to citalopram. Neuropsychobiology, 2005. 52:155–162.
Choi, M.J., et al., Brain-derived neurotrophic factor gene polymorphism (Val66Met) and citalopram response in major depressive disorder. Brain Res, 2006. 1118(1):176–182.
Clark, J.A., et al., Differential hormonal regulation of tryptophan hydroxylase-2 mRNA in the murine dorsal raphe nucleus. Biol Psychiatr, 2005. 57(8):943–946.
Courtet, P., et al., Association between violent suicidal behavior and the low activity allele of the serotonin transporter gene. Mol Psychiatr, 2001. 6(3):338–341.
Craddock, N., M.C. O'Donovan, and M.J. Owen, Genes for Schizophrenia and Bipolar Disorder? Implications for Psychiatric Nosology. Schizophr Bull, 2006. 32(1):9–16.
Craig, S.P., et al., Localization of human tryptophan hydroxylase (TPH) to chromosome 11p15.3–p14 by in situ hybridization. Cytogenet Cell Genet, 1991. 56(3–4):157–159.
Cramer, J.A. and R. Rosenheck, Compliance with medication regimens for mental and physical disorders. Psychiatr Serv, 1998. 49:196–201.
Crissman, A.M., M.M. Makhay, and J.M. O'Donnell, Discriminative stimulus effects of centrally administered isoproterenol in rats: mediation by beta-1 adrenergic receptors. Psychopharmacology (Berl), 2001. 154(1):70–75.
Crowley, J.J., et al., Pharmacogenomic evaluation of the antidepressant citalopram in the mouse tail suspension test. Neuropsychopharmacology, 2006. 31(11):2433–2442.
Cusin, C., et al., Influence of 5-HTTLPR and TPH Variants on Illness Time Course in Mood Disorders. J Psychiatric Res, 2001. 35(4):217–223.
Cusin, C., et al., Influence of monoamine oxydase A and serotonin receptor 2A polymorphisms in SSRIs antidepressant activity. Int Neuropsychopharmacol, 2002. 5:27–35.
Dalen, P., et al.,10-Hydroxylation of nortriptyline in white persons with 0, 1, 2, 3, and 13 functional CYP2D6 genes. Clin Pharmacol Ther, 1998. 63(4):444–452.
Danish University Antidepressant, G., Clomipramine dose-effect study in patients with depression: clinical end points and pharmacokinetics. Danish University Antidepressant Group (DUAG). Clin Pharmacol Ther, 1999. 66(2):152–165.
de Lara, C.L., et al., Effect of tryptophan hydroxylase-2 gene variants on suicide risk in major depression. Biol Psychiatr, 2007. 62(1):72–80.
de Leon, J., The crucial role of the therapeutic window in understanding the clinical relevance of the poor versus the ultrarapid metabolizer phenotypes in subjects taking drugs metabolized by CYP2D6 or CYP2C19. J Clin Psychopharmacol, 2007. 27(3):241–245.
De Luca, V., et al., Investigation of polymorphism in the MDR1 gene and antidepressant-induced mania. Pharmacogenomics J, 2003. 3(5):297–299.
Dentino, A.N., et al., Association of interleukin-6 and other biologic variables with depression in older people living in the community. J Am Geriatr Soc, 1999. 47(1):6–11.
Di Giovine, F., E. Takhsh, and A. Blakemore, Single base polymorphism at -511 in the human interleukin-1 gene (IL1). Hum Mol Genet, 1992. 1:450.
Dick, F., et al., Gene-environment interactions in parkinsonism and Parkinson's disease: the Geoparkinson study. Occup Environ Med, 2007. 64(10):673–680.
Dunn, A.J., Systemic interleukin-1 administration stimulates hypothalamic norepinephrine metabolism parallelling the increased plasma corticosterone. Life Sci, 1988. 43(5):429–435.

Durham, L.K., et al., The serotonin transporter polymorphism, 5HTTLPR, is associated with a faster response time to sertraline in an elderly population with major depressive disorder. Psychopharmacology (Berl), 2004. 174(4):525–529.

Duric, G., et al. [Polymorphisms in the genes of cytochrome oxidase P450 2D6 (CYP2D6), paraoxonase 1 (PON1) and apolipoprotein E (APOE) as risk factors for Parkinson's disease]. Vojnosanit Pregl, 2007. 64(1):25–30.

Ebstein, R.P., et al., Dopamine D4 receptor (D4DR) exon III polymorphism associated with the human personality trait of Novelty Seeking. Nat Genet, 1996. 12(1):78–80.

Eichelbaum, M., M.F. Fromm, and M. Schwab, Clinical aspects of the MDR1 (ABCB1) gene polymorphism. Ther Drug Monit, 2004. 26(2):180–185.

El-Omar, E.M., et al., Interleukin-1 polymorphisms associated with increased risk of gastric cancer. Nature, 2000. 404(6776):398–402.

Entsuah, A.R., H. Huang, and M.E. Thase, Response and remission rates in different subpopulations with major depressive disorder administered venlafaxine, selective serotonin reuptake inhibitors, or placebo. J Clin Psychiatr, 2001. 62(11):869–77.

Eyada, T.K., et al., Study of genetic polymorphism of xenobiotic enzymes in acute leukemia. Blood Coagul Fibrinolysis, 2007. 18(5):489–495.

Feinn, R., M. Nellissery, and H.R. Kranzler, Meta-analysis of the association of a functional serotonin transporter promoter polymorphism with alcohol dependence. Am J Med Genet B Neuropsychiatr Genet, 2005. 133(1):79–84.

Fojo, A.T., et al., Expression of a multidrug-resistance gene in human tumors and tissues. Proc Natl Acad Sci U S A, 1987. 84(1):265–269.

Francke, U. and B.E. Foellmer, The glucocorticoid receptor gene is in 5q31-q32 [corrected]. Genomics, 1989. 4(4):610–612.

Fuke, S., et al., The VNTR polymorphism of the human dopamine transporter (DAT1) gene affects gene expression. Pharmacogenomics J, 2001. 1(2):152–156.

Fukui, N., et al., Dose-dependent effects of the 3435 C>T genotype of ABCB1 gene on the steady-state plasma concentration of fluvoxamine in psychiatric patients. Ther Drug Monit, 2007. 29(2):185–189.

Garriock, H., et al., Lack of association of TPH2 exon XI polymorphisms with major depression and treatment resistance. Mol Psychiatr, 2005. 10(11):976–977.

Garriock, H.A., et al., Number of risk genotypes is a risk factor for major depressive disorder: a case control study. Behav Brain Funct, 2006. 2:24.

Gaysina, D., et al., The serotonin transporter gene: polymorphism and haplotype analysis in Russian suicide attempters. Neuropsychobiology, 2006. 54(1):70–74.

Gelernter, J., et al., D4 dopamine-receptor (DRD4) alleles and novelty seeking in substance-dependent, personality-disorder, and control subjects. Am J Hum Genet, 1997. 61(5):1144–1152.

Gemma, C., et al., Interleukin-1 induces changes in sleep, brain temperature, and serotonergic metabolism. Am J Physiol, 1997. 272(Pt 2):R601–R606.

Gemma, C., L. Imeri, and M.R. Opp, Serotonergic activation stimulates the pituitary-adrenal axis and alters interleukin-1 mRNA expression in rat brain. Psychoneuroendocrinology, 2003. 28(7):875–884.

Geracitano, R., et al., On the effects of psychostimulants, antidepressants, and the antiparkinsonian drug levodopa on dopamine neurons. Ann N Y Acad Sci, 2006. 1074:320–329.

Giros, B., et al., Cloning, pharmacological characterization, and chromosome assignment of the human dopamine transporter. Mol Pharmacol, 1992. 42(3):383–390.

Golab-Janowska, M., et al., CYP2D6 gene polymorphism as a probable risk factor for Alzheimer's disease and Parkinson's disease with dementia. Neurol Neurochir Pol, 2007. 41(2):113–121.

Gold, P.W. and G.P. Chrousos, Organization of the stress system and its dysregulation in melancholic and atypical depression: high vs. low CRH/NE states. Mol Psychiatr, 2002. 7(3):254–275.

Grasmader, K., et al., Impact of polymorphisms of cytochrome-P450 isoenzymes 2C9, 2C19 and 2D6 on plasma concentrations and clinical effects of antidepressants in a naturalistic clinical setting. Eur J Clin Pharmacol, 2004. 60(5):329–336.

Grossman, M.H., B.S. Emanuel, and M.L. Budarf, Chromosomal mapping of the human catechol-O-methyltransferase gene to 22q11.1–q11.2. Genomics, 1992. 12(4):822–825.

Guasch, J.F., R.M. Bertina, and P.H. Reitsma, Five novel intragenic dimorphisms in the human interleukin-1 genes combine to high informativity. Cytokine, 1996. 8(8):598–602.

Gutierrez, B., et al., Variability in the serotonin transporter gene and increased risk for major depression with melancholia. Hum Gen, 1998. 103(3):319–322.

Gutknecht, L., et al., Tryptophan hydroxylase-2 gene variation influences personality traits and disorders related to emotional dysregulation. Int J Neuropsychopharmacol, 2007. 10(3):309–320.

Hahn, M.K., M.S. Mazei-Robison, and R.D. Blakely, Single nucleotide polymorphisms in the human norepinephrine transporter gene affect expression, trafficking, antidepressant interaction, and protein kinase C regulation. Mol Pharmacol, 2005. 68(2):457–466.

Ham, B., et al., No association between the tryptophan hydroxylase gene polymorphism and major depressive disorders and antidepressant response in a Korean population. Psychiatr Genet, 2005. 15(4):229–301.

Ham, B.J., et al., Association between the tryptophan hydroxylase-1 gene A218C polymorphism and citalopram antidepressant response in a Korean population. Prog Neuropsychopharmacol Biol Psychiatr, 2007. 31(1):104–107.

Heils, A., et al., Allelic variation of human serotonin trasporter gene expression. J Neurochem, 1996. 66(6):2621–2624.

Hemeryck, A. and F.M. Belpaire, Selective serotonin reuptake inhibitors and cytochrome P-450 mediated drug-drug interactions: an update. Curr Drug Metab, 2002. 3(1):13–37.

Hemrick-Luecke, S.K., H.D. Snoddy, and R.W. Fuller, Evaluation of nefazodone as a serotonin uptake inhibitor and a serotonin antagonist in vivo. Life Sci, 1994. 55(7):479–483.

Herbert, T.B. and S. Cohen, Depression and immunity: a meta-analytic review. Psychol Bull, 1993. 113(3):472–486.

Holmes, A., D.L. Murphy, and J.N. Crawley, Abnormal behavioral phenotypes of serotonin transporter knockout mice: parallels with human anxiety and depression. Biol Psychiatr, 2003a. 54(10):953–959.

Holmes, A., et al., Mice lacking the serotonin transporter exhibit 5-HT(1A) receptor-mediated abnormalities in tests for anxiety-like behavior. Neuropsychopharmacology, 2003b. 28(12):2077–2088.

Holsboer, F., The corticosteroid receptor hypothesis of depression. Neuropsychopharmacology, 2000. 23(5):477–501.

Hong, C.J., Y.C. Wang, and S.J. Tsai, Association study of angiotensin I-converting enzyme polymorphism and symptomatology and antidepressant response in major depressive disorders. J Neural Transm, 2002. 109(9):1209–1214.

Hong, C.J., et al., Response to fluoxetine and serotonin 1A receptor (C-1019G) polymorphism in Taiwan Chinese major depressive disorder. Pharmacogenomics J, 2006. 6(1):27–33.

Hranilovic, D., et al., Serotonin transporter promoter and intron 2 polymorphisms: relationship between allelic variants and gene expression. Biol Psychiatr, 2004. 55(11):1090–1094.

Hrdina, P.D. and T.B. Vu, Chronic fluoxetine treatment upregulates 5-HT uptake sites and 5-HT2 receptors in rat brain: an autoradiographic study. Synapse, 1993. 14(4):324–331.

Hu, X., et al., An expanded evaluation of the relationship of four alleles to the level of response to alcohol and the alcoholism risk. Alcohol Clin Exp Res, 2005. 29(1):8–16.

Hu, X.Z., et al., Association between a functional serotonin transporter promoter polymorphism and citalopram treatment in adult outpatients with major depression. Arch Gen Psychiatr, 2007. 64(7):783–792.

Ito, K., et al., A variable number of tandem repeats in the serotonin transporter gene does not affect the antidepressant response to fluvoxamine. Psychiatr Res, 2002. 111(2–3):235–239.

Itokawa, M., et al., A structural polymorphism of human dopamine D2 receptor, D2(Ser311 -->Cys). Biochem Biophys Res Commun, 1993. 196(3):1369–1375

Jezova, D., et al., Brain angiotensin II modulates sympathoadrenal and hypothalamic pituitary adrenocortical activation during stress. J Neuroendocrinol, 1998. 10(1):67–72.

Jonsson, E.G., et al., Tryptophan hydroxylase and catechol-O-methyltransferase gene polymorphisms: relationships to monoamine metabolite concentrations in CSF of healthy volunteers. Eur Arch Psychiatr & Clin Neurosci, 1997. 247(6):297–302.

Joyce, P.R., et al., Age-dependent antidepressant pharmacogenomics: polymorphisms of the serotonin transporter and G protein beta3 subunit as predictors of response to fluoxetine and nortriptyline. Int J Neuropsychopharmacol, 2003. 6(4):339–346.

Kabiersch, A., et al., Interleukin-1 induces changes in norepinephrine metabolism in the rat brain. Brain Behav Immun, 1988. 2(3):267–274.

Kang, R.H., et al., Association study of the serotonin transporter promoter polymorphism and mirtazapine antidepressant response in major depressive disorder. Prog Neuropsychopharmacol Biol Psychiatr, 2007. 31(6):1317–1321.

Kato, M., et al., Controlled clinical comparison of paroxetine and fluvoxamine considering the serotonin transporter promoter polymorphism. Int Clin Psychopharmacol, 2005. 20(3):151–156.

Kato, M., et al. Effects of the serotonin type 2A, 3A and 3B receptor and the serotonin transporter genes on paroxetine and fluvoxamine efficacy and adverse drug reactions in depressed Japanese patients. Neuropsychobiology, 2006. 53(4):186–195.

Katzenberg, D., et al., A CLOCK polymorphism associated with human diurnal preference. Sleep, 1998. 21(6):569–576.

Kehne, J.H., The CRF1 receptor, a novel target for the treatment of depression, anxiety, and stress-related disorders. CNS Neurol Disord Drug Targets, 2007. 6(3):163–182.

Kiecolt-Glaser, J.K. and R. Glaser, Depression and immune function: central pathways to morbidity and mortality. J Psychosom Res, 2002. 53(4):873–876.

Kim, C.H., et al., A polymorphism in the norepinephrine transporter gene alters promoter activity and is associated with attention-deficit hyperactivity disorder. Proc Natl Acad Sci U S A, 2006. 103(50):19164–19169.

Kim, D.K., et al., Serotonin transporter gene polymorphism and antidepressant response. Neuroreport, 2000. 11(1):215–219.

Kim, H., eta l., Monoaminet ransporterge nepol ymorphismsa nda ntidepressantr esponsei nk oreans with late-life depression. Jama, 2006. 296(13):1609–1618.

Kim, S.W., S.Y. Park, and O. Hwang, Up-regulation of tryptophan hydroxylase expression and serotonin synthesis by sertraline. Mol Pharmacol, 2002. 61(4):778–785.

Kirchheiner, J., et al., CYP2D6 and CYP2C19 genotype-based dose recommendations for antidepressants: a first step towards subpopulation-specific dosages. Acta Psychiatr Scand, 2001. 104(3):173–192.

Kirchheiner, J., et al., Impact of the CYP2D6 ultrarapid metabolizer genotype on mirtazapine pharmacokinetics and adverse events in healthy volunteers. J Clin Psychopharmacol, 2004. 24(6):647–652.

Kirchheiner, J., D. Grundemann, and E. Schomig, Contribution of allelic variations in transporters to the phenotype of drug response. J Psychopharmacol, 2006a. 20(Suppl 4):27–32.

Kirchheiner, J., et al., A 40-basepair VNTR polymorphism in the dopamine transporter (DAT1) gene and the rapid response to antidepressant treatment. Pharmacogenomics J, 2006b.

Kirchheiner, J., et al., A 40-basepair VNTR polymorphism in the dopamine transporter (DAT1) gene and the rapid response to antidepressant treatment. Pharmacogenomics J, 2007. 7(1):48–55.

Kirigiti, P., et al., Rat beta 1-adrenergic receptor regulatory region containing consensus AP-2 elements recognizes novel transactivator proteins. Mol Cell Biol Res Commun, 2000. 3(3):181–192.

Kishimoto, J., et al., Localization of brain nitric oxide synthase (NOS) to human chromosome 12. Genomics, 1992. 14(3):802–804.

Kobilka, B.K., et al., An intronless gene encoding a potential member of the family of receptors coupled to guanine nucleotide regulatory proteins. Nature, 1987. 329(6134):75–79.

Kohen, R., et al., Cloning, characterization, and chromosomal localization of a human 5-HT6 serotonin receptor. J Neurochem, 1996. 66(1):47–56.

Kraft, J., et al., Sequence analysis of the serotonin transporter and associations with antidepressant response. Biol Psychiatr, 2005. 58(374–381).

Kraft, J.B., et al., Analysis of association between the serotonin transporter and antidepressant response in a large clinical sample. Biol Psychiatr, 2007. 61(6):734–742.

Kramer, M.S., et al., Distinct mechanism for antidepressant activity by blockade of central substance P receptors. Science, 1998. 281(5383):1640–1645.

Kremer, I., et al., Association of the serotonin transporter gene with smoking behavior. Am J Psychiatr, 2005. 162(5):924–930.

Lachman, H.M., et al., Association of codon 108/158 catechol-O-methyltransferase gene polymorphism with the psychiatric manifestations of velo-cardio-facial syndrome. Am J Med Genet, 1996. 67(5):468–472.

Lachman, H.M., et al., Linkage studies suggest a possible locus for bipolar disorder near the velo-cardio-facial syndrome region on chromosome 22. Am J Med Genet, 1997. 74:121–128.

Laika, B., S. Leucht, and W. Steimer, ABCB1 (P-glycoprotein/MDR1) gene G2677T/a sequence variation (polymorphism): lack of association with side effects and therapeutic response in depressed inpatients treated with amitriptyline. Clin Chem, 2006. 52(5):893–895.

Landen, M. and M.E. Thase, A model to explain the therapeutic effects of serotonin reuptake inhibitors: the role of 5-HT2 receptors. Psychopharmacol Bull, 2006. 39(1):147–166.

Lavandera, J.V., et al., CYP2D6 polymorphisms in patients with porphyrias. Mol Med, 2006. 12(9–10):259–263.

Le Beau, M.M. and J.D. Rowley, Chromosomal abnormalities in leukemia and lymphoma: clinical and biological significance. Adv Hum Genet, 1986. 15:1–54.

Lee, H.J., et al., Association between a G-protein beta3 subunit gene polymorphism and the symptomatology and treatment responses of major depressive disorders. Pharmacogenomics J, 2004. 4(1):29–33.

Lee, M.S., et al., Serotonin transporter promoter gene polymorphism and long-term outcome of antidepressant treatment. Psychiatr Genet, 2004. 14(2):111–115.

Lee, S., et al., Association between the 5-HT6 receptor C267T polymorphism and response to antidepressant treatment in major depressive disorder. Psychiatr Clin Neurosci, 2005. 59:140–145.

Lee, S.Y., et al., Sequence-based CYP2D6 genotyping in the Korean population. Ther Drug Monit, 2006. 28(3):382–387.

Lemonde, S., et al., Impaired trans-repression at a 5-HT1A receptor gene polimorphism associated with major depression and suicide. J Neurosci, 2003. 23(25):8788–8799.

Lemonde, S., et al., Association of the C(1019)G 5-HT1A functional promoter polymorphism with antidepressant response. Psychopharmacology (Berl), 2004. 24:24.

Lemos, M.C., et al., Genetic polymorphism of CYP2D6 influences susceptibility to papillary thyroid cancer. Clin Endocrinol (Oxf), 2007. 67(2):180–183.

Lesch, K.P., et al., Organization of the human serotonin transporter gene. J Neural Transm Gen Sect, 1994. 95(2):157–162.

Levine, M.A., et al.., Molecular cloning of beta-3 subunit, a third form of the G protein beta-subunit polypeptide. Proc. Nat. Acad. Sci U S A, 1990. 87:2329–2333.

Li, Z., K. Inenaga, and H. Yamashita, GABAergic inputs modulate effects of interleukin-1 beta on supraoptic neurones in vitro. Neuroreport, 1993. 5(2):181–183.

Licinio, J. and M.L. Wong, The role of inflammatory mediators in the biology of major depression: central nervous system cytokines modulate the biological substrate of depressive symptoms, regulate stress-responsive systems, and contribute to neurotoxicity and neuroprotection. Mol Psychiatr, 1999. 4(4):317–327.

Licinio, J.,et al., Association of a corticotropin-releasing hormone receptor 1 haplotype and antidepressant treatment response in Mexican-Americans. Mol Psychiatr, 2004. 9: 1075–1082.

Lim, J.E.,et al., Tryptophan hydroxylase 2 (TPH2) haplotypes predict levels of TPH2 mRNA expression in human pons. Mol Psychiatr, 2007. 12(5):491–501.

Lin, J.H. and A.Y. Lu, Inhibition and induction of cytochrome P450 and the clinical implications. Clin Pharmacokinet, 1998. 35(5):361–390.
Linthorst, A.C., et al., Local administration of recombinant human interleukin-1 beta in the rat hippocampus increases serotonergic neurotransmission, hypothalamic-pituitary-adrenocortical axis activity, and body temperature. Endocrinology, 1994. 135(2):520–532.
Liu, Z., et al., Association study of corticotropin-releasing hormone receptor1 gene polymorphisms and antidepressant response in major depressive disorders. Neurosci Lett, 2007. 414(2):155–158.
Liu, Z.C., X.N. Luo, and G.H. Wang, Corticotropin-releasing factor andmajor depression. Foreign Med Sci: Section of Psychiatry , 2002. 2:156–158.
Lopez de Lara, C., et al., STin2 variant and family history of suicide as significant predictors of suicide completion in major depression. Biol Psychiatr, 2006. 59(2):114–120.
MacKenzie, A. and J. Quinn, A serotonin transporter gene intron 2 polymorphic region, correlated with affective disorders, has allele-dependent differential enhancer-like properties in the mouse embryo. Proc Natl Acad Sci U S A, 1999. 96(26):15251–15255.
Maes, M. and H.Y. Meltzer, The serotonin hypothesis of major depression, in Psychopharmacology: The fourth generation of progress, F.E. Bloom and D.J. Kupfer, Editors. 1995, Raven Press: New York, pp 933–944.
Maes, M., Interleukin-1 beta: a putative mediator of HPA axis hyperactivity in major depression? Am J Psychiatr, 1993. 150(8):1189–1193.
Maj, J., et al., The effects of paroxetine given repeatedly on the 5-HT receptor subpopulations in the rat brain. Psychopharmacology (Berl), 1996. 127(1):73–82.
Malhotra, A.K. and D. Goldman, The dopamine D(4) receptor gene and novelty seeking. Am J Psychiatr, 2000. 157(11):1885–1886.
Manev, H. and T. Uz, Clock genes: influencing and being influenced by psychoactive drugs. Trends Pharmacol Sci, 2006. 27(4):186–189.
Mannisto, and S. Kaakkola, Catechol-O-methyltransferase (COMT): biochemistry, molecular biology, pharmacology, and clinical efficacy of the new selective COMT inhibitors. Pharmacol Rev, 2005. 51(4):593–628.
Manuck, S.B., et al., A regulatory polymorphism of the monoamine oxidase-A gene may be associated with variability in aggression, impulsivity, and central nervous system serotonergic responsivity. Psychiatr Res, 2000. 95(1):9–23.
Marsh, S. and H.L. McLeod, Pharmacogenetics and oncology treatment for breast cancer. Expert Opin Pharmacother, 2007. 8(2):119–127.
Masand, P.S., Tolerability and adherence issues in antidepressant therapy. Clin Ther, 2003. 25(8):2289–2304.
Mason, D.A., et al., A gain-of-function polymorphism in a G-protein coupling domain of the human beta1-adrenergic receptor. J Biol Chem, 1999. 274(18):12670–12674.
Matsushita, S., et al., Serotonin transporter regulatory region polymorphism is associated with anorexia nervosa. Am J Med Genet B Neuropsychiatr Genet, 2004. 128(1):114–117.
McMahon, F.J., et al., Variation in the gene encoding the serotonin 2A receptor is associated with outcome of antidepressant treatment. Am J Hum Genet, 2006. 78(5):804–814.
Meyer, J.H., et al., The effect of paroxetine on 5-HT2A receptors in depression: an [18F]Setoperone PET imaging study. Am J Psychiatr, 2001. 158:78–85.
Michelson, D., et al., Chronic imipramine is associated with diminished hypothalamic-pituitary-adrenal axis responsivity in healthy humans. J Clin Endocrinol Metab, 1997. 82(8):2601–2606.
Mignone, F., et al., Untranslated regions of mRNAs. Genome Biol, 2002. 3(3):REVIEWS0004.
Miller, L.G., et al., Interleukin-1 augments gamma-aminobutyric acidA receptor function in brain. Mol Pharmacol, 1991. 39(2):105–108.
Minov, C., et al., Serotonin-2A-receptor and -transporter polymorphisms: lack of association in patients with major depression. Neurosci Lett, 2001. 303(2):119–122.
Mitchell, A.J., Two-week delay in onset of action of antidepressants: new evidence. Br J Psychiatr, 2006. 188:105–106.

Mitchell, E.S., T. Sexton, and J.F. Neumaier, Increased expression of 5-HT6 receptors in the rat dorsomedial striatum impairs instrumental learning. Neuropsychopharmacology, 2007. 32(7):1520–1530.
Moncrieff, J. and I. Kirsch, Efficacy of antidepressants in adults. BMJ, 2005. 331(7509): 551–557.
Mossner, R., et al., Aberrant accumulation of serotonin in dopaminergic neurons. Neurosci Lett, 2006. 401(1–2):49–54.
Muller, D.J., . Moclobemide response in depressed patients: association study with a functional polymorphism in the monoamine oxidase-A promoter. in Eighth World Congress on Psychiatric Genetics. 2000. Versailles, France: Neuropsychiatric Genetics, Wiley-Liss.
Muller, M. and R. Renkawitz, The glucocorticoid receptor. Biochim Biophys Acta, 1991. 1088(2):171–182.
Mundo, E., et al., The Role of Serotonin Transporter Protein Gene in Antidepressant-Induced Mania in Bipolar Disorder. Arch Gen Psychiatr, 2001. 58:539–544.
Murphy, G., et al. Effects of the serotonin transporter promoter polymorphism on paroxetine and mirtazapine efficacy and side effects in geriatric major depression. in Pharmacogenetics in psychiatry meeting. 2003. New York.
Murphy, G.M., Jr., et al., Pharmacogenetics of Antidepressant Medication Intolerance. Am J Psychiatr, 2003. 160(10):1830–1835.
Murphy, G.M., Jr., et al.., Effects of the serotonin transporter gene promoter polymorphism on mirtazapine and paroxetine efficacy and adverse events in geriatric major depression. Arch Gen Psychiatr, 2004. 61(11):1163–1169.
Myers, R. L., Airey, D. C., Manier, D. H., Shelton, R. C., Sanders-Bush, E. Polymorphisms in the regulatory region of the human serotonin 5-HT2A receptor gene (HTR2A) influence gene expression. Biol Psychiatry, 2007. 61(2):167–173.
Nebert, D. and M. Dieter, The evolution of drug metabolism. Pharmacology, 2002. 61(3):124–135.
Nemeroff, C.B., et al., Elevated concentrations of CSF corticotropin-releasing factor-like immunoreactivity in depressed patients. Science, 1984. 226(4680):1342–1344.
Nemeroff, C.B., et al., Reduced corticotropin releasing factor binding sites in the frontal cortex of suicide victims. Arch Gen Psychiatr, 1988. 45(6):577–579.
Newman, T.K., et al., Monoamine oxidase A gene promoter variation and rearing experience influences aggressive behavior in rhesus monkeys. Biol Psychiatr, 2005. 57(2):167–172.
Newton, R.A., et al., Characterisation of human 5-hydroxytryptamine2A and 5-hydroxytryptamine2C receptors expressed in the human neuroblastoma cell line SH-SY5Y: comparative stimulation by hallucinogenic drugs. J Neurochem, 1996. 67(6):2521–2531.
Niculescu, A.B., et al., Identifying a series of candidate genes for mania and psychosis: a convergent functional genomics approach. Physiol Genom, 2000. 4(1):83–91.
Nierenberg, A.A., Predictors of response to antidepressants general principals and clinical implications. Psychiatr Clin North Am, 2003. 26:345–352.
Nobile, M., et al., A case-control and family-based association study of the 5-HTTLPR in pediatric-onset depressive disorders. Biol Psychiatr, 2004. 56(4):292–295.
Nutt, D., Substance-P antagonists: a new treatment for depression? Lancet, 1998. 352(9141):1644–1646.
O'Reilly, R.L., L. Bogue, and S.M. Singh, Pharmacogenetic response to antidepressants in a multicase family with affective disorder. Biol Psychiatr, 1994. 36(7):467–471.
Ogilvie, A.D., et al., Polymorphism in serotonin transporter gene associated with susceptibility to major depression. Lancet, 1996. 347:731–733.
Ohara, K., et al., CYP2D6*10 alleles do not determine plasma fluvoxamine concentration/dose ratio in Japanese subjects. Eur J Clin Pharmacol, 2003. 58(10):659–661.
Okamura, K., et al., Lack of an association between 5-HT receptor gene polymorphisms and suicide victims. Psychiatr Clin Neurosci, 2005. 59(3):345–349.
Olsavsky, K.M., et al., Gene expression profiling and differentiation assessment in primary human hepatocyte cultures, established hepatoma cell lines, and human liver tissues. Toxicol Appl Pharmacol, 2007. 222(1):42–56.

Overstreet, D. and G. Griebel, Antidepressant-like effects of CRF1 receptor antagonist SSR125543 in an animal model of depression. Eur J Pharmacol, 2004. 497(1):49–53.
Ozdemir, V., R. Tyndale, and K. Reed, Paroxetine steady-state plasma concentration in relation to CYP2D6 genotype in extensive metabolizers. J Clin Psychopharmacol, 1999. 19(5):472–475.
Pae, C.U., et al., Dysbindin associated with selective serotonin reuptake inhibitor antidepressant efficacy. Pharmacogenetics Genomics, 2007. 17(1):69–75.
Park, J.W., et al., Serotonin transporter polymorphism and harm avoidance personality in chronic tension-type headache. Headache, 2004. 44(10):1005–1009.
Parsian, A., Sequence analysis of exon 8 of MAO-A gene in alcoholics with antisocial personality and normal controls. Genomics, 1999. 55(3):290–295.
Paul, I.A., Antidepressant activity and calcium signaling cascades. Hum Psychopharmacol, 2001. 16(1):71–80.
Perez, V., et al.., Randomised, double-blind, placebo-controlled trial of pindolol in combination with fluoxetine antidepressant treatment. Lancet, 1997. 349:1594–1597.
Perlis, R.H., et al., Serotonin transporter polymorphisms and adverse effects with fluoxetine treatment. Biol Psychiatr, 2003. 54(9):879–883.
Perlis, R.H., Pharmacogenetic studies of antidepressant response: how far from the clinic? Psychiatr Clin North Am, 2007. 30(1): 125–138.
Peters, E.J., et al., Investigation of serotonin-related genes in antidepressant response. Mol Psychiatr, 2004. 9(9):879–889.
Pociot, F., et al., A TaqI polymorphism in the human interleukin-1 beta (IL-1 beta) gene correlates with IL-1 beta secretion in vitro. Eur J Clin Invest, 1992. 22(6):396–402.
Pollock, B.G., et al., Allelic variation in the serotonin transporter promoter affects onset of paroxetine treatment response in late-life depression. Neuropsychopharmacology, 2000. 23(5):587–590.
Popp, J., et al., Serotonin transporter polymorphisms and side effects in antidepressant therapy–a pilot study. Pharmacogenomics, 2006. 7(2):159–166.
Preisig, M., F. Ferrero, and A. Malafosse, Monoamine oxidase a and tryptophan hydroxylase gene polymorphisms: are they associated with bipolar disorder? Am J Pharmacogenomics, 2005. 5(1):45–52.
Raadsheer, F.C., et al., Increased numbers of corticotropin-releasing hormone expressing neurons in the hypothalamic paraventricular nucleus of depressed patients. Neuroendocrinology, 1994. 60(4):436–444.
Rada, P., et al., Interleukin-1 beta decreases acetylcholine measured by microdialysis in the hippocampus of freely moving rats. Brain Res, 1991. 550(2):287–290.
Ramamoorthy, S., et al.., Antidepressant- and cocaine-sensitive human serotonin transporter: molecular cloning, expression, and chromosomal localization. Proc Natl Acad Sci U S A, 1993. 90(6):2542–2546.
Rao, N., The clinical pharmacokinetics of escitalopram. Clin Pharmacokinet, 2007. 46(4):281–290.
Rau, T., et al., CYP2D6 genotype: impact on adverse effects and nonresponse during treatment with antidepressants-a pilot study. Clin Pharmacol Ther, 2004. 75(5):386–393.
Rausch, J.L., et al., Initial conditions of serotonin transporter kinetics and genotype: influence on ssri treatment trial outcome. Biol Psychiatr, 2002. 51(9):723–732.
Reppert, S.M. and D.R. Weaver, Coordination of circadian timing in mammals. Nature, 2002. 418(6901):935–941.
Rigat, B., et al., An insertion/deletion polymorphism in the angiotensin I-converting enzyme gene accounting for half the variance of serum enzyme levels. J Clin Invest, 1990. 86(4):1343–1346.
Roberts, R.L., et al., A common P-glycoprotein polymorphism is associated with nortriptyline-induced postural hypotension in patients treated for major depression. Pharmacogenomics J, 2002. 2(3):191–196.
Rosskopf, D., ., G protein beta 3 gene: structure, promoter, and additional polymorphisms. Hypertension, 2000. 36(1):33–41.

Rousseva, A., et al., Antidepressant-induced mania, rapid cycling and the serotonin transporter gene polymorphism. Pharmacogenomics J, 2003. 3(2):101–104.

Ruiz-Velasco, V. and S.R. Ikeda, A splice variant of the G protein beta 3-subunit implicated in disease states does not modulate ion channels. Physiol Genomics, 2003. 13(2):85–95.

Rujescu, D., et al., Genetic variations in tryptophan hydroxylase in suicidal behavior: analysis and meta-analysis. Biol Psychiatr, 2003. 54(4):465–473.

Sabol, S.Z., S. Hu, and D. Hamer, A functional polymorphism in the monoamine oxidase A gene promoter. Human Genetic, 1998. 103:273–279.

Sakai, K., et al., The genomic organization of the human corticotropin-releasing factor type-1 receptor. Gene, 1998. 219(1–2):125–130.

Sakowski, S.A., et al., Differential tissue distribution of tryptophan hydroxylase isoforms 1 and 2 as revealed with monospecific antibodies. Brain Res, 2006. 1085(1):11–18.

Sato, K., et al., Association between -1438G/A promoter polymorphism in the 5-HT(2A) receptor gene and fluvoxamine response in Japanese patients with major depressive disorder. Neuropsychobiology, 2002. 46(3):136–140.

Scheuch, K., et al., Characterization of a functional promoter polymorphism of the human tryptophan hydroxylase 2 gene in serotonergic raphe neurons. Biol Psychiatr, 2007. 62(11):1288–1294.

Scordo, M.G., et al., Influence of CYP2C9, 2C19 and 2D6 genetic polymorphisms on the steady-state plasma concentrations of the enantiomers of fluoxetine and norfluoxetine. Basic Clin Pharmacol Toxicol, 2005. 97(5):296–301.

Sei, H., et al., Clock mutant mice with Jcl/ICR background shows an impaired learning ability in water maze, but not in passive avoidance, at the beginning of dark phase. Congenit Anom (Kyoto), 2006. 46(2):81–85.

Serretti, A., et al., Mode of inheritance in mood disorders families according to fluvoxamine response. Acta Psychiatrica Scandinavica, 1998. 98(6):443–450.

Serretti, A., et al., Dopamine receptor D4 is not associated with antidepressant activity of sleep deprivation. Psychiatr Res, 1999. 89:107–114.

Serretti, A., et al., Influence of tryptophan hydroxylase and serotonin transporter genes on fluvoxamine antidepressant activity. Mol Psychiatr, 2001a. 6:586–592.

Serretti, A., et al., No association between dopamine D2 and D4 receptor gene variants and antidepressant activity of two selective serotonin reuptake inhibitors. Psychiatr Res, 2001b. 104(3):195–203.

Serretti, A., et al., Tryptophan hydroxylase gene associated with paroxetine antidepressant activity. Eur Neuropsychopharmacol, 2001c. 11(5):375–380.

Serretti, A., et al.Genetic dissection of psychopathological symptoms: Insomnia in mood disorders and CLOCK gene polymorphism. Am J Med Genet, 2003a. 121B(1):39–43.

Serretti, A., et al., SSRIs antidepressant activity is influenced by Gbeta3 variants. Eur Neuropsychopharmacol, 2003b. 13(2):117–122.

Serretti, A., et al., Further evidence of a combined effect of SERTPR and TPH on SSRIs response in mood disorders. Am J Med Genet B Neuropsychiatr Genet, 2004a. 129(1):36–40.

Serretti, A., et al., Genetic features of antidepressant induced mania and hypo-mania in bipolar disorder. Psychopharmacology (Berl), 2004b. 174(4):504–511.

Serretti, A., et al., Pharmacogenetics of selective serotonin reuptake inhibitor response: a 6-month follow-up. Pharmacogenetics, 2004c. 14(9):607–613.

Serretti, A., et al., The C(-1019)G polymorphism of the 5-HT1A gene promoter and antidepressant response in mood disorders: preliminary findings. Int J Neuropsychopharmacol, 2004d. 7(4):453–460.

Serretti, A., et al., Insomnia improvement during antidepressant treatment and CLOCK gene polymorphism. Am J Med Genet B Neuropsychiatr Genet, 2005a. 10:10.

Serretti, A., et al., The influence of Serotonin Transporter Promoter Polymorphism (SERTPR) and other polymorphisms of the serotonin pathway on the efficacy of antidepressant treatments. Prog Neuropsychopharmacol Biol Psychiatr, 2005b. 29(6):1074–1084.

Serretti, A., et al., Serotonin transporter gene variants and behavior: a comprehensive review. Curr Drug Targets, 2006a. 7(12):1659–1669.

Serretti, A., et al.Temperament and character in mood disorders: influence of DRD4, SERTPR, TPH and MAO-A polymorphisms. Neuropsychobiology, 2006b. 53(1):9–16.

Serretti, A., et al., Meta-analysis of serotonin transporter gene promoter polymorphism (5-HTTLPR) association with selective serotonin reuptake inhibitor efficacy in depressed patients. Mol Psychiatr, 2007a. 12(3):247–257.

Serretti, A., et al., Serotonin transporter gene influences the time course of improvement of "core" depressive and somatic anxiety symptoms during treatment with SSRIs for recurrent mood disorders. Psychiatr Res, 2007b. 149(1–3):185–193.

Seymour, P., A. Schmidt, and D. Schulz, The pharmacology of CP-154,526, a non-peptide antagonist of the CRH1 receptor: a review. CNS Drug Rev, 2003. 9(1):57–96.

Shams, M.E., et al., CYP2D6 polymorphism and clinical effect of the antidepressant venlafaxine. J Clin Pharm Ther, 2006. 31(5):493–502.

Shang, Y., et al., Displacement of serotonin and dopamine transporters by venlafaxine extended release capsule at steady state: a [123I]2beta-carbomethoxy-3beta-(4-iodophenyl)-tropane single photon emission computed tomography imaging study. J Clin Psychopharmacol, 2007. 27(1):71–75.

Shintani, F., et al., Interleukin-1 beta augments release of norepinephrine, dopamine, and serotonin in the rat anterior hypothalamus. J Neurosci, 1993. 13(8):3574–3581.

Shirayama, Y., et al., Reduction of substance P after chronic antidepressants treatment in the striatum, substantia nigra and amygdala of the rat. Brain Res, 1996. 739(1–2):70–78.

Siffert, W., et al., Association of a human G-protein beta3 subunit variant with hypertension. Nat Genet, 1998. 18(1):45–48.

Sims, K.B., et al., Monoamine oxidase deficiency in males with an X chromosome deletion. Neuron, 1989. 2(1):1069–1076.

Skrebuhhova, T., L. Allikmets, and V. Matto, Effects of anxiogenic drugs in rat forced swimming test. Meth Find Exp Clin Pharmacol, 1999. 21(3):173–178.

Smeraldi, E., eta l., Polymorphismw ithint hepr omoterof t hes erotonint ransporterge nea nd antidepressant efficacy of fluvoxamine. Mol Psychiatr, 1998. 3(6):508–511.

Smeraldi, E., et al., Serotonin transporter gene-linked polymorphic region: possible pharmacogenetic implications of rare variants. Psychiatr Genet, 2006. 16(4):153–158.

Smith, D.A., et al., Human cytochrome P450s: selectivity and measurement in vivo. Xenobiotica, 1998. 28(12):1095–1128.

Smith, G., et al., Molecular genetics of the human cytochrome P450 monooxygenase superfamily. Xenobiotica, 1998. 28(12):1129–1165.

Smits, K., et al., Serotonin transporter polymorphisms and the occurrence of adverse events during treatment with selective serotonin reuptake inhibitors. Int Clin Psychopharmacol, 2007. 22(3):137–143.

Smits, K.M., et al., Influence of SERTPR and STin2 in the serotonin transporter gene on the effect of selective serotonin reuptake inhibitors in depression: a systematic review. Mol Psychiatr, 2004. 9(5):433–441.

Smits, K.M., et al., Does pretreatment testing for serotonin transporter polymorphisms lead to earlier effects of drug treatment in patients with major depression? A decision-analytic model. Clin Ther, 2007. 29(4):691–702.

Spurlock, G., et al., A family based association study of T102C polymorphism in 5HT2A and schizophrenia plus identification of new polymorphisms in the promoter. Mol Psychiatr, 1998. 3(1):42–49.

Steeves, T.D., et al., Molecular cloning and characterization of the human clock gene: mexpression in the suprachiasmatic nuclei. Genomics, 1999. 57:198–200.

Steiger, H., et al.The 5HTTLPR polymorphism, psychopathologic symptoms, and platelet [3H-] paroxetine binding in bulimic syndromes. Int J Eat Disord, 2005. 37(1):57–60.

Steimer, W., et al., Amitriptyline or not, that is the question: pharmacogenetic testing of CYP2D6 and CYP2C19 identifies patients with low or high risk for side effects in amitriptyline therapy. Clin Chem, 2005. 51(2):376–385.

Stockmeier, C.A., et al., Increase in serotonin-1A autoreceptors in the midbrain of suicide victims with major depression-postmortem evidence for decreased serotonin activity. J Neurosci, 1998. 18(18):7394–7401.
Sugai, T., et al., The effect of 5-hydroxytryptamine 3A and 3B receptor genes on nausea induced by paroxetine. Pharmacogenomics J, 2006. 6(5):351–356.
Suzuki, E., et al., Long-term imipramine treatment increases nitrate levels in the rat hypothalamus. Cell Mol Neurobiol, 2003. 23(6):953–962.
Suzuki, Y., K. Sawamura, and T. Someya, The effects of a 5-hydroxytryptamine 1A receptor gene polymorphism on the clinical response to fluvoxamine in depressed patients. Pharmacoeconomics J, 2004. 4(4):283–286.
Suzuki, Y., K. Sawamura, and T. Someya, Polymorphisms in the 5-hydroxytryptamine 2A receptor and CytochromeP4502D6 genes synergistically predict fluvoxamine-induced side effects in japanese depressed patients. Neuropsychopharmacology, 2006. 31(4):825–831.
Svenningsson, P., et al., Biochemical and behavioral evidence for antidepressant-like effects of 5-HT6 receptor stimulation. J Neurosci, 2007. 27(15):4201–4209.
Szegedi, A., et al., The catechol-O-methyltransferase Val108/158Met polymorphism affects short-term treatment response to mirtazapine, but not to paroxetine in major depression. Pharmacogenomics J, 2005. 5(1):49–53.
Tadic, A., et al., The MAOA T941G polymorphism and short-term treatment response to mirtazapine and paroxetine in major depression. Am J Med Genet B Neuropsychiatr Genet, 2007. 144(3):325–331.
Takahashi, H., et al., No association between the serotonergic polymorphisms and incidence of nausea induced by fluvoxamine treatment. Eur Neuropsychopharmacol, 2002. 12(5):477–481.
Thuerauf, N. and J. Lunkenheimer, The impact of the CYP2D6-polymorphism on dose recommendations for current antidepressants. Eur Arch Psychiatr Clin Neurosci, 2006. 256(5):287–293.
Trent, J.M. and C.M. Witkowski, Clarification of the chromosomal assignment of the human P-glycoprotein/mdr1 gene: possible coincidence with the cystic fibrosis and c-met oncogene. Cancer Genet Cytogenet, 1987. 26(1):187–190.
Tringali, G., et al., Evidence for the neuronal origin of immunoreactive interleukin-1 beta released by rat hypothalamic explants. Neurosci Lett, 1996. 219(3):143–146.
Tsai, S.J., et al., Association study of serotonin 1B receptor (A-161T) genetic polymorphism and suicidal behaviors and response to fluoxetine in major depressive disorder. Neuropsychobiology, 2004. 50(3):235–238.
Usiello, A., et al., Distinct functions of the two isoforms of dopamine D2 receptors. Nature, 2000. 408(6809):199–203.
Uz, T., et al., Effect of fluoxetine and cocaine on the expression of clock genes in the mouse hippocampus and striatum. Neuroscience, 2005. 134(4):1309–1316.
van der Weide, J., E.H. van Baalen-Benedek, and J.E. Kootstra-Ros, Metabolic ratios of psychotropics as indication of cytochrome P450 2D6/2C19 genotype. Ther Drug Monit, 2005. 27(4):478–483.
Van Pett, K., et al., Distribution of mRNAs encoding CRF receptors in brain and pituitary of rat and mouse. J Comp Neurol, 2000. 428(2):191–212.
van Rossum, E.F., et al., Polymorphisms of the glucocorticoid receptor gene and major depression. Biol Psychiatr, 2006. 59(8):681–688.
Van Tol, H.H., et al., Cloning of the gene for a human dopamine D4 receptor with high affinity for the antipsychotic clozapine. Nature, 1991. 350(6319):610–614.
Veefkind, A., Haffmans, and E. Hoencamp, Venlafaxine serum levels and CYP2D6 genotype. Ther Drug Monit, 2000. 22(2):202–208.
Vilar, R., et al., Association of A313 G polymorphism (GSTP1*B) in the glutathione-S-transferase P1 gene with sporadic Parkinson's disease. Eur J Neurol, 2007. 14(2):156–161.
Wegener, G., et al., Local, but not systemic, administration of serotonergic antidepressants decreases hippocampal nitric oxide synthase activity. Brain Res, 2003. 959(1):128–134.

Weinshilboum, R.M., D.M. Otterness, and C.L. Szumlanski, Methylation pharmacogenetics: catechol O-methyltransferase, thiopurine methyltransferase, and histamine N-methyltransferase. Annu Rev Pharmacol Toxicol, 1999. 39:19–52.

Wendland, J.R., Simultaneous genotyping of four functional loci of human SLC6A4, with a reappraisal of 5-HTTLPR and rs25531. Mol Psychiatr, 2006. 11(3):224–226.

Wendland, J.R., et al., A Large Case-Control Study of Common Functional SLC6A4 and BDNF Variants in Obsessive-Compulsive Disorder. Neuropsychopharmacology, 2007.

Wesolowska, A. and A. Nikiforuk, Effects of the brain-penetrant and selective 5-HT6 receptor antagonist SB-399885 in animal models of anxiety and depression. Neuropharmacology, 2007. 52(5):1274–1283.

Wess, J., Molecular basis of receptor/G-protein-coupling selectivity. Pharmacol Ther, 1998. 80(3):231–264.

Wilkie, M.J., et al., A splice site polymorphism in the G-protein beta subunit influences antidepressant efficacy in depression. Pharmacogenet Genomics, 2007. 17(3):207–215.

Willner, P., A.S. Hale, and S. Argyropoulos, Dopaminergic mechanism of antidepressant action in depressed patients. J Affect Disord, 2005. 86(1):37–45.

Wu, W.H., et al., Association study of the 5-HT(6) receptor polymorphism (C267T) and symptomatology and antidepressant response in major depressive disorders. Neuropsychobiology, 2001. 44(4):172–175.

Xu, J., et al., Heterodimerization of alpha 2A- and beta 1-adrenergic receptors. J Biol Chem, 2003. 278(12):10770–10777.

Xu, W., et al., Regional localization of the gene coding for human brain nitric oxide synthase (NOS1) to 12q24.2-->24.31 by fluorescent in situ hybridization. Cytogenet Cell Genet, 1993. 64(1):62–63.

Yamauchi, M., et al., Desensitization of 5-HT2A receptor function by chronic administration of selective serotonin reuptake inhibitors. Brain Res, 2005:164–169.

Yan, Z. and Y. Wu, [Relation between cytochrome P450 2D6 and lung cancer susceptibility caused by smoking]. Wei Sheng Yan Jiu, 2007. 36(1):112–3, 116.

Yang, K., et al., Levels of serum interleukin (IL)-6, IL-1beta, tumour necrosis factor-alpha and leptin and their correlation in depression. Aust N Z J Psychiatr, 2007. 41(3):266–273.

Yang-Feng, T.L., et al., Chromosomal organization of adrenergic receptor genes. Proc Natl Acad Sci U S A, 1990. 87(4):1516–1520.

Yatham, L.N., et al., Decrease in brain serotonin 2 receptor binding in patients with major depression following desipramine treatment – A positron emission tomography study with fluorine-18-labeled setoperone. Arch Gen Psychiatr, 1999. 56(8):705–711.

Yeo, A., et al., Association between a functional polymorphism in the serotonin transporter gene and diarrhoea predominant irritable bowel syndrome in women. Gut, 2004. 53(10):1452–1458.

Yin, O.Q., et al., Phenotype-genotype relationship and clinical effects of citalopram in Chinese patients. J Clin Psychopharmacol, 2006. 26(4):367–372.

Yoshida, K., et al., Influence of the serotonin transporter gene-linked polymorphic region on the antidepressant response to fluvoxamine in Japanese depressed patients. Prog Neuropsychopharmacol Biol Psychiatr, 2002. 26(2):383–386.

Yoshida, K., et al. Monoamine oxidase: A gene polymorphism, tryptophan hydroxylase gene polymorphism and antidepressant response to fluvoxamine in Japanese patients with major depressive disorder. Prog Neuropsychopharmacol Biol Psychiatr, 2002. 26(7–8):1279–1283.

Yoshida, K., et al., Monoamine oxidase A gene polymorphism, 5-HT 2A receptor gene polymorphism and incidence of nausea induced by fluvoxamine. Neuropsychobiology, 2003. 48(1):10–13.

Yoshida, K., et al., Prediction of antidepressant response to milnacipran by norepinephrine transporter gene polymorphisms. Am J Psychiatr, 2004. 161(9):1575–1580.

Yu, Y., et al., Association study of a monoamine oxidase a gene promoter polymorphism with major depressive disorder and antidepressant response. Neuropsychopharmacology, 2005. 30(1719–1723).

Yu, Y.W., et al., Association study of the serotonin transporter promoter polymorphism and symptomatology and antidepressant response in major depressive disorders. Mol Psychiatr, 2002. 7(10):1115–1119.

Yu, Y.W., et al., Association analysis for neuronal nitric oxide synthase gene polymorphism with major depression and fluoxetine response. Neuropsychobiology, 2003a. 47(3):137–140.

Yu, Y.W., et al., Association study of the interleukin-1beta (C-511T) genetic polymorphism with major depressive disorder, associated symptomatology, and antidepressant response. Neuropsychopharmacology, 2003b. 28(6):1182–1185.

Yu, Y.W., et al., Association study of two serotonin 1A receptor gene polymorphisms and fluoxetine treatment response in Chinese major depressive disorders. Eur Neuropsychopharmacol, 2006. 16(7):498–503.

Zanardi, R., et al., Efficacy of paroxetine in depression is influenced by a functional polymorphism within the promoter of serotonin transporter gene. J Clin Psychopharmacol, 2000. 20(1):105–107.

Zanardi, R., et al., Factors affecting fluvoxamine antidepressant activity: influence of pindolol and 5-HTTLPR in delusional and nondelusional depression. Biol Psychiatr, 2001. 50(5):323–330.

Zangen, A., et al., Association between depressive behavior and absence of serotonin-dopamine interaction in the nucleus accumbens. Psychopharmacology (Berl), 2001. 155(4):434–439.

Zateyshchikov, D.A., et al., Association of CYP2D6 and ADRB1 genes with hypotensive and antichronotropic action of betaxolol in patients with arterial hypertension. Fundam Clin Pharmacol, 2007. 21(4):437–443.

Zeise, M.L., S. Madamba, and G.R. Siggins, Interleukin-1 beta increases synaptic inhibition in rat hippocampal pyramidal neurons in vitro. Regul Pept, 1992. 39(1):1–7.

Zhang, X., et al., Tryptophan hydroxylase-2 controls brain serotonin synthesis. Science, 2004. 305(5681):217.

Zhang, X., et al., Loss-of-Function Mutation in Tryptophan Hydroxylase-2 Identified in Unipolar Major Depression. Neuron, 2005. 45(11–16).

Zhou, Z., et al., Haplotype-based linkage of tryptophan hydroxylase 2 to suicide attempt, major depression, and cerebrospinal fluid 5-hydroxyindoleacetic acid in 4 populations. Arch Gen Psychiatr, 2005. 62(10):1109–1118.

Zill, P., et al., Evidence for an association between a G-protein beta3-gene variant with depression and response to antidepressant treatment. Neuroreport, 2000. 11(9):1893–1897.

Zill, P., et al., Beta-1-adrenergic receptor gene in major depression: Influence on antidepressant treatment response. Am J Med Genet, 2003. 120B(1):85–89.

Zill, P., et al., Single nucleotide polymorphism and haplotype analysis of a novel tryptophan hydroxylase isoform (TPH2) gene in suicide victims. Biol Psychiatr, 2004a. 56(8):581–586.

Zill, P., et al., The dysbindin gene in major depression: an association study. Am J Med Genet B Neuropsychiatr Genet, 2004b. 129(1):55–58.

Zill, P., Buttner, A., Eisenmenger, W., Moller, H. J., Ackenheil, M., Bondy, B., Analysis of tryptophan hydroxylase I and II mRNA expression in the human brain: a post-mortem study. J Psychiatr Res, 2007. 41(1–2):168–173.

Zorrilla, E.P., et al., The relationship of depression and stressors to immunological assays: a meta-analytic review. Brain Behav Immun, 2001. 15(3):199–226.

Chapter 15
Perspectives for an Integrated Biomarker Approach to Drug Discovery and Development

Irina Antonijevic(✉), Roman Artymyshyn, Carlos Forray, Sylvia Rabacchi, Kelli Smith, Chad Swanson, Joseph Tamm, Wiktor Mazin, and Christophe Gerald

Abstract Today's psychopharmacological drugs remain focused on targets that were identified serendipitously more than half a century ago. As these targets have not proven to be at the core of the pathophysiology of the major psychiatric disorders, a better understanding of the disease biology seems a crucial step to identify more efficacious treatments. The tools to realize this goal include neuroendocrine, protein, transcription and genetic markers, neuroimaging and neurophysiological approaches. Obviously, the benefit for psychiatric patients of identifying a pattern of blood-based markers that combine information on the disease biology and treatment response would be enormous. Our transcription data from human blood cells suggest that this is a realistic possibility for the future. Ideally, markers identified in patients would be translated into distinct, hypothesis-driven animal models to facilitate conclusions on the potential therapeutic utility of novel compounds. The combined use of disease state and mechanistic models may characterize cellular and molecular mechanisms of various aspects of psychiatric disorders. This information, in turn, could help establish *in vitro* models that link cellular targets to (novel) pharmacological approaches. With the discussions around DSM-V, it can be hoped that the ambitious research agenda will guide and stimulate systematic research into the biology and biological markers of psychiatric disorders.

Abbreviations Akt: Serine/threonine-specific protein kinase; ANOVA: Analysis of Variance; BA: Broadman Area; Bcl-2: B-cell CLL/lymphoma 2; BDNF: Brain derived neurotrophic factor; CMS: Chronic mild stress; CRF: Corticotropin-releasing factor; DISC-1: Disrupted-in-schizophrenia 1; DNA: Deoxyribonucleic acid; DSM: Diagnostic and Statistical Manual; DST: Dexamethasone suppression test; FST: Forced swim test; GR: Glucocorticoid receptor; HAB rats: High anxiety bred rats; HPA: axis Hypothalamic–pituitary–adrenal axis; 5-HT: Serotonin; HUPO: Human Proteome Organization; ICD: International Classification of Diseases;

I. Antonijevic
Lundbeck Research USA, Paramus, NJ 07652 USA
iant@lundbeck.com

C.W. Turck (ed.) *Biomarkers for Psychiatric Disorders*,
doi: 10.1007/978-0-387-79251-4_15, © Springer Science+Business Media, LLC 2008

LAB rats: Low anxiety bred rats; LC: Locus coeruleus; LH: Learned helplessness; mGlu2/3: Metabotropic glutamate 2/3 receptor; MPfc: Medial prefrontal cortex; MR: Mineralocorticoid receptor; NCE: New Chemical Entity; NIMH: National Institute of Mental Health; NRG1: Neuregulin-1; PBMC: Peripheral blood mononuclear cells; Pelora: Penalized logistic regression; ProDaC: Proteomics Data Collection; qPCR: Quantitative Polymerase-chain reaction; RNA: Ribonucleic acid; SGZ: Subgranular zone; siRNA: Small interfering ribonucleic acid; SLR: Stepwise logistic regression; SVZ: Subventricular zone; TST: Tail suspension test; VEGF: Vascular endothelial growth factor

15.1 Introduction

According to the National Institute of Mental Health (NIMH), mental disorders affect an estimated 26.2 percent of Americans aged 18 and older in a given year. Unlike many other chronic and disabling disorders, mental illnesses strike early in life, with unipolar depression and substance abuse accounting for 28% and 20%, respectively, of the disability from all medical causes in people aged 15–44 years (Insel and Scolnick, 2006). These data are in line with the Global Burden of Disease study, reporting that mental illness, including suicide, accounts for over 15% of the burden of disease in established market economies. This is more than the disease burden caused by all cancers.

Despite these alarming observations, biomedical research in the field of psychiatry has remained focused on treatment targets that were identified serendipitously more than half a century ago. Almost all available drugs target primarily monoamine transporter and receptors, in various combinations, leading to slightly different profiles. However, these differences rarely have a clinically relevant impact in terms of efficacy or safety. Moreover, the targets have not proven to be at the core of the pathophysiology of the major psychiatric disorders to this day, which may explain the modest efficacy of all available drugs when tested in not well defined patient populations. This is in contrast to research into other major chronic diseases, such as cancer and heart disease that has shed light on the biology, which has resulted in the successful development of new treatment targets.

However, it is tempting to look at the glass as half full in view of the ambitious research agenda for DSM-V, clearly stating the aim to develop ‘an etiologically based, scientifically sound (diagnostic) classification system’ (2002). Biomedical research that focuses on the disease rather than on treatment may improve our understanding of core biological alterations associated with psychiatric disorders. A better understanding of the disease biology, and the biological differences among patients, should advance the diagnostic classification, which today is entirely descriptive. In doing this, one would expect to also identify new targets that may yield more efficacious treatments, at least for subgroups of patients who share core biological disturbances.

A range of biological read-outs are mentioned in the DSM-V research agenda that may advance our understanding of the biology of mental disorders, including neuroendocrine, protein, transcription and genetic markers, neuroimaging and neurophysiological approaches. Not surprisingly, the biological markers were and still are focused on the brain, where the pathophysiology of mental disorders is thought to occur. Although the brain certainly is a critical site to study the biology of mental disorders, there is increasing evidence for peripheral changes associated with mental disorders *per se* as well as with the response to treatment (Gold and Charney, 2002; Iga et al., 2007a; Maino et al., 2007). However, so far the majority of studies as well as the DSM-V research agenda focus on post-mortem assessment of brain tissue to discover molecular markers of psychiatric disorders, such as proteins and gene transcripts. While these studies can yield interesting molecular targets associated with a clinical syndrome and/or treatment response, several notes of caution are warranted: Firstly, one critical aspect related to biomarkers in psychiatry is their predictive value for either course of disease or treatment response. Obviously, markers identified in post-mortem brain tissue cannot address this need. Secondly, several variables can affect results of gene expression studies in post-mortem brain tissue and need to be carefully controlled for, as recently discussed in a publication by the members of the NIMH Conte Center and the Pritzker Neuropsychiatric Disorders Research Consortium (Atz et al., 2007).

Recently, multiple forms of blood markers as alternatives to brain markers have received significant attention (Avissar et al., 2004; Iga et al., 2007a, b; Ray et al., 2007b; Segman et al., 2005; Zieker et al., 2007).

Though quality control remains a very critical issue for all biomarker investigations, relatively simple procedures are available to ensure standardized blood collection with good preservation of RNA and reproducible transcription data in multi-center clinical trials. Obviously, the advantage of blood-based biomarkers that combine information on the disease biology and treatment response for psychiatric patients would be enormous.

In fact, the search for peripheral (blood) markers for affective states and treatment response is not new, but dates back several decades when publications showed differences in monoamines and related receptor levels in platelets from patients suffering from depression or schizophrenia, and effects of treatment (Garcia-Sevilla et al., 1981; Stahl et al., 1983). Though studies have not yielded a robust marker to date, advances in technology allow the study of multiple markers simultaneously and quantitatively at the transcriptional level in human peripheral leukocytes. Though the same is in principle true for protein biomarkers, there is no standardized procedure for sampling of plasma or serum in large clinical trials. Therefore, the detection and subsequent validation of protein biomarkers for psychiatric disorders will have to wait for agreed upon standards and guidelines by initiatives such as the EU funded 'Proteomics Data Collection' (ProDaC) workshops.

Meanwhile, data on gene transcription profiles from peripheral leukocytes showed interesting changes in patients suffering from an acute psychiatric disorder, suggesting that this approach is an encouraging route for the discovery of disease biomarkers in psychiatry (Segman et al., 2005; Zieker et al., 2007). As some of the very targeted transcription data also showed normalization with treatment, peripheral leukocyte transcription profiles could provide both disease as well as treatment-related biomarkers (Iga et al., 2007a). In this chapter we discuss the use of gene transcription patterns as objective markers to identify biologically distinct subgroups of patients, to help understand the biology within these subgroups and to assess the treatment response. We will illustrate, using some of our own early data and experience, the possibilities transcription profiles from peripheral leukocytes offer in terms of add-ons to psychiatric diagnostic and efficacy markers.

As gene transcription analysis allows one to evaluate entire signaling pathways, this approach could discover patterns that distinguish between patients and controls as well as between subgroups of patients with high specificity and sensitivity. This, in turn, is a prerequisite for a biological marker to become a clinically relevant tool. Moreover, gene transcription related to cellular signaling and metabolism shows a high correlation between blood leukocytes and the brain in humans (Ray et al., 2007; Sullivan et al., 2006). Therefore, signaling alterations identified in human blood leukocytes can also guide hypotheses on the CNS pathophysiology of psychiatric disorders and provide critical information for the discovery of novel treatment targets.

The focus on patterns rather than individual markers should also help to establish complex relationships between biological markers and intermediate phenotypes. The importance of intermediate phenotypes is emphasized by the discussion put forward in the DSM-V research agenda on a dimensional vs. categorical classification system (2002). The dimensional approach includes an aspect often disregarded in psychiatric biomedical research, namely the examination of control populations. In this chapter we will provide data in support of the relevance of intermediate phenotypes by showing an association between transcription patterns and psychiatrically relevant clinical variables in a control population. Such data can be used to provide criteria for clinically relevant dimensions in a future classification system. Moreover, intermediate endophenotypes seem particularly relevant for drug development, as examination of drug effects in such ‘control’ subjects could provide early signs for efficacy in a patient population.

Another change in paradigm, put forward in the research agenda for DSM-V, is the importance of exploring biomarkers, including genetic markers, in smaller but biologically homogenous subject groups (2002). Therefore, a smaller group of subjects who share a distinct transcription profile could provide a stronger basis for detecting genetic markers than large but very heterogeneous populations, typically examined in pivotal clinical trials.

15.2 Clinical Drug Development

15.2.1 General Considerations

Due to the descriptive and purposefully non-etiological nature of today’s psychiatric diagnostic classification (DSM-IV-TR and ICD-10), patients included in clinical trials are heterogeneous in terms of the clinical presentation and, as far as we know, the underlying biology. One possibly extreme example of this unsatisfactory situation is the diagnosis of major depression, which encompasses patients with melancholic and atypical features, which are almost opposite in their clinical presentation and biology (Carroll et al., 2007; Gold and Chrousos, 2002). Moreover, there is evidence for a differential treatment response for these two subtypes (Quitkin et al., 1993; Stewart et al., 1993).

Another clinical discriminator likely to reflect biological differences is psychiatric co-morbidity. Many patients in clinical practice meet criteria for more than one disorder, particularly within the spectrum of affective disorders (Rush et al., 2005; Zimmerman et al., 2005). However, clinical trials examining safety and efficacy of a 'new chemical entity' (NCE) typically exclude patients with obvious psychiatric co-morbidity. This conundrum leads to an overrepresentation of patients with milder syndromes, but not necessarily a more homogenous patient population in terms of biology or treatment response (Khan et al., 2007; Souery et al., 2007). Support for this hypothesis comes from a recent analysis of antidepressant treatment trials performed between 1981 and 2000, showing an increase in the placebo response as well as highly variable responses to both placebo and active treatment (Walsh et al., 2002).

Hopefully, with further biological research, this unsatisfactory situation will be improved by identifying the biological signatures of disorders, including common co-morbid disorders, which may be re-classified as one disorder if a common biological basis for the symptoms is uncovered. One approach is to identify biological markers associated with certain features of psychiatric disorders both within and across toady's diagnostic entities. Thus, it can be presumed that distinct subtypes of major depression, such as depression with melancholic and atypical features, are distinguishable with regard to the (peripheral) biomarker profile (Gold and Chrousos, 2002). In fact, peripheral biomarker patterns may help to delineate core clinical features for these two depression subtypes, which remain a matter of debate (Parker, 2000; Parker et al., 2005; Posternak and Zimmerman, 2002). This would be an example of how biological markers contribute to a scientifically sound classification system.

An extension of the above approach is to address the biology of distinct clinical features across the boundaries of current diagnoses. The analysis of such complex relationships should help to characterize multiple intermediate phenotypes, which in turn may predispose to develop certain mental illnesses, e.g. when exposed to environmental stressors. Examples in psychiatry include impaired cognitive executive function, which can occur in schizophrenia, some forms of depression and in substance abuse. Another example is fatigue that can occur in different psychiatric disorders such as depression and anxiety, but also in disorders associated with a high incidence of depressive disorders, such as Parkinson's disease, Multiple Sclerosis and obesity (Heesen et al., 2006; Vgontzas et al., 2006; Weintraub and Stern, 2005). A focus on biological markers of distinct clinical features is in line with the DSM-V research agenda, which stresses the importance of studying (a) complex relationships between biological and clinical variables and (b) intermediate phenotypes (Charney et al., 2002). A better understanding of the biological basis of certain clinical symptoms that are associated with different neuropsychiatric (and other) disorders will also help to improve treatment strategies for these complex disorders.

A prerequisite for the identification of such complex relationships is a thorough characterization of subjects, including clinical, environmental and biological factors. If taken seriously, this will markedly affect the information (clinical and biological) collected in clinical trials. The emphasis on biological markers and structured clinical information is rooted in the observation that information gathered through routine clinical methods is not reliable, leading to missing critical information such as previous diagnoses, that can markedly affect the treatment outcome (Ramirez et al., 2000).

Ultimately, the understanding of the biological underpinnings of core psychiatric syndromes will also advance the development of animal disease state models as well as cellular models to mimic specific aspects of a human disorder. This, in turn, will help profile new drug targets across today's diagnostic classification and should improve the predictive validity for novel drug targets in terms of efficacy (and safety) in humans.

15.2.2 Specific Biological Considerations

Ideally, by the time a NCE is first tested in man, preclinical data should have been collected that guide the clinical trials in terms of which target symptoms and biological features could be addressed with the potential new drug. While more specific clinical profiles are sometimes addressed as part of the life cycle management, leading to a broadening of the approved indication(s), reliance on biological read-outs is hampered by the lack of consensus parameters.

Though it is true that today no biological read-outs can be considered validated and of regulatory relevance, biological markers can (and should) be used for internal decision making. With some systematic guidance from the efforts around the DSM-V research agenda, biological read-outs can be selected from an abundance of (more ore less confirmed) data from decades of biological research in the field of psychiatry. This chapter will highlight some possibilities of how biological markers could address (biological) efficacy and selection of target populations in clinical development.

A starting point is the convergence of targets across current diagnostic entities. Thus, atypical antipsychotics are efficacious in treating depressive syndromes and recent drug development efforts target modulation of glutamate neurotransmission and inflammation for both affective disorders and schizophrenia. Also, there is growing evidence that risk factors, as well as alterations in neuronal circuitries, e.g. the limbic system and the basal ganglia, are shared between major psychiatric disorders such as schizophrenia and affective disorders. Moreover, some of the abnormalities observed in acutely ill patients can be detected in normal subjects who are or may be at risk to develop a disorder. Such intermediate phenotypes can provide a useful enrichment strategy to support biologically based efficacy end-points of NCEs in exploratory clinical trials.

In this regard, identification of biological markers which can provide reliable associations with distinct symptom clusters, treatment responses and safety profiles are needed to increase the probability of showing superior efficacy of novel drug treatments, and ultimately to improve patient care by hastening the process of identifying optimal treatment options for individual patients. The subsequent sections will focus on the opportunities that the disease biology provides for the discovery of diagnostic and efficacy biomarkers.

15.2.3 Disease Biology

Several biological alterations, including hyperactivity of the hypothalamic–pituitary–adrenal (HPA) axis, altered sleep EEG architecture, and metabolic and volumetric changes in limbic and prefrontal cortical brain areas have been identified in several major psychiatric disorders. Some of these changes also seem to confer a risk factor to develop a disorder rather than being themselves markers of the disease. The expansion of our knowledge into the molecular basis of these changes offers the chance to better understand the pathophysiology of the disorders and also to discover new treatment targets, including prophylactic treatments. As some biological alterations seem relevant for several disorders, systematic research into biomarker read-outs holds the promise to develop drugs that target specific biological changes across the current diagnostic boundaries. This approach should ultimately help to fulfill the goal of developing a scientifically sound classification system. The following sections will briefly review biological changes associated with major psychiatric disorders that could be used as early read-outs in drug development.

15.2.3.1 Stress and the Hypothalamic–Pituitary–Adrenal Axis

The HPA axis controls the release of glucocorticoids and is an essential component of an individual's response to stress. Several changes related to the HPA axis have been reported in psychiatric disorders, with most consistent changes described in patients with major depression. In general, changes related to the HPA axis can be divided into hyper- and hypoactive states, with the former being most consistently seen in patients with melancholic or psychotic features (Carroll et al., 1976, 2007; Gold and Chrousos, 2002; Holsboer, 1999; Kunzel et al., 2003). Several challenge tests have been developed, including the dexamethasone suppression test (DST), first employed more than 30 years ago, the refined dexamethasone-corticotropin-releasing factor (CRF) test, CRF and insulin tests. While none of these tests have reached the level of sensitivity and specificity to be considered as a diagnostic marker, some data suggest that the dexamethasone-CRF test could be an early indicator of the treatment response (Ising et al., 2007).

Non-pharmacological challenge tests have also been explored and have shown that early childhood experiences have a marked and long-lasting impact on the stress responses in adulthood (Heim et al., 2000, 2004; Pruessner et al., 2004). These studies are interesting for several reasons: (1) They highlight that differences in the stress response among patients with depression also impact the treatment response, thereby providing a tool to select patients for a specific treatment trial. (2) They offer the opportunity to study non-depressed subjects with an exaggerated stress response in early clinical development as a way to profile the impact of a new compound on this clinically very relevant biological system. The importance of this possibility is underscored by the lack of beneficial effects of most psychopharmacological treatments in a super healthy population, typically recruited when first testing a NCE in humans. By enriching the study population to exhibit some biological changes relevant for the indication, information of a NCE is obtained earlier and is more relevant for further profiling of its target population. In light of the fact that recent data have demonstrated HPA overactivity in first-onset psychosis and in patients at risk for schizotypal personality disorders (Pariante et al., 2005; Walker et al., 2001), a new drug effective at attenuating exaggerated HPA activity could be considered beyond an indication for affective disorders.

While HPA axis tests can be part of small clinical trials, these tests are not suitable for large multi-center trials and are not likely to be used as decision points in clinical practice. However, a modified HPA activity should alter transcription patterns in circulating blood leukocytes. So, if blood biomarkers are explored in parallel during early clinical development, one could correlate easily accessible blood markers with neuroendocrine markers. This approach could bridge the gap between complex markers explored in small and exploratory trials and the need for simple markers that help select patients for large pivotal trials.

While stress tests have been used by some companies in early clinical development, their value to correctly predict clinical efficacy, at least for a segment of the patient population, remains to be shown.

15.2.3.2 Sleep

Analogous to neuroendocrine markers, sleep EEG recordings are not easily conducted in large multi-center clinical trials. But, as in the case for neuroendocrine markers, sleep EEG changes in psychiatric disorders have been studied for more than three decades and have yielded some promising insights into disease biology, in particular for depressive syndromes.

While subjective sleep disturbances are present in many patients with psychiatric (and other) disorders, only a much more restricted group of patients show consistent and objective changes of the sleep architecture. Common sleep disturbances reported by many patients include prolonged sleep latency, early morning awakenings with inability to return to sleep and intermittent awakenings. While none of these symptoms is specific for a psychiatric (or other) disorder, patients with both prolonged sleep latency and increased nocturnal awakenings show greater symptoms of depression when compared to patients with only one of these features (Taylor et al., 2005). Also, the presence of early morning awakening, increased REM density, short REM latency and reduced non-REM sleep is indicative of severe and recurrent depression (Kupfer et al., 1982, 1986). However, a short REM latency has also been noticed in schizophrenia, schizoaffective illness (Kupfer, 1976) and in elderly subjects (Lauer et al., 1991).

Therefore, recent data on the microarchitecture of sleep may be more promising in terms of specificity. Spectral sleep EEG analysis is considered state-of-the art today and using this method more subtle sleep EEG changes can be detected, which can help profile patients and treatment responses. Recent spectral EEG data have shown increased beta power in non-REM sleep in major depression. This sleep EEG marker of arousal was correlated with altered brain glucose metabolism in cortical areas (Nofzinger et al., 2000). As beta power in non-REM sleep was inversely correlated with subjective sleep quality, a NCE that reduces this beta activity and hence arousal may lead to both subjective and objective sleep improvements. Moreover, hyperarousal, thought to involve increased activity of afferent noradrenergic neurons, is a feature of some forms of depression, PTSD, and schizophrenia, which would be rational target indications for a drug that attenuates beta activity in non-REM sleep.

In addition, lower temporal coherence of left to right EEG synchronization has been reported in depression (Armitage et al., 1999, 2006; Robert et al., 2006) and seems to be a trait-like feature. Furthermore, an altered pattern of delta activity during sleep, namely a reduction of delta activity in the first sleep cycle, has been associated with depression, chronic course and vulnerability for recurrence if not normalized at clinical remission (Armitage et al., 2000, 2001; Kupfer, 1995). As few drugs increase delta activity, particularly during its physiological peak in the first sleep cycle, and commonly used hypnotics decrease delta activity but increase higher frequency activity, NCEs that positively affect delta activity could be a welcome add-on to the current portfolio of drugs to improve sleep (Walsh et al., 2007).

In summary, clinically relevant sleep markers can be identified from the existing data. If further explored systematically and confirmed, and linked to molecular markers, these could yield novel relevant treatment targets. Though the practical application of sleep EEG is restricted to studies with small numbers of patients, the similar physiology of sleep regulation in rodents and humans suggests that changes in sleep microarchitecture could be used as 'translatable' biomarkers for profiling of novel treatment targets.

15.2.3.3 Structural and Functional Brain Imaging

Considerable progress has been made in identifying the brain regions and neural circuits that underlie normal and abnormal cognitive-processing, 'emotion-processing' and mood regulation. Resting cerebral blood flow, glucose metabolism and structural neuroimaging studies point to a widespread network of frontal, limbic, and subcortical areas underlying the cardinal feature of negative emotionality, flat affect and cognitive deficits observed in affective disorders and schizophrenia (Gur et al., 2007; Phillips et al., 2003). Interestingly, these studies also indicate some pathophysiological convergence at the functional neuroanatomical level, and particularly at the level of the hippocampus, for primary thought disorders, such as schizophrenia, and primary emotional disorders such as major depression and even primary cognitive disorders (Gur et al., 2007; Kitayama et al., 2005; Nemeroff et al., 2006; Vermetten et al., 2003). As reduced hippocampal volume has also been considered a marker of increased vulnerability to develop depression or PTSD (Gilbertson et al., 2002; Macmaster et al., 2007), small hippocampal volume could be another marker that confers a risk to develop a mental illness. As early treatment to prevent long-term disability is increasingly recognized, further longitudinal research in large populations of young subjects at risk, identified through a combination of clinical, e.g. family history, and biological, e.g. small hippocampi, factors, is needed.

However, differences are also noted: Unlike depression, schizophrenia is more consistently associated with reduced whole brain and amygdala volumes. In contrast, affective disorders seem to be associated with no change or an increase in amygdala volumes (Frodl et al., 2004; McDonald et al., 2004; Zetzsche et al., 2006). Another approach is to elucidate functional brain circuitries contributing to psychiatric syndromes. A recent elegant example provided by Milak and colleagues (Milak et al., 2005) demonstrated that regional changes in brain glucose metabolism can be linked to specific symptom clusters in patients with major depression. It will be interesting to see if these correlations hold true across today's diagnostic boundaries.

Furthermore, longitudinal studies in depression also provided evidence of the normalization of amygdala activation in response to negative facial expressions after successful antidepressant treatment, suggesting the feasibility of functional MRI studies to identify predictors of treatment outcome (Davidson et al., 2003; Fu et al., 2004; Sheline et al., 2001).

So far, drawbacks with available data from neuroimaging technology include (1) its applicability to small groups of subjects (and hence the question about confirmation) and (2) the lack of standardized data acquisition, that makes comparison (and hence confirmation) between studies often impossible. On the other hand, standardization, systematic extension and confirmation of existing knowledge can be used to evaluate NCEs in exploratory clinical trials and to explore correlations between neuroimaging and clinical findings. This would aid not only in optimal dose selection, but also in the identification of the optimal target population(s). With regard to dose finding, current trends in the development of novel antidepressants rely heavily on the translation of brain target occupancy derived from pharmacodynamic and pharmacokinetic relationships from animal models into the doses required to achieve similar occupancy of the drug target in brains of normal volunteers using selective PET ligands in Phase 1 studies (Frank and Hargreaves, 2003). Results from these studies are used to guide the therapeutic dose selection generally in the absence of information about efficacy. The validity of these assumptions is challenged by the fact that drugs that have no effect on mood in normal people relieve depression in those who are ill. The combination of target occupancy with functional read-outs such as fMRI responses to emotional stimuli in depressed patients could provide an improved and reliable assessment of the relation between occupancy and target engagement.

Whether a detailed correlation of behavioral symptoms and neuroanatomical activation patterns can also be used reliably in rodents, remains to be confirmed. However, some early data support the hope that localized brain glucose metabolism in rodents can be linked to behavior as well as localized neurotransmitter release (Marsteller et al., 2007).

15.2.3.4 Transcription Markers

In evaluating neurochemical or molecular changes that occur in animal models and in the human disease state, the most commonly considered biomarkers to examine are proteins, metabolites or RNA transcripts. Each of these approaches offers specific advantages and limitations. Analysis of RNA transcripts is often a first choice because of the broad range of tools available for transcription profiling. Proteomic and metabolomic approaches offer great promise for identification and characterization of functional biomarkers for psychiatric disease, as they are able to detect patterns of changes that may be more directly linked to a disease state than transcription analysis. At the present time however, limitations in the technology (and cost), sampling standards and knowledge base limit the utility of these approaches for large scale biomarker identification projects.

The ideal marker set would be one that is altered in a disease state (diagnostic) and would be normalized following successful drug treatment and concomitant remission of symptoms (surrogate marker). Biomarkers for other aspects of treatment of psychiatric disorders, such as drug metabolism and toxicology, have been used successfully and are important for developing drugs and understanding drug metabolism (Searfoss et al., 2005; Suzuki et al., 2006). However, they are beyond the scope of this discussion.

One commonly used method for identification of biomarkers relevant to psychiatric disorders is RNA transcription analysis. RNA transcripts can be analyzed individually with high anatomical resolution using in situ hybridization, quantified in relatively small numbers (but with very high sensitivity) using quantitative PCR (qPCR) or analyzed on a genome wide scale using DNA microarrays. It can be successfully argued that analysis of protein distribution or function is more functionally relevant to the disease state than to the analysis of RNA levels, as transcript levels are not always closely coupled to the expression levels of a protein. Thus, smaller, more rapidly turned over proteins and enzymes, such as CRF and tyrosine hydroxylase (Ganguly et al., 2002; Wong and Tank, 2007; Xu et al., 2005) tend to have better correlation between RNA and protein levels compared to G-protein coupled receptors (Duncan et al., 1993; Frohna et al., 1995), where there are many examples of protein regulation without a concomitant increase (or decrease) of the RNA encoding that protein. Also, transcription analysis does not provide information about the functional state of the protein, e.g. phosphorylation of receptors or of signaling cascade elements. However, RNA transcription patterns provide valuable insights about distribution and regulation of RNA (and to some measure the proteins that they encode) in the tissue of interest. Tools are available for examining (essentially) every gene expressed, whereas the same is not true for proteins at this time.

In the past decade vast strides have been made in improving methodology for assessing transcripts, especially real time qPCR and DNA microarrays. Initially, there were questions regarding replication of data across different array platforms (Jarvinen et al., 2004). However, improvements in arrays and analysis techniques have minimized these issues. Recent studies measuring expression across platforms and comparison with quantitative techniques confirm that current arrays have acceptable reliability, although confirmation of data by another method such as qPCR is still considered to be essential (Wang et al., 2006; Bosotti et al., 2007; Canales et al., 2006). Further validation and a reference data set come from the MicroArray Quality Control Project data set (Shi et al., 2006), as well as improved methods of analysis (Stafford and Brun, 2007). In an effort to establish reporting standards the FDA (in August 2007) issued a draft guidance for industry with regard to handling, preparation and analysis of RNA and microarray analysis. Despite the improvements in microarray platforms and sample preparation, microarrays cannot be used as the sole method for transcription analysis. qPCR, which is more accurate and precise, is still essential for the identification and confirmation of biomarker candidates.

Another very recent and attractive possibility with regard to transcript analysis is to examine micro RNA transcription. Micro RNAs fine-tune posttranscriptional expression changes of a large number of genes in mammals and have been suggested to significantly influence phenotypic variations (Georges et al., 2007; Schratt et al., 2006). Though the work today is focused on cancer indications, this approach holds great promise for complex traits common in psychiatric disorders.

Brain vs Blood Leukocyte Transcription Markers

The primary source of tissue for transcription profiling in psychiatric disorders has been postmortem brain samples. These have proved to be, and will likely continue to be, invaluable in collecting information about psychiatric disorders; however they have serious limitations. Using postmortem tissue introduces a large number of technical variables including varying post mortem intervals, agonal time prior to death and the resultant change in pH and subsequent alteration in RNA quality (Atz et al., 2007). The greatest limitation however, is the inability to assess changes in patients relative to symptoms and to assess the effects of treatment. Transcription profiling of blood or blood fractions is more desirable as samples are accessible and can be collected at multiple time points from the same individual, e.g. at different stages in the course of the disease. One approach is to isolate peripheral blood mononuclear cells (PBMCs). Although it is a commonly used methodology, it has drawbacks for multi-center clinical trials in that it is difficult to isolate PBMCs at the point of collection and there is good evidence for post collection changes in transcription (Tanner et al., 2002). An alternate approach is to draw whole blood from donors directly into tubes containing a chaotropic agent (example: PaxGene™, Becton Dickenson), which eliminates post collection changes in transcription, preserves RNA during transportation and storage and allows isolation of high quality RNA. Though the transcription profile will be different in

PBMCs as compared to whole blood, either will provide valid data (Feezor et al., 2004). As there are no standardized protocols available for protein or peptide preservation, transcription analysis is still the best available method for assessing regulation of multiple factors in clinical trials for comparison with animals and cellular systems.

As exemplified below, it seems reasonable to pursue a translational approach by examining transcription patterns identified in blood leukocytes from distinct patient populations in relevant animal brain areas, both in disease states and after pharmacological interventions.

Hypothesis-Driven vs Hypothesis-Free Approaches

The development of high throughput expression analysis methods using microarrays (Brenner et al., 2000; Schena et al., 1995, 1996; Velculescu et al., 1995), allowing a hypothesis-free data analysis, has had a profound impact on assessing gene expression changes. Microarrays will yield considerable information if the experiments have been designed and analyzed properly, and have been used successfully to probe CNS function and to assess changes in psychiatric disorders (Luo and Geschwind, 2001; Marcotte et al., 2003; Mirnics and Pevsner, 2004; Mirnics et al., 2006). A major benefit of using arrays is that patterns of change, or pathways affected, can be identified without prior knowledge or hypotheses. Careful experimental design considerations are key to any successful experiment, but are particularly important in array experiments. There is a tremendous differential expression of both RNA and proteins across regions of the brain, and altered regulation may be occurring in a small population of cells only. This can make detection of subtle but relevant changes difficult. Although arrays have been successfully used to profile changes in the brain in both animal models and in psychiatric disorders, it should be noted that a number of findings in array studies have not been replicated. The most extensively studied psychiatric disorder using microarrays and post-mortem CNS tissue is probably schizophrenia (Konradi, 2005; Mirnics et al., 2006). Changes were identified in genes related to presynaptic markers, mitochondria/energy homeostasis, myelination related genes and the GABA-Glutamate system. These have led to an increased understanding of the disorder, but have not yielded a usable biomarker or novel therapy up to this point.

Major depression, another heterogeneous psychiatric disorder, has been profiled using microarrays with mixed results. Evans et al. (Evans et al., 2004)], using arrays to profile BA46, have identified the FGF system as a factor in major depression. Sibille (Sibille et al., 2004) examined transcription in BA9 and BA47 and did not find changes in the FGF system, pointing either to regional differences in the cortex or lack of a strong association between unspecified major depression and the FGF system. Other studies (Altar et al., 2004; Newton et al., 2003; Shi et al., 2006; Sun et al., 2005) have identified that NPY and TRH RNA are increased in the CNS after electroconvulsive therapy. This is consistent with data that show that NPY administration has an antidepressant effect in the forced swim test (Husum et al., 2000). To date, there have been no studies on postmortem brain to examine generalized anxiety in humans.

An alternate method is to use a hypothesis driven approach and focus on a subset of genes that are likely to be involved in pathways related to the disease. Much of the data currently in the literature on psychiatric disorders has come from experiments looking at specific pathways or a few individual genes and not from genome-wide array studies. In fact, many of the gene classes that were found to be involved in schizophrenia using microarrays were predicted in earlier studies using technology other than large scale transcription profiling: e.g. changes in genes with a presynaptic function such as synapsin (Browning et al., 1993), synaptophysin (Glantz and Lewis, 1997), synaptic proteins (Karson et al., 1999), and decreased neuropil (Selemon and Goldman-Rakic, 1999). There are two major methodologies that can be implemented for a focused approach: Focused arrays and qPCR for multiple relevant genes have been used in various studies including expression profiling of schizophrenia (Lehrmann et al., 2003) as well as for identifying stress related genes relevant to depression (Ohmori et al., 2005; Rokutan et al., 2005b).

The throughput and data quality of qPCR methodologies have improved dramatically in recent years. Also, at present transcription data from microarrays are supposed to be confirmed using qPCR, as the latter offers greater sensitivity, dynamic range, accuracy and precision as compared to arrays. These can be critical factors as the expected changes in transcription levels are small. Using this approach, our early data show that it is possible to segment normal control populations as well as to identify distinct changes in different untreated patient segments. Our data also show that care must be exercised when evaluating changes in transcription levels, as even significant changes between two groups may not be a clinically relevant biomarker if variance of the readout is sufficiently high in a normal population.

Biomarkers that can be identified and measured in peripheral tissues such as blood have greater utility than those identified in CNS regions, as clinical samples are readily available. The underlying assumption is that changes seen in the brain will be reflected in the periphery and lymphocytes have been proposed as suitable neural probes (Gladkevich et al., 2004). Although not universally true for all genes, there is a good correlation between expression of many genes in the CNS and blood, especially genes related to cell signaling and metabolism (Ray et al., 2007; Sullivan et al., 2006). In previous studies of blood samples from schizophrenia patients a set of eight genes was identified that segregated populations of schizophrenic and bipolar patients from controls with high specificity (Tsuang et al., 2005). However, these studies must be interpreted with caution as other studies have identified different genes associated with schizophrenia (Middleton et al., 2002; Vawter et al., 2004a, b). There have also been several studies that examined expression patterns in peripheral blood in PTSD patients (Segman et al., 2005; Zieker et al., 2007) and patients with major depression (Kalman et al., 2005; Ohmori et al., 2005; Rokutan et al., 2005a). These studies have identified gene sets that are involved in the disease, and with additional confirmation, many may become clinically useful biomarkers. Taken as a whole these studies provide solid evidence that analysis of peripheral blood can be useful for the assessment of psychiatric disorders.

Regardless of which technology is used to identify a transcription (or, for that matter, a protein or metabolite) biomarker set, it is important to note that it is unlikely to find an association between complex psychiatric disorders and a single marker. A more likely scenario is that there will be changes in multiple genes that must be examined as a pattern to achieve acceptable sensitivity and specificity.

Real World Data Using Focused Transcription Analysis from Blood Leukocytes

Both microarrays and qPCR have been widely used to measure gene expression changes in a variety of experimental systems and each method has its own benefits and challenges (Brazeau, 2004; Morey et al., 2006). However, given the larger dynamic range and sensitivity of qPCR (Mir, 2006), in addition to the speed and lower cost per sample, we have selected it as the platform of choice for our first studies. Blood samples are collected in the clinic with PaxGene™ RNA tubes.

While using qPCR to measure gene expression is more sensitive than using microarrays, one significant limitation is the restriction in the number of genes that can be assayed. We have addressed this limitation by using reports from the literature to enrich our list of target genes with those that have a high probability of being linked to the biochemical pathways that are affected in affective disorders.

Data Normalization

In order to effectively compare gene expression patterns between different samples, it is necessary to control for variables that could mask the underlying biological changes related to the psychiatric syndrome. The best way to minimize the influence of these variables is through the use of normalization genes (Andersen et al., 2004; Huggett et al., 2005; Jin et al., 2004; Vandesompele et al., 2002). Because it is unlikely that a "universal" normalization gene exists, we have identified several genes which, collectively, are stably expressed in blood samples derived from multiple groups of control subjects and different depressed populations.

Data Analysis and Statistics

Currently there is no consensus on what denotes a significant change in expression, since measurements are being made in different tissues using different techniques. Furthermore, while some expression changes might prove to be large between healthy and depressed individuals, others are likely to be more subtle, but not less important. For these reasons, especially in the early stages of the project, we believe that it is necessary to analyze the expression data using multiple approaches in order to have the best chance of uncovering patterns that are associated with a specific disease state, and/or correlate with response to treatment. Classical univariate methods such as *t* tests and ANOVA (or their non-parametric equivalents) are useful for comparing expressions between different groups, one gene at a time, or between one gene and one clinical variable within a group. When interpreting such analyses, it is prudent not to be too stringent with respect to p value cut-offs for fear of missing an important contributing gene during the exploratory phase. Confirmatory studies have to employ prospectively defined hypotheses and more rigorous analysis methods. We are also actively investigating the use of multivariate approaches, such as MANOVA, stepwise logistic regression (SLR) and Pelora (penalized logistic regression) (Dettling and Buhlmann, 2004) to facilitate the identification of combinations of gene expression values that allow the discrimination between control and patient populations.

The combination of gene expression data and clinical information provides the opportunity to identify biomarkers that differentiate between the healthy and disease state and/or between different segments of diseased as well as healthy individuals. As described below for the analysis of a large set of control subjects, the latter has made it possible to examine intermediate phenotypes, which is an important step towards an improved understanding of the biology of psychiatric disorders. Whenever patients are being treated with medication, the opportunity exists to identify biomarkers that relate to treatment response and safety.

Analysis of Control Subjects

Many clinical trials, by design, do not contain a control group. Because our intention is to compare the gene expression patterns in healthy controls to those seen in depressed patients, it was important to generate gene expression data on a large number of control subjects. In addition, we speculated that the control group itself may be a valuable subject pool to investigate for intermediate phenotypes, since there is likely to be a continuum between completely healthy individuals and those with a clinically manifest psychiatric disorder. This approach is in line with suggestions raised in the context of DSM-V, emphasizing a dimensional rather than a categorical approach. Table 15.1 shows some combinations of gene expression and clinical parameters that produce the most significant p values ($p < 0.001$) in *t* tests or ANOVAs. Note that two genes, "G" and "Q", produce significant correlations with multiple parameters, making them strong candidates for genes whose expression may be linked to a clinically relevant feature of depression.

Analysis of Patients with Disease State 3

Comparison of gene expression patterns between matched control subjects and 21 patients with Disease state 3 demonstrates that 13 genes out of the 25 tested are differentially expressed between the two groups ($p < 0.01$)[Table 15.2]. Similar results are obtained if a larger number of unmatched controls are used for comparison. The expression data have also been analyzed using two multivariate approaches, stepwise logistic regression and Pelora. These techniques have begun to provide insight into combinations of gene expression values that allow the two groups to be differentiated.

Summary of Analysis of Multiple Patient Populations

Depression, as a complex disorder, will in all likelihood be associated with multiple, potentially subtle, gene expression alterations that may differ between subtypes of the disease. The heatmap shown in Table 15.2 summarizes the gene expression differences between controls and patients, derived from several patient populations involving different disease subtypes. The results depicted in this figure can be viewed in two

Table 15.1 Association of clinical parameters and gene expression in blood samples from healthy control subjects

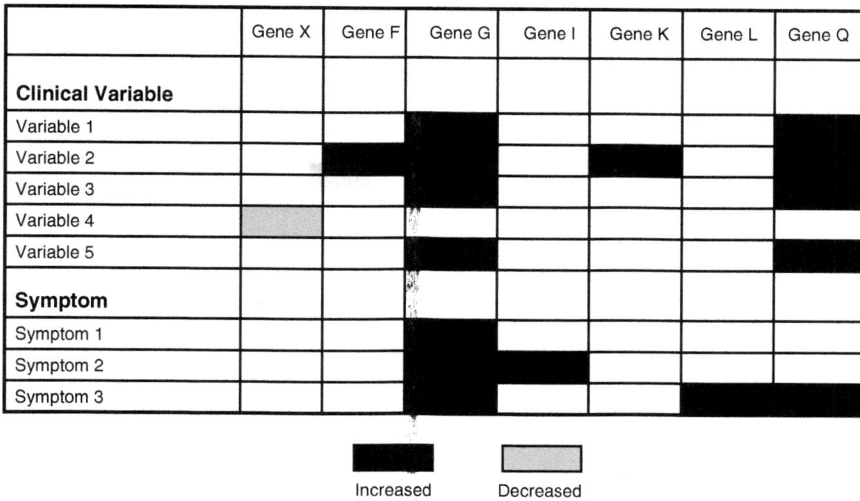

The expression of 29 genes in 299 control subjects was measured in blood using qPCR. ANOVAs or *t* tests, as appropriate, were used to correlate the expression values with responses to 29 items from the self-assessed questionnaires provided by the subjects. In some cases responses to multiple items were grouped to create additional composite scores for testing. For example, the scores for a combination of items related to symptoms of depression were summed to create a new category (Symptom 3). All of the combinations marked with a filled box in the table yielded a p value of $p < 0.001$ in the statistical test.

The heatmap shows gene expression changes between controls and depressed patients measured by quantitative PCR. The expression for each gene is compared between controls and patients using a *t* test. The direction and magnitude of gene expression change is shown by the colored bar and statistical significance is indicated by the number of stars (2 stars $p < 0.01$, 3 stars $p < 0.001$). Disease state one includes 172 patients and 59 control subjects. Disease state two contains 24 patients and 24 matched control subjects. Disease state three contains 21 patients and 21 matched controls. No correction was made for multiple testing.

different ways to extract different information. If viewed by rows across the different disease states, one can identify genes that display similar directional changes across multiple patient populations. For example, relative to the controls, the expression of genes "A", "G", "K", and "P" are increased while genes "C", "H", and "J" tend to be decreased. These genes potentially represent common biomarkers for all depressed patients, regardless of disease subtype. Viewing the table by columns reveals that patients in each of the disease states display very different patterns of gene expression changes. For example, patients in disease state one have approximately twice as many genes that are increased in expression as are decreased. Disease states two and three almost exclusively display patterns of increased expression. However, even these two disease states exhibit different patterns and can be distinguished from one another based on several genes. It is our intention to utilize differences such as these to segment the patient populations in our studies.

Table 15.2 Summary of gene expression changes in different patient groups relative to control subjects (*See Color Plates*)

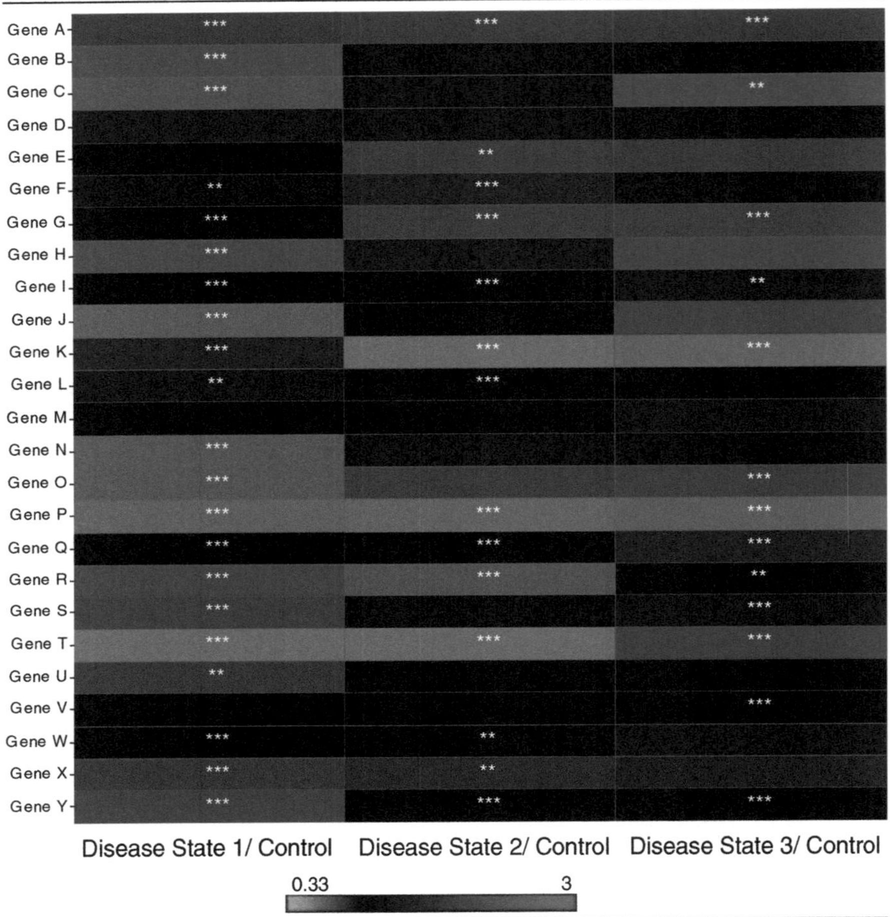

Comparison to Previous Reports

Publications have begun to appear in the literature in recent years reporting changes in peripheral gene expression in patients suffering from affective disorders (Iga et al., 2005, 2006, 2007a; Lima et al., 2005; Ohmori et al., 2005; Rocc et al., 2002; Segman et al., 2005; Tsao et al., 2006; Vogel et al., 2004; Zieker et al., 2007). However, meaningful comparisons between any of these studies and our work are very difficult for a variety of reasons. Each lab has used a different combination of experimental platforms (microarrays vs qPCR), normalization genes, source of tissue (PBMC vs whole blood), and ethnic background of the study group. This highlights the importance of standardizing the technological and methodological aspects as much as possible to lay the foundation for a systematic approach in the future.

In addition, because the previous studies have relied on a small sample size, usually 20 patients or less, all conclusions await confirmation in larger populations. Indeed, this last point also applies to some of our data and is actively being corrected by the analysis of more patients. But similar profiles from different world regions point to the value of this approach. Ultimately, more experiments will clarify which observations reflect the real world.

15.2.3.5 Other Plasma Markers

Methods for characterizing changes in large numbers of proteins have improved in recent years. The advent of proteomics, including protein arrays, improved mass spectrometry technology (SELDI™, Ciphergen Inc) and better analysis of 2D gel electrophoresis have greatly augmented the quality and speed of protein analysis. It is however, far more difficult to assess large numbers of samples for changes in multiple proteins than it is to assess RNA transcripts. The need for standardization in proteomics has been recognized, as evidenced by initiatives such as the one of the Human Proteome Organization (HUPO). HUPO is seeking to characterize the human CNS proteome (HUPO brain proteome project) and to establish collection and analysis standards (HUPO Proteome Standards Initiative). Despite great strides in the field, there are still more tools for examining RNA transcription than currently exist for proteins. One other limitation of proteomic analysis is that samples collected in multi-center studies cannot be immediately processed, making RNA collection still the best available method for assessing regulation of multiple factors in blood samples in multi-center clinical trials.

On the other hand, if a focused approach is pursued, protein markers may provide a valuable translational bridge between human peripheral blood and animal CNS data. Thus, beta arrestin-1 protein levels have been shown to be reduced in leukocytes of depressed patients and this decrease is proportional to the severity of the symptoms (Avissar et al., 2004). In the same study it was shown that antidepressants raise beta arrestin-1 levels in relevant areas of the rodent brain, suggesting that markers identified in human blood could be the basis for studying therapeutic approaches in animal models. The relevance of plasma protein markers for CNS disorders has been recently illustrated in an elegant study using 18 signaling proteins that not only classified patients with Alzheimer's dementia with 90% accuracy, but also predicted progression to Alzheimer's disease over 2–6 years in patients with mild cognitive impairment.

In a similar vein, Kromer et al., (Kromer et al., 2005a) have identified glyoxalase as a marker of anxiety in mice bred for high or low anxiety. Animals with low anxiety compared to high anxiety animals had higher levels of glyoxalase 1 protein in the hypothalamus and amygdala. These changes were also seen in blood from these animals and the amount of glyoxalase 1 was associated with the degree of anxiety in the animals. The regulation of glyoxalase 1 as an anxiety marker has not been confirmed in human studies, although there is evidence for the linkage of the glyoxalase locus with unipolar depression (Tanna et al., 1989).

Also, biochemical markers derived from endogenous metabolic pathways that can be measured in plasma samples have the potential to be used in large patient groups. In this regard the application of a metabolomics approach to study geriatric depression illustrates the power of this technology, and shows great promise to aid in the elucidation of the biochemical pathways perturbed in disease and treatment (Paige et al., 2006). Besides its potential to yield diagnostic and efficacy markers, this approach also holds promise related to safety, as shown by recently published metabolomics data related to lipid metabolism changes in schizophrenic patients treated with different atypical antipsychotics (Kaddurah-Daouk et al., 2007).

In addition, the assessment of peripheral monoamines and their metabolites met with little success in the past. However, metabolomics, by virtue of its ability to measure simultaneously multiple metabolites in biological fluids, shows great promise for the development of biomarkers, related to monoamines and beyond, for many psychiatric disorders. One example is inflammation-induced changes in serotonin metabolism, such as a reduction in plasma tryptophan and an increase in plasma kynurenine, which have been associated with incidence of depressive symptoms during interferon-α treatment (Bonaccorso et al., 2002; Capuron et al., 2003a). A more detailed exploration of plasma metabolites related to monoamine pathways in these patients could be a fruitful approach to identify robust biomarkers, at least for cytokine-induced depression (see also the following section).

15.2.4 Perspectives for Biomarkers in Drug Development

The widely accepted official NIH definition of a biomarker is 'a characteristic that is objectively measured and evaluated as an indicator of normal biologic processes, pathogenic processes, or pharmacologic responses to a therapeutic intervention' (Biomarkers Definitions Working Group, 2001). Based on this definition, all of the above mentioned technologies have the potential to yield a biomarker, but with differences in their applicability to clinical practice.

Obviously, a biomarker that can be measured in a single blood sample and which yields good specificity and sensitivity is more practical than a marker that requires MRI or PET scans. Blood transcription markers also offer the advantage of being able to explore multiple pathways and hence use patterns rather than single markers, which, as our early data suggest, seem capable of differentiating segments of patients with a high predictive value.

On the other hand, a conclusion on the pathophysiological relevance of such transcriptional markers would be greatly advanced by showing a correlation with functional alterations in brain circuitries. This, in turn, would also provide important guidance for animal studies and would facilitate translation between transcriptional changes in human blood leukocytes and animal brain areas. Finally, transcriptional changes (or, for that matter, protein, peptide or metabolite markers), that signal a relevant pathophysiological change and/or effect of treatment can also be explored in cellular assays. As the next sections will illustrate, such reductionist models could help profile novel drug targets early on, particularly if they are designed to address a specific (molecular) aspect of a disorder and if they use molecular read-outs, such as transcriptional changes, that are derived from human (patient) data.

Although it remains to be shown that such an approach is valid, the lack of substantial progress in treating psychiatric disorders for several decades also emphasizes the need to find alternative approaches. It has been recognized that the main reasons for the delay in the introduction of new antidepressants are the notoriously small effect sizes and large placebo effects (Walsh et al., 2002). An alternative approach is needed that closes the gap between the patient symptoms and the pharmacodynamic profile of the drugs that are being developed. To achieve this goal the initial studies should aim to characterize the efficacy profile of the drug, by taking advantage of trait-markers for patient enrollment as well as the broad assessment of biomarkers that inform on the engagement of pathways relevant to the disease biology, and in the process enable the identification of relevant state-markers which predict treatment response and could be used in successive trials for enrichment of the trial with patients that respond to the treatment. In addition the assessment of biomarkers in Phase 3, the confirmatory phase of clinical development, is an important step towards the validation and qualification. Besides the necessary specificity and sensitivity of a biomarker an important feature is the easy access to source in order to enable their application to large clinical studies. The development of biochemical biomarkers that can be measured in blood samples have the potential to be used in large patient groups. This, in turn, is a prerequisite as the utility of biomarkers is directly related to their validation and qualification for decision-making. The qualification process involves a graded evidentiary process that links a biomarker with biological and clinical end points. The extensive efforts that such a process requires can only be achieved with a consortium approach that distributes the cost and risk and effectively drives the research and ultimately the regulatory acceptance of biomarkers (Wagner et al., 2007).

In summary, existing data and technologies allow for the identification of biological markers that can be used in exploratory clinical trials with NCEs. Although none of the markers can be considered validated today, they can be used for internal decision making, on dose, optimal target population(s) and target symptom(s). Importantly, the selection of biomarkers should be based on the preclinical data and the mechanism of action of a NCE, but should also take into account the discussions that have been initiated by several public-private partnerships as well as health authorities. In particular, consortia in both Europe and the USA have been set up, some of which include participation from health authorities, with the aim to establish standards and prioritize biomarkers, specifically for psychiatric disorders.

15.2.4.1 Examples Where Biomarkers Could Be Used Today

A recent example is major depression during cytokine therapy for cancer or hepatitis C (Capuron and Dantzer, 2003; Capuron and Miller, 2004; Wichers et al., 2005). It has been proposed that exaggerated production of inflammatory cytokines further stimulates the hypothalamic–pituitary–adrenal (HPA) axis and thereby contributes to the development of major depression (Leonard, 2005). The high incidence of depression under interferon-α treatment allows to prospectively collect samples and identify vulnerability markers (in subjects who do develop depression) and markers of resilience (in subjects who don't develop depression) (Capuron et al., 2003b). Such paradigms could aid in the understanding of the neurobiology of at least some forms of major depression. The clinical relevance of this approach is underscored by the high incidence of depression in disorders with known CNS inflammation, such as Multiple Sclerosis and Alzheimer's dementia (Ghaffar and Feinstein, 2007; Leonard, 2007; Lyketsos and Olin, 2002).

Another example is melancholic major depression. This old (fashioned) concept has recently received more attention, as it is a disorder with definable clinical signs and can be verified by laboratory tests and treatment responses (Taylor and Fink, 2007). The cardinal clinical manifestations of major depression with melancholic features include sustained and pervasive ahedonia and vegetative features, particularly early morning awakening, loss of appetite, weight loss and loss of libido. Also, there is evidence of physiological hyperarousal, which is thought to arise from the sustained stress system activation evidenced by hypercortisolism and sympathetic nervous system activation, and accompanied by suppression of the growth hormone and reproductive axes, insomnia, and loss of appetite. Another feature of melancholia is a diurnal variation in the severity of depressed mood, with greater severity in the morning than in the evening (Gold and Chrousos, 2002). Although only 25–30% of patients with major depression present with pure melancholic features, this subtype of major depression has been repeatedly associated with several laboratory findings, including dexamethasone non-suppression, high nocturnal cortisol, short REM latency, high REM density and reduced SWS and delta activity (Taylor and Fink, 2007). The biological features of melancholia should facilitate the identification of biomarkers related to the underlying pathophysiology. Furthermore, modeling of these abnormalities in animals provides ample opportunities for translational approaches to assess pathophysiological and etiological mechanisms as well as potential novel therapeutic targets.

15.3 Animal Models

There are a multitude of animal models that have been employed to study particular aspects of major depression. A comprehensive overview of these models has been covered elsewhere and is beyond the scope of this text. However, a selection of these models will be highlighted here in relation to their practical use for identifying novel compounds with therapeutic antidepressant efficacy. It is important to keep in mind that the ultimate goal of any animal model is to achieve construct validity by sharing homologous etiology and pathophysiology to the disease state in humans. Unlike neurological diseases in which the disease biology is well-established, the biological underpinnings of major psychiatric disorders are varied and still relatively poorly understood, thus making the development of such models, e. g. for major depression, rather rare.

15.3.1 Animal Models of Depression

Most of the animal models for major depression in use today are designed to simulate some aspect of the disorder itself by exposing the animal to a single or series of stressful conditions. While this approach has proven useful, caution must be used in interpreting results from these studies since the biological substrate(s) involved in producing the response may not reflect the human condition. Still other models base their utility on predictive validity, or the ability of the model to distinguish compounds that are clinically useful, while often times having no direct relation to depression per se.

The choice of an animal model for any drug discovery research program presents a number of unique technical and theoretical challenges. Most notably, routine use of a model requires that it be reliable, well-suited to high-throughput testing and offer predictive validity. To this end, the most widely used assays employ modifications of a rodent model based on 'learned helplessness' (LH). In its original form, the LH paradigm described the inability of animals to escape a stressor when given the option following a session in which the same stressor was presented in an inescapable fashion (Seligman et al., 1980). Subsequent iterations of the LH model involve stress-induced coping strategies expressed as the eventual acquisition of an immobile posture brought about by acute inescapable stress. These models are among the most widely used in animal research for detecting antidepressant-like activity and they include the forced swim test (FST) and the tail suspension test (TST).

15.3.1.1 Forced Swim Test

The FST is the most widely used animal model in depression research (Cryan and Holmes, 2005). The utility of the FST has been outlined extensively (Borsini, 1995; Cryan and Holmes, 2005b; Nestler et al., 2002; Petit-Demouliere et al., 2005). In short, the FST involves placing a rodent in an inescapable cylindrical water bath where it swims around the container searching for a means of escape. Eventually, the animal assimilates an immobile posture with intermittent bouts of specific activities (swimming and struggling/climbing). The test measure focuses on the time spent immobile, which is thought to reflect a change in coping strategy akin to a state of behavioral despair (Cryan and Mombereau, 2004; Porsolt et al., 1977), an alteration from active to passive [(Holmes, 2003), a learned adaptation (De Pablo et al., 1989; West, 1990) and/or a means to conserve energy (Holmes, 2003; Petit-Demouliere et al., 2005).

The FST represents an obvious advantage in discovery research due to its ability to detect acute and chronic antidepressant-like activity over a large range of clinically relevant antidepressant drug treatments including tricyclics, monoamine oxidase inhibitors and atypical antidepressants, as well as non-drug based paradigms such as electroconvulsive shock and exercise. (Cryan et al., 2005b). More recently, modifications to the FST have enabled the reliable detection of selective serotonin reuptake inhibitors and monitoring specific behaviors associated with serotonergic and/or noradrenergic activities (Detke et al., 1995; Lucki, 1997). Thus, the FST offers good predictive validity and is used routinely as a 'frontline' assay for screening compounds for antidepressant-like activity. The ability to use this test in both mouse and rat rodent species has also added to the utility of this assay.

15.3.1.2 Tail Suspension Test

The tail suspension test (TST) is based on the same theoretical coping strategy as the FST that is thought to come about in the advent of an animal being presented with an inescapable stressor. The utility of the TST has been described elsewhere, and detailed comparisons between the TST and the FST have been discussed (Cryan et al., 2005a). In this test, a mouse is hung upside down by the tail, where it initially engages in escape-related behavior and eventually adopts an immobile posture. Like the FST, the TST is sensitive to a number of clinically used antidepressant compounds, although specific differences have been noted between TST and FST with respect to sensitivity and efficacy among classes of antidepressant agents (Cryan and Mombereau, 2004; Cryan et al., 2005a; Porsolt et al., 1987). The TST is almost always carried out in mice, but other rodent species have been used, and it is generally considered to offer good predictive validity for antidepressant-like activity.

15.3.1.3 Limitations of Current Animal Models

As two of the most widely used animal model tests in use today to detect antidepressant-like activity, the FST and the TST share some common limitations. Notably, these tests have been developed largely based on monoamine-based therapeutics. While monoamine-based drugs have been successful in the clinic, it is becoming increasingly apparent that novel targets must be identified to improve efficacy over currently used therapies. It seems unlikely that future iterations of monoaminergic drugs will sufficiently lack unwanted side-effects in addition to providing superior efficacy to those compounds already in use. To this end, experimental paradigms must be employed that detect novel mechanisms of action.

An important limitation of the FST and TST models stems from the lack of understanding of the immobility behavior itself. For instance, one must consider whether the immobility behavior has direct relevance to the human symptoms of depression, and whether the actions of antidepressant drugs in these models is reflective of their therapeutic actions per se. Perhaps the major criticism of the FST and TST arises from the acute actions of antidepressant compounds in these models. It is well known that clinically effective antidepressant drugs require approximately 3–4 weeks of administration to achieve their therapeutic effect. From this standpoint, it seems unlikely that acute effects of these compounds in the FST or TST are reflective of potential 'therapeutic' efficacy. These questions highlight the concern that non-monoaminergic targets will be adequately identified using these models.

Another limitation of FST and TST is that the measured behavioral endpoint for these models is confounded by potential locomotor effects of the test compound. Thus, false positives may be identified based on a compound's ability to nonspecifically increase spontaneous locomotor activity. While this has not generally been an issue for monoamine-based drugs, this possibility raises a valid issue for using these models to assess potential novel targets for potential antidepressant-like activity. Ultimately, the value for models like the FST and TST in demonstrating antidepressant-like effects of compounds with novel mechanisms awaits the clinical validation of the target in question. In the meantime, one approach for preclinical research on depression is to begin to address the need to improve efficacy read-outs by employing mechanistic and disease-state models of the disorder.

15.3.1.4 Future Directions for Depression Research

Appropriate diagnoses and corresponding successful treatment of specific depression segments has been hampered by our lack of clear understanding of the biological underpinnings that are responsible for the various symptoms of depressive disorders. However, research in depression has recently emphasized the advantage of focusing on these individual segments of depression to parse out specific phenotypes, rather than on depression as a whole (Cryan and Slattery, 2007). It is well known that the experience of significant life stressors is a clear risk factor for depression, and anxiety is often co-morbid with depression. In this context, it is proposed that efforts to utilize disease state and mechanistic models that specifically focus on stress and stress/HPA neurocircuitry (as it relates to depression) may serve as a valuable way forward in the drug development process. By taking this focused approach on the stress axis, it is hoped that depression-related neuronal pathways will be characterized and that molecular aspects of this circuitry can be more readily identified through translational methods.

A convenient framework with which to conceptualize this notion for the following discussion is to define the stress axis according to a modified version of the Gould and Chrousos model (Gold and Chrousos, 2002). In this framework, the stress axis consists of the complex and integrated neuronal interplay between the CRF-containing neurons of the hypothalamus (HPA axis), the amygdala and the noradrenergic containing locus coeruleus (LC). Emphasis for potential mechanistic studies will be placed on terminal fields of the LC such as the medial prefrontal cortex (mPfc) and the ventral hippocampus. The specific segment of depression that will be addressed by such mechanistic models is melancholia, which is characterized by a hyperarousal, associated with hyperactivity of the HPA axis and the LC.

Disease State Models

A number of tests have employed chronic stress paradigms to model certain aspects of depression and, more recently, schizophrenia. One of the depression models, chronic mild stress (CMS), is induced by exposing animals to intermittent, unpredictable and uncontrollable stressors over a period of several weeks to achieve the desired behavioral endpoint (Willner, 1997). Animals exposed to CMS exhibit a reduction in the consumption of a normally palatable sweet solution, and this is considered to model the ahedonia that is a hallmark feature of major depressive disorders. Additional changes are noted in a number of physiological and neuroendocrine parameters that are thought to model depressive symptoms in humans. Moreover, exposure to CMS increases immobility in the FST (Molina et al., 1994). An attractive feature of the CMS model is the fact that the effects of repeated stress are reversed by antidepressant agents over a time-course that mimics their time-course to clinical efficacy. However, a major hindrance to the widespread use of the CMS model is the fact that between-laboratory reproducibility is poor, possibly related to differences in frequency, duration and stressors employed across experimental protocols.

A different approach to exploiting the stress axis as it relates to depression is through the use of genetic models via selective breeding. For instance, high anxiety bred (HAB) rats and mice exhibit enhanced immobility in FST, HPA hyperactivity and increased anxiety compared to their low anxiety bred (LAB) counterparts (Kromer et al., 2005b; Landgraf et al., 1999; Liebsch et al., 1998a, b). The observation of elevated levels of glyoxalase- 1 in a number of brain regions has been identified as a potential marker in LAB mice, suggesting that low expression of this protein may identify trait anxiety in HAB animals. HAB rats display HPA dysregulation (Keck et al., 2002, 2003) that is accompanied by an over expression of hypothalamic vasopressin (Landgraf and Wigger, 2002) and LC hyperactivity (Salchner et al., 2006). Interestingly, the HPA dysregulation noted in HAB rats is normalized by a V1 receptor antagonist, suggesting that the HAB model of anxiety and depressive-like behavior may lend itself to identifying activity of novel therapeutic targets.

Disease state models offer the unique advantage of modeling specific (endo-) phenotypes of depression. In addition to detecting the activity of potential novel therapeutics, it is desirable to gain knowledge about the molecular underpinnings of the response being measured. To this end, it is of particular interest to investigate the responders and non-responders to treatment in these paradigms and to translate those findings to the human condition as they may relate to the clinical observations that note differences between treatment responsive and non-responsive patients. Such an understanding would aid in identifying candidate genes, at the genetic and expression level, that may confer a response to treatment.

Mechanistic Models of Stress Axis Activity

Mechanistic models are valuable in discovery research as they allow for the desired aspects of a particular model to be relatively isolated for in-depth study. While portions of the neurochemical approach outlined below are rather broad in nature, the scope of mechanistic models is mostly limited to the LC hyperactivity component and the subsequent effect on cortical function. As is the case for disease state models, the goal for these mechanism-based assays is to allow for the detection of novel compound activity while simultaneously facilitating an understanding of the molecular mechanisms involved in the behavioral endpoint of the model. An understanding of the molecular mechanisms involved in the behavioral and neurochemical sequelae associated with isolated preparations may begin to bridge the gap between preclinical and clinical outcomes and will help establish cellular models to begin to address the causal molecular factors contributing to psychiatric disorders.

Adult Neurogenesis

Adult neurogenesis involves the turnover and maturation of progenitor cells into new neurons in the dendate gyrus of the hippocampus (Gross, 2000). Neurogenesis has been demonstrated to occur in rodent and human dentate gyrus (Eriksson et al., 1998), and much attention has been given to the relevance of the interplay between stress and neurogenesis in rodent models (Malberg and Duman, 2003; McEwen, 2001). Activation of the HPA axis decreases neurogenesis in rodents and clinically effective antidepressant therapies have been shown to reverse these deleterious effects on cellular proliferation (Malberg and Duman, 2003; Van Praag, 1999). From the drug discovery perspective, the potential utility of the neurogenesis approach is buttressed by its apparent ability to detect novel compounds of potential therapeutic relevance (David et al., 2007). However, it should be cautioned that neurogenesis does not seem a requisite for behavioral effects of all antidepressant-like compounds (David et al., 2007).

At the molecular level, chronic HPA activation has been associated with alterations in genes known to regulate cell proliferation, and these alterations can be prevented with tricyclic antidepressant treatment (Alfonso et al., 2004). These findings are consistent with antidepressant-mediated reversal of chronic stress effects on features of neurogenesis and are of particular importance for supporting a translational research approach in attempts to link alterations in gene expression to disease states.

Neurochemistry

The notion that monoaminergic neurotransmission is implicated in the regulation of mood is a decades-old concept. Moreover, the fact that monoaminergic transporter blockers represent the most widely prescribed antidepressant treatment underscores the importance of enhanced monoamine transmission in the treatment of depression. The fact that antidepressant drugs enhance monoamine levels acutely while their clinical efficacy takes weeks to develop precludes the temptation to overestimate the value of data from acute neurochemical studies. Nonetheless, the fact that protracted monoamine enhancement is thought to mediate the plasticity necessary to achieve therapeutic effects suggests that acute neurochemical data can be useful in estimating a drug's antidepressant potential following long-term treatment. Specifically, the assessment of monoamine release in corticolimbic structures with *in vivo* microdialysis following acute administration offers a valid approach to compound assessment (Millan, 2004; Millan et al., 2000). While the utility of this approach for monoamine-based therapies is obvious, it can also aid in evaluating a number of novel targets since it has been demonstrated that several of these can affect monoamine transmission either directly or indirectly. For instance, NMDA receptor blockers (Millan et al., 2000), metabotropic glutamate 2/3 receptor ($mGlu_{2/3}$) agonists and antagonists (Cartmell and Schoepp, 2000), CRF_1 receptor antagonists and melatonin receptor agonists (Millan, 2003) all seem to increase corticolimbic monoamine release in some capacity. A number of studies also support a role for additional neurotransmitter

(Swanson et al., 2005b) and peptidergic systems in mood disorders (David et al., 2007; Swanson et al., 2005a). The current need for novel therapeutics underscores the potential utility of an integrated drug discovery approach that includes *in vivo* microdialysis assessment of glutamatergic, GABAergic and peptidergic function in colimbic areas, in addition to the traditional survey of monoamine transmitter systems. Furthermore, evaluation of compound effects on corticolimbic acetylcholine transmission is also of great interest given the supposed pro-cognitive benefits of enhancing this neurotransmitter system in the cortex.

In keeping within the framework of the stress axis defined earlier, the *in vivo* microdialysis approach can be useful for assessing mechanistic read-outs of novel target function in the LC-mPfc pathway. mPfc, which receives noradrenergic input from LC (Ungerstedt, 1971), is an important brain structure implicated in mood disorders (Glavin, 1985a, b). A consequence of aberrant LC hyperactivity and the subsequent noradrenergic elevation that results in melancholia is the deleterious inhibition of mPfc function (Gold and Chrousos, 2002), and it has been well demonstrated that various acute stressors increase norepinephrine turnover/release in this structure (Glavin, 1985a). It is proposed that the mechanistic effects of novel compounds may be assayed by exploiting the effects of acute stress on increased mPfc norepinephrine. For example, $mGlu_{2/3}$ receptor agonists have recently been identified as useful agents of therapeutic potential in anxiety disorders (Swanson et al., 2005b). In this light, it has been demonstrated that $mGlu_{2/3}$ receptor agonists can reduce immobilization- (Swanson et al., 2004) and elevated platform stress-induced enhancement (Lorrain et al., 2005) of mPfc norephinephrine in rats. $mGlu_{2/3}$ agonist-mediated reversal of the NE increase in response to elevated platform stress has also been demonstrated in a mouse elevated platform exposure model [Fig. 15.1]. It is important to note that Lorrain et al. (Lorrain et al., 2005) also demonstrated a similar reversal in platform-induced increases in mPfc norephinephrine with the clinically effective anxiolytic diazepam, and a putative anxiolytic-like CRF_1 receptor antagonist (Koob, 1999). Interestingly, a recent study showed clinical efficacy of $mGlu_{2/3}$ agonists to treat positive and negative symptoms of schizophrenia (Patil et al., 2007).

Collectively these data suggest that mechanistic models that effectively isolate some aspect of the stress axis may be useful in identifying brain regions of interest for translational studies that are useful to cross current diagnostic boundaries. Thus, it is hoped that mechanistic assays may eventually guide the identification of novel biomarkers relevant to disease biology.

In Vivo Electrophysiology

Mechanistic models at the cellular level are also potentially valuable for parsing out mechanisms related to endophenotypes of depression. Again turning to the LC-NE system, a number of stressful stimuli increase LC neuronal activity (Abercrombie and Jacobs, 1987) and exogenous stimulation of LC activity has been associated with stress responses in non-human primates and humans (Charney et al., 1984). Interestingly, acute administration of known (Draper et al., 2007; Salchner and Singewald, 2002) antidepressant agents has been reported to elicit anxiety-like

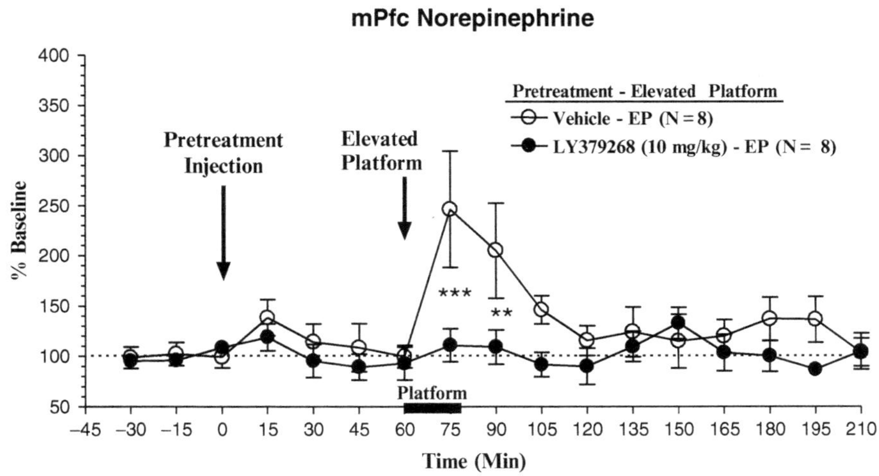

Fig. 15.1 Effects of the $mGlu_{2/3}$ receptor agonist, LY379268, on stress-induced increases in extracellular medial prefrontal cortex norepinephrine levels.Following baseline sample collection, freely moving CD-1 mice were injected with saline vehicle or LY379268 (10 mg/kg, IP; 10 mg/kg) 1 h prior to exposure to elevated platform exposure. The elevated platform consisted of an open, elevated platform made of clear plexiglass top (15 × 7.5 cm) supported 18 cm from the test cage floor by an aluminum tripod. Microdialysis samples were collected every 15 min by an automated fraction collector and norepinephrine concentrations were measured from the dialysate via HPLC with electrochemical detection. A within subjects design was used to generate percent baseline data by expressing each time point for each animal as a percentage of the average of its last 3 baseline samples. Data are then expressed as the mean ± SEM for all animals at each time point within each treatment group.Two-way ANOVA reveals significant effect of time (stress) [$F = 2.505$; $P < 0.0015$] and treatment [$F = 17.05$; $P < 0.0001$] with a significant interaction [$F = 1.897$; $P < 0.0216$]. ** $P < 0.01$; *** $P < 0.001$ compared to vehicle-EP with Bonferroni Post-test. $N = 8$ per group.

behaviors in rodents and this effect has been proposed to arise from enhanced LC activity (Salchner and Singewald, 2002).

Precipitated withdrawal from morphine administration is a model system for studying the effects of compounds on single unit activity of LC neurons and the associated somatic (Frenois et al., 2002) and behavioral (Schulteis et al., 1998) anxiety it produces. In this model, morphine can be administered to rodents over 1–2 days at different doses and a withdrawal syndrome elicited by treating them with an opiate antagonist. The withdrawal syndrome varies in intensity depending on the dose of morphine administered, duration of treatment and dose of opiate antagonist, and this behavioral syndrome is accompanied by enhanced single unit LC firing (Rasmussen et al., 1990). It has been determined that enhanced LC firing upon morphine withdrawal is the result of increased glutamatergic input to this region (Akaoka and Aston-Jones, 1991). In line with microdialysis results mentioned above and the proposed presynaptic mechanism of $mGlu_{2/3}$ receptor function (Cartmell and Schoepp, 2000), $mGlu_{2/3}$ receptor agonists reduce both the enhanced LC firing and behavioral effects of morphine withdrawal (Vandergriff and Rasmussen, 1999). Again, consistent with microdialysis results using platform stress, CRF_1 antagonists block morphine withdrawal-enhanced norepinephrine release in cortex (Funada et al., 2001).

15.3.1.5 Conclusion on Animal Models

The above data highlight the synergistic utility of utilizing distinct, hypothesis-driven mechanistic models based on a conceptual framework of segmentation to draw parallel and supporting conclusions on the potential therapeutic utility of novel compounds. Thus, the combined use of disease state and mechanistic models may offer an advantage to identifying and characterizing cellular and molecular mechanisms involved in various aspects of major psychiatric disorders, including depression, anxiety and schizophrenia.

15.4 Cellular Models

A comprehensive approach to the identification of predictive biomarkers in neuropsychiatric disorders (for both disease and treatment) should include *in vitro* cellular models to augment the clinical and preclinical approaches already discussed here. Cellular models by virtue of their simplicity may seem to represent a (detrimentally) reductionist interpretation of psychiatric disorders and the real human suffering they entail. However it is therein that their promise lies —for out of simplicity can emerge clarity, if the right components of the system are maintained (and unnecessary components abandoned). As symptom clusters and their underlying biochemistries become better defined, it is possible to envision *in vitro* models that could capture key elements of the morphological and molecular processes that emerge in the organism as dimensions of psychiatric illness. In the best embodiment, *in vitro* models can serve as abridged versions of the larger system, displaying emergent properties but providing the practical advantages of working at the cellular rather than at the organismal level. Moreover, cellular models do not need to recapitulate all aspects of a tissue/network/pathway/illness to be both instructive and predictive. When a model has been shown to provide outputs consonant with inputs (such as treatment benchmarks), it may be used to test novel treatments and generate hypotheses for testing in more complex systems. For example, changes in gene expression in response to treatment in clinical samples can be investigated *in vitro*. Similarly, dysregulated biochemical pathways associated with pathology *in vivo* (such as hypercortisolemia in melancholic depression) are more easily defined *in vitro* and may yield novel biomarkers or therapeutic targets. *In vitro* systems are also highly amenable to hypothesis testing since tools for dissecting association from causation, such as selective blockers and siRNAs, are practical to use. Finally, *in vitro* models may help diminish the gap between preclinical and clinical data by providing early confirmation in human cells of expected/desired outcomes. We believe that development of cellular models aimed at common symptom clusters and their underlying biochemistries (representing disease segments of the future) will augment and complement the larger effort to identify biomarkers that is the primary focus of this chapter. An overview of reported efforts in this direction is provided below, with a focus on cell culture models rather than acute *ex vivo* slices. While the following sections are not a comprehensive review, they illustrate the possibilities that cellular assays hold in terms of shedding light onto molecular aspects of disease-related processes, and particularly HPA axis (dys-) regulation and neuronal proliferation, survival and remodeling.

15.4.1 In Vitro Models of HPA Axis Dysregulation in Neuronal Cells

Many of the effects of chronic stress are believed to be mediated by the sustained elevation in glucocorticoid levels. Groups working on *in vitro* models of HPA axis dysfunction have primarily used hippocampal neurons, hippocampal-derived cell lines, or neural stem cells and have focused on the respective roles of the two corticosteroid receptors, the mineralocorticoid receptor (MR) and glucocorticoid receptor (GR): Treatment with a GR agonist triggers neuronal apoptosis, which is counteracted by a MR agonist (Crochemore et al., 2005) hence reflecting the *in vivo* situation (Rozeboom et al., 2007; Sousa et al., 2007).

Another regulator associated with stress and depression is CRF (Charmandari et al., 2005; Claes, 2004). CRF regulation and signaling pathways have been studied in primary amygdalar cultures (Kasckow et al., 1999), on a cell line derived from primary amygdalar cultures (Mulchahey et al., 1999) as well as in hypothalamic primary cultures (Emanuel et al., 1990; Hillhouse and Milton, 1989; Kasckow et al., 2003). Altogether, these *in vitro* studies have indicated that CRF levels in the amygdala are affected by some of the same stimuli that regulate hypothalamic CRF.

Recently, the neuroblastoma cell line Neuro 2A has been used to show that a variety of antidepressants inhibit CRF gene transcription (Budziszewska et al., 2004). Interestingly, similar results were obtained when the same cells were exposed to some antipsychotic treatments (Basta-Kaim et al., 2005).

Physiological stress is also associated with increased firing rate of LC neurons (Bremner et al., 1996), possibly mediated by CRF (Smagin et al., 1996). Mechanistic studies of the modulatory effects that corticosteroids and CRF exert on receptor signaling have been done by Bundley and collaborators (Bundey et al., 1997; Bundey and Kendall, 1999) using a LC-like neuronal cell line (Suri et al., 1993). In the same cell line, Thiel et al (Thiel and Cibelli, 1999) have shown that CRF triggers a PKA-dependent signaling cascade that leads to activation of gene transcription.

15.4.2 In Vitro Models of Monoaminergic Function

The brainstem raphe nuclei comprise the major source of serotonin (5-HT) to the rest of the brain. Primary cultures of raphe neurons have been employed to study the regulation of serotonin release and of presynaptic receptors on serotoninergic neurons (Birthelmer et al., 2007). These cells were also recently used by Scheuch (Scheuch et al., 2007) to link specific tryptophan hydroxylase 2 promoter polymorphisms observed in humans to gene expression abnormalities.

In addition, a raphe-derived cell line was developed and used to study regulation mechanisms of serotonin synthesis, uptake and 5-HT1A coupling mechanisms (Koldzic-Zivanovic et al., 2006; Kushwaha and Albert, 2005). The same cell line has been used to study the polymorphism of 5-HT1A receptor as well as of the serotonin transporter gene promoter (Czesak et al., 2006; Mortensen et al., 1999). In addition, Zhu (Zhu et al., 2006) showed that proinflammatory cytokines activate serotonin transporters, providing an *in vitro* model to study molecular mechanisms related to inflammation-induced depression (Capuron and Miller, 2004; Wichers et al., 2005).

The septal nuclei rich with cholinergic neurons represent a component of the limbic system that regulates emotions and impulses. In a co-culture system of septal with raphe neurons it was shown that the regulation of presynaptic receptor function strongly depends on the concentrations of endogenous transmitter in the neuronal environment (Ehret et al., 2007). These findings open up the possibility that pathological conditions or therapeutic drug treatment could be modeled using this *in vitro* approach.

15.4.3 Neurogenesis and Neuroplasticity

It is now well established that antidepressants increase hippocampal neurogenesis in experimental animals (D'Sa and Duman, 2002). Moreover, cell proliferation has been shown to be necessary for the action of many antidepressant drugs in animal models (Sahay and Hen, 2007).

Though antipsychotics can increase neurogenesis in the subventricular zone (SVZ) as well as the subgranular zone (SGZ) of the hippocampus, conflicting results have been reported (Newton and Duman, 2007).

15.4.3.1 Antidepressants and Neurogenesis

Numerous studies focused on understanding the molecular mechanisms of antidepressants have employed models of *in vitro* neurogenesis mostly using neural stem cells derived from the adult hippocampus. These studies have indicated that in addition to increasing proliferation, antidepressants also promote survival and/or differentiation of neural stem cells (Chiou et al., 2006a, b; Peng et al., 2007). Intracellular signaling underlying these processes has also been analyzed in these studies.

RNA interference *in vitro* studies demonstrated that both neuroprotective and differentiation effects were dependent on BDNF (brain derived neurotrophic factor) expression and Bcl-2 (B-cell leukemia/lymphoma 2) activation (Huang et al., 2007; Peng et al., 2007), confirming *in vivo* findings (D'Sa and Duman, 2002; Schmidt and Duman, 2007). Besides BDNF, vascular endothelial growth factor (VEGF) has also been found to exert direct mitogenic effects on neural progenitors *in vitro* (Warner-Schmidt and Duman, 2007). VEGF is induced by multiple classes of antidepressants at time points consistent with the induction of cell proliferation and therapeutic action of these treatments.

An alternative *in vitro* model to study the effects of antidepressants on cell proliferation has been developed by Zusso (Zusso et al., 2004) using cerebellar granule neurons in culture. The authors reported an increase in proliferation after treatment with the antidepressant fluoxetine. The newly generated cells can be expanded and induced to differentiate into neurons, astrocytes, and oligodendrocytes.

15.4.3.2 Antipsychotics and Neurogenesis

Antipsychotic drugs increase the proliferation of non-neuronal cell types in the prefrontal cortex (Newton and Duman, 2007), consistent with the emerging idea that neurogenesis may be disturbed in schizophrenia (Toro and Deakin, 2007). There is growing evidence that the action of antipsychotic drugs could be mediated at least in part by increased proliferation of neuronal as well as glial cells (Newton and Duman, 2007; Wang et al., 2004). However, while effects for antidepressants are consistent, the results obtained with antipsychotics so far are conflicting (Halim et al., 2004; Kodama et al., 2004; Wang et al., 2004).

15.4.3.3 Neuroplasticity in Mood Disorders and Schizophrenia

In vitro models have been shown to be valuable for clarifying the biological function of specific genes that are found to be linked to psychiatric disorders. Thus, multiple genes such as neuregulin-1 (NRG1), Akt (a serine/threonine-specific protein kinase), disrupted-in-schizophrenia 1 (DISC-1) and dysbindin-1, have recently been associated with schizophrenia although their role in the pathophysiology of the disease is not clear (Owen et al., 2005). A variety of *in vitro* neuronal assays showed that some of these genes are implicated in the process of neurite formation. A comprehensive review of these *in vitro* cellular assays can be found in Bellon et al, (Bellon, 2007). These data lend support to clinical findings of abnormal cytoarchitecture and disconnectivity in several regions of the brains of schizophrenic patients (Gur et al., 2007; Harrison and Weinberger, 2005). Interestingly, some of the genes reported above, namely DISC-1 and NRG, have also been described as genetic susceptibility factors for depression and bipolar disorders (Blackwood et al., 2007; Maier et al., 2006), pointing to a common neurobiological process that contributes to major psychiatric disorders.

15.4.4 Non-Neuronal Cells

In vitro cellular models from peripheral cells (e.g. blood cells, fibroblasts) have been used to probe for gene expression abnormalities that might be present in patients with psychiatric disorders. The identification of altered genes or signaling pathways in readily available human cells is particularly useful in the context of biomarkers to facilitate diagnosis and segmentation of patients as well as to monitor response to medication. Signaling pathways show a correlation between brain and peripheral cells, and hence peripheral cell read-outs in humans may be used to monitor treatment response (Ray et al., 2007; Sullivan et al., 2006). Peripheral blood cells are of obvious interest as biomarkers and have been discussed elsewhere in the chapter. Skin fibroblasts are also relatively accessible and show promise as "windows" into the brain. As several biochemical changes observed in neural tissue of depressed patients have also been observed in fibroblasts, they may be useful for *in vitro* models. Fibroblasts from patients with melancholic major depression confirmed the abnormalities in 5-HT2A receptor-stimulated signaling previously observed in the brain (Shelton et al., 2004). Furthermore, it has been reported that patients with melancholia show distinctly different gene expression compared to controls and patients with depression but without melancholic features (Shelton et al., 2004). In conclusion, non-neuronal cell models can also begin to provide data with which to segment patients for appropriate treatment and track therapeutic efficacy.

15.4.5 Conclusion on Cellular Models

True cellular models of psychiatric disorders do not (yet) exist. The progress described here represents the first stages in developing *in vitro* models relevant to human neuropsychiatric disorders and treatments. Indeed, *in vitro* models have proved useful to unravel some of the molecular mechanisms and to attribute direct cellular targets to specific therapeutic treatments. Furthermore, *in vitro* models have provided functional information on human gene polymorphisms, and also hints on the potential function of some genes, i.e. by knocking down or over-expressing the gene of interest. As clinicians increasingly seek to define biologically-driven endophenotypes in schizophrenia and mood disorders, cellular models may begin to play a larger role in deciphering the underlying biochemistries as well as enabling the discovery of novel targets, treatments, and biomarkers for these disorders.

15.5 Summary and Perspectives

In order to achieve the ambitious goal of the DSM-V research agenda, namely to develop a scientifically sound classification system, a better understanding of the biological underpinnings of psychiatric disorders seems to be a crucial first step. In this regard biomarkers should play a pivotal role.

The previous sections have outlined perspectives on how research into biomarkers, if pursued systematically, could advance our understanding of the biology of psychiatric disorders. Besides the obvious advantages of objective markers for the determination of a diagnosis in psychiatry, biomarkers could also become pivotal for the choice of treatment for a particular patient. In fact, biomarkers that could help make a diagnosis, could relate to the pathophysiology, and could be used to monitor treatment effects could have a major impact on drug development. Animal models could be constructed that mimic a specific biomarker profile of a human disease state and allow testing of novel drug effects on such a circumscribed clinically relevant biomarker read-outs. This approach provides an opportunity to assess effects of new drug targets on multiple distinct biological abnormalities relevant to one or several patient segments. Such a biologically focused evaluation could also help identify multiple therapeutic indications for a novel compound. In addition, these mechanistic constructs could form the basis for cellular assays to assess drug targets very early.

Beforehand, however, the available biomarker data should be critically reviewed and prioritized for future research. This seems important, as there are no confirmed and validated biomarkers available despite the abundant literature on biological findings in psychiatric disorders. This unsatisfying situation reflects the lack of both a systematic approach and use of standardized evaluation methods for clinical and biological factors. Also, if biomarkers are to address diagnostic uncertainty and support treatment selection and novel drug development, they should be assessable in humans and rodents and, at least in humans, at multiple time-points for the same individual. In addition, the biomarker data must yield a consistent result if (practical) standardized procedures are followed and should show high specificity and sensitivity. The ideal biomarker would also show, in individuals at risk, an intermediate level between that seen in acutely ill and super-healthy subjects.

Although many biomarkers explored in scientific publications are not useful for general psychiatric practice, e.g. functional neuroimaging markers or neuroendocrine challenge tests, this could change if they can be correlated with highly accessible markers from blood samples. In this regard, transcription profiling from peripheral leukocytes seems more valuable today than proteomics and metabolomics approaches, as standards exist with regard to sampling, preservation, storage and analysis. In addition, transcription profiling, particularly if a pattern based on a limited number of genes with satisfactory predictive value is identified, can be applied to animal and cellular models. Such mechanistic constructs will also facilitate the discovery of new molecular targets. Once a transcription pattern has been confirmed, it seems reasonable to expect that related protein and metabolite markers will be identified and turned into a commercial biomarker kit.

Implicit in the above paragraph are at least two paradigm shifts, which appear needed to progress from several decades of small and incremental improvements of drug development in psychiatry: Firstly, biological findings used for mechanistic constructs in animals should be based on objectively measured data in patients. Secondly, the link between biological findings and clinical characteristics should not be confined to the current diagnostic entities.

With regard to the first point, some pharmaceutical companies have followed this route by using human genetic mutations in the target populations to discover new drug targets. Whether this effort will prove fruitful for polygenetic disorders, such as psychiatric disorders, remains to be seen. This approach may, in fact, become more useful once biological findings help to delineate homogenous patient segments which are then used for the discovery of genetic polymorphisms.

The second point is addressed in the DSM-V conferences and in some recent publications. Thus, specific symptom clusters rather than an entire disorder are linked to biological findings and can help uncover the biology of distinct phenotypes. This point may be of particular importance when one aims to address early onset of disorders or prophylactic treatments. As the specific symptoms an individual develops depend on the individual's genetic makeup and the environmental context, objective markers that allow recognition of endophenotypes associated with increased vulnerability will help select individuals for prophylactic treatment.

With the discussions around DSM-V, it can be hoped that the process and the discussions will continue to guide and stimulate systematic research into the biology and biological markers of psychiatric disorders long after DSM-V has been realized.

Acknowledgments The authors thank Erik Mosekilde from the Technical University of Denmark for his contribution to the exploratory statistical analyses of the transcription data generated by Lundbeck Research USA. The work done by Erik Mosekilde and Wiktor Mazin was supported by the European Union through the Network of Excellence BioSim, Contract No. LSHB-CT-2004-005137.

References

A Research Agenda for DSM-V™. Washington, D.C. American Psychiatric Association (2002)

Abercrombie, E. D., Jacobs, B. L.: Single-unit response of noradrenergic neurons in the locus coeruleus of freely moving cats. I. Acutely presented stressful and nonstressful stimuli. J Neurosci 7, 2837–2843 (1987)

Akaoka, H., Aston-Jones, G.: Opiate withdrawal-induced hyperactivity of locus coeruleus neurons is substantially mediated by augmented excitatory amino acid input. J Neurosci 11, 3830–3839 (1991)

Alfonso, J., Pollevick, G. D., van der Hart, M. G., Flugge, G., Fuchs, E., Frasch, A. C.: Identification of genes regulated by chronic psychosocial stress and antidepressant treatment in the hippocampus. Eur J Neurosci 19, 659–666 (2004)

Altar, C. A., Laeng, P., Jurata, L. W., Brockman, J. A., Lemire, A., Bullard, J., Bukhman, Y. V., Young, T. A., Charles, V., Palfreyman, M. G.: Electroconvulsive seizures regulate gene expression of distinct neurotrophic signaling pathways. J Neurosci 24, 2667–2677 (2004)

Andersen, C. L., Jensen, J. L., Orntoft, T. F.: Normalization of real-time quantitative reverse transcription-PCR data: a model-based variance estimation approach to identify genes suited for normalization, applied to bladder and colon cancer data sets. Cancer Res 64, 5245–5250 (2004)

Armitage, R., Hoffmann, R. F., Rush, A. J.: Biological rhythm disturbance in depression: temporal coherence of ultradian sleep EEG rhythms. Psychol. Med. 29, 1435–1448 (1999)

Armitage, R., Hoffmann, R., Trivedi, M., Rush, A. J.: Slow-wave activity in NREM sleep: sex and age effects in depressed outpatients and healthy controls. Psychiatry Res. 95, 201–213 (2000)

Armitage, R., Emslie, G. J., Hoffmann, R. F., Rintelmann, J., Rush, A. J.: Delta sleep EEG in depressed adolescent females and healthy controls. J. Affect. Disord. 63, 139–148 (2001)

Armitage, R., Hoffmann, R., Emslie, G., Rintelmann, J., Robert, J.: Sleep microarchitecture in childhood and adolescent depression: temporal coherence. Clin. EEG. Neurosci. 37, 1–9 (2006)

Atz, M., Walsh, D., Cartagena, P., Li, J., Evans, S., Choudary, P., Overman, K., Stein, R., Tomita, H., Potkin, S., Myers, R., Watson, S. J., Jones, E. G., Akil, H., Bunney, W. E., Jr., Vawter, M. P.: Methodological considerations for gene expression profiling of human brain. J. Neurosci. Methods 163, 295–309 (2007)

Avissar, S., Matuzany-Ruban, A., Tzukert, K., Schreiber, G.: Beta-arrestin-1 levels: reduced in leukocytes of patients with depression and elevated by antidepressants in rat brain. Am. J. Psychiatry 161, 2066–2072 (2004)

Basta-Kaim, A., Budziszewska, B., Jaworska-Feil, L., Tetich, M., Kubera, M., Leskiewicz, M., Lason, W.: Inhibitory effect of imipramine on the human corticotropin-releasing-hormone gene promoter activity operates through a PI3-K/AKT mediated pathway. Neuropharmacology 49, 156–164 (2005)

Bellon, A.: New genes associated with schizophrenia in neurite formation: a review of cell culture experiments. Mol Psychiatry 12, 620–629 (2007)

Biomarkers Definitions Working Group: Biomarkers and surrogate endpoints: preferred definitions and conceptual framework. Clin. Pharmacol. Ther. 69, 89–95 (2001)

Birthelmer, A., Ehret, A., Riegert, C., Rothmaier, A. K., Leemhuis, J., Jackisch, R.: Modulation of electrically evoked serotonin release in cultured rat raphe neurons. J Neurochem 100, 1613–1625 (2007)

Blackwood, D. H., Pickard, B. J., Thomson, P. A., Evans, K. L., Porteous, D. J., Muir, W. J.: Are some genetic risk factors common to schizophrenia, bipolar disorder and depression? Evidence from DISC1, GRIK4 and NRG1. Neurotox. Res 11, 73–83 (2007)

Bonaccorso, S., Marino, V., Puzella, A., Pasquini, M., Biondi, M., Artini, M., Almerighi, C., Verkerk, R., Meltzer, H., Maes, M.: Increased depressive ratings in patients with hepatitis C receiving interferon-alpha-based immunotherapy are related to interferon-alpha-induced changes in the serotonergic system. J. Clin. Psychopharmacol. 22, 86–90 (2002)

Borsini, F.: Role of the serotonergic system in the forced swimming test. Neurosci Biobehav. Rev 19, 377–395 (1995)

Bosotti, R., Locatelli, G., Healy, S., Scacheri, E., Sartori, L., Mercurio, C., Calogero, R., Isacchi, A.: Cross platform microarray analysis for robust identification of differentially expressed genes. BMC Bioinformatics 8 Suppl 1, S5 (2007)

Brazeau, D. A.: Combining genome-wide and targeted gene expression profiling in drug discovery: microarrays and real-time PCR. Drug Discov. Today. 9, 838–845 (2004)

Bremner, J. D., Krystal, J. H., Southwick, S. M., Charney, D. S.: Noradrenergic mechanisms in stress and anxiety: I. Preclinical studies. Synapse. 23, 28–38 (1996)

Brenner, S., Johnson, M., Bridgham, J., Golda, G., Lloyd, D. H., Johnson, D., Luo, S., McCurdy, S., Foy, M., Ewan, M., Roth, R., George, D., Eletr, S., Albrecht, G., Vermaas, E., Williams, S.R ., Moon, K., Burcham, T., Pallas, M., DuBridge, R. B., Kirchner, J., Fearon, K., Mao, J., Corcoran, K.: Gene expression analysis by massively parallel signature sequencing (MPSS) on microbead arrays. Nat. Biotechnol. 18, 630–634 (2000)

Browning, M. D., Dudek, E. M., Rapier, J. L., Leonard, S., Freedman, R.: Significant reductions in synapsin but not synaptophysin specific activity in the brains of some schizophrenics. Biol Psychiatry 34, 529–535 (1993)

Budziszewska, B., Jaworska-Feil, L., Tetich, M., Basta-Kaim, A., Kubera, M., Leskiewicz, M., Lason, W.: Regulation of the human corticotropin-releasing-hormone gene promoter activity by antidepressant drugs in Neuro-2A and AtT-20 cells. Neuropsychopharmacology 29, 785–794 (2004)

Bundey, R. A., Iredale, P. A., Kendall, D. A.: Corticosteroid modulation of signal transduction in the CATH.a cell line. J Neurochem 68, 362–369 (1997)

Bundey, R. A., Kendall, D. A.: Inhibition of receptor-mediated calcium responses by corticotrophin-releasing hormone in the CATH.a cell line. Neuropharmacology 38, 39–47 (1999)

Canales, R. D., Luo, Y., Willey, J. C., Austermiller, B., Barbacioru, C. C., Boysen, C., Hunkapiller, K., Jensen, R. V., Knight, C. R., Lee, K. Y., Ma, Y., Maqsodi, B., Papallo, A., Peters, E. H., Poulter, K., Ruppel, P. L., Samaha, R. R., Shi, L., Yang, W., Zhang, L., Goodsaid, F. M.: Evaluation of DNA microarray results with quantitative gene expression platforms. Nat. Biotechnol. 24, 1115–1122 (2006)

Capuron, L., Dantzer, R.: Cytokines and depression: the need for a new paradigm. Brain Behav. Immun. 17 Suppl 1, S119–S124 (2003)

Capuron, L., Miller, A. H.: Cytokines and psychopathology: lessons from interferon-alpha. Biol. Psychiatry 56, 819–824 (2004)

Capuron, L., Neurauter, G., Musselman, D. L., Lawson, D. H., Nemeroff, C. B., Fuchs, D., Miller, A. H.: Interferon-alpha-induced changes in tryptophan metabolism. relationship to depression and paroxetine treatment. Biol. Psychiatry 54, 906–914 (2003a)

Capuron, L., Raison, C. L., Musselman, D. L., Lawson, D. H., Nemeroff, C. B., Miller, A. H.: Association of exaggerated HPA axis response to the initial injection of interferon-alpha with development of depression during interferon-alpha therapy. Am. J. Psychiatry 160, 1342–1345 (2003b)

Carroll, B. J., Curtis, G. C., Mendels, J.: Neuroendocrine regulation in depression. II. Discrimination of depressed from nondepressed patients. Arch. Gen. Psychiatry 33, 1051–1058 (1976)

Carroll, B. J., Cassidy, F., Naftolowitz, D., Tatham, N. E., Wilson, W. H., Iranmesh, A., Liu, P. Y., Veldhuis, J. D.: Pathophysiology of hypercortisolism in depression. Acta Psychiatrica Scandinavica 115, 90–103 (2007)

Cartmell, J., Schoepp, D. D.: Regulation of neurotransmitter release by metabotropic glutamate receptors. J Neurochem 75, 889–907 (2000)

Charmandari, E., Tsigos, C., Chrousos, G.: Endocrinology of the stress response. Annu. Rev Physiol 67, 259–284 (2005)

Charney, D. S., Heninger, G. R., Breier, A.: Noradrenergic function in panic anxiety. Effects of yohimbine in healthy subjects and patients with agoraphobia and panic disorder. Arch. Gen. Psychiatry 41, 751–763 (1984)

Charney DS, Barlow DH, Botteron K, Cohen JD, Goldman D, Gur RE, Lin K-M, Lopez JF, Meador-Woodruff JH, Moldin SO, Nestler EJ, Watson SJ, Zalcman SJ. Neuroscience Research Agenda to Guide Development of a Pathophysiologically Based Classification System. In: Kupfer DJ, First MB, Regier DA, editors. A Research Agenda for DSM-V™. Washington, D.C.: American Psychiatric Association, 2002. p. 31–84.

Chiou, S. H., Chen, S. J., Peng, C. H., Chang, Y. L., Ku, H. H., Hsu, W. M., Ho, L. L., Lee, C. H.: Fluoxetine up-regulates expression of cellular FLICE-inhibitory protein and inhibits LPS-induced apoptosis in hippocampus-derived neural stem cell. Biochem Biophys. Res Commun. 343, 391–400 (2006a)

Chiou, S. H., Ku, H. H., Tsai, T. H., Lin, H. L., Chen, L. H., Chien, C. S., Ho, L. L., Lee, C. H., Chang, Y. L.: Moclobemide upregulated Bcl-2 expression and induced neural stem cell differentiation into serotoninergic neuron via extracellular-regulated kinase pathway. Br. J Pharmacol 148, 587–598 (2006b)

Claes, S. J.: Corticotropin-releasing hormone (CRH) in psychiatry: from stress to psychopathology. Ann Med 36, 50–61 (2004)

Crochemore, C., Lu, J., Wu, Y., Liposits, Z., Sousa, N., Holsboer, F., Almeida, O. F.: Direct targeting of hippocampal neurons for apoptosis by glucocorticoids is reversible by mineralocorticoid receptor activation. Mol Psychiatry 10, 790–798 (2005)

Cryan, J. F., Holmes, A.: The ascent of mouse: advances in modelling human depression and anxiety. Nat. Rev Drug Discov. 4, 775–790 (2005)

Cryan, J. F., Mombereau, C.: In search of a depressed mouse: utility of models for studying depression-related behavior in genetically modified mice. Mol Psychiatry 9, 326–357 (2004)

Cryan, J. F., Slattery, D. A.: Animal models of mood disorders: Recent developments. Curr. Opin. Psychiatry 20, 1–7 (2007)

Cryan, J. F., Mombereau, C., Vassout, A.: The tail suspension test as a model for assessing antidepressant activity: review of pharmacological and genetic studies in mice. Neurosci Biobehav. Rev 29, 571–625 (2005a)

Cryan, J. F., Valentino, R. J., Lucki, I.: Assessing substrates underlying the behavioral effects of antidepressants using the modified rat forced swimming test. Neurosci Biobehav. Rev 29, 547–569 (2005b)

Czesak, M., Lemonde, S., Peterson, E. A., Rogaeva, A., Albert, P. R.: Cell-specific repressor or enhancer activities of Deaf-1 at a serotonin 1A receptor gene polymorphism. J Neurosci 26, 1864–1871 (2006)

David, D. J., Klemenhagen, K. C., Holick, K. A., Saxe, M. D., Mendez, I., Santarelli, L., Craig, D. A., Zhong, H., Swanson, C. J., Hegde, L. G., Ping, X. I., Dong, D., Marzabadi, M. R., Gerald, C. P., Hen, R.: Efficacy of the MCHR1 antagonist N-[3-(1-{[4-(3,4-difluorophenoxy) phenyl]methyl}(4-piperidyl))-4-methylphen yl]-2-methylpropanamide (SNAP 94847) in mouse models of anxiety and depression following acute and chronic administration is independent of hippocampal neurogenesis. J Pharmacol Exp Ther 321, 237–248 (2007)

Davidson, R. J., Irwin, W., Anderle, M. J., Kalin, N. H.: The neural substrates of affective processing in depressed patients treated with venlafaxine. Am. J. Psychiatry 160, 64–75 (2003)

De Pablo, J. M., Parra, A., Segovia, S., Guillamon, A.: Learned immobility explains the behavior of rats in the forced swimming test. Physiol Behav 46, 229–237 (1989)

Detke, M. J., Rickels, M., Lucki, I.: Active behaviors in the rat forced swimming test differentially produced by serotonergic and noradrenergic antidepressants. Psychopharmacology. (Berl.) 121, 66–72 (1995)

Dettling, M., Buhlmann, P.: Finding predictive gene groups from microarray data. J. Multivariate Analysis 90, 106–131 (2004)

Draper, I., Kurshan, P. T., McBride, E., Jackson, F. R., Kopin, A. S.: Locomotor activity is regulated by D2-like receptors in Drosophila: an anatomic and functional analysis. Dev Neurobiol 67, 378–393 (2007)

D'Sa, C., Duman, R. S.: Antidepressants and neuroplasticity. Bipolar Disord. 4, 183–194 (2002)

Duncan, G. E., Breese, G. R., Criswell, H. E., Johnson, K. B., Schambra, U. B., Mueller, R. A., Caron, M. G., Fremeau, R. T., Jr.: D1 dopamine receptor binding and mRNA levels are not altered after neonatal 6-hydroxydopamine treatment: evidence against dopamine-mediated induction of D1 dopamine receptors during postnatal development. J Neurochem 61, 1255–1262 (1993)

Ehret, A., Birthelmer, A., Rutz, S., Riegert, C., Rothmaier, A. K., Jackisch, R.: Agonist-mediated regulation of presynaptic receptor function during development of rat septal neurons in culture. J Neurochem 102, 1071–1082 (2007)

Emanuel, R. L., Girard, D. M., Thull, D. L., Majzoub, J. A.: Second messengers involved in the regulation of corticotropin-releasing hormone mRNA and peptide in cultured rat fetal hypothalamic primary cultures. Endocrinology 126, 3016–3021 (1990)

Eriksson, P. S., Perfilieva, E., Bjork-Eriksson, T., Alborn, A. M., Nordborg, C., Peterson, D. A., Gage, F. H.: Neurogenesis in the adult human hippocampus. Nat. Med 4, 1313–1317 (1998)

Evans, S. J., Choudary, P. V., Neal, C. R., Li, J. Z., Vawter, M. P., Tomita, H., Lopez, J. F., Thompson, R. C., Meng, F., Stead, J. D., Walsh, D. M., Myers, R. M., Bunney, W. E., Watson, S. J., Jones, E. G., Akil, H.: Dysregulation of the fibroblast growth factor system in major depression. Proc Natl Acad Sci U S A 101, 15506–15511 (2004)

Feezor, R. J., Baker, H. V., Mindrinos, M., Hayden, D., Tannahill, C. L., Brownstein, B. H., Fay, A., MacMillan, S., Laramie, J., Xiao, W., Moldawer, L. L., Cobb, J. P., Laudanski, K., Miller-Graziano, C. L., Maier, R. V., Schoenfeld, D., Davis, R. W., Tompkins, R. G.: Whole blood and leukocyte RNA isolation for gene expression analyses. Physiol Genomics 19, 247–254 (2004)

Frank, R., Hargreaves, R.: Clinical biomarkers in drug discovery and development. Nat. Rev. Drug Discov. 2, 566–580 (2003)

Frenois, F., Cador, M., Caille, S., Stinus, L., Le, M. C.: Neural correlates of the motivational and somatic components of naloxone-precipitated morphine withdrawal. Eur J Neurosci 16, 1377–1389 (2002)

Frodl, T., Meisenzahl, E. M., Zetzsche, T., Hohne, T., Banac, S., Schorr, C., Jager, M., Leinsinger, G., Bottlender, R., Reiser, M., Moller, H. J.: Hippocampal and amygdala changes in patients with major depressive disorder and healthy controls during a 1-year follow-up. J. Clin. Psychiatry 65, 492–499 (2004)

Frohna, P. A., Neal-Beliveau, B. S., Joyce, J. N.: Neonatal 6-hydroxydopamine lesions lead to opposing changes in the levels of dopamine receptors and their messenger RNAs. Neuroscience 68, 505–518 (1995)

Fu, C. H., Williams, S. C., Cleare, A. J., Brammer, M. J., Walsh, N. D., Kim, J., Andrew, C. M., Pich, E. M., Williams, P. M., Reed, L. J., Mitterschiffthaler, M. T., Suckling, J., Bullmore, E. T.: Attenuation of the neural response to sad faces in major depression by antidepressant treatment: a prospective, event-related functional magnetic resonance imaging study. Arch. Gen. Psychiatry 61, 877–889 (2004)

Funada, M., Hara, C., Wada, K.: Involvement of corticotropin-releasing factor receptor subtype 1 in morphine withdrawal regulation of the brain noradrenergic system. Eur J Pharmacol 430, 277–281 (2001)

Ganguly, S., Coon, S. L., Klein, D. C.: Control of melatonin synthesis in the mammalian pineal gland: the critical role of serotonin acetylation. Cell Tissue Res 309, 127–137 (2002)

Garcia-Sevilla, J. A., Zis, A. P., Hollingsworth, P. J., Greden, J. F., Smith, C. B.: Platelet alpha 2-adrenergic receptors in major depressive disorder. Binding of tritiated clonidine before and after tricyclic antidepressant drug treatment. Arch. Gen. Psychiatry 38, 1327–1333 (1981)

Georges, M., Coppieters, W., Charlier, C.: Polymorphic miRNA-mediated gene regulation: contribution to phenotypic variation and disease. Curr. Opin. Genet. Dev. 17, 166–176 (2007)

Ghaffar, O., Feinstein, A.: The neuropsychiatry of multiple sclerosis: a review of recent developments. Curr. Opin. Psychiatry 20, 278–285 (2007)

Gilbertson, M. W., Shenton, M. E., Ciszewski, A., Kasai, K., Lasko, N. B., Orr, S. P., Pitman, R. K.: Smaller hippocampal volume predicts pathologic vulnerability to psychological trauma. Nat. Neurosci. 5, 1242–1247 (2002)

Gladkevich, A., Kauffman, H. F., Korf, J.: Lymphocytes as a neural probe: potential for studying psychiatric disorders. Prog. Neuropsychopharmacol. Biol Psychiatry 28, 559–576 (2004)

Glantz, L. A., Lewis, D. A.: Reduction of synaptophysin immunoreactivity in the prefrontal cortex of subjects with schizophrenia. Regional and diagnostic specificity. Arch. Gen. Psychiatry 54, 943–952 (1997)

Glavin, G. B.: Selective noradrenaline depletion markedly alters stress responses in rats. Life Sci 37, 461–465 (1985a)

Glavin, G. B.: Selective noradrenaline depletion markedly alters stress responses in rats. Life Sci 37, 461–465 (1985b)

Glavin, G. B.: Stress and brain noradrenaline: a review. Neurosci Biobehav. Rev 9, 233–243 (1985c)

Gold, P. W., Charney, D. S.: Diseases of the mind and brain: depression: a disease of the mind, brain, and body. Am. J. Psychiatry 159, 1826 (2002)

Gold, P. W., Chrousos, G. P.: Organization of the stress system and its dysregulation in melancholic and atypical depression: high vs low CRH/NE states. Mol. Psychiatry 7, 254–275 (2002)

Gross, C. G.: Neurogenesis in the adult brain: death of a dogma. Nat. Rev Neurosci 1, 67–73 (2000)

Gur, R. E., Keshavan, M. S., Lawrie, S. M.: Deconstructing psychosis with human brain imaging. Schizophr. Bull. 33, 921–931 (2007)

Halim, N. D., Weickert, C. S., McClintock, B. W., Weinberger, D. R., Lipska, B. K.: Effects of chronic haloperidol and clozapine treatment on neurogenesis in the adult rat hippocampus. Neuropsychopharmacology 29, 1063–1069 (2004)

Harrison, P. J., Weinberger, D. R.: Schizophrenia genes, gene expression, and neuropathology: on the matter of their convergence. Mol Psychiatry 10, 40–68 (2005)

Heesen, C., Nawrath, L., Reich, C., Bauer, N., Schulz, K. H., Gold, S. M.: Fatigue in multiple sclerosis: an example of cytokine mediated sickness behaviour? J. Neurol. Neurosurg. Psychiatry 77, 34–39 (2006)

Heim, C., Newport, D. J., Heit, S., Graham, Y. P., Wilcox, M., Bonsall, R., Miller, A. H., Nemeroff, C. B.: Pituitary–adrenal and autonomic responses to stress in women after sexual and physical abuse in childhood. JAMA 284, 592–597 (2000)

Heim, C., Plotsky, P. M., Nemeroff, C. B.: Importance of studying the contributions of early adverse experience to neurobiological findings in depression. Neuropsychopharmacology 29, 641–648 (2004)

Hillhouse, E. W., Milton, N. G.: Effect of acetylcholine and 5-hydroxytryptamine on the secretion of corticotrophin-releasing factor-41 and arginine vasopressin from the rat hypothalamus *in vitro*. J Endocrinol. 122, 713–718 (1989)

Holmes, P. V.: Rodent models of depression: reexamining validity without anthropomorphic inference. Crit. Rev Neurobiol 15, 143–174 (2003)

Holsboer, F.: The rationale for corticotropin-releasing hormone receptor (CRH-R) antagonists to treat depression and anxiety. J. Psychiatr. Res. 33, 181–214 (1999)

Huang, Y. Y., Peng, C. H., Yang, Y. P., Wu, C. C., Hsu, W. M., Wang, H. J., Chan, K. H., Chou, Y. P., Chen, S. J., Chang, Y. L.: Desipramine activated Bcl-2 expression and inhibited lipopolysaccharide-induced apoptosis in hippocampus-derived adult neural stem cells. J Pharmacol Sci 104, 61–72 (2007)

Huggett, J., Dheda, K., Bustin, S., Zumla, A.: Real-time RT-PCR normalisation; strategies and considerations. Genes Immun. 6, 279–284 (2005)

Husum, H., Mikkelsen, J. D., Hogg, S., Mathe, A. A., Mork, A.: Involvement of hippocampal neuropeptide Y in mediating the chronic actions of lithium, electroconvulsive stimulation and citalopram. Neuropharmacology 39, 1463–1473 (2000)

Iga, J., Ueno, S., Yamauchi, K., Motoki, I., Tayoshi, S., Ohta, K., Song, H., Morita, K., Rokutan, K., Ohmori, T.: Serotonin transporter mRNA expression in peripheral leukocytes of patients with major depression before and after treatment with paroxetine. Neurosci Lett. 389, 12–16 (2005)

Iga, J., Ueno, S., Yamauchi, K., Numata, S., Motoki, I., Tayoshi, S., Kinouchi, S., Ohta, K., Song, H., Morita, K., Rokutan, K., Tanabe, H., Sano, A., Ohmori, T.: Gene expression and association analysis of LIM (PDLIM5) in major depression. Neurosci Lett. 400, 203–207 (2006)

Iga, J., Ueno, S., Yamauchi, K., Numata, S., Kinouchi, S., Tayoshi-Shibuya, S., Song, H., Ohmori, T.: Altered HDAC5 and CREB mRNA expressions in the peripheral leukocytes of major depression. Prog. Neuropsychopharmacol. Biol. Psychiatry 31, 628–632 (2007a)

Iga, J., Ueno, S., Yamauchi, K., Numata, S., Tayoshi-Shibuya, S., Kinouchi, S., Nakataki, M., Song, H., Hokoishi, K., Tanabe, H., Sano, A., Ohmori, T.: Gene expression and association analysis of vascular endothelial growth factor in major depressive disorder. Prog. Neuropsychopharmacol. Biol. Psychiatry 31, 658–663 (2007b)

Insel, T. R., Scolnick, E. M.: Cure therapeutics and strategic prevention: raising the bar for mental health research. Mol. Psychiatry 11, 11–17 (2006)

Ising, M., Horstmann, S., Kloiber, S., Lucae, S., Binder, E. B., Kern, N., Kunzel, H. E., Pfennig, A., Uhr, M., Holsboer, F.: Combined dexamethasone/corticotropin releasing hormone test predicts treatment response in major depression - a potential biomarker? Biol. Psychiatry 62, 47–54 (2007)

Jarvinen, A. K., Hautaniemi, S., Edgren, H., Auvinen, P., Saarela, J., Kallioniemi, O. P., Monni, O.: Are data from different gene expression microarray platforms comparable? Genomics 83, 1164–1168 (2004)

Jin, P., Zhao, Y., Ngalame, Y., Panelli, M. C., Nagorsen, D., Monsurro, V., Smith, K., Hu, N., Su, H., Taylor, P. R., Marincola, F. M., Wang, E.: Selection and validation of endogenous reference genes using a high throughput approach. BMC Genomics 5, 55 (2004)

Kaddurah-Daouk, R., McEvoy, J., Baillie, R. A., Lee, D., Yao, J. K., Doraiswamy, P. M., Krishnan, K. R.: Metabolomic mapping of atypical antipsychotic effects in schizophrenia. Mol. Psychiatry 1–12 (2007)

Kalman, J., Palotas, A., Juhasz, A., Rimanoczy, A., Hugyecz, M., Kovacs, Z., Galsi, G., Szabo, Z., Pakaski, M., Feher, L. Z., Janka, Z., Puskas, L. G.: Impact of venlafaxine on gene expression profile in lymphocytes of the elderly with major depression–evolution of antidepressants and the role of the "neuro-immune" system. Neurochem Res 30, 1429–1438 (2005)

Karson, C. N., Mrak, R. E., Schluterman, K. O., Sturner, W. Q., Sheng, J. G., Griffin, W. S.: Alterations in synaptic proteins and their encoding mRNAs in prefrontal cortex in schizophrenia: a possible neurochemical basis for 'hypofrontality'. Mol Psychiatry 4, 39–45 (1999)

Kasckow, J. W., Regmi, A., Sheriff, S., Mulchahey, J., Geracioti, T. D., Jr.: Regulation of corticotropin-releasing factor messenger RNA by nicotine in an immortalized amygdalar cell line. Life Sci 65, 2709–2714 (1999)

Kasckow, J. W., Aguilera, G., Mulchahey, J. J., Sheriff, S., Herman, J. P.: *In vitro* regulation of corticotropin-releasing hormone. Life Sci 73, 769–781 (2003)

Keck, M. E., Wigger, A., Welt, T., Muller, M. B., Gesing, A., Reul, J. M., Holsboer, F., Landgraf, R., Neumann, I. D.: Vasopressin mediates the response of the combined dexamethasone/CRH test in hyper-anxious rats: implications for pathogenesis of affective disorders. Neuropsychopharmacology 26, 94–105 (2002)

Keck, M. E., Welt, T., Muller, M. B., Uhr, M., Ohl, F., Wigger, A., Toschi, N., Holsboer, F., Landgraf, R.: Reduction of hypothalamic vasopressinergic hyperdrive contributes to clinically relevant behavioral and neuroendocrine effects of chronic paroxetine treatment in a psychopathological rat model. Neuropsychopharmacology 28, 235–243 (2003)

Khan, A., Schwartz, K., Kolts, R. L., Ridgway, D., Lineberry, C.: Relationship between depression severity entry criteria and antidepressant clinical trial outcomes. Biol. Psychiatry 62, 65–71 (2007)

Kitayama, N., Vaccarino, V., Kutner, M., Weiss, P., Bremner, J. D.: Magnetic resonance imaging (MRI) measurement of hippocampal volume in posttraumatic stress disorder: a meta-analysis. J. Affect. Disord. 88, 79–86 (2005)

Kodama, M., Fujioka, T., Duman, R. S.: Chronic olanzapine or fluoxetine administration increases cell proliferation in hippocampus and prefrontal cortex of adult rat. Biol Psychiatry 56, 570–580 (2004)

Koldzic-Zivanovic, N., Seitz, P. K., Cunningham, K. A., Thomas, M. L., Hughes, T. K.: Serotonin regulation of serotonin uptake in RN46A cells. Cell Mol Neurobiol 26, 979–987 (2006)

Konradi, C.: Gene expression microarray studies in polygenic psychiatric disorders: applications and data analysis. Brain Res Brain Res Rev 50, 142–155 (2005)

Koob, G. F.: Corticotropin-releasing factor, norepinephrine, and stress. Biol Psychiatry 46, 1167–1180 (1999)

Kromer, S. A., Kessler, M. S., Milfay, D., Birg, I. N., Bunck, M., Czibere, L., Panhuysen, M., Putz, B., Deussing, J. M., Holsboer, F., Landgraf, R., Turck, C. W.: Identification of glyoxalase-I as a protein marker in a mouse model of extremes in trait anxiety. J Neurosci 25, 4375–4384 (2005a)

Kromer, S. A., Kessler, M. S., Milfay, D., Birg, I. N., Bunck, M., Czibere, L., Panhuysen, M., Putz, B., Deussing, J. M., Holsboer, F., Landgraf, R., Turck, C. W.: Identification of glyoxalase-I as a protein marker in a mouse model of extremes in trait anxiety. J Neurosci 25, 4375–4384 (2005b)

Kunzel, H. E., Binder, E. B., Nickel, T., Ising, M., Fuchs, B., Majer, M., Pfennig, A., Ernst, G., Kern, N., Schmid, D. A., Uhr, M., Holsboer, F., Modell, S.: Pharmacological and nonpharmacological factors influencing hypothalamic–pituitary–adrenocortical axis reactivity in acutely depressed psychiatric in-patients, measured by the Dex-CRH test. Neuropsychopharmacology 28, 2169–2178 (2003)

Kupfer, D. J.: REM latency: a psychobiologic marker for primary depressive disease. Biol. Psychiatry 11, 159–174 (1976)
Kupfer, D. J.: Sleep research in depressive illness: clinical implications--a tasting menu. Biol. Psychiatry 38, 391–403 (1995)
Kupfer, D. J., Targ, E., Stack, J.: Electroencephalographic sleep in unipolar depressive subtypes: support for a biological and familial classification. J. Nerv. Ment. Dis. 170, 494–498 (1982)
Kupfer, D. J., Reynolds, C. F., III, Grochocinski, V. J., Ulrich, R. F., McEachran, A.: Aspects of short REM latency in affective states: a revisit. Psychiatry Res. 17, 49–59 (1986)
Kushwaha, N., Albert, P. R.: Coupling of 5-HT1A autoreceptors to inhibition of mitogen-activated protein kinase activation via G beta gamma subunit signaling. Eur J Neurosci 21, 721–732 (2005)
Landgraf, R., Wigger, A.: High vs low anxiety-related behavior rats: an animal model of extremes in trait anxiety. Behav. Genet. 32, 301–314 (2002)
Landgraf, R., Wigger, A., Holsboer, F., Neumann, I. D.: Hyper-reactive hypothalamo-pituitary–adrenocortical axis in rats bred for high anxiety-related behaviour. J. Neuroendocrinol. 11, 405–407 (1999)
Lauer, C. J., Riemann, D., Wiegand, M., Berger, M.: From early to late adulthood. Changes in EEG sleep of depressed patients and healthy volunteers. Biol. Psychiatry 29, 979–993 (1991)
Lehrmann, E., Hyde, T. M., Vawter, M. P., Becker, K. G., Kleinman, J. E., Freed, W. J.: The use of microarrays to characterize neuropsychiatric disorders: postmortem studies of substance abuse and schizophrenia. Curr. Mol Med 3, 437–446 (2003)
Leonard, B. E.: The HPA and immune axes in stress: the involvement of the serotonergic system. Eur. Psychiatry 20 Suppl 3, S302–S306 (2005)
Liebsch, G., Linthorst, A. C., Neumann, I. D., Reul, J. M., Holsboer, F., Landgraf, R.: Behavioral, physiological, and neuroendocrine stress responses and differential sensitivity to diazepam in two Wistar rat lines selectively bred for high- and low-anxiety-related behavior. Neuropsychopharmacology 19, 381–396 (1998a)
Liebsch, G., Montkowski, A., Holsboer, F., Landgraf, R.: Behavioural profiles of two Wistar rat lines selectively bred for high or low anxiety-related behaviour. Behav Brain Res 94, 301–310 (1998b)
Lima, L., Mata, S., Urbina, M.: Allelic isoforms and decrease in serotonin transporter mRNA in lymphocytes of patients with major depression. Neuroimmunomodulation. 12, 299–306 (2005)
Lorrain, D. S., Baccei, C. S., Correa, L. D., Bristow, L. J.: Comparison of the effects of diazepam, the CRF1 antagonist CP-154,526 and the group II mGlu receptor agonist LY379268 on stress-evoked extracellular norepinephrine levels. Neuropharmacology 48, 927–935 (2005)
Lucki, I.: The forced swimming test as a model for core and component behavioral effects of antidepressant drugs. Behav Pharmacol 8, 523–532 (1997)
Luo, Z., Geschwind, D. H.: Microarray applications in neuroscience. Neurobiol Dis 8, 183–193 (2001)
Lyketsos, C. G., Olin, J.: Depression in Alzheimer's disease: overview and treatment. Biol. Psychiatry 52, 243–252 (2002)
Macmaster, F. P., Mirza, Y., Szeszko, P. R., Kmiecik, L. E., Easter, P. C., Taormina, S. P., Lynch, M., Rose, M., Moore, G. J., Rosenberg, D. R.: Amygdala and Hippocampal Volumes in Familial Early Onset Major Depressive Disorder. Biol. Psychiatry (2007)
Maier, W., Zobel, A., Wagner, M.: Schizophrenia and bipolar disorder: differences and overlaps. Curr. Opin. Psychiatry 19, 165–170 (2006)
Maino, K., Gruber, R., Riedel, M., Seitz, N., Schwarz, M., Muller, N.: T- and B-lymphocytes in patients with schizophrenia in acute psychotic episode and the course of the treatment. Psychiatry Res. 152, 173–180 (2007)
Malberg, J. E., Duman, R. S.: Cell proliferation in adult hippocampus is decreased by inescapable stress: reversal by fluoxetine treatment. Neuropsychopharmacology 28, 1562–1571 (2003)
Marcotte, E. R., Srivastava, L. K., Quirion, R.: cDNA microarray and proteomic approaches in the study of brain diseases: focus on schizophrenia and Alzheimer's disease. Pharmacol Ther 100, 63–74 (2003)

Marsteller, D. A., Barbarich-Marsteller, N. C., Patel, V. D., Dewey, S. L.: Brain metabolic changes following 4-week citalopram infusion: Increased (18)FDG uptake and gamma-amino butyric acid levels. Synapse 61, 877–881 (2007)
McDonald, C., Zanelli, J., Rabe-Hesketh, S., Ellison-Wright, I., Sham, P., Kalidindi, S., Murray, R. M., Kennedy, N.: Meta-analysis of magnetic resonance imaging brain morphometry studies in bipolar disorder. Biol. Psychiatry 56, 411–417 (2004)
McEwen, B. S.: Plasticity of the hippocampus: adaptation to chronic stress and allostatic load. Ann N Y Acad Sci 933, 265–277 (2001)
Middleton, F. A., Mirnics, K., Pierri, J. N., Lewis, D. A., Levitt, P.: Gene expression profiling reveals alterations of specific metabolic pathways in schizophrenia. J Neurosci 22, 2718–2729 (2002)
Milak, M. S., Parsey, R. V., Keilp, J., Oquendo, M. A., Malone, K. M., Mann, J. J.: Neuroanatomic correlates of psychopathologic components of major depressive disorder. Arch. Gen. Psychiatry 62, 397–408 (2005)
Millan, M. J.: The neurobiology and control of anxious states. Prog. Neurobiol 70, 83–244 (2003)
Millan, M. J.: The role of monoamines in the actions of established and "novel" antidepressant agents: a critical review. Eur J Pharmacol 500, 371–384 (2004)
Millan, M. J., Gobert, A., Bervoets, K., Rivet, J. M., Veiga, S., Brocco, M.: Induction of spontaneous tail-flicks in rats by blockade of transmission at N-methyl-D-aspartate receptors: roles of multiple monoaminergic receptors in relation to the actions of antipsychotic agents. J Pharmacol Exp Ther 292, 672–683 (2000)
Mir, K. U.: Ultrasensitive RNA profiling: counting single molecules on microarrays. Genome. Res 16, 1195–1197 (2006)
Mirnics, K., Pevsner, J.: Progress in the use of microarray technology to study the neurobiology of disease. Nat. Neurosci 7, 434–439 (2004)
Mirnics, K., Levitt, P., Lewis, D. A.: Critical appraisal of DNA microarrays in psychiatric genomics. Biol Psychiatry 60, 163–176 (2006)
Molina, V. A., Heyser, C. J., Spear, L. P.: Chronic variable stress enhances the stimulatory action of a low dose of morphine: reversal by desipramine. Eur J Pharmacol 260, 57–64 (1994)
Morey, J. S., Ryan, J. C., Van Dolah, F. M.: Microarray validation: factors influencing correlation between oligonucleotide microarrays and real-time PCR. Biol Proced. Online. 8, 175–193 (2006)
Mortensen, O. V., Thomassen, M., Larsen, M. B., Whittemore, S. R., Wiborg, O.: Functional analysis of a novel human serotonin transporter gene promoter in immortalized raphe cells. Brain Res Mol Brain Res 68, 141–148 (1999)
Mulchahey, J. J., Regmi, A., Sheriff, S., Balasubramaniam, A., Kasckow, J. W.: Coordinate and divergent regulation of corticotropin-releasing factor (CRF) and CRF-binding protein expression in an immortalized amygdalar neuronal cell line. Endocrinology 140, 251–259 (1999)
Nemeroff, C. B., Bremner, J. D., Foa, E. B., Mayberg, H. S., North, C. S., Stein, M. B.: Posttraumatic stress disorder: a state-of-the-science review. J. Psychiatr. Res. 40, 1–21 (2006)
Nestler, E. J., Gould, E., Manji, H., Buncan, M., Duman, R. S., Greshenfeld, H. K., Hen, R., Koester, S., Lederhendler, I., Meaney, M., Robbins, T., Winsky, L., Zalcman, S.: Preclinical models: status of basic research in depression. Biol Psychiatry 52, 503–528 (2002)
Newton, S. S., Collier, E. F., Hunsberger, J., Adams, D., Terwilliger, R., Selvanayagam, E., Duman, R. S.: Gene profile of electroconvulsive seizures: induction of neurotrophic and angiogenic factors. J Neurosci 23, 10841–10851 (2003)
Newton, S. S., Duman, R. S.: Neurogenic actions of atypical antipsychotic drugs and therapeutic implications. CNS. Drugs 21, 715–725 (2007)
Nofzinger, E. A., Price, J. C., Meltzer, C. C., Buysse, D. J., Villemagne, V. L., Miewald, J. M., Sembrat, R. C., Steppe, D. A., Kupfer, D. J.: Towards a neurobiology of dysfunctional arousal in depression: the relationship between beta EEG power and regional cerebral glucose metabolism during NREM sleep. Psychiatry Res. 98, 71–91 (2000)

Ohmori, T., Morita, K., Saito, T., Ohta, M., Ueno, S., Rokutan, K.: Assessment of human stress and depression by DNA microarray analysis. J Med Invest. 52 Suppl, 266–271 (2005)
Owen, M. J., Craddock, N., O'donovan, M. C.: Schizophrenia: genes at last? Trends Genet 21, 518–525 (2005)
Pariante, C. M., Dazzan, P., Danese, A., Morgan, K. D., Brudaglio, F., Morgan, C., Fearon, P., Orr, K., Hutchinson, G., Pantelis, C., Velakoulis, D., Jones, P. B., Leff, J., Murray, R. M.: Increased pituitary volume in antipsychotic-free and antipsychotic-treated patients of the AEsop first-onset psychosis study. Neuropsychopharmacology 30, 1923–1931 (2005)
Parker, G.: Classifying depression: should paradigms lost be regained? Am. J. Psychiatry 157, 1195–1203 (2000)
Parker, G., Malhi, G., Mitchell, P., Wilhelm, K., Austin, M. P., Crawford, J., Hadzi-Pavlovic, D.: Progressing a spectrum model for defining non-melancholic depression. Acta Psychiatr. Scand. 111, 139–143 (2005)
Patil, S. T., Zhang, L., Martenyi, F., Lowe, S. L., Jackson, K. A., Andreev, B. V., Avedisova, A. S., Bardenstein, L. M., Gurovich, I. Y., Morozova, M. A., Mosolov, S. N., Neznanov, N. G., Reznik, A. M., Smulevich, A. B., Tochilov, V. A., Johnson, B. G., Monn, J. A., Schoepp, D. D.: Activation of mGlu2/3 receptors as a new approach to treat schizophrenia: a randomized Phase 2 clinical trial. Nat. Med. (2007)
Peng, C. H., Chiou, S. H., Chen, S. J., Chou, Y. C., Ku, H. H., Cheng, C. K., Yen, C. J., Tsai, T. H., Chang, Y. L., Kao, C. L.: Neuroprotection by Imipramine against lipopolysaccharide-induced apoptosis in hippocampus-derived neural stem cells mediated by activation of BDNF and the MAPK pathway. Eur Neuropsychopharmacol. (2007)
Petit-Demouliere, B., Chenu, F., Bourin, M.: Forced swimming test in mice: a review of antidepressant activity. Psychopharmacology. (Berl.) 177, 245–255 (2005)
Phillips, M. L., Drevets, W. C., Rauch, S. L., Lane, R.: Neurobiology of emotion perception II: Implications for major psychiatric disorders. Biol. Psychiatry 54, 515–528 (2003)
Porsolt, R. D., Le, P. M., Jalfre, M.: Depression: a new animal model sensitive to antidepressant treatments. Nature 266, 730–732 (1977)
Porsolt, R. D., Chermat, R., Lenegre, A., Avril, I., Janvier, S., Steru, L.: Use of the automated tail suspension test for the primary screening of psychotropic agents. Arch. Int Pharmacodyn. Ther 288, 11–30 (1987)
Posternak, M. A., Zimmerman, M.: Partial validation of the atypical features subtype of major depressive disorder. Arch. Gen. Psychiatry 59, 70–76 (2002)
Pruessner, J. C., Champagne, F., Meaney, M. J., Dagher, A.: Dopamine release in response to a psychological stress in humans and its relationship to early life maternal care: a positron emission tomography study using [11C]raclopride. J. Neurosci. 24, 2825–2831 (2004)
Quitkin, F. M., Stewart, J. W., McGrath, P. J., Tricamo, E., Rabkin, J. G., Ocepek-Welikson, K., Nunes, E., Harrison, W., Klein, D. F.: Columbia atypical depression. A subgroup of depressives with better response to MAOI than to tricyclic antidepressants or placebo. Br. J. Psychiatry Suppl 30–34 (1993)
Ramirez, B. M., Bostic, J. Q., Davies, D., Rush, A. J., Witte, B., Hendrickse, W., Barnett, V.: Methods to improve diagnostic accuracy in a community mental health setting. Am. J. Psychiatry 157, 1599–1605 (2000)
Rasmussen, K., Beitner-Johnson, D. B., Krystal, J. H., Aghajanian, G. K., Nestler, E. J.: Opiate withdrawal and the rat locus coeruleus: behavioral, electrophysiological, and biochemical correlates. J Neurosci 10, 2308–2317 (1990)
Ray, S., Britschgi, M., Herbert, C., Takeda-Uchimura, Y., Boxer, A., Blennow, K., Friedman, L. F., Galasko, D. R., Jutel, M., Karydas, A., Kaye, J. A., Leszek, J., Miller, B. L., Minthon, L., Quinn, J. F., Rabinovici, G. D., Robinson, W. H., Sabbagh, M. N., So, Y. T., Sparks, D. L., Tabaton, M., Tinklenberg, J., Yesavage, J. A., Tibshirani, R., Wyss-Coray, T.: Classification and prediction of clinical Alzheimer's diagnosis based on plasma signaling proteins. Nat. Med (2007)
Robert, J. J., Hoffmann, R. F., Emslie, G. J., Hughes, C., Rintelmann, J., Moore, J., Armitage, R.: Sex and age differences in sleep macroarchitecture in childhood and adolescent depression. Sleep 29, 351–358 (2006)

Rocc, P., De, L. C., Eva, C., Marchiaro, L., Milani, A. M., Musso, R., Ravizza, L., Zanalda, E., Bogetto, F.: Decrease of the D4 dopamine receptor messenger RNA expression in lymphocytes from patients with major depression. Prog. Neuropsychopharmacol. Biol Psychiatry 26, 1155–1160 (2002)

Rokutan, K., Morita, K., Masuda, K., Tominaga, K., Shikishima, M., Teshima-Kondo, S., Omori, T., Sekiyama, A.: Gene expression profiling in peripheral blood leukocytes as a new approach for assessment of human stress response. J Med Invest. 52, 137–144 (2005a)

Rokutan, K., Morita, K., Masuda, K., Tominaga, K., Shikishima, M., Teshima-Kondo, S., Omori, T., Sekiyama, A.: Gene expression profiling in peripheral blood leukocytes as a new approach for assessment of human stress response. J Med Invest. 52, 137–144 (2005b)

Rozeboom, A. M., Akil, H., Seasholtz, A. F.: Mineralocorticoid receptor overexpression in forebrain decreases anxiety-like behavior and alters the stress response in mice. Proc Natl Acad Sci U S A 104, 4688–4693 (2007)

Rush, A. J., Zimmerman, M., Wisniewski, S. R., Fava, M., Hollon, S. D., Warden, D., Biggs, M. M., Shores-Wilson, K., Shelton, R. C., Luther, J. F., Thomas, B., Trivedi, M. H.: Comorbid psychiatric disorders in depressed outpatients: demographic and clinical features. J. Affect. Disord. 87, 43–55 (2005)

Sahay, A., Hen, R.: Adult hippocampal neurogenesis in depression. Nat. Neurosci 10, 1110–1115 (2007)

Salchner, P., Singewald, N.: Neuroanatomical substrates involved in the anxiogenic-like effect of acute fluoxetine treatment. Neuropharmacology 43, 1238–1248 (2002)

Salchner, P., Sartori, S. B., Sinner, C., Wigger, A., Frank, E., Landgraf, R., Singewald, N.: Airjet and FG-7142-induced Fos expression differs in rats selectively bred for high and low anxiety-related behavior. Neuropharmacology 50, 1048–1058 (2006)

Schena, M., Shalon, D., Davis, R. W., Brown, P. O.: Quantitative monitoring of gene expression patterns with a complementary DNA microarray. Science 270, 467–470 (1995)

Schena, M., Shalon, D., Heller, R., Chai, A., Brown, P. O., Davis, R. W.: Parallel human genome analysis: microarray-based expression monitoring of 1000 genes. Proc Natl Acad Sci U S A 93, 10614–10619 (1996)

Scheuch, K., Lautenschlager, M., Grohmann, M., Stahlberg, S., Kirchheiner, J., Zill, P., Heinz, A., Walther, D. J., Priller, J.: Characterization of a Functional Promoter Polymorphism of the Human Tryptophan Hydroxylase 2 Gene in Serotonergic Raphe Neurons. Biol Psychiatry (2007)

Schmidt, H. D., Duman, R. S.: The role of neurotrophic factors in adult hippocampal neurogenesis, antidepressant treatments and animal models of depressive-like behavior. Behav Pharmacol 18, 391–418 (2007)

Schratt, G. M., Tuebing, F., Nigh, E. A., Kane, C. G., Sabatini, M. E., Kiebler, M., Greenberg, M. E.: A brain-specific microRNA regulates dendritic spine development. Nature 439, 283–289 (2006)

Schulteis, G., Yackey, M., Risbrough, V., Koob, G. F.: Anxiogenic-like effects of spontaneous and naloxone-precipitated opiate withdrawal in the elevated plus-maze. Pharmacol Biochem Behav 60, 727–731 (1998)

Searfoss, G. H., Ryan, T. P., Jolly, R. A.: The role of transcriptome analysis in pre-clinical toxicology. Curr. Mol Med 5, 53–64 (2005)

Segman, R. H., Shefi, N., Goltser-Dubner, T., Friedman, N., Kaminski, N., Shalev, A. Y.: Peripheral blood mononuclear cell gene expression profiles identify emergent post-traumatic stress disorder among trauma survivors. Mol. Psychiatry 10, 500–13, 425 (2005)

Selemon, L. D., Goldman-Rakic, P. S.: The reduced neuropil hypothesis: a circuit based model of schizophrenia. Biol Psychiatry 45, 17–25 (1999)

Seligman, M. E., Weiss, J., Weinraub, M., Schulman, A.: Coping behavior: learned helplessness, physiological change and learned inactivity. Behav Res Ther 18, 459–512 (1980)

Sheline, Y. I., Barch, D. M., Donnelly, J. M., Ollinger, J. M., Snyder, A. Z., Mintun, M. A.: Increased amygdala response to masked emotional faces in depressed subjects resolves with antidepressant treatment: an fMRI study. Biol. Psychiatry 50, 651–658 (2001)

Shelton, R. C., Liang, S., Liang, P., Chakrabarti, A., Manier, D. H., Sulser, F.: Differential expression of pentraxin 3 in fibroblasts from patients with major depression. Neuropsychopharmacology 29, 126–132 (2004)

Shi, L., Reid, L. H., Jones, W. D., Shippy, R., Warrington, J. A., Baker, S. C., Collins, P. J., de, L. F., Kawasaki, E. S., Lee, K. Y., Luo, Y., Sun, Y. A., Willey, J. C., Setterquist, R. A., Fischer, G. M., Tong, W., Dragan, Y. P., Dix, D. J., Frueh, F. W., Goodsaid, F. M., Herman, D., Jensen, R. V., Johnson, C. D., Lobenhofer, E. K., Puri, R. K., Schrf, U., Thierry-Mieg, J., Wang, C., Wilson, M., Wolber, P. K., Zhang, L., Amur, S., Bao, W., Barbacioru, C. C., Lucas, A. B., Bertholet, V., Boysen, C., Bromley, B., Brown, D., Brunner, A., Canales, R., Cao, X. M., Cebula, T. A., Chen, J. J., Cheng, J., Chu, T. M., Chudin, E., Corson, J., Corton, J. C., Croner, L. J., Davies, C., Davison, T. S., Delenstarr, G., Deng, X., Dorris, D., Eklund, A. C., Fan, X. H., Fang, H., Fulmer-Smentek, S., Fuscoe, J. C., Gallagher, K., Ge, W., Guo, L., Guo, X., Hager, J., Haje, P. K., Han, J., Han, T., Harbottle, H. C., Harris, S. C., Hatchwell, E., Hauser, C. A., Hester, S., Hong, H., Hurban, P., Jackson, S. A., Ji, H., Knight, C. R., Kuo, W. P., LeClerc, J. E., Levy, S., Li, Q. Z., Liu, C., Liu, Y., Lombardi, M. J., Ma, Y., Magnuson, S. R., Maqsodi, B., McDaniel, T., Mei, N., Myklebost, O., Ning, B., Novoradovskaya, N., Orr, M. S., Osborn, T. W., Papallo, A., Patterson, T. A., Perkins, R. G., Peters, E. H., Peterson, R., Philips, K. L., Pine, P. S., Pusztai, L., Qian, F., Ren, H., Rosen, M., Rosenzweig, B. A., Samaha, R. R., Schena, M., Schroth, G. P., Shchegrova, S., Smith, D. D., Staedtler, F., Su, Z., Sun, H., Szallasi, Z., Tezak, Z., Thierry-Mieg, D., Thompson, K. L., Tikhonova, I., Turpaz, Y., Vallanat, B., Van, C., Walker, S. J., Wang, S. J., Wang, Y., Wolfinger, R., Wong, A., Wu, J., Xiao, C., Xie, Q., Xu, J., Yang, W., Zhang, L., Zhong, S., Zong, Y., Slikker, W., Jr.: The MicroArray Quality Control (MAQC) project shows inter- and intraplatform reproducibility of gene expression measurements. Nat. Biotechnol. 24, 1151–1161 (2006a)

Sibille, E., Arango, V., Galfalvy, H. C., Pavlidis, P., Erraji-Benchekroun, L., Ellis, S. P., John, M. J.: Gene expression profiling of depression and suicide in human prefrontal cortex. Neuropsychopharmacology 29, 351–361 (2004)

Smagin, G. N., Harris, R. B., Ryan, D. H.: Corticotropin-releasing factor receptor antagonist infused into the locus coeruleus attenuates immobilization stress-induced defensive withdrawal in rats. Neurosci Lett. 220, 167–170 (1996)

Souery, D., Oswald, P., Massat, I., Bailer, U., Bollen, J., Demyttenaere, K., Kasper, S., Lecrubier, Y., Montgomery, S., Serretti, A., Zohar, J., Mendlewicz, J.: Clinical factors associated with treatment resistance in major depressive disorder: results from a European multicenter study. J. Clin. Psychiatry 68, 1062–1070 (2007)

Sousa, N., Cerqueira, J. J., Almeida, O. F.: Corticosteroid receptors and neuroplasticity. Brain Res Rev (2007)

Stafford, P., Brun, M.: Three methods for optimization of cross-laboratory and cross-platform microarray expression data. Nucleic. Acids Res 35, e72 (2007)

Stahl, S. M., Woo, D. J., Mefford, I. N., Berger, P. A., Ciaranello, R. D.: Hyperserotonemia and platelet serotonin uptake and release in schizophrenia and affective disorders. Am. J. Psychiatry 140, 26–30 (1983)

Stewart, J. W., McGrath, P. J., Quitkin, F. M., Rabkin, J. G., Harrison, W., Wager, S., Nunes, E., Ocepek-Welikson, K., Tricamo, E.: Chronic depression: response to placebo, imipramine, and phenelzine. J. Clin. Psychopharmacol. 13, 391–396 (1993)

Sullivan, P. F., Fan, C., Perou, C. M.: Evaluating the comparability of gene expression in blood and brain. Am. J. Med. Genet. B Neuropsychiatr. Genet. 141, 261–268 (2006)

Sun, W., Park, K. W., Choe, J., Rhyu, I. J., Kim, I. H., Park, S. K., Choi, B., Choi, S. H., Park, S. H., Kim, H.: Identification of novel electroconvulsive shock-induced and activity-dependent genes in the rat brain. Biochem Biophys. Res Commun. 327, 848–856 (2005)

Suri, C., Fung, B. P., Tischler, A. S., Chikaraishi, D. M.: Catecholaminergic cell lines from the brain and adrenal glands of tyrosine hydroxylase-SV40 T antigen transgenic mice. J Neurosci 13, 1280–1291 (1993)

Suzuki, Y., Sawamura, K., Someya, T.: Polymorphisms in the 5-hydroxytryptamine 2A receptor and CytochromeP4502D6 genes synergistically predict fluvoxamine-induced side effects in japanese depressed patients. Neuropsychopharmacology 31, 825–831 (2006)

Swanson, C. J., Perry, K. W., Schoepp, D. D.: The mGlu2/3 receptor agonist, LY354740, blocks immobilization-induced increases in noradrenaline and dopamine release in the rat medial prefrontal cortex. J Neurochem 88, 194–202 (2004)

Swanson, C. J., Blackburn, T. P., Zhang, X., Zheng, K., Xu, Z. Q., Hokfelt, T., Wolinsky, T. D., Konkel, M. J., Chen, H., Zhong, H., Walker, M. W., Craig, D. A., Gerald, C. P., Branchek, T. A.: Anxiolytic- and antidepressant-like profiles of the galanin-3 receptor (Gal3) antagonists SNAP 37889 and SNAP 398299. Proc Natl Acad Sci U S A 102, 17489–17494 (2005a)

Swanson, C. J., Bures, M., Johnson, M. P., Linden, A. M., Monn, J. A., Schoepp, D. D.: Metabotropic glutamate receptors as novel targets for anxiety and stress disorders. Nat. Rev Drug Discov. 4, 131–144 (2005b)

Tanna, V. L., Wilson, A. F., Winokur, G., Elston, R. C.: Linkage analysis of pure depressive disease. J Psychiatr. Res 23, 99–107 (1989)

Tanner, M. A., Berk, L. S., Felten, D. L., Blidy, A. D., Bit, S. L., Ruff, D. W.: Substantial changes in gene expression level due to the storage temperature and storage duration of human whole blood. Clin Lab. Haematol. 24, 337–341 (2002)

Taylor, D. J., Lichstein, K. L., Durrence, H. H., Reidel, B. W., Bush, A. J.: Epidemiology of insomnia, depression, and anxiety. Sleep 28, 1457–1464 (2005)

Taylor, M. A., Fink, M.: Restoring melancholia in the classification of mood disorders. J. Affect. Disord. (2007)

Thiel, G., Cibelli, G.: Corticotropin-releasing factor and vasoactive intestinal polypeptide activate gene transcription through the cAMP signaling pathway in a catecholaminergic immortalized neuron. Neurochem Int 34, 183–191 (1999)

Toro, C. T., Deakin, J. F.: Adult neurogenesis and schizophrenia: a window on abnormal early brain development? Schizophr. Res 90, 1–14 (2007)

Tsao, C. W., Lin, Y. S., Chen, C. C., Bai, C. H., Wu, S. R.: Cytokines and serotonin transporter in patients with major depression. Prog. Neuropsychopharmacol. Biol Psychiatry 30, 899–905 (2006)

Tsuang, M. T., Nossova, N., Yager, T., Tsuang, M. M., Guo, S. C., Shyu, K. G., Glatt, S. J., Liew, C. C.: Assessing the validity of blood-based gene expression profiles for the classification of schizophrenia and bipolar disorder: a preliminary report. Am. J Med Genet B Neuropsychiatr. Genet 133, 1–5 (2005)

Ungerstedt, U.: Stereotaxic mapping of the monoamine pathways in the rat brain. Acta Physiol Scand. Suppl. 367, 1–48 (1971)

Vandergriff, J., Rasmussen, K.: The selective mGlu2/3 receptor agonist LY354740 attenuates morphine-withdrawal-induced activation of locus coeruleus neurons and behavioral signs of morphine withdrawal. Neuropharmacology 38, 217–222 (1999)

Vandesompele, J., De, P. K., Pattyn, F., Poppe, B., Van, R. N., De, P. A., Speleman, F.: Accurate normalization of real-time quantitative RT-PCR data by geometric averaging of multiple internal control genes. Genome. Biol 3, RESEARCH0034 (2002)

Van Praag, H. M.: On what evidence should education concerning depression be based? Int Clin Psychopharmacol. 14, Suppl 3, S35–S39 (1999)

Vawter, M. P., Ferran, E., Galke, B., Cooper, K., Bunney, W. E., Byerley, W.: Microarray screening of lymphocyte gene expression differences in a multiplex schizophrenia pedigree. Schizophr. Res 67, 41–52 (2004a)

Vawter, M. P., Shannon, W. C., Ferran, E., Matsumoto, M., Overman, K., Hyde, T. M., Weinberger, D. R., Bunney, W. E., Kleinman, J. E.: Gene expression of metabolic enzymes and a protease inhibitor in the prefrontal cortex are decreased in schizophrenia. Neurochem Res 29, 1245–1255 (2004b)

Velculescu, V. E., Zhang, L., Vogelstein, B., Kinzler, K. W.: Serial analysis of gene expression. Science 270, 484–487 (1995)

Vermetten, E., Vythilingam, M., Southwick, S. M., Charney, D. S., Bremner, J. D.: Long-term treatment with paroxetine increases verbal declarative memory and hippocampal volume in posttraumatic stress disorder. Biol. Psychiatry 54, 693–702 (2003)

Vgontzas, A. N., Bixler, E. O., Chrousos, G. P.: Obesity-related sleepiness and fatigue: the role of the stress system and cytokines. Ann. N. Y. Acad. Sci. 1083, 329–344 (2006)

Vogel, M., Pfeifer, S., Schaub, R. T., Grabe, H. J., Barnow, S., Freyberger, H. J., Cascorbi, I.: Decreased levels of dopamine D3 receptor mRNA in schizophrenic and bipolar patients. Neuropsychobiology 50, 305–310 (2004)

Wagner, J. A., Williams, S. A., Webster, C. J.: Biomarkers and surrogate end points for fit-for-purpose development and regulatory evaluation of new drugs. Clin. Pharmacol. Ther. 81, 104–107 (2007)

Walker, E. F., Walder, D. J., Reynolds, F.: Developmental changes in cortisol secretion in normal and at-risk youth. Dev. Psychopathol. 13, 721–732 (2001)

Walsh, B. T., Seidman, S. N., Sysko, R., Gould, M.: Placebo response in studies of major depression: variable, substantial, and growing. JAMA 287, 1840–1847 (2002)

Walsh, J. K., Deacon, S., Dijk, D. J., Lundahl, J.: The selective extrasynaptic GABAA agonist, gaboxadol, improves traditional hypnotic efficacy measures and enhances slow wave activity in a model of transient insomnia. Sleep 30, 593–602 (2007)

Wang, H. D., Dunnavant, F. D., Jarman, T., Deutch, A. Y.: Effects of antipsychotic drugs on neurogenesis in the forebrain of the adult rat. Neuropsychopharmacology 29, 1230–1238 (2004)

Wang, Y., Barbacioru, C., Hyland, F., Xiao, W., Hunkapiller, K. L., Blake, J., Chan, F., Gonzalez, C., Zhang, L., Samaha, R. R.: Large scale real-time PCR validation on gene expression measurements from two commercial long-oligonucleotide microarrays. BMC Genomics 7, 59 (2006)

Warner-Schmidt, J. L., Duman, R. S.: VEGF is an essential mediator of the neurogenic and behavioral actions of antidepressants. Proc Natl Acad Sci U S A 104, 4647–4652 (2007)

Weintraub, D., Stern, M. B.: Psychiatric complications in Parkinson disease. Am. J. Geriatr. Psychiatry 13, 844–851 (2005)

West, A. P.: Neurobehavioral studies of forced swimming: the role of learning and memory in the forced swim test. Prog. Neuropsychopharmacol. Biol Psychiatry 14, 863–877 (1990)

Wichers, M. C., Koek, G. H., Robaeys, G., Verkerk, R., Scharpe, S., Maes, M.: IDO and interferon-alpha-induced depressive symptoms: a shift in hypothesis from tryptophan depletion to neurotoxicity. Mol. Psychiatry 10, 538–544 (2005)

Willner, P.: Validity, reliability and utility of the chronic mild stress model of depression: a 10-year review and evaluation. Psychopharmacology. (Berl.) 134, 319–329 (1997)

Wong, D. L., Tank, A. W.: Stress-induced catecholaminergic function: transcriptional and post-transcriptional control. Stress 10, 121–130 (2007)

Xu, J. F., Chen, X. Q., Du, J. Z., Wang, T. Y.: CRF receptor type 1 mediates continual hypoxia-induced CRF peptide and CRF mRNA expression increase in hypothalamic PVN of rats. Peptides. 26, 639–646 (2005)

Zetzsche, T., Frodl, T., Preuss, U. W., Schmitt, G., Seifert, D., Leinsinger, G., Born, C., Reiser, M., Moller, H. J., Meisenzahl, E. M.: Amygdala volume and depressive symptoms in patients with borderline personality disorder. Biol. Psychiatry 60, 302–310 (2006)

Zhu, C. B., Blakely, R. D., Hewlett, W. A.: The proinflammatory cytokines interleukin-1beta and tumor necrosis factor-alpha activate serotonin transporters. Neuropsychopharmacology 31, 2121–2131 (2006)

Zieker, J., Zieker, D., Jatzko, A., Dietzsch, J., Nieselt, K., Schmitt, A., Bertsch, T., Fassbender, K., Spanagel, R., Northoff, H., Gebicke-Haerter, P. J.: Differential gene expression in peripheral blood of patients suffering from post-traumatic stress disorder. Mol Psychiatry 12, 116–118 (2007)

Zimmerman, M., Chelminski, I., Posternak, M. A.: Generalizability of antidepressant efficacy trials: differences between depressed psychiatric outpatients who would or would not qualify for an efficacy trial. Am. J. Psychiatry 162, 1370–1372 (2005)

Zusso, M., Debetto, P., Guidolin, D., Giusti, P.: Cerebellar granular cell cultures as an *in vitro* model for antidepressant drug-induced neurogenesis. Crit. Rev Neurobiol 16, 59–65 (2004)

Chapter 16
Hunting for Peripheral Biomarkers to Support Drug Development in Psychiatry

Enrico Domenici(✉), Pierandrea Muglia, and Emilio Merlo-Pich

Abstract The development of novel treatments in psychiatry is suffering from the lack of objective criteria to select patients on the basis of measurable diagnostic parameters and biological markers to predict drug response. Clinical trials are usually long-term investigations, involving a large number of patients across multiple sites in the attempt to obtain sufficient statistical power, where drug efficacy is assessed by rather subjective and poorly sensitive psychopathologic rating scales. There is a clear need for hard read-outs that could complement the clinician's assessment at different stages of the trial, creating the premises for more effective "proof of concept" studies. Biological or molecular markers have long been investigated in psychiatry mainly on the basis of mechanistic hypotheses. Although the results are encouraging, there is substantial lack of replication for most putative biomarkers investigated so far. Owing to the heterogeneity of psychiatric disorders, it appears unlikely that markers based on single mechanisms might prove to be useful in the clinics, and approaches aimed at detecting biomarker profiles rather than single markers are warranted. The search for peripheral biomarkers is supported by the evidence of multiple interactions between the brain and the periphery through complex neuroendocrine and neuroinflammatory networks. Recent developments in the application of genomic, proteomic, and metabonomic approaches are creating novel opportunities for the search of multiple biomarkers in psychiatry, opening up the chance to explore novel pathways in an unbiased manner. A series of studies documenting the possibility of detecting molecular signatures for psychiatric diseases and drug response in blood by large-scale approaches are reported. Integration of biomarker profiling approaches with neuroimaging strategies and genetic approaches on a genome-wide scale is expected to increase the chance of identifying novel markers which could assist modern drug development in psychiatry.

E. Domenici
Biology, Psychiatry CEDD, GlaxoSmithKline Medicines Research Centre,
Via Fleming 4, 37135, Verona, Italy
enrico.h.domenici@gsk.com

C.W. Turck (ed.) *Biomarkers for Psychiatric Disorders*,
doi: 10.1007/978-0-387-79251-4_16,

Abbreviations BDNF: Brain-derived neurotrophic factor; CNS: Central nervous system; CRH: Corticotropin releasing hormone; CSF: Cerebrospinal fluid; FDA: Food and Drug Administration; HAM-D: Hamilton rating scale for depression; HPA: Hypothalamus–pituitary–adrenocortical; MRI: Magnetic resonance imaging; PBMC: Peripheral blood mononuclear cell; PoC: Proof of concept; PTSD: Post-traumatic stress disorder

16.1 Introduction

An important milestone in the modern drug discovery process is the achievement of "Proof of Concept" (PoC), i.e., to establish the confidence that a novel candidate compound can be developed into an effective drug in the clinics. PoC studies consist of efficacy trials conducted in relatively small sets of patients representing the target population of the novel treatment. Modern trials with adaptive design have been successfully implemented in the oncology area, strengthening the rationale for the progression to development of novel therapeutics (Capdeville et al., 2002; Kaelin et al., 2007). The success can be further increased by making use of biological markers to identify subjects that can benefit from the novel treatment (Coudert et al., 2007), or to improve the evaluation of the biological effect of the drug treatment (DePrimo and Bello, 2007; DePrimo et al., 2007).

In psychiatry, the design and implementation of PoC studies are more difficult, owing to the convergence of a series of factors that can be summarized as follows: (1) poor knowledge of the disease etiology; (2) lack of biological markers that can help classify the patients into more homogeneous disorder groups; (3) unavailability of instrumental or molecular markers as surrogates for efficacy measures or with a prognostic value; (4) subjectivity of the clinician-rated scales used as efficacy endpoints, and, last but not least, (5) occurrence of an important placebo response (for a thorough discussion on the implications of placebo response in depression, see reviews by Fava et al. (Fava et al., 2003) and Dworkin et al.(Dworkin et al., 2005)).

With these premises, combined with the delayed onset of action of several weeks for most drugs used in psychiatry, PoC studies are bound to be long-term investigations, involving large numbers of patients across multiple sites in the attempt to obtain sufficient statistical power, and with uncertain output even for well-established drugs. Khan et al. have recently analyzed the FDA Summary Basis of Approval for all antidepressants investigated between 1985 and 2000, showing that more than half of all recent clinical trials of commonly used antidepressants failed to show statistical superiority of the drug over placebo (Khan et al., 2003). The inability of these trials to detect clinically relevant effects of otherwise active drugs is worrisome and, owing to relevant placebo effects (Benedetti et al., 2005) cannot be considered unique to depression. Nonetheless, because of the high variability of the clinical parameters observed, and the heterogeneity of the patient population, the failure to show statistical superiority in an individual trial cannot be taken as a definitive indication of the lack of psychotropic efficacy.

The above observations dictate the need to explore alternative study designs and more appropriate endpoints for the development of effective PoC studies. In this regard, biological markers, or biomarkers, can potentially represent an important support to drug development, and more generally to clinical practice, in psychiatry (Domenici and Muglia, 2007). They have been defined as "characteristics that can be objectively measured and evaluated as an indicator of normal biologic processes, or pharmacologic responses to a therapeutic intervention" (The Biomarker Definitions Working Group, 2001). Conceptually, they could help to characterize the clinical study population from a biological standpoint, leading to the identification of homogeneous sets of patients who will benefit most from the treatment, and placebo responders. Biomarkers could also complement the clinical assessment of the psychiatrists and highlight changes in biological parameters occurring in parallel or ahead of the clinical symptoms, thereby enabling early response detection. Finally, when validated they could serve as read-outs for treatment efficacy with improved sensitivity over the clinical rating scales.

All these features would improve drug discovery and development in psychiatry and increase the rate of success of PoCs, leading to the selection of more efficacious and tolerated treatment in clinical practice. The Food Drug Administration (FDA) has recently analyzed the challenges and opportunities on the critical path to new medical product discovery, reaching a consensus on biomarkers as one of the most important areas for improving drug development (U.S. Department of Health and Human Services, Food and Drug Administration, 2006a). Accordingly, diagnostic markers for neuropsychiatry conditions now appear specifically in the List of Opportunities aimed at speeding the development and approval of medical products (U.S. Department of Health and Human Services, Food and Drug Administration, 2006b).

16.2 The Rationale for the Search of Peripheral Biomarkers in Psychiatry

The selection of biomarkers to investigate the treatment effects of psychoactive compounds has been guided so far by the knowledge of the disease biology and pathophysiological mechanisms, mainly derived from preclinical investigations. In the last 20 years, the outstanding advancements in the field of neuroscience about the mechanisms that regulate behavior have driven the concept that psychiatric disorders are mostly brain diseases, involving specific neural circuits and structures (Hyman, 2007). Large investments in technology and scientific investigation have been made to map brain activity using neuroimaging or electro-magnetoencephalograpy in patients. This research has generated a growing body of evidence that central "imaging" markers can be used to measure changes in the disease state and potentially to predict treatment efficacy. Recent reviews describe the initial advancements and highlight the possible application to drug discovery (Matthews et al., 2006; Paulus and Stein, 2007).

The present chapter focuses on the development of peripheral markers in psychiatry, and in particular on the chances offered by the detection of molecular markers that can be easily, and rather noninvasively, measured in the blood district.

The approach is mainly based on the evidence of a cross-talk between the brain and the periphery in both directions. The brain is indeed in continuous interaction with peripheral signals generated in other organs via blood, cerebrospinal fluid (CSF), the lymphatic system, and local neuroendocrine networks (Gladkevich et al., 2004). The possibility of bodily or peripheral factors to produce changes of psychological states and induce psychiatric disorders has been proposed by several authors in the past (e.g., James–Lange theory of emotion). In turn, changes in brain activity and psychological states have been associated with an impact on the levels of circulating factors involved in the homeostasis of inflammatory, endocrine, and metabolic pathways, which might affect their functioning and contribute to enhance a disease state. The above concepts are an integral component of the stress theory proposed by Hans Seyle during 1950s and 1960s and can account for the observed comorbidity between systemic and psychiatry disorders (Plante, 2005). Epidemiologic and clinical findings collected in diabetes, metabolic syndrome, and cardiovascular diseases show high incidence of comorbidity with depression (Arbelaez et al., 2007; Carnethon et al., 2007; Vaccarino et al., 2007). Accordingly, changes in circulating glucocorticoids (Holsboer, 2000), C-reactive protein (CRP) and pro-inflammatory cytokines (Raison et al., 2006) have been commonly reported in subjects with depressive disorders. These findings are not unique to depression and therefore unlikely to be developed into specific biomarkers; nonetheless they support the investigation of common pathological mechanisms underlying systemic disorders as risk factors for psychiatric diseases (Krishnan, 2002; Naarding et al., 2005; Tiemeier, 2003).

Changes in the brain metabolism for hormones and neurotransmitters can therefore impact through direct and indirect mechanisms on the levels of a number of circulating factors that can have potential value as biomarkers. A parallel and prolific line of research has developed on the basis of the evidence of expression of a number of genes coding for brain-relevant proteins and receptors at the level of peripheral blood cells. Accordingly, neurotransmitter receptors or transporters in blood cells have been investigated as putative disease trait or state markers, in particular for depression and schizophrenia, as described in the next paragraph.

More recently, the commonality between gene expression in peripheral blood and in the central nervous system (CNS) has been addressed on a genome-wide scale by looking at data derived from DNA microarray investigations. Nicholson et al. (Nicholson et al., 2004) have examined the expression of genes belonging to the synthetic, biochemical, and regulatory psycho–neuroendocrine–immune (PNI) pathways in peripheral blood mononuclear cells (PBMCs) and found a higher number of expressed genes than anticipated. About 65% of roughly 1,600 genes belonging to PNI pathways were also expressed by peripheral blood cells, and among them more than 50% classified as "nervous" were positive for PBMC expression by microrrays. In a subsequent study, the microarray gene expression data from a number of human tissues, including 16 different CNS tissues and peripheral blood, have been directly compared, and a significant similarity between the transcriptional profile of whole blood and multiple brain regions was found (Sullivan et al., 2006). By cluster analysis of microarray data from blood and brain tissues, it can be observed that whole-blood expression profile appears in close proximity to pituitary expression profiles and relatively close to several regions of relevance for neuropsychiatry disorders, such as amygdala and hypothalamus. In schizophrenia, for example, the direct comparison of gene expression profiles from postmortem brain tissues with those obtained from peripheral cells of schizophrenics has allowed the identification of putative disease biomarkers having corresponding differential expression in brain and blood (Glatt et al., 2005). The above examples, together with the observed degree of similarity between the blood and brain transcriptome, provide a further rationale for the periphery as an useful surrogate for central gene expression and warrant the application of genome-wide profiling techniques to the identification of biomarker signatures for response to psychotropic drugs. Taking into account that the genetic makeup of a given individual is the same in every tissue, it would also be possible to identify trait markers by considering gene polymorphisms associated with psychiatry diseases, which may have a similar functional impact in the CNS and in the periphery.

16.3 Biomarker Findings in Psychiatry by Hypothesis-Based Approaches

The search of peripheral markers reflecting psychiatric disease states and traits has been under constant scrutiny for a few decades, and numerous experimental attempts to generate reliable blood-derived markers based on several disease pathogenesis hypotheses have been made. Much attention has been focused on depression and schizophrenia, for which efforts have been made to identify specific biomarkers based on the hypothesis of monoamine dysfunction, the immuno-inflammatory hypothesis, as well as the neuroendocrine and the neuroplasticity hypothesis. We will briefly discuss the areas where suggestive evidences have been provided for the development of disease markers with potential impact on clinical trials.

16.3.1 The Monoaminergic Hypothesis

The identification of the central monoaminergic neural systems as substrates of the activity of antidepressant and antipsychotic drugs and the evidence of commonalities between central neurochemical mechanisms and peripheral neuroendocrine systems have stimulated a line of investigation on monoamine-related marker in periphery. Several authors have investigated the peripheral level of neurotransmitter metabolites as well as the expression of monoaminergic receptors in both lymphocytes and platelets from psychiatric patients, either by radioligand binding or expression studies.

Overall, the literature suggests a systemic dysregulation that might reflect, or mimic, the alteration of the central mechanism in psychiatric disorders. However, most investigations have produced variable results. For example, in depression, most of the studies focusing on the blood or CSF level of metabolites for noradreline (MHPG and VMA), serotonin (5-HIAA), and dopamine (HVA and DOPC) have yielded inconsistent findings or lack of replication (Connor and Leonard, 2004). Chatecholaminergic receptors in peripheral cells (such as β-adrenoceptor in lymphocytes and α_2-adrenoceptor in platelets) have been largely investigated in depression (Gurguis et al., 1999a,b). The results have been often equivocal or contradictory, most likely due to the heterogeneity of the patients, the use of different radioligands for binding studies, and perhaps different wash-out periods in patients under medication (Connor and Leonard, 2004). More convincing data have been collected for the serotonin transporter in platelets, for which imipramine binding has been proposed as putative biological marker for depression (see Mossner et al.,(2007) and reference therein).

In schizophrenia, plasma concentrations of the dopamine metabolite homovanillic acid (HVA) have been used as an indirect tool to assess changes in dopamine turnover, and efforts to correlate its level with clinical outcome have resulted with limited success (see Davila et al., (2007)). Indeed, while correlations between plasma HVA levels and clinical improvement have generally been found for positive symptoms, inconsistent results have been produced when attempting to correlate clinical data for negative symptoms. Dopamine receptors have been deeply investigated in schizophrenia, but the results are contradictory. For dopamine D3 receptor (DRD3), an increased expression was found in lymphocytes from two different cohorts of schizophrenic patients (Ilani et al., 2001; Kwak et al., 2001); however, a subsequent study reported a reduction of D3 expression in schizophrenics (with higher mRNA level correlated to negative symptoms), which was normalized upon therapy (Vogel et al., 2004). In the same study, a decreased lymphocyte DRD3 expression was found also in bipolar patients. In addition, measurement of mRNA expression for DRD3 has been found to correlate with other potentially psychiatric-related conditions, such as personality traits and smoking (Czermak et al., 2004a,b), making it a likely nonspecific marker.

16.3.2 The Immuno-Inflammatory Pathway and Mood Disorders

Multiple lines of evidence suggest a link between inflammatory pathways and psychiatric disorders, in particular stress-related disorders, leading to the development of the long-standing "cytokine hypothesis for depression" (Dantzer and Kelley, 2007; Irwin and Miller, 2007; Schiepers et al., 2005). According to this theory, the dysregulation of the immuno-inflammatory system is associated with the etiology and pathophysiology of major depressive disorders (MDD). Increased circulating levels of pro-inflammatory cytokines, acute phase proteins, and chemokines are known to be associated with symptoms of depression and fatigue in humans and preclinical species (Raison et al., 2006). To date, most studies on cytokines in MDD have investigated a single cytokine subset, especially monocytic proinflammatory cytokines such as interleukin-6 (IL-6) and tumor necrosis factor-α (TNF-α). Increased circulating levels of IL-6 and TNF-α were described in subpopulations of patients suffering from MDD during the symptomatic episode (Hestad et al., 2003; Kim et al., 2007a; Mikova et al., 2001), and correlation between the high morning levels of IL-6 and depressive symptoms were found in MDD patients by Alesci et al. (Alesci et al., 2005). Successful antidepressant treatments of MDD episodes with selective serotonin reuptake inhibitors (SSRIs) or tricyclic antidepressants (TCAs) have been associated with the reduction of circulating cytokine levels, in particular TNF-α (Narita et al., 2006; Tuglu et al., 2003) and IL-6 (Kim et al., 2007a; Lanquillon et al., 2000), suggesting their potential use as surrogate markers for response. Interestingly, many reports have shown that, when administered as therapeutic treatment, cytokines can elicit mood-like symptoms. Treatment with interferon-α induces depressive symptoms, including fatigue, motor retardation, and depression in more than 50% individuals (Raison et al., 2006; Wichers et al., 2006). While the main evidences are on the activation of monocytic proinflammatory cytokines, the involvement of both Th1 and Th2 cytokines has been reported in MDD. A series of studies have shown that the Th1/Th2 ratio is increased in depressed patients and can be decreased by antidepressant treatment (Kim et al., 2007a; Kubera et al., 2001; Myint et al., 2005). Overall, the literature has generated a series of converging data on a number of potential cytokine markers, somewhat inconsistently showing their modulation by using different techniques and approaches. The above findings would need to be replicated on a larger scale (both in terms of study population and panel of measured cytokines). Recently, a larger panel of pro- and anti-inflammatory cytokines has been measured in a case/control MDD population (49 cases and 49 controls), showing elevation of a number of additional cytokines not previously implicated in MDD, as well as of some chemokines (Simon et al., 2007). We have measured a wider range of cytokines, chemokines, and neurotrophins represented by the Multi-Analyte Profiles (RBM Inc.) in a large case–control collection of psychiatric patients (250 depressed and 250 schizophrenic patients, plus 250 controls), and have found a number of significant signals exceeding the one expected under the null hypothesis (Domenici et al., 2006). These rather promising data call for the application of wider chemokine/cytokine assay technologies as biomarker discovery platforms in well-designed longitudinal studies with psychotropic drugs.

16.3.3 Neuroendocrine Markers: HPA Axis in Depression

In mood disorders, there is a large body of evidence for neuroendocrine abnormalities, including changes in the secretion of prolactin and growth hormones and dysregulation of the hypothalamic–pituitary–adrenal (HPA) and thyroid axes, which have been studied as potential biological markers by many investigators.

As the neuroendocrine system is strongly regulated by neurochemical pathways, the analysis of neuroendocrine parameters as peripheral markers is appealing, as it can provide indirect information on changes in neurotransmitter modulation occurring at the CNS level. The impaired regulation of the HPA system in depression is likely one of the most consistent laboratory findings in psychiatry, which led to the proposal of the neuroendocrine hypothesis of depression. According to this hypothesis, an impaired feedback of the HPA axis in depressed patients is at the basis of the hypercortisolemia observed during acute episodes (Holsboer, 2000). Along these lines, a neuroendocrine test measuring the impaired suppression of cortisol secretion after dexamethasone (DEX) challenge was developed and applied to the characterization of depressed patients with different severities (Holsboer, 1983). More recently, an evolution of the test has been proposed, based on the analysis of the stimulating effect of corticotrophin-releasing hormone (CRH) under the suppressive action of DEX. This test (also called the combined DEX/CRH test) has been shown to have great potential as a predictor of response to antidepressant treatment in selected cases of depressed patients (Ising et al., 2005, 2006).

Studies focusing on the genetic of the HPA axis in depression have additionally shown the potential of neuroendocrine-related markers as potential predictors for response in clinical trial. On the basis of a candidate gene association study on genes regulating the HPA axis and antidepressant response, a significant association was found for a polymorphism in FKBP5, a GR-receptor-regulating co-chaperone (Binder et al., 2004). Overall, the duration to response (as defined by a reduction of the HAM-D scores from admission by 50%) was reduced by about 2 weeks (from 5 to 3), in TT homozygotes with respect to CC and CT genotypes in a single nucleotide polymorphism (SNP) marker in the FKBP5 gene. Interestingly, altered levels of FKBP5 (at both mRNA and protein levels) in lymphocytes were found in TT homozygotes as compared with individuals with the two other genotypes. This observation suggests a possible functional effect of the FKBP5 polymorphisms, pointing to the possibility of identifying peripheral biomarkers driven by genetic association studies. For the CRH receptor type 1, the potential association with antidepressant response has also been reported (Licinio et al., 2004; Liu et al., 2007), and similar alterations of circulating CRH in depression have been described (Catalan et al., 1998; Galard et al., 2002).

16.3.4 Neuroplasticity and Neurotrophins as Markers in Psychiatry

Recent lines of evidence suggest the involvement of neuroplasticity-related pathways in the pathophysiology of psychiatric disorders, and a body of data support the impairment in neurotrophin signaling in mood disorders (Pittenger and Duman, 2008) and schizophrenia (Buckley et al., 2007a). In addition, psychotropic agents have been suggested to exert their long-term effect through induction of neuroplastic changes such as changes in synaptic plasticity, neurogenesis, and synaptogenesis, which in the case of antidepressants may partially explain the delayed-onset action of antidepressants (Castren et al., 2007; Castren, 2004; Duman et al., 1997; Manji et al., 2001; Nestler et al., 2002).

Much effort has been devoted to the understanding of the function of neurotrophins in diseases of the brain, and in particular on the role of brain-derived neurotrophic factor (BDNF) in the etiology of depression. BDNF has been largely studied in humans as a potential genetic factor (Gratacos et al., 2007). BDNF gene variants, shown to have functional effect in vitro (Egan et al., 2003), have been associated with mood disorders (see e.g., Neves-Pereira et al., 2005; Sklar et al., 2002; Strauss et al., 2005), depression-related personality (Sen et al., 2003), and schizophrenia (Muglia et al., 2003). Preclinically, a number of studies in rodents have shown that antidepressants are able to modulate BDNF function, and that BDNF modulation by genetic and nongenetic approaches can exert antidepressant-like properties (Angelucci et al., 2005). The possibility of measuring BDNF in the periphery has prompted a number of clinical investigations to assess its potential as a putative biomarker for depression and other psychiatric disorders. A large number of studies have focused on its levels in serum or plasma in different psychiatric diseases, including schizophrenia (Buckley et al., 2007b; Gama et al., 2007; Grillo et al., 2007; Hori et al., 2007; Huang and Lee, 2006; Jockers-Scherubl et al., 2004; Palomino et al., 2006; Pirildar et al., 2004; Shimizu et al., 2003a,b; Tan et al., 2005a,b; Toyooka et al., 2002; Zhang et al., 2007), depression (Aydemir et al., 2006; Bocchio-Chiavetto et al., 2006; Gervasoni et al., 2005; Gonul et al., 2005; Huang et al., 2007; Karege et al., 2002, 2005 Kim et al., 2007b; Lang et al., 2004; Lee et al., 2006; Piccinni et al., 2007; Yoshimura et al., 2007), or autism (Hashimoto et al., 2006; Katoh-Semba et al., 2007; Miyazaki et al., 2004).

The data appear to be more consistent in mood disorders and in particular in bipolar patients, where BDNF appears to be consistently downregulated. As far as peripheral BDNF as a marker for antidepressant efficacy is concerned, most studies report an increase of BDNF after several weeks of treatment, but no conclusive evidence exists for BDNF as a state marker for depression. For both schizophrenia and autism, an increase, a decrease, or no change in serum BDNF level have been documented. A recent study has investigated the relationship between peripheral BDNF concentration and brain metabolite levels measured by magnetic resonance spectroscopy in healthy volunteers. Interestingly, the data show a correlation with the level of *N*-acetylaspartate in the cerebral cortex, a putative marker of brain function and plasticity, making BDNF further attractive as a peripheral surrogate marker (Lang et al., 2007). However, the overall lack of specificity and somewhat conflicting findings produced around peripheral BDNF make it still questionable as a specific marker in psychiatry.

16.4 Novel Large-Scale Approaches to Biomarkers in Psychiatry

The above examples illustrate the potential application of peripheral markers to classify psychiatric patients or characterize them in terms of severity and drug response. In many cases, there is mixed or conflicting evidence that can be ascribed to a number of issues, such as small number of participants to the study, difference in the population origin, or differences in sampling schedules and assay procedures. Although some methodological issues might be resolved, it appears unlikely that a restricted number of biomarkers could help in the diagnosis, prognosis, and response assessment, given the heterogeneity of the psychiatric disorders both in terms of symptoms and the underlying etiology (see Hasler et al., (2004) for an overview on depression). Rather, a combination of biomarkers belonging to the same or different pathways looks more promising. The approach is largely justified by the high level of interconnection between the different systems and the complexity of psychiatric disorders at various levels. Several encouraging attempts to combine different markers have been reported, such as the integration of cytokine with neurotransmitter-related (Tsao et al., 2006) or HPA (Himmerich et al., 2006) markers, or BDNF with HPA markers (Schule et al., 2006). We will now discuss how approaches on a larger scale might revolutionize this area.

16.4.1 Disease Biomarker Signatures

Large-scale approaches by "omic" technologies (Stoughton and Friend, 2005) have given a novel impetus in biomarker discovery, allowing search in an unbiased manner the genome, the proteome, and the metabonome for novel molecular disease markers. The large-scale approaches are (1) expanding the number of potentially testable hypothesis by few orders of magnitude; (2) giving access to yet untapped pathways; and (3) creating the premises for the identification of biomarker signatures (i.e., combination of multiple markers instead of single biomarkers).

Their application in psychiatry is quite recent, in spite of the lack of understanding of the disease pathophysiology at the molecular level. Nonetheless, genomic and proteomic analysis of preclinical models or postmortem brain tissues from psychiatric patients have already opened new prospects for the identification of novel CNS pathways involved in psychiatric disorders (Mirnics et al., 2006). The approach, which is based on the detection of differences in the expression profiles of tissues from psychiatric patients vs. healthy controls, is now being applied also to peripheral tissues, showing that molecular signatures can be used as potential disease classification markers.

Encouraging data are coming from the application of microrrays to the transcriptional profiling of peripheral blood. One of the first examples of the potential of microarray analysis of peripheral blood in dissecting psychiatric phenotypes is the work by Tsuang et al. (Tsuang et al., 2005), who have identified a specific genomic signature for diseases on the basis of 89 genes differentially expressed in blood cells of schizophrenics, bipolar patients, and healthy controls. More recently, a number of papers have been published where genomic or proteomic technologies have been applied to peripheral blood samples from psychiatric patients, including schizophrenics, and patients with bipolar disorders, depression, and autism. For a comprehensive analysis of the potential of the above approaches in delivering diagnostic or disease classification biomarkers in psychiatry, the reader is referred to our recent review (Domenici and Muglia, 2007).

Attempts to create a relationship between transcriptional profiles from peripheral samples and disease outcome have also been described. Segman et al. (Segman et al., 2005) have run a longitudinal microarray study on peripheral samples from patients surviving from trauma and later developing post-traumatic stress disorders (PSTD). They were able to identify gene-expression signatures that can predict whether a subject will develop PTSD 1 or 4 months later, providing evidence that transcriptional profiles can determine symptom/disease outcome in psychiatry when they are thoroughly assessed at their onset.

16.4.2 Detection of Drug-Response Biomarkers

The promising examples arising from baseline assessment of transcriptional or proteomic profiles from psychiatric patients warrant further application of genome-wide approaches to the identification of biomarker signatures for response in longitudinal clinical studies.

So far, only a few papers have described the application of microarrays to the detection of peripheral signatures for response in patients undergoing a pharmacological treatment with a psychotropic drug. A few examples have come from Puskas and coworkers, who have applied printed DNA microarrays on lymphocytes samples from patients treated with different antidepressants. In one study, they have compared the microarray profile of lymphocytes from elderly depressed patients (Kalman et al., 2005a) at baseline and after 4 weeks of treatment with venlafaxine, and found a number of differentially expressed genes. However, likely because of the small number of participants, the authors did not address the relationship between gene expression and treatment outcome. The differentially expressed genes (mainly associated with cell survival, neural plasticity, signal transduction, and general metabolism) are therefore of restricted clinical value, being most likely the pharmacological peripheral signature of venlafaxine. The same group has analyzed the effect of another antidepressant drug, citalopram, on the lymphocyte gene expression profile from patients with Alzheimer's disease and controls (Kalman et al., 2005b; Palotas et al., 2004). Again, the study focused on the identification of genes differentially regulated by citalopram treatment, which were then categorized on the basis of similarities in biological functions. Their analysis revealed that citalopram affects differently healthy controls and Alzheimer lymphocytes, suggesting a possible link to the underlying disease pathophysiology.

Interestingly, Gene Logic has recently reported a microarray study on Affymetrix DNA chips conducted in blood from depressed patients treated with citalopram, where the relationship with treatment outcome was investigated. The authors claim to have identified gene subsets that are possibly implicated in the response to the antidepressant, as well as genes that may allow the prediction of this response on a gender-specific basis prior to the start of the treatment (Mamdani et al., 2007).

As far as proteomics is concerned, probably the only example reported so far in the longitudinal assessment of the effect of a psychotropic drug on peripheral samples from human patients is the study by Patil et al. (Patil et al., 2007). The authors have investigated the effect of 2 weeks of treatment with a centrally active norepinephrine reuptake inhibitor (atomoxetine) on the plasma and CSF proteome of healthy volunteers enrolled in a clinical trial. However, following drug treatment a change in the CSF protein profile was detected, but no significant change in plasma proteins was identified.

The most encouraging examples of large-scale profiling of peripheral tissues in longitudinal clinical studies in psychiatry so far have come from the application of metabonomic or metabolomic technologies, which allow the quantification of several different metabolites in tissues or fluids (Dettmer et al., 2007; German et al., 2007; Oresic et al., 2006). Kaddurah-Daouk and colleagues have applied a combined gas chromatography/mass spectrometry platform approach to the analysis of the metabolomic profile of plasma from psychiatric patients who were followed longitudinally during therapeutic treatment. A metabolomic analysis of blood plasma from depressed, remitted, and never-depressed older adults has been carried out, in which approximately 800 metabolites were analyzed and compared among the three groups (Paige et al., 2007). The study resulted in the identification of alterations in lipid metabolism and neurotransmitters in the depressed state, which might be modulated by antidepressant treatment. The same authors have also conducted a metabolomic study in the plasma of schizophrenic patients submitted to different antipsychotic drugs (Kaddurah-Daouk et al., 2007). Using a "lipomic" platform, able to quantify over 300 polar and non-polar lipid metabolites, the authors investigated the effects of three commonly used atypical antipsychotic drugs after 2–3 weeks' treatment. Interestingly, they were able to identify baseline lipid alterations in specific lipid classes which correlated with acute treatment response. The data have set the ground for the identification of patterns of lipid changes in schizophrenic patients specifically related to metabolic side effects or to therapeutic efficacy of atypical antipsychotics. The above examples make a more realistic expectation of the possibility to derive peripheral biomarkers for drug-response phenotypes by linking metabolic or expression profiles with clinical outcome.

16.4.3 Novel Biomarker Discovery in Clinical Investigations

Genomic, proteomic, or metabonomic approaches for the search of peripheral biomarkers in psychiatry might help to overcome some of the theoretical obstacles of hypothesis-driven biomarker investigations. As previously noted, they could increase the chance of identifying useful biomarkers by expanding the number of pathways being monitored and driving to the identification of biomarker signatures rather than stand-alone biomarkers. However, as genome- or proteome-wide approaches can generate a large volume of data, usually derived from a limited number of subjects, additional issues need to be considered when searching for markers with a prospective clinical utility. Peripheral blood is a highly dynamic tissue, in contact with nearly every tissue of the body, therefore reflecting a number of external factors that need to be controlled to ensure sufficient sensitivity and reproducibility of any profiling approach (Baird, 2006). The task appears to be more challenging in psychiatric disorders, where postmortem expression profiling studies reveal relatively modest molecular changes at the CNS level when compared to nonpsychiatric controls (Mirnics et al., 2006), and likely to manifest in the periphery with even more subtle changes.

The study design can be fundamental to maximize the chance of finding a reliable biomarker signature in the periphery. The choice of the most sensitive and reliable platform for genomic, proteomic, or metabolomic profiling and the downstream data analysis are also important factors that can contribute to the success of the approach. However, to minimize variability it is of paramount importance to accurately define or select the clinical population by using restrictive demographic and clinical criteria. In addition, the sample size theoretically should allow sufficient statistical power even after stratification for the main potential confounders. Finally, the sampling process should be established with accuracy. Time of sampling, diet, smoking, alcohol use, and exercise should all be monitored during a biomarker investigation, and procedures for sample processing accurately standardized. This can be achieved in well-designed randomized, placebo-controlled clinical trials, which could represent an opportunity to validate preliminary findings obtained in disease collections for their value as disease-state markers.

16.5 Integration with Pharmacogenetic and Imaging Markers

The advent of genome-, proteome- and metabonome-wide approaches are setting the premises for a paradigm shift in the search and validation of biomarkers in many fields, and psychiatry is gradually taking advantage of this technological evolution. Data originating from preliminary studies applied to psychiatry patients and healthy controls are encouraging. A number of methodological issues need to be resolved before large-scale approaches could become readily applicable in the clinical context; however, the application of the emerging "omic" technologies suggest that biological signatures for psychiatric disorders may exist and can be identified.

The field can take great advantage of two areas that are in similar rapid evolution.

As previously discussed, the results obtained so far by neuroimaging approaches have testified their great potential, especially as predictive markers for response to psychotropic drug treatment (Chen et al., 2007; Whalen et al., 2007). Even though instrumental approaches such as magnetic resonance imaging (MRI) or positron emission tomography (PET) are probably less widely applicable owing to cost constraints, their integration with peripheral biomarker studies in small and focused trials could provide further incentives to the application of biomarker strategies in clinical psychiatry. The evidence of a significant degree of correlation between peripheral BDNF levels and a brain spectroscopy signal related to cortical plasticity should encourage further attempts to integrate different methodological approaches to support the central–peripheral link for a given biomarker (Lang et al., 2007).

Another important area of development is the integration of biomarker approaches with pharmacogenetic approaches. The efficacy of psychotropic agents has been shown to be, to some extent, under genetic control, and a number of polymorphisms that may determine subjective response to drugs have been identified over the last few years. So far, pharmacogenetic studies of psychiatric drugs have focused on the candidate gene approach based on a mechanistic hypothesis. Besides metabolizing enzymes, most of the variants are related to genes coding for proteins in the putative pathways of the drug mechanism of action, such as the serotonin system (e.g., serotonin transporter, 5HT1a and 5HT2a, tryptophan hydroxylase, see (Serretti et al., 2005)) and the HPA axis (Binder and Holsboer, 2006) for antidepressants, or the dopamine and serotonin system for antipsychotics (e.g., dopamine receptors, 5HT2a and 2c, COMT, see (Malhotra et al., 2004; Prathikanti and Weinberger, 2005)). Larger data sets and wider gene panels are starting to produce encouraging data, such as, for example, the pharmacogenetic investigation on the STAR*D cohort ((Paddock et al., 2007). In this study, a list of 68 candidate genes was assessed and a significant and reproducible association between citalopram treatment and response was found in HTR2A and in genes beyond the monoamine pathway (such as GRIK4). Similar to biomarker profiling studies, the possibility to carry out genetic association studies on a genome-wide scale is now providing the opportunity to uncover the contribution of genes beyond the usual list of candidates. The Foundation of NIH is considering within the Biomarker Consortium (see http://www.biomarkersconsortium.org/) a key program for neurosciences for the identification of genomic biomarkers of treatment response in Major Depressive Disorder based on the extension of the whole genome association study of the STAR*D cohort. A similar approach is currently being followed by pharmaceutical companies. Pfizer has just announced the results from a whole genome association study run on eight depression trials and shown that the largest proportion of the associated genes are novel and not known to be CNS-relevant so far (Sakul et al., 2006).

To increase the chances of success, wide research efforts are being established through consortia or large projects, in an effort to combine multiple genetic and biomarker data sets across a large number of study centers. A recent example is the GENDEP (Genome-based Therapeutic Drugs for Depression) Project, an EC-funded integrated project aimed at the validation of pharmacogenomic methods for symptom improvement, the prediction of response to psychiatric drug treatment, and the reduction of adverse effects (Uher et al., 2007). The project is probably the first multidisciplinary large-scale multicenter human pharmacogenomic study focused on the prediction of therapeutic response to antidepressants, which includes an integrated analysis of the results with biomarker data originating from blood transcriptomic and proteomic approaches (see http://gendep.iop.kcl.ac.uk/).

16.6 Summary and Perspectives

The identification of a reliable peripheral marker or signature to assist drug development and allow decision making in a clinical trial is still far from being achieved. The recent progress in the application of genetic, genomic, and proteomic approaches in psychiatry, and the support from regulatory agencies, should, however, encourage further investments in the field. From a technological standpoint, we are witnessing a rapid evolution both in terms of decreasing costs and improvement of the standardization procedure, which will make possible the application of "omic" approaches on a wider scale. This is fundamental in psychiatry, where, owing to the heterogeneity of the patients and the subtle changes expected in the periphery, large populations are likely to be required to reach sufficient statistical power.

An additional spin for a renewed interest in the quest for peripheral biomarkers in psychiatric disorders is expected to derive from the parallel explosion of genotyping technologies, which are offering unprecedented opportunity to perform association studies at the whole-genome level.

Whole-genome association studies using high-throughput DNA chips are opening up the chance to search for risk factors at the genome-wide level (Barrett and Cardon, 2006; Manolio et al., 2007), and studies conducted on large and thoroughly assessed populations are leading to the identification of novel causative gene variants for common diseases (Couzin and Kaiser, 2007) including psychiatric ones (Baum et al., 2007; Wellcome Trust Case Control Consortium, 2007). The identification of causative gene variants, on one hand, could dictate more focused approaches to the search of validation of peripheral markers, as the gene can also be expressed in the periphery and undergo a dysregulation in the context of the disease, such as in the case of FKBP5. On the other hand, the integration of expression data from an individual with information on the genetic makeup based on whole genome scan will be essential, as there are lines of evidence showing that gene expression traits can be substantially explained by genotype differences. Gene expression traits in lymphocytes have been shown to be inheritable in multiplex pedigrees (Vawter et al., 2006), and points of convergence between gene expression changes in peripheral cells and previously known loci have been identified (Middleton et al., 2005). There are increasing examples of databases for allele-specific expression at the genome-wide level in human cell lines (Dixon et al., 2007; Morley et al., 2004; Stranger et al., 2005, 2007) and in brain tissues (Myers et al., 2007), which confirm the large impact of genotypic differences on the transcriptional machinery at the cellular level. This notion should be factored in when analyzing expression data sets, knowing that at global functional levels, gene variants can impact also on brain structure and function, as shown by combinations of brain imaging with genotype data (imaging genomics (Hariri and Weinberger, 2003)). As the cellular function of many genes can be common to both neural and peripheral tissues, or can be detected indirectly in peripheral tissues, the integration of risk factor data with combined genotype and expression data will increase the chance to identify peripheral biomarkers with closer relevance for disorders of the brain, an otherwise inaccessible "organ".

Acknowledgments The authors thank the members of the psychiatry biomarker team at GSK for their valuable support.

References

Alesci S, Martinez PE, Kelkar S et al. (2005) Major depression is associated with significant diurnal elevations in plasma interleukin-6 levels, a shift of its circadian rhythm, and loss of physiological complexity in its secretion: clinical implications. J Clin Endocrinol Metab 90:2522–2530

Angelucci F, Brene S, Mathe AA (2005) BDNF in schizophrenia, depression and corresponding animal models. Mol Psychiatr 10:345–352

Arbelaez JJ, Ariyo AA, Crum RM et al. (2007) Depressive symptoms, inflammation, and ischemic stroke in older adults: a prospective analysis in the cardiovascular health study. J Am Geriatr Soc 55:1825–1830

Aydemir C, Yalcin ES, Aksaray S et al. (2006) Brain-derived neurotrophic factor (BDNF) changes in the serum of depressed women. Prog Neuropsychopharmacol Biol Psychiatr 30:1256–1260

Baird AE (2006) The blood option: transcriptional profiling in clinical trials. Pharmacogenomics 7:141–144

Barrett JC and Cardon LR (2006) Evaluating coverage of genome-wide association studies. Nat Genet 38:659–662

Baum AE, Akula N, Cabanero M et al (2007) A genome-wide association study implicates diacylglycerol kinase eta (DGKH) and several other genes in the etiology of bipolar disorder. Mol Psychiatr doi: 10.1038/sj.mp.4002012

Benedetti F, Mayberg HS, Wager TD et al. (2005) Neurobiological mechanisms of the placebo effect. J Neurosci 25:10390–10402

Binder EB and Holsboer F (2006) Pharmacogenomics and antidepressant drugs. Ann Med 38:82–94

Binder EB, Salyakina D, Lichtner P et al. (2004) Polymorphisms in FKBP5 are associated with increased recurrence of depressive episodes and rapid response to antidepressant treatment. Nat Genet 36:1319–1325

Bocchio-Chiavetto L, Zanardini R, Bortolomasi M et al. (2006) Electroconvulsive Therapy (ECT) increases serum Brain Derived Neurotrophic Factor (BDNF) in drug resistant depressed patients. Eur Neuropsychopharmacol, 16:620–624

Buckley PF, Mahadik S, Pillai A, Terry A Jr (2007a) Neurotrophins and schizophrenia. Schizophr Res 94:1–11

Buckley PF, Pillai A, Evans D et al. (2007b) Brain derived neurotropic factor in first-episode psychosis. Schizophr Res 91:1–5

Capdeville R, Buchdunger E, Zimmermann J et al. (2002) Glivec (STI571, imatinib), a rationally developed, targeted anticancer drug. Nat Rev Drug Discov, 1:493–502

Carnethon MR, Biggs ML, Barzilay JI et al. (2007) Longitudinal association between depressive symptoms and incident type 2 diabetes mellitus in older adults: the cardiovascular health study. Arch Intern Med 167:802–807

Castren E (2004) Neurotrophic effects of antidepressant drugs. Curr Opin Pharmacol 4:58–64

Castren E, Voikar V, Rantamaki T (2007) Role of neurotrophic factors in depression. Curr Opin Pharmacol 7:18–21

Catalan R, Gallart JM, Castellanos JM, Galard R (1998) Plasma corticotropin-releasing factor in depressive disorders. Biol Psychiatr 44:15–20

Chen CH, Ridler K, Suckling J et al. (2007) Brain imaging correlates of depressive symptom severity and predictors of symptom improvement after antidepressant treatment. Biol Psychiatr, 62:407–414

Connor TJ, Leonard BE.(2004) Biological markers for Depression. In: Preskorn SH et al. (eds.) Handbook of Experimental Pharmacology (Volume 157). Antidepressants. Past, Present and Future, Springer, Berlin Heidelberg New York

Coudert BP, Largillier R, Arnould L et al. (2007) Multicenter phase II trial of neoadjuvant therapy with trastuzumab, docetaxel, and carboplatin for human epidermal growth factor receptor-2-overexpressing stage II or III breast cancer: results of the GETN(A)-1 trial. J Clin Oncol 25:2678–2684

Couzin J and Kaiser J (2007) Genome-wide association. Closing the net on common disease genes. Science 316:820–822

Czermak C, Lehofer M, Wagner EM et al. (2004a) Reduced dopamine D3 receptor expression in blood lymphocytes of smokers is negatively correlated with daily number of smoked cigarettes: a peripheral correlate of dopaminergic alterations in smokers. Nicotine Tob Res 6:49–54

Czermak C, Lehofer M, Renger H et al. (2004b) Dopamine receptor D3 mRNA expression in human lymphocytes is negatively correlated with the personality trait of persistence. J Neuroimmunol 150:145–149

Dantzer R and Kelley KW (2007) Twenty years of research on cytokine-induced sickness behavior. Brain Behav Immun 21:153–160

Davila R, Zumarraga M, Basterreche N et al. (2007) Plasma homovanillic acid levels in schizophrenic patients: correlation with negative symptoms. Psychiatr Res 151:163–168

DePrimo SE and Bello C (2007) Surrogate biomarkers in evaluating response to anti-angiogenic agents: focus on sunitinib. Ann Oncol 18(Suppl 10):x11–x19

DePrimo SE, Bello CL, Smeraglia J et al. (2007) Circulating protein biomarkers of pharmacodynamic activity of sunitinib in patients with metastatic renal cell carcinoma: modulation of VEGF and VEGF-related proteins. J Transl Med 5:32

Dettmer K, Aronov PA, Hammock BD (2007) Mass spectrometry-based metabolomics. Mass Spectrom Rev 26:51–78

Dixon AL, Liang L, Moffatt MF et al. (2007) A genome-wide association study of global gene expression. Nat Genet 39:1202–1207

Domenici E and Muglia P (2007) The search for peripheral markers in psychiatry by genomic and proteomic approaches. Exp Opin Med Diagn, 1:235–251

Domenici E, Prokopenko I, Wille D et al. (2006) Identification of biomarkers for depression by multi analyte profiling of peripheral samples from case/control collections. Am.J.Med.Genet. B Neuropsychiatr.Genet. 141B: 790

Duman RS, Heninger GR, Nestler EJ (1997) A molecular and cellular theory of depression. Arch Gen Psychiatr 54:597–606

Dworkin RH, Katz J, Gitlin MJ (2005) Placebo response in clinical trials of depression and its implications for research on chronic neuropathic pain. Neurology, 65:S7–S19

Egan MF, Kojima M, Callicott JH et al. (2003) The BDNF val66met polymorphism affects activity-dependent secretion of BDNF and human memory and hippocampal function. Cell 112:257–269

Fava M, Evins AE, Dorer DJ, Schoenfeld DA (2003) The problem of the placebo response in clinical trials for psychiatric disorders: culprits, possible remedies, and a novel study design approach. Psychother Psychosom 72:115–127

Galard R, Catalan R, Castellanos JM, Gallart JM (2002) Plasma corticotropin-releasing factor in depressed patients before and after the dexamethasone suppression test. Biol Psychiatr 51:463–468

Gama CS, Andreazza AC, Kunz M et al (2007) Serum levels of brain-derived neurotrophic factor in patients with schizophre-nia and bipolar disorder. Neurosci Lett 420:45–48

German JB, Gillies LA, Smilowitz JT et al. (2007) Lipidomics and lipid profiling in metabolomics. Curr Opin Lipidol 18:66–71

Gervasoni N, Aubry JM, Bondolfi G et al. (2005) Partial normalization of serum brain-derived neurotrophic factor in remitted patients after a major depressive episode. Neuropsychobiology 51:234–238

Gladkevich A, Kauffman HF, Korf J (2004) Lymphocytes as a neural probe: potential for studying psychiatric disorders. Prog Neuropsychopharmacol Biol Psychiatr 28:559–576
Glatt SJ, Everall IP, Kremen WS et al. (2005) Comparative gene expression analysis of blood and brain provides concurrent validation of SELENBP1 up-regulation in schizophrenia. Proc Natl Acad Sci USA 102:15533–15538
Gonul AS, Akdeniz F, Taneli F et al. (2005) Effect of treatment on serum brain-derived neurotrophic factor levels in depressed patients. Eur Arch Psychiatry Clin Neurosci 255:381–386
Gratacos M, Gonzalez JR, Mercader JM et al. (2007) Brain-derived neurotrophic factor Val66Met and psychiatric disorders: meta-analysis of case-control studies confirm association to substance-related disorders, eating disorders, and schizophrenia. Biol Psychiatr 61:911–922
Grillo RW, Ottoni GL, Leke R et al. (2007) Reduced serum BDNF levels in schizophrenic patients on clozapine or typical antipsychotics. J Psychiatr Res 41:31–35
Gurguis GN, Vo SP, Griffith JM, Rush AJ (1999a) Platelet alpha2A-adrenoceptor function in major depression: Gi coupling, effects of imipramine and relationship to treatment outcome. Psychiatr Res 89:73–95
Gurguis GN, Vo SP, Griffith JM, Rush AJ (1999b) Neutrophil beta(2)-adrenoceptor function in major depression: G(s) coupling, effects of imipramine and relationship to treatment outcome. Eur J Pharmacol 386:135–144
Hariri AR and Weinberger DR (2003) Imaging genomics. Br Med Bull 65:259–270
Hashimoto K, Iwata Y, Nakamura K et al. (2006) Reduced serum levels of brain-derived neurotrophic factor in adult male patients with autism. Prog Neuropsychopharmacol Biol Psychiatr 30:1529–1531
Hasler G, Drevets WC, Manji HK, Charney DS (2004) Discovering endophenotypes for major depression. Neuropsychopharmacology 29:1765–1781
Hestad KA, Tonseth S, Stoen CD et al. (2003) Raised plasma levels of tumor necrosis factor alpha in patients with depression: normalization during electroconvulsive therapy. J ECT 19:183–188
Himmerich H, Binder EB, Kunzel HE et al. (2006) Successful antidepressant therapy restores the disturbed interplay between TNF-alpha system and HPA axis. Biol Psychiatr 60:882–888
Holsboer F (1983) The dexamethasone suppression test in depressed patients: clinical and biochemical aspects. J Steroid Biochem 19:251–257
Holsboer F (2000) The corticosteroid receptor hypothesis of depression. Neuropsychopharmacology 23:477–501
Hori H, Yoshimura R, Yamada Y et al. (2007) Effects of olanzapine on plasma levels of catecholamine metabolites, cytokines, and brain-derived neurotrophic factor in schizophrenic patients. Int Clin Psychopharmacol 22:21–27
Huang TL and Lee CT (2006) Associations between serum brain-derived neurotrophic factor levels and clinical phenotypes in schizophrenia patients. J Psychiatr Res 40:664–668
Huang TL, Lee CT, Liu YL (2007) Serum brain-derived neurotro-phic factor levels in patients with major depression: Effects of antidepressants. J Psychiatr Res
Hyman SE (2007) Can neuroscience be integrated into the DSM-V? Nat Rev Neurosci 8:725–732
Ilani T, Ben Shachar D, Strous RD et al. (2001) A peripheral marker for schizophrenia: Increased levels of D3 dopamine receptor mRNA in blood lymphocytes. Proc Natl Acad Sci USA 98:625–628
Irwin MR and Miller AH (2007) Depressive disorders and immunity: 20 years of progress and discovery. Brain Behav Immun 21:374–383
Ising M, Kunzel HE, Binder EB et al. (2005) The combined dexamethasone/CRH test as a potential surrogate marker in depression. Prog Neuropsychopharmacol Biol Psychiatr 29:1085–1093
Ising M, Horstmann S, Kloiber S et al. (2006) Combined Dexamethasone/Corticotropin Releasing Hormone Test Predicts Treatment Response in Major Depression-A potential Biomarker? Biol Psychiatr 62:400–406

Jockers-Scherubl MC, Danker-Hopfe H, Mahlberg R et al. (2004) Brain-derived neurotrophic factor serum concentrations are increased in drug-naive schizophrenic patients with chronic cannabis abuse and multiple substance abuse. Neurosci Lett 371:79–83

Kaddurah-Daouk R, McEvoy J, Baillie RA et al. (2007) Metabolomic mapping of atypical antipsychotic effects in schizophrenia. Mol Psychiatr 12:934–945

Kaelin WG, Sawyers CL, Mihich E (2007) Nineteenth Annual Pezcoller Symposium: hypothesis-driven clinical investigation in cancer. Cancer Res 67:11102–11105

Kalman J, Palotas A, Juhasz A et al. (2005a) Impact of venlafaxine on gene expression profile in lymphocytes of the elderly with major depression--evolution of antidepressants and the role of the "neuro-immune" system. Neurochem Res 30:1429–1438

Kalman J, Kitajka K, Pakaski M et al. (2005b) Gene expression profile analysis of lymphocytes from Alzheimer's patients. Psychiatr Genet, 15:1–6

Karege F, Perret G, Bondolfi G et al. et al. (2002) Decreased serum brain-derived neurotrophic factor levels in major depressed patients. Psychiatry Res 109:143–148

Karege F, Bondolfi G, Gervasoni N et al. (2005) Low brain-derived neurotrophic factor (BDNF) levels in serum of depressed patients probably results from lowered platelet BDNF release unrelated to platelet reactivity. Biol Psychiatr 57:1068–1072

Katoh-Semba R, Wakako R, Komori T et al. (2007) Age-related changes in BDNF protein levels in human serum: differences between autism cases and normal controls. Int J Dev Neurosci 25:367–372

Khan A, Detke M, Khan SR, Mallinckrodt C (2003) Placebo response and antidepressant clinical trial outcome. J Nerv Ment Dis 191:211–218

Kim YK, Na KS, Shin KH et al. (2007a) Cytokine imbalance in the pathophysiology of major depressive disorder. Prog Neuropsychopharmacol Biol Psychiatr, 31:1044–1053

Kim YK, Lee HP, Won SD et al. (2007b) Low plasma BDNF is associated with suicidal behavior in major depression. Prog Neuropsychopharmacol Biol Psychiatr 31:78–85

Krishnan KR (2002) Biological risk factors in late life depression. Biol Psychiatr, 52:185–192

Kubera M, Lin AH, Kenis G et al. (2001) Anti-Inflammatory effects of antidepressants through suppression of the interferon-gamma/interleukin-10 production ratio. J Clin Psychopharmacol, 21:199–206

Kwak YT, Koo MS, Choi CH, Sunwoo I (2001) Change of dopamine receptor mRNA expression in lymphocyte of schizophrenic patients. BMC Med Genet 2:3

Lang UE, Hellweg R, Gallinat J (2004) BDNF serum concentrations in healthy volunteers are associated with depression-related personality traits. Neuropsychopharmacology, 29:795–798

Lang UE, Hellweg R, Seifert F et al. (2007) Correlation between serum brain-derived neurotrophic factor level and an in vivo marker of cortical integrity. Biol Psychiatr 62:530–535

Lanquillon S, Krieg JC, Bening-Abu-Shach U, Vedder H (2000) Cytokine production and treatment response in major depressive disorder. Neuropsychopharmacology 22:370–379

Lee BH, Kim H, Park SH, Kim YK (2006) Decreased plasma BDNF level in depressive patients. J Affect Disord 101:239–244

Licinio J, O'Kirwan F, Irizarry K et al. (2004) Association of a corticotropin-releasing hormone receptor 1 haplotype and antidepressant treatment response in Mexican-Americans. Mol Psychiatr 9:1075–1082

Liu Z, Zhu F, Wang G et al. (2007) Association study of corticotropin-releasing hormone receptor1 gene polymorphisms and antidepressant response in major depressive disorders. Neurosci Lett 414:155–158

Malhotra AK, Murphy GM, Jr., Kennedy JL (2004) Pharmacogenetics of psychotropic drug response. Am J Psychiatr 161:780–796

Mamdani F, Sequeira PA, ffrench-Mullen J et al. (2007) Gene Expression in Peripheral Lymphocyte Cells of Patients with Major Depressive Disorder Treated with Citalopram. 62nd Annual Scientific Convention of the Society of Biological Psychiatry, San Diego, USA

Manji HK, Drevets WC, Charney DS (2001) The cellular neurobiology of depression. Nat Med 7:541–547

Manolio TA, Rodriguez LL, Brooks L et al. (2007) New models of collaboration in genome-wide association studies: the Genetic Association Information Network. Nat Genet 39:1045–1051
Matthews PM, Honey GD, Bullmore ET (2006) Applications of fMRI in translational medicine and clinical practice. Nat Rev Neurosci 7:732–744
Middleton FA, Pato CN, Gentile KL et al. (2005) Gene expression analysis of peripheral blood leukocytes from discordant sib-pairs with schizophrenia and bipolar disorder reveals points of convergence between genetic and functional genomic approaches. Am J Med Genet B Neuropsychiatr Genet 136:12–25
Mikova O, Yakimova R, Bosmans E et al. (2001) Increased serum tumor necrosis factor alpha concentrations in major depression and multiple sclerosis. Eur Neuropsychopharmacol 11:203–208
Mirnics K, Levitt P, Lewis DA (2006) Critical appraisal of DNA microarrays in psychiatric genomics. Biol Psychiatr 60:163–176
Miyazaki K, Narita N, Sakuta R et al. (2004) Serum neurotrophin concentrations in autism and mental retardation: a pilot study. Brain Dev 26:292–295
Morley M, Molony CM, Weber TM et al. (2004) Genetic analysis of genome-wide variation in human gene expression. Nature 430:743–747
Mossner R, Mikova O, Koutsilieri E et al. (2007) Consensus paper of the WFSBP Task Force on Biological Markers: biological markers in depression. World J Biol Psychiatr 8:141–174
Muglia P, Vicente AM, Verga M et al. (2003) Association between the BDNF gene and schizophrenia. Mol Psychiatr 8:146–147
Myers AJ, Gibbs JR, Webster JA et al. (2007) A survey of genetic human cortical gene expression. Nat Genet 39:1494–1499
Myint AM, Leonard BE, Steinbusch HW, Kim YK (2005) Th1, Th2, and Th3 cytokine alterations in major depression. J Affect Disord 88:167–173
Naarding P, Schoevers RA, Janzing JG et al. (2005) A study on symptom profiles of late-life depression: the influence of vascular, degenerative and inflammatory risk-indicators. J Affect Disord 88:155–162
Narita K, Murata T, Takahashi T et al. (2006) Plasma levels of adiponectin and tumor necrosis factor-alpha in patients with remitted major depression receiving long-term maintenance antidepressant therapy. Prog Neuropsychopharmacol Biol Psychiatr 30:1159–1162
Nestler EJ, Barrot M, DiLeone RJ et al. (2002) Neurobiology of depression. Neuron 34:13–25
Neves-Pereira M, Cheung JK, Pasdar A et al. (2005) BDNF gene is a risk factor for schizophrenia in a Scottish population. Mol Psychiatr 10:208–212
Nicholson AC, Unger ER, Mangalathu R et al. (2004) Exploration of neuroendocrine and immune gene expression in peripheral blood mononuclear cells. Brain Res Mol Brain Res 129:193–197
Oresic M, Vidal-Puig A, Hanninen V (2006) Metabolomica pproachest oph enotypec haracterization and applications to complex diseases. Expert Rev Mol Diagn 6:575–585
Paddock S, Laje G, Charney D et al. (2007) Association of GRIK4 with outcome of antidepressant treatment in the STAR*D cohort. Am J Psychiatr 164:1181–1188
Paige LA, Mitchell MW, Krishnan KR et al. (2007) A preliminary metabolomic analysis of older adults with and without depression. Int J Geriatr Psychiatr 22:418–423
Palomino A, Vallejo-Illarramendi A, Gonzalez-Pinto A et al. (2006) Decreased levels of plasma BDNF in first-episode schizophrenia and bipolar disorder patients. Schizophr Res 86:321–322
Palotas A, Puskas LG, Kitajka K et al. (2004) The effect of citalopram on gene expression profile of Alzheimer lymphocytes. Neurochem Res 29:1563–1570
Patil ST, Higgs RE, Brandt JE, Knierman MD (2007) Identifying pharmacodynamic protein markers of centrally active drugs in humans: a pilot study in a novel clinical model. J Proteome Res 6:955–966
Paulus MP and Stein MB (2007) Role of functional magnetic resonance imaging in drug discovery. Neuropsychol Rev 17:179–188

Piccinni A, Marazziti D, Catena M et al (2007) Plasma and serum brain-derived neurotrophic factor (BDNF) in depressed patients during 1 year of antidepressant treatments. J Affect Disord 105:279–283

Pirildar S, Gonul AS, Taneli F et al. (2004) Low serum levels of brain-derived neurotrophic factor in patients with schizophrenia do not elevate after antipsychotic treatment. Prog Neuropsychopharmacol Biol Psychiatr 28:709–713

Pittenger C and Duman RS (2008) Stress, depression, and neuroplasticity: a convergence of mechanisms. Neuropsychopharmacology 33:88–109

Plante GE (2005) Depression and cardiovascular disease: a reciprocal relationship. Metabolism 54:45–48

Prathikanti S and Weinberger DR (2005) Psychiatric genetics–the new era: genetic research and some clinical implications. Br Med Bull 73–74:107–122

Raison CL, Capuron L, Miller AH (2006) Cytokines sing the blues: inflammation and the pathogenesis of depression. Trends Immunol 27:24–31

Sakul H, Hinds DA, Frazer KA et al. (2006) Results of a whole genome association study for major depressive disorders (MDD). Am.J.Med.Genet.B Neuropsychiatr. Genet. 141B:693

Schiepers OJ, Wichers MC, Maes M (2005) Cytokines and major depression. Prog Neuropsychopharmacol Biol Psychiatr 29:201–217

Schule C, Zill P, Baghai TC et al. (2006) Brain-derived neurotrophic factor Val66Met polymorphism and dexamethasone/CRH test results in depressed patients. Psychoneuroendocrinology 31:1019–1025

Segman RH, Shefi N, Goltser-Dubner T et al. (2005) Peripheral blood mononuclear cell gene expression profiles identify emergent post-traumatic stress disorder among trauma survivors. Mol Psychiatr 10:500–13, 425

Sen S, Nesse RM, Stoltenberg SF et al. (2003) A BDNF coding variant is associated with the NEO personality inventory domain neuroticism, a risk factor for depression. Neuropsychopharmacology 28:397–401

Serretti A, Benedetti F, Zanardi R, Smeraldi E (2005) The influence of Serotonin Transporter Promoter Polymorphism (SERTPR) and other polymorphisms of the serotonin pathway on the efficacy of antidepressant treatments. Prog Neuropsychopharmacol Biol Psychiatr 29:1074–1084

Shimizu E, Hashimoto K, Watanabe H et al. (2003a) Serum brain-derived neurotrophic factor (BDNF) levels in schizophrenia are indistinguishable from controls. Neurosci Lett 351:111–114

Shimizu E, Hashimoto K, Okamura N et al. (2003b) Alterations of serum levels of brain-derived neurotrophic factor (BDNF) in depressed patients with or without antidepressants. Biol Psychiatr 54:70–75

Simon NM, McNamara K, Chow CW et al (2007) A detailed examination of cytokine abnormalities in Major Depressive Disorder. Eur Neuropsychopharmacol 18:230–233

Sklar P, Gabriel SB, McInnis MG et al. (2002) Family-based association study of 76 candidate genes in bipolar disorder: BDNF is a potential risk locus. Brain-derived neutrophic factor. Mol Psychiatr 7:579–593

Stoughton RB and Friend SH (2005) How molecular profiling could revolutionize drug discovery. Nat Rev Drug Discov 4:345–350

Stranger BE, Forrest MS, Clark AG et al. (2005) Genome-wide associations of gene expression variation in humans. PLoS Genet 1:e78

Stranger BE, Forrest MS, Dunning M et al. (2007) Relative impact of nucleotide and copy number variation on gene expression phenotypes. Science 315:848–853

Strauss J, Barr CL, George CJ et al. (2005) Brain-derived neurotrophic factor variants are associated with childhood-onset mood disorder: confirmation in a Hungarian sample. Mol Psychiatr 10:861–867

Sullivan PF, Fan C, Perou CM (2006) Evaluating the comparability of gene expression in blood and brain. Am J Med Genet B Neuropsychiatr Genet 141:261–268

Tan YL, Zhou DF, Cao LY et al. (2005a) Decreased BDNF in serum of patients with chronic schizophrenia on long-term treatment with antipsychotics. Neurosci Lett 382:27–32
Tan YL, Zhou DF, Zhang XY (2005b) Decreased plasma brain-derived neurotrophic factor levels in schizophrenic patients with tardive dyskinesia: association with dyskinetic movements. Schizophr Res 74:263–270
The Biomarker Definitions Working Group. (2001) Biomarkers and surrogate endpoints: preferred definitions and conceptual framework. Clin Pharmacol Ther 69:89–95
Tiemeier H (2003) Biological risk factors for late life depression. Eur J Epidemiol 18:745–750
Toyooka K, Asama K, Watanabe Y et al. (2002) Decreased levels of brain-derived neurotrophic factor in serum of chronic schizophrenic patients. Psychiatr Res 110:249–257
Tsao CW, Lin YS, Chen CC et al. (2006) Cytokines and serotonin transporter in patients with major depression. Prog Neuropsychopharmacol Biol Psychiatr 30:899–905
Tsuang MT, Nossova N, Yager T et al. (2005) Assessing the validity of blood-based gene expression profiles for the classification of schizophrenia and bipolar disorder: a preliminary report. Am J Med Genet B Neuropsychiatr Genet 133:1–5
Tuglu C, Kara SH, Caliyurt O et al. (2003) Increased serum tumor necrosis factor-alpha levels and treatment response in major depressive disorder. Psychopharmacology (Berl) 170:429–433
Uher R, Farmer A, Maier W et al (2007) Measuring depression: comparison and integration of three scales in the GENDEP study. Psychol Med1–12
US Department of Health and Human Services, Food and Drug Administration (2006a) Innovation or Stagnation: Critical Path Opportunities Report. http://www.fda.gov/oc/initiatives/criticalpath/reports/opp_report.pdf
US Department of Health and Human Services, Food and Drug Administration (2006b) Innovation or Stagnation: Critical Path Opportunities List. http://www.fda.gov/oc/initiatives/criticalpath/reports/opp_list.pdf
Vaccarino V, Johnson BD, Sheps DS et al. (2007) Depression, inflammation, and incident cardiovascular disease in women with suspected coronary ischemia: the National Heart, Lung, and Blood Institute-sponsored WISE study. J Am Coll Cardiol 50:2044–2050
Vawter MP, Atz ME, Rollins BL et al. (2006) Genome scans and gene expression microarrays converge to identify gene regulatory loci relevant in schizophrenia. Hum Genet, 119:558–570
Vogel M, Pfeifer S, Schaub RT et al. (2004) Decreased levels of dopamine D3 receptor mRNA in schizophrenic and bipolar patients. Neuropsychobiology, 50:305–310
Wellcome Trust Case Control Consortium (2007) Genome-wide association study of 14,000 cases of seven common diseases and 3,000 shared controls. Nature 447:661–678
Whalen PJ, Johnstone T, Somerville LH et al (2007) A Functional Magnetic Resonance Imaging Predictor of Treatment Response to Venlafaxine in Generalized Anxiety Disorder. Biol Psychiatr
Wichers MC, Kenis G, Leue C et al. (2006) Baseline immune activation as a risk factor for the onset of depression during interferon-alpha treatment. Biol Psychiatr 60:77–79
Yoshimura R, Mitoma M, Sugita A et al. (2007) Effects of paroxetine or milnacipran on serum brain-derived neurotrophic factor in depressed patients. Prog Neuropsychopharmacol Biol Psychiatr, 31:1034–1037
Zhang XY, Tan YL, Zhou DF et al (2007) Serum BDNF levels and weight gain in schizophrenic patients on long-term treatment with antipsychotics. J Psychiatr Res 41:997–1004

Chapter 17
Biomarkers for the Development of Antidepressant and Anxiolytic Drugs

Marcus Ising(✉) and Florian Holsboer

Abstract Depression and anxiety disorders are by far the most prevalent mental disorders. Even though current treatments are effective overall, a large number of patients treated with antidepressants do not benefit sufficiently from therapy. To improve this unsatisfactory situation, drugs with a diverse profile of action are required to provide a more specific treatment to patients not sufficiently responding to standard therapy. Biomarkers sensitive for drug action are the ideal tools to identify compounds with a specific profile of action at an early stage of drug development. This chapter summarizes the suitability of neuroendocrine tests, sleep and other EEG markers, imaging techniques, gene expression, and protein markers for serving as clinical biomarkers in depression and anxiety, and discusses their potential for improving drug discovery and development.

List of abbreviations α_2: Alpha 2-adrenergic receptor; ACTH: Corticotropin; AVP: Vasopressin; CRH: Corticotropin releasing hormone; CSF: Cerebrospinal fluid; DEX: Dexamethasone; DEX/CRH: Combined dexamethasone/CRH test; DST: Dexamethasone suppression test; EEG: Electroencephalography; EMG: Electromyography; EOG: Electrooculography; fMRI: Functional magnetic resonance imaging; $GABA_A$: Gamma-aminobutyric acid receptor A; GH: Growth hormone; GHRH: Growth hormone releasing hormone; GLX1: Glyoxalase 1; H_1: Histamine 1 receptor; HPA: Hypothalamic–pituitary–adrenocortical; HPT: Hypothalamic–pituitary–thyroid; $5\text{-}HT_{2/3}$: Serotonin 2, 3 receptors; IGF-1: Insuline-like growth factor 1; MARS: Munich Antidepressant Response Signature project; mRNA: Messenger ribonucleic acid; PET: Positron emission tomography; PRL: Prolactin; REM: Rapid eye movement; SPECT: Single photon emission computed tomography; T3: Triiodothyronine; T4: Thyroxine; TRH: Thyreotropin-releasing hormone; TSH: Thyroidea-stimulating hormone; WHO: World Health Organisation

---

M. Ising, PhD
Max Planck Institute of Psychiatry, Kraepelinstr. 2-10, 80804 Munich, Germany
ising@mpipsykl.mpg.de

C.W. Turck (ed.) *Biomarkers for Psychiatric Disorders*,
doi: 10.1007/978-0-387-79251-4_17,

17.1 Introduction

Depression and anxiety disorders are by far the most frequent mental disorders with lifetime prevalence rates in Europe and the USA approaching 20% for depression (Jacobi et al., 2004; Kessler et al., 1994) and ranging between 20 (Bijl et al., 1998) and 25% (Kessler et al., 1994) for anxiety disorders that frequently precede or co-occur with depression. Of the 870 million people living in the European Region, about 100 million people are estimated to suffer from anxiety or depression at any time (WHO European Ministerial Conference, 2005). Depression and the related disorders are not only debilitating for the sufferers, their relatives, and friends, but also pose a huge burden upon health care systems: depression is a leading cause for early retirement and represents an independent risk factor for cardiovascular disease, diabetes, and neurodegenerative disorders including Alzheimer's and Parkinson's diseases (Ishihara and Brayne, 2006; Modrego and Ferrandez, 2004). Finally, depression-related suicides are trailing only fatal accidents as the cause of death among young adults (Bertolote and Fleischmann, 2002).

Pharmacotherapy of depression today relies entirely upon drugs that modulate the monoaminergic neurotransmission in the brain, a pharmacological principle discovered serendipitously in the 1950s in Switzerland. Major progress has been made regarding the adverse effects of these drugs, and treatment efficacy of currently available antidepressants is acceptable. According to the Munich Antidepressant Response Signature (MARS) project (http://www.mars-depression.de), a naturalistic longitudinal study evaluating antidepressant treatment outcome in depression, almost 80% of inpatients suffering from depressive disorders show response to antidepressant treatment, i.e., an improvement of at least 50% in terms of symptom severity. Full remission (i.e., near absence of residual depressive symptoms), however, is achieved only in about 60% of the patients. Depression also poses a major risk factor for an unfavorable outcome in a number of other medical and neurological diseases. For instance, comorbid depression increased the risk for rehospitalization up to 3 times in an unselected sample of inpatients with cardiovascular, respiratory, or metabolic disorders (Kartha et al., 2007).

We perceive a situation in which patients are diagnosed on verbally transmitted information in addition to features such as psychomotor changes or facial expression of sadness and distress. The selection of drug treatment is based upon the doctor's choice and there is yet no antidepressant available that has a clinical profile that is clearly distinct from another. The only difference, which is also clinically deployed, is the profile of potential side effects. Twenty percent of the patients do not sufficiently benefit from antidepressant treatment even if combined with psychotherapy and other supportive treatments. These patients are treatment-resistant, and preventing such an unfavorable and desperate clinical condition is one of the yet most challenging unmet needs in depression research.

We envisage several developments that are capable of driving the progress in the management of depression: the discovery of novel targets emerging from human genetics and the discovery of compounds that specifically modulate these target structures. As is well perceived from the field, genes do not act alone; external factors that accumulate over a lifespan have major repercussions upon gene expression, protein processing, and other biological parameters. Biomarker alterations reflecting disturbances in biological systems will be an important avenue to identify new drug targets for mood disorders. Even more importantly, biomarkers can serve as screening tools to judge whether new compounds developed as antidepressants provide sufficient efficacy to normalize the disturbances in biological systems reflected by the biomarkers. And finally, on the way to administer the right drug to the right patient at the right time, biomarkers will be required to match specific drugs with individual patients to maximize benefit (Holsboer, 2008). We will summarize the evidence on selected candidates for clinical biomarkers in depression and anxiety and discuss their suitability for drug discovery and development.

17.2 Clinical Biomarkers in Depression and Anxiety

In 2001, the Biomarkers Workgroup of the U.S. National Institute of Health suggested defining biomarkers as characteristics that are objectively measured and evaluated as indicators of normal biological processes, pathogenic processes, or pharmacologic responses to a therapeutic intervention (Biomarkers Definitions Working Group, 2001). The latter part of this definition, biomarkers as indicators of a pharmacological response to a therapeutic intervention, is in the focus of this chapter. Indicators satisfying this definition should be most appropriate to identify compounds effectively targeting the pathological process related to the biomarker.

Most of the biomarkers proposed for depression and anxiety disorders are derived from neuroendocrinology, electroencephalographic (EEG) studies, brain imaging, gene expression, and proteomics. Recent advances in genetic research have generated several candidates for genetic markers in mental diseases including depression and anxiety disorders. As genetic markers are invariant and static, they are important sources for identifying new drug targets and enhancing individual diagnostics, for instance, by characterizing subgroups of patients with specific vulnerabilities requiring special treatment. Genetic markers, however, do not indicate processes and, therefore, are not biomarkers in a narrow sense; they are not discussed in this chapter.

17.2.1 Neuroendocrine Tests

17.2.1.1 HPA Axis

A number of profound endocrine abnormalities can be observed in acute depression and (less predominantly) also in anxiety disorders. The most consistent neuroendocrine finding is an impaired regulation of the neuroendocrine stress hormone system, the hypothalamic–pituitary–adrenocortical (HPA) axis (Holsboer, 2000 ; Pariante et al., 2004; Raison and Miller, 2003). In response to any kind of stress, the parvocellular neurons of the hypothalamus are stimulated to secrete the neuropeptides corticotropin releasing hormone (CRH) and vasopressin (AVP) into the portal vessel system to activate the synthesis and release of corticotropin (ACTH) from the anterior pituitary. ACTH, in turn, stimulates the adrenal cortex to synthesize and release glucocorticoids, in particular cortisol (in humans). These hormones have a multiplicity of functions, which are necessary for adaptation to the acute stress but can be pathogenic when the organism is persistently overexposed (de Kloet et al., 2005).

Depressed patients show increased plasma and urinary free cortisol levels (Sachar, 1967) and an elevated number of cortisol secretory episodes (Rubin et al., 1987) suggesting impaired negative feedback inhibition of the HPA axis. These observations are in agreement with findings of increased concentrations of CRH in the cerebrospinal fluid (Nemeroff et al., 1984; Roy et al., 1987), which, if extrapolated to the situation in the brain, is consistent with reduced CRH binding in forebrains of depressed suicide victims (Merali et al., 2004; Nemeroff et al., 1988) and elevated numbers of CRH-producing neurons in the paraventricular hypothalamic nucleus of patients with depression (Raadsheer et al., 1994).

The dexamethasone suppression test (DST) is a laboratory test evaluating the pituitary negative feedback inhibition of the HPA axis. In this test, the suppressive effect of 1–2 mg of the synthetic glucocorticoid dexamethasone (dex) upon plasma cortisol concentration (Carroll et al., 1981) is examined. The DST was advocated as a diagnostic test for the subgroup of melancholic depression, but it did not prove to be the case, as the same neuroendocrine alterations occur also among patients meeting other diagnostic criteria (Holsboer et al., 1986b). In fact, a considerable number of patients with mood disorders show elevated plasma cortisol levels following dex administration, presumably due to an impaired function of corticosteroid receptors at the pituitary level (Holsboer, 2000), which normalizes after antidepressant treatment (Holsboer et al., 1982). Persistent cortisol nonsuppression after dex is associated with early relapse or poor clinical outcome in depression (Holsboer et al., 1982, Ribeiro et al., 1993). In the CRH stimulation test, CRH is intravenously administered as a bolus. Most depressed patients show a blunted ACTH response but normal cortisol secretion (Gold et al., 1984; Holsboer et al., 1984). This divergent response pattern is interpreted as a result of desensitized pituitary CRH receptors due to hypothalamic CRH oversecretion combined with a functional hyperplasia of the adrenal cortex in these patients (Gold et al., 1986; Holsboer et al., 1986a). Both the dex suppression and the CRH stimulation tests have been extensively studied in affective disorders. However, they suffer from a limited sensitivity to detect HPA alterations in depressed patients (Arana and Ornsteen, 1985; Chrousos et al., 1985; Gold and Chrousos, 1985).

To overcome the problem of sensitivity, a combined dex suppression/CRH stimulation test was proposed by Holsboer and coworkers (Holsboer et al., 1987). Elevated cortisol responses to the dex/CRH test could be consistently observed in patients suffering from an acute major depressive episode (Ising et al., 2005), but also in panic disorder (Erhardt et al., 2006; Schreiber et al., 1996). Superior sensitivity of the dex/CRH test compared to the regular DST was repeatedly confirmed (Deuschle et al., 1998; Heuser et al., 1994; Rybakowski and Twardowska, 1999; Watson et al., 2002). During antidepressant treatment, neuroendocrine response to the dex/CRH test attenuates, irrespective of the primary pharmacological mode of action of the drugs (tricyclic antidepressants: Deuschle et al., 1997; Frieboes et al., 2003; Heuser et al., 1987, 1996, Holsboer-Trachsler et al., 1991; selective serotonin re-uptake inhibitors: Nickel et al., 2003, Rinne et al., 2003; serotonin re-uptake enhancer tianeptine: Nickel et al., 2003; combined α_2/5-$HT_{2/3}$/H_1 receptor antagonist mirtazapine: Schüle et al., 2003). This is explained by the common effects of antidepressants on GR gene expression resulting in a restoration of the GR sensitivity (Herr et al., 2003; Holsboer, 2000, Raison and Miller, 2003; Reul et al., 1993). These and related findings led to the hypothesis that a restoration of corticosteroid receptor function resulting in an improvement of the HPA axis regulation is the mediator for the clinical action of antidepressant drugs (Holsboer, 2000). Consequently, a restoration of the HPA system regulation should precede clinical amelioration. Likewise, the persistency of altered HPA system function during treatment is prognostically less favorable. A number of studies examined the association between the hormonal response to a single dex/CRH test during depression and subsequent treatment outcome. Even though the results are inconsistent, in most studies they are negative (Deuschle et al., 1997; Kunugi et al., 2006; Nickel et al., 2003). A possible explanation for these negative findings is the inadequacy of a single test to evaluate whether the dysregulation of the HPA axis is improving under the actual medication. If a second dex/CRH test is applied, change in neuroendocrine response predicts antidepressant treatment outcome. This was demonstrated in a naturalistic longitudinal study, where reduced cortisol response to a second dex/CRH test after 2–3 weeks was associated with future antidepressant treatment response as well as with achieving remission (Ising et al., 2007). Zobel and coworkers (Zobel et al., 1999, 2001) could show that patients remitted from a major depressive episode who displayed still elevated cortisol responses to a second dex/CRH test had an almost sixfold higher risk for relapse during a 6-month follow-up period than remitted patients with an attenuated cortisol response. This has been confirmed by recent studies in outpatients (Appelhof et al., 2006; Aubry et al., 2007).

17.2.1.2 Other Neuroendocrine Systems

Another neuroendocrine candidate system for a biomarker in depression is the hypothalamic–pituitary–thyroid (HPT) axis. Thyreotropin-releasing hormone (TRH) is produced in the paraventricular nucleus of the hypothalamus activating thyroidea-stimulating hormone (TSH), and also prolactin (PRL), both released from the anterior pituitary gland. TSH in turn stimulates the release of the thyroid hormones, mostly thyroxine (T4), but also triiodothyronine (T3), with T4 being convertible to the more active T3. The HPT axis is a major regulator of the metabolism of the organism and influences the general activation level and drive in the individual. In acute depression, slightly elevated T4 can be observed in combination with a loss of the nocturnal rise of the circadian TSH levels. In the TRH test, plasma TSH secretion in response to intravenous administration of 200 μg TRH is measured. In acute depression (Jackson, 1998), as also in panic disorder (Corrigan et al., 1992; Tukel et al., 1999), the TSH response to this test is attenuated. The evaluation of the morning (or evening) serum TSH in response to the TRH test has been suggested as a predictor for antidepressant treatment outcome in depression (Duval et al., 1996), and antidepressants have been discussed to normalize altered HPT function (Bauer et al., 2002), which, however, was contradicted by other studies (Brambilla et al., 1982; Rosenbaum et al., 1993).

Further potential biomarker candidates are neuroendocrine tests embarking on growth hormone (GH) and prolactin (PRL). GH is secreted from the anterior part of the pituitary gland and predominantly stimulated by growth hormone releasing hormone (GHRH) from the hypothalamic nucleus arcuatus; the peptide ghrelin secreted from epithelial cells of the stomach and other intestinal organs, and also from the pituitary and the hypothalamus, additionally stimulates GH secretion. Inhibitory effects on GH are induced by the hypothalamic peptide somatostatin (periventricular nucleus) and via feedback inhibition by circulating GH and insulin-like growth factor 1 (IGF-1), primarily secreted from the liver and other target tissues, which in turn is stimulated by GH. The GHRH-containing region in the medial basal hypothalamus receives massive noradrenergic and also dopaminergic innervations, and injection of noradrenaline strongly stimulates GH. GH stimulation can also be observed after administration of clonidine, a partial α_2-adrenoceptor agonist. PRL is also secreted from the anterior pituitary gland, predominantly regulated by dopamine from the arcuate nucleus of the hypothalamus exhibiting inhibitory effects and by TRH that stimulates also PRL besides TSH. Apomorphine, which is a nonselective dopamine agonist, strongly inhibits PRL secretion and, similar to clonidine, stimulates GH, but also ACTH and cortisol.

A blunted GH response to oral clonidine can be observed in depression (Valdivieso et al., 1996), and also in generalized anxiety disorder (Abelson et al., 1991), panic disorder (Brambilla et al., 1995; Uhde et al., 1989), and social phobia (Tancer et al., 1993). These findings have been interpreted as evidence for a decreased sensitivity of α_2-adrenergic receptors. The clonidine stimulation test is not sensitive with respect to antidepressant treatment; patients with depression as well as panic disorders treated with different types of antidepressants did not restore the GH response to clonidine stimulation (Ansseau et al., 1988; Coplan et al., 1995). Inconsistent findings have also been observed for GH and PRL levels in response to the dopaminergic agonist apomorphine (Lal, 1988; McPherson et al., 2003), which also exhibits a number of side effects, limiting its applicability as a biomarker.

Table 17.1 summarizes the suitability of the proposed neuroendocrine markers as potential biomarkers in terms of their sensitivity towards antidepressant treatment and prediction of treatment outcome.

Table 17.1 Neuroendocrine tests in depression

Systems	Markers	Sensitivity to antidepressant treatment	Prediction of antidepressant treatment outcome
HPA axis	Dex suppression test	+	(+)
	CRH test	(+)	+/–
	Dex/CRH test	++	+
HPT axis	TRH test	+/–	+/–
Monoaminergic stimulation	Clonidine test	+/–	+/–
	Apomorphine test	+/–	+/–

Note: ++ = excellent, + = moderate, (+) = limited, +/– = inconsistent evidence

17.2.2 Night Sleep and Other EEG Markers

17.2.2.1 Night Sleep

Disturbed sleep is a cardinal symptom of depression. For the diagnosis of sleep disturbances, polysomnographic recordings including electroencephalography (EEG), eye movement measurements (electrooculography, EOG), and assessment of muscle tension (musculus mentalis or submentalis in the chin/mouth region; electromyography, EMG) are performed during night sleep. Using these data, five sleep stages can be defined, one stage of rapid eye movement (REM) sleep characterized by bursts of REM, low tonic muscle tension, and mixed frequency EEG; and four non-REM sleep stages with higher muscle tension and decreasing EEG frequencies. Alpha and theta wave band activity is predominant in non-REM sleep stage 1; theta activity combined with the occurrence of a typical EEG waveform called K-complex and high frequency bursts (sleep spindles) appear in stage 2; increasing delta activity appears in stage 3, and predominance of delta activity in stage 4. Non-REM stages 3 and 4 are also summarized as the slow-wave sleep stages. The sleep stages occur in cycles, starting with the non-REM stages and terminated by REM sleep. One cycle typically lasts between 1.5 and 2 h with about four to five cycles occurring during normal night sleep. While slow-wave sleep decreases during the cycles, REM sleep increases, suggesting reciprocal regulation of both sleep types. While the secretion of GHRH triggers slow-wave sleep, REM sleep is predominantly regulated by CRH (Steiger, 2007).

Sleep in depression is characterized by decreased sleep efficacy (difficulty falling asleep, nocturnal awakenings, early-morning awakening), decreased slow-wave sleep, shortened REM latency, and increased REM density, a measure of relative REM intensity (Thase et al., 1997; Tsuno et al., 2005). REM latency and REM density have been proposed as biomarkers for antidepressant treatment response (Kimura and Steiger, 2008; Kupfer et al., 1976, 1980). However, the interpretation of REM sleep parameters under antidepressant treatment is impaired by the fact that most of these drugs also have outcome-independent effects on REM sleep. Slow-wave sleep is not, or only marginally, affected by antidepressant treatment

(Sharpley and Cowen, 1995; Tsuno et al., 2005). Further details about the suitability of sleep parameters as biomakers in depression have been reported by Kimura and Steiger (2008).

17.2.2.2 Other EEG Markers

Until brain imaging techniques became available, EEG was a unique and easily accessible method to investigate brain synaptic activity in vivo. Several EEG markers including resting EEG parameters (e.g., frontal alpha activity) or event-related EEG potentials (e.g., odd/ball P300 component, loudness dependence) have been proposed as diagnostic markers for depression and anxiety. Despite several significant findings, sensitivity and specificity are limited (Bruder, 1992; Pollock and Schneider, 1990). Recently, quantitative EEG cordance has been introduced as a measure combining the relative and absolute power of a specific wave band at a given electrode assumed to correlate with regional cerebral perfusion (Leuchter et al., 1994). Change in frontal theta band cordance after 1 week of fluoxetine or venlafaxine treatment was demonstrated to be associated with treatment response (Cook et al., 2002), which was replicated in two further studies (Bares et al., 2007; Cook et al., 2005). These findings are promising, but require independent replications in larger samples.

In Table 17.2, the properties of night sleep and EEG markers are summarized in terms of their sensitivity towards antidepressant treatment and prediction of treatment outcome, prerequisites for biomarkers in depression.

17.2.3 Brain Imaging

Preclinical imaging methods such as autoradiography or in vitro/in vivo target occupancy studies are an established part of drug design and development. The increasing availability and technical progress in functional imaging methods such as single photon emission computed tomography (SPECT), positron emission tomography (PET), or functional magnetic resonance imaging (fMRI) provide new promising tools also for human studies. While SPECT and PET use radioactive

Table 17.2 Night sleep and other EEG markers in depression

Systems	Markers	Sensitivity to antidepressant treatment	Prediction of antidepressant treatment outcome
Night sleep	REM sleep	++	(+)
	Slow-wave sleep	(+)	+/–
Other EEG Markers	Resting EEG, event-related potentials	(+)	+/–
	Frontal theta cordance	+	+

Note: ++ = excellent, += moderate, (+)= limited, +/– = inconsistent evidence

Table 17.3 Properties of brain imaging techniques

Imaging markers	Properties
Single photon emission computed tomography (SPECT)	• Low effort and costs • Limitations concerning test repetitions • Low anatomical resolution • Moderate signal-to-noise ratio
Positron emission tomography (PET)	• High effort and costs • Limitations concerning test repetitions • Moderate anatomical resolution • Good signal-to-noise ratio
Functional magnetic resonance imaging (fMRI)	• Low costs • No limitations for test repetitions • Good anatomical resolution • Low signal-to-noise ratio

ligands (tracers) for revealing biochemical and physiological processes in the brain, fMRI applies magnetic resonance pulses to evaluate local blood-oxygen-level-dependent (BOLD) contrasts in response to external trigger events that can be embedded in neuropsychological tasks or emotional tests, addressing specific deficits characteristic for depression and anxiety. All three methods have advantages and disadvantages. A significant limitation of SPECT is its low resolution, while the availability of inexpensive tracers with long half-lives is an important advantage. PET provides excellent signal quality and a moderate resolution, but development of PET-based biomarkers is prohibited by the cost factor. Performing fMRI is less expensive and provides a good anatomical resolution, but its sensitivity is limited. The advantages and disadvantages of the methods are summarized in Table 17.3.

A well-established impairment in depression is a mood-congruent processing bias such that ambiguous or positive stimuli tend to be perceived as negative. Such impairment is reflected as reduced capacity for activation in several brain areas deemed to be implicated in the neuropathology of depression, which tends to resolve under successful antidepressant treatment (Harmer et al., 2004; Norbury et al., 2007).

17.2.4 Gene Expression and Protein Markers

Gene expression profiling (microarrays) and protein arrays (e.g., antibody arrays), which are described in detail elsewhere (de Souza and Dias-Neto, 2008; Turck et al., 2008), are probably the most promising approaches to identify effective biomarkers for drug discovery and development. These approaches are unbiased, i.e., they can be used in an explorative manner and without limiting the potential outcome by previous knowledge or assumptions. Unbiased approaches are particularly valuable for psychotropic drug discovery, as our understanding of the brain and its interaction with drugs is still poor. They provide the opportunity to identify unexpected biomarkers, which in turn unravel unknown pathomechanisms providing new targets for antidepressant and anxiolytic drug action.

The most important limitation of such approaches is presumably the restricted access to proper target tissues, which are, in case of mental disorders, brain tissue and cerebrospinal fluid (CSF). Human gene expression studies with the brain tissue of depressives can only be performed postmortem, which in turn is connected to other limitations, e.g., the restriction to cross-sectional studies and the problem of the unknown impact of terminal suffering while passing away. CSF protein array studies can be conducted in in vivo samples and represent the most promising source for protein-based biomarkers. This approach can be complemented by animal models carrying a disease-related phenotype. Such an extrapolation from animal models to human tissue has recently been reported to be feasible: the protein glyoxalase 1 (Glx 1), an enzyme involved in the detoxification of oxidative stress metabolites, reliably distinguished two mouse strains selectively bred for high and low anxiety-related behavior (Ditzen et al., 2006; Kromer et al., 2005). Altered mRNA levels for Glx 1 have also been observed in white blood cells from patients suffering from a depressive episode (Fujimoto et al., 2008). Notably, this result emerged from an unbiased approach.

17.3 Biomarkers in Drug Discovery and Development

Despite intensive psychopharmacological research, antidepressants entirely rely on modulation of monoaminergic neurotransmission in the brain, while most anxiolytics act at the $GABA_A$ receptor complex, both principles serendipitously discovered in the 1950s. As already mentioned, at least 20% of the patients treated with antidepressants do not sufficiently benefit from their treatment. To improve this unsatisfactory situation, more courageous decision making by pharmaceutical companies regarding the selection of compounds entering drug development is required. We need more drugs with diverse profiles of action in order to provide more specific treatment to those patients not sufficiently responding to standard therapy. This is not affordable for the pharmaceutical industry if the actual rigid process of drug development lasting years and consuming hundreds of millions of dollars is maintained. However, if reliable criteria for continuation or discontinuation of a compound become available already at a very early stage of phase 1 or phase 2 of drug development, a strategic shift might become a reality: from a conservative approach focusing on drugs with established mechanisms of action and general efficacy to an innovative approach testing drug candidates with diverse mechanisms of action and specific efficacy.

Biomarkers sensitive to drug action are the ideal tools to accelerate the time until making a reliable decision in favor of continuation or discontinuation of a candidate compound. They can be applied in healthy volunteers participating in a phase 1 proof-of-concept study simulating the target symptoms by experimental procedures

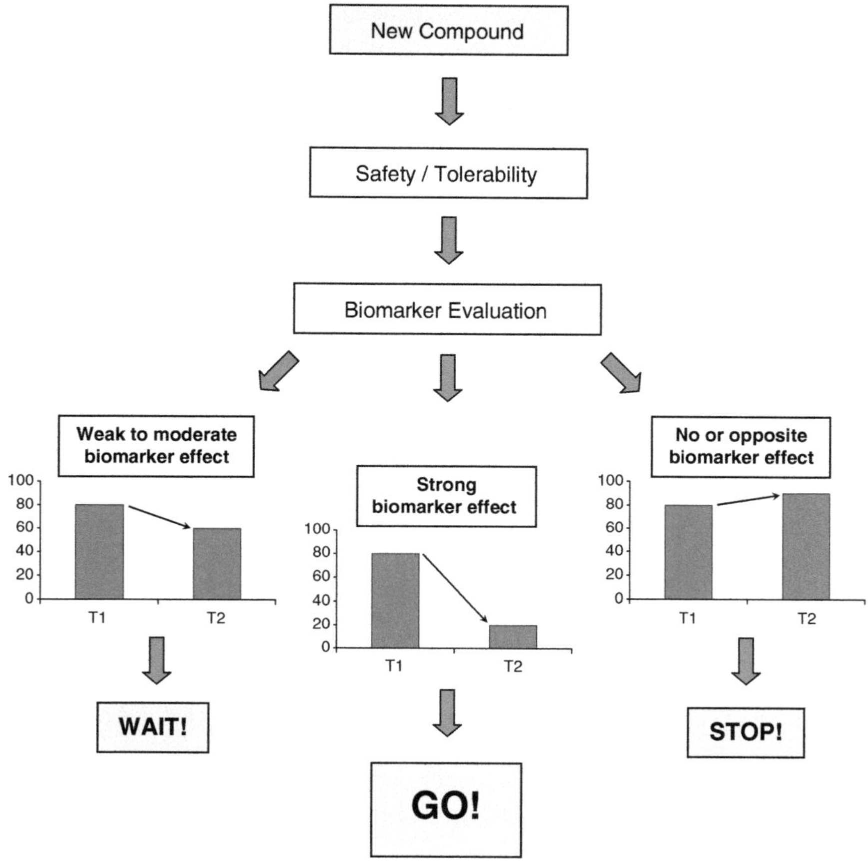

Fig. 17.1 Biomarker-based decision-making process about continuation, halting, or discontinuation of a new compound during phase 1 or early phase 2 development

or in clinical samples of early phase 2 studies. If a compound does not produce the expected biomarker effect, its development can be immediately halted, while those showing the expected results should be continued, as they are likely to provide sufficient efficacy. Compounds without a clear positive or negative effect might be put on hold until additional evidence is available. This is illustrated in Fig. 17.1.

17.4 Summary and Perspectives

Research over the past 5 years has demonstrated that detailed knowledge about genetic variances will not suffice to discover new drugs. Genomic regulation is much too complex to allow identification of specific targets. We are only beginning to understand how genes interact and how environmental factors impact upon gene activity. Epigenetic modulation is one way in which gene activation can be altered over long periods of time, sometimes lifelong, and these epigenetic marks can even be heritable. Other genomic adaptations to environmental influences are variations in gene copy numbers. Keeping all these possibilities in mind, it becomes quite obvious that we need measures that integrate all these effects. Biomarkers, obtained from gene expression profiles that closely reflect the activity of genomic systems, will become important tools to identify signaling pathways activated by a specific intervention. Such biomarkers will become increasingly valuable for two reasons: (1) they may guide us to unprecedented targets; and (2) they will help us to identify proteome-derived biomarkers. The latter field is the most straightforward research area, as most drugs act on proteins. In fact, there are 25,000 genes, 150,000 transcripts, and an estimated 1,000,000 proteins. These figures make it clear that the discovery of proteomic biomarkers is a particularly daunting task.

All these genetic and biochemical effects, as evidenced by genetics, gene expression profiles, and proteomics, interact throughout the organism and converge in clinical biomarkers, which in the case of a disease will reflect the individual's systems pathology. The value of such biomarkers is obvious: they may serve as tools for making clinical decisions ranging from prognostic considerations to targeted drug treatments. A combination of genotypes with biomarkers will create new diagnostic subgroups, for which more specific treatments will become available. Clearly, this is the beginning of a departure from classic psychiatric diagnosis, which in the near future will experience a renaissance if knowledge from clinical and basic neuroscience is integrated.

Another great opportunity is biomarker-assisted drug discovery and development. Neuroendocrine tests and structural or fMRI as well as EEG, all have high potential to deliver biomarkers. These will help us to sort out the most promising drug candidates for neurobiologically defined subgroups. Thus, biomarkers will be drivers for the development of the new era of personalized treatment of mood disorders.

References

Abelson JL, Glitz D, Cameron OG, Lee MA, Bronzo M, Curtis GC (1991) Blunted growth hormone response to clonidine in patients with generalized anxiety disorder. Arch Gen Psychiatr 48:157–162

Ansseau M, von Freckell R, Maassen D, Cerfontaine JL, Papart P, Timsit-Berthier M, Legros JJ, Franck G (1988) Prediction of treatment response to selective antidepressants from clonidine and apomorphine neuroendocrine challenges. In: Briley M, Fillion G (eds) New concepts in depression. McMillan, London, pp 269–276

Appelhof BC, Huyser J, Verweij M, Brouwer JP, van Dyck R, Fliers E, Hoogendijk WJ, Tijssen JG, Wiersinga WM, Schene AH (2006) Glucocorticoids and relapse of major depression (dexamethasone/corticotropin-releasing hormone test in relation to relapse of major depression). Biol Psychiatr 59:696–701

Arana GWB, Ornsteen M (1985) The dexamethasone suppression test for diagnosis and prognosis in psychiatry. Arch Gen Psychiatr 42:1193–1204

Aubry JM, Gervasoni N, Osiek C, Perret G, Rossier MF, Bertschy G, Bondolfi G (2007) The DEX/CRH neuroendocrine test and the prediction of depressive relapse in remitted depressed outpatients. J Psychiatr Res 41:290–294

Bares M, Brunovsky M, Kopecek M, Stopkova P, Novak T, Kozeny J, Hoschl C (2007) Changes in QEEG prefrontal cordance as a predictor of response to antidepressants in patients with treatment resistant depressive disorder: a pilot study. J Psychiatr Res 41:319–325

Bauer M, Heinz A, Whybrow PC (2002) Thyroid hormones, serotonin and mood: of synergy and significance in the adult brain. Mol Psychiatr 7:140–156

Bertolote JM, Fleischmann S (2002) Suicide and psychiatric diagnosis: a world-wide perspective. World Psychiatr 1:181–186

Bijl RV, Ravelli A, van Zessen G (1998) Prevalence of psychiatric disorder in the general population: results of The Netherlands Mental Health Survey and Incidence Study (NEMESIS). Soc Psychiatry Psychiatr Epidemiol 33:587–595

Biomarkers Definitions Working Group (2001) Biomarkers and surrogate endpoints: preferred definitions and conceptual framework. Clin Pharmacol Ther 69:89–95

Brambilla F, Scarone S, Massironi R, Nobile P, Pugnetti L (1982) Abnormal anterior pituitary responsiveness to hypothalamic hormones in primary affective disorders. Effect of desipramine therapy. Neuropsychobiology 8:269–275

Brambilla F, Perna G, Garberi A, Nobile P, Bellodi L (1995) Alpha 2-adrenergic receptor sensitivity in panic disorder: I. GH response to GHRH and clonidine stimulation in panic disorder. Psychoneuroendocrinology 20:1–9

Bruder GE (1992) P300 findings for depressive and anxiety disorders. Ann NY Acad Sci 658:205–222

Carroll BJ, Feinberg M, Greden JF (1981) A specific laboratory test for the diagnosis of melancholia. Arch Gen Psychiatr 38:15–22

Chrousos GP, Schuermeyer TH, Doppman J, Oldfield EH, Schulte HM, Gold PW, Loriaux DL (1985) NIH conference. Clinical applications of corticotropin-releasing factor. Ann Intern Med 102:344–358

Cook IA, Leuchter AF, Morgan M, Witte E, Stubbeman WF, Abrams M, Rosenberg S, Uijtdehaage SH (2002) Early changes in prefrontal activity characterize clinical responders to antidepressants. Neuropsychopharmacology 27:120–131

Cook IA, Leuchter AF, Morgan ML, Stubbeman W, Siegman B, Abrams M (2005) Changes in prefrontal activity characterize clinical response in SSRI nonresponders: a pilot study. J Psychiatr Res 39:461–466

Coplan JD, Papp LA, Martinez J, Pine D, Rosenblum LA, Cooper T, Liebowitz MR, Gorman JM (1995) Persistence of blunted human growth hormone response to clonidine in fluoxetine-treated patients with panic disorder. Am J Psychiatr 152:619–622

Corrigan MH, Gillette GM, Quade D, Garbutt JC (1992) Panic, suicide, and agitation: independent correlates of the TSH response to TRH in depression. Biol Psychiatr 31:984–992

de Kloet ER, Joels M, Holsboer F (2005) Stress and the brain: from adaptation to disease. Nat Rev Neurosci 6:463–475

de Souza DM, Dias-Neto E (2008) RNA biomarkers in schizophrenia. In: Turck CW (ed) Biomarkers for psychiatric disorders. Springer, Berlin Heidelberg New York

Deuschle M, Schmider J, Weber B, Standhardt H, Korner A, Lammers CH, Schweiger U, Hartmann A, Heuser I (1997) Pulse-dosing and conventional application of doxepin: effects on psychopathology and hypothalamus-pituitary-adrenal (HPA) system. J Clin Psychopharmacol 17:156–160

Deuschle M, Schweiger U, Gotthardt U, Weber B, Korner A, Schmider J, Standhardt H, Lammers CH, Krumm B, Heuser IJ (1998) The combined dexamethasone/corticotropin-releasing hormone stimulation test is more closely associated with features of diurnal activity of the hypothalamo-pituitary-adrenocortical system than the dexamethasone suppression test. Biol Psychiatr 43:762–766

Ditzen C, Jastorff AM, Kessler MS, Bunck M, Teplytska L, Erhardt A, Kromer SA, Varadarajulu J, Targosz BS, Sayan-Ayata EF, Holsboer F, Landgraf R, Turck CW (2006) Protein biomarkers in a mouse model of extremes in trait anxiety. Mol Cell Proteomics 5:1914–1920

Duval F, Mokrani MC, Crocq MA, Jautz M, Bailey P, Diep TS, Macher JP (1996) Effect of antidepressant medication on morning and evening thyroid function tests during a major depressive episode. Arch Gen Psychiatr 53:833–840

Erhardt A, Ising M, Unschuld PG, Kern N, Lucae S, Putz B, Uhr M, Binder EB, Holsboer F, Keck ME (2006) Regulation of the hypothalamic-pituitary-adrenocortical system in patients with panic disorder. Neuropsychopharmacology 31:2515–2522

Frieboes RM, Sonntag A, Yassouridis A, Eap CB, Baumann P, Steiger A (2003) Clinical outcome after trimipramine in patients with delusional depression – a pilot study. Pharmacopsychiatry 36:12–17

Fujimoto M, Uchida S, Watanuki T, Wakabayashi Y, Otsuki K, Matsubara T, Suetsugi M, Funato H, Watanabe Y (2008) Reduced expression of glyoxalase-1 mRNS in mood disorder patients. Neurosci Lett 438:196–199

Gold PW, Chrousos GP (1985) Clinical studies with corticotropin releasing factor: implications for the diagnosis and pathophysiology of depression, Cushing's disease, and adrenal insufficiency. Psychoneuroendocrinology 10:401–419

Gold PW, Chrousos G, Kellner C, Post R, Roy A, Augerinos P, Schulte H, Oldfield E, Loriaux DL (1984) Psychiatric implications of basic and clinical studies with corticotropin-releasing factor. Am J Psychiatr 141:619–627

Gold PW, Loriaux DL, Roy A, Kling MA, Calabrese JR, Kellner CH, Nieman LK, Post RM, Pickar D, Gallucci W (1986) Responses to corticotropin-releasing hormone in the hypercortisolism of depression and Cushing's disease. Pathophysiologic and diagnostic implications. New Engl J Med 314:1329–1335

Harmer CJ, Shelley NC, Cowen PJ, Goodwin GM (2004) Increased positive versus negative affective perception and memory in healthy volunteers following selective serotonin and norepinephrine reuptake inhibition. Am J Psychiatr 161:1256–1263

Herr AS, Tsolakidou AF, Yassouridis A, Holsboer F, Rein T (2003) Antidepressants differentially influence the transcriptional activity of the glucocorticoid receptor in vitro. Neuroendocrinology 78:12–22

Heuser IJ, Yassouridis A, Holsboer F (1994) The combined dexamethasone/CRH test: a refined laboratory test for psychiatric disorders. J Psychiatr Res 28:341–356

Heuser IJ, Schweiger U, Gotthardt U, Schmider J, Lammers CH, Dettling M, Yassouridis A, Holsboer F (1996) Pituitary-adrenal-system regulation and psychopathology during amitriptyline treatment in elderly depressed patients and normal comparison subjects. Am J Psychiatr 153:93–99

Holsboer F (2000) The corticosteroid receptor hypothesis of depression. Neuropsychopharmacology 23:477–501

Holsboer F (2008) How can we realize the promise of personalized antidepressant medicines? Nat Rev Neurosci 9:638–646

Holsboer F, Liebl R, Hofschuster E (1982) Repeated dexamethasone suppression test during depressive illness. Normalisation of test result compared with clinical improvement. J Affect Disord 4:93–101

Holsboer F, von Bardeleben U, Gerken A, Stalla GK, Müller OA (1984) Blunted corticotropin and normal cortisol response to human corticotropin-releasing factor in depression. N Engl J Med 311:1127

Holsboer F, Gerken A, von Bardeleben U, Grimm W, Beyer H, Müller OA, Stalla GK (1986a) Human corticotropin-releasing hormone in depression – correlation with thyrotropin secretion following thyrotropin-releasing hormone. Biol Psychiatr 21:601–611

Holsboer F, Philipp M, Steiger A, Gerken A (1986b) Multisteroid analysis after DST in depressed patients–a controlled study. J Affect Disord 10:241–249

Holsboer F, von Bardeleben U, Wiedemann K, Müller OA, Stalla GK (1987) Serial assessment of corticotropin-releasing hormone response after dexamethasone in depression. Implications for pathophysiology of DST nonsuppression. Biol Psychiatr 22:228–234

Holsboer-Trachsler E, Stohler R, Hatzinger M (1991) Repeated administration of the combined dexamethasone-human corticotropin releasing hormone stimulation test during treatment of depression. Psychiatr Res 38:163–171

Ishihara L, Brayne C (2006) A systematic review of depression and mental illness preceding Parkinson's disease. Acta Neurol Scand 113:211–220

Ising M, Künzel HE, Binder EB, Nickel T, Modell S, Holsboer F (2005) The combined dexamethasone/CRH test as a potential surrogate marker in depression. Prog Neuropsychopharmacol Biol Psychiatr 29:1085–1093

Ising M, Horstmann S, Kloiber S, Lucae S, Binder EB, Kern N, Künzel HE, Pfennig A, Uhr M, Holsboer F (2007) Combined Dexamethasone/Corticotropin Releasing Hormone test predicts treatment response in major depression – a potential biomarker? Biol Psychiatr 62:47–54

Jackson IM (1998) The thyroid axis and depression. Thyroid 8:951–956

Jacobi F, Wittchen HU, Holting C, Hofler M, Pfister H, Müller N, Lieb R (2004) Prevalence, co-morbidity and correlates of mental disorders in the general population: results from the German Health Interview and Examination Survey (GHS). Psychol Med 34:597–611

Kartha A, Anthony D, Manasseh CS, Greenwald JL, Chetty VK, Burgess JF, Culpepper L, Jack BW (2007) Depression Is a Risk Factor for Rehospitalization in Medical Inpatients. Prim Care Companion J Clin Psychiatr 9:256–262

Kessler RC, McGonagle KA, Zhao S, Nelson CB, Hughes M, Eshleman S, Wittchen HU, Kendler KS (1994) Lifetime and 12-month prevalence of DSM-III-R psychiatric disorders in the United States. Results from the National Comorbidity Survey. Arch Gen Psychiatr 51:8–19

Kimura M, Steiger A (2008) Sleep EEG provides biomarker in depression. In: Turck CW(ed) Biomarkers for psychiatric disorders. Springer, Berlin Heidelberg New York

Kromer SA, Kessler MS, Milfay D, Birg IN, Bunck M, Czibere L, Panhuysen M, Putz B, Deussing JM, Holsboer F, Landgraf R, Turck CW (2005) Identification of glyoxalase-I as a protein marker in a mouse model of extremes in trait anxiety. J Neurosci 25:4375–4384

Kunugi H, Ida I, Owashi T, Kimura M, Inoue Y, Nakagawa S, Yabana T, Urushibara T, Kanai R, Aihara M, Yuuki N, Otsubo T, Oshima A, Kudo K, Inoue T, Kitaichi Y, Shirakawa O, Isogawa K, Nagayama H, Kamijima K, Nanko S, Kanba S, Higuchi T, Mikuni M (2006) Assessment of the dexamethasone/CRH test as a state-dependent marker for hypothalamic-pituitary-adrenal (HPA) axis abnormalities in major depressive episode: a Multicenter Study. Neuropsychopharmacology 31:212–220

Kupfer DJ, Foster FG, Reich L, Thompson SK, Weiss B (1976) EEG sleep changes as predictors in depression. Am J Psychiatr 133:622–626

Kupfer DJ, Spiker DG, Coble PA, Neil JF, Ulrich R, Shaw DH (1980) Depression, EEG sleep, and clinical response. Compr Psychiatr 21:212–220

Lal S (1988) Apomorphine in the evaluation of dopaminergic function in man. Prog Neuropsychopharmacol Biol Psychiatr 12:117–164

Leuchter AF, Cook IA, Lufkin RB, Dunkin J, Newton TF, Cummings JL, Mackey JK, Walter DO (1994) Cordance: a new method for assessment of cerebral perfusion and metabolism using quantitative electroencephalography. Neuroimage 1:208–219

McPherson H, Walsh A, Silverstone T (2003) Growth hormone and prolactin response to apomorphine in bipolar and unipolar depression. J Affect Disord 76:121–125

Merali Z, Du L, Hrdina P, Palkovits M, Faludi G, Poulter MO, Anisman H (2004) Dysregulation in the suicide brain: mRNA expression of corticotropin-releasing hormone receptors and GABA(A) receptor subunits in frontal cortical brain region. J Neurosci 24:1478–1485

Modrego PJ, Ferrandez J (2004) Depression in patients with mild cognitive impairment increases the risk of developing dementia of Alzheimer type: a prospective cohort study. Arch Neurol 61:1290–1293

Nemeroff CB, Widerlov E, Bissette G, Walleus H, Karlsson I, Eklund K, Kilts CD, Loosen PT, Vale WW (1984) Elevated concentrations of CSF corticotropin-releasing factor-like immunoreactivity in depressed patients. Science 226:1342–1344

Nemeroff CB, Owens MJ, Bissette G, Andorn AC, Stanley M (1988) Reduced corticotropin releasing factor binding sites in the frontal cortex of suicide victims. Arch Gen Psychiatr 45:577–579

Nickel T, Sonntag A, Schill J, Zobel AW, Ackl N, Brunnauer A, Murck H, Ising M, Yassouridis A, Steiger A, Zihl J, Holsboer F (2003) Clinical and neurobiological effects of tianeptine and paroxetine in major depression. J Clin Psychopharmacol 28:155–168

Norbury R, Mackay CE, Cowen PJ, Goodwin GM, Harmer CJ (2007) Short-term antidepressant treatment and facial processing. Functional magnetic resonance imaging study. Br J Psychiatr 190:531–532

Pariante CM, Thomas SA, Lovestone S, Makoff A, Kerwin RW (2004) Do antidepressants regulate how cortisol affects the brain? Psychoneuroendocrinology 29:423–447

Pollock VE, Schneider LS (1990) Quantitative, waking EEG research on depression. Biol Psychiatr 27:757–780

Raadsheer FC, Hoogendijk WJ, Stam FC, Tilders FJ, Swaab DF (1994) Increased numbers of corticotropin-releasing hormone expressing neurons in the hypothalamic paraventricular nucleus of depressed patients. Neuroendocrinology 60:436–444

Raison CL, Miller AH (2003) When not enough is too much: the role of insufficient glucocorticoid signaling in the pathophysiology of stress-related disorders. Am J Psychiatr 160:1554–1565

Reul JM, Stec I, Soeder M, Holsboer F (1993) Chronic treatment of rats with the antidepressant amitriptyline attenuates the activity of the hypothalamic-pituitary-adrenocortical system. Endocrinology 133:312–320

Ribeiro SC, Tandon R, Grunhaus L, Greden JF (1993) The DST as a predictor of outcome in depression: a meta-analysis. Am J Psychiatr 150:1618–1629

Rinne T, de Kloet ER, Wouters L, Goekoop JG, de Rijk RH, van den Brink W (2003) Fluvoxamine reduces responsiveness of HPA axis in adult female BPD patients with a history of sustained childhood abuse. Neuropsychopharmacology 28:126–132

Rosenbaum JF, Fava M, Pava JA, McCarthy MK, Steingard RJ, Bouffides E (1993) Anger attacks in unipolar depression, Part 2: Neuroendocrine correlates and changes following fluoxetine treatment. Am J Psychiatr 150:1164–1168

Roy A, Pickar D, Paul S, Doran A, Chrousos GP, Gold PW (1987) CSF corticotropin-releasing hormone in depressed patients and normal control subjects. Am J Psychiatr 144:641–645

Rubin RT, Poland RE, Lesser IM, Winston RA, Blodgett ALN (1987) Neuroendocrine aspects of primary endogenous depression. Arch Gen Psychiatr 44:336

Rybakowski JK, Twardowska K (1999) The dexamethasone/corticotropin-releasing hormone test in depression in bipolar and unipolar affective illness. Journal of Psychiatr Res 33:363–370

Sachar EJ (1967) Corticosteroids in depressive illness. II. A longitudinal psychoendocrine study. Arch Gen Psychiatr 17:554–567

Schreiber W, Lauer CJ, Krumrey K, Holsboer F, Krieg JC (1996) Dysregulation of the hypothalamic-pituitary-adrenocortical system in panic disorder. Neuropsychopharmacology 15:7–15

Schüle C, Baghai T, Zwanzger P, Ella R, Eser D, Padberg F, Möller HJ, Rupprecht R (2003) Attenuation of hypothalamic-pituitary-adrenocortical hyperactivity in depressed patients by mirtazapine. Psychopharmacology 166:271–275

Sharpley AL, Cowen PJ (1995) Effect of pharmacologic treatments on the sleep of depressed patients. Biol Psychiatr 37:85–98

Steiger A (2007) Neuroendocrinology of Sleep. In: Lajtha A, Blaustein J (eds) Handbook of Neurochemestry and Molecular Neurobiology – Behavioral Neurochemistry, Neuroendocrinology and Molecular Neurobiology. Springer, Berlin Heidelberg New York, pp 897–932

Tancer ME, Stein MB, Uhde TW (1993) Growth hormone response to intravenous clonidine in social phobia: comparison to patients with panic disorder and healthy volunteers. Biol Psychiatr 34:591–595

Thase ME, Kupfer DJ, Fasiczka AJ, Buysse DJ, Simons AD, Frank E (1997) Identifying an abnormal electroencephalographic sleep profile to characterize major depressive disorder. Biol Psychiatr 41:964–973

Tsuno N, Besset A, Ritchie K (2005) Sleep and depression. J Clin Psychiatr 66:1254–1269

Tukel R, Kora K, Hekim N, Oguz H, Alagol F (1999) Thyrotropin stimulating hormone response to thyrotropin releasing hormone in patients with panic disorder. Psychoneuroendocrinology 24:155–160

Turck CW, Ditzen C, Sayan-Ayata EF (2008) Proteomic strategies for biomarker discovery: from differential expression to isoforms to pathways. In: Turck CW (ed) Biomarkers for psychiatric disorders. Springer, Berlin Heidelberg New York

Uhde TW, Stein MB, Vittone BJ, Siever LJ, Boulenger JP, Klein E, Mellman TA (1989) Behavioral and physiologic effects of short-term and long-term administration of clonidine in panic disorder. Arch Gen Psychiatr 46:170–177

Valdivieso S, Duval F, Mokrani MC, Schaltenbrand N, Castro JO, Crocq MA, Macher JP (1996) Growth hormone response to clonidine and the cortisol response to dexamethasone in depressive patients. Psychiatr Res 60:23–32

Watson S, Gallagher P, Del-Estal D, Hearn A, Ferrier IN, Young AH (2002) Hypothalamic – pituitary – adrenal axis function in patients with chronic depression. Psychol Med32:1021–1028

WHO European Ministerial Conference (2005) Mental health: fac-ing the challenges, building solutions. WHO Regional Office for Europe, Copenhagen

Zobel AW, Yassouridis A, Frieboes RM, Holsboer F (1999) Prediction of medium-term outcome by cortisol response to the combined dexamethasone-CRH test in patients with remitted depression. Am J Psychiatr 156:949–951

Zobel AW, Nickel T, Sonntag A, Uhr M, Holsboer F, Ising M (2001) Cortisol response in the combined dexamethasone/CRH test as predictor of relapse in patients with remitted depression. a prospective study. J Psychiatr Res 35:83–94

Chapter 18
DNA Biomarkers for Pharmacogenomics and Personalized Medicine

John Raelson(✉) and Abdelmajid Belouchi

Abstract Genome-wide association studies are expected soon to provide a significant increase in disease associated DNA sequences that can serve as biomarkers for diagnosis and treatment for psychiatric illnesses. These biomarkers have the potential to identify precisely defined disease phenotypes and to predict effective individually specific therapies. This chapter reviews the current knowledge of the association between DNA markers and disease phenotypes and discusses some of the anticipated problems that need to be overcome before DNA biomarkers can provide specific and sensitive diagnostic tests for both the presence of disease and the prediction of individual response to specific drugs.

Abbreviations ADHD: Attention deficit hyperactivity disorder; ADME Genes: Genes whose products are involved in absorption, distribution, metabolism and excretion of drugs; BPD: Bipolar disorder; CATIE: Clinical Antipsychotic Trials of Intervention Effectiveness; $\chi 2$: Chi-square statistic for significance of counts; DNA: Deoxyribonucleic acid; FDA: Food and Drug Administration USA; GAD: Generalized anxiety disorder; GAIN: Genetic Association Information Network; GWAS: Genome-wide association study(ies) (=WGAS); LD: Linkage disequilibrium; MDD: Major depressive disorder; NIH: National Institute of Health USA; NPV: Negative predictive value of a diagnostic test; PPV: Positive predictive value of a diagnostic test; Pr: Prevalence of a disease or phenotype in a population; RNA: Ribonucleic acid; Se: Sensitivity of a diagnostic test; SNP: Single nucleotide polymorphism; Sp: Specificity of a diagnostic test; STAR*D: Sequenced Treatment Alternatives for Depression Study; SZ: Schizophrenia; WGAS: Whole genome association study(ies) (=GWAS); WTCCC: Wellcome Trust Case-Control Consortium

J. Raelson
Genizon Biosciences, 880, McCaffrey, Ville St-Laurent, Canada QCH4T 2C7
John.raelson@teksavvy.com

C.W. Turck (ed.) *Biomarkers for Psychiatric Disorders*,
doi: 10.1007/978-0-387-79251-4_18, © Springer Science+Business Media, LLC 2008

Gene Abbreviations ABCB1: Adenosine tri-phosphate-binding cassette sub-family B member 1, chromosome 7q21.12; ADH1A: Class I alcohol dehydrogenase 1 alpha subunit, chromosome 4q23; ADH1B: Class I alcohol dehydrogenase 1 beta subunit, chromosome 4q23; ADH4: Class II alcohol dehydrogenase 4 pi subunit, chromosome 4q23; ALDH2: Mitochondrial aldehyde dehydrogenase 2, chromosome 12q24.12; AOX: Aldehyde oxidase, chromosome 2p33.1; BDNF: Brain derived neurotrophic factor, chromosome 11p14.1; BRCA1: Breast cancer 1, chromosome 17q21.311; CARD15 (=NOD2): Caspase recruitment domain 15 gene, chromosome 16q12.1; CHRM2: Cholinergic receptor muscarinic 2, chromosome 7q33; COMT: Catechol-*O*-methyltransferase, chromosome 22q11.21; CYP (=P450): Cytochrome heme protein with pigment at 450 nm absorption; very large family of heme containing proteins involved in multi-component electron transfer chains. Many are involved in drug metabolism.; CYP2E1: Cytochrome P450, family 2, subfamily E, polypeptide 1 variant, chromosome 10q26.3; DAO: D-amino acid oxidase, chromosome 12q24.11; DAT1 (SLC6A3): Dopamine transporter, (=solute carrier family 6) chromosome 5p15.33; DBH: Dopamine β-hydroxylase precursor, chromosome 9q34.2; DISC1: Disrupted in schizophrenia 1, chromosome 1q42.2; DRD1: Dopamine receptor D1, chromosome 5q35.2; DRD2: Dopamine receptor D2, chromosome 11q23.1-q23.2; DRD3: Dopamine receptor D3, chromosome 3q13.31; DRD4: Dopamine receptor D4, chromosome 11p15.5; DRD5: Dopamine receptor D5, chromosome 4p16.1; DTNBP1: Distrobrevin binding protein 1, (=dysbindin), chromosome 6p22.3; ERRB4 (=HER4): The genomic homologue of erythroblastic leukemia viral oncogene, chromosome 2q34; FMO3: Flavin containing mono-oxygenase 3 isoform, chromosome 1q24.3; GABRA2: Gamma aminobutyric acid A receptor, chromosome 4p12; GAD2: Glutamate decarboxylase 2, chromosome 10p12.1; HTR1B: 5-Hydroxy tryptamine (serotonin) receptor 1B, chromosome 6q14; HTR2A: 5-Hydroxy tryptamine (serotonin) receptor 2A, chromosome 13q14.2; 5-HTT (=SLC6A4, =hSERT): 5-Hydroxy tryptamine transporter (=serotonin symporter), chromosome 17q11.1-q12; 5-HTTLPR: 5-HTT linked polymorphic region, a repeat length polymorphism in the promoter of the serotonin transporter gene; LG72 (=G72=D AOA): D-amino acid oxidase activator, chromosome 13q33.2; MAOA: Monoamine oxidase A, chromosome Xp11.3; NOD2 (=CARD15): Nucleotide binding oligomerization domain, chromosome 16q12.1; NRG1: Neuregulin 1, chromosome 8p12; PI3K/AKT: Phosphatidylinositol 3-kinase-AKT protein substrate, cell signaling pathway, multiple genes; PPARG: Peroxisome proliferative activator receptor gamma, chromosome 3p25.2-p25.1; SLC6A4 (=5-HTT = hSER): Solute carrier family 6 neurotransmitter transporter (=5- hydroxy tryptamine transporter = serotonin symporter), chromosome 17q11.1-q12; SNAP25: Synaptosome associated protein, chromosome 20p12.2; TAS2R16: Taste receptor 16, chromosome 7q31.32; TPH1: Tryptophan hydroxylase 1, chromosome 11p15.1; UGT1A4: Uridine di-phosphate glycosyltransferase 1 family, polypeptide A4, chromosome 2q37.1

18.1 Introduction

Biomarkers are biochemicals whose presence or quantity has been correlated with some trait (phenotype), usually disease status. They may be DNA sequences, RNA molecules, proteins or secondary metabolites. The detection of a specific biomarker assists in the diagnosis and treatment of the disease. A classic example is the use of biomarkers for cardiac necrosis in guiding emergency room decisions concerning the reality of suspected cases of cardiac arrest. Blood protein levels of troponin I and C, creatin kinase-MB, and myoglobin can all indicate the presence of actual damage to cardiac muscles and confirm that infarct has in fact occurred (Ginsburg et al., 2005).

DNA sequences are unique among the four classes of biomarkers in that they are likely to be directly implicated in the etiology of the disease, whereas the other types of molecules, while possibly causative in some instances, are more likely to be consequences of the disease state. Because genetic cause and effect is complex, modified by gene interaction and environment, DNA markers may be less useful than other types of biomarkers for monitoring disease progression; however, their identification offers the possibility of understanding the biochemical basis of the disease and provides specific targets for novel effective therapies. Moreover, DNA biomarkers offer the possibility of personalized medicine, providing specific treatments for patients with appropriate genotypes; as an alternative to the "one drug fits all" model whose significant limitations are decreased efficacy or serious adverse reactions in some groups of patients (Vizirianakis, 2005; Hitchen, 2006).

Establishing the correlation between DNA sequences, or genotypes, and specific disease phenotypes has two goals: it allows the identification of novel targets for therapeutics and it can provide the basis for more exact disease diagnosis. It is in this diagnostic role that DNA sequences serve as biomarkers. DNA based diagnostics, in turn, has two aspects: diagnosis of specific disease phenotype and prediction of probable patient responses to a specific therapy.

The efficiency of DNA markers for disease diagnosis depends upon the percentage of the phenotype variation that is collectively explained by all known markers, which at the time of writing is not large for most psychiatric diseases but which promises to dramatically increase as the results for several soon to be completed genome-wide association studies (GWAS) become available. With respect to the diagnostic efficiency of DNA markers, a distinction must be made between simple genetic diseases such as fragile X associated mental retardation and complex genetic diseases such as attention deficit hyperactivity disorder (ADHD), bipolar disorder (BPD) schizophrenia (SZ), generalized anxiety disorder (GAD) or major depressive disorder (MDD). A simple genetic disease is directly caused by the presence of specific DNA polymorphisms found within a single gene in either one or two copies depending upon the disease. These disease-causing alleles are said to have a high penetrance. If a patient has the appropriate genotype then they are very likely to have the disease. Consequently, the presence of the specific disease associated alleles is a very good predictor of disease. The link between DNA polymorphisms and complex genetic disease is less direct. Multiple interacting alleles in multiple distinct genes and alternative disease pathways are likely to be involved in the etiology of these diseases and, in addition to the individual gene effects, there are very strong effects of gene-gene interaction, gene-environment interaction and purely environmental effects. For example we have identified multiple disease associated DNA markers for Crohn disease, yet we have some patients who have none of the identified alleles and some controls who have one or more of them.

The feasibility of using DNA polymorphisms as diagnostic biomarkers for the presence of complex genetic psychiatric diseases and for prediction of treatment response will be examined in this chapter. Current methods for discovery of disease correlated DNA sequences are reviewed. Some of the known DNA associations for several psychiatric diseases are discussed and possible strategies for combining individual gene-disease associations into useful diagnostic tools are also examined. Finally, strategies for pharmacogenomic studies designed to determine correlations between DNA markers and drug response are discussed.

18.2 Methods for Discovery of Disease Associated DNA Biomarkers

The currently known susceptibility genes for psychiatric diseases have been found using a number of techniques. For example, the DISC1 gene association with schizophrenia was discovered using a cytogenetic approach. A chromosomal translocation with a break point located within the gene was observed to co-segregate with schizophrenic and affective phenotypes in a large Scottish pedigree (Blackwood et al., 2001). Other discoveries have resulted from the candidate gene approach in which polymorphisms in possibly functionally relevant genes are tested for association with disease in case-control studies (for example, Shifman et al., 2002). Case-control association studies assess the statistical significance of differences in allele or genotype frequency between case and control samples for DNA polymorphisms located within candidate genes. If other extraneous factors such as population structure, age, and other phenotype differences between the cases and the controls have been carefully controlled, then these frequency differences can be attributed to genetic contribution to disease status. However, the major tool used for the discovery of disease susceptibility genes for both simple and complex genetic diseases over the last 25 years has been linkage-based genome-wide scans.

Linkage genome wide scans use family based samples, either large extended pedigrees with multiple affected members or else many small families each consisting of a pair of siblings where either both (concordant) or only one (discordant) is affected by the disease. The co-segregation of alleles for specific DNA polymorphisms with disease phenotype within families is systematically examined over a set of markers spaced across the genome. Typically, hundreds of sequence repeat polymorphisms (microsatellite markers) were tested. The location of the linked marker indicates the approximate location of the disease gene. Dozens of linkage scans have been reported for SZ, BPD, AHDH, MDD, and GAD. Candidate linkage regions have been reported on every chromosome except Y for schizophrenia, with some candidate regions mapping both to schizophrenia and to bipolar disorder (Maziade et al., 2005). Unfortunately, very few of these reported linkage regions have been translated into known disease susceptibility genes.

The major reason for the failure of linkage studies to identify complex disease genes is that family samples do not provide a sufficient number of meiotic events to produce enough genetic recombination to finely resolve the candidate region to a size small enough to contain either a single or at most just a few genes. Generally, linkage studies identify disease linked candidate regions that average several megabases of DNA in size. Such regions are large enough to contain dozens to hundreds of genes. Consequently, there has usually been a long lag time, often many years, between identification of a linkage region and identification of the disease gene. The discovery of the familial breast cancer gene, BRCA1, provides a good example. The linkage between family breast cancer and a microsatellite locus on chromosome 17q21 was reported in 1990 (Hall et al., 1990) but, despite intense international efforts, the gene was not identified until 1994. The process involved international collaboration to assemble ever larger samples for better meiotic resolution (for example, Simard et al., 1993). Eventually the gene was isolated by a brute force search for obvious mutations among the multiple candidate genes within the remaining unresolved chromosomal fragment (Miki et al., 1994). A series of probably deleterious mutations, a deletion, an insertion, a nonsense mutation and a missense mutation, were found within a previously unidentified expressed DNA sequence located within the unresolved candidate region that co-segregated with disease status in Utah kindreds. Familial breast cancer is a simple genetic disease. This approach is less successful for complex genetic diseases because the susceptibility polymorphisms involved in genes with small genetic effects tend not to be obviously disruptive mutations such as frame shifts deletions and stop codons, all of which have major genetic effects. It is much more difficult to distinguish between a disease causing mutation and a neutral polymorphism in many complex disease susceptibility genes.

A refined approach designed for faster resolution of linkage candidate regions involves saturation of the linkage region with either microsatellite or single nucleotide polymorphism (SNP) DNA markers in a sample of independent cases and controls followed by case-control association analysis across the linkage region. Resolution for association analysis is approximately 100 times greater than for linkage analysis because individuals within the same population samples are separated by many more historical meiotic events than are family members. The effect of genetic recombination in case control samples is measured as linkage disequilibrium (LD) which extends over much shorter distances than familial linkage effects providing a corresponding increase in genetic resolution. A good example of the identification of a complex genetic disease gene using this approach is the discovery of the CARD15/NOD2 gene association with Crohn disease (Hugot et al., 2001). The success of this approach for finding complex disease genes is limited, however, by the fact that the initial discovery depends upon linkage analysis which has low power to detect genes that impart a low relative risk for a disease.

Relative risk, the probability of developing a disease given the fact that one carries a specific genotype or allele, is a measure of the strength of a genetic effect. Very moderate relative risks in the range of 1.25–2.0 are typical of known susceptibility genes for complex diseases. Neil Risch and Katherine Merikangas (Risch and Merikangas, 1996) in a seminal paper, demonstrated that thousands of sib-pairs (linkage analysis) are required to detect a gene with a relative risk of only 2 whereas association studies can detect this relative risk using only a few hundred subjects, an order of magnitude fewer. The converse is that, for a given fixed number of subjects, association studies will detect smaller relative risk genes than linkage analysis.

It has long been recognized that an entire genome-wide scan based upon association analysis would be more powerful than a linkage based genome-wide scan for discovery of genes for complex diseases (Carlson et al., 2004; Lawrence et al., 2005); however, sufficiently dense mapping marker sets were not available until very recently. A new mapping paradigm, genome-wide association studies (GWAS) or whole genome association studies (WGAS) has recently emerged to address the problems of complex disease gene discovery. Hundreds of thousands (soon to be 1 million at the time of writing) of SNP markers, arranged across the genome, are tested for significant differences in allele frequency between case and control samples. The detection of a statistically significant association implies that the associated marker is either causative of disease or else is correlated through LD with the true nearby causative polymorphism.

Very high marker densities are required for GWAS because the effect of genetic recombination among population samples LD extends over much shorter distances than linkage. The human HapMap project (International HapMap Consortium, 2005) has mapped the patterns of LD across the genome in Caucasian, Asian and African populations. This has allowed the development of genome wide SNP maps designed to adequately test known LD regions for association with disease (Magi et al., 2007).

The results of the first systematic GWAS are just being reported in late 2006 and 2007. The most advanced mapping has occurred for Crohn disease (Duerr et al., 2006; Hampe et al., 2007; Libioulle et al., 2007; Rioux et al., 2007; Parkes et al., 2007). Prior to these recent publications there were two unequivocally accepted Crohn disease loci, CARD15/NOD2 and IBD5 on 5q31.1. These associations took decades of work to discover. Subsequent to the GWAS publications, there are now at least five additional accepted replicated genes reported over a period of a few months.

Genome-wide association studies are also being undertaken for psychiatric diseases The Genetic Association Information Network (GAIN) Initiative, http://www.fnih.org/GAIN2/home_new.shtml/) is a public/private collaboration sponsored by NIH/NGRI for GWAS of several complex diseases including Schizophrenia (PV Gejman, Northwestern University); ADHD (SV Faraone, State University of New York) and bipolar disorder (JR Kelsoe UCSD). The results for ADHD are expected to be released in July 2007.

The Wellcome Trust Case Control Consortium (WTCCC http://www.wtccc.org.uk/) has just released (at the time of writing) the results of a GWAS for Bipolar disorder (The Wellcome Trust Case Control Consortium, 2007). An association with a *p* value of significance of 6.29×10^{8} was observed for SNP, rs420259, located within an intron of the "partner and localizer of BRCA2" (PALB2) gene. This finding has not yet been replicated in a second population. Pamela Sklar from Harvard University described an ongoing collaboration, between US and UK groups, for a GWAS of BPD at the 15th World Congress of Psychiatric Genetic Meetings in 2006, as did Michael Boehnke from the University of Michigan (abstracts, 2006). No results from either study have yet been published, at the date of this writing.

The published results for GWAS for psychiatric genetic diseases are still too few to indicate whether these studies will achieve the same level of success as those for Crohn disease. Our group at Genizon Biosciences has completed seven unpublished GWAS to date, including studies for ADHD and schizophrenia which have identified multiple associations significant at genome-wide levels. In the course of these studies we have learned several lessons on increasing the power to detect true positively associated genes. For example, we have learned that using haplotype association in addition to single tag SNP association results in increased power for gene detection in low LD regions, even when using as many as 300,000 tag SNPs. We have also discovered that some associations with modest *p* values for significance of association that do not meet strict Bonferroni requirements for genome-wide significance may nevertheless be replicated in second populations and can dramatically increase in significance in combined population samples We have also observed that the effect of regional population stratification, that is, mismatch in geographic origins between cases and controls, not just differences in ethnicity, can be a serious problem that is generally underestimated in GWAS.

18.3 Some Known Associations Between Psychiatric Disease and DNA Sequences

Perhaps on an average, a half dozen or fewer susceptibility genes have been unequivocally identified for each of the major psychiatric diseases. There are many reports of candidate gene association studies in the literature, the majority of them from under-powered studies that report very modest *p* values for significance of association and that are often not replicated in subsequent studies. Recently, meta-analyses of multiple studies have begun to appear that confirm or fail to confirm certain reported associations (for example, Barnett et al., 2007; Naoe et al., 2007).

Many more genes than will be mentioned here have been reported to be associated with each disease in the literature; however many of these reported associations have not been well replicated by other groups. The lists given below represent a conservative view of the numbers of genes that have been well replicated and that are generally accepted as truly associated genes by the research community. Whether others should be added to this list as well is certainly a valid debatable point. The essential point of this review is that the currently known genes associated with the major complex diseases do not account for the majority of the genetic risk factor for the diseases and, thus, are insufficient for disease diagnosis. There are many additional susceptibility genes that remain to be discovered, perhaps by the emerging GWAS approach.

18.3.1 Schizophrenia

Many genes have been reported to be associated with schizophrenia; however many of these associations have been poorly replicated in other studies. Six well-replicated genes that are generally accepted as being true positive associations are reported here:

1. DISC1 (disrupted in schizophrenia 1, chromosome 1q42.2) This gene encodes a peri-nuclear protein that participates with multiple partners such as the NUDEL proteins to form complexes involved in membrane localization of receptors, in maintenance of cytoskelatal structure and in intracellular transport during neurite growth and cortical development. The DISC1 association was identified through a chromosomal translocation, t(1:11)(p42.1;q14.3) that co-segregates with schizophrenia and affective phenotypes in a Scottish kindred (Blackwood et al., 2001)
2. NRG1 (neuregulin 1, chromosome 8p12): This gene encodes a glycoprotein ligand for the NEU/ERBB2 tyrosine kinase receptor which, among other functions, is involved in neuronal development. This gene was discovered through linkage analysis, using Icelandic families, followed by association analysis to resolve the location (Stefansson et al., 2002).
3. COMT (catechol-*O*-methyltransferase, chromosome 22q11.21): This gene encodes a protein that transfers a methyl group from s-adenosylmethionine to the catecholamine neurotransmitters, dopamine, adrenaline and noradrenaline, leading to their enhanced degradation. The association was discovered using an association test of a candidate gene (Shifman et al., 2002).
4. DTNBP1 (distrobrevin binding protein 1, dysbindin, chromosome 6p22.3): This gene encodes an ubiquitously expressed protein that binds to members of the dystrophin associated protein complex involved in trafficking lysosomal related organelles to localized positions within presynaptic vesicle membranes and adjacent microtubules. The association was discovered by candidate gene testing within a previously reported linkage region (Straub et al., 2002).
5. LG72 (G72 or DAOA) and DAO (D-amino acid oxidase activator chromosome 13q33.2 and D-amino acid oxidase chromosome 12q24.11): LG72 encodes a protein that activates D-amino acid oxidase, which in turn degrades D-amino acids including D-serine which is a co-agonist of the NMDA class of glutamate receptors. The association was discovered in a sample of schizophrenic patients

from the Quebec founder population using association analysis across a linkage region on chromosome 13 that had been reported by multiple studies. A yeast two hybrid system search for interacting proteins identified DAO which was also found to be associated with schizophrenia (Chumakov et al., 2002).
6. ERRB4 (HER4) (the genomic homologue of erythroblastic leukemia viral oncogene, chromosome 2q34): This gene encodes a tyrosone kinase receptor that is a member of the epidermal growth factor receptor family. This association was discovered as a result of a candidate gene association test based upon the known fact that ERRB4 is regulated by neuregulin (Norton et al., 2006; Silberberg et al., 2006; Nicodemus et al., 2006).

A summary of the details of the associations for these genes is presented in Table 18.1, which provides information on the populations that were tested, the specific SNPs studied, the odds ratios or relative risks for strength of association and P_{excess} values (Bengtsson and Thomson, 1981). Note that if non-phenotyped controls are used in the case control study, which is typically the case for low prevalence diseases such as schizophrenia, then the allele frequencies in controls become an estimate of the population frequency of the disease allele so that the odds ratio becomes:

$$[P(\text{allele 1} \mid \text{cases})/\text{freq allele 1}]/[P(\text{allele 2} \mid \text{cases})/\text{freq allele 2}] \quad (18.1)$$

which is an estimate of the true population relative risk and that P_{excess} becomes:

$$\frac{[(\text{Freq allele 1} \mid \text{cases}) - (\text{Population Freq allele 1})]}{(1 - \text{Population Freq allele 1})} \quad (18.2)$$

which is an estimate of the population attributable risk (PAR) or the percentage of cases in which the gene is involved (Levin and Bertell, 1978). PAR is not additive due to gene interaction.

DISC1 and NRG1, particularly in the Chinese population, have fairly strong relative risks, which is consistent with their discovery in family studies, while COMT and dysbindin have more modest relative risks, typical of many genes associated with complex genetic diseases. Most of the other associated genes for psychiatric diseases impart similarly modest relative risks for disease, typical of complex disease genes in general. The message provided by Table 18.1 is that the known schizophrenia associations do not explain a large amount of the genetic variation of the disease.

18.3.2 Bipolar Disorder

Several detailed reviews of the genetics of bipolar disorders have recently been published. Carter (2007) provides an exhaustive overview of putative bipolar genetics using a functional systems approach. He has organized the one hundred or so genes

Table 18.1 Some characteristics of SZ disease gene associations

Gene	Population	#cases #controls	Poly-morphism	OR or RR	$P_{exces\ \%s}$	References
DISC1	Scotland	677–648	2-marker haplotype (CA) for rs751229 rs3738401	RR 5.04	21.6	Zhang et al. (2006)
	Japan	532–519	−274 G > C = rs3738398	NS	NS	Kockelkorn et al. (2004)
NRG1	Scotland	609–618	7-marker SNP haplotype	RR 1.8	4.6	Stefansson et al. (2003)
	Iceland	402–394	Same 7-Marker Haplotype	RR 2.1	7.4	
	China	231–271	Micros atellite markers	RR 3.1	8.0	Li et al. (2004)
COMT	Ashkenazi Jewish	708–2849	3-marker haplotype SNPS rs165688 rs737865 rs165599	OR 1.46	32.0 F 13.5 M	Shifman et al. (2002)
	Korea	320–379	rs6267 A22/72S	OR 1.8 Ser	9.7	Lee et al. (2005)
DTNBP1	Central Europe	125 trios	6-marker Haplotype	Trans/Non Trans ratio 1.24		Schwab et al. (2003)
	UK	708–711	3-marker haplotype	OR 1.4	5.9	Williams et al. (2004)
LG72	French Canadian	213–241	Multliple SNPs	OR 1.46	20.0	Chumakov et al. (2002)
	Russia	183–183		1.42	18.4	
ERRB4	Ashkenazi Jewish	59–130	G-G-A haplotype for rs707284 rs 839523 rs7598440	OR 2.18	28.0	Silberberg et al. (2006)

that have been reported by at least one group as being associated with bipolar disorders into a few relevant biological pathways such the PI3K/AKT signaling network, the brain derived neurotrophic factor (BDNF) signaling pathway and the endoplasmic reticulum stress pathway. Carter's review is intentionally inclusive for the sake of building a comprehensive overview of the systems biology of bipolar disorders that stresses the argument that all reported genes fall into relatively few pathways; however, this argument may be, at least in part, circular as many of the candidate genes were chosen for testing based upon prior knowledge of the biochemical pathways in which they occur. Most researchers accept far fewer than 100 genes as replicated BD associations. Two other recent reviews (Kato, 2007; Farmer et al., 2007) were much more conservative in accepting what is an established BD association. Indeed Kato states that, "To date, no causative gene or genetic risk factor has been identified for bipolar disorder or depression."

Several of the genes associated with schizophrenia have been reported to be also associated with BD. The association of the LG72 gene with BD has been reported by multiple studies (for example, Schumacher et al., 2004; Chen et al., 2004). While the association with LG72 is the most robust of all reported BD associations, associated haplotypes and locations of maximum association are not always consistently located within the G72-G30 region (Detera-Wadleigh and McMahon, 2006). DISC1 has also been reported to be associated with bipolar disorders (Thomson et al., 2005). This association has been replicated, albeit at a very modest p value, and a statistically significant reduction in DISC1 mRNA levels was demonstrated in lymphoblasts from affected individuals (Maeda et al., 2006). A modest association of NRG1 with bipolar 1 has also been reported (Green et al., 2005).

Among the hundreds of other genes reported to be associated with BD, the best known associations include the dopamine receptors, DRD1, 2, 3 and 4, the dopamine transporter gene, DAT1, the 5-HTTLPR polymorphism in the serotonin transporter gene, SLC6A4, the serotonin receptor gene, HTR2A, the tryptophan hydroxylase 1 gene, TPH1, the monoamine oxidase A gene, MAOA, and the brain derived neurotrophic factor gene, BDNF; however, none of these associations have been consistently replicated.

18.3.3 Attention Deficit Hyperactivity Disorder

Dozens of gene associations with ADHD have been reported in the literature, the most consistent being associations with the dopamine transporter gene, DAT1 (SLC6A3) (Cook et al., 1995), the dopamine receptor genes, DRD4 and DRD5 (Barr et al., 2001; Payton et al., 2001; Daly et al., 1999), the dopamine β-hydrolyase gene, DBH (Smith et al., 2003), the serotonin transporter gene, SLC6A4, (5-HTTLPR promoter mutation) (Manor et al., 2001), the serotonin receptor 1B gene, HTR1B (Li et al., 2005), and the synaptosome associated protein gene, SNAP25 (Barr et al., 2000). The SNAP25 protein forms part of the protein complex attached to the presynaptic plasma membrane that is involved in the regulation of neurotransmitter release. Several meta-analyses of these associations have been undertaken (Faraone et al., 2001, 2005; Lowe et al., 2004). Only these eight genes were found to be significantly associated with ADHD across these meta-analyses.

18.3.4 Major Depression Disorder (MDD) and Generalized Anxiety Disorder (GAD)

MDD and GAD are highly prevalent psychiatric disorders likely to be more heterogeneous than schizophrenia, BD or ADHD. Correspondingly research into genetic associations is not as advanced. Several associations, the S allele of 5-HTTLPR (SLC6A4) (Caspi et al., 2003), BDNF (Frodl et al., 2007) and plexin-A, PLXNA2, a semaphorin receptor involved in axon path finding during neuron development (Wray et al., 2007) have been reported for MDD but none of them are well replicated. The 5-HTTLPR polymorphism has also been reported to be associated with GAD (Battaglia et al., 2005) as has COMT (McGrath et al., 2004) and mouse functional studies have implicated the hormone neuropeptide Y. Again these associations are not well replicated.

18.3.5 Alcoholism

The major genetic research on alcohol dependence has been conducted within the Collaborative Study on the Genetics of Alcoholism (COGA). Several gene associations with alcoholism have been replicated. These are:

1. ADH1A, ADH1B, and ADH4 (alcohol dehydrogenase, chromosome 4q23) and ALDH2 (aldehyde dehdyrogenase, chromosome 12q24.12). Alcohol is first converted to acetaldehyde in the liver by both alcohol dehydrogenase and by cytochrome P450 (CYP2E1) (Gemma et al., 2006) and then acetaldehyde is very quickly converted to acetate by aldehyde dehydrogenase. Some very low activity ALDH2 and ADH1B mutations result in aldehyde build up making drinking of alcohol very unpleasant and are therefore protective against alcoholism (Crabb et al., 1989). These variants are rare in Caucasians and Africans but more frequent in Asians. More recently, several SNPs within ADH4, ADH1A and ADH1B have been found to be associated with alcoholism in Caucasians and African Americans (Edenberg et al., 2006).
2. GABRA2 (GABA A receptor, chromosome 4p12). This is a receptor for the neurotransmitter, γ-aminobutyric acid A. It is a ligand gated chloride channel which is a target for the benzodiazepine drugs. This association has been replicated in several studies (Edenberg et al., 2004; Covault et al., 2004; Lappalainen et al., 2005; Fehr et al., 2006).
3. CHRM2 (Cholinergic receptor muscarinic 2, chromosome 7q33). The protein product of this gene is a G protein coupled receptor for acetylcholine. This

association has been reported in two independent studies (Wang et al., 2004; Luo et al., 2005).
4. TAS2R16 (Taste receptor 16 chromosome 7q31.32). The protein encoded by this gene is a G protein coupled receptor found in the tongue that is responsible for bitter taste. An association between this gene and alcoholism has been replicated (Hinrichs et al., 2006; Wang et al., 2007). The risk allele is more frequent in African than Caucasian populations.
5. GAD2 (Glutamate decarboxylase 2, chromosome 10p12.1). This gene encodes an enzyme that is involved in the production of the GABA neurotransmitter. It has been reported to be associated with alcoholism in two independent populations (Lappalainen et al., 2007).

DRD2 and 5-HTTLPR have also been reported to be associated with alcoholism but these have failed to replicate in many studies (Dick and Bierut, 2006).

18.3.6 Conclusions

The current state of disease gene identification for complex psychiatric diseases is characterized by hundreds of reported associations identified from studies, which have mostly used too small a sample size for adequate statistical power. Nevertheless, a consensus across multiple studies has emerged accepting perhaps two dozen genes, involved in neurotransmission or neuron development that are generally accepted as true positive associations. It is anticipated that this number will dramatically increase as the results of GWAS become available in the near future. The GWAS studies are expected to produce de novo associations and will also offer the possibility of replication of already published results. This is an additional major advantage as each systematic genome-wide association study will serve as a systematic large sample replication test for the plethora of poorly replicated, under powered, candidate gene associations reported in the literature. It is expected that these studies will help assemble an expanded picture of consistently associated disease genes.

18.4 Diagnostic Tests Using DNA Biomarkers

For disease associated DNA polymorphisms to be useful as biomarkers, they must be efficient in predicting disease phenotypes, disease sub-phenotypes and response to therapy. The spectrum of psychotic disorders is, in many respects, a continuum (Lake, 2007) as is indicated by both the shared susceptibility genes and by co-morbidity within families, for schizophrenia schizoaffective disorder and bipolar disorder. Once a comprehensive identification of susceptibility genes is achieved, combinations of specific associated polymorphisms could be useful for predicting organically distinct sub-phenotypes of psychosis; however, this may be mainly of academic interest. The practical goal of such fine molecular based diagnoses is to serve as a guide for appropriate effective treatment that is designed to correct the root biochemical imbalance leading to the disease.

For a diagnostic test to be useful as a predictor, it has to have a relatively high level of specificity and sensitivity (Altman and Bland, 1994). Unfortunately, single associated genes for complex genetic diseases have neither. The diagram below illustrates the concepts of specificity and sensitivity for a simple diagnostic test. The test is the presence or absence of a known associated disease gene polymorphism or a combination of known polymorphisms in a second case control test sample that is independent of the sample in which the association was discovered. In the case control diagram below, *A* is the number of positive results (associated polymorphisms) within the cases, *B* is the number of positive results within controls, *C* is the number of negative test results (other non-risk polymorphisms of the gene) in cases and *D* is the number of negative results in the controls.

	Counts	
Test value	Cases	Controls
+	*A*	*B*
	C	*D*
Total	*A* + *C*	*B* + *D*

The sensitivity of the test is the frequency of a positive test result within the cases (=*A*/(*A* + *C*)) and specificity is the frequency of a negative test result within the controls (=*D*/(*B* + *D*)). Formally, sensitivity is defined as the probability of a positive test result given a disease (=P(+ | case)) and sensitivity is the probability of a negative result given no disease (=P(− | control)). The results of the association study by Sagiv Shifman and co-workers (Shifman et al., 2002) for the association of two marker haplotypes for SNPs rs737865 and rs165688 found within the COMT gene with schizophrenia can serve as an example.

	% Frequency	
Test value	Cases	Controls
+=G-G haplotype	43.61	37.71
−=other haplotypes	56.39	62.29
Total	100.00	100.00

A test consisting of the presence or absence of the G-G haplotype for these two SNPs would have a sensitivity of 43.61% and a specificity of 62.29%; however sensitivity and specificity are not the primary measures of interest for a diagnostic test. Sensitivity is the probability of a positive result given a disease. We really want to know the positive predictive value of the test (PPV) which is the probability of disease given a positive result. Likewise the negative predictive value (NPV) is the probability of no disease given a negative test result. Formally, PPV = P (disease (case) | + test result) and NPV = P(no disease | − test result). Bayes theorem can be applied to solve these probabilities using the sensitivity and specificity of the test and the known prevalence of the disease (Pr).

The formula for calculating PPV is:

$$PPV = SePr/[(SePr) + (1 - Sp)(1 - Pr)] \quad (18.3)$$

The formula for NPV is:

$$NPV = Sp(1 - Pr)/[(1 - Se)Pr + Sp(1 - Sp)] \quad (18.4)$$

where Se is sensitivity, Sp is specificity and Pr is disease prevalence. Using a value of 1% prevalence for schizophrenia and the Shifman data, we get a positive predictive value of just 1.15% (compared to 1.00% if we just guessed whether someone had schizophrenia using 1:99 odds) and a negative predictive value of 99.09.

Such a test has no practical value; however, it is not likely that a diagnostic test would be used in this way to randomly test people off the street for the presence or absence of schizophrenia. It is more likely that a diagnostic test would be used on suspected cases of disease in order to increase the odds of accurately confirming or ruling out disease. The prevalence of disease among this group of people may be more in the order of 50%. For 50% prevalence, the positive predictive value of the Shifman data test becomes 53.62% and the negative predictive value becomes 52.48%. This is still not a useful improvement over the 50:50 odds of just guessing. It is evident that higher rates of sensitivity and specificity must be achieved in order to produce a useful diagnostic test from disease gene associations.

The way to improve specificity and sensitivity for the diagnosis of a complex genetic disease is to identify all the major genes associated with the disease and to then determine the effects of their interactions on the incidence of the disease. The soon to be completed genome-wide association studies offer the best promise for achieving this but a simple list of most of the truly associated genes is not, by itself, sufficient for a diagnostic test. Further analyses would be required to elucidate the effect of gene interaction in determining a phenotype.

A phenotype has several etiological components, the major single gene effects being just one of these. It is also likely that there are very strong gene interaction effects (epistasis), epigenetic (gene expression) effects, gene by environment interactions and purely environmental effects. Perhaps the most underappreciated of these are the gene by gene interactions. It is relatively straight forward to test gene interaction among genes with known individual effects but it is likely that there will be susceptibility genes that are only detectable in their interactions with no apparent individual effects. These will be very difficult to identify, as the number of possible interactions among the markers in a GWAS map is enormous. Recent mouse experiments (Hoover-Plow et al., 2006) have clearly indicated the importance of genetic background in determining the effects of specific genes associated with thrombosis. These varied considerably in their effect, depending upon the genetic background in which they occurred in chromosomal substitution lines and their crosses.

One classic tool for looking at multiple genetic predictors of a phenotype has been logistic regression (McCullagh and Nelder, 1999) in which a regression of a set of independent variables (presence of identified gene polymorphisms) and their interactions is used as a predictor of the dependent variable, case or control status. The problem with this approach is that the interaction terms become cumbersome and the analysis time becomes untenable for more than a few primary variables (genes). Also the primary effects (individual gene effects) must be known before the interactions can be modeled and tested

Examples of alternative methods for addressing the difficult problem of detecting gene interactions include Random Forests (Breiman, 2001), neural networks (Haykin, 1999) and multivariate correspondence analysis (Benzecri, 1992; Greenacre, 1993). We are currently exploring all of these approaches.

The Random Forests algorithm uses a random collection of classification trees where N individuals are classed into either case or control status, for example, or into finer sub-phenotype classes, based upon a set of m input variables (SNP polymorphisms) out of a total of M (all SNP alleles in the GWAS) possible input variables. The number of m input variables used for each classification tree is set by the user. A very large number of random classification trees are generated and tested for accuracy of classification against the remaining ("out of bag") individuals who were not used to construct the specific tree. The overall best predictors are voted on over the entire random forest of classification trees. The method is unbiased and powerful for detection interactions between SNPs that need not have been identified for their individual effects; however, the computational load is very heavy, especially for the hundreds of thousands of SNP markers used in current GWAS, but possible if sufficient resources are devoted to it.

Artificial neural networks (ANN) are computing systems consisting of a collection of processing elements that are multiply connected in parallel with an adaptive interaction between elements. There are multiple inputs (genotypes) into the network of elements with only a few output states (case or control status for example). A learning data set is used to train the network on the appropriate weights to give to the input variables to produce the given output states. The exact details of multiple input interactions are determined implicitly by the system which does not require explicit formulations by the user. The trained network is then employed, to predict case control status on an independent data set. The trained network analysis is the basis of the diagnostic test. The analysis predicts case or control status (disease or no disease) but the details of specific genes and gene interaction remain unknown. Software programs for running ANN are available from The MathWorks, http://www.mathworks.com/products/neuralnet/description1.html. We have experimented with neural networks for gene mapping in simulated data sets but have not yet attempted their use on a full dense GWAS.

Correspondence analysis is a multivariate technique that is analogous to multivariate factor analysis but which is adapted to discrete variables (case control status and marker genotype) rather than to continuous variables. The case control data matrix for all SNPs is formulated so that the cumulative frequency of all cells of the case control table sum to 1. Singular value decomposition is used to calculate the Eigen vector and Eigen values of the case control matrix and the principal components for the marker variables. Markers are then clustered according to their contribution to the total inertia (explanation of case control status) over all principal components in order to produce groupings of markers that together best discriminate between case and control status. Case and control status can also be further subdivided into specific sub-phenotypes. Similar to the other methods, specific marker interactions do not need to be specified *a priori*. We have experimented with correspondence analysis using hundreds of thousands of input genotypes and have had promising results for building a predictor of Crohn disease.

It may prove very difficult to completely deduce gene interactions within population based samples. In this respect, the large pedigree disease collections formerly used for linkage analysis are likely to become very important again, once the individual gene mutations are identified using population-based samples. These families could be used to study the patterns of simultaneously co-segregating known multiple disease alleles and their relation to disease phenotype within the pedigrees.

Complex genetic diseases such as schizophrenia and bipolar disorder have a heritability of approximately 70–80%. This implies that, even if the entire genetic component of the disease were to be determined, there would still be a large effect due to environment and environment and gene interaction that is not predictable using genetic information alone. It is improbable that both sensitivity and specificity of over 95% can be achieved using DNA biomarkers as predictors of complex disease status. Seventy-five to 80% sensitivity and specificity seems like a more reasonable target. In the example above, if Shifman and coworkers had been able to advise a multi-genic predictor of disease that achieved both 75% sensitivity and 75% specificity, they would have a test with 75% positive and negative predictive value for a partially phenotyped patient with only a 50% chance of being affected. This represents a definite improvement over guessing and illustrates how such tests might be used as methods to significantly increase the odds of correct disease diagnosis. This discussion has attempted to indicate the scope of the effort that will be required for the construction of a diagnostic test for disease status. Fortunately, we now have the necessary tools in the form of GWAS and advanced analyses to begin to address this problem.

18.5 Pharmacogenomic Studies

There exists no optimal therapy for schizophrenia or other psychiatric disorders that controls psychotic symptoms and reduces cognitive impairment without any adverse side effects in all patients. It is generally felt that antipsychotic drug treatment has improved with the advent of the second generation antipsychotics but this view has been recently challenged by the results of the Clinical Antipsychotic Trials of Intervention Effectiveness (CATIE) study (Keefe et al., 2007). The CATIE trials were designed to assess the long term efficacy of the newer atypical antipsychotic drugs, including olanzapine, quetiapine, risperidone and ziprasidone compared to the older typical antipsychotic perphenazine; however, the trials failed to detect a major difference in discontinuation rate. The discontinuation rate of the different drugs varied between 62% for olanzapine to 82% for quetiapine. The discontinuation rate for perphenazine was 75%, due mostly to the presence of extrapyramidal symptoms. This rate was not significantly different than the discontinuation rate for olanzapine, which was due mostly to excess weight gain (Lieberman et al., 2005). The second generation drugs also have other serious side effects besides excessive weight gain and progression to type 2 diabetes in some patients, including somnolence, postural hypotension, dizziness, and a sense of sedation (Gorwood, 2006; Muller and Kennedy, 2005). The promise of whole genome association studies to provide targets leading to more effective drugs with fewer side effects is not likely to bear fruit for several more years; however, association studies can be used immediately to detect the genetic variation that is involved in determining drug response to current drugs and to those in the process of being approved.

The study of the interaction between specific genotype and treatment response has been given the terms pharmacogenetics or pharmacogenomics. Generally the distinction is that pharmacogenetics examines the interaction between drugs and genotypes for well established drug metabolizing enzymes and has a longer history, whereas the term pharmacogenomics implies using newer genomic approaches to discovery of additional genes affecting drug response (Kalow, 2005).

The problem of pharmacogenomics is similar to that of disease gene discovery, the main difference being simply one of phenotype. In one case the phenotype is the disease status and in the other case it is the reaction to a specific drug; however, there remain other important differences. In both cases two approaches are possible. One approach tests a set of suspected candidate genes for association with response phenotype, the second systematically looks for unknown associations without any prior assumptions concerning the genes likely to be involved. The major difference between discovery of disease genes and discovery of drug response genes is that the candidate gene approach has proven far more successful for discovery of drug response genes (Goldstein, 2005). This is because the target biochemical pathway of the drug is usually known and because the genetic variation of various drug metabolizing enzymes is well characterized. There is a very good chance that a positive association to drug response will be found in one of these two categories. Goldstein and his co-workers (Goldstein et al., 2003) examined the literature for successfully replicated associations between genes and drug response. Twenty-one of a total of 42 identified genes included either the specific drug target gene, another gene within the same biochemical pathway, or a gene involved in the drug's metabolism. Thirteen of the 21 associations were located within drug metabolizing enzymes such as cytochrome P450 genes. Accordingly, specialized DNA genotyping chips designed to test an array of known gene polymorphisms in the so called ADME genes (involved in absorption, distribution, metabolism and excretion of drugs) are now or will very soon be (at the time of writing) commercially available for pharmacogenomic studies (Ahmadi et al., 2005; Illumina, 2007, de Leon, 2006).

Despite the successful identification of multiple associations for ADME genes, these genes do not completely explain the genetics of drug response. The CATIE study included a pharmacogenomic aspect (Goldstein and Sullivan, 2006). Subjects were genotyped for a panel of candidate genes including the dopamine, serotonin, glutamate, γ-amino butyric acid, and acetyl CoA receptors, a panel of CYP P450 genes, and the AOX, FMO3 ABCB1 and UGT1A4 drug metabolizing genes. No significant associations with efficacy were found to any of these ADME genes.

The Sequenced Treatment Alternatives for Depression (STAR*D) study is another recently conducted psychiatric clinical trial with a pharmacogenomic aspect. Over a thousand subjects suffering from MDD were treated with the selective serotonin uptake inhibitor, citalopram. A collection of 768 SNP markers found within 68 candidate genes that are plausibly involved with monoamine neuron transmission but not involved in drug metabolism were analyzed for an association with treatment response. Genetic polymorphisms within the serotonin transporter SLC6A4, believed to be the target of citalopram, were not associated with a positive response (Kraft et al., 2007); however, an intronic SNP, but not two coding SNPs, within the post-synaptic Serotonin 2A, receptor gene (HTR2A) was highly associated with a positive response. None of the other 68 candidate genes were significantly associated with drug response and the association with HTR2A did not account for all the variability in response (McMahon et al., 2006).

The failure of the CATIE pharmacogenomic candidate gene study to identify the genes involved in the efficacy of antipsychotic drug treatment and the only limited success of the STAR*D candidate gene study suggest that there may be real value in performing whole genome association studies to identify additional drug response genes to provide biomarkers for psychiatric drug therapy. The major consideration in planning such a study is the sample size required to achieve adequate power for gene discovery. In the absence of a full empirical power study we can provide a rough answer to this question by examining the effect of sample size on $-\log_{10}p$ value for significance of association in an allelic 2 × 2 χ^2 association table of the form:

Allele	Cases	Controls
1	A	B
2	C	D
Total	A + C	B + D

Figure 18.1 shows the effect of sample size and relative risk imparted by the associated gene allele on the $-\log_{10}p$ value for association significance of this type of test. The x-axis of Fig. 18.1 shows total sample size for an equal number of cases and controls. For example, a value of 500 corresponds to 250 cases and 250 controls. The y-axis corresponds to the corresponding $-\log_{10}p$ value for association if the tested marker is the causative polymorphism. The genetic model employed to produce the data in Fig. 18.1 implicitly assumes additive allelic interaction and explicitly assumes a disease prevalence of 1% and the use of non-phenotyped, general population samples as controls. The population frequency for the disease allele was set at 25%.

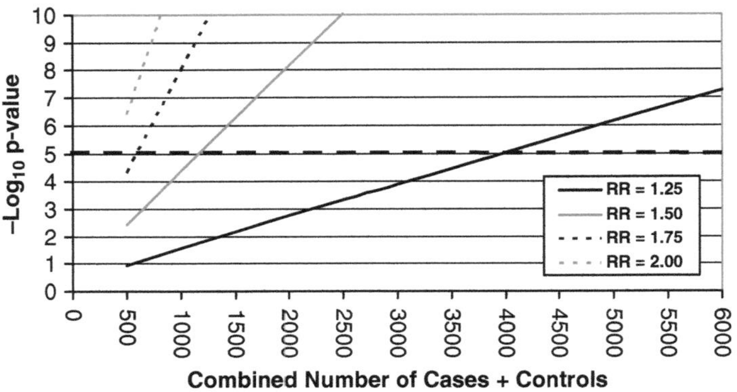

Fig. 18.1 Effect of sample size on $-\text{Log}_{10}p$ value for association in a 2 × 2 allelic case-control association table

These data do not consider the significance of association of an SNP that is only in partial LD with the causative SNP and do not consider correction for multiple testing. We have observed that using conservative Bonferroni cutoffs for genome wide significance can result in excessive type 2 error, that is, false rejection of true associations. One reason for this is that the Bonferroni cutoff assumes independence of all tests which is not the case due to LD between markers. We prefer to use a genome wide significance test that permutes case control status multiple times, each time repeating the entire genome wide scan under the null hypothesis, in order to determine the distribution of the ordered ranked *p* values observed under the null hypothesis to which observed nominal ranked *p* values can be compared. For example the Bonferroni cut off would be $-\log_{10}$ p = 7.00 for a 500 thousand SNP scan but we have observed genome-wide significant nominal *p* values at 10^{-4} for this map density, depending upon the ordered nominal *p* values that were observed in the random permutation tests. We have replicated a locus in a second independent population for Crohn disease that displayed a nominal *p* value of only 10^{-4} in the initial genome wide scan. The nominal *p* value for association at this locus dropped to below 10^{-8} in the combined population samples. Generally false negatives are more serious than false positives in genome-wide association studies. False positives can always be eliminated in the back up replication study; however, falsely rejected associations tend to be lost from the study and from further consideration.

Accordingly the cutoff line for rejection of the null hypothesis has been set at 10^{-5} in Fig. 18.1. It is apparent from Fig. 18.1 that relative risks as low as 1.5 can readily be detected at $-\log_{10}p$ values above this nominal log *p* value using a reasonable sample size 600 cases and controls and relative risks above 1.75 can be detected using 250 cases and controls; however, detection of lower relative risks would require much larger sample sizes. Relative risks of 1.25 or lower have been observed for associations with complex genetic disease phenotypes, (for example PPARG with diabetes, Altshuler et al., 2000). It is entirely possible, but not certain, that relative risks for drug response might be generally higher than those for disease phenotype. If the goal of the study is to detect all genes that impart moderate relative risks (as low as 1.25), which seems like a prudent, achievable goal, then a sample size of at least 2,000 cases (negative responders) and 2,000 controls (positive responders) would be required. The CATIE study examined a total of 1,493 schizophrenia patients but these were randomly assigned five drugs. Therefore there were approximately only 300 patients for each drug. Of these approximately 75% were poor responders (~225) so that there were, at most, 75 good responders for each drug. These sample sizes are definitely too small for a powered GWAS for drug response genes. It would be difficult, but certainly possible, to assemble sufficiently large sample sizes for strictly controlled randomized pharmacogenomic GWAS clinical trials. Perhaps such trials could be organized by the large publicly funded research organizations such as NIH or FDA working with the pharmaceutical industry. On the other hand, genotyping subjects for smaller sample sizes in current clinical trials may still be worth-while to detect any higher relative risk (≥1.5–1.75) genes that may be present.

18.6 Conclusion

DNA polymorphisms have a potential to serve as predictive biomarkers for disease phenotype and drug response; however the current state of knowledge of gene-phenotype association is not adequate for this purpose. Fortunately the tools to achieve the knowledge of gene-disease association necessary for this task have become available. It is anticipated that the next two years will witness a great advance in this field.

References

Abstracts for 15th World Congress of Psychiatric Genetics (2006) Am J Med Genet B Neuropsychiatr Genet141:683–824

Ahmadi KR, Weale ME, Xue ZY, Soranzo N, Yarnall DP, Briley JD, Maruyama Y, Kobayashi M, Wood NW, Spurr NK, Burns DK, Roses AD, Saunders AM, Goldstein DB (2005) A single-nucleotide polymorphism tagging set for human drug metabolism and transport. Nat Genet 37:84–89

Altman DG, Bland JM (1994) Statistic notes: diagnostic tests 1: sensitivity and specificity. BMJ 308:1552

Altshuler D, Hirschhorn JN, Klannemark M, Lindgren CM, Vohl MC, Nemesh J, Lane CR, Schaffner SF, Bolk S, Brewer C, Tuomi T, Gaudet D, Hudson TJ, Daly M, Groop L, Lander ES (2000) The common PPARgamma Pro12Ala polymorphism is associated with decreased risk of type 2 diabetes. Nat Genet 26:76–80

Barnett JH, Jones PB,Robbins TW, Muller U (2007) Effects of the catechol-O-methyltransferase Val158Met polymorphism on executive function: a meta-analysis of the Wisconsin Card Sort Test in schizophrenia and healthy controls. Mol Psychiatr 12:502–509

Barr CL, Feng Y, Wigg K, Bloom S, Roberts W, Malone M, Schachar R, Tannock R, Kennedy JL (2000) Identification of DNA variants in the SNAP-25 gene and linkage study of these polymorphisms and attention-deficit hyperactivity disorder. Mol Psychiatr 5:405–409

Barr CL, Feng Y, Wigg KG, Schachar R, Tannock R, Roberts W, Malone M, Kennedy JL (2001) 5'-untranslated region of the do-pamine D4 receptor gene and attention-deficit hyperactivity dis-order. Am J Med Genet 105:84–90

Battaglia M, Ogliari A, Zanoni A, Citterio A, Pozzoli U, Giorda R, Maffei C, Marino C (2005) Influence of the serotonin trans-porter promoter gene and shyness on children's cerebral re-sponses to facial expressions. Arch Gen Psychiatr 62:85–94

Bengtsson BO, Thomson G (1981) Measuring the strength of asso-ciations between HLA antigens and diseases. Tissue Antigens 18:365–363

Benzecri J-P (1992) Correspondence analysis handbook. Dekker, New York

Blackwood DH, Fordyce A, Walker MT, St Clair DM, Porteous DJ, Muir WJ (2001) Schizophrenia and affective disorders--cosegregation with a translocation at chromosome 1q42 that di-rectly disrupts brain-expressed genes: clinical and P300 findings in a family. Am J Hum Genet 69:428–433

Breiman L (2001) Random forests. Mach Learn 45:5–32

Carlson CS, Eberle MA, Kruglyak L, Nickerson DA (2004) Map-ping complex disease loci in whole-genome association studies. Nature 429:446–452

Carter JC (2007) Multiple genes and factors associated with bipo-lar disorder converge on growth factor and stress activated kinase pathways controlling translation initiation: implications for oligodendrocyte viability. Neurochem Int 50:461–490

Caspi A, Sugden K, Moffitt TE, Taylor A, Craig IW, Harrington H, McClay J, Mill J, Martin J, Braithwaite A, Poulton R (2003) Influence of life stress on depression: moderation by a poly-mor-phism in the 5-HTT gene. Science 301:291–293

Covault J, Gelernter J, Hesselbrock V, Nellissery M, Kranzler HR (2004) Allelic and haplotypic association of GABRA2 with al-cohol dependence. Am J Med Genet B Neuropsychiatr Genet 129:104–109

Chen YS, Akula N, Detera-Wadleigh SD, Schulze TG, Thomas J, Potash JB, DePaulo JR, McInnis MG, Cox NJ, McMahon FJ (2004) Findings in an independent sample support an association between bipolar affective disorder and the G72/G30 locus on chromosome 13q33. Mol Psychiatr 9:87–92

Chumakov I, Blumenfeld M, Guerassimenko O, Cavarec L, Palicio M, Abderrahim H, Bougueleret L, Barry C, Tanaka H, La Rosa P, Puech A, Tahri N, Cohen-Akenine A, Delabrosse S, Lissarrague S, Picard F-P, Maurice K, Essioux L, Millasseau P, Grel P, Debailleul V, Simon A-M, Caterina D, Dufaure I, Maleksadeh K, Belova M, Luan J-J, Bouilolot M, Sambucy J-L, Primas G, Saumier M, Boubkiri N, Martin-Saumier S, Nasroune M, Peixoto H, Delaye A, Pinchot V, Bastucci M, Guillou S, Chevillon M, Sainz-Fuertes R, Meguenni S, Aurich-Costa J, Cherif D, Gimalac A, Van Duijn C, Gauvreau D, Ouellette G, Fortier I, Raelson J, Sherbatich T, Riazanskaia N, Rogaev E, Raeymaekers P, Aerssens J, Konings F, Luyten W, Macciardi F, Sham PC, Straub R, Weinberger, Cohen N, Cohen D (2002) Genetic and physiological data implicating the new human gene G72 and the gene for D-amino acid oxidase in schizophrenia. Proc Nat Acad Sci USA 99:13675–13680

Cook EH, Stein MA, Krasowski MD, Cox NJ, Olkon DM, Kieffer JE, Leventhal BL (1995) Association of attention-deficit disorder and the dopamine transporter gene. Am J Hum Genet 56:993–998

Crabb DW, Edenberg HJ, Bosron WF, Li TK (1989) Genotypesf ora ldehyded ehydrogenase deficiency and alcohol sensitivity. The inactive ALDH2(2) allele is dominant. J Clin Invest 83:314–316

de Leon J (2006) AmpliChip CYP450 test: personalized medicine has arrived in psychiatry. Exp Rev Mol Diagn 6:277–286

Detera-Wadleigh SD, McMahon FJ (2006) G72/G30 in schizo-phrenia and bipolar disorder: review and meta-analysis. Biol Psychiatr 60:106–114

Dick DM Bierut LJ (2006) The genetics of alcohol dependence. Curr Psychiatr Rep 8:151–157

Duerr RH, Taylor KD, Brant SR, Rioux JD, Silverberg MS, Daly MJ, Steinhart H, Abraham C, Regueiro M, Griffiths A, Das-sopoulos T, Bitton A, Yang H, Targan S, Datta LW, Kistner EO, Schumm LP, Lee A, Gregersen PK, Barmada MM, Rotter JI Nicolae DL, Cho JH (2006)

A genome-wide association study Identifies IL23R as an inflammatory bowel disease gene. Sci-ence 314:1461–1463
Edenberg HJ, Dick DM, Xuei X, Tian H, Almasy L, Bauer LO, Crowe RR, Goate A, Hesselbrock V, Jones K, Kwon J, Li TK, Nurnberger JI, O'Connor SJ, Reich T, Rice J, Schuckit MA, Porjesz B, Foroud T, Begleiter H (2004) Variations in GABRA2, encoding the alpha 2 subunit of the GABA(A) receptor, are associated with alcohol dependence and with brain oscillations. Am J Hum Genet 74:705–714
Edenberg HJ, Xuei X, Chen HJ, Tian H, Wetherill LF, Dick DM, Almasy L, Bierut L, Bucholz KK, Goate A, Hesselbrock V, Ku-perman S, Nurnberger J, Porjesz B, Rice J, Schuckit M, Tisch-field J, Begleiter H, Foroud T (2006) Association of alcohol de-hydrogenase genes with alcohol dependence: a comprehensive analysis. Hum Mol Genet 15:1539–1549
Faraone SV, Doyle AE, Mick E, Biederman J (2001) Meta-analysis of the association between the 7-repeat allele of the dopamine D(4) receptor gene and attention deficit hyperactivity disorder. Am J Psychiatr 1052–1057
Faraone SV, Perlis RH, Doyle AE, Smoller JW, Goralnick JJ, Holmgren MA, Sklar P (2005) Molecular genetics of attention-deficit/hyperactivity disorder. Biol Psychiatr 57:1313–1323
Farmer A, Elkin A, McGuffin P (2007) The genetics of bipolar af-fective disorder. Curr Opin Psychiatr 20:8–12
Fehr C, Sander T, Tadic A, Lenzen KP, Anghelescu I, Klawe C, Dahmen N, Schmidt LG, Szegedi A (2006) Confirmation of as-sociation of the GABRA2 gene with alcohol dependence by sub-type-specific analysis. Psychiatr Genet 16:9–17
Frodl T, Schöle C, Schmitt G, Born C, Baghai T, Zill P, Bottlender R, Rupprecht R, Bondy B, Reiser M, Müller HJ, Meisenzahl EM (2007) Association of the brain-derived neurotrophic factor Val66Met polymorphism with reduced hippocampal volumes in major depression. Arch Gen Psychiatr 64:410–416
Gemma S, Vichi S, Testai E (2006) Individual susceptibilitiy and alcohol effectes: biochemical and genetic aspects. Ann 1st Suk-per Sanita 42:8–16
Goldstein DB (2005) The genetics of human drug response. Phil Trans R Soc B 360:1571–1572
Goldstein DB, Sullivan P (2006) Session 45 Pharmacogenomics ASHG 56th Annual Meeting New Orleans Louisiana
Goldstein DB, Tate SK, Sisodiya SM (2003) Pharmacogenetics goes genomic. Nat Rev Genet 4:937–947
Gorwood P (2006) Meeting everyday challenges: antipsychotic therapy in the real world. Eur Neuropsychopharmacol 16:S156–S162
Ginsburg GS, Donahue MP, Newby LK (2005) Cardiovascular ge-nomic medicine: Prospects for personalized cardiovascular medicine. J Am Col Cardiol 46:1615–1627
Green EK, Raybould R, Macgregor S, Gordon-Smith K, Heron J, Hyde S, Grozeva D, Hamshere M, Williams N, Owen MJ, O'Donovan MC, Jones L, Jones I, Kirov G, Craddock N (2005) Operation of the schizophrenia susceptibility gene, neuregulin 1, across traditional diagnostic boundaries to increase risk for bipo-lar disorder. Arch Gen Psychiatr 62:642–648
Greenacre MJ (1993) Correspondence analysis in practice. Academic, London
Hall JM, Lee MK, Newman B, Morrow JE, Anderson LA, Huey B, King MC (1990) Linkage of early-onset familial breast cancer to chromosome 17q21. Science 250:1684–1689
Hampe J Franke A, Rosentiel P, Till A, Teuber M, Huse K Albrecht M, Mayr G, De la Vega FM, Briggs J, Gunter S, Pres-cott NJ, Onnie CM, Hasler R, Sipos B, Folsch UR, Lengauer T, Platzer M, Mathew CG, Krawczack M, Schreiber S (2007) A genome-wide association scan of nonsynonymous SNPs identi-fies a susceptibility variant for Crohn disease in ATG16L1. Nat Genet 39:207–211
Haykin S (1999) Neural networks 2nd edition. Prentice Hall, New York
Hinrichs AL, Wang JC, Bufe B, Kwon JM, Budde J, Allen R, Bertelsen S, Evans W, Dick D, Rice J, Foroud T, Nurnberger J, Tischfield JA, Kuperman S, Crowe R, Hesselbrock V, Schuckit M, Almasy L, Porjesz B, Edenberg HJ, Begleiter H, Meyerhof W, Bierut LJ, Goate AM (2006) Functional variant in a bitter-taste receptor (hTAS2R16) influences risk of alcohol dependence. Am J Hum Genet 78:103–111

Hitchen L (2006) Adverse drug reactions result in 250 000 UK admissions a year. BMJ 332:1109

Hoover-Plow J, Shchurin A, Hart E, Sha J, Hill AE, Singer JB, Nadeau JH (2006) Genetic background determines response to hemostasis and thrombosis. BMC Blood Disord 6:6

Hugot JP, Chamaillard M, Zouali H, Lesage S, Cezard JP, Belaiche J, Almer S, Tysk C, O'Morain CA, Gassul M, Binder V, Finkel Y, Cortot A, Modigliani R, Laurent-Puig P, Gower-Rousseau, C Macry J, Colombel JF, Shabatou M, Thomas G (2001) Association of NOD2 leucine-rich repeat variants with susceptibility to Crohn's disease. Nature 411:537–539

Illumina Inc (2007) New Illumina human 1M and human 450S genotyping bead chips. Pre Release Information Sheet. techsup-port@illumina.com

International HapMap Consortium (2005) A haplotype map of the human genome. Nature 437:1299–1320

Kalow W (2005) Pharmacogenomics: historical perspective and current status. MethMol Biol 311:3–15

Kato D (2007) Molecular genetics of bipolar disorder and depres-sion. Psychiatr Clin Neurosci 61:3–19

Keefe RS, Bilder RM, Davis SM, Harvey PD, Palmer BW, Gold JM, Meltzer HY, Green MF, Capuano G, Stroup TS, McEvoy JP, Swartz MS, Rosenheck RA, Perkins DO, Davis CE, Hsiao JK, Lieberman JA; CATIE Investigators; Neurocognitive Work-ing Group (2007) Neurocognitive effects of antipsychotic medi-cations in patients with chronic schizophrenia in the CATIE Trial. Arch Gen Psychiatr 64:633–647

Kockelkorn TTJP, Arai M, Matsumoto H, Fukuda N, Yamada K, Minabe Y, Toyota T, Ujike H, Sora I, Mori N, Yoshikawa T, Itokawa M (2004) Association study of polymorphisms in the 5' upstream region of human DISC1 gene with schizophrenia. Neurosci Lett 368:41–45

Kraft JB, Peters EJ, Slager SL, Jenkins GD, Reinalda MS, McGrath PJ Hamilton SP (2007) Analysis of association between the serotonin transporter and antidepressant response in a large clinical sample. Biol Psychiatr 61:734–742

Lake CR (2007) Disorders of thought are severe mood disorders: the selective attention defect in mania challenges the Kraepelinian dichotomy–A review. Schizophr Bulln May 21 Epublication ahead of print

Lappalainen J, Krupitsky E, Remizov M, Pchelina S, Taraskina A, Zvartau E, Somberg LK, Covault J, Kranzler HR, Krystal JH, Gelernter J (2005) Association between alcoholism and gamma-amino butyric acid alpha2 receptor subtype in a Russian population. Alcohol Clin Exp Res 29:493–498

Lappalainen J, Krupitsky E, Kranzler HR, Luo X, Remizov M, Pchelina S, Taraskina A, Zvartau E, Räsanen P, Makikyro T, Somberg LK, Krystal JH, Stein MB, Gelernter J (2007) Muta-tion screen of the GAD2 gene and association study of alcohol-ism in three populations. Am J Med Genet B Neuropsychiatr Genet 144:183–192

Lawrence RW, Evans DM, Cardon LR (2005) Prospects and pit-falls in whole genome association studies. *Philos Trans R Soc Lond B Biol Sci* 360:1589–95

Lee S-G, Joo Y, Kim B, Chung S,, Kim H-L, Lee I, Choi B, Kim C, Song K (2005) Association of ala72ser polymorphism with COMT enzyme activity and the risk of schizophrenia in Koreans. Hum Genet 116:319–328

Levin M, Bertell R (1978) Simple estimation of population attrib-utable risk from case control studies. Am J Epidemiol 108:78–79

Li T, Stefansson H, Gudfinnsson E, Cai G, Liu X, Murray RM, Steinthorsdottir V, Januel D, Gudnadottir VG, Petursson H, Ingason A, Gulcher JR, Stefansson K, Collier DA (2004) Identification of a novel neuregulin 1 at-risk haplotype in Han schizophrenia Chinese patients, but no association with the Icelandic/Scottish risk haplotype. Mol Psychiatr 9:698–704

Li J, Wang Y, Zhou R, Zhang H, Yang L, Wang B, Khan S, Faraone SV (2005) Serotonin 5-HT1B receptor gene and atten-tion deficit hyperactivity disorder in Chinese Han subjects. Am J Med Genet B Neuropsychiatric Genet 132:59–63

Libioulle C, Louis E, Hansoul S, Sandor C, Farnir F, Franchimont D, Vermeire S, Dewit O, de Vos M, Dixon A, Demarche B, Gut I, Heath S, Foglio M, Liang L, Laukens D, Mni M, Zelenika

D, Van Gossum A, Rutgeerts P, Belaiche J, Lathrop M, Georges M (2007) Novel Crohn disease locus identified by genome-wide association maps to a gene desert on 5p13.1 and modulates ex-pression of PTGER4. PLOS Genet 3:538–543

Lieberman JA, Stroup TS, McEvoy JP, Swartz MS, Rosenheck RA, Perkins DO, Keefe RS, Davis SM, Davis CE, Lebowitz BD, Severe J, Hsiao JK, Clinical Antipsychotic Trials of Intervention Effectiveness (CATIE) Investigators (2005) Effectiveness of antipsychotic drugs in patients with chronic schizophrenia. N Engl J Med 353:1209–23

Lowe N, Kirley A, Hawi Z, Sham P, Wickham H, Kratochvil CJ, Smith SD, Lee SY, Levy F, Kent L, Middle F, Rohde LA, Ro-man T, Ta hir E, Yazgan Y, Asherson P, Mill J, Thapar A, Payton A, Todd RD, Stephens T, Ebstein RP, Manor I, Barr CL, Wigg KG, Sinke RJ, Buitelaar JK, Smalley SL, Nelson SF, Biederman J, Faraone SV, Gill M (2004) Joint analysis of the DRD5 marker concludes association with attention-deficit/hyperactivity disorder confined to the predominantly in-attentive and combined subtypes. Am J Hum Genet 74:348–356

Luo X, Kranzler HR, Zuo L, Wang S, Blumberg HP, Gelernter J (2005) CHRM2 gene predisposes to alcohol dependence, drug dependence and affective disorders: results from an extended case-control structured association study. Hum Mol Genet 14:2421–2432

Maeda K, Nwulia E, Chang J, Balkissoon R, Ishizuka K, Chen H, Zandi P, McInnis MG, Sawa (2006) A Differential expression of disrupted-in-schizophrenia (DISC1) in bipolar disorder. Biol Psychiatr 60:929–935

Magi R, Pfeufer A, Nelis M, Montpetit A, Metspalu A, Remm M (2007) Evaluating the performance of commercial whole-genome marker sets for capturing common genetic variation. BMC Genomics 8:159

Manor I, Eisenberg J, Tyano S, Sever Y, Cohen H, Ebstein RP, Kotler M (2001) Family-based association study of the serotonin transporter promoter region polymorphism (5-HTTLPR) in attention deficit hyperactivity disorder. Am J Med Genet 105:91–95

Maziade M, Roy MA, Chagnon YC, Cliche D, Fournier JP, Mont-grain N, Dione C, Lavallee JC, Garneau Y, Gingras N, Nicole L, Pires A, Ponton AM, Potvin A, Wallot H, Merette C (2005) Shared and specific susceptibility loci for schizophrenia and bi-polar disorder: a dense genome scan in Eastern Quebec families. Mol Psychiatr 10:486–499

McCullagh P, Nelder JA (1999) Generalized linear models second edition. Chapman and Hall/ CRC, Boca Raton

McGrath M, Kawachi I, Ascherio A, Colditz GA, Hunter DJ, De Vivo I (2004) Association between catechol-O-methyltransferase and phobic anxiety. Am J Psychiatr 161:1703–1705

McMahon FJ, Buervenich S, Charney D, Lipsky R, Rush AJ, Wilson AF, Sorant AJM, Papanicolaou GJ, Laje G, Fava M, Trivedi MH, Wisniewski SR, Manji H (2006) Variation in the gene encoding the serotonin 2A receptor is associated with outcome of antipressant treatment. Am J Hum Genet 78:804–812

Miki Y, Swensen J, Shattuck-Eidens D, Futreal PA, Harshman K, Tav tigian S, Liu Q, Cochran C, Bennett LM, Ding W. et al. (1994) A strong candidate for the breast and ovarian cancer susceptibility gene BRCA1. Science 266:66–71

Müller DJ, Kennedy JL (2005) Genetics of antipsychotic treatment emergent weight gain in schizophrenia. Pharmacogenomics 7:863–887

Naoe Y, Shinkai T, Hori H, Fukunaka Y, Utsunomiya K Sakata S, Matsumoto C, Shimizu K, Hwang R, Ohmori O, Nakamura J (2007) No association between the brain-derived neurotrophic factor (BDNF) Val66Met polymorphism and schizophrenia in Asian populations: Evidence from a case-control study and meta-analysis. Neurosci Lett 415:108–112

Nicodemus KK, Luna A, Vakkalanka R, Goldberg T, Egan M, Straub RE, Weinberger DR (2006) Further evidence for associa-tion between ErbB4 and schizophrenia and influence on cognitive intermediate phenotypes in healthy controls. Mol Psychiatr 11:1062–1069

Norton N, Moskvina V, Morris DW, Bray NJ, Zammit S, Williams NM, Williams HJ, Preece AC, Dwyer S, Wilkinson JC, Spurlock G, Kirov G, Buckland P, Waddington JL, Gill M, Cor-vin AP, Owen MJ, O,Donovan MC (2006) Evidence that inter-action between neuregulin 1 and its receptor erbB4 increases susceptibility to schizophrenia. Am J Med Genet B Neuropsychiatr Genet 141:96–101

Parkes M, Barett JC, Presscott NJ, Tremelling M, Anderson CA, Fisher SA Roberts RG, Nimmo ER, Cummings FR, Soars D, Drummond H, Lees CW, Khawaja SA, Bagnall R, Burke DA, Todhunter CE, Ahmad T, Onnie CM, McArdle W, Strachan D, Bethel G, Bryan C, Lewis CM, Deloukas P, Forbes A, Sanderson J, Jewell DP, Sat sangi J, Mansfield JC, The Wellcome Trust Case Control Consortium, Cardon L, Mathew CG (2007) Sequence variants in the autophagy gene IRGM and multiple other replicating loci contrib.ute to Crohn's disease susceptibility. Nat Genet 39:830–832

Payton A, Holmes J, Barrett JH, Hever T, Fitzpatrick H, Trumper AL, Harrington R, McGuffin P, O'Donovan M, Owen M, Ollier W, Worthington J, Thapar A (2001) Examining for association between candidate gene polymorphisms in the dopamine path-way and attention-deficit hyperactivity disorder: a family-based study. Am J Med Genet 105:464–470

Rioux JD, Xavier RJ, Taylor KD Silverberg MS, Goyette P, Huett A, Green T, Kuballa P, Barmada MM, Datta LW, Shugart YY, Griffiths Am, Targan SR, Ippoliti AF, Bernard E-J, Mei L, Nicolae DL, Regueiro M, Schumm LP, Steinhard AH, Rotter JI, Duerr RH, Cho JH, Daly MJ Brant SR (2007) Genome-wide as-sociation study identifies new susceptibility loci for Crohn dis-ease and implicates autophagy in disease pathogenesis. Nat Ge-net 39:596–634

Risch N, Merikangas K (1996) The future of genetic studies of complex human diseases. Science 273:1516–1517

Schumacher J, Jamra RA, Freudenberg J, Becker T, Ohlraun S,Otte AC, Tullius M, Kovalenko S, Bogaert AV, Maier W, Rietschel M, Propping P, Noethen MM, Cichon S (2004) Examination of G72 and D-amino-acid oxidase as genetic risk factors for schizophrenia and bipolar affective disorder. Mol Psychiatr 9:203–207

Shifman S, Bronstein M, Sternfeld M, Pisante-Shalom A, Lev-Lehman E, Weizman A, Reznik I, Spivak B, Grisaru N, Karp L, Schiffer R, Kotler M, Strous RD, Swartz-Vanetik M, Knobler HY, Shinar E, Beckmann JS, Yakir B, Risch N, Zak NB, Dar-vasi A (2002) A highly significant association between a COMT haplo type and schizophrenia. Am J Hum Genet 71:1296–1302

Silberberg G, Darvasi A, Pinkas-Kramarski R, Navon R (2006) The involvement of erbB4 with schizophrenia. Am J Med Genet B Neruopsychiatr Genet 141B:142–148

Simard J, Jeunteun J, Lenoir G, Tonin P, Normand T, The VL, Vivier A, Laska D, Morgan K, Rouleau GA, Lynch H, Labrie F, Narod SA (1993) Gentic mapping of the breast-ovarian cancer syndrome to a small interval on chromosome 17q12–21: A ex-clusion of candidate genes EDH17B2 and RARA. Hum Mol Genet 2:1193–1199

Smith KM, Daly M, Fischer M, Yiannoutsos CT, Bauer L, Barkley R, Navia BA (2003) Association of the dopamine beta hydroxy-lase gene with attention deficit hyperactivity disorder: genetic analysis of the Milwaukee longitudinal study. Am J Med Genet B Neuropsychiatr Genet 119:778–85

Stefansson H, Sigurdsson E, Steinthorsdottir V, Bjornsdottir S, Sigmundsson T, Ghosh S, Brynjolfsson J, Gunnarsdottir S, Bjornsson A, Ingvarsson B, Ingason A, Sigfusson S, Hardardottir H, Harvey RP, Lai D, Zhou M, Brunner D, Mutel V, Gonzalo A, Lemke G, Sainz J, Johannesson G, Andersson T, Gudbjartsson D, Manolescu A, Frigge ML, Gurney ME, Kong A, Gulcher JR, Pertusson H, Steffanson Kal (2002) Neuregulin 1 and susceptibility to schizophrenia. Am J Hum Genet 71:877–892

Stefansson H, Sarginson J, Kong A, Yates P, Steinhorsdottir V, Gudfinnsson E, Gunnarsdottir S, Walker N, Petursson H, Crombie C, Ingason A, Gulcher J, Stefansson K, St Clair D (2003) Association of Neuregulin 1 with Schizophrenia confirmed in a Scottish population. Am J Hum Genet 72:83–87

Straub RE, Jiang Y, MacLean CJ, Ma Y, Webb BT, Myakishev MV, Harris-Kerr C, Wormley B, Sadek H, Kadambi B, Cesare AJ, Givverman A, Wang X O'Neill FA, Walsh D, Kendler KS (2002) Genetic variation in the 6p22.3 gene DTNBP1, the human ortholog of t he mouse dysbindin gene, is associated with schizophrenia. Am J Hum Genet 71:337–348

Schwab SG, Knapp M, Mondabon S, Hallmayer J, Borrmann-Hassenbach M, Albus M, Lerer B, Rietschel M, Trixler M, Maier W, Wilde nauer DB (2003) Support for association of schizophrenia with genetic variation in the 6p22.3 gene, dys-bindin, in sib-pair families with linkage and in an additional sample of triad families. Am J Hum Genet 72:185–190

Thomson PA, Wray NR, Millar JK, Evans KL, Hellard SL, Condie A, Muir WJ, Blackwood DH, Porteous DJ (2006) Association between the TRAX/DISC locus and both bipolar disorder and schizophrenia in the Scottish population. Mol Psychiatr 657–668

Vizirianakis I (2005) Improving pharmacotherapy outcomes by pharmacogenomics: from expectation to reality. Pharmacoge-nomics 6:701–711

Wang JC, Hinrichs AL, Stock H, Budde J, Allen R, Bertelsen S, Kwon JM, Wu W, Dick DM, Rice J, Jones K, Nurnberger JI, Tischfield J, Porjesz B, Edenberg HJ, Hesselbrock V, Crowe R, Schuckit M, Begleiter H, Reich T, Goate AM, Bierut LJ (2004) Evidence of common and specific genetic effects: association of the muscarinic acetylcholine receptor M2 (CHRM2) gene with alcohol dependence and major depressive syndrome. Hum Mol Genet 13:1903–1911

Wang JC, Hinrichs AL, Bertelsen S, Stock H, Budde JP, Dick DM, Bucholz KK, Rice J, Saccone N, Edenberg HJ, Hesselbrock V, Kuperman S, Schuckit MA, Bierut LJ, Goate AM (2007) Functional variants in TAS2R38 and TAS2R16 influence alcohol consumption in high-risk families of African-American origin. Alcohol Clin Exp Res 31:209–215

Wellcome Trust Case Control Consortium (2007) Genome-wide association study of 14,000 cases of seven common diseases and 3,000 shared controls. Nature 447:661–678

Williams NM, Preece A, Morris DW, Spurlock G, Bray NJ, Stephens M, Norton N, Williams H, Clement M, Dwyer S, Curran C, Wilkinson J, Voskvina B, Waddington JL, Gill M, Corvin P, Zammit S, Kirov G, Owen MJ, O'Donovan MC (2004) Identification in 2 independent samples of a novel schizophrenia risk haplotype of the dystrobrevin binding protein gene (DTNBP1). Arch Gen Psychiatr 61:336–344

Wray NR, James MR, Mah SP, Nelson M, Andrews G, Sullivan PF, Montgomery GW, Birley AJ, Braun A, Martin NG (2007) Anxiety and co-morbid measures associated with PLXNA2. Arch Gen Psychiatr 64:318–326

Zhang F, Sarginson J, Crombie C, Walker N, St Clair D, Shaw D (2006) Genetic association between Schizophrenia and DISC1 gene in the Scottish population. Am J Med Genet B Neuropsychiatr Genet 141:155–159

Chapter 19
Biological Modeling in the Discovery and Validation of Cognitive Dysfunctions Biomarkers

François Iris

Abstract Cognitive disorders are highly heterogenous in terms of symptoms, clinical etiologies, disease progression and therapeutic responses. Furthermore, their potential biological causes remain largely unknown. Progress at these different levels is currently mired in a vicious circle.

The identification of coherent biomarkers, essential for clinical and therapeutic progress, requires an understanding of either the relevant pathogenic processes or, at the very least, of the parameters that need to be monitored. But, syndrome-dominated conceptual thinking has become a barrier to understanding the biological processes linked to diseases characterized by clinical and therapeutic heterogeneity. As a result, current biomarkers of cognitive disorders are much too numerous, too heterogenous, and too variable to serve useful purposes. This leads to an untenable situation that precludes coherent therapeutic developments since it effectively prevents defining what could constitute valid biological, clinical and therapeutic approaches. How to escape from this situation? The problem could be partly resolved by adopting the much wider views allowed by "system-wide" approaches: in effects, by constructing predictive theoretical models of what could constitute pathological cognitive processes. This, naturally, requires the integration of massive amounts of highly heterogenous and often conflicting information. This chapter aims to provide a necessarily brief overview of the concepts, the breadth of data and the variety of network dynamics that will have to be considered while proposing a functional, experimentally validated model-building approach that could be fruitfully utilized.

F. Iris
Bio-Modeling Systems, 26 Rue St Lambert, 75017 Paris, France
francois.iris@bmsystems.net
francois.iris@laposte.net

C.W. Turck (ed.) *Biomarkers for Psychiatric Disorders*,
doi: 10.1007/978-0-387-79251-4_19, © Springer Science+Business Media, LLC 2008

Abbreviations ADHD: Attention-deficit/hyperactivity disorder; ACM: Astrocyte-conditioned medium; ADP: After-depolarization membrane potential; AHP: Ca2+-dependent, K+-mediated after-hyperpolarization; Akt: Protein kinase (for "rac-family serine/threonine-protein kinase homolog"); Akt-P: Phosophorylated form of Akt; AMPA: Alpha-amino-3-hydroxy-5-methylisoxazole-4-propionic acid; BDNF: Brain-derived neurotrphic factor; BDWG: Biomarkers Definitions Working Group; BOLD: Blood oxygenation level-dependent (neurophysiology); BST: Biochemical systems theory; CA1/3: Cornu ammonis, field 1/3; CaMKII: Protein kinase, (for "Calmodulin kinase type II"); Ca^{2+}: Calcium ion; CCK: Cholecystokinin; Cdc42: GTP-binding protein/GTPase (for "cell division cycle protein 42"); c-fos: Transcription factor (for "F-type osteosarcoma viral oncogene homolog"); CNS: Central nervous system; Cx: Connexins (isoforms 26, 30, 43, etc.); EC: Entorhinal cortex; EEG: Electroencephalogram; EPSC: Electric potential generated by spontaneous slow cation conductence; EPSP: Electric potential generated by spontaneous patterns of activity; ERM: Ezrin-Moesin-Radixin protein family; FD: Fascia dentata; FDA: Food & Drugs Administration (USA); fra-2: Transcription factor (for "fos-related antigen 2"); GABA: Gamma-aminobutyric acid; GAP-43: Memory-associated protein (for "growth & plasticity-associated protein 43"); GAT1: GABA transporter, isoform I; GluR: Glutamate & AMPA receptors (isoforms 1 to 4); GTPase: Guanydine triphosphate catabolic enzyme; Hz: Hertz (wave frequencies measurement unit); 5-HT4R: 5-Hydroxytryptamine type 4 receptor; iEEG: Intra-cranial electroencephalogram; IPSPs: Inhibitory post-synaptic potentials; JunB: Transcription factor (for "J-type unnamed osteosarcoma oncogene B"); LTD: Long-term depression (neurophysiology); LTP: Long-term potentiation (neurophysiology); M2: Muscarinic receptor type II; MAPK: Protein kinase (for "mitogen-activated protein kinase"); MAPK-P:Phos phorylatedformof MAPK; MEK: Protein kinase kinase (for "MAPK/Erk kinase"); mIPSC: Miniature inhibitory post-synaptic conductivity; MMPs: Matrix metalloproteases; MRI: Magnetic resonance imaging; mRNA: Messenger ribonucleic acid; MS-DB: Medial septal-diagonal band nuclei; NGF: Nerve growth factor; NHE1: Na+/H+ exchanger type I; NHERF: Na+/H+ exchanger regulatory factor; NIH: National Institute of Health (USA); NMDA: N-methyl-D-aspartate; NPY: Neuro-peptide Y; ORF: Open reading frame (coding genetic sequence); PET: Positron emission tomography; PI3K: Phosphoinositol triphosphate kinase; PKC: Protein kinase C; PTM: Post-translational modification; PV: Parvalbumin; p75NTR: Neurotrophin receptor, (for "protein of 75 kilo-dalton in molecular mass); Rac1: GTP-binding protein/GTPase (for "ras-related C3 botulinum toxin substrate 1"); Raf: Serine/threonine-protein kinase (for "ras-associated factor"); Ras: GTP-binding protein/GTPase (for "rat sarcoma oncogene"); RF: Reticular formation; RhoA/B: GTP-binding proteins/GTPases (for "ras homolog gene family, member A/B"); RTCs: Receptor-laden transport carriers; Shc: Adaptor protein, (for "src homology 2 domain-containing protein"); Tgfbr2: Transforming growth factor beta receptor II, isoform 2; TrkA/B: Neural receptor protein-tyrosine kinase, typeA/B; TTX: Tetrodotoxin; WHO: World Health Organization.

19.1 Introduction

Mental disorders are highly prevalent, heterogeneous, and of multifactorial etiologies. Collectively, they are associated with significant morbidity, mortality, and economic cost. However, major difficulties are encountered in translating and quantifying critical end-points into therapeutic activity that benefit patient outcome. In the absence of a validated biomarker for psychiatric illness activity, symptomatic remission and functional restoration are the only available markers of wellness in psychiatry (McIntyre et al., 2006). Furthermore, it is increasingly accepted that the imprecision of categorical psychiatric diagnoses can be a limiting factor in understanding both the genetic and functional basis of human behavioral abnormalities. Each of the current clinical forms of mental disorders, be it depressive disorder, bipolar disorder, schizophrenia, anxiety disorders, or attention-deficit/hyperactivity disorder (ADHD), is defined by a number of symptoms that differ considerably between affected individuals with respect to their presence, frequency, severity, and topography. This heterogeneity in symptoms has complicated the search for the etiology of the diseases and the mechanisms for their treatment. For more than a century, it has been uncertain whether or not the major diagnostic categories of psychosis, namely schizophrenia and bipolar disorder, are distinct disease entities with specific genetic causes and neuro-anatomical substrates. Indeed, the results of studies directed toward etiologies and the interpretation of the complex relationships between genes and behavior have shown very limited levels of reproducibility. So much so that the reputation of the field of psychiatric genetics has become tarnished in the view of many human geneticists. Too many linked loci have been claimed and withdrawn, too many association studies published and not confirmed and too many new and different chromosomal regions have been implicated for the same disorder.

It is becoming generally accepted that the current diagnostic system often guarantees, rather than diminishes, disease heterogeneity. In effect, syndrome-dominated conceptual thinking has become a barrier to understanding the biological causes of diseases such as psychiatric disorders, characterized by clinical and therapeutic heterogeneity.

This leads to an untenable situation that precludes coherent therapeutic developments as it effectively prevents defining what could constitute valid biological, clinical and therapeutic biomarkers.

It must be said however, that in the context of psychiatric disorders, the current definitions of biomarkers leave much to be desired. There are three broadly accepted definition addressing the issue of what a biomarker should be.

The world Health Organization (WHO) provided the first definition in 1993. The WHO defined a biomarker as "any parameter that can be used to measure an interaction between a biological system and an environmental agent, which may be chemical, physical or biological." The broadness of this definition raised so many problems that in 1999 the FDA proposed a new definition. According to the FDA, a biomarker is "a characteristic that is objectively measured and evaluated as an indicator of normal biological processes, pathogenic processes, or pharmacological responses to a therapeutic intervention." This definition in effect presupposes a knowledge of what functionally distinguishes a "normal" biological processes from "pathogenic" processes and it further presupposes relative homogeneity in terms of responses to a given drug treatment. While certainly valid for many infective and metabolic diseases, this definition turned out to be inapplicable in a plethora of other medical domains, and in particular in psychiatry. In 2001, in an attempt to accommodate the evident functional complexities attached to the distinctions between "normal" and "pathogenic" biological processes, the Biomarkers Definitions Working Group (BDWG) at the NIH proposed that a biomarker should be viewed as "a molecular indicator of a specific biological property, a biochemical feature or facet that can be used to measure the progress of disease or the responses to a therapeutic intervention"(Group, 2001). Taking into account the state of our current understanding of pathological processes, this definition literally opens a 'Pandora's Box'. In the context of functionally very heterogeneous disorders, there might be as many biomarkers as there are affected individuals. Furthermore, this definition does not distinguish between biological markers and clinical markers. Yet the goals of experimental therapies in psychiatric, and in other CNS disorders, are twofold: (1) better symptomatic therapies, and (2) treatments that slow disease progression or delay disease onset (Ravina et al., 2003). In the latter case, clinical endpoints are used as biomarkers (clinical measures such as blood pressure have been used in the past), and they are not measured for the purpose of detecting clinical benefit but for their reflection of the underlying pathological process.

Thus, since a "biomarker" is typically defined as a laboratory measurement that reflects the activity of a disease process, there are many such markers identified for many diseases of the nervous system, such as, for example, various magnetic resonance imaging (MRI) measures in multiple sclerosis and Alzheimer's disease treatments, positron emission tomographic (PET) scanning of dopamine transporters in Parkinson's disease, etc. (Couderc, 2000; Bohnen and Frey, 2003; Nurmi et al., 2003; Rovaris et al., 2003). Essentially, in all cases, these markers quantitatively correlate, either directly or inversely, with disease progression. A "surrogate marker" however, can be defined as "a laboratory measurement or physical sign that is used in therapeutic trials as a substitute for a clinically meaningful endpoint that is a direct measure of how a patient feels, functions, or survives and is expected to predict the effect of the therapy." (Tardif et al., 2006). Hence, the major difference between a biomarker and a surrogate marker is that a biomarker is a "candidate" surrogate marker, whereas a surrogate marker is a test used, and taken, as a measure of the effects of a specific treatment.

Nevertheless, it must be admitted that, although certainly imperfect, both the FDA and the NIH definitions hold some inescapable truths. Furthermore, it must also be admitted that biomarkers are at the roots of evidence-based medicine (who should be treated, how and with what) and that without valid biomarkers, advances in better targeted therapies will remain limited and treatments will remain largely empirical.

19.2 The Need for Biological Models and the Role of Systems Biology

How to escape from this vicious circle? There seem to be very few choices. We must imperatively obtain a functional understanding of the pathological processes associated with these disorders. But this primarily means that we must first and foremost understand what we call "normal" processes so as to distinguish what could constitute potentially pathological deviations or events.

In other words, we have to develop predictive functional models sufficiently detailed so as to enable the precise identification, in mechanistic terms, of events leading to pathological consequences, hence identifying the markers associated with these events together with the modes of intervention most likely to prevent or alleviate the problems.

However, psychiatric disorders arise from adverse events, both internal and external, affecting cognitive processes. Could we really have the pretension to "model" the cognitive processes in the brain? And assuming that anyone could dare hold such pretension, could we even start doing this? To have the least chance of success, one would have to integrate massive amounts of information, pertaining to literally every domain of functional biology, if not into a coherent whole, then at least in coherent pathways, leading to defined mechanisms, leading in turn to defined networks and interactions and finally to defined processes and functional modes. Could this be seriously contemplated?

Not only can it be seriously contemplated, but it must be undertaken.

Psychiatric conditions result from a complex interaction of genetic susceptibility and environmental effects. None of the many psychiatric conditions investigated to date has been shown to present a purely genetic background. Furthermore, studies in molecular biology have indicated that in the CNS gene expression is influenced by several environmental factors, including early experiences, traumas, learning, and memory processes (Mundo, 2006). However, the heterogeneity of available information together with the entire absence of functional models of pathogenesis and pathological evolution leads to a situation where not only psychiatric illness activity cannot be coherently approached but where therapy also becomes highly problematic as unidentified or evolving medical conditions may precipitate rapid and unanticipated changes in the status (Carpenter et al., 2004; Belgamwar and Fenton, 2005; Cheung et al., 2005; Gillies et al., 2005; Bogenschutz et al., 2006; Segura-Bruna et al., 2006).

Acquisition of the necessary knowledge can be obtained, in parts, using in-silico models produced through analytical approaches and processes collectively known as "Systems Biology."

19.2.1 What is "Systems Biology"?

Systems biology is the discipline that specifically addresses the fundamental properties of the complexity that living systems represent. Living organisms present systems complexities that span at least five dimensions: (1) molecular complexity; (2) structural complexity; (3) temporal complexity; (4) abstraction and emergence; and (5) algorithmic complexity (Huang and Wikswo, 2006).

In effect, systems biology addresses the need to shift from a component-based reductionist view of biology to a system-wide perspective. It can be described as a global quantitative analysis of how all components in a biological system interact to determine its phenotype. Although the definitions may vary, systems biology can usually be characterized as interdisciplinary, iterative, computationally intensive, and information greedy. An explanation of these terms is beneficial to set the stage for the remainder of this discussion:

Interdisciplinary. Although it is sometimes possible to answer a reductionist question by a collaboration of scientists from the same discipline, a global view requires a multidisciplinary team. Scientists from multiple quantitative disciplines (mathematics, computer science, statistics), medical specialties (clinicians, radiologists, epidemiologists), and life sciences disciplines (molecular and cell biologists, geneticists, chemists, physiologists) need to work together for a real system-wide model to be developed.

Iterative. An intriguing aspect of systems approaches is the iterative nature of the process, in which scientific publications provide information required for modeling; based on this information, a model is created and tested by experiments. The results of these experiments are in turn used to refine the model.

Computationally intensive. The integration of multiple levels of information, multiple datasets from high-throughput technologies, the use of advanced mathematical and computational tools, and the iterative nature of the analysis require high-powered and parallel computing.

Information greedy. In contrast to reductionist approaches, where only recognized, directly relevant information is used, in systems biology all information is potentially relevant.

A recent review (Aderem, 2005) described three concepts that are also worth introducing in any discussion of systems biology:

Emergence. The most basic principle in systems biology. Biological systems will express emergent characteristics that cannot be predicted from the sole knowledge of their parts. The colloquial equivalent of this concept is "the whole is bigger than the sum of its parts." For example, the presence of collagen I and III and matrix metalloproteases (MMPs) in the lung does not necessarily result in fibrosis; it is the combination of multiple factors and their deregulation that leads to the abnormal phenotype.

Robustness. This concept describes how stable the system is in response to environmental stresses and genetic variation. This includes understanding of internal regulatory loops and positive and negative feedback mechanisms. In general, negative feedback loops increase the system's robustness and positive feedback loops reduce it. It is worth considering that similar organ phenotypes may differ in their robustness (Kitano, 2007). Understanding the system's robustness and the factors that increase its sensitivity may be important in design of therapeutic interventions.

Modularity. Complex biological systems are organized by functional modules, where multiple components are co-regulated by a process and, when activated, lead to a similar outcome. A module can be a set of genes located in a similar genomic region that is activated by the same transcription factors, or a set of chemokines that are bound to a similar glycoprotein. An increase in the peptidase that degrades this protein will cause predictable response through all the associated molecules. The benefit of modular analysis is that it reduces dependence on the relative variability in levels of single molecules. Modularity often helps to resolve conflicting pieces of information by allowing for spatial, temporal, and contextual organization of the information.

19.2.2 The Current State of Systems Biology

A generic approach to biological systems modeling and analysis, called *biochemical systems theory* (BST) (Savageau, 1969a, b) was originally designed for studying the dynamics and other features of biochemical and gene regulatory systems, but is not restricted to these application areas in terms of its mathematical foundation. BST is almost forty years old, and its development, expansions, and applications have been documented in several books and hundreds of journal articles, proceedings, and chapters in books (see for example (Goel et al., 2006). However, the real obstacle to fast progress in bio-mathematical modeling is the determination of unknown parameters from biological information. Even within the same modeling framework, this task may be attacked in distinctly different ways, leading to entirely different results (Gilbert et al., 2006) because one is never short of possible mathematical manipulations.

Furthermore, despite the identification of a number of novel disease-predisposing genes (Dwyer et al., 2004; Edwards et al., 2005; Florez et al., 2006; Herbert et al., 2006), progress in uncovering the mechanisms by which these genes lead to disease has been far slower. Even for cases in which genes validated as causal for disease are known to operate in what are thought to be well-understood pathways, it is often unclear whether the connection to disease regarding such genes involves the known pathways, whether these "known" pathways are more general than is presently known, or whether the disease-associated genes operate in multiple pathways and some of which are yet to be defined (Lander, 2007). An example is the gene Tgfbr2, a key component of the transforming growth factor-ß signaling pathway, that involves only a modest number of proteins, but whose expression in the liver of mice from an F2 inter-cross population was shown to associate with thousands of other genes ostensibly unrelated to this classic signaling pathway. The gene was subsequently identified and validated as causal for obesity in a segregating mouse population (Schadt et al., 2005), but how variations in this gene can lead to obesity is not yet understood.

There currently exists a vast diversity of systems biology programs with a wide variety of goals (Huang and Wikswo, 2006). However, in spite of numerous technological advances (Albeck et al., 2006; Hua et al., 2006; Janes and Yaffe, 2006; Ng et al., 2006; Stromback et al., 2006; Wolkenhauer et al., 2007), systems biology remains in its early infancy and models accurately describing complex pathophysiological processes remain extremely rare, with the notable exception of those published in 2003 and 2005 by one European group (Gadal et al., 2003; Gadal et al., 2005). While these models, addressing breast cancer, were experimentally validated a posteriori using cell lines, their latest model, describing the mechanisms leading to neuronal death and vacuolation in Creutzfeld-Jakob disease, has now been experimentally validated in vivo (J-P Deslys, personal communication).

Hence, efficient data appropriation and data integration across the "five-dimensional" manifold leads to the production of highly detailed predictive models, that, although entirely theoretical in nature, indicate very precisely what should be experimentally investigated, where, when, how and, most of all, why.

Although the models arising from these integrative approaches cannot, by any means, be regarded as biologically true in the absolute, they do represent a "least biased" and detailed view of the mechanisms potentially associated with a given physiological state and/or governed by the biological components under consideration, together with precise indications of the means whereby these could be manipulated.

Thus, although such a model only represents an approximation of biological reality, it has the considerable advantage of providing the investigators with precise indications of what should be scrutinized and why. The new data arising from these experimentations can then be re-injected into the model, rapidly leading to a clear and factual understanding of the biological processes under investigation.

19.2.3 "Information" is a Double-Bladed Tool to be Manipulated with Caution

Nevertheless, it remains certain that, if predictive models of psychiatric pathogenesis and progression are to ever be constructed, enormous masses of information originating from a multitude of biological investigations and encompassing an enormous functional complexity will have to be integrated.

While we might be suffering from too much data complexity, we are certainly not suffering from lack of available information or from lack of redundancies in this information. Thus, although daunting in amplitude, if approached coherently, the flood of highly heterogeneous, and often conflicting information that currently hampers most biological fields, can become an invaluable tool.

This tool however must be approached and manipulated with utmost caution.

The information shockwave is of such amplitude that the scientific literature has become complex almost beyond measure. But to make matter worse, most of that information arose from reductionist approaches that, in attempts to compensate for the enormous, and often insurmountable, experimental difficulties presented by in vivo systems, utilized in-vitro experimentation on material far removed from functional physiological reality. Furthermore, such studies usually focus on a single factor or pathway of a given biological phenomenon. Although in some simple cases linear connections between signals, effecter molecules, and physiological consequences may provide general explanations of a biological process, in many other situations it is quite difficult to draw, from data obtained with reductionist approaches, universal schemes, which describe the biological phenomenon beyond the restrictive context of the experiment. Hence, the enormous diversity of information obtained in association with most physiological networks poses a particular challenge in that it represents, per se, a highly distorted compounded view of the various modulations that can potentially affect each such network, albeit without any distinctions of means or of actual effectors. Thus, due to the enormous diversity of experimental systems that gave rise to this information, and the often complete lack of physiological compatibilities between experimental systems, the information thereby generated is characterized by three properties that cannot be controlled by either the producers of this information or the potential users. All information thus produced is always (1) incomplete, to an unknown extent; (2) biased, in unknown manners and to an unknown extent; and (3) erroneous, to an unknown extent.

As a consequence, without analytical approaches that specifically incorporate the facts that all that is called "information" is not necessarily useful nor utilizable and that all information should be considered as a priori suspect, modeling attempts will fail because of the much too numerous conflicting and, although correct in molecular terms, physiologically invalid reports. This is particularly true in the domain of psychiatric disorders since these are characterized by functionally fuzzy concepts such as behavioral and cognitive 'dysfunctions.'

A rapid survey of the types of data and concepts that will need to be integrated to generate models applicable to these disorders will demonstrate the fact.

19.3 The Concept of Endophenotypes and Functional Genetic Reality

Since 1972, a conceptual approach has slowly emerged around the idea of "endophenotypes." Here, reducing complex behaviors into components, whether they are neurophysiological, biochemical, endocrine, neuroanatomical, cognitive or neuropsychological, is described as an "endophenotype strategy" (Gottesman and Shields, 1973). Symptoms and clinical sub-typing (i.e., depression with or without psychosis) generally are not considered endophenotypes. However, sub-typing in this manner amounts to little more than altering the defining observations of a complex behavior. Decades of applying this approach have resulted in only slightly greater reproducibility than with the broad definition of the disorders themselves, whether in schizophrenia (Gottesman and Erlenmeyer-Kimling, 2001; Weinberger et al., 2001; Hariri and Weinberger, 2003; Heinrichs, 2005), bipolar disorder (Ahearn et al., 2002; Lenox et al., 2002; Glahn et al., 2004), depression (Niculescu and Akiskal, 2001; Hasler et al., 2004), ADHD (Gould et al., 2001; Castellanos and

Tannock, 2002; Doyle et al., 2005; Waldman, 2005) or obsessive-compulsive disorder (Chamberlain et al., 2005; Miguel et al., 2005). It is increasingly obvious that there exist an overwhelming number of potential biological markers associated with psychiatric diseases. However, these often solitary findings frequently have limited reproducibility, both among and within patients, and may only represent state-dependent results. Even with a better definition of what might constitute a credible endophenotype, namely that it must be (1) heritable, (2) associated with illness in the population, (3) manifest in an individual whether or not illness is active (state independent) but age-normed, (4) co-segregate with illness within families, and (5) found in unaffected relatives of probands at a higher rate than in the general population (Gould and Gottesman, 2006), the results remain far below expectations (Braff et al., 2007; Broekman et al., 2007; Calkins et al., 2007). This may not be surprising since the biology of psychiatric disorders is not only complex but further complicated by epigenetic and stochastic contributing factors together with various gene/gene and gene/environment interactions and co-actions. Many gene products interact at many levels, leading to activation of multiple neuronal circuits, which results in behavioral variations. This is further complicated by the fact that there can be more than one pathway to a given behavior.

19.3.1 Biological Functions Result from Interactions Between Integrative and Discontinuous Mechanisms

In effect, the idea that endophenotypes could represent defined and quantifiable measures may be somewhat over-simplistic. The concept is clearly dependent upon the argument that endophenotypes involve few genes, and therefore fewer interacting levels, ultimately affecting only a few, if not a single, set of neuronal circuits.

This is reminescent of the views that were put forward in other medical fields, including gerontology, with respect to the benefits to be derived from a knowledge of the human genome sequence (Blumenthal, 2001; Futterman and Lemberg, 2001; Hata, 2002; Kannabiran and Panicker, 2002; Tanaka, 2002; Gerling et al., 2003).

It simply does not take into account the fact that biological functions result from integrative and non-linear processes subject to discontinuities.

The human genome contains much fewer open reading frames (ORFs; about 30,000 ORFs), encoding functional proteins, than was generally predicted. Like all other completed genomes, it contains many "novel" genes with no ascribed functions. Moreover, because of processes such as alternative mRNA splicing, RNA editing, and post-translational protein modifications, one gene can encode several functionally different proteins. Therefore, the functional complexity of an organism far exceeds that indicated by its genome alone. There is often a poor correlation between mRNA abundance and the quantity of the corresponding functional protein present within a cell (Anderson and Seilhamer, 1997; Gygi et al., 1999). Additionally, co-translational and post-translational modification (PTM) events result in a diversity of protein products from a single ORF. These modifications include phosphorylation, sulfation, glycosylation, hydroxylation, *N*-methylation, carboxymethylation, acetylation, prenylation, and *N*-myristohylation. These dynamic processes, together with protein maturation and degradation, control the amount of functionally active protein within a cell (McGregor and Dunn, 2006). However, most proteins carry out their physiological functions by interacting with other proteins. And this is particularly the case in the CNS where, for many receptor sub-units, the variety of potential interaction partners can be bewildering with very significant functional consequences (Collinson et al., 2002; Qiu et al., 2003; Jovanovic et al., 2004; Mody, 2005; Sinagra et al., 2005; Enz, 2007). Hence, it is not merely a matter of what genes are expressed and to what level, but rather of what other potential interaction partners are present and in what state. It is the qualitative aspects that are important here, and not the quantitative considerations. A protein deemed "physiologically important" may be entirely absent or constitutively non-functional without producing deleterious phenotypic effects. This is amply demonstrated by the numerous such instances observed in knockout mice (Scarff et al., 2004; Knobeloch et al., 2005; Bungartz et al., 2006; Hosl et al., 2006; Leaf et al., 2006; Sentman et al., 2006; Spazierer et al., 2006; Kalkonde et al., 2007). Furthermore, the deleterious effects of inactivating mutations affecting a given protein can often be compensated by inactivating mutations simultaneously affecting another protein (Hashimoto et al., 2004; Puolakkainen et al., 2005; Kobsar et al., 2006; Parlato et al., 2006) or by the corrective mechanisms, involving many genes, that its functional absence induces (de Caprona et al., 2004; Gottsch et al., 2005; Coppola et al., 2006; Xu et al., 2006). Ultimately, what appear most likely to be at play in the realm of psychiatric disorders could be the effects of complex and extensive haplotypes, involving numerous genetic loci, with positive as well as negative complementation effects. While providing a coherent explanation for the numerous apparent conflicts associated with psychiatric genetics (Fortune et al., 2003; Tsutsumi et al., 2004; Abou Jamra et al., 2006; Detera-Wadleigh and McMahon, 2006; Williams et al., 2006), this view is also in agreement with numerous puzzling and so far unexplained observations, including discordance among monozygotic twins (Cote and Gyftodimou, 1991; Singh et al., 2002; Petronis et al., 2003; Fraga et al., 2005; Mill et al., 2006; Oates et al., 2006) and non-mendelian segregation within probands families (Kendler and Zerbin-Rudin, 1996; Maher et al., 2002; Zubenko et al., 2004; Camp et al., 2005).

19.3.2 The Effects of Time, Gene Expression Switching and Complex Haplotypes

With the exception of clear congenital malformations or maternal drug abuse, there are few reported cases of "constitutive" mental disorders (onset in early infancy) (Cannon et al., 2003) and most of these are various forms of autism (Courchesne et al., 2004; Hazlett et al., 2005; Larsson et al., 2005). While this certainly could be due to diagnostic difficulties, it could, just as equally, be due to the mechanisms associated with cognitive development. The CNS, much more so than many other organs, is eminently affected by time (the act of being actively alive over a period of significant duration), in terms of both structural anatomy as well as networks interactions (Shenton et al., 2001; Pantelis et al., 2005; Rapoport et al., 2005). Functional evolution over time or, to use the term under which this is better known, ageing, is associated with numerous switches in gene expression patterns (Lochner et al., 2006; Pearce et al., 2007; Shames et al., 2007). This is evident in many organs, the CNS included (Shen et al., 2006; Popesco et al., 2007). Interpreted in terms of functional genetics, this leads to the inescapable conclusion that, depending upon complex haplotypes, the functional situation implemented when time-point "B" will be reached may be radically different from what it was at time-point "A." The numerous reported cases of spontaneous remission in a variety of severe psychiatric conditions (Klingemann and Efionayi-Mader, 1994; Weiss et al., 2000; Lambert and Bickman, 2004; Bischof et al., 2005; Sekine, 2005) may well be striking examples of this.

A rapid survey of the known mechanisms that constitute "cognition" may be useful to obtain a clear idea of the actual scope.

19.4 The Nature of Cognitive Processes

Working memory may be viewed as a multi-component system involving the active maintenance and manipulation of stored information in the service of planning/guiding behavior. This necessarily implicates the subjective perception of time. Impaired spatial working memory and disturbed experience of time are consistent findings in schizophrenia patients and has been related to impairment in frontostriatal connectivity (Bruder et al., 2004; Goodman et al., 2005; Reilly et al., 2007; Vogeley and Kupke, 2007). It is evident that patients with schizophrenia are consistently more impaired on spatial working memory measures than healthy controls. These impairments may be related to social disability and explain some cognitive deficits that characterize the clinical presentation of schizophrenia (Piskulic et al., 2007). However, patients presenting either schizophrenia or bipolar disorder with psychotic features share overlapping neuropsychological impairments. Both are impaired on the spatial span tasks (storage capacity), which require the maintenance and retrieval of stored information. In contrast, only schizophrenia patients show a significant deficit in working memory (search errors), which requires both maintenance and manipulation of information. They exhibit both a mnemonic and an executive dysfunction. Spatial span is particularly important to accurate planning ability in bipolar patients. In contrast, in patients with schizophrenia poor spatial working memory is a significant predictor of planning impairments (accuracy and latency), consistent with failures in goal selection, evaluation and/or execution. Here, initial planning time is positively correlated with the latency to complete a planning sequence. This pattern of slow cognitive processing in schizophrenia patients, only, resembled that reported in patients with basal ganglia disorders. Hence there is a possible common disturbance in fronto-parietal circuitry in the two disorders together with a specific disturbance of fronto-striatal circuitry in schizophrenia that is not present in bipolar disorder (Pantelis et al., 1997; Badcock et al., 2005; Pirkola et al., 2005; Glahn et al., 2006).

19.4.1 Cognition is Driven by Neural-Network Oscillations at Distinct Frequencies

But the cerebral mechanisms associated with cognitive functions can very rapidly become bewilderingly complex. Given a variety of evidence implicating the prefrontal cortex and its dopaminergic circuits in cognition, most of the research conducted to date has focused on genes regulating dopaminergic function. However, the specific neurocognitive processes involved continue to be a matter of debate (Li et al., 2006; Green et al., 2007; Schumacher et al., 2007).

Part of the difficulty is in distinguishing between false positives, pleiotropy and the influence of a general intelligence factor, g. Also at issue is the complexity of the relevant neuromolecular pathways, which make the inference of simple causal relationships difficult (Goldberg and Weinberger, 2004; Diaz-Asper et al., 2006; Savitz et al., 2006).

The involvement of the hippocampus and the prefrontal cortex in cognitive processes, and particularly in learning and memory, has been known for a long time. There is a direct monosynaptic pathway along the projection that connects the ventral CA1 region of the hippocampus and subiculum to specific areas of the prefrontal cortex and the hippocampal to prefrontal cortex synapses are modifiable and can express different forms of plasticity, including long-term potentiation (LTP), long-term depression (LTD), and depotentiation. The available evidence suggests that functional interactions between the hippocampus and prefrontal cortex in cognition (the consolidation of information and working memory) are more complex than previously anticipated, with bi-directional regulation of synaptic strength as a function of the specific demands of tasks. The hippocampal-medial prefrontal cortex pathway apparently functionally integrates discrete sources of hippocampal information via cooperativity between short- and long-term plasticity (Laroche et al., 2000; Izaki et al., 2002; Jay et al., 2004; Craig and Commins, 2005; Kawashima et al., 2006). But, cognitive processes, although critically dependent upon hippocampal and entorhinal cortex integrity, involve intense, long range signaling traffic between many cerebral structures.

Human scalp electroencephalograms (EEG) have demonstrated that global coherence among distant areas increases during cognitive tasks, suggesting that oscillating neural activities work to generate global neuronal assemblies for cognitive functions. During declarative memory operations, oscillatory activity occurs in the gamma (60–90 Hz) and theta (4.5–8.5 Hz) ranges of frequencies (Osipova et al., 2006). Theta oscillations with large amplitudes, which emerge during mental tasks around the frontal midline region, are called "fm" theta. If theta oscillation concerns the global neuronal assemblies, fm theta should be associated with regional activities that depend on task conditions. EEG-related negative blood oxygenation level dependent (BOLD) signal is dominant over anterior medial regions, suggesting a major contribution of negative BOLD to fm theta. But negative and positive BOLD are found over distant regions (Micheloyannis et al., 2005; Mizuhara et al., 2005). Functional connectivity analyses reveal that the connectivity varies remarkably according to the number of cognitive operations required to perform a task and also to mental conditions. In the 'rest' condition, the connectivity is localized, whereas in the 'task' condition, a long-range coherent network is formed by the anterior midline, posterior cingulate and right middle temporal cortices with linking between the right middle temporal and left lateral cortices during numerical processing. Further EEG analyses indicate that the long-range coherent network executing cognitive functions is coordinated in the time window of theta oscillations (Mizuhara et al., 2004).

Long-range phase synchronization is often observed during cognitive tasks and is considered to provide a possible neural principle for the functional integration of distributed neural activities. Synchronization could be reflected at the neuron firing level or at the local field potential and could appear in the scalp EEG under certain conditions on neural spatial and temporal coherence. There is a dominant task-dependent increase of phase synchronization around 14 Hz (in beta frequency) across bilateral parietal sites that are associated with both negative and positive BOLD responses. Functional connectivity analyses of these regions demonstrate that an increase in hemispheric beta synchronization is associated with a linking between the cross-hemispheric regions (left angular gyrus and right superior parietal gyrus) and also among the anterior-posterior regions (right dorso-lateral prefrontal cortex, putamen, and right superior temporal gyrus) (Mizuhara et al., 2005). Thus, the positive BOLD regions (dorso-lateral prefrontal cortex and superior parietal lobule) are linked with other negative BOLD regions. In addition, multi-electrode intra-cranial EEG (iEEG) recordings have provided unequivocal evidence that at many cortical locations, theta oscillations are gated. That is, theta power rises sharply when working memory becomes required, is maintained throughout the memory task, and decreases when working memory is no longer required. Furthermore, in experiments where successful retrieval depends on the long-term retention of items, theta power during encoding predicts subsequent recall (Sederberg et al., 2003). A substantial fraction of electrode sites in occipital/parietal and temporal cortices are gated by the task. Surprisingly, this aspect of working-memory function is virtually absent in the frontal cortex. Furthermore, whereas nearby gated sites (<20 mm) are often, but not always, coherent, distant gated sites are almost never coherent (Raghavachari et al., 2006). This implies that there are local mechanisms for the generation of cortical theta. Indeed, following electrical stimulations, individual CA3 pyramidal cell can activate the CA1 neuronal network in vivo in rat hippocampus by producing simultaneous intracellular gamma and extracellular theta and slow (0.5–1 Hz) frequencies (Mikkonen et al., 2006). Hence, an individual pyramidal cell can contribute to the self-organization of a neuronal small-scale network.

During declarative memory operations, both theta and gamma activities are stronger for the "later remembered" compared with the "later forgotten" items (subsequent memory). In memory retrieval tasks, theta and gamma activity is stronger for recognized items compared with correctly rejected new items (old/new effect). Gamma activity is also stronger for recognized compared with forgotten old items (recognition effect). The effects in the theta band occur over the right parieto-temporal areas, whereas the sources of the effects in the gamma band were identified in the Brodmann area 18/19 (Osipova et al., 2006). The increase in neuronal synchronization in the gamma band in occipital areas may result in a stronger drive to subsequent areas, thus facilitating both memory encoding and retrieval. Alternatively, the gamma synchronization might reflect representations being reinforced by top-down activity from higher-level memory areas.

19.4.2 Dual Oscillations and Directional Firing Patterns Encode Multiple Units of Information

Hence, in the hippocampus, interacting oscillations in the theta and gamma frequency range are part of a common functional system. Gamma activity arises when pools of interneurons receive a tonic or slowly varying excitation. The frequency of the oscillation depends upon the strength of this excitation and on the parameters regulating the inhibitory coupling between the interneurons. The interneuron network output is then imposed upon pyramidal neurons in the form of rhythmic synchronized inhibitory post-synaptic potentials (IPSPs) (Traub et al., 1996). These oscillations apparently form a coding scheme that is used in the hippocampus to organize the readout from long-term memory of the discrete sequence of upcoming places, as cued by current position (Jensen and Lisman, 1996a, 1996b; Bose et al., 2000; Hok et al., 2007). Plots of the theta phase of spikes vs. position on a track show a systematic progression of phase as rats run through a place field. This is termed the phase precession (Kamondi et al., 1998; Maurer et al., 2006a, b). Two cells with nearby place fields show a systematic difference in phase, as indicated by a cross-correlation having a peak with a temporal offset that is a significant fraction of a theta cycle (Skaggs et al., 1996). Several different decoding algorithms demonstrate the information content of theta phase in predicting the animal's position. It appears that small phase differences corresponding to jitter within a gamma cycle do not carry information (Traub et al., 1996). This evidence, together with the finding that principal cells fire preferentially at a given gamma phase (Osipova et al., 2006), supports the concept of theta/gamma coding: a given place is encoded by the spatial pattern of neurons that fire in a given gamma cycle (the exact timing within a gamma cycle being unimportant) while sequential places are encoded in sequential gamma sub-cycles of the theta cycle, albeit with different discrete theta phase (Lisman and Otmakhova, 2001; Lisman, 2005; Smith and Mizumori, 2006). CA3 place fields maintain a greater degree of population coherence than CA1 place fields after a rearrangement of salient landmarks in an environment. CA3 place field appears to store information about the spatiotemporal sequences of place fields, starting with the first exposure to a cue-altered environment, while CA1 place fields store this information only on a temporary basis. The lateral and medial entorhinal cortex (EC) appears to convey fundamentally different representations to the hippocampus, with spatial information conveyed by the medial EC and non-spatial information conveyed by the lateral EC. The dentate gyrus and CA3 regions may create configural "object-plus-place" (item-plus-context) representations that provide the spatiotemporal context of an episodic memory (Knierim et al., 2006).

It appears that this general form of coding is not restricted to readout of information from long-term memory in the hippocampus because similar patterns of theta/gamma oscillations have been observed in multiple brain regions, including regions involved in working memory and sensory integration (Vinogradova, 2001; Lisman et al., 2005). Hence, dual oscillations appear to encode multiple units of information (items) in a way that preserves their serial order. Here, theta would provide the absolute phase reference needed for the encoding order. Theta/gamma coding therefore bears some relationship to the concept of "word" in digital computers, with word length corresponding to the number of gamma cycles within a theta cycle, and discrete phase corresponding to the ordered "place" within a word.

A fundamental property of many associative memory networks is the ability to decorrelate overlapping input patterns before information is stored. In the hippocampus, this neuronal pattern separation is expressed as the tendency of ensembles of place cells to undergo extensive 'remapping' in response to changes in the sensory or motivational inputs to the hippocampus (Knierim and McNaughton, 2001; Yoganarasimha et al., 2006). Remapping is expressed under some conditions as a change of firing rates in the presence of a stable place code ('rate remapping'), and under other conditions as a complete reorganization of the hippocampal place code in which both place and rate of firing take statistically independent values ('global remapping') (Yoganarasimha and Knierim, 2005; Leutgeb et al., 2006). The nature of hippocampal remapping can be predicted by ensemble dynamics in place-selective grid cells in the medial entorhinal cortex, one synapse upstream of the hippocampus. Whereas rate remapping is associated with stable grid fields, global remapping is always accompanied by a coordinate shift in the firing vertices of the grid cells. Grid fields of co-localized medial entorhinal cortex cells move and rotate in concert during this realignment. In contrast to the multiple environment-specific representations coded by place cells in the hippocampus, local ensembles of grid cells thus maintain a constant spatial phase structure, allowing position to be represented and updated by the same translation mechanism in all environments encountered (Fyhn et al., 2007).

19.4.3 CNS Anatomy and the Mechanisms of Information Gathering and Storage

One of the basic obstacles in explaining why cortical neurons produce action potentials in a particular pattern is the lack of knowledge of the identity and number of input neurons in the required detail. In most cortical areas many populations of input axons and several distinct populations of recipient neurons are mixed in space, making synaptic connections difficult to predict. The cortical area with the least heterogeneous neuronal population and the smallest number of extrinsic inputs is probably the CA1 area, one reason for its popularity for studying the cortical network. The alignment of the somata and dendrites of pyramidal cells into defined layers and the laminar segregation of much of the extrinsic and intrinsic inputs provide the best chance for defining the synaptic relationships of distinct cell types and the basic cortical circuit.

The pyramidal cells are generally considered to form a single population, but there may be at least three distinct groups, which do not necessarily share the same inputs and response properties. Pyramidal cells in the compact layer of stratum pyramidale next to stratum radiatum are weakly immunopositive for calbindin and are smaller than pyramidal cells more loosely arranged towards stratum oriens, which are calbindin immunonegative (Baimbridge and Miller, 1982). Soma size usually correlates with the size of the axonal arborization, but, to our knowledge, the difference in axonal projections between calbindin negative and positive neurons has not been tested. A third population of pyramidal cells is located in stratum radiatum (Gulyas et al., 1998), some of them being at the border with stratum lacunosum-moleculare. These cells are distinct from the other two populations as they project uniquely to the accessory olfactory bulb (van Groen and Wyss, 1990). Unlike the other two populations they may have local axon collaterals also within stratum radiatum, in addition to striatus oriens.

There are five known significant glutamatergic inputs to CA1 pyramidal cells: from CA3 pyramidal cells, entorhinal cortical pyramidal cells, the thalamus, CA1 pyramidal cells and the amygdala. Their spatio-temporal interactions and modulation hold the key to explaining the role of this cortical area. Numerically the largest extrinsic glutamatergic input to the CA1 area is from the ipsi- and contralateral CA3 pyramidal cells terminating in stratum radiatum and oriens, with a sharp cut off at the radiatum–lacunosum-moleculare border. This input has a topographical organization with as yet unknown consequences, as the CA3 pyramids closest to the CA1 area innervate only stratum oriens, whereas those closest to the dentate hilus innervate only stratum radiatum (Ishizuka et al., 1990; Li et al., 1994). Therefore, the basal and apical dendrites of pyramidal cells and interneurons with dendrites restricted to striatus oriens or radiatum receive input from different individual CA3 pyramidal cells.

Integration of sensory and motor processing underlies social behavior. This requires processing of multimodal sensory information by the hippocampus, and in particular, by its input and output structures.

In the hippocampus, CA3 neurons show uniform multimodal, mainly inhibitory, rapidly habituating sensory responses (Abenavoli et al., 2002), while a substantial part of CA1-subicular neurons have phasic reactions and patterned on-responses, depending on the characteristics of the stimuli (Klausberger et al., 2003; Klausberger et al., 2004).

These differences result from the organization of the afferent inputs to CA1 and CA3. The brainstem reticular input is largely responsible for the characteristic tonic functional responses of CA3, concurrently with suppression of CA1 synaptic responses (Jiang and Khanna, 2004). Before entering the hippocampus, these inputs are additionally preprocessed at the medial septal nucleus and the nucleus of diagonal band (MS-DB) relay, where they are smoothed and frequency-modulated in the range of theta oscillations (Jackson and Bland, 2006). Here, new sensory stimuli produce inhibitory reset, after which synchronized theta-modulation is triggered (Bland et al., 2006; Jackson and Bland, 2006). Other stimuli, appearing in the background of the ongoing theta oscillation, do not evoke any responses from the hippocampal neurons. Thus, theta-modulation can be regarded as a mechanism of attention arousing, which prolongs response to a selected stimulus while simultaneously protecting its processing against interference (Vinogradova et al., 1998; Watanabe et al., 2006). The cortical inputs to the hippocampus deliver highly differentiated information for analysis at the highest levels of the neocortex, through the intermediary of the entorhinal cortex and presubiculum (Sommer and Wennekers, 2001). However, only CA1-subiculum receives this information directly (Apostolova et al., 2006; Wang et al., 2006). Before entering into CA3, it is additionally pre-processed at the fascia dentata (FD) relay, where secondary signal simplifications occur (Reid et al., 2001; Wohrl et al., 2007). As a result, CA3 only receives messages indicating the presence and levels of incoming signals in each of its two input pathways (MS-DB and FD), and performs the relatively simple function of determining their degree of match/mismatch and relative weights. For this comparator system, the presence of signal in the reticulo-septal input is equivalent to the quality of novelty (Redish and Touretzky, 1998; Martin et al., 2005). The cortical signal appears with some delay, following its analysis in the neocortex and shaping in the prehippocampal structures (Floresco et al., 1997; Jackson, 2004). Its amplitude is gradually increased due to LTP-like incremental changes in pyramidal (PP) and mossy fiber synapses (Krucker et al., 2002; Huang et al., 2004). The CA3 neurons with synapses potentiated by cortical input do not respond to sensory stimuli (Wiebe et al., 1997; Moxon et al., 2003). Hence, increased transmission of the cortical signals can be regarded as signifying "familiarity," terminating the reactive state of the CA3 neurons (Freedman et al., 1996; Ghisolfi et al. 2002). The integrity of both inputs is necessary for gradual habituation of sensory responses in the hippocampus. The output signals of CA3 flowing through the precommissural fornix to the output-relay locus subcoeruleus (LS) nucleus and to the brain-stem structures have strong regulatory influences over the level of brain activity attached to "arousal," an important condition for processing and registration of information (Flach et al., 1996; Mavanji and Datta, 2003). The primary targets of this output signal are the subcoeruleus and raphe nuclei, which suppress activity of the ascending excitatory reticular formation (RF) (Datta et al., 1998; Sugaya et al., 1998; Jones, 2004; Rudomin et al., 2004). In the background state, activity of CA3 neurons through the intermediary of raphe keeps RF under tonic inhibitory control. Inhibition of the majority of CA3 pyramidal neurons during a novel stimulus decreases the volume of its output signal to raphe and releases RF from tonic inhibition (increase in level of forebrain activity signifying "arousal") (Jiang and Khanna, 2004). When the responses of CA3 neurons habituate, the initial high background activity, as well as tonic suppression of RF is reinstated (Vinogradova and Brazhnik, 1977). Activity in the second output pathway of CA3 (via Schaffer's collaterals to CA1) can block access of cortical signals from PP to CA1 neurons by activating the local system of inhibitory neurons, or by shunting the propagation of signals to apical dendrites (Jiang and Khanna, 2006). Thus, CA3 can act as a filter controlling information transmission by CA1 (Csicsvari et al., 2000). At any given moment, such transmission is allowed only in those CA1 neurons that receive signals from CA3 neurons, and are responding to the sensory stimulus by suppression of their activity (Vinogradova et al., 1999). Disconnection of the CA3 output fibers results in disappearance of habituation in all its target structures (raphe, RF, CA1) (Vinogradova and Brazhnik, 1977). The output signal of CA1-subiculum flows, via the postcommissural fornix, to the chain of structures of the main limbic circuit: mammillary bodies (medial nucleus), anterior thalamic nuclei (mainly antero-ventral nucleus), and cingulate limbic cortex (mainly posterior area). In each of these links, the signal is additionally processed. Habituation is nearly absent in these structures. Instead, strong incremental dynamics are observed under the form of various types of signal shaping, often accompanied by changes in levels and structures of background activities (Vinogradova et al., 1998; Wong, 1997). Within this output circuit, the farther the output structure is from the hippocampus, the more repetitive the stimulus needs to be for shaping sensory responses. This is why this system is regarded as both a chain of integrators, where each link starts to respond only after the previous link has initiated a response, and, as a delay line, preventing premature fixation of spurious, irrelevant, low probability signals (Kaut and Bunsey, 2001; Heckers et al., 2002; Moses et al., 2005; Steffenach et al., 2005). The responses in the higher link of this system, the posterior limbic cortex, may be regarded as the ultimate signal for information fixation in the non-primary areas of the neocortex (Barbas and Blatt, 1995; Barbas, 2000; Ghashghaei and Barbas, 2001). In this way, the two morpho-functional modules of the hippocampus, the regulatory circuits (based on CA3) and the informational circuits (based on CA1), perform the unified functions of attention arousing and initial stages of memory trace fixation.

However, all this is dependent upon complex receptor trafficking as well as dynamic synaptic remodelling at both neuronal and glial levels.

19.5 Neurotransmitter Networks and Glial Populations

GABA is the major inhibitory neurotransmitter in the mammalian brain. Its fast actions are mediated through ligand-gated anion channels, GABA type A ($GABA_A$) receptors, which are distributed throughout the brain. Receptor activation at cell membrane alters the conformation of the receptor normally leading to opening of the ionophore and anion flux, hyperpolarization of the cell, and inhibition of neuronal excitability. The $GABA_A$ receptor subtypes are composed of five subunits and are formed by temporal and spatial regulation of subunit $\alpha1$–6, $\beta1$–3, $\gamma1$–3, δ, ε, θ, $\rho1$–3, and π (Barnard et al., 1998; Bonnert et al., 1999; Sinkkonen et al., 2000) expression in brain regions and/or by cellular regulation of assembly to pentameric receptor complexes. $GABA_A$ ergic drugs, such as benzodiazepines, barbiturates, and volatile anesthetics, enhance the actions of GABA (Sieghart, 1995) and are used to treat anxiety, insomnia, and epilepsy, and in general anesthesia. Many drug effects on $GABA_A$ receptor function have been shown to depend on subunit combinations (subtypes) and even on critical amino acids in specific subunits (Korpi et al., 2002).

19.5.1 Regional Distribution and Molecular Structures of $GABA_A$ Receptors

There is an extraordinary heterogeneity in the distribution of $GABA_A$-receptor subunits, as evidenced by abrupt changes in immunoreactivity along well-defined cytoarchitectonic boundaries and by pronounced differences in the cellular distribution of subunits among various types of neurons. Thus, functionally and morphologically diverse neurons are characterized by a distinct $GABA_A$-receptor subunit repertoire. Multiple staining experiments identified 12 subunit combinations in defined neurons. The most prevalent combination is the triplet α1/β2,3/γ2, detected in numerous cell types throughout the brain. An additional subunit (α2, α3, or δ) sometimes is associated with this triplet, pointing to the existence of receptors containing four subunits. The triplets α2/β2,3/γ2, α3/β2,3/γ2, and α5/β2,3/γ2 are also identified in discrete cell populations. The prevalence of these seven combinations suggests that they represent major $GABA_A$-receptor subtypes. Five combinations also apparently lack the β2,3-subunits, including one devoid of γ2-subunit: α1/α2/γ2; α2/γ2, α3/γ2; α2/α3/γ2 and α2/α5/δ (Fritschy and Mohler, 1995).

The γ2 subunit is necessary for the expression of the full benzodiazepine pharmacology of $GABA_A$ receptors and is one of the major subunits in the brain. The γ2 subunit is present in type 2, "symmetrical" synapses in most of areas the brain, with the exception of the granule cell layer of the cerebellum. It is frequently co-localized in the same synaptic junction with the α1 and β2/3 subunits, often occurring in multiple clusters in the synapses, coincident with the junctional membrane specialization of the active zone. In the hippocampus, the γ2 subunit is present in basket cell synapses on the somata and proximal dendrites and in axo-axonic cell synapses on the axon initial segment of pyramidal and granule cells. Some synapses on the dendrites of GABAergic interneurones have dense clusters of γ2, α1 and β2/3 subunits. In the cerebellum, the γ2 subunit is present in both distal and proximal Purkinje cell dendritic synapses established by the stellate and basket cell, respectively, while weakly so on basket cell synapses in the soma of Purkinje cells. Synapses on interneuron dendrites are more densely labeled for the γ2, α1 and β2/3 subunits than synapses on Purkinje or granule cells, Golgi synapses being very weakly labeled. In the globus pallidus, many type 2 synapses are labeled for the γ2 subunit together with α1 and β2/3 subunits. Thus, the γ2 and β2/3 subunits receptor channels are highly concentrated in GABAergic synapses that also contain the α1 and β2/3 subunits. Channels containing the γ2 subunit are expressed in synapses on functionally distinct domains of the same neuron receiving GABA from different presynaptic sources. There are quantitative differences in the density of $GABA_A$ receptors at synapses on different cell types in the same brain area (Somogyi et al., 1996).

Pyramidal cells, expressing at least 14 subunits of the heteropentameric $GABA_A$ receptor, receive GABAergic input on their soma and proximal dendrites from basket cells, activating $GABA_A$ receptors and containing either parvalbumin or cholecystokinin and vasoactive intestinal polypeptide. The properties of $GABA_A$ receptors are determined by the subunit composition, and synaptic receptor content governs the effect of the presynaptic neuron. Synapses made by parvalbumin-negative terminals exhibit five times more α2 subunits than synapses made by parvalbumin-positive basket cells, whose synapses were frequently negative. This difference is likely to be due to specific $GABA_A$ receptor α subunit composition, because neither synaptic size nor density of the β2/3 subunits, indicating total receptor content, are different in these two synapse populations. Synapses established by axo-axonic cells on axon initial segments show an intermediate number of α2 subunit compared to those made by basket cells but, due to their smaller size, the density of the α2 subunit is higher in synapses on the axon. Because the two basket cell types innervate the same domain of the pyramidal cell, pyramidal cells have mechanisms to target $GABA_A$ receptors, under presynaptic influence, preferentially to distinct synapses and the two basket cell types act via partially distinct $GABA_A$ receptor populations (Nyiri et al., 2001).

Within the rat striatum, the θ subunit (50.5% sequence identity with the β1 subunit) co-assembles with α2, β1, and γ1, suggesting that $GABA_A$ receptors consisting of arrangements other than α + β + γ, δ, or ε do exist. Expression of α2β1γ1θ in transfected mammalian cells leads to the formation of receptors with a 4-fold decrease in the affinity for GABA compared with α2β1γ1. This subunit has a unique distribution, with studies so far suggesting significant expression within monoaminergic neurons of both the human and monkey brain (Bonnert et al., 1999).

19.5.2 Interactions with Other Neurotransmitter Networks

Hippocampal pyramidal cells receive GABAergic innervation from several distinct interneurons (Freund and Buzsaki, 1996). For example, axo-axonic cells innervate only axon initial segments, basket cells innervate mainly somata and the proximal dendrites, and other interneurons innervate only dendrites. The same postsynaptic domain of pyramidal cells may be targeted by more than one class of interneuron. Thus, pyramidal cell somata are innervated by two distinct basket cells expressing either parvalbumin (PV) or cholecystokinin (CCK). These distinct basket cells differ in their soma position, local and subcortical innervation, and in the presynaptic control of transmitter release (Hajos et al., 1998); (Katona et al., 1999), predicting distinct roles in the hippocampal network.

Interneurons expressing M2 muscarinic acetylcholine receptor and various types of M2-positive axon terminals have been described in the hippocampal formation. In the CA1 subfield, neurons immunoreactive for M2 have horizontal dendrites; they are located at the stratum oriens/alveus border and have an axon that project to the dendritic region of pyramidal cells. In the CA3 sub-field and the hilus, M2-positive neurons are multipolar and are scattered in all layers except stratum lacunosum-moleculare. In stratum pyramidale of the CA1 and CA3 regions, striking axon terminal staining for M2 are observed, surrounding the somata and axon initial segments of pyramidal cells in a basket-like manner.

GABA is also present in the somata of most M2-immunoreactive interneurons, as well as in the majority of M2-positive terminals in all layers. The calcium-binding protein parvalbumin is absent from practically all M2-immunoreactive cell bodies and dendrites. In contrast, many of the terminals synapsing on pyramidal cell somata and axon initial segments co-localise with parvalbumin and M2, suggesting a differential distribution of M2 receptors on the axonal and somadendritic membrane of parvalbumin-containing basket and axo-axonic cells. However, the co-existence of m2 receptors with the calcium-binding protein calbindin and the neuropeptides cholecystokinin and vasoactive intestinal polypeptide is rare throughout the hippocampal formation. Only calretinin and somatostatin show an appreciable degree of co-localization with M2 (20% and 15%, respectively). Some of the M2-positive cells in stratum oriens project to the medial septum, accounting for 38% of all projection neurons. Thus, there is a differential distribution of M2 receptor immunoreactivity on the axonal vs. the somadendritic membranes of distinct interneuron types and acetylcholine, via M2 receptors, which may reduce GABA release presynaptically from the terminals of perisomatic inhibitory cells, while acting to increase the activity of another class of interneuron, which innervates the dendritic region of pyramidal cells (Hajos et al., 1998).

Cortical excitatory and inhibitory amino acid receptors are regulated by the action of phosphorylating/dephosphorylating enzymes and the regulatory effects induced by cell depolarization and agonist stimulation are based on similar mechanisms. Alkaline phosphatase increases binding levels for $GABA_A$ and AMPA receptors while PKA has the opposite effect. Increases in cell depolarizations by veratridine leads to an increase in labeled $GABA_A$ receptors, but to decreases in labeled AMPA receptors. Increases in binding are differentially blocked by the phosphatase inhibitors sodium β-D-glycerol phosphate and sodium vanadate, while decreases in binding are blocked by protein kinase inhibitor. Agonist stimulation of $GABA_A$ and AMPA receptors leads to a decrease in receptor binding which can be blocked in both cases by protein kinase inhibitor (Pasqualotto et al., 1993).

The multiple sources of GABA, released by distinct interneurons, and the large variety of distinct $GABA_A$ receptors raise the possibility that the segregation of inputs is supported by molecular specializations in postsynaptic receptors. The α_1, $\beta_{2/3}$, and γ_2 subunits have been demonstrated in many GABAergic synapses on pyramidal cells (Nusser et al., 1996; Somogyi et al., 1996). However, the α_2 subunit was found more frequently in synapses on axon-initial segments than on somata (Nusser et al., 1996), and, on the latter, α_2 subunit immunoreactivity was present at much higher levels in synapses formed by PV-negative (presumably CCK-positive) basket cells than in synapses formed by PV-positive cells (Nyiri et al., 2001). There is thus an input-specific enrichment of α_2 subunit-containing receptors and the ratio of α_2 vs. α_1 containing receptors is important, because the α_1 and α_2 subunit-containing receptors are responsible for different behavioral and pharmacological responsiveness in mice (McKernan et al., 2000; Rudolph et al., 2001).

Oscillations in neuronal population activity within the gamma frequency band (>25 Hz) are correlated with cognition: Gamma oscillations could bind together features of a sensory stimulus by generating synchrony between discrete cortical areas (Eckhorn et al., 1988; Singer and Gray, 1995). Morphine and beta-endorphin disrupt this long-range synchrony of gamma oscillations while leaving the synchrony of local oscillations relatively intact. The effect is caused by a decrease in type $GABA_A$ receptor-mediated inhibition of both excitatory pyramidal cells and inhibitory interneurons. The effects of morphine on gamma oscillations are blocked by mu-opioid receptor antagonists but not by antagonists of delta or kappa receptors. Synaptic excitation from pyramidal cells is no longer balanced by synchronous inhibitory postsynaptic potentials allowing morphine to also produce burst firing in interneurons; because synaptic excitation from pyramidal cells is no longer balanced by synchronous inhibitory postsynaptic potentials (Whittington et al., 1998). The loss of synchrony of gamma oscillations induced by morphine may constitute one mechanism involved in producing the cognitive deficits that this drug causes clinically.

19.5.3 $GABA_A$ Receptor Endocytosis

Insulin treatment of neurons in culture (Wan et al., 1997) and the kindling model of epileptogenesis (Nusser et al., 1998) have both been shown to increase $GABA_A$ receptor surface number. In contrast to these observations, $GABA_A$ receptors have also been found to be down-regulated by an agonist-dependent mechanism (Tehrani and Barnes, 1997). Both recombinant and neuronal $GABA_A$ receptors can constitutively recycle between the cell surface and an intracellular endosomal compartment (Connolly et al., 1999b). Furthermore, $GABA_A$ receptor levels are reduced upon protein kinase C (PKC) activation (Chapell et al., 1998; Connolly et al., 1999a). Dynamin-dependent endocytosis has been shown to be important in the regulation of cell surface levels of a number of integral membrane proteins (Schmid, 1997), including opioid receptors (Chu et al., 1997), the β-adrenergic receptor (Pitcher et al., 1998), and ionotropic glutamate receptors (Carroll et al., 1999; Luscher et al., 1999; Man et al., 2000). Endocytosis of such membrane proteins involves their recruitment into clathrin-coated pits by adaptor proteins. The target protein-adaptor complex is then capable of interacting with other binding partners, including clathrin, the GTPase dynamin, and its binding partner amphiphysin (Marsh and McMahon, 1999), which are key elements of the endocytotic machinery. Internalisation of $GABA_A$ receptors is mediated by clathrin-dependent endocytosis. $GABA_A$ receptors associate with the adaptin complex AP2 and colocalize with AP2 in cultured hippocampal neurons. Inhibition of endocytosis dramatically affects the miniature IPSC (mIPSC) amplitude, resulting in an increase of function of synaptic $GABA_A$ receptors in cultured hippocampal neurons (Kittler et al., 2000). Thus, the removal of $GABA_A$ receptors from synaptic sites plays a critical role in controlling the efficacy of inhibitory synaptic transmission.

19.5.4 Homeostatic Plasticity of Inhibitory Activity

Neurons can respond to changing activity patterns by altering the array or properties of their voltage-dependent conductances or by adjusting the level of synaptic transmission by controlling the number or properties of ionotropic or metabotropic neurotransmitter receptors (Turrigiano and Nelson, 2000). Adjustments of voltage-dependent conductance is usually referred to as '-intrinsic homeostatic plasticity' as opposed to the 'synaptic homeostatic plasticity' that involves the fine-tuning of synaptic strength (Turrigiano, 1999; Turrigiano and Nelson, 2000). The intrinsic homeostatic plasticity has been known to exist for some time in invertebrate central pattern generators (Golowasch et al., 1999), and has been recently demonstrated in mammalian cortical neurones (Desai et al., 1999; Stemmler and Koch, 1999; Poolos et al., 2002). Homeostatic synaptic plasticity, or synaptic scaling, was discovered in cultured cortical neurons as the adjustment of the quantal amplitude of AMPA receptor-mediated miniature EPSCs following blockade of action potential firing by TTX or GABAergic inhibitory activity by bicuculline (Turrigiano et al., 1998). Such plasticity has been widely demonstrated at central synapses and at the neuromuscular junction (Turrigiano, 1999; Davis and Bezprozvanny, 2001). Driven by the need to extend Hebbian correlative mechanisms of plasticity, the study of synaptic scaling, to date, has mainly focused on GluRs (Turrigiano, 1999; Abbott and Nelson, 2000). The homeostatic plasticity of inhibition has received considerably less attention. Deprivation of neuronal activity by TTX in cultured neurones produces a downsizing of mIPSC amplitudes presumably by a loss of $GABA_A$Rs from the synapses (Davis and Bezprozvanny, 2001). Another study, done at the network level in the amygdala, reported the balance of synaptic weight to be conserved by adjusting the synaptic excitation of principal cells and of GABAergic interneurones such that in intercalated neurons at least, inverse heterosynaptic plasticity tends to compensate for homosynaptic long-term potentiation and depression, thus stabilizing total synaptic weight (Royer and Pare, 2003).

However, the loss of a specific inhibitory input can generate a remarkable adjustment in the control of excitability of cerebellar granule cells. Recent studies have shown direct relationships between an increased tonic inhibition and a presumably homeostatic down-regulation of phasic inhibition. When inserted throughout the forebrain, α6 subunit-containing $GABA_A$Rs become inserted in the membrane at extrasynaptic sites resulting in an enhanced tonic inhibition in CA1 PC while phasic inhibition is down-regulated as the average amplitude of mIPSCs becomes smaller (Wisden et al., 2002).

After its release from interneurons in the CNS, the major inhibitory neurotransmitter GABA is taken up by GABA transporters (GATs). The predominant neuronal GABA transporter GAT1 is localized in GABAergic axons and nerve terminals, where it is thought to influence GABAergic synaptic transmission. The presence of a highly augmented tonic inhibition in GABA transporter GAT-$1^{-/-}$ in animals results in fewer mIPSCs. GAT1 deficiency leads to enhanced extracellular GABA levels resulting in an overactivation of $GABA_A$ receptors responsible for a postsynaptic tonic conductance. Chronically elevated GABA levels also downregulate phasic GABA release and reduce presynaptic signaling via $GABA_B$ receptors thus causing an enhanced tonic and a diminished phasic inhibition. (Jensen et al., 2003). The alterations in the number and properties of $GABA_A$Rs found during development (Hollrigel et al., 1998; Cohen et al., 2000), in pathological conditions such as epilepsy (Coulter, 2001), or the hyperexcitability seen following withdrawal from drugs initially potentiating $GABA_A$R function (De Witte et al., 2003) should really be considered as part of global homeostatic plasticity mechanisms. In the case of pathological alterations, these mechanisms may have ultimately tried, but failed, to readjust neuronal excitability to the levels before the pathogenic disturbance. In this regard, it is interesting to consider that the brain may possess several endogenous 'homeostatic' agents. Through combined effects on different neurones, synapses, receptors and voltage-dependent mechanisms, such agents may achieve a global effect on neuronal excitability in a much shorter time than the equivalent effect that could be accomplished by the activation of homeostatic mechanisms. Neuropeptides may satisfy these criteria, as recently demonstrated by the multitude of effects of NPY on cortical neurones resulting in the global dampening of excitability in the network (Bacci et al., 2002).

19.5.5 Neurotrophins, Synaptic Plasticity and Cognitive Processes

Neurotrophins also exert antagonistic effects during activity-dependent synaptic plasticity processes. For example, BDNF and its receptor TrkB play a crucial role in the induction and maintenance of hippocampal long-term potentiation (LTP) (Poo, 2001). In contrast, deletion of $p75^{NTR}$ leads to a significant impairment in long-term depression (LTD) without affecting LTP. $p75^{NTR}$ localizes in dendritic spines, in addition to afferent terminals, of CA1 neurons and deletion of $p75^{NTR}$ in mice selectively impairs N-methyl-D-aspartate (NMDA) receptor functions, through a decrease in the expression of NR2B, an NMDA receptor subunit uniquely involved in LTD (Woo et al., 2005). Furthermore, expression of the AMPA receptor subunits GluR2 and GluR3, but not GluR1 or GluR4, are significantly altered in the hippocampus of $p75^{NTR}$ -deficient mice, the overall expression level of GluR2 being reduced by about half while that of GluR3 is increased about 3.5-fold (Rosch et al., 2005).

Deletion of $p75^{NTR}$ has been shown to improve spatial learning (Greferath et al., 2000). Furthermore, TNF-α binding to the TNF receptor, which, like $p75^{NTR}$, is a member of the TNFR superfamily, enhances synaptic efficacy by increasing surface expression of AMPA receptors (Beattie et al., 2002). $p75^{NTR}$ was also shown to mediate a rapid switch in neurotransmitter release in individual sympathetic neurons acting on cardiac myocytes (Yang et al., 2002). Stimulating neurons with NGF promotes the release of norepinephrine, resulting in an increased twitching frequency of myocytes, whereas application of BDNF triggers the release of acetylcholine, which has the opposite effect. The NGF-induced release of the excitatory transmitter is mediated by TrkA (Lockhart et al., 1997), whereas the BDNF-induced switch to cholinergic inhibitory transmission depends on $p75^{NTR}$ (Yang et al., 2002). Here, activation of presynaptic CaMKII is both necessary and sufficient for a shift from excitatory to inhibitory transmission (Slonimsky et al., 2006).

This switching mechanism has direct consequences upon the processes associated with working memory (cholinergic neurons in the basal nuclei of Meynert) and decision-taking (noradrenergic neurons in the locus ceruleus). Disorganizing either of these two processes leads, in rhesus macaques, to Alzheimer's disease-like neurological dysfunctions. However, while impairment of the cholinergic aspect manifests itself in delayed visual differentiation and as a significant decrease in correct responses, dependent upon both the duration of the delay and the type of visual information, impairment of the noradrenergic component is characterized by an increase in refusal to take decisions that is independent of the duration of delays and the type of visual information (Dudkin et al., 2005).

19.5.6 The Critical Roles of Astrocytes and Glial Cells

Astrocytes constitute the largest glial population in the mammalian brain. Evidence suggest that they are involved in provision of metabolic substrates for neurons (Voutsinos-Porche et al., 2003), maintenance of the extracellular ionic environment and pH (Fields and Stevens-Graham, 2002), uptake of neurotransmitters (Danbolt, 2001), regulation of synaptic strength and plasticity and provide a pathway for synaptic cross-talk (Pascual et al., 2005). They interact with one another to form large networks (John et al., 1999), and associate with neurons, oligodendrocytes and endothelia (Janzer and Raff, 1987). Astrocytes extend ezrin and radixin-containing thin lamellate processes into the neuropil, in particular around synapses, where they can modulate synaptic function or mediate glial-neuronal communication (Derouiche and Frotscher, 2001). The structural and functional properties of these processes suggest that they represent a separate astroglial compartment (Derouiche et al., 2002). Thus, parts of the brain (such as the hippocampus) are divided by astrocytes into separate compartments, each one the sole domain of an individual astrocyte (Bushong et al., 2002).

The association of synaptic plasticity with the RhoA, RhoB, Rac1 and Cdc42 molecular switches regulating the actin cytoskeleton is well documented (Olenik et al.,

1997; Tashiro and Yuste, 2004; Weernink et al., 2004; Zhang et al., 2005). In neurons, moesin is concentrated in specialized microdomains such as filopodia, microvilli, microspikes, and retraction fibres and is apparently involved in their dynamic restructuring (Amieva and Furthmayr, 1995, Sousa et al., 2006). The axonal cell adhesion molecule L1 binds directly to members of the ERM (ezrin-moiesin-radixin) family and this association appears to coordinate axonal morphogenesis, in particular the elaboration of membrane protrusions and axon branching (Dickson et al., 2002). Significant manifold decrease of moesin is associated with developmental impairment of Down syndrome brain, including deteriorated arborisation, neuritic outgrowth, and neuronal migration (Lubec et al., 2001). Hence, the ERM proteins play a key role in both the formation and maintenance of afferent connections and neuronal networks (Paglini et al., 1998; Majores et al., 2005; Kitajiri et al., 2004).

Besides these structural functions, ezrin and moesin are also involved in neuroreceptor cellular dynamics and functions such as receptor-laden transport carriers (RTCs) tethering and fusion with the plasma membrane (Deretic et al., 2004), membrane-cytoskeleton ligation of neurotransmitter receptors in the down-regulation of NMDA receptor activity and in NMDA-induced dendrite remodeling (Penzes et al., 2001; Park et al., 2003; Norenberg et al., 1999; (Bretscher et al., 2002). Furthermore, The Na^+/H^+ exchanger regulatory factor (NHERF) interacts specifically with and recruits 5-hydroxytryptamine type 4 receptor (5-HT4R), involved in learning, feeding and respiratory control, to microvilli where it localizes with activated ezrin (Joubert et al., 2004). In addition to mediating Na(+)/H(+) transport, the NHE1 Na(+)/H(+) exchanger interacts with ERM proteins, which tethers NHE1 to cortical actin cytoskeleton, to regulate cell shape, adhesion, motility, and resistance to apoptosis. There, NHE1 acts as a scaffold for recruitment of a signaling complex that includes ERM, PI3K, and Akt (Wu et al., 2004).

Slow NMDA channels (deactivation approximately 150 ms) form networks that multiplex different memories in different gamma subcycles of a low frequency theta oscillation. The NMDA channels are in the synapses of recurrent collaterals and govern synaptic modification in accord with known physiological properties. Because slow NMDA channels have a time constant that spans several gamma cycles, synaptic connections will form between cells that represent different memories. This enables brain structures that have slow NMDA channels to store hetero-associative sequence information in long-term memory. Recall of this stored sequence information can be initiated by presentation of initial elements of the sequence. The remaining sequence is then recalled at a rate of one memory every gamma cycle. Recall at gamma frequency works well if slow NMDA channels provide the dominant component of the EPSP at the synapse of recurrent collaterals: The slow onset of these channels and their long duration allows the firing of one memory during one gamma cycle to trigger the next memory during the subsequent gamma cycle (Jensen et al., 1996). An interesting feature of the readout mechanism is that the activation of a given memory is due to cumulative input from multiple previous memories in the stored sequence, not just the previous one. The network thus stores sequence information in a doubly redundant way: Activation of a memory depends on the strength of synaptic inputs from multiple cells of multiple previous memories.

The cumulative property of sequence storage has support from the psychophysical literature. Cumulative learning also provides a solution to the disambiguation problem that occurs when different sequences have a region of overlap. Furthermore, coupling an auto-associative network to a hetero-associative network allows the storage of episodic memories (a unique sequence of briefly occurring known items). The auto-associative network (cortex) captures the sequence in short-term memory and provides the accurate, time-compressed repetition required to drive synaptic modification in the hetero-associative network (hippocampus) (Jensen and Lisman, 1996a). Hence, network oscillations, recurrent collaterals, AMPA channels, NMDA channel subtypes, the ADP, and the AHP can act together to accomplish memory storage and recall.

19.5.7 Astrocyte Coupling, Synaptic Remodeling and Neurotrophins

In virtually all regions of the CNS, gap junctions couple astrocytes into extensive networks (Rash et al., 2000). Astrocytes have been reported to express connexins (Cx) 26, 30, and 43 (Giaume et al., 1991; Kunzelmann et al., 1999; Nagy et al., 2001) and to form, in vivo, two classes of gap junctions with each other, one containing Cx26 and the other containing Cx43 and Cx30 (Altevogt and Paul, 2004). Cx hemichannels permit the rapid exchange of ions and of small molecules between the cytoplasm and the extracellular space (Kondo et al., 2000; Cotrina et al., 2000; Bruzzone et al., 2001) and have been implicated in the regulation of various physiological and signal transduction processes, including calcium wave propagation (Fahrenfort et al., 2005; Plotkin et al., 2002; Goodenough and Paul, 2003; Giaume, 1996; Naus et al., 1997) as well as in the pathogenesis of certain neurological disorders (Contreras et al., 2002; Castro et al., 1999; Liang et al., 2005). In basal condition, Cx30, reported to be subject to tissue specific splicing (Essenfelder et al., 2005; Dahl et al., 1996), is expressed only in grey matter astrocytes with distinct regional patterns in the developing and adult brain (Rash et al., 2001; Princen et al., 2001). Neuronal-glial interactions play an important role in information processing in the CNS (Alvarez-Maubecin et al., 2000) and neurotoxic kainate treatment induces strong and region-specific changes in astroglial Cx30 mRNA levels, suggesting that neuronal activity may influence the level of intercellular communication between astrocytes through gap junctions channels (Condorelli et al., 2002). Glial pathways of junctional communication appear to be determined by the connexin composition and conductance regulation of junctional channels (Nagy and Rash, 2000) and exacerbated hemichannel opening, which contributes to the loss of chemical gradients across the plasma membrane, is reported to occur in metabolically inhibited cells, including cortical astrocytes (Saez et al., 2003). Electrophysiological studies indicate that hemichannels have a low open probability under physiological conditions but may have a much higher open probability under certain pathological conditions (Pfahnl and Dahl, 1999, Muller et al., 2002).

Cx hemichannels, some of which which incorporate calmodulin as an integral regulatory subunit (Peracchia et al., 2000), tend to be closed by negative membrane potentials, high concentrations of extracellular $Ca2^+$ and intracellular H^+ ions and by protein phosphorylation. They are opened by positive membrane potentials and low extracellular Ca^{2+} (Gomez-Hernandez et al., 2003; Saez et al., 2005). Cx 30 is reported to gate at significantly lower voltage than Cx26, its closest homologue (Dahl et al., 1996), and there is evidence that Cx30 channels are involved in the secretion of enzymes (Levin and Mercola, 2000).

Astrocytes promote the formation and function of excitatory synapses in the CNS. Neuronal coculture with astrocytes or treatment with astrocyte-conditioned medium (ACM) increases the number of inhibitory presynaptic terminals, the frequency of miniature IPSCs, and the number and synaptic localization of $GABA_AR$ clusters during the first 10 days in vitro. ACM from BDNF- or TrkB-deficient astrocytes increases inhibitory pre-synaptic terminals and postsynaptic $GABA_AR$ clusters in wild-type neurons, suggesting that BDNF and TrkB expression in astrocytes is not required for these effects. In contrast, although the increase in the number of inhibitory presynaptic terminals is persistent, no increase is observed in postsynaptic $GABA_AR$ clusters after ACM treatment of hippocampal neurons lacking BDNF or TrkB (Elmariah et al., 2005). This suggests that neurons, not astrocytes, are the relevant source of BDNF and are the site of TrkB activation required for postsynaptic $GABA_AR$ modulation. Furthermore, astrocytes may modulate postsynaptic development indirectly by stimulating Trk signaling between neurons. Hence, astrocytes modulate inhibitory synapse formation via distinct presynaptic and postsynaptic mechanisms.

A significant body of evidence indicates an important role for BDNF in the hippocampal synaptic plasticity (Black, 1999; McAllister et al., 1999; Poo, 2001). In vivo, TrkB signaling directly controls neuronal plasticity associated with learning and long term-memory.

In transgenic mice, over-expression of the full-length TrkB receptor leads to its continuous activation. This is associated with improved learning and memory but an attenuated LTP (Koponen et al., 2004b). The most prominent increase in full-length TrkB expression is found in the cortical layer V pyramidal neurons and dentate gyrus of the hippocampus. This is associated with up-regulated expression of c-fos, fra-2 and junB. The mRNA expression of growth-associated protein 43 (GAP-43) is induced in both the hippocampus and parietal cortex, and that of α-Ca2+/calmodulin-dependent kinase II (CaMKIIα) is reduced in the corresponding regions. However, while the mRNA expression of the transcription factor CREB is not altered in TrkB.TK+ mice, the density of neuropeptide Y (NPY)-expressing cells is increased in the dentate hilus (Koponen et al., 2004a). Furthermore, production of BDNF mRNA in the amygdala is regulated by neuronal activity during fear conditioning. Fear conditioning results in an increase in BDNF protein levels and in TrkB phosphorylation in the amygdala. In addition, TrkB receptor blockade or expression of a dominant-negative TrkB receptor impairs amygdala-dependent learning and memory (Rattiner et al., 2004). Fear conditioning induces the association of Shc and TrkB, Shc and Ras, increase in Ras activity and phosphorylation of

MAPK. Treatment of amygdala slices with BDNF for 15 min increases the levels of active Ras, MAPK-P and Akt-P. BDNF-induced MAPK phosphorylation is completely abolished by MEK inhibitors, and only partially inhibited by farnesyltransferase or PI3K inhibitors. On the other hand, BDNF-induced Akt phosphorylation is unaffected by farnesyltransferase or MEK inhibitors, but can be blocked by PI3K inhibitors (Ou and Gean, 2006). Thus, BDNF is required for the learning of fear and the memory-enhancing effect of BDNF involves the activation of Ras, MAPK and PI3K. BDNF-induced MAPK phosphorylation in the amygdala is mediated via TrkB and the Shc-binding site. Shc binding to TrkB leads to activation of Ras, Raf, and MEK. In addition, BDNF can induce phosphorylation of MAPK via activation of PI3K.

19.6 Systems Biology Applied to Mental Disorders

19.6.1 The Model Building Rationale

Phenotypes and behavior depend on the integrated effects of multiple signaling pathways and molecules, genetic polymorphism and environmental stimuli. This is true for all biological systems, from individual cells all the way to organisms. The brief survey above provides some indication of the density and extent of the integrative effects that constitute the main mechanisms associated with cognitive processes.

In the classical approach to systems biology, it is indispensable to construct databases that are as exhaustive as possible. Furthermore, it is just as vital that the contents of these databases be exploited with the highest possible flexibility (Bono et al., 2003; Nicholson et al., 2004). This results in strategies for updating, archiving, indexing, text-mining, information mapping, etc. of ever increasing complexity.

But in actual fact, how were these databases constructed? It was first necessary to impose filters at the data collection stage. Can one have the least idea of what could constitute a good filter, as opposed to a bad one, when, at this stage, it is impossible to define the types of information that will be required in fine? Indexing strategies had then to be implemented. Can one have the least idea of what could constitute a good strategy, as opposed to a bad one, when, at this stage, it is impossible to define the types of relationships that will have to be eliminated? But that is not all. The information entered in such a database is always incomplete, biased and partly erroneous. The "true" is thus intermixed with the "uncertain,", without possibility to eliminate the "false" and even less to determine in which context the "true" may become "false" (Fig. 19.1A).

But what happens if all information is regarded as suspect? Not only the databases need not be exhaustive anymore, but, their contents, which are largely subject to caution, do not present the least analytical importance. What becomes analytically important are the indices attached to these databases. Not because these indices

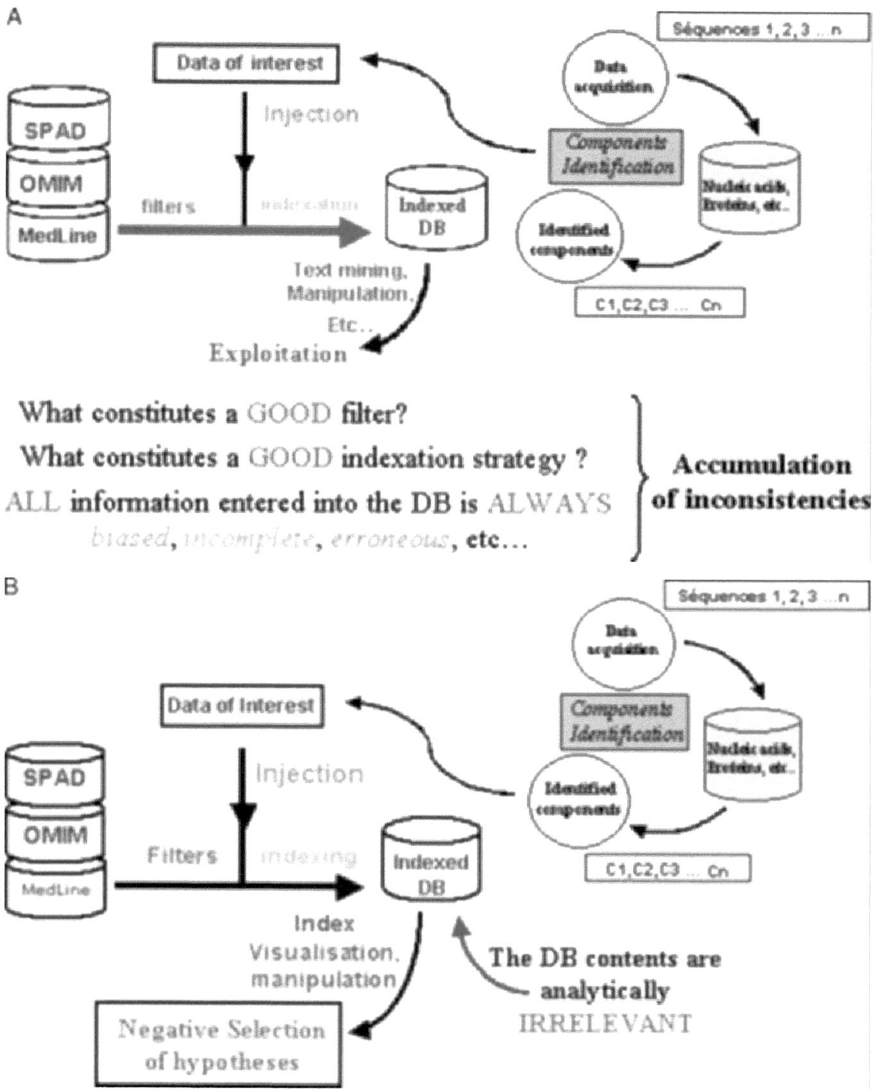

Fig 19.1. (*See Color Plates*)

could be credible *per se*, but because they will allow the very rapid production of a large number of hypotheses that will systematically be subjected to destruction attempts, using the multiple crosschecks allowed by the scientific literature as archived in the public databases. A false hypothesis is very easy to destroy and the

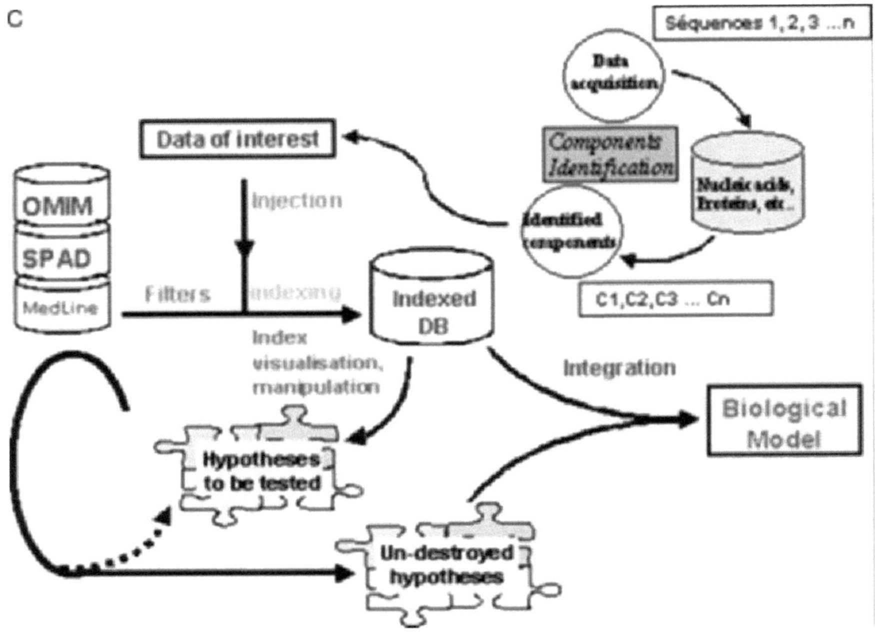

Fig. 19.1 (continued) (*See Color Plates*)

elements that allow the destruction of a hypothesis can then be used to build a new, more solid hypothesis, and so on iteratively, until a hypothesis that cannot be refuted has been obtained. This does not mean that this hypothesis is correct. It merely means that this hypothesis, supported by multiple information-intersects, must be seriously considered (Fig. 19.1B). Undestroyed hypotheses are then merged into meta-hypotheses and the iterative destructive process is applied anew, eventually leading to a very detailed biological model that clearly suggests what should be investigated, how, where, when and why.

This model-building process, functioning on the basis of negative selection of working hypotheses, embodies three major operating principles. First, biological systems are characterized by integrative, non-linear mechanisms (human genome = less than 3×10^4 genes, transcriptome = over 2×10^5 entities, proteome = about 10^6 individual components, etc.). This implies that **any analytical approach tending to linearity by approximations will only generate artifacts**. Second, the functions of biological components are context-dependent (Chapman and Asthagiri, 2004; Grant et al., 2006; Pullikuth et al., 2006). This implies that, within biological systems, events dictate to contexts how to distort, contexts dictate to components how to behave and components dictate to events how to arise. This, in turn, means that **analytical treatments in terms of components must be abandoned**. It becomes necessary to adopt a relativistic and event-driven analytical approach.

Thirdly, published information, be it under the form of text, images, charts, sequence, etc. is always incomplete (to an unknown extent), biased (in an unknown manner and to an unknown extent) and erroneous (to an unknown extent). This implies that **analytical approaches based upon positive selection** (information is considered valid) **will necessarily generate incoherencies**. This approach allows to very rapidly reconstruct events that define context distortions, which, in turn, define component-associated behaviors and functions, leading to the inhibition of specific events while favoring the arousal of other specific events and so on, iteratively. The net result is a biological model, entirely supported by a very large body of well identified published information, that very clearly suggests coherent mechanisms for hitherto unexplained observations and that describes in detail the conditions allowing the development of a biological phenomenon, the mechanisms associated with its progression and the means whereby it could be prevented or alleviated (Fig. 19.1C).

However, it is important to realize that such a model can only be an approximation of biological reality. Furthermore, the more complex the reality attached to the model, the coarser the model will be. It is therefore indispensable that such a model be confronted by the biological reality it is supposed to represent. The data arising from these experimental verifications can then be re-injected into the model, correcting errors and mis-directions. This approach results in a situation where biological investigations proceed at a rapid pace with a hitherto unachievable success rate.

19.6.2 The Model Building Process

In the present case, the aim of the models would be to propose novel avenues of investigation into complex mechanisms associated with behavioral & cognitive disorders. The role of the models being to suggest what should be investigated, where and why. Thus, the model building procedure has to be initiated directly from published information. The procedure to be followed is schematically depicted in Fig. 19.2.

The process is initiated from a query-building interface linked to a database (DB DH). The purpose of this database, which is initially empty, is threefold. First, to record all queries sent to external databases; second, to harbor both the queries and the retrieved information attached to working hypotheses demonstrated as incorrect, and third, to allow filtering of all new query to avoid unnecessary redundancies. Following this filtration step, the queries are then dispatched (processing module 1) to small machines linked to public databases via a web information retrieval interface (Web IR). The information retrieved in answer to a query, largely under the form of published literature and images, is then processed (Processing module 2) to determine whether this information could support the hypothesis attached to the query, refute this hypothesis, or neither support nor refute the hypothesis but provide material for a new formulation of this hypothesis. If this information is in support of the working hypothesis, it is, together with the query, directed to a dedicated

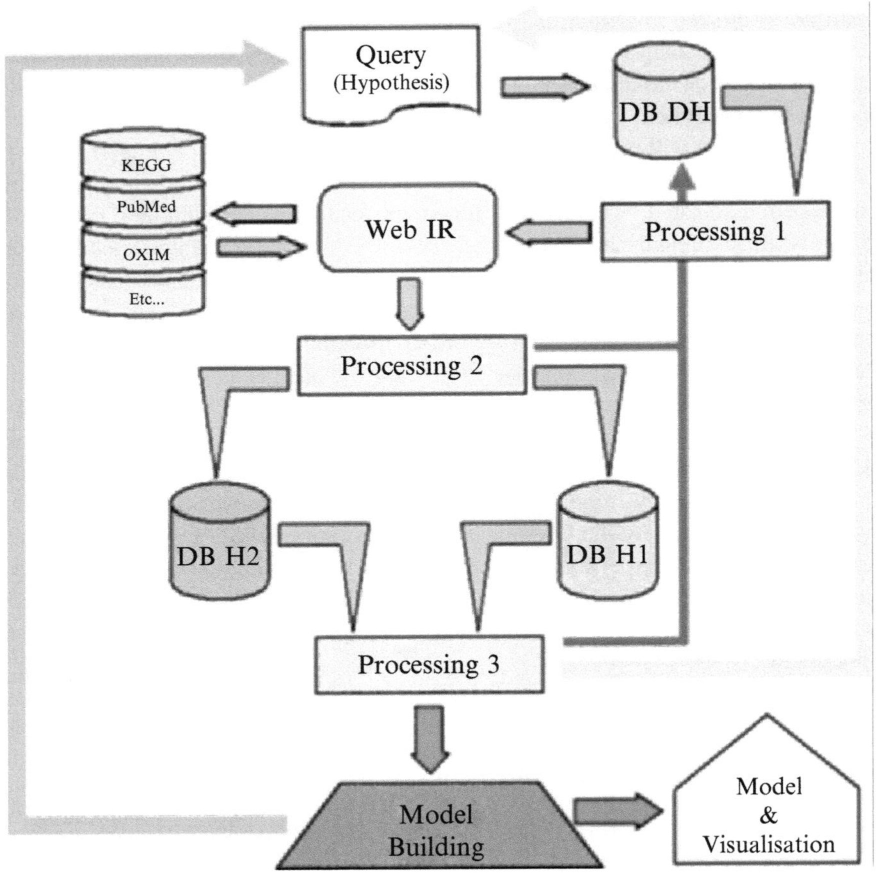

Fig. 19.2 (*See Color Plates*)

database (DB H1). The fact that a working hypothesis finds support in the published literature does not mean that this hypothesis is correct. It merely means that it is not in contradiction with publicly accessible information. If the hypothesis is refuted, the retrieved information and the query are directed to the DB DH database. If the hypothesis is neither refuted nor supported, the retrieved information and the query are directed to the DB H2 database. Specialized biologists, assisted by information processing software, carry out this complex procedure.

This first level of iterative querying procedures reaches an end when most new queries lead to material being directed to DB H2, resulting in a growth rate of the contents of this database over four times faster than that of the other two databases. At this stage, it can be considered that most available "medium-sized" pieces of the puzzle have been obtained and the model building process itself can now be implemented.

The indices of the databases DB H1 and H2 are visualized in order to generate "meta-hypotheses" (Processing module 3) constructed from the merging of either already supported hypotheses (DB H1) or of supported hypotheses associated with neither supported nor destroyed hypotheses (DB H1 + DB H2). These meta-hypotheses are in turn subjected to the testing mechanism described above. Meta-hypotheses, which find support in the literature enter the model building module, while those that are demonstrated incorrect enter the DB DH database and those neither supported nor destroyed enter a new sector in DB H2. Once again, the process is ended when the contents of the new sector in DB H2 grows much faster then that of either DB DH or of that in the model building module. At this stage, most of the large pieces of the puzzle that can be reconstructed using published information have been obtained. But numerous gaps and uncertainties still remain. Thus, during the model-building phase, numerous questions do arise and these are in turn processed to the querying interface in order to find supported solutions or to propose possible answers. The building process ends when the querying process mostly generates uncertainties.

The predictive models arising from this procedure are thus entirely theoretical and can only be viewed as an approximation, albeit supported by a very large body of precisely identified published information, of both the biological mechanisms attached to a complex pathological state and of the physiological events giving rise to this phenomenon. Such a model, although certainly imperfect, presents the substantial advantage of clearly indicating what should be experimentally verified, what should be investigated, both in vitro and in vivo, and for what reason.

19.7 Conclusion

Most behavioral dysfunctions appear primarily as disorders of brain functional and structural connectivity. GABA neurotransmission appears to play a prominent role in schizophrenia. A recent analysis revealed high probability candidate genes involved in GABA neurotransmission (GABRA1, GABBR1, and GAD2), glutamate neurotransmission (GRIA2), neuropeptide signaling (TAC1), synaptic functions (SYN2 and KCNJ4), myelin/glial function (CNP, MAL, MBP, PLP1, MOBP and GFAP), and lipid metabolism (LPL) (Le-Niculescu et al., 2007). These findings may explain the EEG gamma band abnormalities detected in schizophrenia. Bipolar (manic-depressive) and related disorders, on the other hand, appear strongly associated with DARPP-32 (dopamine- and cAMP-regulated phosphoprotein of 32 kDa), PENK (preproenkephalin), and TAC1 (tachykinin 1, substance P), suggesting that more primitive molecular mechanisms involved in pleasure and pain may have been recruited by evolution to play a role in higher mental functions such as mood (Ogden et al., 2004).

Unfortunately, there is no convincing data that a crucial druggable molecular target exists which, if targeted, would yield medications with efficacies greater than those currently available and are thought to exert their main antipsychotic effect through antagonism of dopamine D2 receptors (Stone and Pilowsky, 2006). It appears, instead, that drugs which interact with a multiplicity of molecular targets are likely to show greater efficacy in treating the core symptoms of schizophrenia (Roth, 2006) and some of the above pathways suggest possible avenues for the augmentation of the pharmacotherapy of schizophrenia with other existing agents, such as benzodiazepines, anticonvulsants and lipid modulating agents.

However, it remains amply evident that without predictive functional models sufficiently detailed so as to enable the precise identification, in mechanistic terms, of events leading to pathological consequences, hence identifying the markers associated with these events together with the modes of intervention most likely to prevent or alleviate the problems, the above suggestions shall remain just that - mere suggestions.

References

Abbott, L. F. and S. B. Nelson (2000). Synaptic plasticity: taming the beast. *Nat Neurosci* 3(Suppl): 1178–83.

Abenavoli, A., L. Forti, et al. (2002). Multimodal quantal release at individual hippocampal synapses: evidence for no lateral inhibition. *J Neurosci* 22(15): 6336–46.

Abou Jamra, R., C. Schmael, et al. (2006). The G72/G30 gene locus in psychiatric disorders: a challenge to diagnostic boundaries? *Schizophr Bull* 32(4): 599–608.

Aderem, A. (2005). Systems biology: its practice and challenges. *Cell* 121(4): 511–3.

Ahearn, E. P., M. C. Speer, (2002). Investigation of Notch3 as a candidate gene for bipolar disorder using brain hyperintensities as an endophenotype. *Am J Med Genet* 114(6): 652–8.

Albeck, J. G., G. MacBeath, et al. (2006). Collecting and organizing systematic sets of protein data. *Nat Rev Mol Cell Biol* 7(11): 803–12.

Altevogt, B. M. and D. L. Paul (2004). Four classes of intercellular channels between glial cells in the CNS. *J Neurosci* 24(18): 4313–23.

Alvarez-Maubecin, V., F. Garcia-Hernandez, (2000). Functional coupling between neurons and glia. *J Neurosci* 20(11): 4091–8.

Amieva, M. R. and H. Furthmayr (1995). Subcellular localization of moesin in dynamic filopodia, retraction fibers, and other structures involved in substrate exploration, attachment, and cell-cell contacts. *Exp Cell Res* 219(1): 180–96.

Anderson, L. and J. Seilhamer (1997). A comparison of selected mRNA and protein abundances in human liver. *Electrophoresis* 18(3–4): 533–7.

Apostolova, L. G., R. A. Dutton, (2006). Conversion of mild cognitive impairment to Alzheimer disease predicted by hippocampal atrophy maps. *Arch Neurol* 63(5): 693–9.

Bacci, A., J. R. Huguenard, et al. (2002). Differential modulation of synaptic transmission by neuropeptide Y in rat neocortical neurons. *Proc Natl Acad Sci U S A* 99(26): 17125–30.

Badcock, J. C., P. T. Michiel, et al. (2005). Spatial working memory and planning ability: contrasts between schizophrenia and bipolar I disorder. *Cortex* 41(6): 753–63.

Baimbridge, K. G. and J. J. Miller (1982). Immunohistochemical localization of calcium-binding protein in the cerebellum, hippocampal formation and olfactory bulb of the rat. *Brain Res* 245(2): 223–9.

Barbas, H. (2000). Connections underlying the synthesis of cognition, memory, and emotion in primate prefrontal cortices. *Brain Res Bull* 52(5): 319–30.

Barbas, H. and G. J. Blatt (1995). Topographically specific hippocampal projections target functionally distinct prefrontal areas in the rhesus monkey. *Hippocampus* 5(6): 511–33.

Barnard, E. A., P. Skolnick, et al. (1998). International Union of Pharmacology. XV. Subtypes of gamma-aminobutyric acidA receptors: classification on the basis of subunit structure and receptor function. *Pharmacol Rev* 50(2): 291–313.
Beattie, E. C., D. Stellwagen, et al. (2002). Control of synaptic strength by glial TNFalpha. *Science* 295(5563): 2282–5.
Belgamwar, R. B. and M. Fenton (2005). Olanzapine IM or velotab for acutely disturbed/agitated people with suspected serious mental illnesses. Cochrane Database Syst Rev (2): CD003729.
Bischof, G., H. J. Rumpf, et al. (2005). Influence of psychiatric comorbidity in alcohol-dependent subjects in a representative population survey on treatment utilization and natural recovery. *Addiction* 100(3): 405–13.
Black, I. B. (1999). Trophic regulation of synaptic plasticity. *J Neurobiol* 41(1): 108–18.
Bland, B. H., J. Bird, et al. (2006). Medial septal modulation of the ascending brainstem hippocampal synchronizing pathways in the freely moving rat. *Hippocampus* 16(1): 11–9.
Blumenthal, H. T. (2001). Milestone or genomania? The relevance of the Human Genome Project to biological aging and the age-related diseases. *J Gerontol A Biol Sci Med Sci* 56(9): M529–37.
Bogenschutz, M. P., C. M. Geppert, et al. (2006). The role of twelve-step approaches in dual diagnosis treatment and recovery. *Am J Addict* 15(1): 50–60.
Bohnen, N. I. and K. A. Frey (2003). The role of positron emission tomography imaging in movement disorders. *Neuroimaging Clin N Am* 13(4): 791–803.
Bonnert, T. P., R. M. McKernan, et al. (1999). Theta, a novel gamma-aminobutyric acid type A receptor subunit. *Proc Natl Acad Sci U S A* 96(17): 9891–6.
Bono, H., I. Nikaido, et al. (2003). Comprehensive analysis of the mouse metabolome based on the transcriptome. *Genome Res* 13(6B): 1345–9.
Bose, A., V. Booth, et al. (2000). A temporal mechanism for generating the phase precession of hippocampal place cells. *J Comput Neurosci* 9(1): 5–30.
Braff, D. L., R. Freedman, et al. (2007). Deconstructing schizophrenia: an overview of the use of endophenotypes in order to understand a complex disorder. *Schizophr Bull* 33(1): 21–32.
Bretscher, A., K. Edwards, et al. (2002). ERM proteins and merlin: integrators at the cell cortex. *Nat Rev Mol Cell Biol* 3(8): 586–99.
Broekman, B. F., M. Olff, et al. (2007). The genetic background to PTSD. *Neurosci Biobehav Rev* 31(3): 348–62.
Bruder, G. E., B. E. Wexler, et al. (2004). Verbal memory in schizophrenia: additional evidence of subtypes having different cognitive deficits. *Schizophr Res* 68(2–3): 137–47.
Bruzzone, S., L. Guida, et al. (2001). Connexin 43 hemi channels mediate Ca2+-regulated transmembrane NAD+ fluxes in intact cells. *Faseb J* 15(1): 10–12.
Bungartz, G., S. Stiller, et al. (2006). Adult murine hematopoiesis can proceed without beta1 and beta7 integrins. *Blood* 108(6): 1857–64.
Bushong, E. A., M. E. Martone, et al. (2002). Protoplasmic astrocytes in CA1 stratum radiatum occupy separate anatomical domains. *J Neurosci* 22(1): 183–92.
Calkins, M. E., D. J. Dobie, et al. (2007). The Consortium on the Genetics of Endophenotypes in Schizophrenia: model recruitment, assessment, and endophenotyping methods for a multisite collaboration. *Schizophr Bull* 33(1): 33–48.
Camp, N. J., M. R. Lowry, et al. (2005). Genome-wide linkage analyses of extended Utah pedigrees identifies loci that influence recurrent, early-onset major depression and anxiety disorders. *Am J Med Genet B Neuropsychiatr Genet* 135(1): 85–93.
Cannon, T. D., T. G. van Erp, et al. (2003). Early and late neurodevelopmental influences in the prodrome to schizophrenia: contributions of genes, environment, and their interactions. *Schizophr Bull* 29(4): 653–69.
Carpenter, S., M. Berk, et al. (2004). Clotiapine for acute psychotic illnesses. Cochrane Database Syst Rev(4): CD002304.
Carroll, R. C., E. C. Beattie, et al. (1999). Dynamin-dependent endocytosis of ionotropic glutamate receptors. *Proc Natl Acad Sci U S A* 96(24): 14112–7.

Castellanos, F. X. and R. Tannock (2002). Neuroscience of attention-deficit/hyperactivity disorder: the search for endophenotypes. *Nat Rev Neurosci* 3(8): 617–28.

Castro, C., J. M. Gomez-Hernandez, et al. (1999). Altered formation of hemichannels and gap junction channels caused by C-terminal connexin-32 mutations. *J Neurosci* 19(10): 3752–60.

Chamberlain, S. R., A. D. Blackwell, et al. (2005). The neuropsychology of obsessive compulsive disorder: the importance of failures in cognitive and behavioural inhibition as candidate endophenotypic markers. *Neurosci Biobehav Rev* 29(3): 399–419.

Chapell, R., O. F. Bueno, et al. (1998). Activation of protein kinase C induces gamma-aminobutyric acid type A receptor internalization in Xenopus oocytes. *J Biol Chem* 273(49): 32595–601.

Chapman, S. and A. R. Asthagiri (2004). Resistance to signal activation governs design features of the MAP kinase signaling module. *Biotechnol Bioeng* 85(3): 311–22.

Cheung, E., V. Wong, et al. (2005). Topiramate-valproate-induced hyperammonemic encephalopathy syndrome: case report. *J Child Neurol* 20(2): 157–60.

Chu, P., S. Murray, et al. (1997). Delta and kappa opioid receptors are differentially regulated by dynamin-dependent endocytosis when activated by the same alkaloid agonist. *J Biol Chem* 272(43): 27124–30.

Cohen, A. S., D. D. Lin, et al. (2000). Protracted postnatal development of inhibitory synaptic transmission in rat hippocampal area CA1 neurons. *J Neurophysiol* 84(5): 2465–76.

Collinson, N., F. M. Kuenzi, et al. (2002). Enhanced learning and memory and altered GABAergic synaptic transmission in mice lacking the alpha 5 subunit of the GABAA receptor. *J Neurosci* 22(13): 5572–80.

Condorelli, D. F., G. Mudo, et al. (2002). Connexin-30 mRNA is up-regulated in astrocytes and expressed in apoptotic neuronal cells of rat brain following kainate-induced seizures. *Mol Cell Neurosci* 21(1): 94–113.

Connolly, C. N., J. T. Kittler, et al. (1999a). Cell surface stability of gamma-aminobutyric acid type A receptors. Dependence on protein kinase C activity and subunit composition. *J Biol Chem* 274(51): 36565–72.

Connolly, C. N., J. M. Uren, et al. (1999b). Subcellular localization and endocytosis of homomeric gamma2 subunit splice variants of gamma-aminobutyric acid type A receptors. *Mol Cell Neurosci* 13(4): 259–71.

Contreras, J. E., H. A. Sanchez, et al. (2002). Metabolic inhibition induces opening of unapposed connexin 43 gap junction hemichannels and reduces gap junctional communication in cortical astrocytes in culture. *Proc Natl Acad Sci U S A* 99(1): 495–500.

Coppola, G., S. H. Choi, et al. (2006). Gene expression profiling in frataxin deficient mice: microarray evidence for significant expression changes without detectable neurodegeneration. *Neurobiol Dis* 22(2): 302–11.

Cote, G. B. and J. Gyftodimou (1991). Twinning and mitotic crossing-over: some possibilities and their implications. *Am J Hum Genet* 49(1): 120–30.

Cotrina, M. L., J. H. Lin, et al. (2000). ATP-mediated glia signaling. *J Neurosci* 20(8): 2835–44.

Couderc, R. (2000). Search of biological markers of Alzheimer's disease. *Ann Biol Clin (Paris)* 58(5): 581–93.

Coulter, D. A. (2001). Epilepsy-associated plasticity in gamma-aminobutyric acid receptor expression, function, and inhibitory synaptic properties. *Int Rev Neurobiol* 45: 237–52.

Courchesne, E., E. Redcay, et al. (2004). The autistic brain: birth through adulthood. *Curr Opin Neurol* 17(4): 489–96.

Craig, S. and S. Commins (2005). Interaction between paired-pulse facilitation and long-term potentiation in the projection from hippocampal area CA1 to the entorhinal cortex. *Neurosci Res* 53(2): 140–6.

Csicsvari, J., H. Hirase, et al. (2000). Ensemble patterns of hippocampal CA3-CA1 neurons during sharp wave-associated population events. *Neuron* 28(2): 585–94.

Dahl, E., D. Manthey, et al. (1996). Molecular cloning and functional expression of mouse connexin-30, a gap junction gene highly expressed in adult brain and skin. *J Biol Chem* 271(30): 17903–10.

Danbolt, N. C. (2001). Glutamate uptake. *Prog Neurobiol* 65(1): 1–105.

Datta, S., D. F. Siwek, et al. (1998). Localization of pontine PGO wave generation sites and their anatomical projections in the rat. *Synapse* 30(4): 409–23.
Davis, G. W. and I. Bezprozvanny (2001). Maintaining the stability of neural function: a homeostatic hypothesis. *Annu Rev Physiol* 63: 847–69.
de Caprona, M. D., K. W. Beisel, et al. (2004). Partial behavioral compensation is revealed in balance tasked mutant mice lacking otoconia. *Brain Res Bull* 64(4): 289–301.
Deretic, D., V. Traverso, et al. (2004). Phosphoinositides, ezrin/moesin, and rac1 regulate fusion of rhodopsin transport carriers in retinal photoreceptors. *Mol Biol Cell* 15(1): 359–70.
Derouiche, A. and M. Frotscher (2001). Peripheral astrocyte processes: monitoring by selective immunostaining for the actin-binding ERM proteins. *Glia* 36(3): 330–41.
Derouiche, A., E. Anlauf, et al. (2002). Anatomical aspects of glia-synapse interaction: the perisynaptic glial sheath consists of a specialized astrocyte compartment. *J Physiol Paris* 96(3–4): 177–82.
Desai, N. S., L. C. Rutherford, et al. (1999). Plasticity in the intrinsic excitability of cortical pyramidal neurons. *Nat Neurosci* 2(6): 515–20.
Detera-Wadleigh, S. D. and F. J. McMahon (2006). G72/G30 in schizophrenia and bipolar disorder: review and meta-analysis. *Biol Psychiatr* 60(2): 106–14.
De Witte, P., E. Pinto, et al. (2003). Alcohol and withdrawal: from animal research to clinical issues. *Neurosci Biobehav Rev* 27(3): 189–97.
Diaz-Asper, C. M., D. R. Weinberger, et al. (2006). Catechol-O-methyltransferase polymorphisms and some implications for cognitive therapeutics. *NeuroRx* 3(1): 97–105.
Dickson, T. C., C. D. Mintz, et al. (2002). Functional binding interaction identified between the axonal CAM L1 and members of the ERM family. *J Cell Biol* 157(7): 1105–12.
Doyle, A. E., E. G. Willcutt, et al. (2005). Attention-deficit/hyperactivity disorder endophenotypes. *Biol Psychiatr* 57(11): 1324–35.
Dudkin, K. N., V. Chueva, et al. (2005). Impairments in working memory and decision-taking processes in monkeys in a model of Alzheimer's disease. *Neurosci Behav Physiol* 35(3): 281–9.
Dwyer, J. H., H. Allayee, et al. (2004). Arachidonate 5-lipoxygenase promoter genotype, dietary arachidonic acid, and atherosclerosis. *N Engl J Med* 350(1): 29–37.
Eckhorn, R., R. Bauer, et al. (1988). Coherent oscillations: a mechanism of feature linking in the visual cortex? Multiple electrode and correlation analyses in the cat. *Biol Cybern* 60(2): 121–30.
Edwards, A. O., R. Ritter, 3rd, et al. (2005). Complement factor H polymorphism and age-related macular degeneration. *Science* 308(5720): 421–4.
Elmariah, S. B., E. J. Oh, et al. (2005). Astrocytes regulate inhibitory synapse formation via Trk-mediated modulation of postsynaptic GABAA receptors. *J Neurosci* 25(14): 3638–50.
Enz, R. (2007). The trick of the tail: protein-protein interactions of metabotropic glutamate receptors. *Bioessays* 29(1): 60–73.
Essenfelder, G. M., G. Larderet, et al. (2005). Gene structure and promoter analysis of the human GJB6 gene encoding connexin 30. *Gene* 350(1): 33–40.
Fahrenfort, I., J. Klooster, et al. (2005). The involvement of glutamate-gated channels in negative feedback from horizontal cells to cones. *Prog Brain Res* 147: 219–29.
Fields, R. D. and B. Stevens-Graham (2002). New insights into neuron-glia communication. *Science* 298(5593): 556–62.
Flach, K. A., L. E. Adler, et al. (1996). Sensory gating in a computer model of the CA3 neural network of the hippocampus. *Biol Psychiatr* 40(12): 1230–45.
Floresco, S. B., J. K. Seamans, et al. (1997). Selective roles for hippocampal, prefrontal cortical, and ventral striatal circuits in radial-arm maze tasks with or without a delay. *J Neurosci* 17(5): 1880–90.
Florez, J. C., K. A. Jablonski, et al. (2006). TCF7L2 polymorphisms and progression to diabetes in the Diabetes Prevention Program. *N Engl J Med* 355(3): 241–50.
Fortune, M. T., J. L. Kennedy, et al. (2003). Anticipation and CAG*CTG repeat expansion in schizophrenia and bipolar affective disorder. *Curr Psychiatr Rep* 5(2): 145–54.
Fraga, M. F., E. Ballestar, et al. (2005). Epigenetic differences arise during the lifetime of monozygotic twins. *Proc Natl Acad Sci U S A* 102(30): 10604–9.

Freedman, R., L. E. Adler, et al. (1996). Inhibitory gating of an evoked response to repeated auditory stimuli in schizophrenic and normal subjects. Human recordings, computer simulation, and an animal model. *Arch Gen Psychiatr* 53(12): 1114–21.

Freund, T. F. and G. Buzsaki (1996). Interneurons of the hippocampus. *Hippocampus* 6(4): 347–470.

Fritschy, J. M. and H. Mohler (1995). GABAA-receptor heterogeneity in the adult rat brain: differential regional and cellular distribution of seven major subunits. *J Comp Neurol* 359(1): 154–94.

Futterman, L. G. and L. Lemberg (2001). The mysteries of the human genome uncovered--medicine is changed forever. *Am J Crit Care* 10(2): 125–32.

Fyhn, M., T. Hafting, et al. (2007). Hippocampal remapping and grid realignment in entorhinal cortex. *Nature* 446(7132): 190–4.

Gadal, F., C. Bozic, et al. (2003). Integrated transcriptome analysis of the cellular mechanisms associated with Ha-ras-dependent malignant transformation of the human breast epithelial MCF7 cell line. *Nucleic Acids Res* 31(19): 5789–804.

Gadal, F., A. Starzec, et al. (2005). Integrative analysis of gene expression patterns predicts specific modulations of defined cell functions by estrogen and tamoxifen in MCF7 breast cancer cells. *J Mol Endocrinol* 34(1): 61–75.

Gerling, I. C., S. S. Solomon, et al. (2003). Genomes, transcriptomes, and proteomes: molecular medicine and its impact on medical practice. *Arch Intern Med* 163(2): 190–8.

Ghashghaei, H. T. and H. Barbas (2001). Neural interaction between the basal forebrain and functionally distinct prefrontal cortices in the rhesus monkey. *Neuroscience* 103(3): 593–614.

Giaume, C. (1996). Communicating junctions of glial cells in the central nervous system. *Ann Endocrinol (Paris)* 57(6): 487–91.

Giaume, C., C. Fromaget, et al. (1991). Gap junctions in cultured astrocytes: single-channel currents and characterization of channel-forming protein. *Neuron* 6(1): 133–43.

Gilbert, D., H. Fuss, et al. (2006). Computational methodologies for modelling, analysis and simulation of signalling networks. *Brief Bioinform* 7(4): 339–53.

Gillies, D., A. Beck, et al. (2005). Benzodiazepines alone or in combination with antipsychotic drugs for acute psychosis. Cochrane Database Syst Rev (4): CD003079.

Glahn, D. C., C. E. Bearden, et al. (2004). The feasibility of neuropsychological endophenotypes in the search for genes associated with bipolar affective disorder. *Bipolar Disord* 6(3): 171–82.

Glahn, D. C., C. E. Bearden, et al. (2006). Differential working memory impairment in bipolar disorder and schizophrenia: effects of lifetime history of psychosis. *Bipolar Disord* 8(2): 117–23.

Goel, G., I. C. Chou, et al. (2006). Biological systems modeling and analysis: a biomolecular technique of the twenty-first century. *J Biomol Tech* 17(4): 252–69.

Goldberg, T. E. and D. R. Weinberger (2004). Genes and the parsing of cognitive processes. *Trends Cogn Sci* 8(7): 325–35.

Golowasch, J., M. Casey, et al. (1999). Network stability from activity-dependent regulation of neuronal conductances. *Neural Comput* 11(5): 1079–96.

Gomez-Hernandez, J. M., M. de Miguel, et al. (2003). Molecular basis of calcium regulation in connexin-32 hemichannels. *Proc Natl Acad Sci U S A* 100(26): 16030–5.

Goodenough, D. A. and D. L. Paul (2003). Beyond the gap: functions of unpaired connexon channels. *Nat Rev Mol Cell Biol* 4(4): 285–94.

Goodman, C., G. Knoll, et al. (2005). Insight into illness in schizophrenia. *Compr Psychiatr* 46(4): 284–90.

Gottesman, II and L. Erlenmeyer-Kimling (2001). Family and twin strategies as a head start in defining prodromes and endophenotypes for hypothetical early-interventions in schizophrenia. *Schizophr Res* 51*(1)*: 93–102.

Gottesman, II and J. Shields (1973). Genetic theorizing and schizophrenia. *Br J Psychiatr* 122(566): 15–30.

Gottsch, M. L., H. Zeng, et al. (2005). Phenotypic analysis of mice deficient in the type 2 galanin receptor (GALR2). *Mol Cell Biol* 25(11): 4804–11.

Gould, T. D. and Gottesman, II (2006). Psychiatric endophenotypes and the development of valid animal models. *Genes Brain Behav* 5(2): 113–9.

Gould, T. D., T. M. Bastain, et al. (2001). Altered performance on an ocular fixation task in attention-deficit/hyperactivity disorder. *Biol Psychiatr* 50(8): 633–5.

Grant, M. R., K. E. Mostov, et al. (2006). Simulating properties of in vitro epithelial cell morphogenesis. *PLoS Comput Biol* 2(10): e129.

Green, M. J., C. M. Cahill, et al. (2007). The cognitive and neurophysiological basis of emotion dysregulation in bipolar disorder. J Affect Disord. 103(1–3):29–42.

Greferath, U., A. Bennie, et al. (2000). Enlarged cholinergic forebrain neurons and improved spatial learning in p75 knockout mice. *Eur J Neurosci* 12(3): 885–93.

Group, B. D. W. (2001). Biomarkers and surrogate endpoints: preferred definitions and conceptual framework. *Clin Pharmacol Ther* 69(3): 89–95.

Gulyas, A. I., K. Toth, et al. (1998). Stratum radiatum giant cells: a type of principal cell in the rat hippocampus. *Eur J Neurosci* 10(12): 3813–22.

Gygi, S. P., Y. Rochon, et al. (1999). Correlation between protein and mRNA abundance in yeast. *Mol Cell Biol* 19(3): 1720–30.

Hajos, N., E. C. Papp, et al. (1998). Distinct interneuron types express m2 muscarinic receptor immunoreactivity on their dendrites or axon terminals in the hippocampus. *Neuroscience* 82(2): 355–76.

Hariri, A. R. and D. R. Weinberger (2003). Functional neuroimaging of genetic variation in serotonergic neurotransmission. *Genes Brain Behav* 2(6): 341–9.

Hashimoto, S., Y. Huang, et al. (2004). Compensation of proximal tubule malabsorption in AQP1-deficient mice without TGF-mediated reduction of GFR. *Acta Physiol Scand* 181(4): 455–62.

Hasler, G., W. C. Drevets, et al. (2004). Discovering endophenotypes for major depression. *Neuropsychopharmacology* 29(10): 1765–81.

Hata, A. (2002). Human genome and made-to-order health care system. *Hokkaido Igaku Zasshi* 77(1): 13–6.

Hazlett, H. C., M. Poe, et al. (2005). Magnetic resonance imaging and head circumference study of brain size in autism: birth through age 2 years. *Arch Gen Psychiatr* 62(12): 1366–76.

Heckers, S., A. P. Weiss, et al. (2002). Hippocampal and brain stem activation during word retrieval after repeated and semantic encoding. *Cereb Cortex* 12(9): 900–7.

Heinrichs, R. W. (2005). The primacy of cognition in schizophrenia. *Am Psychol* 60(3): 229–42.

Herbert, A., N. P. Gerry, et al. (2006). A common genetic variant is associated with adult and childhood obesity. *Science* 312(5771): 279–83.

Hok, V., P. P. Lenck-Santini, et al. (2007). Goal-related activity in hippocampal place cells. *J Neurosci* 27(3): 472–82.

Hollrigel, G. S., S. T. Ross, et al. (1998). Temporal patterns and depolarizing actions of spontaneous GABAA receptor activation in granule cells of the early postnatal dentate gyrus. *J Neurophysiol* 80(5): 2340–51.

Hosl, K., H. Reinold, et al. (2006). Spinal prostaglandin E receptors of the EP2 subtype and the glycine receptor alpha3 subunit, which mediate central inflammatory hyperalgesia, do not contribute to pain after peripheral nerve injury or formalin injection. *Pain* 126(1–3): 46–53.

Hua, F., S. Hautaniemi, et al. (2006). Integrated mechanistic and data-driven modelling for multivariate analysis of signalling pathways. *J R Soc Interface* 3(9): 515–26.

Huang, S. and J. Wikswo (2006). Dimensions of systems biology. *Rev Physiol Biochem Pharmacol* 157: 81–104.

Huang, K. P., F. L. Huang, et al. (2004). Neurogranin/RC3 enhances long-term potentiation and learning by promoting calcium-mediated signaling. *J Neurosci* 24(47): 10660–9.

Ishizuka, N., J. Weber, et al. (1990). Organization of intrahippocampal projections originating from CA3 pyramidal cells in the rat. *J Comp Neurol* 295(4): 580–623.

Izaki, Y., M. Takita, et al. (2002). Local properties of CA1 region in hippocampo-prefrontal synaptic plasticity in rats. *Neuroreport* 13(4): 469–72.

Jackson, B. S. (2004). Including long-range dependence in integrate-and-fire models of the high interspike-interval variability of cortical neurons. *Neural Comput* 16(10): 2125–95.

Jackson, J. and B. H. Bland (2006). Medial septal modulation of the ascending brainstem hippocampal synchronizing pathways in the anesthetized rat. *Hippocampus* 16(1): 1–10.
Janes, K. A. and M. B. Yaffe (2006). Data-driven modelling of signal-transduction networks. *Nat Rev Mol Cell Biol* 7(11): 820–8.
Janzer, R. C. and M. C. Raff (1987). Astrocytes induce blood-brain barrier properties in endothelial cells. *Nature* 325(6101): 253–7.
Jay, T. M., C. Rocher, et al. (2004). Plasticity at hippocampal to prefrontal cortex synapses is impaired by loss of dopamine and stress: importance for psychiatric diseases. *Neurotox Res* 6(3): 233–44.
Jensen, O. and J. E. Lisman (1996a). Hippocampal CA3 region predicts memory sequences: accounting for the phase precession of place cells. *Learn Mem* 3(2–3): 279–87.
Jensen, O. and J. E. Lisman (1996b). Theta/gamma networks with slow NMDA channels learn sequences and encode episodic memory: role of NMDA channels in recall. *Learn Mem* 3(2–3): 264–78.
Jensen, O., M. A. Idiart, et al. (1996). Physiologically realistic formation of autoassociative memory in networks with theta/gamma oscillations: role of fast NMDA channels. *Learn Mem* 3(2–3): 243–56.
Jensen, K., C. S. Chiu, et al. (2003). GABA transporter-1 (GAT1)-deficient mice: differential tonic activation of GABAA versus GABAB receptors in the hippocampus. *J Neurophysiol* 90(4): 2690–701.
Jiang, F. and S. Khanna (2004). Reticular stimulation evokes suppression of CA1 synaptic responses and generation of theta through separate mechanisms. *Eur J Neurosci* 19(2): 295–308.
Jiang, F. and S. Khanna (2006). Microinjection of carbachol in the supramammillary region suppresses CA1 pyramidal cell synaptic excitability. *Hippocampus* 16(10): 891–905.
John, G. R., E. Scemes, et al. (1999). IL-1beta differentially regulates calcium wave propagation between primary human fetal astrocytes via pathways involving P2 receptors and gap junction channels. *Proc Natl Acad Sci U S A* 96(20): 11613–8.
Jones, B. E. (2004). Paradoxical REM sleep promoting and permitting neuronal networks. *Arch Ital Biol* 142(4): 379–96.
Joubert, L., B. Hanson, et al. (2004). New sorting nexin (SNX27) and NHERF specifically interact with the 5-HT4a receptor splice variant: roles in receptor targeting. *J Cell Sci* 117(Pt 22): 5367–79.
Jovanovic, J. N., P. Thomas, et al. (2004). Brain-derived neurotrophic factor modulates fast synaptic inhibition by regulating GABA(A) receptor phosphorylation, activity, and cell-surface stability. *J Neurosci* 24(2): 522–30.
Kalkonde, Y. V., W. W. Morgan, et al. (2007). Chemokines in the MPTP model of Parkinson's disease: absence of CCL2 and its receptor CCR2 does not protect against striatal neurodegeneration. *Brain Res* 1128(1): 1–11.
Kamondi, A., L. Acsady, et al. (1998). Theta oscillations in somata and dendrites of hippocampal pyramidal cells in vivo: activity-dependent phase-precession of action potentials. *Hippocampus* 8(3): 244–61.
Kannabiran, C. and S. G. Panicker (2002). Of genes and disease. *Indian J Ophthalmol* 50(1): 2–3.
Katona, I., B. Sperlagh, et al. (1999). Presynaptically located CB1 cannabinoid receptors regulate GABA release from axon terminals of specific hippocampal interneurons. *J Neurosci* 19(11): 4544–58.
Kaut, K. P. and M. D. Bunsey (2001). The effects of lesions to the rat hippocampus or rhinal cortex on olfactory and spatial memory: retrograde and anterograde findings. *Cogn Affect Behav Neurosci* 1(3): 270–86.
Kawashima, H., Y. Izaki, et al. (2006). Cooperativity between hippocampal-prefrontal short-term plasticity through associative long-term potentiation. *Brain Res* 1109(1): 37–44.
Kendler, K. S. and E. Zerbin-Rudin (1996). Abstract and review of "Studien über Vererbung und Entstehung Geistiger Störungen. I. Zur Vererbung und Neuentstehung der Dementia praecox."

(Studies on the inheritance and origin of mental illness: I. To the problem of the inheritance and primary origin of dementia praecox). 1916. *Am J Med Genet* 67(4): 338–42.
Kitajiri, S., K. Fukumoto, et al. (2004). Radixin deficiency causes deafness associated with progressive degeneration of cochlear stereocilia. *J Cell Biol* 166(4): 559–70.
Kitano, H. (2007). The theory of biological robustness and its implication in cancer. Ernst Schering Res Found Workshop (61): 69–88.
Kittler, J. T., P. Delmas, et al. (2000). Constitutive endocytosis of GABAA receptors by an association with the adaptin AP2 complex modulates inhibitory synaptic currents in hippocampal neurons. *J Neurosci* 20(21): 7972–7.
Klausberger, T., P. J. Magill, et al. (2003). Brain-state- and cell-type-specific firing of hippocampal interneurons in vivo. *Nature* 421(6925): 844–8.
Klausberger, T., L. F. Marton, et al. (2004). Spike timing of dendrite-targeting bistratified cells during hippocampal network oscillations in vivo. *Nat Neurosci* 7(1): 41–7.
Klingemann, K. H. and D. Efionayi-Mader (1994). How much treatment does a person need? Addiction, spontaneous remission and "family" as biographical as leitmotiv. *Schweiz Rundsch Med Prax* 83(34): 937–49.
Knierim, J. J. and B. L. McNaughton (2001). Hippocampal place-cell firing during movement in three-dimensional space. *J Neurophysiol* 85(1): 105–16.
Knierim, J. J., I. Lee, et al. (2006). Hippocampal place cells: parallel input streams, subregional processing, and implications for episodic memory. *Hippocampus* 16(9): 755–64.
Knobeloch, K. P., O. Utermohlen, et al. (2005). Reexamination of the role of ubiquitin-like modifier ISG15 in the phenotype of UBP43-deficient mice. *Mol Cell Biol* 25(24): 11030–4.
Kobsar, I., C. Oetke, et al. (2006). Attenuated demyelination in the absence of the macrophage-restricted adhesion molecule sialoadhesin (Siglec-1) in mice heterozygously deficient in P0. *Mol Cell Neurosci* 31(4): 685–91.
Kondo, R. P., S. Y. Wang, et al. (2000). Metabolic inhibition activates a non-selective current through connexin hemichannels in isolated ventricular myocytes. *J Mol Cell Cardiol* 32(10): 1859–72.
Koponen, E., M. Lakso, et al. (2004a). Overexpression of the full-length neurotrophin receptor trkB regulates the expression of plasticity-related genes in mouse brain. *Brain Res Mol Brain Res* 130(1–2): 81–94.
Koponen, E., V. Voikar, et al. (2004b) .Transgenic mice overexpressing the full-length neurotrophin receptor trkB exhibit increased activation of the trkB-PLCgamma pathway, reduced anxiety, and facilitated learning. *Mol Cell Neurosci* 26(1): 166–81.
Korpi, E. R., G. Grunder, et al. (2002). Drug interactions at GABA(A) receptors. *Prog Neurobiol* 67(2): 113–59.
Krucker, T., G. R. Siggins, et al. (2002). Targeted disruption of RC3 reveals a calmodulin-based mechanism for regulating metaplasticity in the hippocampus. *J Neurosci* 22(13): 5525–35.
Kunzelmann, P., W. Schroder, et al. (1999). Late onset and increasing expression of the gap junction protein connexin30 in adult murine brain and long-term cultured astrocytes. *Glia* 25(2): 111–9.
Lambert, W. and L. Bickman (2004). Child & adolescent psychiatry: the "clock-setting" cure: how children's symptoms might improve after ineffective treatment. *Psychiatr Serv* 55(4): 381–2.
Lander, A. D. (2007). Morpheus unbound: reimagining the morphogen gradient. *Cell* 128(2): 245–56.
Laroche, S., S. Davis, et al. (2000). Plasticity at hippocampal to prefrontal cortex synapses: dual roles in working memory and consolidation. *Hippocampus* 10(4): 438–46.
Larsson, H. J., W. W. Eaton, et al. (2005). Risk factors for autism: perinatal factors, parental psychiatric history, and socioeconomic status. *Am J Epidemiol* 161(10): 916–25; discussion 926–8.
Leaf, I., J. Tennessen, et al. (2006). Sfrp5 is not essential for axis formation in the mouse. *Genesis* 44(12): 573–8.
Le-Niculescu, H., Y. Balaraman, et al. (2007). Towards understanding the schizophrenia code: an expanded convergent functional genomics approach. *Am J Med Genet B Neuropsychiatr Genet* 144(2): 129–58.

Lenox, R. H., T. D. Gould, et al. (2002). Endophenotypes in bipolar disorder. *Am J Med Genet* 114(4): 391–406.

Leutgeb, S., J. K. Leutgeb, et al. (2006). Fast rate coding in hippocampal CA3 cell ensembles. *Hippocampus* 16(9): 765–74.

Levin, M. and M. Mercola (2000). Expression of connexin 30 in Xenopus embryos and its involvement in hatching gland function. *Dev Dyn* 219(1): 96–101.

Li, X. G., P. Somogyi, et al. (1994). The hippocampal CA3 network: an in vivo intracellular labeling study. *J Comp Neurol* 339(2): 181–208.

Li, S. C., Y. Brehmer, et al. (2006). Neuromodulation of associative and organizational plasticity across the life span: empirical evidence and neurocomputational modeling. *Neurosci Biobehav Rev* 30(6): 775–90.

Liang, G. S., M. de Miguel, et al. (2005). Severe neuropathy with leaky connexin32 hemichannels. *Ann Neurol* 57(5): 749–54.

Lisman, J. (2005). The theta/gamma discrete phase code occuring during the hippocampal phase precession may be a more general brain coding scheme. *Hippocampus* 15(7): 913–22.

Lisman, J. E. and N. A. Otmakhova (2001). Storage, recall, and novelty detection of sequences by the hippocampus: elaborating on the SOCRATIC model to account for normal and aberrant effects of dopamine. *Hippocampus* 11(5): 551–68.

Lisman, J. E., L. M. Talamini, et al. (2005). Recall of memory sequences by interaction of the dentate and CA3: a revised model of the phase precession. *Neural Netw* 18(9): 1191–201.

Lochner, K., A. Gaemlich, et al. (2006). Expression of decorin and collagens I and III in different layers of human skin in vivo: a laser capture microdissection study. Biogerontology. 8(3):269–82.

Lockhart, S. T., G. G. Turrigiano, et al. (1997). Nerve growth factor modulates synaptic transmission between sympathetic neurons and cardiac myocytes. *J Neurosci* 17(24): 9573–82.

Lubec, B., R. Weitzdoerfer, (2001). Manifold reduction of moesin in fetal Down syndrome brain. *Biochem Biophys Res Commun* 286(5): 1191–4.

Luscher, C., H. Xia, et al. (1999). Role of AMPA receptor cycling in synaptic transmission and plasticity. *Neuron* 24(3): 649–58.

Maher, B. S., M. L. Marazita, et al. (2002). Genetic segregation analysis of alcohol and other substance-use disorders in families with recurrent, early-onset major depression. *Am J Drug Alcohol Abuse* 28(4): 711–31.

Majores, M., V. Schick, et al. (2005). Mutational and immunohistochemical analysis of ezrin-, radixin-, moesin (ERM) molecules in epilepsy-associated glioneuronal lesions. *Acta Neuropathol (Berl)* 110(6): 537–46.

Man, H. Y., J. W. Lin, et al. (2000). Regulation of AMPA receptor-mediated synaptic transmission by clathrin-dependent receptor internalization. *Neuron* 25(3): 649–62.

Marsh, M. and H. T. McMahon (1999). The structural era of endocytosis. *Science* 285(5425): 15–20.

Martin, S. J., L. de Hoz, et al. (2005). Retrograde amnesia: neither partial nor complete hippocampal lesions in rats result in preferential sparing of remote spatial memory, even after reminding. *Neuropsychologia* 43(4): 609–24.

Maurer, A. P., S. L. Cowen, et al. (2006a). Organization of hippocampal cell assemblies based on theta phase precession. *Hippocampus* 16(9): 785–94.

Maurer, A. P., S. L. Cowen, et al. (2006b). Phase precession in hippocampal interneurons showing strong functional coupling to individual pyramidal cells. *J Neurosci* 26(52): 13485–92.

Mavanji, V. and S. Datta et al. (2003). Activation of the phasic pontine-wave generator enhances improvement of learning performance: a mechanism for sleep-dependent plasticity. *Eur J Neurosci* 17(2): 359–70.

McAllister, A. K., L. C. Katz, et al. (1999). Neurotrophins and synaptic plasticity. *Annu Rev Neurosci* 22: 295–318.

McGregor, E. and M. J. Dunn (2006). Proteomics of the heart: unraveling disease. *Circ Res* 98(3): 309–21.

McIntyre, R. S., A. Fallu, et al. (2006). Measurable outcomes in psychiatric disorders: remission as a marker of wellness. *Clin Ther* 28(11): 1882–91.

McKernan, R. M., T. W. Rosahl, et al. (2000). Sedative but not anxiolytic properties of benzodiazepines are mediated by the GABA(A) receptor alpha1 subtype. *Nat Neurosci* 3(6): 587–92.

Micheloyannis, S., V. Sakkalis, et al. (2005). Neural networks involved in mathematical thinking: evidence from linear and non-linear analysis of electroencephalographic activity. *Neurosci Lett* 373(3): 212–7.

Miguel, E. C., J. F. Leckman, et al. (2005). Obsessive-compulsive disorder phenotypes: implications for genetic studies. *Mol Psychiatr* 10(3): 258–75.

Mikkonen, J. E., J. Huttunen, et al. (2006). Contribution of a single CA3 neuron to network synchrony. *Neuroimage* 31(3): 1222–7.

Mill, J., E. Dempster, et al. (2006). Evidence for monozygotic twin (MZ) discordance in methylation level at two CpG sites in the promoter region of the catechol-O-methyltransferase (COMT) gene. *Am J Med Genet B Neuropsychiatr Genet* 141(4): 421–5.

Mizuhara, H., L. Q. Wang, et al. (2004). A long-range cortical network emerging with theta oscillation in a mental task. *Neuroreport* 15(8): 1233–8.

Mizuhara, H., L. Q. Wang, et al. (2005). Long-range EEG phase synchronization during an arithmetic task indexes a coherent cortical network simultaneously measured by fMRI. *Neuroimage* 27(3): 553–63.

Mody, I. (2005). Aspects of the homeostaic plasticity of GABAA receptor-mediated inhibition. *J Physiol* 562(Pt 1): 37–46.

Moses, S. N., C. Cole, et al. (2005). Relational memory for object identity and spatial location in rats with lesions of perirhinal cortex, amygdala and hippocampus. *Brain Res Bull* 65(6): 501–12.

Moxon, K. A., G. A. Gerhardt, et al. (2003). Inhibitory control of sensory gating in a computer model of the CA3 region of the hippocampus. *Biol Cybern* 88(4): 247–64.

Muller, D. J., G. M. Hand, et al. (2002). Conformational changes in surface structures of isolated connexin 26 gap junctions. *Embo J* 21(14): 3598–607.

Mundo, E. (2006). Neurobiology of dynamic psychotherapy: an integration possible? *J Am Acad Psychoanal Dyn Psychiatr* 34(4): 679–91.

Nagy, J. I. and J. E. Rash (2000). Connexins and gap junctions of astrocytes and oligodendrocytes in the CNS. *Brain Res Brain Res Rev* 32(1): 29–44.

Nagy, J. I., X. Li, et al. (2001). Connexin26 in adult rodent central nervous system: demonstration at astrocytic gap junctions and colocalization with connexin30 and connexin43. *J Comp Neurol* 441(4): 302–23.

Naus, C. C., J. F. Bechberger,et al. (1997). Altered gap junctional communication, intercellular signaling, and growth in cultured astrocytes deficient in connexin43. *J Neurosci Res* 49(5): 528–40.

Ng, A., B. Bursteinas, et al. (2006). Resources for integrative systems biology: from data through databases to networks and dynamic system models. *Brief Bioinform* 7(4): 318–30.

Nicholson, J. K., E. Holmes, et al. (2004). The challenges of modeling mammalian biocomplexity. *Nat Biotechnol* 22(10): 1268–74.

Niculescu, A. B., 3rd and H. S. Akiskal (2001). Proposed endophenotypes of dysthymia: evolutionary, clinical and pharmacogenomic considerations. *Mol Psychiatr* 6(4): 363–6.

Norenberg, W., F. Hofmann, et al. (1999). Rundown of somatodendritic N-methyl-D-aspartate (NMDA) receptor channels in rat hippocampal neurones: evidence for a role of the small GTPase RhoA. *Br J Pharmacol* 127(5): 1060–3.

Nurmi, E., J. Bergman, et al. (2003). Progression of dopaminergic hypofunction in striatal subregions in Parkinson's disease using [18F]CFT PET. *Synapse* 48(3): 109–15.

Nusser, Z., W. Sieghart, et al. (1996). Differential synaptic localization of two major gamma-aminobutyric acid type A receptor alpha subunits on hippocampal pyramidal cells. *Proc Natl Acad Sci USA* 93(21): 11939–44.

Nusser, Z., N. Hajos, et al. (1998). Increased number of synaptic GABA(A) receptors underlies potentiation at hippocampal inhibitory synapses. *Nature* 395(6698): 172–7.

Nyiri, G., T. F. Freund, et al. (2001). Input-dependent synaptic targeting of alpha(2)-subunit-containing GABA(A) receptors in synapses of hippocampal pyramidal cells of the rat. *Eur J Neurosci* 13(3): 428–42.

Oates, N. A., J. van Vliet, et al. (2006). Increased DNA methylation at the AXIN1 gene in a monozygotic twin from a pair discordant for a caudal duplication anomaly. *Am J Hum Genet* 79(1): 155–62.

Ogden, C. A., M. E. Rich, et al. (2004). Candidate genes, pathways and mechanisms for bipolar (manic-depressive) and related disorders: an expanded convergent functional genomics approach. *Mol Psychiatry* 9(11): 1007–29.

Olenik, C., H. Barth, et al. (1997). Gene expression of the small GTP-binding proteins RhoA, RhoB, Rac1, and Cdc42 in adult rat brain. *Brain Res Mol Brain Res* 52(2): 263–9.

Osipova, D., A. Takashima, et al. (2006). Theta and gamma oscillations predict encoding and retrieval of declarative memory. *J Neurosci* 26(28): 7523–31.

Ou, L. C. and P. W. Gean (2006). Regulation of amygdala-dependent learning by brain-derived neurotrophic factor is mediated by extracellular signal-regulated kinase and phosphatidylinositol-3-kinase. *Neuropsychopharmacology* 31(2): 287–96.

Paglini, G., P. Kunda, et al. (1998). Suppression of radixin and moesin alters growth cone morphology, motility, and process formation in primary cultured neurons. *J Cell Biol* 143(2): 443–55.

Pantelis, C., T. R. Barnes, et al. (1997). Frontal-striatal cognitive deficits in patients with chronic schizophrenia. *Brain* 120(Pt 10): 1823–43.

Pantelis, C., M. Yucel, et al. (2005). Structural brain imaging evidence for multiple pathological processes at different stages of brain development in schizophrenia. *Schizophr Bull* 31(3): 672–96.

Park, E., M. Na, et al. (2003). The Shank family of postsynaptic density proteins interacts with and promotes synaptic accumulation of the beta PIX guanine nucleotide exchange factor for Rac1 and Cdc42. *J Biol Chem* 278(21): 19220–9.

Parlato, R., C. Rieker, et al. (2006). Survival of DA neurons is independent of CREM upregulation in absence of CREB. *Genesis* 44(10): 454–64.

Pascual, O., K. B. Casper, et al. (2005). Astrocytic purinergic signaling coordinates synaptic networks. *Science* 310(5745): 113–6.

Pasqualotto, B. A., R. A. Lanius, et al. (1993). Regulation of GABAA and AMPA receptors by agonist and depolarizing stimulation requires phosphatase or kinase activity. *Neuroreport* 4(4): 447–50.

Pearce, D. J., F. Anjos-Afonso, et al. (2007). Age-dependent increase in side population distribution within hematopoiesis: implications for our understanding of the mechanism of aging. *Stem Cell* 25(4): 828–35.

Penzes, P., R. C. Johnson, et al. (2001). The neuronal Rho-GEF Kalirin-7 interacts with PDZ domain-containing proteins and regulates dendritic morphogenesis. *Neuron* 29(1): 229–42.

Peracchia, C., A. Sotkis, et al. (2000). Calmodulin directly gates gap junction channels. *J Biol Chem* 275(34): 26220–4.

Petronis, A., Gottesman, II, et al. (2003). Monozygotic twins exhibit numerous epigenetic differences: clues to twin discordance? *Schizophr Bull* 29(1): 169–78.

Pfahnl, A. and G. Dahl (1999). Gating of cx46 gap junction hemichannels by calcium and voltage. *Pflugers Arch* 437(3): 345–53.

Pirkola, T., A. Tuulio-Henriksson, et al. (2005). Spatial working memory function in twins with schizophrenia and bipolar disorder. *Biol Psychiatr* 58(12): 930–6.

Piskulic, D., J. S. Olver, et al. (2007). Behavioural studies of spatial working memory dysfunction in schizophrenia: a quantitative literature review. *Psychiatr Res* 150(2): 111–21.

Pitcher, J. A., N. J. Freedman, et al. (1998). G protein-coupled receptor kinases. *Annu Rev Biochem* 67: 653–92.

Plotkin, L. I., S. C. Manolagas, et al. (2002). Transduction of cell survival signals by connexin-43 hemichannels. *J Biol Chem* 277(10): 8648–57.

Poo, M. M. (2001). Neurotrophins as synaptic modulators. *Nat Rev Neurosci* 2(1): 24–32.

Poolos, N. P., M. Migliore, et al. (2002). Pharmacological upregulation of h-channels reduces the excitability of pyramidal neuron dendrites. *Nat Neurosci* 5(8): 767–74.

Popesco, M. C., S. Lin, et al. (2007). Serial analysis of gene expression profiles of adult and aged mouse cerebellum. Neurobiol Aging. 29(5):774–88.

Princen, F., P. Robe, et al. (2001). Rat gap junction connexin-30 inhibits proliferation of glioma cell lines. *Carcinogenesis* 22(3): 507–13.

Pullikuth, A. K., K. Aimanova, et al. (2006). Molecular characterization of sodium/proton exchanger 3 (NHE3) from the yellow fever vector, Aedes aegypti. *J Exp Biol* 209(Pt 18): 3529–44.

Puolakkainen, P. A., A. D. Bradshaw, et al. (2005). SPARC-thrombospondin-2-double-null mice exhibit enhanced cutaneous wound healing and increased fibrovascular invasion of subcutaneous polyvinyl alcohol sponges. *J Histochem Cytochem* 53(5): 571–81.

Qiu, J., M. A. Bosch, et al. (2003). Rapid signaling of estrogen in hypothalamic neurons involves a novel G-protein-coupled estrogen receptor that activates protein kinase C. *J Neurosci* 23(29): 9529–40.

Raghavachari, S., J. E. Lisman, et al. (2006). Theta oscillations in human cortex during a working-memory task: evidence for local generators. *J Neurophysiol* 95(3): 1630–8.

Rapoport, J. L., A. M. Addington, et al. (2005). The neurodevelopmental model of schizophrenia: update 2005. *Mol Psychiatr* 10(5): 434–49.

Rash, J. E., W. A. Staines, et al. (2000). Immunogold evidence that neuronal gap junctions in adult rat brain and spinal cord contain connexin-36 but not connexin-32 or connexin-43. *Proc Natl Acad Sci U S A* 97(13): 7573–8.

Rash, J. E., T. Yasumura, et al. (2001). Cell-specific expression of connexins and evidence of restricted gap junctional coupling between glial cells and between neurons. *J Neurosci* 21(6): 1983–2000.

Rattiner, L. M., M. Davis, et al. (2004). Brain-derived neurotrophic factor and tyrosine kinase receptor B involvement in amygdala-dependent fear conditioning. *J Neurosci* 24(20): 4796–806.

Ravina, B. M., S. C. Fagan, et al. (2003). Neuroprotective agents for clinical trials in Parkinson's disease: a systematic assessment. *Neurology* 60(8): 1234–40.

Redish, A. D. and D. S. Touretzky (1998). The role of the hippocampus in solving the Morris water maze. *Neural Comput* 10(1): 73–111.

Reid, C. A., R. Fabian-Fine, et al. (2001). Postsynaptic calcium transients evoked by activation of individual hippocampal mossy fiber synapses. *J Neurosci* 21(7): 2206–14.

Reilly, J. L., M. S. Harris, et al. (2007). Antipsychotic drugs exacerbate impairment on a working memory task in first-episode schizophrenia. Biol Psychiatr. 62(7):818–21.

Rosch, H., R. Schweigreiter, et al. (2005). The neurotrophin receptor p75NTR modulates long-term depression and regulates the expression of AMPA receptor subunits in the hippocampus. *Proc Natl Acad Sci U S A* 102(20): 7362–7.

Roth, B. L. (2006). Contributions of molecular biology to antipsychotic drug discovery: promises fulfilled or unfulfilled? *Dialogues Clin Neurosci* 8(3): 303–9.

Rovaris, M., G. Comi, et al. (2003). Short-term correlations between clinical and MR imaging findings in relapsing-remitting multiple sclerosis. *AJNR Am J Neuroradiol* 24(1): 75–81.

Royer, S. and D. Pare (2003). Conservation of total synaptic weight through balanced synaptic depression and potentiation. *Nature* 422(6931): 518–22.

Rudolph, U., F. Crestani, et al. (2001). GABA(A) receptor subtypes: dissecting their pharmacological functions. *Trends Pharmacol Sci* 22(4): 188–94.

Rudomin, P., J. Lomeli, et al. (2004). Differential modulation of primary afferent depolarization of segmental and ascending intraspinal collaterals of single muscle afferents in the cat spinal cord. *Exp Brain Res* 156(3): 377–91.

Saez, J. C., J. E. Contreras, et al. (2003). Gap junction hemichannels in astrocytes of the CNS. *Acta Physiol Scand* 179(1): 9–22.

Saez, J. C., M. A. Retamal, et al. (2005). Connexin-based gap junction hemichannels: gating mechanisms. *Biochim Biophys Acta* 1711(2): 215–24.

Savageau, M. A. (1969a). Biochemical systems analysis. I. Some mathematical properties of the rate law for the component enzymatic reactions. *J Theor Biol* 25(3): 365–9.

Savageau, M. A. (1969b). Biochemical systems analysis. II. The steady-state solutions for an n-pool system using a power-law approximation. *J Theor Biol* 25(3): 370–9.

Savitz, J., M. Solms, et al. (2006). The molecular genetics of cognition: dopamine, COMT and BDNF. *Genes Brain Behav* 5(4): 311–28.

Scarff, K. L., K. S. Ung, et al. (2004). Targeted disruption of SPI3/Serpinb6 does not result in developmental or growth defects, leukocyte dysfunction, or susceptibility to stroke. *Mol Cell Biol* 24(9): 4075–82.

Schadt, E. E., J. Lamb, et al. (2005). An integrative genomics approach to infer causal associations between gene expression and disease. *Nat Genet* 37(7): 710–7.

Schmid, S. L. (1997). Clathrin-coated vesicle formation and protein sorting: an integrated process. *Annu Rev Biochem* 66: 511–48.

Schumacher, E. H., M. W. Cole, et al. (2007). Selection and maintenance of stimulus-response rules during preparation and performance of a spatial choice-reaction task. *Brain Res* 1136(1): 77–87.

Sederberg, P. B., M. J. Kahana, et al. (2003). Theta and gamma oscillations during encoding predict subsequent recall. *J Neurosci* 23(34): 10809–14.

Segura-Bruna, N., A. Rodriguez-Campello, (2006). Valproate-induced hyperammonemic encephalopathy. *Acta Neurol Scand* 114(1): 1–7.

Sekine, Y. (2005). Consideration of the concepts of "remission" and "cured" in schizophrenia: a male case who experienced schizophrenia with psycho-motoric excitation 30 years ago. *Seishin Shinkeigaku Zasshi* 107(11): 1159–68.

Sentman, M. L., M. Granstrom, et al. (2006). Phenotypes of mice lacking extracellular superoxide dismutase and copper- and zinc-containing superoxide dismutase. *J Biol Chem* 281(11): 6904–9.

Shames, D. S., J. D. Minna, et al. (2007). DNA methylation in health, disease, and cancer. *Curr Mol Med* 7(1): 85–102.

Shen, S., A. Liu, et al. (2006). Epigenetic memory loss in aging oligodendrocytes in the corpus callosum. *Neurobiol Aging* 29(3):452–63.

Shenton, M. E., C. C. Dickey, et al. (2001). A review of MRI findings in schizophrenia. *Schizophr Res* 49(1–2): 1–52.

Sieghart, W. (1995). Structure and pharmacology of gamma-aminobutyric acidA receptor subtypes. *Pharmacol Rev* 47(2): 181–234.

Singer, W. and C. M. Gray (1995). Visual feature integration and the temporal correlation hypothesis. *Annu Rev Neurosci* 18: 555–86.

Singh, S. M., B. Murphy, et al. (2002). Epigenetic contributors to the discordance of monozygotic twins. *Clin Genet* 62(2): 97–103.

Sinagra, M., D. Verrier, et al. (2005). Reelin, very-low-density lipoprotein receptor, and apolipoprotein E receptor 2 control somatic NMDA receptor composition during hippocampal maturation in vitro. *J Neurosci* 25(26): 6127–36.

Sinkkonen, S. T., M. C. Hanna, et al. (2000). GABA(A) receptor epsilon and theta subunits display unusual structural variation between species and are enriched in the rat locus ceruleus. *J Neurosci* 20(10): 3588–95.

Skaggs, W. E., B. L. McNaughton, et al. (1996). Theta phase precession in hippocampal neuronal populations and the compression of temporal sequences. *Hippocampus* 6(2): 149–72.

Slonimsky, J. D., M. D. Mattaliano, et al. (2006). Role for calcium/calmodulin-dependent protein kinase II in the p75-mediated regulation of sympathetic cholinergic transmission. *Proc Natl Acad Sci U S A* 103(8): 2915–9.

Smith, D. M. and S. J. Mizumori (2006). Hippocampal place cells, context, and episodic memory. *Hippocampus* 16(9): 716–29.

Sommer, F. T. and T. Wennekers (2001). Associative memory in networks of spiking neurons. *Neural Netw* 14(6–7): 825–34.

Somogyi, P., J. M. Fritschy, et al. (1996). The gamma 2 subunit of the GABAA receptor is concentrated in synaptic junctions containing the alpha 1 and beta 2/3 subunits in hippocampus, cerebellum and globus pallidus. *Neuropharmacology* 35(9–10): 1425–44.

Sousa, A. D., J. S. Berg, et al. (2006). Myo10 in brain: developmental regulation, identification of a headless isoform and dynamics in neurons. *J Cell Sci* 119(Pt 1): 184–94.

Spazierer, D., P. Fuchs, et al. (2006). Epiplakin is dispensable for skin barrier function and for integrity of keratin network cytoarchitecture in simple and stratified epithelia. *Mol Cell Biol* 26(2): 559–68.

Steffenach, H. A., M. Witter, et al. (2005). Spatial memory in the rat requires the dorsolateral band of the entorhinal cortex. *Neuron* 45(2): 301–13.
Stemmler, M. and C. Koch (1999). How voltage-dependent conductances can adapt to maximize the information encoded by neuronal firing rate. *Nat Neurosci* 2(6): 521–7.
Stone, J. M. and L. S. Pilowsky (2006). Antipsychotic drug action: targets for drug discovery with neurochemical imaging. *Expert Rev Neurother* 6(1): 57–64.
Stromback, L., V. Jakoniene, et al. (2006). Representing, storing and accessing molecular interaction data: a review of models and tools. *Brief Bioinform* 7(4): 331–8.
Sugaya, K., Y. Ogawa, et al. (1998). Evidence for involvement of the subcoeruleus nucleus and nucleus raphe magnus in urine storage and penile erection in decerebrate rats. *J Urol* 159(6): 2172–6.
Tanaka, T. (2002). Identification of common disease related genes by means of genome-wide SNP typing. *Gan To Kagaku Ryoho* 29(8): 1479–83.
Tardif, J. C., T. Heinonen, et al. (2006). Vascular biomarkers and surrogates in cardiovascular disease. *Circulation* 113(25): 2936–42.
Tashiro, A. and R. Yuste (2004). Regulation of dendritic spine motility and stability by Rac1 and Rho kinase: evidence for two forms of spine motility. *Mol Cell Neurosci* 26(3): 429–40.
Tehrani, M. H. and E. M. Barnes, Jr. (1997). Sequestration of gamma-aminobutyric acidA receptors on clathrin-coated vesicles during chronic benzodiazepine administration in vivo. *J Pharmacol Exp Ther* 283(1): 384–90.
Traub, R. D., M. A. Whittington, et al. (1996). Analysis of gamma rhythms in the rat hippocampus in vitro and in vivo. *J Physiol* 493(Pt 2): 471–84.
Tsutsumi, T., S. E. Holmes, et al. (2004). Novel CAG/CTG repeat expansion mutations do not contribute to the genetic risk for most cases of bipolar disorder or schizophrenia. *Am J Med Genet B Neuropsychiatr Genet* 124(1): 15–9.
Turrigiano, G. G. (1999). Homeostatic plasticity in neuronal networks: the more things change, the more they stay the same. *Trends Neurosci* 22(5): 221–7.
Turrigiano, G. G. and S. B. Nelson (2000). Hebb and homeostasis in neuronal plasticity. *Curr Opin Neurobiol* 10(3): 358–64.
Turrigiano, G. G., K. R. Leslie, et al. (1998). Activity-dependent scaling of quantal amplitude in neocortical neurons. *Nature* 391(6670): 892–6.
van Groen, T. and J. M. Wyss (1990). Extrinsic projections from area CA1 of the rat hippocampus: olfactory, cortical, subcortical, and bilateral hippocampal formation projections. *J Comp Neurol* 302(3): 515–28.
Vinogradova, O. S. (2001). Hippocampus as comparator: role of the two input and two output systems of the hippocampus in selection and registration of information. *Hippocampus* 11(5): 578–98.
Vinogradova, O. S. and E. S. Brazhnik (1977). Neuronal aspects of septo-hippocampal relations. Ciba Found Symp (58): 145–77.
Vinogradova, O. S., V. F. Kitchigina, et al. (1998). Pacemaker neurons of the forebrain medical septal area and theta rhythm of the hippocampus. *Membr Cell Biol* 11(6): 715–25.
Vinogradova, O. S., V. F. Kitchigina, et al. (1999). Spontaneous activity and sensory responses of hippocampal neurons during persistent theta-rhythm evoked by median raphe nucleus blockade in rabbit. *Neuroscience* 94(3): 745–53.
Vogeley, K. and C. Kupke (2007). Disturbances of time consciousness from a phenomenological and a neuroscientific perspective. *Schizophr Bull* 33(1): 157–65.
Voutsinos-Porche, B., G. Bonvento, et al. (2003). Glial glutamate transporters mediate a functional metabolic crosstalk between neurons and astrocytes in the mouse developing cortex. *Neuron* 37(2): 275–86.
Waldman, I. D. (2005). Statistical approaches to complex phenotypes: evaluating neuropsychological endophenotypes for attention-deficit/hyperactivity disorder. *Biol Psychiatr* 57(11): 1347–56.
Wan, Q., Z.G . Xiong,e ta l. (1997). Recruitmentof f unctionalG ABA(A)r eceptorst opos tsynaptic domains by insulin. *Nature* 388(6643): 686–90.

Wang, L., J. P. Miller, et al. (2006). Abnormalities of hippocampal surface structure in very mild dementia of the Alzheimer type. *Neuroimage* 30(1): 52–60.

Watanabe, H., T. Aihara, et al. (2006). Phase shift of subthreshold theta oscillation in hippocampal CA1 pyramidal cell membrane by excitatory synaptic inputs. *Neuroscience* 140(4): 1189–99.

Weernink, P. A., K. Meletiadis, et al. (2004). Activation of type I phosphatidylinositol 4-phosphate 5-kinase isoforms by the Rho GTPases, RhoA, Rac1, and Cdc42. *J Biol Chem* 279(9): 7840–9.

Weinberger, D. R., M. F. Egan, et al. (2001). Prefrontal neurons and the genetics of schizophrenia. *Biol Psychiatr* 50(11): 825–44.

Weiss, B., T. Catron, et al. (2000). A 2-year follow-up of the effectiveness of traditional child psychotherapy. *J Consult Clin Psychol* 68(6): 1094–101.

Whittington, M. A., R. D. Traub, et al. (1998). Morphine disrupts long-range synchrony of gamma oscillations in hippocampal slices. *Proc Natl Acad Sci U S A* 95(10): 5807–11.

Wiebe, S. P., U. V. Staubli, et al. (1997). Short-term reverberant memory model of hippocampal field CA3. *Hippocampus* 7(6): 656–65.

Williams, N. M., E. K. Green, et al. (2006). Variation at the DAOA/G30 locus influences susceptibility to major mood episodes but not psychosis in schizophrenia and bipolar disorder. *Arch Gen Psychiatr* 63(4): 366–73.

Wisden, W., D. Cope, et al. (2002). Ectopic expression of the GABA(A) receptor alpha6 subunit in hippocampal pyramidal neurons produces extrasynaptic receptors and an increased tonic inhibition. *Neuropharmacology* 43(4): 530–49.

Wohrl, R., D. von Haebler, et al. (2007). Low-frequency stimulation of the direct cortical input to area CA1 induces homosynaptic LTD and heterosynaptic LTP in the rat hippocampal-entorhinal cortex slice preparation. *Eur J Neurosci* 25(1): 251–8.

Wolkenhauer, O., M. Mesarovic, et al. (2007). A plea for more theory in molecular biology. Ernst Schering Res Found Work-shop(61): 117–37.

Wong, C. W. (1997). Two circuits to convert short-term memory into long-term memory. *Med Hypotheses* 49(5): 375–8.

Woo, N. H., H. K. Teng, et al. (2005). Activation of p75NTR by proBDNF facilitates hippocampal long-term depression. *Nat Neurosci* 8(8): 1069–77.

Wu, K. L., S. Khan, et al. (2004). The NHE1 Na+/H+ exchanger recruits ezrin/radixin/moesin proteins to regulate Akt-dependent cell survival. *J Biol Chem* 279(25): 26280–6.

Xu, J., L. Gowen, et al. (2006). Decreased hepatic futile cycling compensates for increased glucose disposal in the Pten heterodeficient mouse. *Diabetes* 55(12): 3372–80.

Yang, B., J. D. Slonimsky, et al. (2002). A rapid switch in sympathetic neurotransmitter release properties mediated by the p75 receptor. *Nat Neurosci* 5(6): 539–45.

Yoganarasimha, D. and J. J. Knierim (2005). Coupling between place cells and head direction cells during relative translations and rotations of distal landmarks. *Exp Brain Res* 160(3): 344–59.

Yoganarasimha, D., X. Yu, et al. (2006). Head direction cell representations maintain internal coherence during conflicting proximal and distal cue rotations: comparison with hippocampal place cells. *J Neurosci* 26(2): 622–31.

Zhang, H., D. J. Webb, et al. (2005). A GIT1/PIX/Rac/PAK signaling module regulates spine morphogenesis and synapse formation through MLC. *J Neurosci* 25(13): 3379–88.

Zubenko, G. S., B. S. Maher, et al. (2004). Genome-wide linkage survey for genetic loci that affect the risk of suicide attempts in families with recurrent, early-onset, major depression. *Am J Med Genet B Neuropsychiatr Genet* 129(1): 47–54.

Index

C

Q

R

S

T

U

V

W

Color Plates

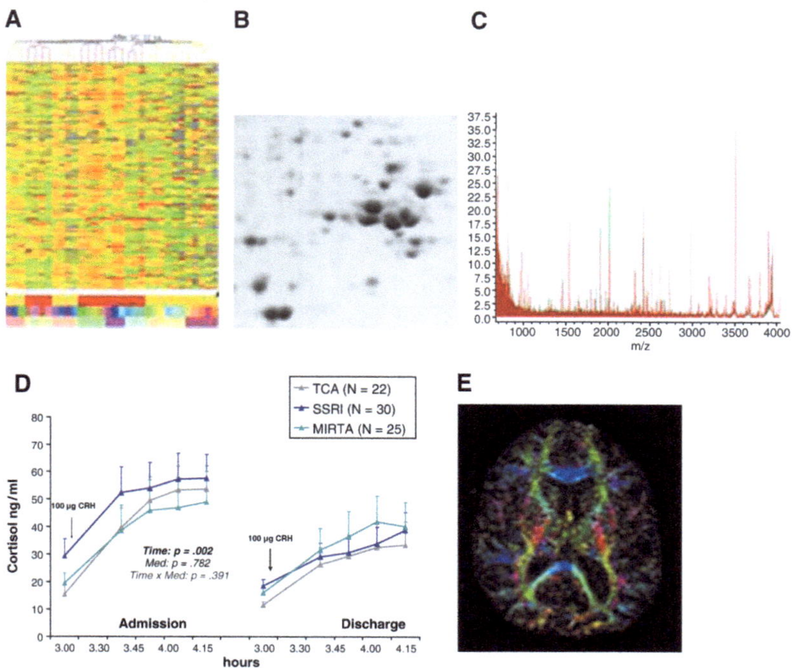

Fig. 3.1 A biomarker is any characteristic that can be objectively measured to reflect physiological, pharmacological, or disease processes in animals or humans (De Gruttola et al., 2001). Various platforms for different types of biomarkers can be used for psychiatric disorders. (**a**) RNA microarray, (**b**) Protein expression analysis by gel electrophoresis, (**c**) Mass spectrometry-based protein and metabolite profiling, (**d**) Neuroendocrine assay, (**e**) Brain MRI scan

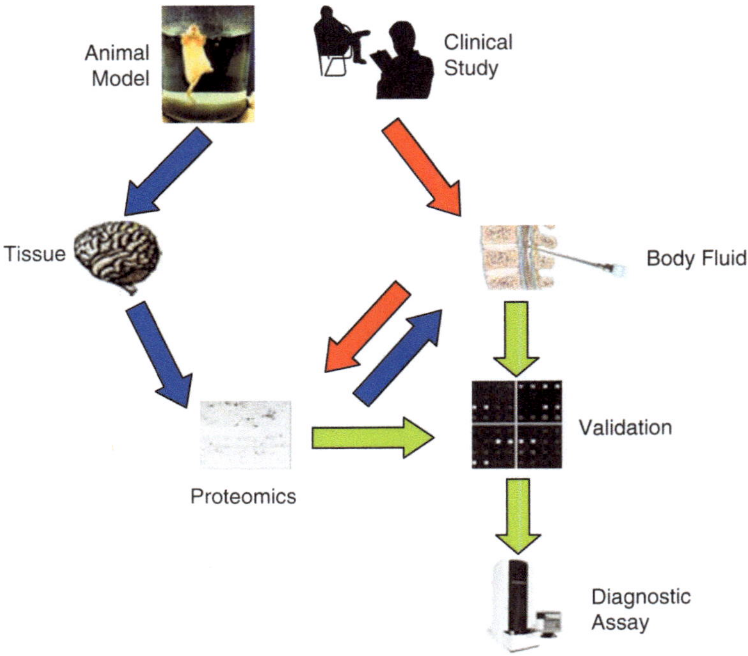

Fig. 3.2 Roads to protein biomarker discovery. On one hand, brains from animal models that reflect certain aspects of the disease phenotype are investigated by the proteomics methods. Once candidate markers have been identified from the animal model, they need to be validated in patient specimens. Alternatively, clinical specimens are used directly for biomarker discovery, validated in a great number of samples, and ultimately used for establishing a diagnostic assay platform

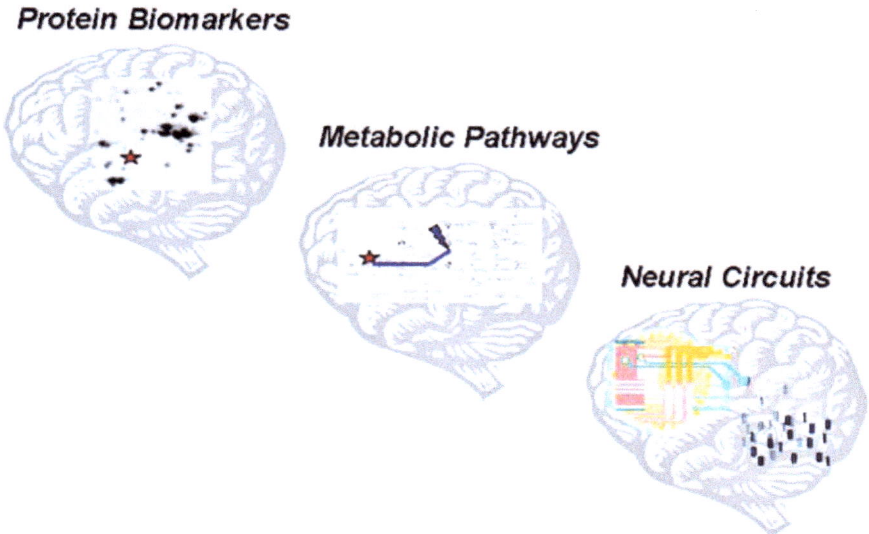

Fig. 3.3 From protein biomarkers to metabolic pathways to neural circuits. Phenotype-related protein expression differences provide information on metabolic pathways that are affected in psychiatric disorders. A combination of altered metabolic pathways may ultimately precipitate disease by distressing critical neural circuits

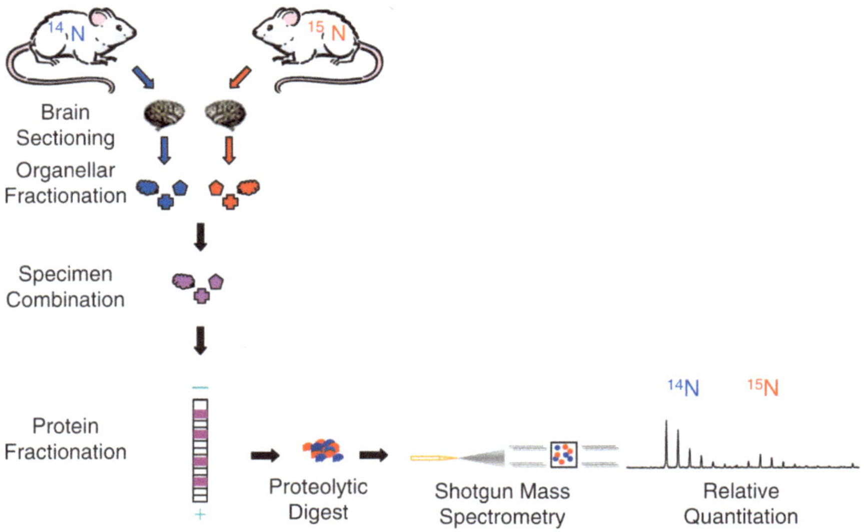

Fig. 3.4 Metabolic labeling approach for sensitive and comprehensive protein analysis and comparison. Mice are fed with either regular ^{14}N or labeled ^{15}N diets. Organ or body fluid specimens are then combined and processed for relative protein quantitation by mass spectrometry

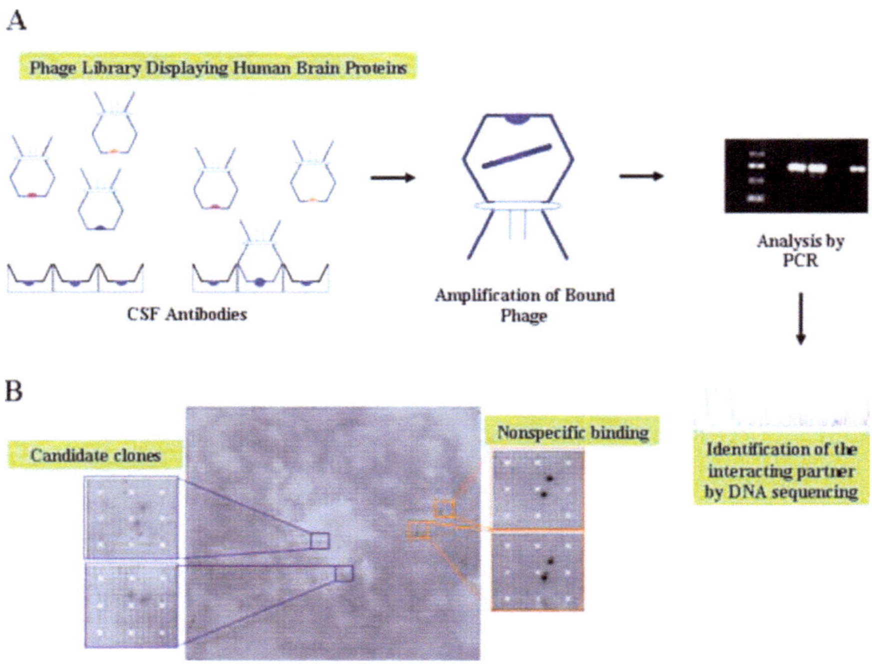

Fig. 3.6 Phage display (**a**) and protein array (**b**) screens for CSF autoantibodies. Positive clones are candidate brain autoantigen biomarkers

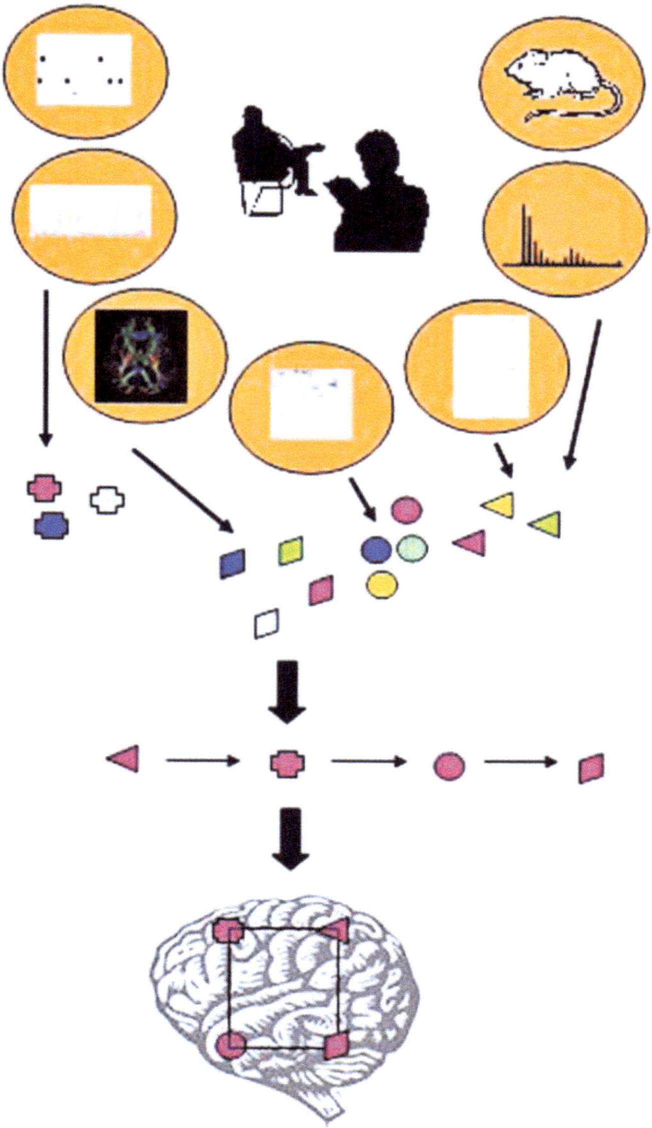

Fig. 3.7 Data integration for insights into psychiatric disorder pathophysiology. Experimental and clinical data obtained from diverse platforms are consolidated with the help of bioinformatics tools in order to identify affected neural pathways

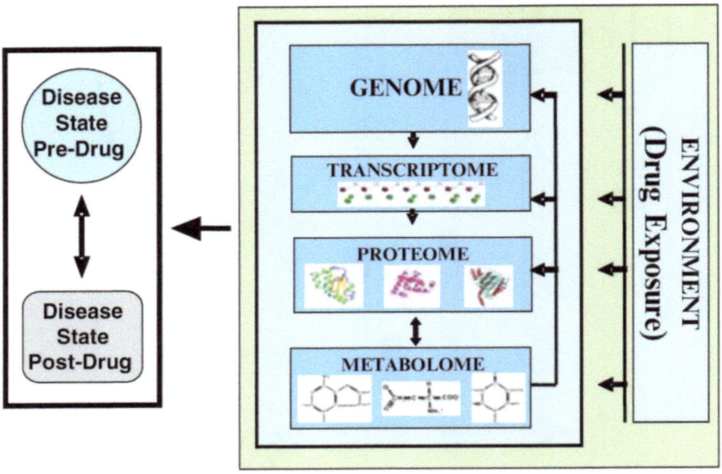

Fig. 6.1 Flow of information from the genetic code (DNA) to proteins, and finally to metabolites. The environment and DNA affect the end products and influence health/disease states. In this case, drug exposure is the environmental alteration being tested

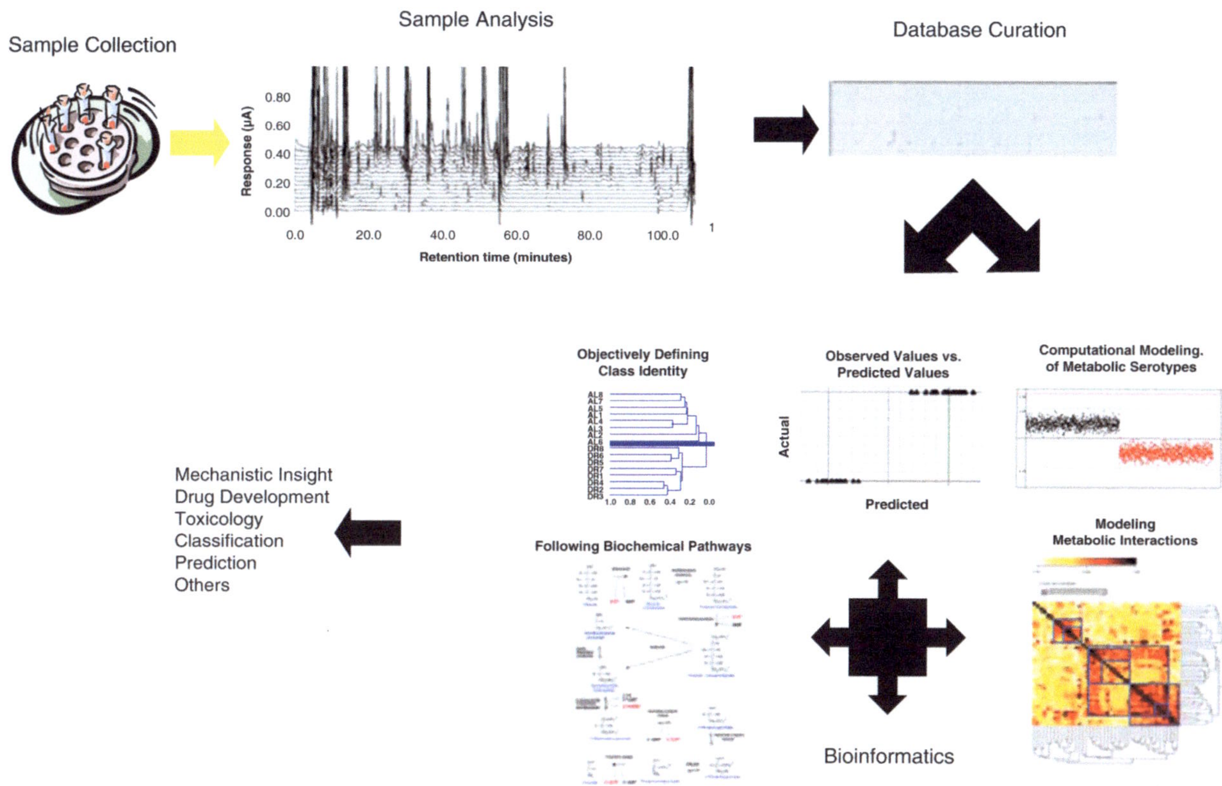

Fig. 6.2 Flow diagram for a typical metabolomics experiment. Samples are collected and individual metabolites are quantitatively analyzed using one or more high data density analytical instruments. These datasets are curated based on the design of the individual experiment, and are then analyzed using a series of high data density informatics approaches. Clockwise from the top left, those shown here are cluster analysis (from Pirouette, infometrix), class prediction (SIMCA-P, Umetrics), principal components analysis of a computationally modeled dataset (SIMCA-P, Umetrics), 2D cluster analysis (GeneLinker Platinum, Improved Outcomes Software), and metabolic pathway analysis (biotech.icmb.utexas.edu)

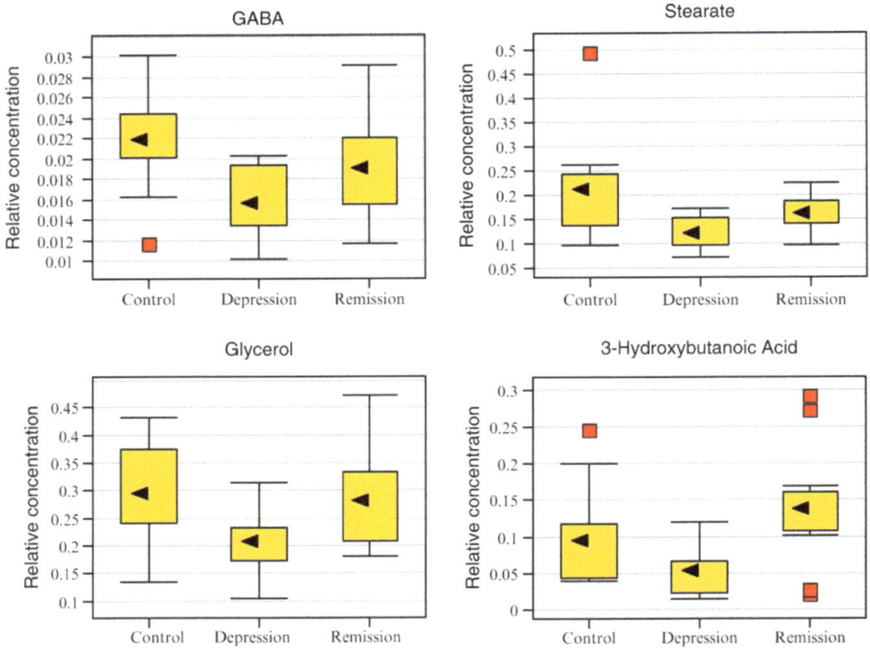

Fig. 6.3 Relative Concentrations of four metabolites in healthy controls and individuals with ongoing or remitted depression. In this box plots, the center line represents the median value, the bottom and top of the box represents the 25th and 75th percentile, respectively, and the dashed lines show the full range for the metabolite in the various groups. When compared with controls and remission groups, the depressed group had overall significant ($p<0.05$) decline in the concentration of GABA and medium chain fatty acids including stearate, palmitate, myristate and oleate. Stearate is shown as a representative fatty acid that is lower in the depressed population

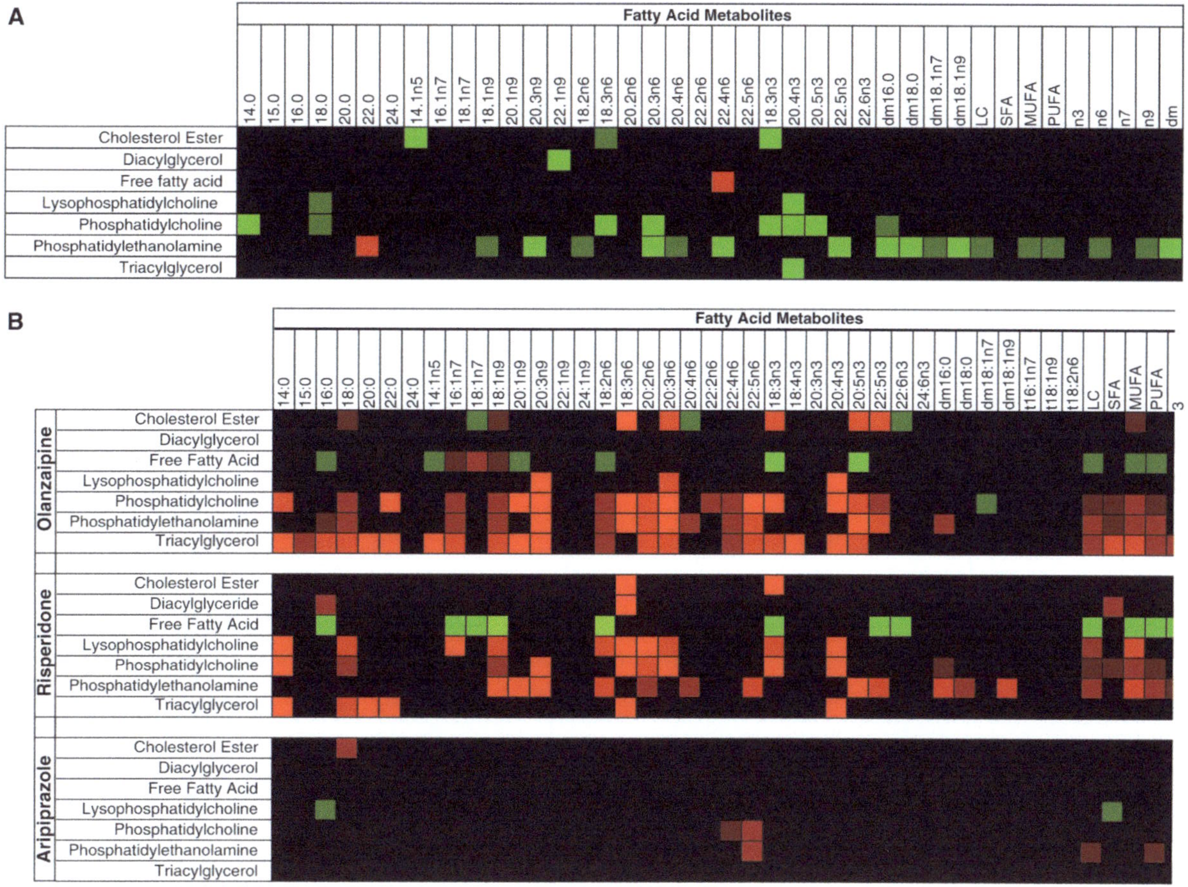
A
Fatty Acid Metabolites
14.0
15.0
16.0
18.0
20.0
22.0
24.0
14.1n5
16.1n7
18.1n7
18.1n9
20.1n9
20.3n9
22.1n9
18.2n6
18.3n6
20.2n6
20.3n6
20.4n6
22.2n6
22.4n6
22.5n6
18.3n3
20.4n3
20.5n3
22.5n3
22.6n3
dm16.0
dm18.0
dm18.1n7
dm18.1n9
LC
SFA
MUFA
PUFA
n3
n6
n7
n9
dm
Cholesterol Ester
Diacylglycerol
Free fatty acid
Lysophosphatidylcholine
Phosphatidylcholine
Phosphatidylethanolamine
Triacylglycerol
B
Fatty Acid Metabolites
14:0
15:0
16:0
18:0
20:0
22:0
24:0
14:1n5
16:1n7
18:1n7
18:1n9
20:1n9
20:3n9
22:1n9
24:1n9
18:2n6
18:3n6
20:2n6
20:3n6
20:4n6
22:2n6
22:4n6
22:5n6
18:3n3
18:4n3
20:3n3
20:4n3
20:5n3
22:5n3
22:6n3
24:6n3
dm16:0
dm18:0
dm18:1n7
dm18:1n9
t16:1n7
t18:1n9
t18:2n6
LC
SFA
MUFA
PUFA
3
Olanzaipine
Cholesterol Ester
Diacylglycerol
Free Fatty Acid
Lysophosphatidylcholine
Phosphatidylcholine
Phosphatidylethanolamine
Triacylglycerol
Risperidone
Cholesterol Ester
Diacylglyceride
Free Fatty Acid
Lysophosphatidylcholine
Phosphatidylcholine
Phosphatidylethanolamine
Triacylglycerol
Aripiprazole
Cholesterol Ester
Diacylglycerol
Free Fatty Acid
Lysophosphatidylcholine
Phosphatidylcholine
Phosphatidylethanolamine
Triacylglycerol

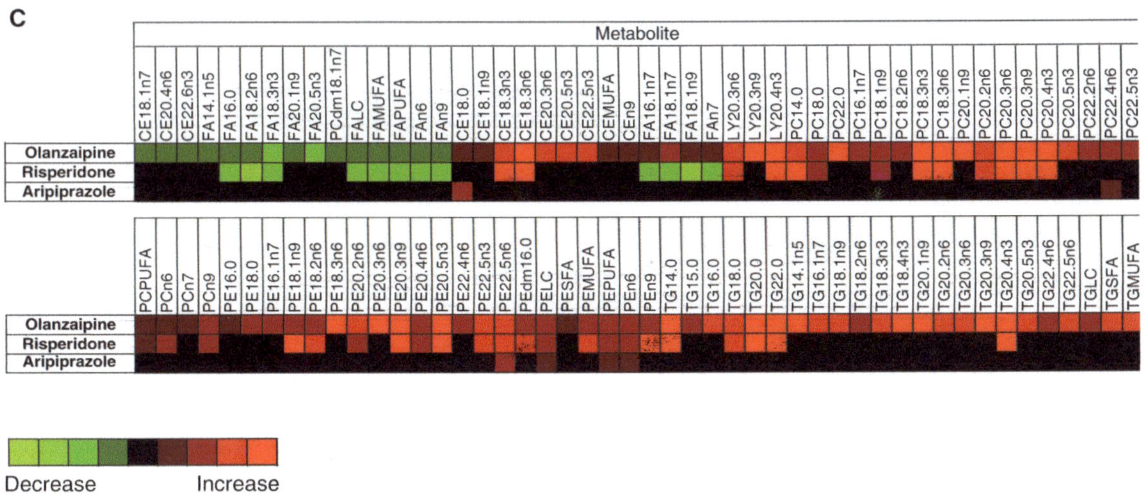

Fig. 6.4 Heat map showing differences in individual lipid metabolites in the plasma of patients with schizophrenia as compared to controls (**a**), in the plasma of schizophrenic patients post treatment as compared to pretreatment (**b**) with olanzapine, risperidone, and aripiprazole. Figure C shows the most significantly modified lipid metabolites in plasma of patients treated with olanzapine and highlights which of these metabolites are also modified upon treatment with risperidone or aripiprazole. The column headers indicate fatty acid metabolites as they appear in each distinct lipid class (rows). In (**a**), lipids whose percent levels were higher in patients vs. controls are shown in red while those with decreased level are shown in green; see methods for details. In (**b**) and (**c**), the percent increase in any lipid upon treatment with drug is shown in red squares and decrease in green squares as described in the methods. The brightness of each color corresponds to the magnitude of the difference in quartiles. The brighter the square is the larger the difference

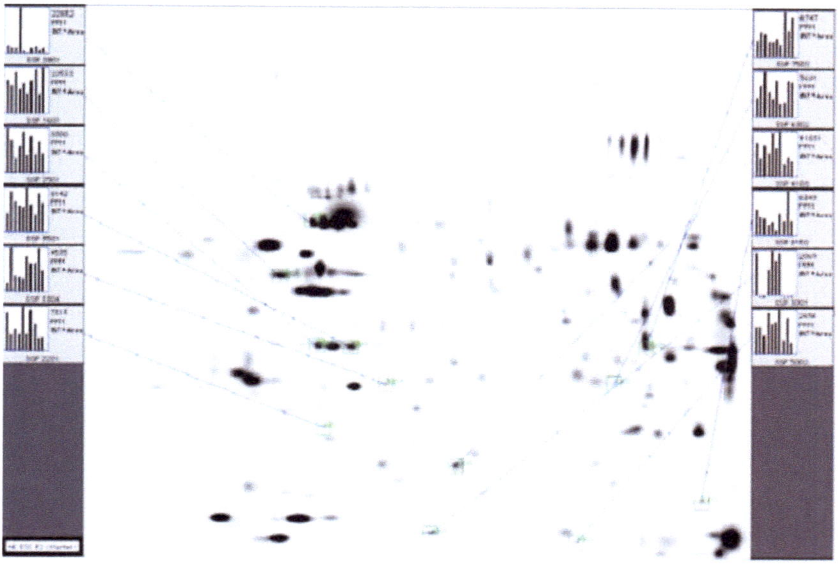

Fig. 8.3 Example of hippocampus master image from synaptosomes of LH–nLH rats. The master is a synthetic image containing the spot data from all the 2D maps in the analysis set. Left and right side bar graphs represent expression of the highlighted proteins in the various comparisons

Fig. 8.4 Example of prefrontal/frontal cortex master image of synaptosomes from FSL/FRL. Left and right side bar graphs represent expression of the highlighted proteins in the various comparisons

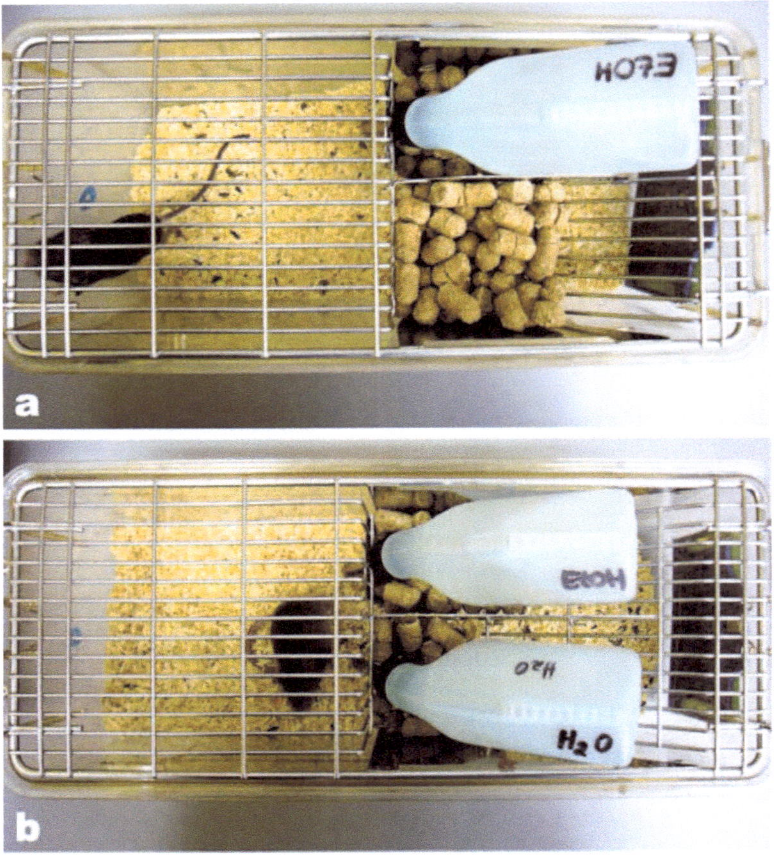

Fig. 10.1 *Forced drinking* (**a**) and *Two bottle-choice paradigm* (**b**). In the *forced drinking paradigm* the animals are forced to consume distinct amounts of drug-solution over an extended period of time. Contrastingly, in the *two bottle-choice paradigm* the animals have the choice between a drug-solution or tap water to evaluate drug preference

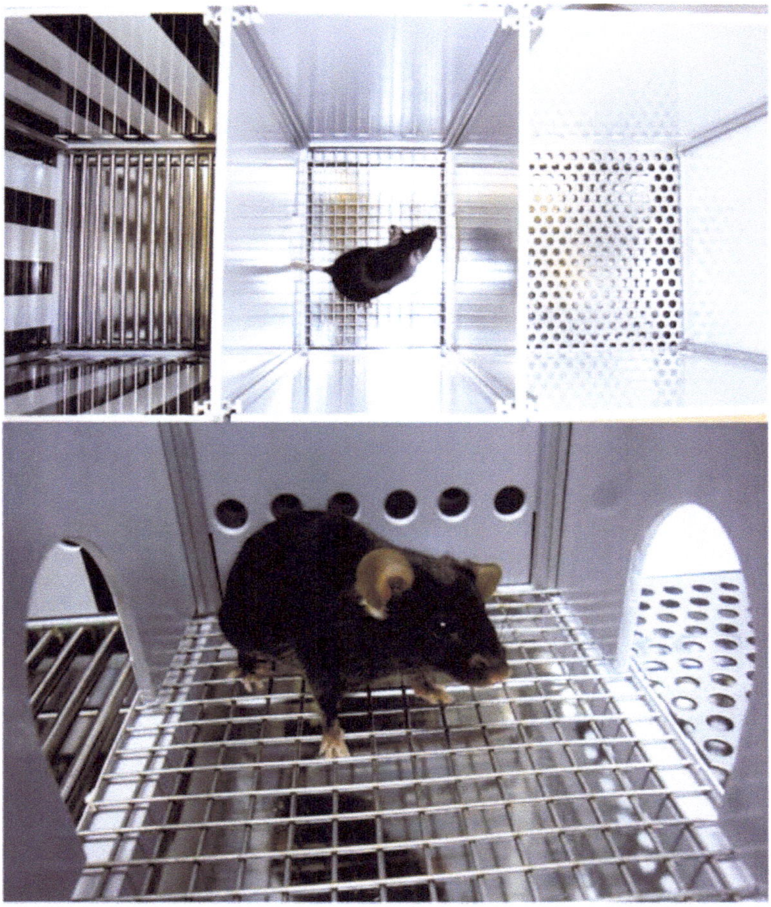

Fig. 10.2 *Conditioned place preference paradigm.* Animals are exposed to two or more distinct neutral environments that are differentiated from each other. After several conditioning sessions, this paradigm provides information about positive and negative reinforcing effects of a drug

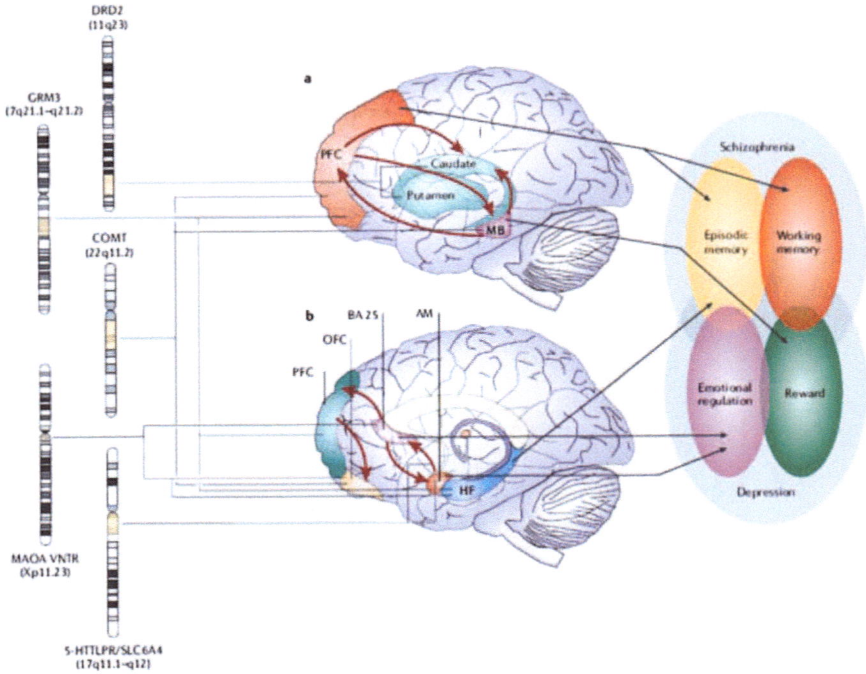

Fig. 11.1 The complex path from genes to behavioral and disease phenotype: mediation through brain circuitry. Multiple genetic risk variants affect, through interaction with each other and the environment, multiple neural systems linked to several neuropsychological and behavioral domains that are impaired, in differing proportions, in psychiatric diseases. As examples, the following genetic variants are depicted (chromosomal variation in parentheses): GRM3 single nucleotide polymorphism 4 (7q21.1–q21.2) [Egan et al., 2004], dopamine receptor D2 (DRD2) Taq 1a (11q23) [Cohen et al., 2005], catechol-O-methyltransferase (COMT) Val66Met (22q11.2) [Egan et al., 2001, Meyer-Lindenberg et al., 2006], serotonin transporter length polymorphism (5-HTTLPR/SLC6A4) (17q11.1–q12) [Hariri et al., 2002, Pezawas et al., 2005] and monoamine oxidase A variable number tandem repeat (MAOA VNTR) (Xp11.23) [Meyer-Lindenberg et al., 2006]. These are shown to affect a circuit that links the prefrontal cortex (PFC) with the midbrain (MB) and striatum (caudate and putamen) (**a**), which is relevant for schizophrenia, and a circuit that connects the amygdala (AM) with regulatory cortical and limbic areas (**b**), which is implicated in depression and anxiety. These circuits, in turn, are shown to mediate risk for schizophrenia and depression and various neuropsychological functions. BA 25, Brodmann's area 25; HF, hippocampal formation; OFC, orbitofrontal cortex. (reprinted with permission from [Meyer-Lindenberg and Weinberger, 2006])

associated with psychiatric disorders or behavioral traits are used to discover neural mechanisms mediating their complex emergent phenotypic associations, implicating these mechanisms in the psychiatric disorders to which they have been linked. Examples include the use of the COMT Val158Met polymorphism to characterize prefrontal function and prefrontal-midbrain interactions linked to risk for schizophrenia [Egan et al., 2001, Meyer-Lindenberg et al., 2005], and the delineation of cingulate circuitry regulating amygdala (AM) function mediating risk for anxiety and depression through an investigation of the MAOA variable number tandem repeat (VNTR) [Hariri et al., 2002, Pezawas et al., 2005]. BA 25, Brodmann's area 25; HF, hippocampal formation; MB, midbrain; OFC, orbitofrontal cortex (reprinted with permission from [Meyer-Lindenberg and Weinberger, 2006])

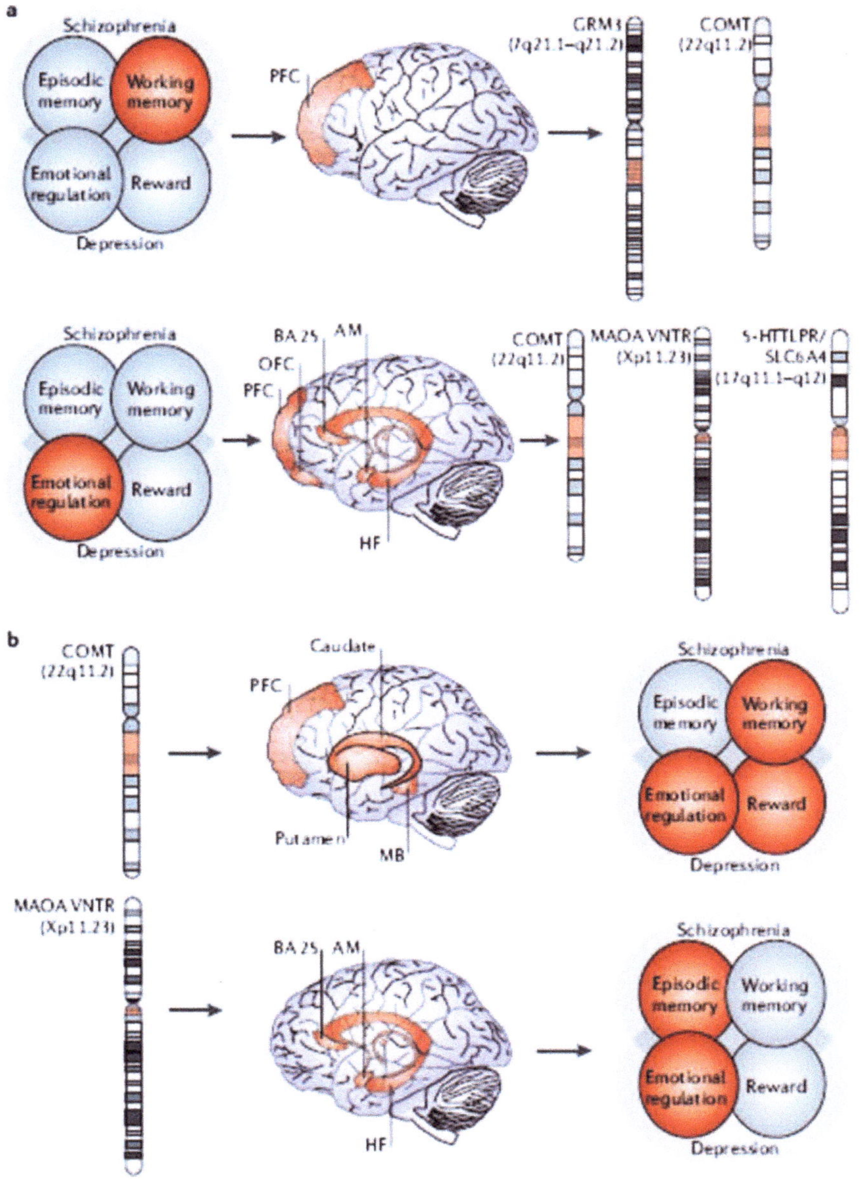

Fig. 11.2 Intermediate phenotypes as tools for gene discovery versus neural mechanism characterization. Examples of two alternative approaches to the identification of genetic variants linked to psychiatric disorders are illustrated, with the relevant genes, neural systems and behavioral phenotypes highlighted in red, and arrows indicating the direction of research inference. (**a**) In the gene discovery approach, behavioral or neural systems phenotypes are used to reduce genetic complexity and increase penetrance to identify genes implicated in psychiatric disorders. In the figure, prefrontal cortex (PFC) dysfunction has been linked to catechol-*O*-methytransferase (COMT) and GRM3 genetic variation [Egan et al., 2004, Meyer-Lindenberg et al., 2006], and emotional regulation has been linked to variation in COMT, monoamine oxidase A (MAOA) and the serotonin transporter length polymorphism (5-HTTLPR/SLC6A4) [Heinz et al., 2005, Meyer-Lindenberg et al., 2006, Pezawas et al., 2005], and so could have been hypothetically employed as a phenotype to identify these genes. (**b**) In the neural mechanism approach, genes known to be

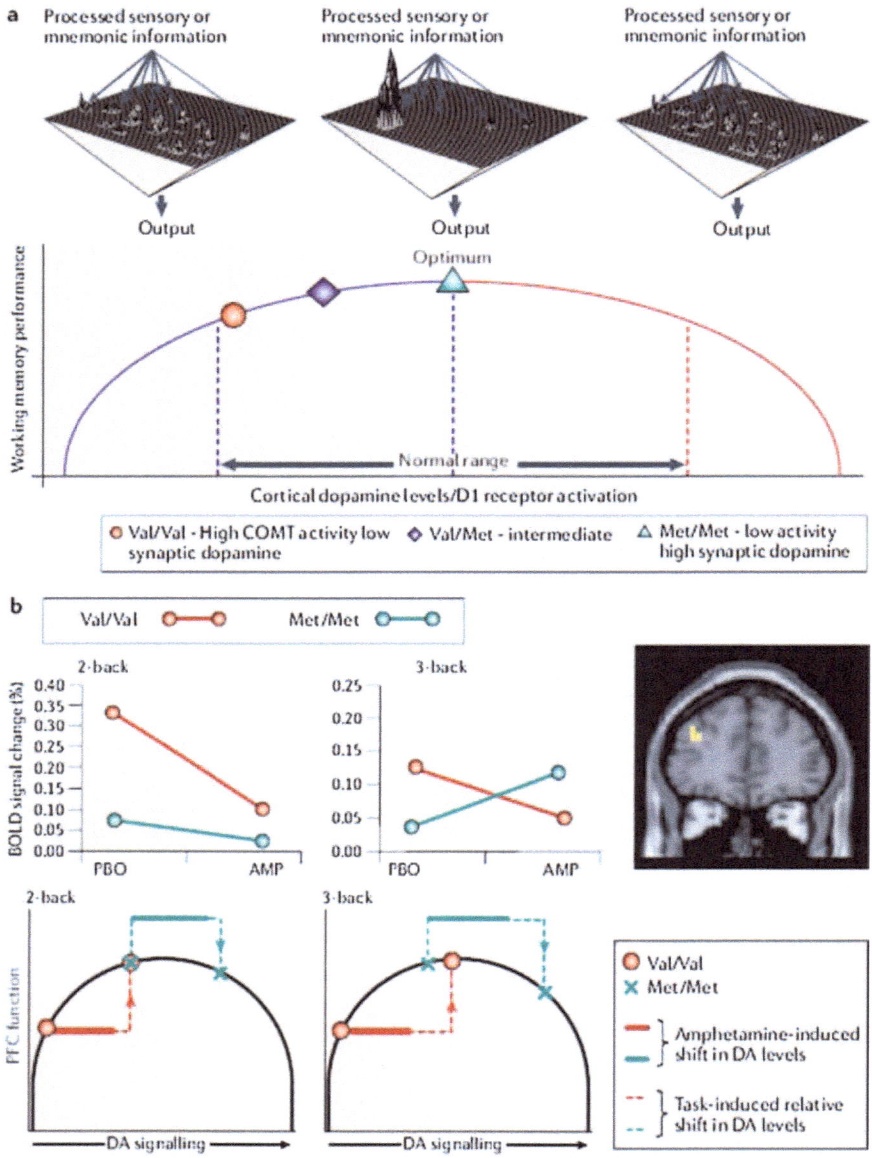

Fig. 11.3 Effects of COMT Val(108/158)Met on prefrontal cortex activity linked to extracellular dopamineConvergent evidence has established an inverted u-shaped relationship between working-memory-related activation of dorsolateral prefrontal cortex (PFC) neurons and dopaminergic tone (especially D1 receptor stimulation) [Williams and Goldman-Rakic, 1995], which has been shown to be essential for optimizing signal-to-noise ratio, or tuning of the neuronal response. This figure demonstrates the use of genetic variation in catechol-O-methyltransferase (COMT) to mechanistically characterize neural circuits relevant for schizophrenia that are dependent on this relationship. (**a**) An inverted u-shaped curve [Goldman-Rakic et al., 2000] links extracellular dopamine to prefrontal signal-to-noise ratio (*top*) and working memory performance (*bottom*): homozygotes for the Val-encoding allele are positioned at the left (suboptimally low dopamine due to high COMT activity) and Met homozygotes are near the optimum (higher dopamine due to reduced COMT activity). Heterozygotes are intermediate. (**b**) Increasing synaptic dopamine (DA) by the administration of

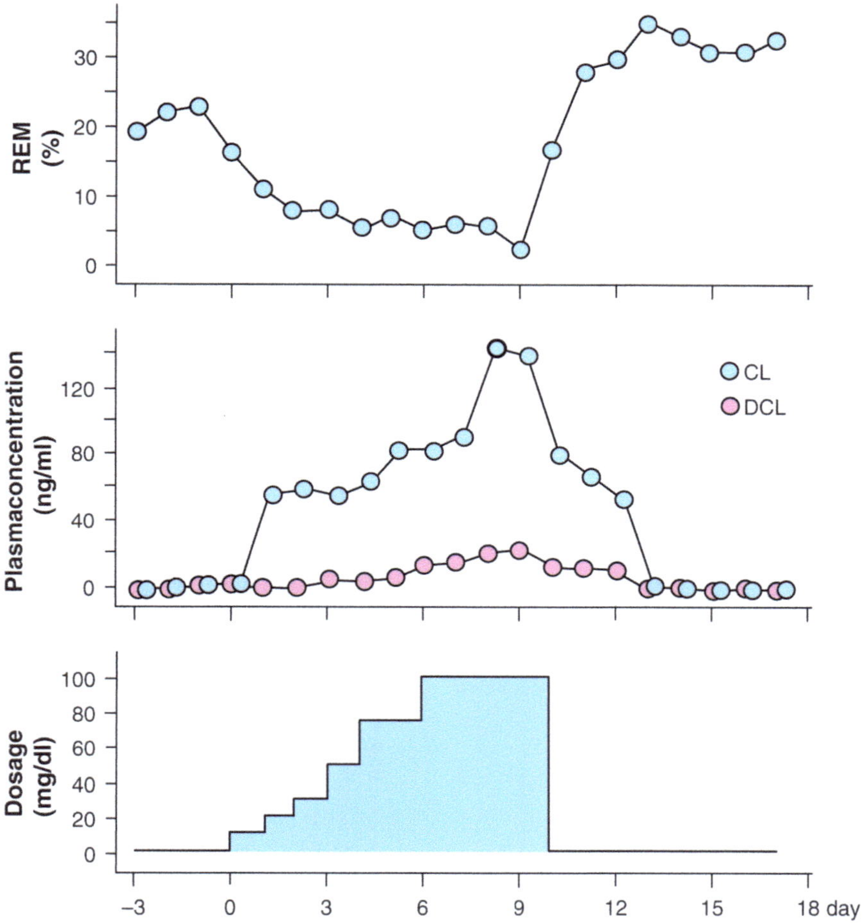

Fig. 12.2 Course of the percentage of REM sleep and the plasma concentration of clomipramine and desmethylclomipramine in a normal subject before, during and after administration of clomipramine. Steiger A: Unterschiede in den Wirkungen von Antidepressiva auf den Schlaf. Psychopharmakotherapie, 3 (1999) 91–95, Fig. 12.1. Copyright Wissenschaftliche Verlagsgesellschaft mbH, Stuttgart. Reproduced with permission

amphetamine (AMP) dissociates the functional states of Val and Met homozygotes [Mattay et al., 2003]. At medium working memory load level (2-back task, where subjects need to remember a stimulus presented 2 trials back), Val homozygotes, on the left of the optimum on the inverted u curve, profit from increased dopamine, whereas Met-homozygotes, near the optimum, show little change (*left*). At high load level (3-back task), dopamine increase by drug intervention now pushes Met homozygotes into the suboptimally high range of dopamine stimulation, leading to reduced prefrontal efficiency (right; localization of activity in the prefrontal cortex is shown on the far right). PBO, placebo (reprinted with permission from [Meyer-Lindenberg and Weinberger, 2006])

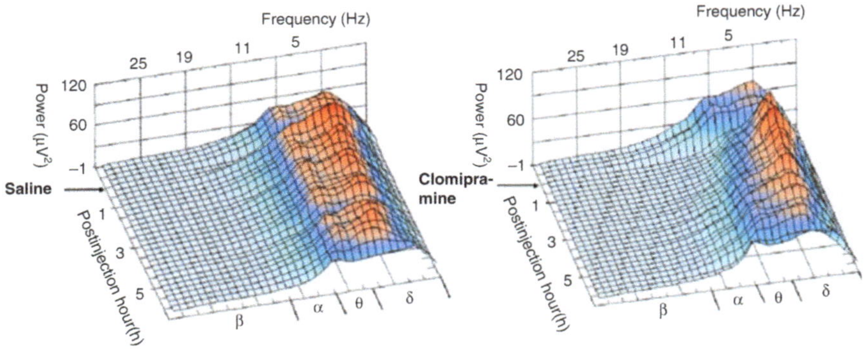

Fig. 12.3 Changes in power spectra of EEGs after saline (*left*) or clomipramine (*right*) injection in the same mice (n = 8). Either vehicle (saline) or clomipramine (30 mg/kg) was given intraperitoneally in the middle of the 12 h-light period, indicated by arrows. Relatively high-intensified power values are illustrated in orange color vs. lower intensity colored in blue. Compared with changes after saline, clomipramine immediately increased delta power but in a much narrower band range, whereas theta power was remarkably diminished for 3 h after injection

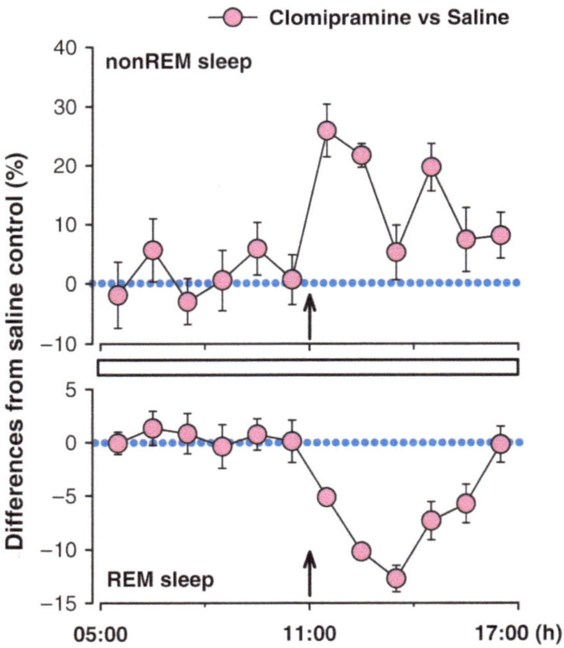

Fig. 12.4 Hourly changes in nonREM and REM sleep before and after clomipramine in mice (indicated with circles, from the same data shown in Fig. 12.3), compared with the saline-treated day (dotted lines). Clomipramine (30 mg/kg) or saline was injected ip in the middle of the 12 h-light period, indicated by arrows. After clomipramine, nonREM sleep increased, but REM sleep was suppressed during the rest of the light period

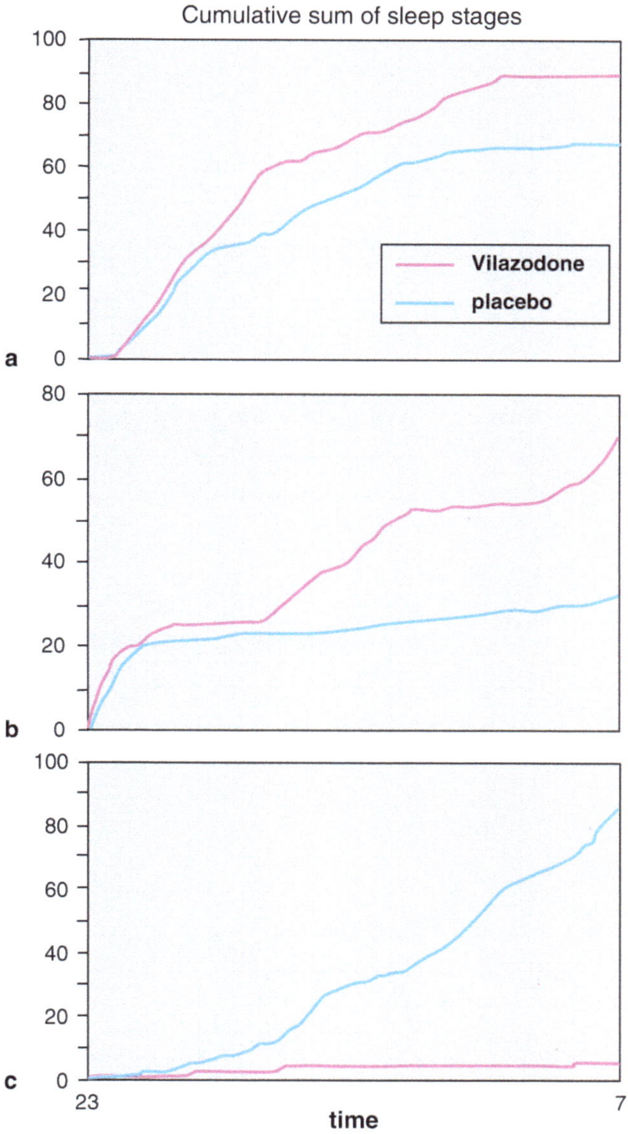

Fig. 12.6 Time course of the accumulation of slow wave sleep (SWS; **a**), wakefulness (**b**), and REM (**c**). Three distinct intervals, which correspond with the thirds of the night, can be distinguished. The increase in SWS after Vilazodone occurs selectively in the first and third one-third, the increase in wakefulness in the second one-third and at the end of the sleep period, whereas REM is suppressed throughout the night. Springer, Psychopharmacology, 155, 2002, p. 189, Distinct temporal pattern of the effects of the combined serotonin-reuptake inhibitor and 5-HT1A agonist EMD 68843 on the sleep EEG in healthy men, Murck H, Frieboes RM, Antonijevic IA, Steiger A, Fig. 2. With kind permission from Springer Science and Business Media

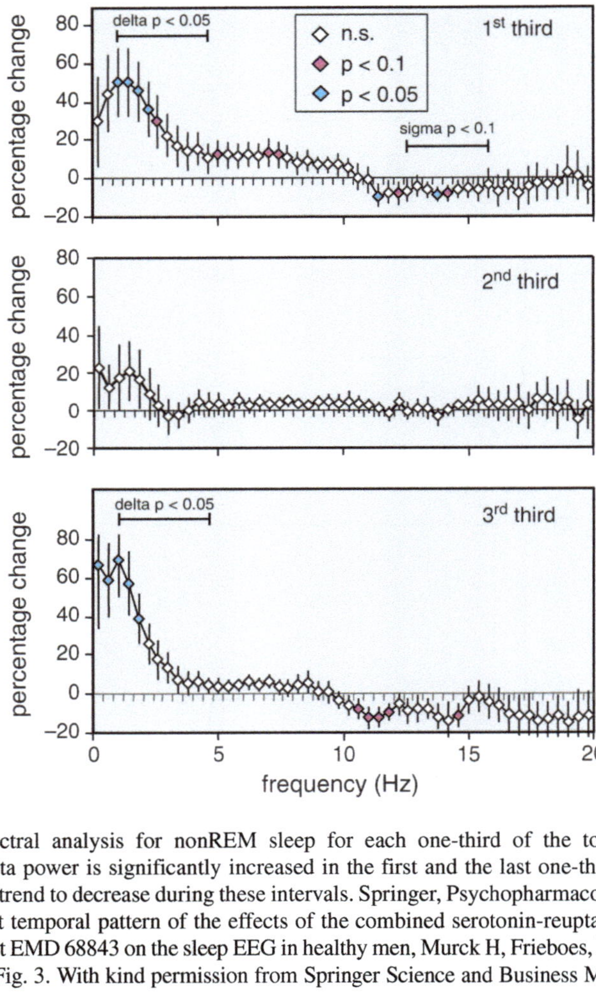

Fig. 12.7 Spectral analysis for nonREM sleep for each one-third of the total night. After Vilazodone delta power is significantly increased in the first and the last one-third, while sigma power shows a trend to decrease during these intervals. Springer, Psychopharmacology, 155, 2002, p. 189, Distinct temporal pattern of the effects of the combined serotonin-reuptake inhibitor and 5-HT1A agonist EMD 68843 on the sleep EEG in healthy men, Murck H, Frieboes, RM, Antonijevic IA, Steiger A, Fig. 3. With kind permission from Springer Science and Business Media

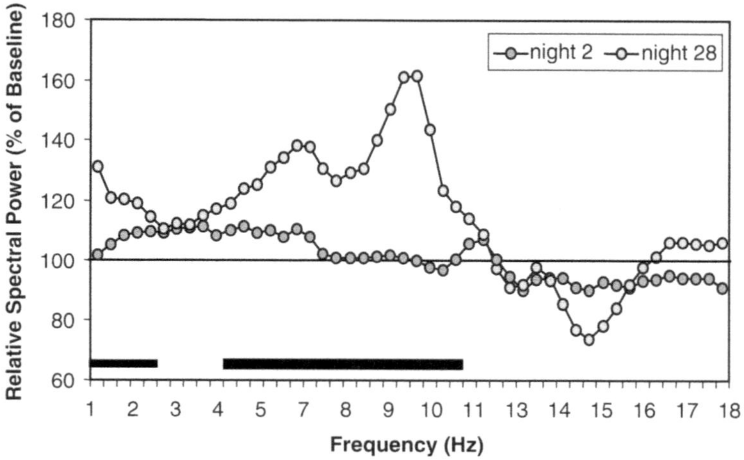

Fig. 12.8 Spectral EEG composition of the largest common amount of nonREM sleep (=155.5 min) 2 days (night 2) and 4 weeks (night 28) after beginning of mirtazapine administration. The values for all frequency bins are plotted as a percentage of baseline night (=100%). The frequency ranges in which the values varied significantly (ANOVA, $p < 0.05$) over all three examination nights are indicated by black bars above the abscissa. Schmid DA, Wichniak A, Uhr M, Ising M, Brunner H, Held K, Weikel J, Sonntag A, Steiger A: Changes of sleep architecture, spectral composition of sleep EEG, the nocturnal secretion of cortisol, ACTH, GH, prolactin, melatonin, ghrelin and leptin and the DEX-CRH-test in depressed patients during treatment with mirtazapine. Neuropsychopharmacology, 31 (2006) 832–844, Fig. 1

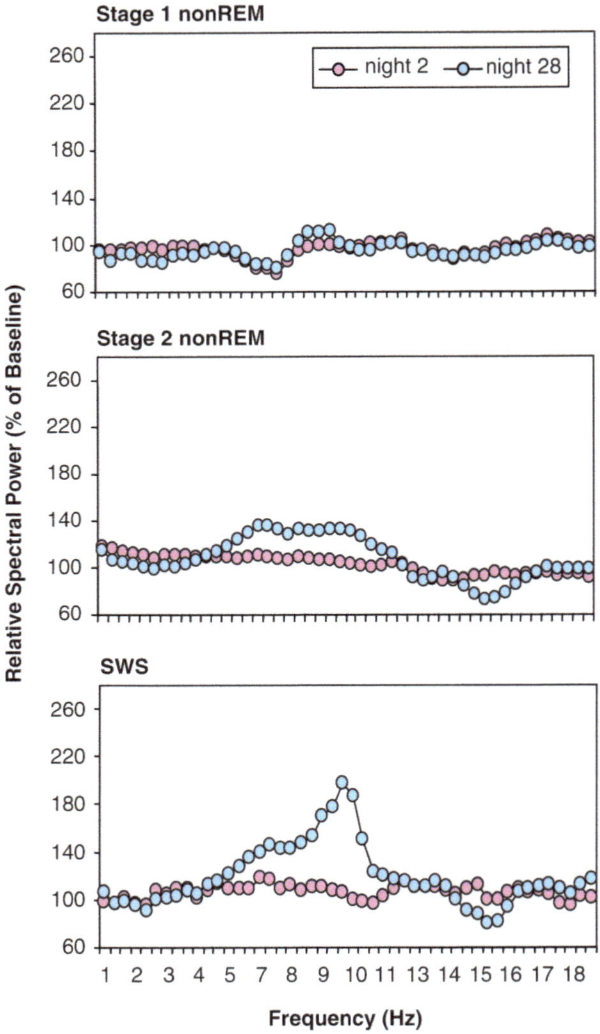

Fig. 12.9 Spectral EEG composition in each of nonREM sleep stages 2 days (night 2) and 4 weeks (night 28) after beginning of mirtazapine administration. The values for all frequency bins are plotted as a percentage of baseline night (= 100%). The increase of alpha activity was apparent only at week 4 and was, in particular, related to SWS. Schmid DA, Wichniak A, Uhr M, Ising M, Brunner H, Held K, Weikel J, Sonntag A, Steiger A: Changes of sleep architecture, spectral composition of sleep EEG, the nocturnal secretion of cortisol, ACTH, GH, prolactin, melatonin, ghrelin and leptin and the DEX-CRH-test in depressed patients during treatment with mirtazapine. Neuropsychopharmacology 31 (2006) 832–844, Fig. 2

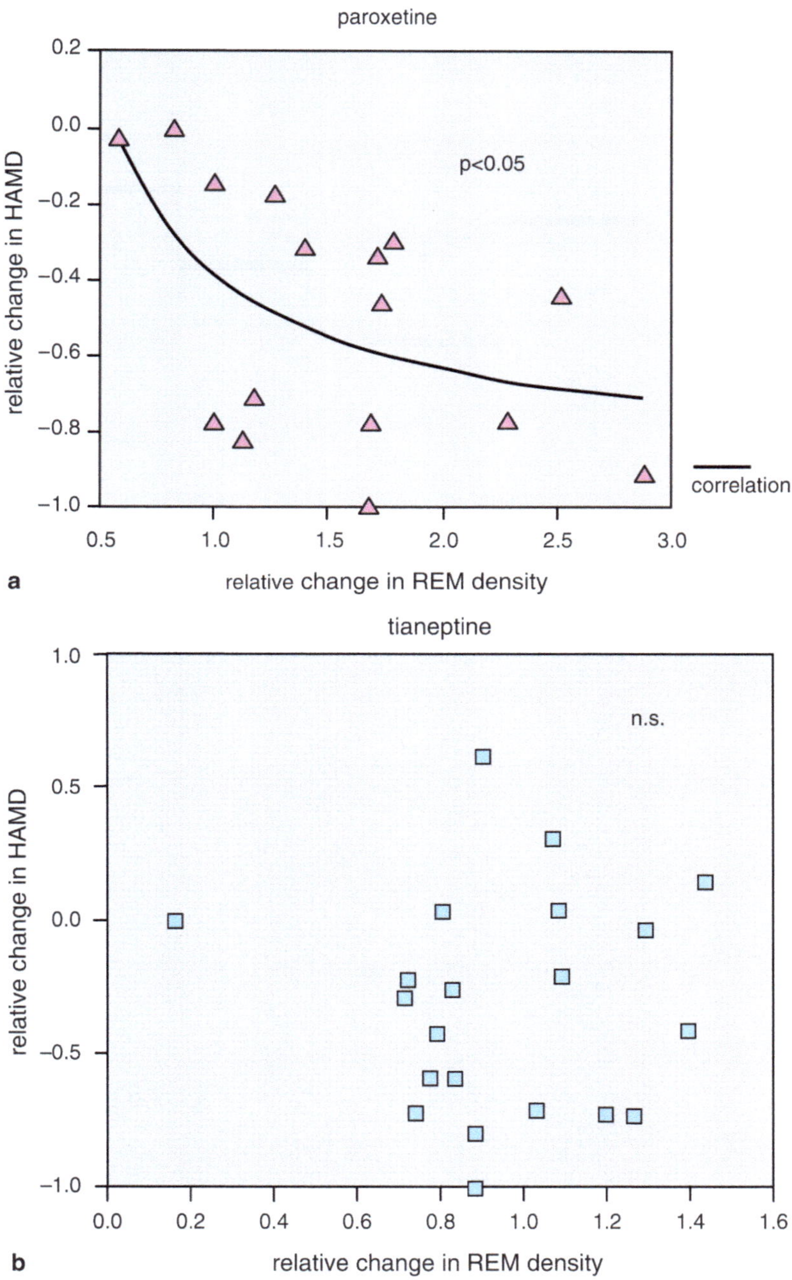

Fig. 12.10 **a** Correlation between the relative change in the Hamilton Depression Score (HAMD) (difference HAMD at day 7 – HAMD at day 42) in the course of treatment with paroxetine in relation to the relative change in REM density. REM density increases with the decline in REM. **b** No correlation is found after tianeptine. Murck H, Nickel T, Künzel H, Antonijevic IA, Schill J, Zobel A, Steiger A, Sonntag A, Holsboer F: State markers of depression in sleep EEG: Dependency on drug and gender in patients treated with tianeptine or paroxetine Neuropsychopharmacology 28 (2003) 348–358, Fig. 1a and 1b

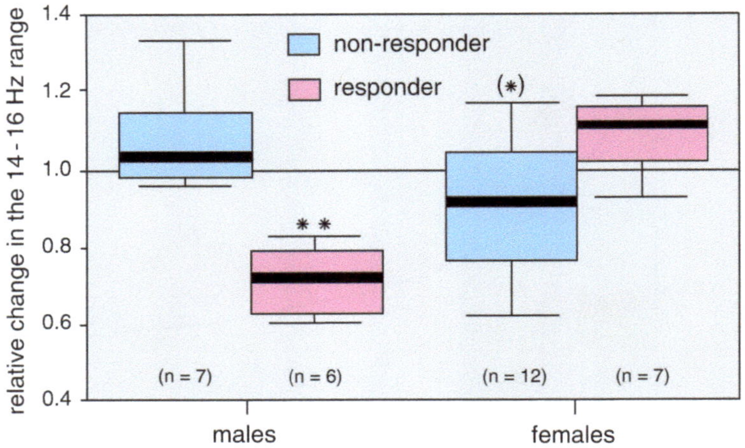

Fig. 12.11 Relative change of EEG power in the 14–16 Hz range divided into gender comparing responders and nonresponders. Male and female subjects behave differentially concerning treatment response (for details, see text). Murck H, Nickel T, Künzel H, Antonijevic IA, Schill J, Zobel A, Steiger A, Sonntag A, Holsboer F: State markers of depression in sleep EEG: Dependency on drug and gender in patients treated with tianeptine or paroxetine. Neuropsychopharmacology 28 (2003) 348–358, Fig. 2

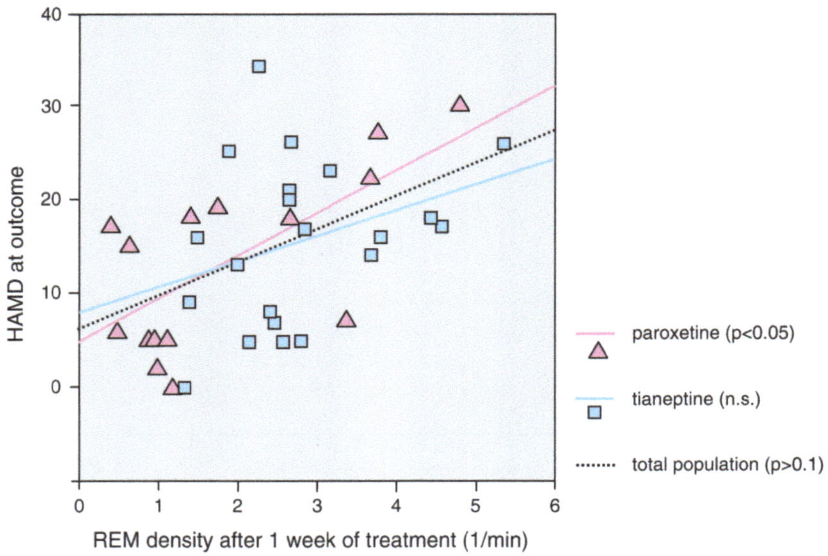

Fig. 12.12 HAMD at day 42 in relation to the REM density at day 7. With paroxetine, the REM density at day 7 is correlated to treatment outcome, whereas with tianeptine no significant correlation existed. Murck H, Nickel T, Künzel H, Antonijevic IA, Schill J, Zobel A, Steiger A, Son ntag A, Holsboer F: State markers of depression in sleep EEG: Dependency on drug and gender in patients treated with tianeptine or paroxetine. Neuropsychopharmacology 28 (2003) 348–358, Fig 3

Table 15.2 Summary of gene expression changes in different patient groups relative to control subjects

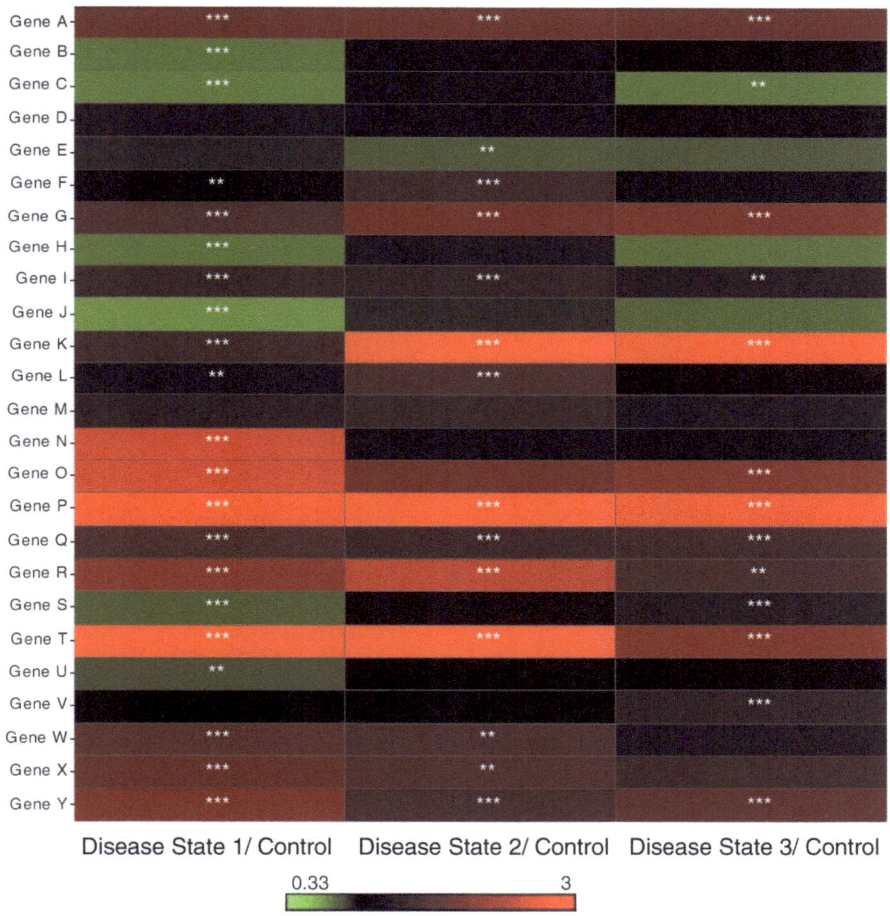

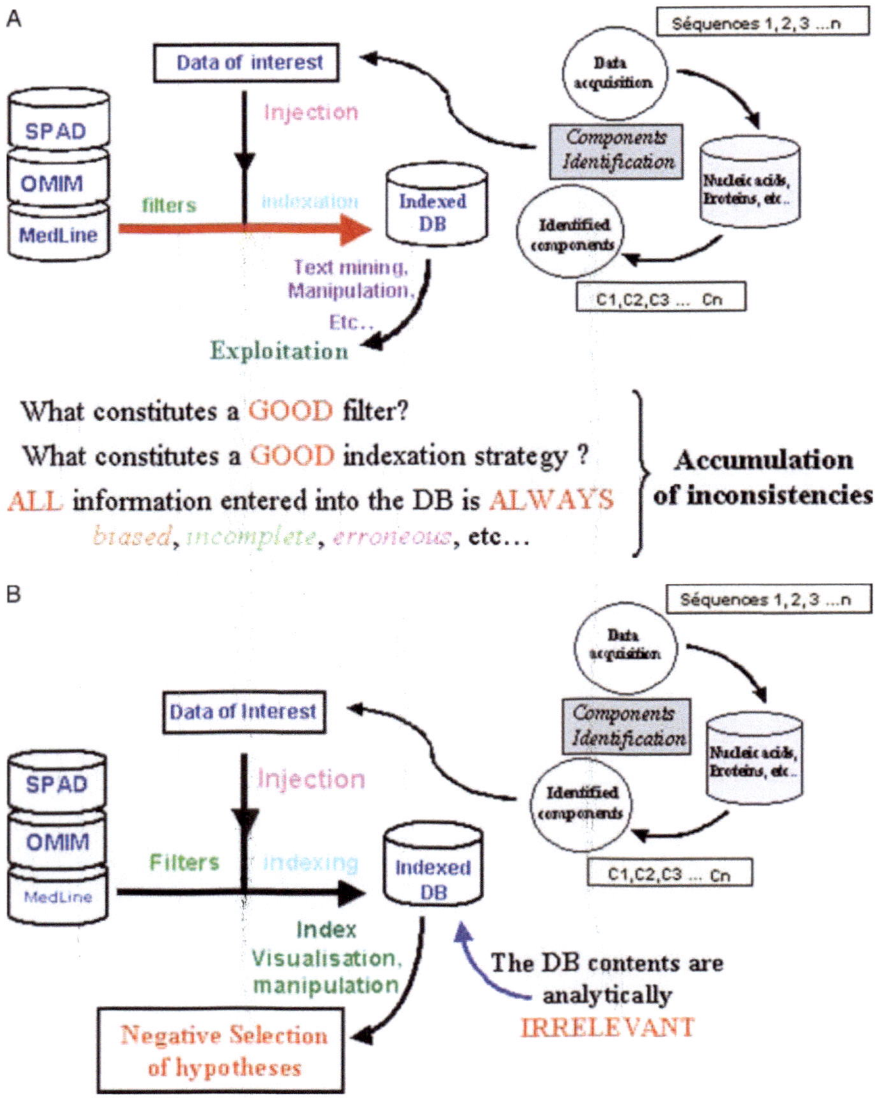
A
Séquences 1,2,3 ...n
Data of interest
Data acquisition
Injection
Components Identification
SPAD
OMIM
MedLine
filters
indexation
Indexed DB
Nucleic acids, Proteins, etc..
Identified components
Text mining, Manipulation, Etc..
C1,C2,C3 ... Cn
Exploitation
What constitutes a GOOD filter?
What constitutes a GOOD indexation strategy ?
ALL information entered into the DB is ALWAYS
biased, incomplete, erroneous, etc…
Accumulation of inconsistencies
B
Séquences 1,2,3 ...n
Data acquisition
Data of Interest
Components Identification
Nucleic acids, Proteins, etc..
SPAD
Injection
Identified components
OMIM
Filters
indexing
Indexed DB
C1,C2,C3 ... Cn
MedLine
Index Visualisation, manipulation
The DB contents are analytically IRRELEVANT
Negative Selection of hypotheses

Fig. 19.1

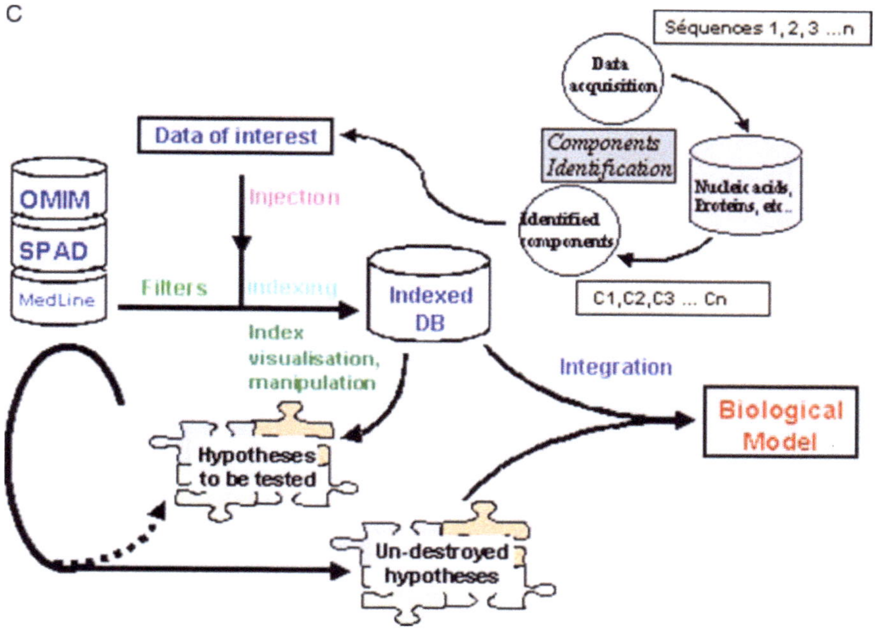

Fig. 19.1 (Continued)

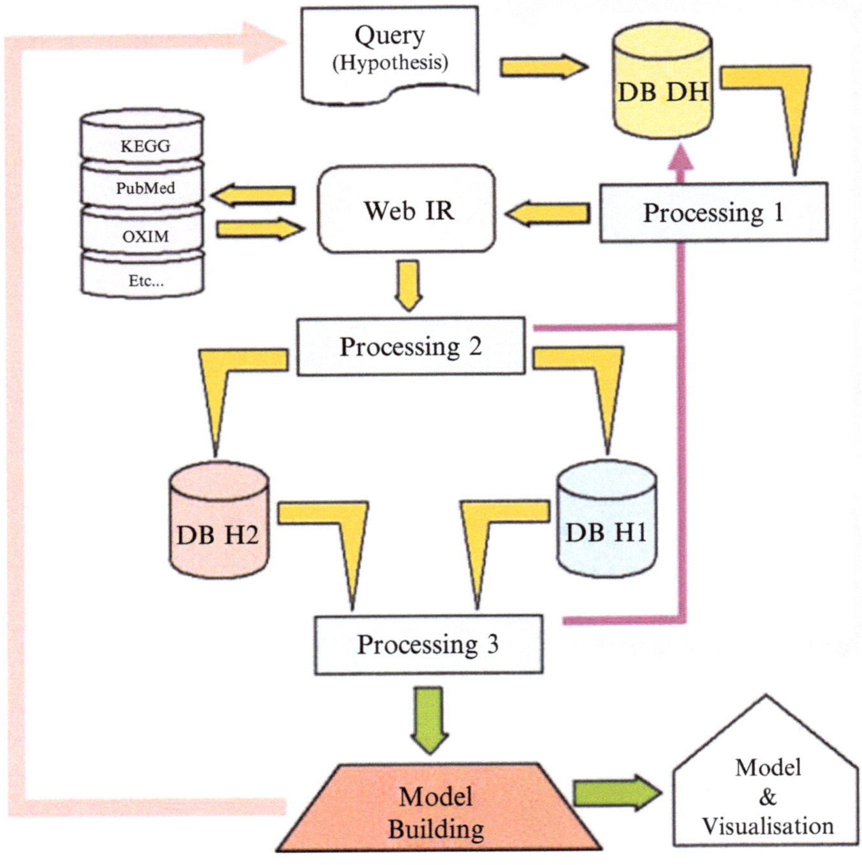
Query
(Hypothesis)
DB DH
KEGG
PubMed
OXIM
Etc...
Web IR
Processing 1
Processing 2
DB H2
DB H1
Processing 3
Model
Building
Model
&
Visualisation

Fig. 19.2

MIX
Papier aus verantwortungsvollen Quellen
Paper from responsible sources
FSC® C105338
FSC
www.fsc.org

If you have any concerns about our products,
you can contact us on
ProductSafety@springernature.com

In case Publisher is established outside the EU,
the EU authorized representative is:
Springer Nature Customer Service Center GmbH
Europaplatz 3, 69115 Heidelberg, Germany

Printed by Libri Plureos GmbH
in Hamburg, Germany